Lecture Notes in Computer Science 12650

Advanced Research in Computing and Software Science

Subline of Lecture Notes in Computer Science

More information about this subseries at http://www.springer.com/series/7407

Stefan Kiefer · Christine Tasson (Eds.)

Foundations of Software Science and Computation Structures

24th International Conference, FOSSACS 2021
Held as Part of the European Joint Conferences
on Theory and Practice of Software, ETAPS 2021
Luxembourg City, Luxembourg, March 27 – April 1, 2021
Proceedings

Springer

Editors
Stefan Kiefer
University of Oxford
Oxford, UK

Christine Tasson
Sorbonne Université - LIP6
Paris, France

ISSN 0302-9743 ISSN 1611-3349 (electronic)
Lecture Notes in Computer Science
ISBN 978-3-030-71994-4 ISBN 978-3-030-71995-1 (eBook)
https://doi.org/10.1007/978-3-030-71995-1

LNCS Sublibrary: SL1 – Theoretical Computer Science and General Issues

This Springer imprint is published by the registered company Springer Nature Switzerland AG
The registered company address is: Gewerbestrasse 11, 6330 Cham, Switzerland

ETAPS Foreword

Welcome to the 24th ETAPS! ETAPS 2021 was originally planned to take place in Luxembourg in its beautiful capital Luxembourg City. Because of the Covid-19 pandemic, this was changed to an online event.

ETAPS 2021 was the 24th instance of the European Joint Conferences on Theory and Practice of Software. ETAPS is an annual federated conference established in 1998, and consists of four conferences: ESOP, FASE, FoSSaCS, and TACAS. Each conference has its own Program Committee (PC) and its own Steering Committee (SC). The conferences cover various aspects of software systems, ranging from theoretical computer science to foundations of programming languages, analysis tools, and formal approaches to software engineering. Organising these conferences in a coherent, highly synchronised conference programme enables researchers to participate in an exciting event, having the possibility to meet many colleagues working in different directions in the field, and to easily attend talks of different conferences. On the weekend before the main conference, numerous satellite workshops take place that attract many researchers from all over the globe.

ETAPS 2021 received 260 submissions in total, 115 of which were accepted, yielding an overall acceptance rate of 44.2%. I thank all the authors for their interest in ETAPS, all the reviewers for their reviewing efforts, the PC members for their contributions, and in particular the PC (co-)chairs for their hard work in running this entire intensive process. Last but not least, my congratulations to all authors of the accepted papers!

ETAPS 2021 featured the unifying invited speakers Scott Smolka (Stony Brook University) and Jane Hillston (University of Edinburgh) and the conference-specific invited speakers Işil Dillig (University of Texas at Austin) for ESOP and Willem Visser (Stellenbosch University) for FASE. Inivited tutorials were provided by Erika Ábrahám (RWTH Aachen University) on analysis of hybrid systems and Madhusudan Parthasarathy (University of Illinois at Urbana-Champaign) on combining machine learning and formal methods.

ETAPS 2021 was originally supposed to take place in Luxembourg City, Luxembourg organized by the SnT - Interdisciplinary Centre for Security, Reliability and Trust, University of Luxembourg. University of Luxembourg was founded in 2003. The university is one of the best and most international young universities with 6,700 students from 129 countries and 1,331 academics from all over the globe. The local organisation team consisted of Peter Y.A. Ryan (general chair), Peter B. Roenne (organisation chair), Joaquin Garcia-Alfaro (workshop chair), Magali Martin (event manager), David Mestel (publicity chair), and Alfredo Rial (local proceedings chair).

ETAPS 2021 was further supported by the following associations and societies: ETAPS e.V., EATCS (European Association for Theoretical Computer Science), EAPLS (European Association for Programming Languages and Systems), and EASST (European Association of Software Science and Technology).

The ETAPS Steering Committee consists of an Executive Board, and representatives of the individual ETAPS conferences, as well as representatives of EATCS, EAPLS, and EASST. The Executive Board consists of Holger Hermanns (Saarbrücken), Marieke Huisman (Twente, chair), Jan Kofron (Prague), Barbara König (Duisburg), Gerald Lüttgen (Bamberg), Caterina Urban (INRIA), Tarmo Uustalu (Reykjavik and Tallinn), and Lenore Zuck (Chicago).

Other members of the steering committee are: Patricia Bouyer (Paris), Einar Broch Johnsen (Oslo), Dana Fisman (Be'er Sheva), Jan-Friso Groote (Eindhoven), Esther Guerra (Madrid), Reiko Heckel (Leicester), Joost-Pieter Katoen (Aachen and Twente), Stefan Kiefer (Oxford), Fabrice Kordon (Paris), Jan Křetínský (Munich), Kim G. Larsen (Aalborg), Tiziana Margaria (Limerick), Andrew M. Pitts (Cambridge), Grigore Roşu (Illinois), Peter Ryan (Luxembourg), Don Sannella (Edinburgh), Lutz Schröder (Erlangen), Ilya Sergey (Singapore), Mariëlle Stoelinga (Twente), Gabriele Taentzer (Marburg), Christine Tasson (Paris), Peter Thiemann (Freiburg), Jan Vitek (Prague), Anton Wijs (Eindhoven), Manuel Wimmer (Linz), and Nobuko Yoshida (London).

I'd like to take this opportunity to thank all the authors, attendees, organizers of the satellite workshops, and Springer-Verlag GmbH for their support. I hope you all enjoyed ETAPS 2021.

Finally, a big thanks to Peter, Peter, Magali and their local organisation team for all their enormous efforts to make ETAPS a fantastic online event. I hope there will be a next opportunity to host ETAPS in Luxembourg.

February 2021

Marieke Huisman
ETAPS SC Chair
ETAPS e.V. President

Preface

This volume contains the papers accepted for the 24th International Conference on Foundations of Software Science and Computation Structures (FoSSaCS). The conference series is dedicated to foundational research with a clear significance for software science. It brings together research on theories and methods to support the analysis, integration, synthesis, transformation, and verification of programs and software systems.

This volume contains 28 contributed papers selected from 88 paper submissions. Each submission was reviewed by at least three Program Committee members, with the help of external reviewers, and the final decisions took into account the feedback from a rebuttal phase. The conference submissions were managed using the EasyChair conference system, which was also used to assist with the compilation of these proceedings.

We wish to thank all the authors who submitted papers to FoSSaCS 2021, the Program Committee members, the Steering Committee members, the external reviewers, and the ETAPS 2021 organizers. Due to the Covid-19 pandemic, ETAPS 2021 was held online.

July 2021

Stefan Kiefer
Christine Tasson

Organization

Program Committee

Sandra Alves	University of Porto
Zena M. Ariola	University of Oregon
Giorgio Bacci	Aalborg University
Nathalie Bertrand	Inria
Véronique Bruyère	University of Mons
Dmitry Chistikov	The University of Warwick
Ugo Dal Lago	Università di Bologna & Inria Sophia Antipolis
Valeria de Paiva	Santa Clara University and Topos Institute
Jacques Garrigue	Nagoya University
Robert Harper	Carnegie Mellon University
Piotr Hofman	University of Warsaw
Stefan Kiefer	University of Oxford
Dexter Kozen	Cornell University
Sebastian Maneth	Universität Bremen
Giulio Manzonetto	Université Sorbonne Paris Nord
Samuel Mimram	École Polytechnique
Peter Selinger	Dalhousie University
Mahsa Shirmohammadi	CNRS
Filip Sieczkowski	University of Wrocław
Jeremy Sproston	University of Turin
Thomas Streicher	TU Darmstadt
Christine Tasson	Sorbonne Université
Nikos Tzevelekos	Queen Mary University of London
Rob van Glabbeek	Data61 - CSIRO

Additional Reviewers

Alcolei, Aurore
Amy, Matthew
Angiuli, Carlo
Avanzini, Martin
Barenbaum, Pablo
Behr, Nicolas
Biernacki, Dariusz
Bjorndahl, Adam
Blumensath, Achim
Bollig, Benedikt
Breuvart, Flavien
Brookes, Steve
Carbone, Marco
Christoff, Zoé
Clemente, Lorenzo
Colcombet, Thomas
Czerwiński, Wojciech
de Frutos-Escrig
de Liguoro, Ugo
de Visme, Marc
Demangeon, Romain
Deng, Yuxin

Derakhshan, Farzaneh
Din, Crystal Chang
Dixon, Alex
Doyen, Laurent
Dubut, Jérémy
Duncan, Ross
Echahed, Rachid
Enea, Constantin
Espírito Santo, José
Exibard, Léo
Fahrenberg, Uli
Falcone, Yliès
Fijalkow, Nathanaël
Finster, Eric
Flesch, János
Fortin, Marie
Frey, Jonas
Garner, Richard
Gavazzo, Francesco
Gay, Simon
Geoffroy, Guillaume
Giacobazzi, Roberto
Graham-Lengrand, Stéphane
Gruber, Hermann
Guerrieri, Giulio
Gusev, Vladimir
Haase, Christoph
Heijltjes, Willem
Hirschkoff, Daniel
Hirschowitz, Tom
Hoffmann, Jan
Hou, Zhe
Husfeldt, Thore
Jansen, David N.
Kaminski, Benjamin Lucien
Kerjean, Marie
Klin, Bartek
Kop, Cynthia
Kopczynski, Eryk
Koutavas, Vasileios
Kura, Satoshi
Kuske, Dietrich
König, Barbara
Laarman, Alfons
Lahav, Ori
Laird, James
Lam, Vitus
Laroussinie, Francois
Lasota, Sławomir
Lazić, Ranko
Le Roux, Stéphane
Lefaucheux, Engel
Leroux, Jérôme
Levy, Jordi
Lin, Yu-Yang
Maillard, Kenji
Malacaria, Pasquale
Malbos, Philippe
Maletti, Andreas
Mansfield, Shane
Mardare, Radu
Marin, Sonia
Markey, Nicolas
Maruyama, Yoshihiro
Maschio, Samuele
Matthes, Ralph
Mayr, Richard
Mazowiecki, Filip
Mazza, Damiano
McCusker, Guy
Michaux, Christian
Mikulski, Lukasz
Milius, Stefan
Moerman, Joshua
Moreira, Nelma
Muscholl, Anca
Møgelberg, Rasmus Ejlers
Najib, Muhammad
Nanevski, Aleksandar
Nordvall Forsberg, Fredrik
O'Connor, Liam
Padovani, Luca
Parys, Paweł
Pasquali, Fabio
Pattathurajan, Mohnish
Patterson, Daniel
Pérez, Guillermo
Pfenning, Frank
Piedeleu, Robin
Pimentel, Elaine
Piróg, Maciej
Pistone, Paolo

Pitts, Andrew
Pous, Damien
Przybyłko, Marcin
Puppis, Gabriele
Purser, David
Péchoux, Romain
Quaas, Karin
Raskin, Jean-François
Riba, Colin
Rowe, Reuben
Ryzhikov, Andrew
Saikawa, Takafumi
Sammartino, Matteo
Sangnier, Arnaud
Schmitz, Sylvain
Schöpp, Ulrich
Seely, Robert
Sobocinski, Pawel
Sofronie-Stokkermans, Viorica
Soudjani, Sadegh
Sprunger, David
Srba, Jiri
Stefanescu, Gheorghe
Sterling, Jonathan
Studer, Thomas
Tamm, Hellis
Tan, Tony
Tang, Qiyi
Tini, Simone
Totzke, Patrick
Tretmans, Jan
Trotta, Davide
Van der Merwe, Brink
van Dijk, Tom
Virtema, Jonni
Vizel, Yakir
Wang, Xinyu
Winkler, Tobias
Winter, Sarah
Wojtczak, Dominik
Zanasi, Fabio
Zeilberger, Noam

Contents

Constructing a universe for the setoid model

Thorsten Altenkirch[1] *, Simon Boulier[2†], Ambrus Kaposi[3] ‡, Christian Sattler[4§], and Filippo Sestini[1]

[1] School of Computer Science, University of Nottingham, Nottingham, UK
{psztxa,psxfs5}@nottingham.ac.uk

[2] Inria, Nantes, France simon.boulier@inria.fr

[3] Eötvös Loránd University, Budapest, Hungary akaposi@inf.elte.hu

[4] Chalmers University of Technology, Gothenburg, Sweden sattler@chalmers.se

Abstract. The setoid model is a model of intensional type theory that validates certain extensionality principles, like function extensionality and propositional extensionality, the latter being a limited form of univalence that equates logically equivalent propositions. The appeal of this model construction is that it can be constructed in a small, intensional, type theoretic metatheory, therefore giving a method to boostrap extensionality. The setoid model has been recently adapted into a formal system, namely Setoid Type Theory (SeTT). SeTT is an extension of intensional Martin-Löf type theory with constructs that give full access to the extensionality principles that hold in the setoid model.

Although already a rich theory as currently defined, SeTT currently lacks a way to internalize the notion of type beyond propositions, hence we want to extend SeTT with a universe of setoids. To this aim, we present the construction of a (non-univalent) universe of setoids within the setoid model, first as an inductive-recursive definition, which is then translated to an inductive-inductive definition and finally to an inductive family. These translations from more powerful definition schemas to simpler ones ensure that our construction can still be defined in a relatively small metatheory which includes a proof-irrelevant identity type with a strong transport rule.

Keywords: type theory · function extensionality · univalence · setoid model · induction-recursion · induction-induction

* Supported by USAF grant FA9550-16-1-0029.

† Supported by ERC Starting Grant CoqHoTT 637339.

‡ Supported by the Bolyai Fellowship of the Hungarian Academy of Sciences (BO/00659/19/3) and by the "Application Domain Specific Highly Reliable IT Solutions" project that has been implemented with support from the National Research, Development and Innovation Fund of Hungary, financed under the Thematic Excellence Programme TKP2020-NKA-06 funding scheme.

§ Supported by USAF grant FA9550-16-1-0029 and Swedish Research Council grant 2019-03765.

S. Kiefer and C. Tasson (Eds.): FOSSACS 2021, LNCS 12650, pp. 1–21, 2021.
https://doi.org/10.1007/978-3-030-71995-1_1

1 Introduction

Intuitionistic type theory is a formal system designed by Per Martin-Löf to be a full-fledged foundation in which to develop constructive mathematics [23,24]. A central aspect of type theory is the coexistence of two notions of equality. On the one hand definitional equality, the computational equality that is built into the formalism. On the other hand "propositional" equality, the internal notion of equality that is actually used to state and prove equational theorems within the system. The precise balance between these two notions is at the center of type theory research; however, it is generally understood that to properly support formalization of mathematics, one should aim for a notion of propositional equality that is as *extensional* as possible.

Two extensionality principles seem particularly desirable, since they arguably constitute the bare minimum for type theory to be comparable to set theory as a foundational system for set-level mathematics, in terms of power and ergonomics. One is function extensionality (or *funext*), according to which functions are equal if point-wise equal. Another is propositional extensionality (or *propext*), that equates all propositions that are logically equivalent.

Type theory with equality reflection, also known as *extensional type theory* (ETT) does support extensional reasoning to some degree, but unfortunately equality reflection makes the problem of type-checking ETT terms computationally unfeasible: it is undecidable.

On the other hand, *intensional type theory* (ITT) has nice computational properties like decidable type checking that can make it more suitable for computer implementation, but as usually defined (for example, in [23]) it severely lacks extensionality. It is known from model constructions that extensional principles like funext are consistent with ITT. Moreover, ITT extended with the principle of *uniqueness of identity proofs* (UIP) and funext is known to be as powerful as ETT [19]. We could recover the expressive power of ETT by adding these principles to ITT as axioms, however destroying some computational properties like canonicity.

What we would like instead is a formulation of ITT that supports extensionality, while retaining its convenient computational behaviour. Unfortunately, canonicity for Martin-Löf's inductively defined identity type says that if two terms are propositionally equal in the empty context, then they are also definitionally equal. This rules out function extensionality. The first step towards a solution is to give up the idea of propositional equality as a single inductive definition given generically for arbitrary types. Instead, equality should be *specific* to each type former in the type theory, or in other words, every type former should be introduced alongside an explanation of what counts as equality for its elements.

This idea of pairing types together with their own equality relation goes back to the notion of *setoid* or *Bishop set*. Setoids provide a quite natural and useful semantic domain in which to interpret type theory. The first setoid model was constructed to justify function extensionality without relying on funext in the metatheory [18]. Moreover, it was shown by Altenkirch [4] that if the model construction is carried out in a type theoretic metatheory with a universe of strict (definitionally proof-irrelevant) propositions, it is possible to define a univalent universe of propositions satisfying propositional extensionality. The setoid model thus satisfies all the extensionality principles that we would like to have in a set-level type theory [5]. The question is whether there exists a version of intensional type theory that supports setoid reasoning, and hence the forms of extensionality enabled by it.

This question was revisited and answered in Altenkirch et al. [5]. In this paper, the authors define Setoid Type Theory (SeTT), an extension of intensional Martin-Löf type theory with constructs for setoid reasoning, where funext and propext hold by definition. SeTT is based on the *strict* setoid model of Altenkirch[6], which makes it possible to show consistency via a syntactic translation. This is in contrast with other type theories based on the setoid model, like Observational Type Theory [9] and XTT [28], which instead rely on ETT for their justification. A major property of SeTT is thus to illustrate how to bootstrap extensionality, by translation into a small intensional core.

SeTT as defined in [5] is already a rich theory, but its introspection capabilities are currently lacking, as its universes are limited to propositions. We would like to internalise the notion of type in SeTT, thus extending the theory with a universe of setoids. This goal brings up several questions, one of which has to do with the notion of equality with which the universe should come equipped: the universe of setoids is itself a setoid (as any type is) so it certainly cannot be univalent, since setoids lack the necessary structure. Another issue is the way such universe can be justified by the setoid model, and in particular what principles are needed in the metatheory to do so.

Contributions This paper documents our work towards the construction of a universe of setoids inside the setoid model, and tries to answer these and other questions related to the design and implementation of this construction. Our main contribution is the construction of the universe in the model; this is given in steps, first as an inductive-recursive definition, which is then translated to an inductive-inductive definition, and subsequently to an inductive type. As a consequence, we show that we only need to assume indexed W-types and proof-irrelevant identity types in the metatheory (along with some obligatory basic tools like Σ and Π types) to construct the universe.

The universe constructions presented in this paper are, to our knowledge, the first examples of two kinds of data type reductions in an intensional metatheory: the first involving an inductive-recursive type which includes strict propositions, and the second involving an infinitary inductive-inductive type.

Finally, the mathematical contents of this paper have been formalized in the proof-assistant Agda (see [10]).

Structure of the paper We begin by describing the metatheory that we will use throughout the paper, in Section 2. In Section 3, after briefly recalling *cate-*

[5] In the sense of HoTT we mean a type theory limited to h-sets.

[6] A *strict* model is one where every equation holds definitionally.

gories with families as an abstract notion of models of type theory, we outline Altenkirch's setoid model as given in [5]. We then briefly discuss the rules of Setoid Type Theory in Section 3.2.

In Section 4 we discuss the setoid model and various design choices related to it. We then recall inductive-recursive universes, and the way they can be equivalently defined as a plain inductive definition, in Section 4.1. We then provide, in Section 4.2, a first complete definition of the setoid universe using a special form of induction-recursion. This form of induction-recursion is not known to be reducible to plain inductive types. Then we describe an alternative definition of the universe in Section 4.3, that does not rely on induction-recursion but instead on infinitary induction-induction. This inductive-inductive encoding of the universe is obtained from the inductive-recursive one, inspired by the method of Section 4.1. We end the series of universe constructions with Section 4.4, where we outline a purely inductive definition of the setoid universe, obtained from the inductive-inductive one.

1.1 Related work

The setoid model was first described in [18] in order to add extensionality principles to Type Theory such as function extensionality and propositional extensionality. A strict variant of the setoid model was given in [4] using a definitionally proof-irrelevant universe of propositions. Recently, support for such a universe was added to the proof-assistants Agda and Coq [17], allowing a full formalization of Altenkirch's setoid model. Setoid Type Theory (SeTT) is a recently developed formal system derived from this model construction [5]. Observational Type Theory (OTT) [9] is a syntax for the setoid model differing from SeTT in the use of a different notion of heterogeneous equality. Moreover, the consistency proof for OTT relies on Extensional Type Theory, whereas for SeTT it is obtained via a syntactic translation. XTT [28] is a cubical variant of OTT where the equality type is defined using an interval pretype [7] . XTT's universes support universe induction, whereas it is left open whether the construction presented here supports this principle. Palmgren and Wilander [27] construct a setoid universe using a translation into constructive set theory. Palmgren [26] constructs an encoding of ETT in ITT through Aczel's encoding of set theory in type theory [3]. He uses type theory as a language for his formalisation but his construction is set-theoretic in nature. Setoids are utilized to encode sets as arbitrarily branching well-founded trees quotiented by bisimulation. His notion of family of setoids does not use strict propositions and it has a weaker form of proof irrelevance which seems to be not enough to obtain a model of SeTT.

The principle of propositional extensionality in the setoid model is an instance of Voevodsky's univalence axiom [29]. The cubical set model is a constructive model justifying this axiom [11]. A type theory extracted from this model is Cubical Type Theory [13]. The relationship between the cubical set

[7] To quote one of the referees: *the fact that the interval is a pretype is but the easiest part of the story.*

model and cubical type theory is similar to that between the setoid model and SeTT. Compared to cubical type theories, SeTT has the advantage that the equality type satisfies more definitional equalities. For instance, whereas in cubical type theory equality of functions is isomorphic to pointwise equality, in SeTT the isomorphism is replaced by a definitional equality. SeTT is also a syntactically straightforward extension of Martin-Löf Type Theory, that does not require exotic objects like the interval pretype. In turn, the obvious advantage of cubical type theory is that it is not limited to setoids.

An exceptional aspect of the metatheory used in this paper is the presence of a proof-irrelevant identity type with a strong transport rule allowing to eliminate into arbitrary types. In [1], Abel gives a proof of normalization for the Logical Framework extended with a similar proof-irrelevant equality type. Abel and Coquand show in [2] that the combination of impredicativity with a strong transport rule results in terms that fail to normalize but this is irrelevant in our setting.

2 $\mathsf{MLTT}^{\mathbf{Prop}}$

This section describes $\mathsf{MLTT}^{\mathbf{Prop}}$, our ambient metatheory. We employ Agda notation to write down $\mathsf{MLTT}^{\mathbf{Prop}}$ terms throughout the paper.

One of the main appeals of Altenkirch's setoid model is that it can justify several useful extensionality principles while being defined in a small intensional metatheory. We tried to stay true to this idea when figuring out the necessary metatheoretical tools for the universe construction in this paper. In particular, we wanted to avoid having to assume strong definition schemas that go beyond inductive families. $\mathsf{MLTT}^{\mathbf{Prop}}$ is thus an intensional type theory in the style of Martin-Löf type theory.

We have sorts $\mathbf{Type}_i$ of types and $\mathbf{Prop}_i$ of strict propositions for $i \in \{0, 1\}$. Here, $i = 0$ means "small" (and we will omit the subscript) and $i = 1$ means "large". We have implicit lifting from $i = 0$ to $i = 1$, but do not assume type formers are preserved. $\mathbf{Type}_1$ has universes for $\mathbf{Type}$ and $\mathbf{Prop}$. We do not distinguish notationally between universes and sorts. We continue to describe only the case $i = 0$; everything introduced has an analogue at level $i = 1$. Propositions lift to types via $\mathsf{Lift} : \mathbf{Prop} \to \mathbf{Type}$, with constructor $\mathsf{lift} : \{P : \mathbf{Prop}\} \to P \to \mathsf{Lift}\ P$ and destructor $\mathsf{unlift} : \{P : \mathbf{Prop}\} \to \mathsf{Lift}\ P \to P$.

We have standard type formers $\Pi, \Sigma, \mathsf{Bool}, \mathbf{0}, \mathbf{1}$ in $\mathbf{Type}$. Σ-types are defined negatively by pairing $-,-$ and projections π_1, π_2. We have definitional η-rules for Π-, Σ-, $\mathbf{1}$-types. We also require indexed W-types, both in $\mathbf{Type}$ and $\mathbf{Prop}$: $W_\square : (S : I \to \mathbf{Type}) \to ((i : I) \to S\ i \to I \to \mathbf{Type}) \to I \to \square$ where $\square \in \{\mathbf{Type}, \mathbf{Prop}\}$. The elimination principle of $W_{\mathbf{Prop}}$ only allows defining functions into elements of $\mathbf{Prop}$. From $W_{\mathbf{Prop}}$ we can define propositional truncation $\|-\| : \mathbf{Type} \to \mathbf{Prop}$, with constructor $|-| : \{A : \mathbf{Type}\} \to A \to \|A\|$ and eliminator $\mathsf{elim}_{\|-\|} : \{P : \mathbf{Prop}\} \to (A \to P) \to \|A\| \to P$.

In addition to type formers in $\mathbf{Type}$, we will need the propositional versions of $\mathbf{0}$, $\mathbf{1}$, Π, and Σ. The latter three can be defined from their $\mathbf{Type}$ counterparts via truncation. That is, given $P : \mathbf{Prop}$ and $Q : P \to \mathbf{Prop}$:

$$\mathbf{1}_{\mathbf{Prop}} :\equiv \|\mathbf{1}\|$$
$$\Pi_{\mathbf{Prop}}\ P\ Q :\equiv \|\Pi\ (\mathsf{Lift}\ P)\ (\mathsf{Lift} \circ Q \circ \mathsf{unlift})\|$$
$$\Sigma_{\mathbf{Prop}}\ P\ Q :\equiv \|\Sigma\ (\mathsf{Lift}\ P)\ (\mathsf{Lift} \circ Q \circ \mathsf{unlift})\|$$

We assume that we have $\mathbf{0}_{\mathbf{Prop}} : \mathbf{Prop}$ together with $\mathsf{exfalso}_{\mathbf{Prop}} : \{A : \mathbf{Type}\} \to \mathbf{0}_{\mathbf{Prop}} \to A$.

Finally, we will assume an identity type in the style of Martin-Löf's inductive identity type. The main difference is that our identity type is a **Prop**-valued relation. We have a transport combinator transp from which J is derivable.

$$\mathsf{Id} : \{A : \mathbf{Type}\} \to A \to A \to \mathbf{Prop}$$
$$\mathsf{refl} : \{A : \mathbf{Type}\}(a : A) \to \mathsf{Id}\ a\ a$$
$$\mathsf{transp} : \{A : \mathbf{Type}\}(C : A \to \mathbf{Type})\{a_0\ a_1 : A\} \to \mathsf{Id}\ a_0\ a_1 \to C\ a_0 \to C\ a_1$$

with $\mathsf{transp}\ C\ \{x\}\ \{x\}\ e\ u \equiv u$. The transp combinator provides a strong elimination principle allowing to eliminate a strict proposition (the identity type) into arbitrary types. We only use this identity type in Section 4.4. For the rest of our constructions, the traditional Martin-Löf's identity type suffices.

2.1 Formalization

A universe of strict propositions has been recently added to the Agda proof assistant [17], making *most* of $\mathsf{MLTT}^{\mathbf{Prop}}$ a subset of Agda, with the exception of the proof-irrelevant identity type. Most of the universe constructions presented here have been formalized and proof-checked using Agda, with the proof-irrelevant identity type and the strong transport rule added via postulates and rewriting. The formalization can be found in [10].

For convenience, we slightly deviate from $\mathsf{MLTT}^{\mathbf{Prop}}$ both in the paper and in the formalization, for instance by relying on pattern matching instead of eliminators, and using primitive versions of **Prop**-valued Π and Σ types instead of deriving them from truncation. We operate under the assumption that everything can be equivalently carried out in $\mathsf{MLTT}^{\mathbf{Prop}}$, although we have not fully checked all the necessary details.

3 Setoid model

By *setoid model* we mean a class of models of type theory where contexts/closed types are interpreted as setoids, i.e. sets with an equivalence relation, and dependent types are interpreted as dependent/indexed setoids. A setoid model was first given for intensional type theory by M. Hofmann [18], in order to provide a semantics for extensionality principles such as function and propositional extensionality.

Here we consider a similar model construction due to Altenkirch [4]. The peculiarity of this model is that it is presented in a type theoretic and intensional metatheory which includes a strict universe of propositions.

The setoid model thus defined validates function extensionality, a universe of propositions with propositional extensionality, and quotient types. Therefore, it provides a way to bootstrap and "explain" extensionality, since the model construction effectively gives an implementation of various extensionality principles in terms of a small, completely intensional theory.

3.1 Setoid model as a CwF

The setoid model can be framed categorically as a category with families (CwF, [14]) with extra structure for the various type and term formers. The core structure of a CwF can be given as the following signature:

$$\begin{aligned}
\mathrm{Con} &: \mathbf{Type} \\
\mathrm{Ty} &: (\Gamma : \mathrm{Con}) \to \mathbf{Type} \\
\mathrm{Sub} &: (\Gamma\ \Delta : \mathrm{Con}) \to \mathbf{Type} \\
\mathrm{Tm} &: (\Gamma : \mathrm{Con}) \to \mathrm{Ty}\ \Gamma \to \mathbf{Type}
\end{aligned}$$

In our presentation of the setoid model, contexts are given by setoids, that is, types together with an equivalence relation. A key point of the model is that the equivalence relation is valued in **Prop** and is thus definitionally proof irrelevant.

$$\frac{\Gamma : \mathrm{Con}}{\begin{aligned}
|\Gamma| &: \mathbf{Type} \\
\Gamma^{\sim} &: |\Gamma| \to |\Gamma| \to \mathbf{Prop} \\
\mathsf{refl}\ \Gamma &: (\gamma : |\Gamma|) \to \Gamma^{\sim}\ \gamma\,\gamma \\
\mathsf{sym}\ \Gamma &: \forall\{\gamma_0\ \gamma_1\} \to \Gamma^{\sim}\ \gamma_0\,\gamma_1 \to \Gamma^{\sim}\ \gamma_1\,\gamma_0 \\
\mathsf{trans}\ \Gamma &: \forall\{\gamma_0\ \gamma_1\ \gamma_2\} \to \Gamma^{\sim}\ \gamma_0\,\gamma_1 \to \Gamma^{\sim}\ \gamma_1\,\gamma_2 \to \Gamma^{\sim}\ \gamma_0\,\gamma_2
\end{aligned}}$$

Types in a context Γ are given by displayed setoids over Γ with a fibration condition given by coe, coh. In the following, we sometimes omit implicit quantifications such as the $\forall\{\gamma_0\ \gamma_1\}$ in the type of $\mathsf{sym}\,\Gamma$.

$$\frac{A : \mathrm{Ty}\ \Gamma}{\begin{aligned}
|A| &: |\Gamma| \to \mathbf{Type} \\
A^{\sim} &: \{\gamma_0\ \gamma_1 : |\Gamma|\} \to \Gamma^{\sim}\ \gamma_0\ \gamma_1 \to |A|\gamma_0 \to |A|\gamma_1 \to \mathbf{Prop} \\
\mathsf{refl}^* &: \{\gamma : |\Gamma|\}(a : |A|\gamma) \to A^{\sim}\ (\mathsf{refl}\ \Gamma\ \gamma)\ a\ a \\
\mathsf{sym}^* &: \forall\{\gamma_0\,\gamma_1\,a_0\,a_1\}\{p : \Gamma^{\sim}\ \gamma_0\ \gamma_1\} \to A^{\sim}\ p\ a_0\ a_1 \to A^{\sim}\ (\mathsf{sym}\ \Gamma\ p)\ a_1\ a_0 \\
\mathsf{trans}^* &: A^{\sim}\ p_0\ a_0\ a_1 \to A^{\sim}\ p_1\ a_1\ a_2 \to A^{\sim}\ (\mathsf{trans}\,\Gamma\,p_0\,p_1)\ a_0\ a_2 \\
\mathsf{coe} &: \Gamma^{\sim}\ \gamma_0\ \gamma_1 \to |A|\gamma_0 \to |A|\gamma_1 \\
\mathsf{coh} &: (p : \Gamma^{\sim}\ \gamma_0\ \gamma_1)(a : |A|\gamma_0) \to A^{\sim}\,p\,a\,(\mathsf{coe}\,A\,p\,a)
\end{aligned}}$$

This definition of types in the setoid model is different from the one in [4], but it is equivalent to it [12, Section 1.6.1]. The main difference here is in the use of a heterogeneous equivalence relation $A^{\sim}$ in the definition of types.

Substitutions are interpreted as functors between the corresponding setoids, whereas terms of type A in context Γ are sections of the type seen as a setoid fibration $\Gamma.A \to \Gamma$. Note that we only need to include components for the functorial action on objects and morphisms, since the functor laws follow from proof-irrelevance in the metatheory, and thus hold definitionally.

$$\frac{\sigma : \mathrm{Sub}\ \Gamma\ \Delta}{\begin{array}{l}|\sigma| : |\Gamma| \to |\Delta| \\ \sigma^{\sim} : \Gamma^{\sim}\ \rho_0\ \rho_1 \to \Delta^{\sim}\ (|\sigma|\rho_0)\ (|\sigma|\rho_1)\end{array}} \qquad \frac{t : \mathrm{Tm}\ \Gamma\ A}{\begin{array}{l}|t| : (\gamma : |\Gamma|) \to |A|\ \gamma \\ t^{\sim} : (p : \Gamma^{\sim}\ \gamma_0\ \gamma_1) \to A^{\sim}\ p\ (|t|\gamma_0)\ (|t|\gamma_1)\end{array}}$$

We can show that the setoid model validates the usual basic type formers (Π, Σ, etc.), function extensionality and a universe of strict propositions with propositional extensionality [4]. Note that we do not need identity types or inductive types (W-types) for this.

3.2 Setoid Type Theory

The setoid model presented in the previous section is *strict*, that is, every equation of a CwF holds by definition in the semantics. One advantage of strict models is that they can be turned into *syntactic translations*, in which syntactic objects of the source theory are interpreted as their counterparts in another *target* theory. In the case of the setoid model, this gives rise to a *setoid translation*, where source contexts are interpreted as target contexts together with a target type representing the equivalence relation, and so on.[8]

A setoid translation is used in [5] to justify Setoid Type Theory (SeTT), an extension of Martin-Löf type theory (+ **Prop**) with equality types for contexts and dependent types that reflect the setoid equality of the model.

We recall the rules of SeTT that extend regular MLTT below, but with a variation: whereas the equality types in [5] are stated as elements of SeTT's internal universe of propositions, here we state the context equalities as elements of the external, metatheoretic universe **Prop**. This generalises the notion of model of SeTT thus making it easier to construct models. Equality on types is defined as before in [5].

We have a universe of propositions Prop defined as follows:

$$\frac{\Gamma : \mathrm{Con}}{\mathsf{Prop} : \mathrm{Ty}\ \Gamma} \qquad \frac{P : \mathrm{Tm}\ \Gamma\ \mathsf{Prop}}{\underline{P} : \mathrm{Ty}\ \Gamma} \qquad \frac{u : \mathrm{Tm}\ \Gamma\ \underline{P} \qquad v : \mathrm{Tm}\ \Gamma\ \underline{P}}{u \equiv v}$$

Equality type constructors for contexts and dependent types internalize the idea that every context and type comes equipped with a setoid equivalence relation. Note that **Prop** is the universe of the metatheory while Prop is the internal

[8] Semantically, this translation corresponds to a model construction, in particular a functor from the category of models of the target theory to the category of models of what will be Setoid Type Theory. Since the setoid translation is structural in the context component, we can work with models in the style of categories with families rather than contextual categories.

one. As in the model, equality for dependent types is indexed over context equality.

$$\frac{\Gamma : \mathrm{Con} \qquad \rho_0, \rho_1 : \mathrm{Sub}\ \Delta\ \Gamma}{\Gamma^{\sim}\ \rho_0\ \rho_1 : \mathbf{Prop}} \qquad \frac{A : \mathrm{Ty}\ \Gamma \qquad \rho_{01} : \Gamma^{\sim}\ \rho_0\ \rho_1 \qquad a_0 : \mathrm{Tm}\ \Delta\ A[\rho_0] \qquad a_1 : \mathrm{Tm}\ \Delta\ A[\rho_1]}{A^{\sim}\ \rho_{01}\ a_0\ a_1 : \mathrm{Tm}\ \Delta\ \mathsf{Prop}}$$

We have rules witnessing that these are indeed equivalence relations. We only recall reflexivity:

$$\frac{\rho : \mathrm{Sub}\ \Delta\ \Gamma}{\mathsf{R}\ \rho : \Gamma^{\sim}\ \rho\ \rho} \qquad \frac{A : \mathrm{Ty}\ \Gamma \qquad \rho : \mathrm{Sub}\ \Delta\ \Gamma \qquad a : \mathrm{Tm}\ \Delta\ A[\rho]}{\mathsf{R}\ a : \mathrm{Tm}\ \Gamma\ \underline{A^{\sim}\ (\mathsf{R}\ \rho)\ a\ a}}$$

In addition, we also have rules representing the fact that every construction in SeTT respects setoid equality, so that we can transport along any such equality:

$$\frac{A : \mathrm{Ty}\ \Gamma \qquad \rho_0, \rho_1 : \mathrm{Sub}\ \Delta\ \Gamma \qquad p : \Gamma^{\sim}\ \rho_0\ \rho_1 \qquad a : \mathrm{Tm}\ \Delta\ A[\rho_0]}{\begin{array}{c}\mathsf{coe}_A\ p\ a : \mathrm{Tm}\ \Delta\ A[\rho_1] \\ \mathsf{coh}_A\ p\ a : \mathrm{Tm}\ \Delta\ \underline{A^{\sim}\ p\ a\ (\mathsf{coe}_A\ p\ a)}\end{array}}$$

Notably, equality types in SeTT compute definitionally on concrete type formers. In particular, they compute to their obvious intended meaning, so that an equality of pairs is a pair of equalities, an equality of functions is a map of equalities, and so on. From this, we get definitional versions of function and propositional extensionality.

We can easily recover the usual Martin-Löf identity type from setoid equality, with transport implemented via coercion.

$$\frac{A : \mathrm{Ty}\ \Gamma \qquad a_0, a_1 : \mathrm{Tm}\ \Gamma\ A}{\mathsf{Id}_A\ a_0\ a_1 :\equiv A^{\sim}\ (\mathsf{R}\ \Gamma)\ a_0\ a_1 : \mathrm{Tm}\ \Gamma\ \mathsf{Prop}}$$

$$\frac{P : \mathrm{Ty}\ (\Gamma.A) \qquad p : \mathrm{Tm}\ \Gamma\ (\mathsf{Id}\ A\ a_0\ a_1) \qquad t : \mathrm{Tm}\ \Gamma\ P[a_0]}{\mathsf{transp}\ P\ p\ t :\equiv \mathsf{coe}\ P\ (\mathsf{R}\ \mathsf{id}, p)\ t : \mathrm{Tm}\ \Gamma\ P[a_1]}$$

We can also derive Martin-Löf's J eliminator for this homogeneous identity type. The only caveat is that transp and the J eliminator do not compute definitionally on reflexivity.

4 Universe of setoids

As pointed out in the introduction, SeTT is seriously limited by the lack of a universes internalizing the notion of setoid. Our goal is to extend SeTT with a universe of setoids; since SeTT is a direct syntactic reflection of the setoid model, this essentially amounts to showing that a universe of setoids with the necessary structure and equations can be constructed within the setoid model. This opens several questions and possible design choices.

A first fundamental consideration has to do with the very definition of the setoid universe: as any type in the setoid model, this universe must be a setoid and thus come equipped with an equivalence relation. However, unlike the universe of propositions, a universe of setoids cannot be univalent, since this would force it to be a groupoid. The obvious choice is therefore to have a non-univalent universe, and instead define the universe's relation so that it reflects a simple syntactic equality of codes rather than setoid equivalence.

Another question has to do with the metatheoretic tools required to carry out the construction of the universe. In fact, one of the main aspects of the setoid model construction recalled in Section 3 and shown originally in [4] is that it can be carried out in a very small type theoretic metatheory, thus providing a way to reduce extensionality to a small intensional core. We would like to stay faithful to this ideal when constructing this setoid universe.

A known and established method for defining universes in type theory relies on induction-recursion (IR), a definition schema developed by Dybjer [15,16]. Inductive-recursive definitions can be found throughout the literature, from the already mentioned type theoretic universes, including the original formulation à la Tarski by Martin-Löf [24], to metamathematical tools like computability predicates.

Although universe constructions in type theory—including our own setoid universe—are naturally presented as inductive-recursive definitions, they may not necessarily require a metatheory with induction-recursion. In fact, it is possible to reduce some instances of induction-recursion to plain induction (more specifically, inductive families), including some universe definitions. We recall this reduction in Section 4.1.

Other design choices on the setoid universe are less essential, but still require careful consideration. For instance, one question is whether the setoid universe should support universe induction, thus exposing the inductive structure of the codes. Such an elimination principle is known to be inconsistent with univalence, although this is not an issue in our case; nevertheless it is not immediately clear if the elimination principle can be justified by the semantics, that is, if our encoding of the setoid universe in the model allows to define such a universe eliminator. The question arises because our final encoding of the setoid universe only supports a weak form of elimination, for reasons that are explained in Section 4.4. Although not currently needed, a stronger eliminator might be necessary to justify universe induction. This problem should not arise in the other encodings of the setoid universe (as given in Section 4.2 and Section 4.3).

Another design choice has to do with how the setoid universe relates to the other universes. One could provide a code for Prop in the setoid universe. Moreover, the setoid universes could form a hierarchy, possibly cumulative.

Yet another choice is whether to have two separate sorts, one for propositions and one for sets (with propositions convertible to sets) or a single sort of types (sets), with propositions given by elements of a universe of propositions, which is a (large) type. We have chosen to present the second option to fit with the standard notion of (unisorted) CwF. However, this has downsides: to even talk about propositions, we need to have a notion of large types. The first option is more symmetric: we can have parallel hierarchies for propositions and sets.

4.1 Inductive-recursive universes

An inductive-recursive universe is given by a type of codes $\mathsf{U} : \mathbf{Type}$, and a family $\mathsf{El} : \mathsf{U} \to \mathbf{Type}$ that assigns, to each code corresponding to some type, the meta-theoretic type of its elements. The resulting definition is inductive-recursive because the inductive type of codes is defined simultaneously with the recursive function El.

An example is the following definition of a small universe with bool and Π.

$$\begin{aligned} &\mathsf{data}\ \mathsf{U}\ : \mathbf{Type} && \mathsf{El}\ : \mathsf{U} \to \mathbf{Type} \\ &\quad \mathsf{bool} : \mathsf{U} && \mathsf{El}\ \mathsf{bool} :\equiv \mathbb{2} \\ &\quad \mathsf{pi} : (A : \mathsf{U}) \to (\mathsf{El}\ A \to \mathsf{U}) \to \mathsf{U} && \mathsf{El}\ (\mathsf{pi}\ A\ B) :\equiv (a : \mathsf{El}\ A) \to \mathsf{El}\ (B\ a) \end{aligned}$$

Induction-recursion is arguably a nice and natural way to define internal universes in type theory, however it is not always strictly required. We can translate basic instances of induction-recursion into inductive families using the equivalence of I-indexed families of types and types over I (that is, $A : \mathbf{Type}$ with $A \to I$) [22].

In our case, we can encode U as an inductive type inU that *carves out* all types in $\mathbf{Type}$ that are in the image of El. In other words, inU is a predicate that holds for any type that would have been obtained via El in the inductive-recursive definition. As El is indexed by the type of codes, the definition of inU quite expectedly reflects the inductive structure of codes.

$$\begin{aligned} &\mathsf{data}\ \mathsf{inU} : \mathbf{Type} \to \mathbf{Type}_1 \\ &\quad \mathsf{inBool} : \mathsf{in\text{-}U}\ \mathbb{2} \\ &\quad \mathsf{inPi}\ \ : \mathsf{inU}\ A \to ((a : A) \to \mathsf{inU}\ (B\ a)) \to \mathsf{inU}\ ((a : A) \to (B\ a)) \end{aligned}$$

U and El can be given by $\mathsf{U} :\equiv \Sigma\ (A : \mathbf{Type})\ (\mathsf{in\text{-}U}\ A)$ and $\mathsf{El} :\equiv \pi_1$.

Note that this construction gives rise to a universe in $\mathbf{Type}_1$, rather than $\mathbf{Type}$, since the definition of U quantifies over all possible types in $\mathbf{Type}$. Hence this kind of construction requires a metatheory with at least one universe.

4.2 Inductive-recursive setoid universe

In this section we give a first definition of the setoid universe, as a direct generalization of the simple inductive-recursive definition just shown. We only consider a very small universe with bool type $\mathbb{2}$ and Π for simplicity; a more realistic universe that includes more type formers can be found in the Agda formalization.

To construct the universe of setoids in the setoid model, we first of all need to define a type $\mathbb{U} : \mathrm{Ty}\ \Gamma$ for every $\Gamma : \mathrm{Con}$, and for every $A : \mathrm{Tm}\ \Gamma\ \mathbb{U}$ a type $\mathbb{E}l\ A : \mathrm{Ty}\ \Gamma$. Recalling Section 3, these are essentially record types made of several components. Since $\mathbb{U}$ is a closed type, it requires the same data of a setoid; in particular, we need a type of codes together with an equivalence relation reflecting equality of codes, in addition to proofs that these are indeed equivalence relations:

$$\begin{aligned} &\mathsf{data}\ \mathcal{U} : \mathbf{Type}_1 && \mathsf{refl}_{\mathcal{U}} : (A : \mathcal{U}) \to A \sim_{\mathcal{U}} A \\ &- \sim_{\mathcal{U}} - : \mathcal{U} \to \mathcal{U} \to \mathbf{Prop}_1 && \mathsf{sym}_{\mathcal{U}} : A \sim_{\mathcal{U}} B \to B \sim_{\mathcal{U}} A \\ & && \mathsf{trans}_{\mathcal{U}} : A \sim_{\mathcal{U}} B \to B \sim_{\mathcal{U}} C \to A \sim_{\mathcal{U}} C \end{aligned}$$

$\mathbb{E}l$ is given by a family of setoids indexed over the universe, that is, a way to assign to each code in the universe a carrier set and an equivalence relation.

$$\mathsf{El} : \mathcal{U} \to \mathbf{Type}$$
$$- \vdash - \sim_{\mathsf{El}} - : \{a\ a' : \mathcal{U}\} \to a \sim_{\mathcal{U}} a' \to \mathsf{El}\ a \to \mathsf{El}\ a' \to \mathbf{Prop}$$

Note that $- \vdash - \sim_{\mathsf{El}} -$ is indexed over equality on the universe, because El is a displayed setoid over $\mathcal{U}$, hence in particular it must respect the setoid equality of $\mathcal{U}$. We also require data and proofs that make sure we get setoids out of El:

$$\begin{aligned}
\mathsf{refl}_{\mathsf{El}} &: (A : \mathcal{U})(x : \mathsf{El}\ A) \to \mathsf{refl}_{\mathcal{U}}\ A \vdash x \sim_{\mathsf{El}} x \\
\mathsf{sym}_{\mathsf{El}} &: p \vdash x \sim_{\mathsf{El}} x' \to \mathsf{sym}_{\mathcal{U}}\ p \vdash x' \sim_{\mathsf{El}} x \\
\mathsf{trans}_{\mathsf{El}} &: p \vdash x \sim_{\mathsf{El}} x' \to q \vdash x' \sim_{\mathsf{El}} x'' \to \mathsf{trans}_{\mathcal{U}}\ p\ q \vdash x \sim_{\mathsf{El}} x'' \\
\mathsf{coe}_{\mathsf{El}} &: A \sim_{\mathcal{U}} B \to \mathsf{El}\ A \to \mathsf{El}\ B \\
\mathsf{coh}_{\mathsf{El}} &: (p : A \sim_{\mathcal{U}} A')\ (x : \mathsf{El}\ A) \to p \vdash x \sim_{\mathsf{El}} \mathsf{coe}_{\mathsf{El}}\ p\ x
\end{aligned}$$

We give an inductive definition of $\mathcal{U}$, mutually with a recursive definition of the 4 functions $- \sim_{\mathcal{U}} -$, $\mathsf{refl}_{\mathcal{U}}$, El and $- \vdash - \sim_{\mathsf{El}} -$. The other functions are then recursively defined: $\mathsf{refl}_{\mathsf{El}}$ alone, $\mathsf{sym}_{\mathcal{U}}$ and $\mathsf{sym}_{\mathsf{El}}$ mutually, $\mathsf{trans}_{\mathcal{U}}$, $\mathsf{trans}_{\mathsf{El}}$, $\mathsf{coe}_{\mathsf{El}}$ and $\mathsf{coh}_{\mathsf{El}}$ mutually. The whole construction is quite long, below we only show the more interesting definitions of $\mathcal{U}$ and El:

$$\begin{aligned}
&\mathsf{data}\ \mathcal{U} : \mathbf{Type}_1 \\
&\quad \mathsf{bool} : \mathcal{U} \\
&\quad \mathsf{pi} : (A : \mathcal{U})(B : \mathsf{El}\ A \to \mathcal{U}) \\
&\qquad \to (\{x\ x' : \mathsf{El}\ A\} \to \mathsf{refl}_{\mathcal{U}}\ A \vdash x \sim_{\mathsf{El}} x' \\
&\qquad \to B\ x \sim_{\mathcal{U}} B\ x') \to \mathcal{U}
\end{aligned}$$

$$\begin{aligned}
&\mathsf{El}\ \mathsf{bool} :\equiv \mathbb{2} \\
&\mathsf{El}\ (\mathsf{pi}\ A\ B\ h) :\equiv \\
&\quad \Sigma\ (f : (a : \mathsf{El}\ A) \to \mathsf{El}\ (B\ a)) \\
&\qquad (\forall \{x\ x'\}(p : \mathsf{refl}_{\mathcal{U}}\ A \vdash x \sim_{\mathsf{El}} x') \\
&\qquad \to h\ p \vdash f\ x \sim_{\mathsf{El}} f\ x')
\end{aligned}$$

Note that in the definition of $\mathcal{U}$ we require that the family $B : \mathsf{El}\ A \to \mathcal{U}$ be a setoid morphism, respecting the setoid equalities involved. This choice is crucial for the definition of El to go through, in particular since we eliminate the code for Π types into the setoid of functions that map equal elements to equal results. To state this mapping property we need to compare elements in different types, coming from applying f to different arguments x and x'. We know that x and x' are equal, but to conclude $B\ x \sim_{\mathcal{U}} B\ x'$ we need to know that B respects setoid equality. This is exactly what we get from our definition of $\mathcal{U}$.

We can now give a full definition of the setoid universe, and of $\mathbb{E}l\ A$ for any $A : \mathrm{Tm}\ \Gamma\ \mathbb{U}$:

$$\begin{aligned}
|\mathbb{U}| &:\equiv \lambda\gamma.\mathcal{U} \\
\mathbb{U}^\sim &:\equiv \lambda p\,x\,y.\, x \sim_{\mathcal{U}} y \\
\mathsf{refl}\ \mathbb{U} &:\equiv \mathsf{refl}_{\mathcal{U}} \\
&\dots \\
\mathsf{coe}\ \mathbb{U} &:\equiv \lambda p\,a.\, a \\
\mathsf{coh}\ \mathbb{U} &:\equiv \lambda p.\, \mathsf{refl}_{\mathcal{U}}
\end{aligned}$$

$$\begin{aligned}
|\mathbb{E}l\ A| &:\equiv \lambda\gamma.\, \mathsf{El}\ (|A|\ \gamma) \\
(\mathbb{E}l\ A)^\sim &:\equiv \lambda p\,x\,y.\, A^\sim\ p \vdash x \sim_{\mathsf{El}} y \\
\mathsf{refl}\ (\mathbb{E}l\ A) &:\equiv \mathsf{refl}_{\mathsf{El}} \\
&\dots \\
\mathsf{coe}\ (\mathbb{E}l\ A) &:\equiv \lambda p.\, \mathsf{coe}_{\mathsf{El}}\ (A^\sim\ p) \\
\mathsf{coh}\ (\mathbb{E}l\ A) &:\equiv \lambda p.\, \mathsf{coh}_{\mathsf{El}}\ (A^\sim\ p)
\end{aligned}$$

We can show that $\mathbb{U}$ is closed under Π types and booleans, and satisfies $\mathbb{E}l\,(\mathsf{pi}\,A\,B) \equiv \Pi\,(\mathbb{E}l\,A)\,(\mathbb{E}l\,B)$ and $\mathbb{E}l\,\mathsf{bool} = \mathsf{Bool}$. The universe can be closed under more constructions if more codes are added to $\mathcal{U}$. This gives a complete definition of a universe of setoids, which is, however, inductive-recursive. Moreover, the kind of recursion involved in this definition is particularly complex, and not obviously reducible to well-understood notions of induction-recursion like the one described in [16]. In any case, we would like to avoid extending the metatheory with any form of induction-recursion in order to keep the metatheory as small and essential as possible.

In the next section we transform our current inductive-recursive definition to one that does not use induction-recursion. The way this is done is inspired by the well-known trick to eliminate induction-recursion described in Section 4.1, but modified in a novel way to account for the presence of **Prop**-valued types. To our knowledge, this is the first time this reduction method is applied to an inductive-recursive type of this kind.

4.3 Inductive-inductive setoid universe

We will follow the method outlined in Section 4.1. In addition to inU for defining U, we also introduce a family $\mathsf{inU}{\sim}$ of binary relations between types in the universe, from which we then define $- \sim_{\mathcal{U}} -$.

$$
\begin{array}{l}
\mathsf{data}\ \mathsf{inU} : \mathbf{Type} \to \mathbf{Type}_1 \\
\quad \mathsf{bool} : \mathsf{inU}\ \mathbb{2} \\
\quad \pi : \mathsf{inU}{\sim}\ a\ a\ A_{\sim} \to (\forall\{x_0\ x_1\}(x_{01} : A_{\sim}\ x_0\ x_1) \to \mathsf{inU}{\sim}\ (b\ x_0)\ (b\ x_1)\ (B_{\sim}\ x_{01})) \\
\qquad \to \mathsf{inU}\ (\Sigma\ (f : (x : A) \to B\ x) \\
\qquad\qquad\qquad ((x_0\ x_1 : A)(x_{01} : A_{\sim}\ x_0\ x_1) \to B_{\sim}\ x_{01}\ (f\ x_0)\ (f\ x_1)))
\end{array}
$$

$$
\begin{array}{l}
\mathsf{data}\ \mathsf{inU}{\sim} : \{A\ A' : \mathbf{Type}\} \to \mathsf{inU}\ A \to \mathsf{inU}\ A' \to (A \to A' \to \mathbf{Prop}) \to \mathbf{Type}_1 \\
\quad \mathsf{bool}_{\sim} : \mathsf{inU}{\sim}\ \mathsf{bool}\ \mathsf{bool}\ (\lambda x_0\ x_1\ .\ x_0 \stackrel{?}{=}_{\mathbb{2}} x_1) \\
\quad \pi_{\sim} : \ \{b_0 : (x_0 : A_0) \to \mathsf{inU}\ (B_0\ x_0)\}\{b_1 : (x_1 : A_1) \to \mathsf{inU}\ (B_1\ x_1)\} \\
\qquad\quad \{a_{0\sim} : \mathsf{inU}{\sim}\ a_0\ a_0\ A_{0\sim}\}\{a_{1\sim} : \mathsf{inU}{\sim}\ a_1\ a_1\ A_{1\sim}\} \\
\qquad\quad \{b_{0\sim} : \forall\{x_0\ x_1\}(x_{01} : A_{0\sim}\ x_0\ x_1) \to \mathsf{inU}{\sim}\ (b_0\ x_0)\ (b_0\ x_1)\ (B_{0\sim}\ x_{01})\} \\
\qquad\quad \{b_{1\sim} : \forall\{x_0\ x_1\}(x_{01} : A_{1\sim}\ x_0\ x_1) \to \mathsf{inU}{\sim}\ (b_1\ x_0)\ (b_1\ x_1)\ (B_{1\sim}\ x_{01})\} \\
\qquad \to \mathsf{inU}{\sim}\ a_0\ a_1\ A_{01\sim} \\
\qquad \to (\forall\{x_0\ x_1\}(x_{01} : A_{01\sim}\ x_0\ x_1) \to \mathsf{inU}{\sim}\ (b_0\ x_0)\ (b_1\ x_1)\ (B_{01\sim}\ x_{01})) \\
\qquad \to \mathsf{inU}{\sim}\ (\pi\ a_0\ a_{0\sim}\ b_0\ b_{0\sim})\ (\pi\ a_1\ a_{1\sim}\ b_1\ b_{1\sim}) \\
\qquad\qquad\qquad (\lambda f_0\ f_1\ .\ \forall (x_0\ x_1) \to A_{01\sim}\ x_0\ x_1 \to B_{01\sim}\ x_{01}\ (\pi_1\ f_0\ x_0)\ (\pi_1\ f_1\ x_1))
\end{array}
$$

Just as the role of inU is, as before, to classify all types that are image of El, in the same way $\mathsf{inU}{\sim}\ a\ a'$ classifies all relations of type $A \to A' \to \mathbf{Prop}$ that are image of $- \vdash - \sim_{\mathsf{El}} -$, given proofs $a : \mathsf{inU}\ A, a' : \mathsf{inU}\ A'$. In particular, this definition of $\mathsf{inU}{\sim}$ states that the appropriate equivalence for boolean elements is the obvious syntactic equality $- \stackrel{?}{=}_{\mathbb{2}} -$, whereas functions are to be compared pointwise. Note that inU appears in the sort of $\mathsf{inU}\!\sim$. Since these types are mutually defined, they form an instance of *induction-induction*, a schema that allows the definition of a type mutually with other types that contain the first one in their signature [25].[9]

As in the universe example in Section 4.1, we now define $\mathcal{U}$ as a Σ type, and El as the corresponding first projection.

$$\mathcal{U} : \mathbf{Type}_1 \qquad \mathsf{El} : \mathcal{U} \to \mathbf{Type}$$
$$\mathcal{U} :\equiv \Sigma\ (X : \mathbf{Type})\ (\mathsf{inU}\ X) \qquad \mathsf{El} :\equiv \pi_1$$

What is left now is to define the setoid equality relation on the universe, as well as the setoid equality relation on $\mathsf{El}\ A$ for any A in $\mathcal{U}$. Two codes A, B in the universe $\mathcal{U}$ are equal when there exists a setoid equivalence relation on their respective sets $\mathsf{El}\ A$ and $\mathsf{El}\ B$. Intuitively, since elements of a setoid are only ever compared to elements of the same setoid, this should only be possible if A and B are codes for the same setoid, that is, if $A \sim_{\mathcal{U}} B$. Existence and well-formedness of such relations is expressed via the type $\mathsf{inU}\!\sim$ just defined, hence we would expect $A \sim_{\mathcal{U}} B$ to be defined as follows:

$$(A, a) \sim_{\mathcal{U}} (B, b) :\equiv \Sigma\ (R : A \to B \to \mathbf{Prop})\ (\mathsf{inU}\!\sim\ a\ b\ R)$$

Unfortunately this definition only manages to capture the idea, but does not actually typecheck. In fact, $- \sim_{\mathcal{U}} -$ should be a $\mathbf{Prop}_1$-valued relation, so $A \sim_{\mathcal{U}} B$ should be a proposition. However, the Σ type shown above clearly is not, since it quantifies over a type of relations, which is not a proposition. One possible solution is actually quite simple, and it just involves truncating the Σ type above to force it to be in $\mathbf{Prop}_1$.

$$- \sim_{\mathcal{U}} - : \mathcal{U} \to \mathcal{U} \to \mathbf{Prop}_1$$
$$(A, a) \sim_{\mathcal{U}} (B, b) :\equiv \|\Sigma\ (R : A \to B \to \mathbf{Prop})\ (\mathsf{inU}\!\sim\ a\ b\ R)\|$$

We are now left to define the indexed equivalence relation on El:

$$- \vdash - \sim_{\mathsf{El}} - : \{A\ B : \mathsf{U}\} \to A\ \sim_{\mathcal{U}}\ B \to \mathsf{El}\ A \to \mathsf{El}\ B \to \mathbf{Prop}$$
$$p \vdash x \sim_{\mathsf{El}} y :\equiv\ ?$$

In the definition above, p has type $\|\Sigma\ (R : \mathsf{El}\, A \to \mathsf{El}\, B \to \mathbf{Prop})\ (\ldots)\|$. If the type was not propositionally truncated, we could define $p \vdash x \sim_{\mathsf{El}} y$ by extracting the relation out of the first component of p, and apply it to x, y. That is, $p \vdash x \sim_{\mathsf{El}} y :\equiv \pi_1\, p\, x\, y$. This would make the definition of $- \sim_{\mathcal{U}} -$ and $- \vdash - \sim_{\mathsf{El}} -$ in line with how we defined $\mathcal{U}$ and El.

However, this does not work in our case, since the type of p *is* propositionally truncated, hence it cannot be eliminated to construct a proof-relevant object. Fortunately, we can work around this limitation by defining $p \vdash x \sim_{\mathsf{El}} y$ by induction on the codes $A\ B : \mathcal{U}$, in a way that ends up being logically equivalent to the proposition we would have obtained by $\pi_1\ p\ x\ y$ if there were no truncation.

[9] The main example of induction-induction is the intrinsic definition of a dependent type theory in type theory [6].

More precisely, we need to construct proofs that for any concrete R and inR, the types $|(R, \mathsf{inR})| \vdash x \sim_{\mathsf{El}} y$ and $R\ x\ y$ are logically equivalent. These in turn need to be defined mutually with $- \vdash - \sim_{\mathsf{El}} -$. We direct the interested reader to the Agda formalization for the full details of these definitions, as they are quite involved.

The full definition of the universe is concluded with the remaining definitions, like $\mathsf{refl}_{\mathcal{U}}$, $\mathsf{refl}_{\mathsf{El}}$, etc., which can be adapted from their IR counterparts more or less straightforwardly. The final result does not use induction-recursion, but it is nevertheless an instance of infinitary induction-induction. The ability to define arbitrary, infinitary inductive-inductive types clashes, again, with our objective of keeping the metatheory as small and simple as possible. The next step is therefore to reduce this inductive-inductive universe to one that does not require (infinitary) induction-induction.

4.4 Inductive setoid universe

This section encodes the inductive-inductive universe of setoids from the previous section without assuming arbitrary inductive-inductive definitions in the metatheory.

Before turning our attention to the setoid universe, we recall the known, systematic method to reduce finitary inductive-inductive types to inductive families.

Reducing finitary induction-induction It is known that finitary inductive-inductive definitions can be reduced to inductive families [8,7,21]. To illustrate the idea, let us consider a well-known example of a finitary inductive-inductive type, the intrinsic encoding of type theory in type theory itself. Actually, we only consider the type of contexts $\mathsf{Con} : \mathbf{Type}$ and the type of types $\mathsf{Ty} : \mathsf{Con} \to \mathbf{Type}$; since the latter is indexed over the former, this is already an example of induction-induction.

Contexts in Con are formed out of empty contexts $\bullet$ and context extension $-,-$. Types in Ty are either the base type ι or Π types.

$$\bullet : \mathsf{Con} \qquad\qquad \iota : (\Gamma : \mathsf{Con}) \to \mathsf{Ty}\ \Gamma$$
$$-,- : (\Gamma : \mathsf{Con}) \to \mathsf{Ty}\ \Gamma \to \mathsf{Con} \qquad\qquad \Pi : \{\Gamma : \mathsf{Con}\}(A : \mathsf{Ty}\ \Gamma) \to \mathsf{Ty}\ (\Gamma, A) \to \mathsf{Ty}\ \Gamma$$

The general method to eliminate induction-induction is to split the original inductive-inductive types into a type of codes and associated well-formedness predicates. In our Con/Ty example, these would be respectively given by codes $\mathsf{Con}_0, \mathsf{Ty}_0 : \mathbf{Type}$ and predicates $\mathsf{Con}_1 : \mathsf{Con}_0 \to \mathbf{Type}, \mathsf{Ty}_1 : \mathsf{Con}_0 \to \mathsf{Ty}_0 \to \mathbf{Type}$.

The definition of the codes and predicate types follows that of the original inductive-inductive type, and can be derived systematically from it. More importantly, they can be defined without induction-induction, since although Con_0 and Ty_0 are defined mutually, their sorts are not indexed.

$$
\begin{aligned}
\bullet_0 &: \mathsf{Con}_0 \\
-,_0- &: \mathsf{Con}_0 \to \mathsf{Ty}_0 \to \mathsf{Con}_0 \\
\iota_0 &: \mathsf{Con}_0 \to \mathsf{Ty}_0 \\
\Pi_0 &: \mathsf{Con}_0 \to \mathsf{Ty}_0 \to \mathsf{Ty}_0 \to \mathsf{Ty}_0
\end{aligned}
\qquad
\begin{aligned}
\bullet_1 &: \mathsf{Con}_1\ \bullet_0 \\
-,_1- &: \forall\{\Gamma_0\ A_0\} \to \mathsf{Con}_1\ \Gamma_0 \to \mathsf{Ty}_1\ \Gamma_0\ A_0 \\
&\quad \to \mathsf{Con}_1\ (\Gamma_0 ,_0 A_0) \\
\iota_1 &: \forall\{\Gamma_0\} \to \mathsf{Con}_1\ \Gamma_0 \to \mathsf{Ty}_1\ \Gamma_0\ (\iota_0\ \Gamma_0) \\
\Pi_1 &: \forall\{\Gamma_0\ A_0\ B_0\} \to \mathsf{Con}_1\ \Gamma_0 \\
&\quad \to \mathsf{Ty}_1\ \Gamma_0\ A_0 \to \mathsf{Ty}_1\ (\Gamma_0 ,_0 A_0)\ B_0 \\
&\quad \to \mathsf{Ty}_1\ \Gamma_0\ (\Pi_0\ \Gamma_0\ A_0\ B_0)
\end{aligned}
$$

We can recover the original inductive-inductive type as $\mathsf{Con} :\equiv \Sigma\,(\Gamma_0 : \mathsf{Con}_0)\,(\mathsf{Con}_1\ \Gamma_0)$ and $\mathsf{Ty}\,\Gamma :\equiv \Sigma\,(A_0 : \mathsf{Ty}_0)\,(\mathsf{Ty}_1\ (\pi_1\ \Gamma)\ A_0)$. Recovering the constructors is straightforward:

$$
\begin{aligned}
&\bullet && :\equiv (\bullet_0, \bullet_1) \\
&(\Gamma_0, \Gamma_1), (A_0, A_1) && :\equiv ((\Gamma_0 ,_0 A_0), (\Gamma_1 ,_1 A_1)) \\
&\iota\ (\Gamma_0, \Gamma_1) && :\equiv (\iota_0\ \Gamma_0, \iota_1\ \Gamma_1) \\
&\Pi\ \{\Gamma_0, \Gamma_1\}(A_0, A_1)(B_0, B_1) && :\equiv (\Pi_0\ \Gamma_0\ A_0\ B_0, \Pi_1\ \Gamma_1\ A_1\ B_1)
\end{aligned}
$$

Finally, we can define eliminators/induction principles for Con and Ty as just defined, by induction on the well-typing predicates.

Following [25], we distinguish two versions of the eliminator: the *simple* and the *general* one. Note that this is orthogonal to the distinction between non-dependent and dependent eliminators, from which we only consider the latter. The motives for the simple eliminator are $C' : \mathsf{Con} \to \mathbf{Type}$, $T' : (\Gamma : \mathsf{Con})(A : \mathsf{Ty}\,\Gamma) \to \mathbf{Type}$ and the eliminators themselves have the following signatures:

$$
\mathsf{elim}'_{\mathsf{Con}} : (\Gamma : \mathsf{Con}) \to C'\ \Gamma \qquad \mathsf{elim}'_{\mathsf{Ty}} : \forall\{\Gamma\}(A : \mathsf{Ty}\ \Gamma) \to T'\ \Gamma\ A
$$

In the case of the general eliminator, the motive for Ty depends on the motive for Con, making the two eliminators *recursive-recursive* functions. For motives $C : \mathsf{Con} \to \mathbf{Type}$ and $T : (\Gamma : \mathsf{Con}) \to \mathsf{Ty}\ \Gamma \to C\ \Gamma \to \mathbf{Type}$ the signatures are:

$$
\mathsf{elim}_{\mathsf{Con}} : (\Gamma : \mathsf{Con}) \to C\ \Gamma \qquad \mathsf{elim}_{\mathsf{Ty}} : \forall\{\Gamma\}(A : \mathsf{Ty}\ \Gamma) \to T\ \Gamma\ A\ (\mathsf{elim}_{\mathsf{Con}}\ \Gamma)
$$

The general eliminators can be derived from our encoding of Con and Ty via untyped codes and well-typing predicates. The way to do it is to first define the graph of the eliminators in the form of inductively-generated relations:

$$
\begin{aligned}
&\mathsf{data\ R\text{-}Con} : (\Gamma : \mathsf{Con}) \to C\ \Gamma \to \mathbf{Type} \\
&\mathsf{data\ R\text{-}Ty}\ \ : \{\Gamma : \mathsf{Con}\}(A : \mathsf{Ty}\ \Gamma)(\gamma : C\ \Gamma) \to T\ \Gamma\ A\ \gamma \to \mathbf{Type}
\end{aligned}
$$

The next step is to prove that these relations are functional, by induction on the untyped codes Con_0 and Ty_0 [21]. From this result, defining the eliminators is immediate.

Reducing the setoid universe The reduction described in the previous section works generically for an arbitrary finitary inductive-inductive type, thus giving a systematic way to reduce finitary inductive-inductive definitions to inductive families. However, it is not clear whether this method extends to *infinitary* induction-induction, of which the setoid universe defined in Section 4.3 is an instance. Of course, the absence of a general reduction method does not mean that we cannot reduce particular concrete instances of infinitary induction-induction, which is exactly what we hope for our universe construction.

The obvious challenge in successfully completing this reduction is to avoid the need for extensionality in the metatheory. In fact, consider the simple infinitary inductive-inductive type obtained from the previous Con/Ty example by replacing the finitary constructor Π with an infinitary one: $\Pi : \{\Gamma : \mathsf{Con}\} \to (\mathbb{N} \to \mathsf{Ty}\ \Gamma) \to \mathsf{Ty}\ \Gamma$. Already with this simple example, we run into problems as soon as we try to define the eliminator. One issue is that the definition of the eliminator relies on a proof that the well-typing predicates $\mathsf{inU}_1, \mathsf{inU}{\sim}_1$ are propositional, that is, any two of their elements are equal. Without further assumptions this proof can only be done by induction, and requires function extensionality since these predicates include higher-order constructors.

One way to get around this is to define the well-typing predicates as **Prop**-valued families, rather than in **Type**:

$$
\begin{aligned}
&\mathsf{data}\ \mathsf{inU}_0\quad : \mathbf{Type} \to \mathbf{Type}_1\\
&\mathsf{data}\ \mathsf{inU}{\sim}_0 : \{A\ A' : \mathbf{Type}\} \to (A \to A' \to \mathbf{Prop}) \to \mathbf{Type}_1\\
&\mathsf{data}\ \mathsf{inU}_1\quad : (A : \mathbf{Type}) \to \mathsf{inU}_0\ A \to \mathbf{Prop}_1\\
&\mathsf{data}\ \mathsf{inU}{\sim}_1 : \{A\ A' : \mathbf{Type}\} \to (R : A \to A' \to \mathbf{Prop}) \to \mathsf{inU}{\sim}_0\ R \to \mathbf{Prop}_1
\end{aligned}
$$

Using **Prop** avoids the issue of proving propositionality altogether, since the predicates are now propositional by definition. However, it introduces a different issue: inU_1 and $\mathsf{inU}{\sim}_1$ give rise to equational constraints on their indices, in the form of proofs of the **Prop**-valued identity type. The definition of the eliminators for inU and $\mathsf{inU}{\sim}$ relies on the ability to transport along these proofs, hence the need to extend our metatheory with a primitive, strong form of transport for Id.[10]

Having **Prop** and a strong transport principle does help to some extent. However, we would still need extensionality to derive the general eliminators for inU and $\mathsf{inU}{\sim}$. In fact, as explained in the previous section, to derive the general recursive-recursive eliminators we need to prove that the corresponding graph relations are functional, which cannot be done without funext.

Luckily, the *simple* elimination principle is sufficient for our purposes: all functions described in Section 4.3 can be defined just using the simple eliminator without recursion-recursion. The simple eliminator itself can be defined by pattern matching on the untyped codes, and does not require extensionality or any extra principles beyond strong transport.

Once the inductive encoding of the inductive-inductive universe is done, the setoid universe can be defined just as in Section 4.3.

[10] Note that this issue cannot be solved by expressing the equational constraints with an identity type in **Type**, since the well-typing predicates force it to necessarily be in **Prop**.

5 Conclusions and further work

We have described the construction of a universe of setoids in the setoid model of type theory; this is given in several steps, first as an inductive-recursive definition, then as an inductive-inductive definition, and finally as an inductive type. Every encoding is obtained from the previous by adapting known data type transformation methods in a novel way that accounts for the peculiarities of our construction. In [5] we present rules for SetTT, clearly these rules need to be extended by the rules for a universe reflecting the semantics presented here.

It is known that finitary IITs can be reduced to inductive types in an extensional setting [21]. In our paper we reduce an infinitary IIT to inductive types in an intensional setting. In the future, we would like to investigate whether this reduction can be generalised to arbitrary infinitary IITs.

In contrast to the inductive-recursive and inductive-inductive versions of the universe, the inductive definition relies on a metatheory with a strong transport rule. As future work, we would like to prove normalization for this metatheory since previous work in this respect [2] seems to suggest that is represents a non-trivial addition.

Another question regards the relationship between SeTT [5] and XTT [28]. Both systems are syntactic representations of the setoid model with similar design choices, like definitional proof-irrelevance. We would like to know whether their respective notions of models are equivalent, that is, if we can obtain an XTT model from a SeTT model, and vice versa. Since XTT universes support universe induction, for one direction we would need to extend our own universe with the same principle (see discussion in Section 3 and the previous paragraph). Thus a related question is whether our encodings of the setoid universe can support universe induction. A further question is whether this mapping of models is functorial.

Groupoids can be regarded as generalized setoids. In the future we would like to design a type theory internalizing the groupoid model of type theory [20], in the same way that SeTT represents a syntax for the setoid model. A further question is whether such "groupoid type theory" can be justified, similarly to SeTT, via a syntactic translation, perhaps with SeTT itself as the target theory.

References

1. Andreas Abel. Extensional normalization in the logical framework with proof irrelevant equality. In Olivier Danvy, editor, *Workshop on Normalization by Evaluation, affiliated to LiCS 2009, Los Angeles, 15 August 2009*, 2009.
2. Andreas Abel and Thierry Coquand. Failure of normalization in impredicative type theory with proof-irrelevant propositional equality, 2019. `arXiv:1911.08174`.
3. Peter Aczel. The type theoretic interpretation of constructive set theory. In Angus Macintyre, Leszek Pacholski, and Jeff Paris, editors, *Logic Colloquium '77*, volume 96 of *Studies in Logic and the Foundations of Mathematics*, pages 55 – 66. Elsevier, 1978. URL: http://www.sciencedirect.com/science/article/pii/S0049237X0871989X, `doi:https://doi.org/10.1016/S0049-237X(08)71989-X`.

4. Thorsten Altenkirch. Extensional equality in intensional type theory. In *Proceedings of the Fourteenth Annual IEEE Symposium on Logic in Computer Science (LICS 1999)*, pages 412–420. IEEE Computer Society Press, July 1999.
5. Thorsten Altenkirch, Simon Boulier, Ambrus Kaposi, and Nicolas Tabareau. Setoid type theory—a syntactic translation. In Graham Hutton, editor, *Mathematics of Program Construction*, pages 155–196, Cham, 2019. Springer International Publishing.
6. Thorsten Altenkirch and Ambrus Kaposi. Type theory in type theory using quotient inductive types. *SIGPLAN Not.*, 51(1):18–29, January 2016. URL: https://doi.org/10.1145/2914770.2837638, doi:10.1145/2914770.2837638.
7. Thorsten Altenkirch, Ambrus Kaposi, András Kovács, and Jakob von Raumer. Reducing inductive-inductive types to indexed inductive types. In José Espírito Santo and Luís Pinto, editors, *24th International Conference on Types for Proofs and Programs, TYPES 2018*. University of Minho, 2018.
8. Thorsten Altenkirch, Ambrus Kaposi, András Kovács, and Jakob von Raumer. Constructing inductive-inductive types via type erasure. In Marc Bezem, editor, *25th International Conference on Types for Proofs and Programs, TYPES 2019*. Centre for Advanced Study at the Norwegian Academy of Science and Letters, 2019.
9. Thorsten Altenkirch, Conor McBride, and Wouter Swierstra. Observational equality, now! In *PLPV '07: Proceedings of the 2007 workshop on Programming languages meets program verification*, pages 57–68, New York, NY, USA, 2007. ACM. doi:http://doi.acm.org/10.1145/1292597.1292608.
10. Thorsten Altenkrich, Simon Boulier, Ambrus Kaposi, Christian Sattler, and Filippo Sestini. Agda formalization of the setoid universe. https://bitbucket.org/taltenkirch/setoid-univ, 2021.
11. Marc Bezem, Thierry Coquand, and Simon Huber. A model of type theory in cubical sets. In Ralph Matthes and Aleksy Schubert, editors, *19th International Conference on Types for Proofs and Programs (TYPES 2013)*, volume 26 of *Leibniz International Proceedings in Informatics (LIPIcs)*, pages 107–128, Dagstuhl, Germany, 2014. Schloss Dagstuhl–Leibniz-Zentrum fuer Informatik. URL: http://drops.dagstuhl.de/opus/volltexte/2014/4628, doi:10.4230/LIPIcs.TYPES.2013.107.
12. Simon Boulier. *Extending Type Theory with Syntactical Models.* PhD thesis, IMT Atlantique, 2018.
13. Cyril Cohen, Thierry Coquand, Simon Huber, and Anders Mörtberg. Cubical type theory: A constructive interpretation of the univalence axiom. In Tarmo Uustalu, editor, *21st International Conference on Types for Proofs and Programs (TYPES 2015)*, volume 69 of *Leibniz International Proceedings in Informatics (LIPIcs)*, pages 5:1–5:34, Dagstuhl, Germany, 2018. Schloss Dagstuhl–Leibniz-Zentrum fuer Informatik. URL: http://drops.dagstuhl.de/opus/volltexte/2018/8475, doi:10.4230/LIPIcs.TYPES.2015.5.
14. Peter Dybjer. Internal type theory. In *International Workshop on Types for Proofs and Programs*, pages 120–134. Springer, 1995.
15. Peter Dybjer. A general formulation of simultaneous inductive-recursive definitions in type theory. *Journal of Symbolic Logic*, 65, 06 2003. doi:10.2307/2586554.
16. Peter Dybjer and Anton Setzer. A finite axiomatization of inductive-recursive definitions. In Jean-Yves Girard, editor, *Typed Lambda Calculi and Applications*, pages 129–146, Berlin, Heidelberg, 1999. Springer Berlin Heidelberg.
17. Gaëtan Gilbert, Jesper Cockx, Matthieu Sozeau, and Nicolas Tabareau. Definitional proof-irrelevance without K. *Proceedings of the ACM on Programming*

Languages, pages 1–28, January 2019. URL: https://hal.inria.fr/hal-01859964, doi:10.1145/329031610.1145/3290316.
18. Martin Hofmann. *Extensional concepts in intensional type theory.* PhD thesis, University of Edinburgh, 1995.
19. Martin Hofmann. Conservativity of equality reflection over intensional type theory. In Stefano Berardi and Mario Coppo, editors, *Types for Proofs and Programs*, pages 153–164, Berlin, Heidelberg, 1996. Springer Berlin Heidelberg.
20. Martin Hofmann and Thomas Streicher. The groupoid interpretation of type theory. In *Twenty-five years of constructive type theory (Venice, 1995)*, volume 36 of *Oxford Logic Guides*, pages 83–111. Oxford Univ. Press, New York, 1998.
21. Ambrus Kaposi, András Kovács, and Ambroise Lafont. For finitary induction-induction, induction is enough. In Marc Bezem and Assia Mahboubi, editors, *25th International Conference on Types for Proofs and Programs (TYPES 2019)*, volume 175 of *Leibniz International Proceedings in Informatics (LIPIcs)*, pages 6:1–6:30, Dagstuhl, Germany, 2020. Schloss Dagstuhl–Leibniz-Zentrum für Informatik. URL: https://drops.dagstuhl.de/opus/volltexte/2020/13070, doi:10.4230/LIPIcs.TYPES.2019.6.
22. Lorenzo Malatesta, Thorsten Altenkirch, Neil Ghani, Peter Hancock, and Conor McBride. Small induction recursion, indexed containers and dependent polynomials are equivalent, 2013. TLCA 2013.
23. Per Martin-Löf. An intuitionistic theory of types: Predicative part. In H.E. Rose and J.C. Shepherdson, editors, *Logic Colloquium '73*, volume 80 of *Studies in Logic and the Foundations of Mathematics*, pages 73 – 118. Elsevier, 1975. URL: http://www.sciencedirect.com/science/article/pii/S0049237X08719451, doi:https://doi.org/10.1016/S0049-237X(08)71945-1.
24. Per Martin-Löf. *Intuitionistic type theory*, volume 1 of *Studies in proof theory.* Bibliopolis, 1984.
25. Fredrik Nordvall Forsberg. *Inductive-inductive definitions.* PhD thesis, Swansea University, 2013.
26. Erik Palmgren. From type theory to setoids and back. *arXiv e-prints*, page arXiv:1909.01414, September 2019. arXiv:1909.01414.
27. Erik Palmgren and Olov Wilander. Constructing categories and setoids of setoids in type theory. *arXiv e-prints*, page arXiv:1408.1364, August 2014. arXiv:1408.1364.
28. Jonathan Sterling, Carlo Angiuli, and Daniel Gratzer. Cubical syntax for reflection-free extensional equality. In Herman Geuvers, editor, *Proceedings of the 4th International Conference on Formal Structures for Computation and Deduction (FSCD 2019)*, volume 131 of *Leibniz International Proceedings in Informatics (LIPIcs)*, pages 31:1–31:25. Schloss Dagstuhl–Leibniz-Zentrum fuer Informatik, 2019. URL: http://drops.dagstuhl.de/opus/volltexte/2019/10538, arXiv:1904.08562, doi:10.4230/LIPIcs.FSCD.2019.31.
29. The Univalent Foundations Program. *Homotopy Type Theory: Univalent Foundations of Mathematics.* https://homotopytypetheory.org/book, Institute for Advanced Study, 2013.

Nominal Equational Problems*

Mauricio Ayala-Rincón[1], Maribel Fernández[2], Daniele Nantes-Sobrinho[1], and Deivid Vale[3]

[1] Departments of Computer Science and Mathematics, Universidade de Brasília, Brasília D.F., Brazil
{ayala, dnantes}@unb.br
[2] Department of Informatics, King's College London, London, UK
maribel.fernandez@kcl.ac.uk
[3] Department of Software Science, Radboud University Nijmegen, Nijmegen, The Netherlands
deividvale@cs.ru.nl

Abstract. We define *nominal equational problems* of the form $\exists\overline{W}\forall\overline{Y} : P$, where P consists of conjunctions and disjunctions of equations $s \approx_\alpha t$, freshness constraints $a\#t$ and their negations: $s \not\approx_\alpha t$ and $a\not\#t$, where a is an atom and s, t nominal terms. We give a general definition of solution and a set of simplification rules to compute solutions in the nominal ground term algebra. For the latter, we define notions of solved form from which solutions can be easily extracted and show that the simplification rules are sound, preserving, and complete. With a particular strategy for rule application, the simplification process terminates and thus specifies an algorithm to solve nominal equational problems. These results generalise previous results obtained by Comon and Lescanne for first-order languages to languages with binding operators. In particular, we show that the problem of deciding the validity of a first-order equational formula in a language with binding operators (i.e., validity modulo α-equality) is decidable.

Keywords: Nominal syntax · Unification · Disunification.

1 Introduction

Nominal unification [23] is the problem of solving equations modulo α-equivalence. A solution consists of a substitution and a freshness context ∇, i.e., a set of primitive constraints of the form $a\#X$ (read: "a is fresh for X"), which intuitively means that a cannot occur free in the instances of X. Nominal unification is decidable and unitary [23], and efficient algorithms exist [5,17], which can be used to solve problems of the form $\exists\overline{X}\,(\bigwedge \Delta_i \vdash s_i \approx_\alpha t_i)$, where s_i, t_i are nominal terms with variables $\overline{X}$ and Δ_i is a freshness context.

* First author partially founded by PrInt MAT-UnB-CAPES and CNPq grant numbers Ed 41/2017 and 07672/2017-4. Third author partially supported by DPI/UnB - 03/2020. Fourth author supported by NWO TOP project "Implicit Complexity through Higher Order Rewriting" (ICHOR), NWO 612.001.803/7571.

S. Kiefer and C. Tasson (Eds.): FOSSACS 2021, LNCS 12650, pp. 22–41, 2021.
https://doi.org/10.1007/978-3-030-71995-1_2

Similarly, nominal disunification is the problem of solving disequations i.e., negated equations of the form $s \not\approx_\alpha t$. An algorithm to solve *nominal constraint problems* of the form

$$\mathcal{P} := \exists \overline{X}\left(\left(\bigwedge \Delta_i \vdash s_i \approx_\alpha t_i\right) \wedge \left(\bigwedge \nabla_j \vdash p_j \not\approx_\alpha q_j\right)\right)$$

is available [1], which finds solutions in the nominal term algebra $\mathcal{T}(\Sigma, \mathbb{A}, \mathbb{X})$ by constructing suitable representation of the witnesses for the variables in $\mathcal{P}$.

Comon and Lescanne [10] investigated a more general version of this problem, called *equational problem*, in their words: "an equational problem is any first-order formula whose only predicate symbol is =", that is, it has the form $\exists w_1, \ldots, w_n \forall y_1, \ldots, y_m : P$ where P is a *system*, i.e., an equation $s = t$, or a disequation $s \neq t$, or a disjunction of systems $\bigvee P_i$, or a conjunction of systems $\bigwedge P_i$, or a failure $\perp$, or success $\top$. The study of such problems was motivated by applications in pattern-matching for functional languages, sufficient completeness for term rewriting systems, negation in logic programming languages, etc.

In order to extend these applications to languages that offer support for binders and α-equivalence following the nominal approach, such as αProlog [6], αKanren [4], αLeanTAP [20], to nominal rewriting [14] and nominal (universal) algebra [15], in this paper we consider *nominal equational problems.*

Based on Comon and Lescanne's work, the nominal extension of a first-order equational problem is a formula $\mathcal{P} ::= \exists W_1 \ldots W_n \forall Y_1 \ldots Y_m : P$ where P is a *nominal system*, i.e., a formula consisting of conjunctions and disjunctions of freshness, equality constraints, and their negations.

Contributions. This paper introduces nominal equational problems (NEPs) and presents simplification rules to find solutions in the ground nominal algebra. The simplification rules are shown to be terminating (by using a measure that strictly decreases with each rule application), and also sound and solution-preserving. The simplification process for NEPs is more challenging than in the syntactic case because it deals with two predicates ($\approx_\alpha$ and $\#$) and needs to consider the interaction between freshness and α-equality constraints, and quantifiers. The elimination of universal quantifiers requires careful analysis since universal variables may occur in freshness constraints and in their negations. To make the process more manageable, we define a set of rules together with a strategy of application (specified by rule conditions) that simplifies the termination proof.

Finally, we show that the irreducible forms are either $\perp$ or problems from which a solution can be easily extracted. In particular, if the NEP consists only of existentially quantified conjunctions of freshness and α-equality constraints, we obtain solved forms consisting of a substitution and a freshness context, as in the standard nominal unification algorithm [23].

Related Work. Comon and Lescanne [10] introduced first-order equational problems and studied their solutions in the algebra of rational trees, the initial term algebra, and the ground term algebra. A restricted version of equational problems, called disunification problems, which do not contain quantified variables, has been extensively studied in the first-order framework [8,3,11,2,22]. More recently, a nominal approach to disunification problems was proposed by Ayala et.al [1], including only conjunctions of equations and disequations and freshness constraints, without quantified variables. Here we generalise this previous work to deal with general formulas including disjunction, conjunction and negation of equations and freshness constraints, as well as existential and universal quantification over variables. To deal with negation of freshness, disjunctive formulas, and quantification we extend the semantic interpretation and design a different set of simplification rules as well as a more elaborated strategy for rule application.

Extensions of first-order equational problems modulo equational theories have also been considered. Although the problem of solving disequations modulo an equational theory is not even semi-decidable in general (as shown by Comon [7]), there are useful decidable and semi-decidable cases. For example, solvability of complement problems (a sub-class of equational problems) is decidable modulo theories with permutative operators (which include commutative theories) [9,13], and for linear complement problems solvability modulo associativity and commutativity is also decidable [16,19,12]. Buntine and Bürckert [3] solve systems of equations and disequations in equational theories with a finitary unification type. Fernández [11] shows that E-disunification is semi-decidable when the theory E is presented by a ground convergent rewrite system, and gives a sound and complete E-disunification procedure based on narrowing. Baader and Schulz [2] show that solvability of disunification problems in the free algebra of the combined theory $E_1 \cup \ldots \cup E_n$ is decidable if solvability of disunification problems with linear constant restrictions in the free algebras of the theories $E_i (1 \leq i \leq n)$ is decidable. Lugiez [18] introduces higher-order disunification problems and gives some decidable cases for which equational problems can be extended to higher-order systems.

Organisation. Section 2 recalls the main concepts of nominal syntax and semantics. Section 3 introduces nominal equational problems and a notion of solution for such problems. Section 4 presents a rule-based procedure for solving `NEP`s, as well as soundness, preservation of solutions, and termination results. Section 5 shows that the simplification rules reach solved forms from which solutions can be easily extracted. Section 6 concludes and discusses future work.

2 Background

We assume the reader is familiar with nominal techniques and recall some concepts and notations that shall be used in the paper; for more details, see [14,21,23].

Nominal Terms. We fix countable infinite pairwise disjoint sets of *atoms* $\mathbb{A} = \{a, b, c, \ldots\}$ and *variables* $\mathbb{X} = \{X, Y, Z, \ldots\}$. Atoms follow the *permutative convention*: names a, b range permutatively over $\mathbb{A}$. Therefore, they represent different objects. Let Σ be a finite set of term-formers disjoint from $\mathbb{A}$ and $\mathbb{X}$ such that for each $f \in \Sigma$, a unique non-negative integer n (the arity of f, written as $f : n$) is assigned. We assume there is at least one $f : n$ such that $n > 0$.

A *permutation* π is a bijection $\mathbb{A} \to \mathbb{A}$ with finite domain, i.e., the set $\mathtt{dom}(\pi) := \{a \in \mathbb{A} \mid \pi(a) \neq a\}$ is finite. We shall represent permutations as lists of *swappings* $\pi = (a_1\ b_1)(a_2\ b_2)\ldots(a_n\ b_n)$. The identity permutation is denoted by $\mathtt{id}$ and $\pi \circ \pi'$ the composition of π and π'. The set $\mathbb{P}$ of all such permutations together with the composition operation form a group $(\mathbb{P}, \circ)$ and it will be denoted simply by $\mathbb{P}$. The *difference set* of π and γ is defined by $\mathtt{ds}(\pi, \gamma) = \{a \in \mathbb{A} \mid \pi(a) \neq \gamma(a)\}$.

Definition 1 (Nominal Terms). *The set $T(\Sigma, \mathbb{A}, \mathbb{X})$ of Nominal Terms, or just terms for short, is inductively defined by the following grammar:*

$$s, t, u ::= a \mid \pi \cdot X \mid [a]t \mid f(t_1, \ldots, t_n),$$

where a is an atom*, $\pi \cdot X$ is a moderated variable, $[a]t$ is the* abstraction *of a in the term t, and $f(t_1, \ldots, t_n)$ is a* function application *with $f \in \Sigma$ and $f : n$. A term is ground if it does not contain variables.*

In an abstraction $[a]t$, t is the scope of the binder $[\cdot]$ and it *binds* all free occurrences of a in t. An occurrence of an atom in a term is *free* if it is not under the scope of a binder. Notice that syntactical equality is not modulo α-equivalence; for example, $[a]a \not\equiv [b]b$. We may denote $s \equiv t$ by $s = t$ with the same intended meaning and $\tilde{t}$ abbreviates an ordered sequence $t_1, \ldots, t_n$ of terms.

Example 1. Let $\Sigma_\lambda := \{\mathtt{lam} : 1, \mathtt{app} : 2\}$ be a signature for the λ-calculus. Using atoms to represent variables, λ-expressions are generated by the grammar:

$$e ::= a \mid \mathtt{lam}([a]e) \mid \mathtt{app}(e, e)$$

As usual, we sugar $\mathtt{app}(s, t)$ to $s\,t$ and $\mathtt{lam}([a]s)$ to $\lambda[a]s$. The following are examples of nominal terms: $(\lambda[a]a)\,X$ and $(\lambda[a](\lambda[b]b\,a)\ c)\,d$.

We inductively extend the action of a permutation π to a term t, denoted as $\pi \cdot t$, by setting: $\pi \cdot a = \pi(a), \pi \cdot (\pi' \cdot X) = (\pi \circ \pi') \cdot X, \pi \cdot ([a]t) = [\pi(a)](\pi \cdot t)$, and $\pi \cdot f(\tilde{t}) = f(\pi \cdot \tilde{t})$.

Substitutions, ranging over $\sigma, \gamma, \tau \ldots$, are maps (with finite domain) from variables to terms. The *action of a substitution* σ on a term t, denoted $t\sigma$, is inductively defined by: $a\sigma = a, (\pi \cdot X)\sigma = \pi \cdot (X\sigma), ([a]t)\sigma = [a](t\sigma)$ and $f(t_1, \ldots, t_n)\sigma = f(t_1\sigma, \ldots, t_n\sigma)$. Notice that $t(\sigma\gamma) = (t\sigma)\gamma$.

Definition 2 (Positions and subterms). *Let s be a nominal term. The set $\mathtt{Pos}(s)$ of positions in s is a set of strings over positive integers defined inductively below. Additionally, $s|_p$ denotes the subterm of s at position p and $s(p)$ denotes the symbol at position p.*

- *If $s = a$ or $s = \pi \cdot X$, then $\mathtt{Pos}(s) = \{\epsilon\}$ and $s|_\epsilon = s$;*
- *if $s = [a]t$ then $\mathtt{Pos}(s) = \{\epsilon\} \cup \{1 \cdot p \mid p \in \mathtt{Pos}(t)\}$, $s|_\epsilon = s$ and $s|_{1 \cdot p} = t|_p$;*
- *if $s = f(s_1, \ldots, s_n)$ then $\mathtt{Pos}(s) = \{\epsilon\} \cup \bigcup_{i=1}^{n}\{i \cdot p \mid p \in \mathtt{Pos}(s_i)\}$, $s|_\epsilon = s$ and $s|_{i \cdot p} = s_i|_p$.*

Freshness and α-equality. A *nominal equation* is the symbol $\top$ or an expression $s \approx_\alpha t$ where s and t are nominal terms. A *trivial equation* is either $s \approx_\alpha s$ or $\top$. *Freshness constraints* have the form $a\#t$ where a is an atom and t a term. A *freshness context* is a finite set of *primitive* freshness constraints of the form $a\#X$, we use Δ, ∇, and Γ to denote them. We extend the notation to sets of atoms: $A\#X$ denotes that $a\#X$ for every $a \in A$.

α-derivability is given by the deduction rules in Figure 1, which define an *equational theory* called CORE.

$$\frac{}{\nabla \vdash a\#b}\ (\#\text{-ax}) \qquad \frac{\pi^{-1}(a)\#X \in \nabla}{\nabla \vdash a\#\pi \cdot X}\ (\#\text{-var}) \qquad \frac{}{\nabla \vdash a\#[a]t}\ (\#\text{-abs-a})$$

$$\frac{\nabla \vdash a\#t}{\nabla \vdash a\#[b]t}\ (\#\text{-abs-b}) \qquad \frac{\nabla \vdash a\#t_1 \quad \cdots \quad \nabla \vdash a\#t_n}{\nabla \vdash a\#f(t_1, \dots t_n)}\ (\#\text{-f})$$

$$\frac{}{\nabla \vdash a \approx_\alpha a}\ (\text{ax}) \qquad \frac{\mathtt{ds}(\pi, \pi')\#X \in \nabla}{\nabla \vdash \pi \cdot X \approx_\alpha \pi' \cdot X}\ (\text{var}) \qquad \frac{\nabla \vdash t \approx_\alpha t'}{\nabla \vdash [a]t \approx_\alpha [a]t'}\ (\text{abs-a})$$

$$\frac{\nabla \vdash t \approx_\alpha (a\,a') \cdot t' \qquad \nabla \vdash a\#t'}{\nabla \vdash [a]t \approx_\alpha [a']t'}\ (\text{abs-b}) \qquad \frac{\nabla \vdash t_1 \approx_\alpha t'_1 \quad \cdots \quad \nabla \vdash t_n \approx_\alpha t'_n}{\nabla \vdash f(t_1, \dots t_n) \approx_\alpha f(t'_1, \dots, t'_n)}\ (\text{f})$$

Fig. 1. CORE freshness and α-equality rules.

- Write $\nabla \vdash a\#t$ when there exists a derivation of $\nabla \vdash a\#t$.
  The judgement $\nabla \vdash a\#t$ intuitively means that using freshness constraints from ∇ as assumptions a does not occur free in t.
- Write $\nabla \vdash s \approx_\alpha t$ when there exists a derivation of $\nabla \vdash s \approx_\alpha t$.
  The judgement $\nabla \vdash s \approx_\alpha t$ intuitively means that using freshness constraints from ∇ as assumptions s is α-equivalent to t.

Semantic Notions. Nominal equational theory has a natural semantic denotation in *nominal sets* since we can easily interpret freshness and abstraction.

A $\mathbb{P}$-set X is an ordinary set equipped with an action in $\mathbb{P} \times X \to X$ (written as $\pi \cdot x$) such that $\mathtt{id} \cdot x = x$ and $\pi \cdot (\pi' \cdot x) = (\pi \circ \pi') \cdot x$. A set of atoms $A \subset \mathbb{A}$ *supports* $x \in X$ iff for all permutations $\pi \in \mathbb{P}$ fixing every element of A $\cdot$ acts trivially on x via π, i.e., if $\pi(a) = a$ for all $a \in A$ then $\pi \cdot x = x$. *Semantic freshness* is defined in terms of support as follows: an atom a is fresh for $x \in X$ iff $a \notin \mathtt{supp}(x)$. We denote this by writing $a\#_{\mathsf{sem}}x$. A *nominal set* is a $\mathbb{P}$-set such that every element is finitely supported.

To build an algebraic ground term-model of CORE, we fix the set G consisting of equivalence classes of provable α-equivalent ground terms. More precisely, given a ground term g, the class $\overline{g}$ is the set of ground terms g' for which there exist a derivation $\vdash g \approx_\alpha g'$. Note that G is a nominal set by defining the natural action: $\pi \cdot \overline{g} = \overline{\pi \cdot g}$. Each function symbol $f \in \Sigma$ is interpreted by an *equivariant function* $f^{\mathcal{I}}$ mapping $(\overline{t_1}, \ldots, \overline{t_n}) \mapsto \overline{f(t_1, \ldots, t_n)}$ and abstractions $[a]t$ are interpreted by an equivariant function $[\_]\_$ in $\mathbb{A} \times G \to G$ such that $a\#_{\mathsf{sem}}[a]g$ always.

Signature interpretation is homomorphically extended to the set of terms as follows: Fix a *valuation function* ς that assigns to every variable $X \in \mathbb{X}$ an element of G. The interpretation of a term t under ς, $[\![t]\!]_\varsigma$, is defined as:

$$[\![a]\!]_\varsigma = \overline{a} \qquad [\![\pi \cdot X]\!]_\varsigma = \pi \cdot \varsigma(X) \qquad [\![[a]t]\!]_\varsigma = [\overline{a}][\![t]\!]_\varsigma$$
$$[\![f(t_1, \ldots, t_n)]\!]_\varsigma = f^{\mathcal{I}}([\![t_1]\!]_\varsigma, \ldots, [\![t_n]\!]_\varsigma)$$

Definition 3 (Validity under ς). *Let $\mathcal{A}$ be any infinite subalgebra of* CORE *with domain A and ς a valuation function assigning for every variable $X \in \mathbb{X}$ an element of A. We say that:*

1. *$[\![a\#t]\!]_\varsigma$ (resp. $[\![t \approx_\alpha u]\!]_\varsigma$) is valid if $a\#_{\mathsf{sem}}\, [\![t]\!]_\varsigma$ (resp. $[\![t]\!]_\varsigma = [\![u]\!]_\varsigma$).*
2. *$[\![\nabla]\!]_\varsigma$ is valid when $a\#_{\mathsf{sem}}\varsigma(X)$ for each $a\#X \in \nabla$.*
3. *$[\![\nabla \vdash a\#t]\!]_\varsigma$ is valid when the validity of $[\![\nabla]\!]_\varsigma$ implies $a\#_{\mathsf{sem}}\, [\![t]\!]_\varsigma$, and*
4. *$[\![\nabla \vdash t \approx_\alpha u]\!]_\varsigma$ is valid when the validity of $[\![\nabla]\!]_\varsigma$ implies $[\![t]\!]_\varsigma = [\![u]\!]_\varsigma$.*

Write $\nabla \models s \approx_\alpha t$ (resp. $\nabla \models a\#t$) when $[\![\nabla \vdash s \approx_\alpha t]\!]_\varsigma$ (resp. $[\![\nabla \vdash a\#t]\!]_\varsigma$) is valid for any valuation ς.

A model of a nominal theory is an interpretation that validates all of its axiomatic judgements $\nabla \vdash s \approx_\alpha t$. It is easy to see that the interpretation we define above is a model of CORE. For the rest of the paper, we slightly abuse notation by calling CORE both the theory and its model making distinctions when necessary.

Remark 1. It is worth noticing the *syntactic* character of CORE: by interpreting atoms as themselves and since there are no equational axioms, we easily connect $\nabla \models a\#t$ and $\nabla \vdash a\#t$. This behaviour is not the rule if equational axioms are considered. For instance, consider the theory LAM that axiomatises β-equality in the λ-calculus. It is a fact that $a\#_{\mathsf{sem}}\overline{(\lambda[a]b)a}$ in LAM but there is no syntactic derivation for $a\#(\lambda[a]b)a$. Furthermore, by completeness for equality derivation, we establish a connection between $\nabla \models s \approx_\alpha t$ and $\nabla \vdash s \approx_\alpha t$.

There are alternative definitions of nominal terms where the syntax is many-sorted. We chose to work with an unsorted syntax for simplicity; all the results below can be extended to the many-sorted case, indeed they are proved for any infinite subalgebra of the ground nominal algebra.

3 Nominal Equational Problems

In this section, we introduce *nominal equational problems* (NEPs) as our main object of study. A NEP is a fist-order formula built only with the predicates $\approx_\alpha$

and $\#$. Their negations, denoted $\not\approx_\alpha$ and $\not\#$, are used to build disequations and non-freshness constraints. A *trivial disequation* is either $s \not\approx_\alpha s$ or $\bot$.

Intuitively, a non-freshness constraint $a\not\#t$ — read *a is not fresh for t* — states that there exists at least one instance of t where a occurs free. Similarly, for disequations: $s \not\approx_\alpha t$ states that s and t are not α-equivalent.

Definition 4. *A* nominal system *is a formula defined by the following grammar:*

$$P, P' ::= \top \mid \bot \mid s \approx_\alpha t \mid s \not\approx_\alpha t \mid a\#t \mid a\not\#t \mid P \wedge P' \mid P \vee P'$$

In the next definition, we make a distinction between the set of variables occurring in a NEP: the mutually disjoint sets $\overline{W} = \{W_1, \dots, W_n\}$ and $\overline{Y} = \{Y_1, \dots, Y_m\}$ denote existentially and universally quantified variables, respectively. The former we call *auxiliary variables* and the latter *parameters.*

Definition 5 (NEP). *A* NEP *is a formula of the form below, where P is a nominal system.*

$$\mathcal{P} ::= \exists W_1 \dots W_n \forall Y_1 \dots Y_m : P$$

The set $\mathtt{Fv}(\mathcal{P})$ contains the free variables occurring in $\mathcal{P}$. For the rest of the paper, we use the following implicit naming scheme for variables: W denotes an auxiliary variable, Y a parameter, X a free variable, and Z an arbitrary variable.

Example 2. Nominal disunification constraints [1] are pairs of the form $\mathcal{P} := \exists \overline{W} \langle E \mid\mid D \rangle$, where E is a finite set of nominal equations-in-context, i.e., $E = \bigcup_{i=0}^{n} \{\Delta_i \vdash s_i \approx_\alpha t_i\}$ and D is a finite set of nominal disequations-in-context, $D = \bigcup_{j=0}^{m} \{\nabla_j \vdash u_j \not\approx_\alpha v_j\}$. This problem is a particular NEP: taking the judgement $\Delta \vdash s \approx_\alpha t$ as $\Delta \Rightarrow s \approx_\alpha t$, or yet as $\neg\Delta \vee s \approx_\alpha t$[4], we obtain the formula:

$$\mathcal{P} := (\bigwedge_{i=0}^{n} (\neg[\Delta_i] \vee s_i \approx_\alpha t_i)) \wedge (\bigwedge_{j=0}^{m} (\neg[\nabla_j] \vee u_j \not\approx_\alpha v_j)),$$

where $[\Delta_i]$, $[\nabla_j]$ are conjunctions of freshness constraints in Δ_i, ∇_j, respectively.

Sufficient completeness, that is, deciding whether a set of pattern (rules) covers all possible cases, is a well-known problem in functional programming. In the next example, we show how to naturally represent such problems as NEPs.

Example 3. Consider the function map which applies a function $[a]F$ to every element of any list L. It may be defined by the rules below:

$$\mathcal{R}_{\mathsf{map}} = \begin{cases} \vdash \mathsf{map}([a]F, \mathsf{nil}) \rightarrow \mathsf{nil} \\ \vdash \mathsf{map}([a]F, \mathsf{cons}(X, L)) \rightarrow \mathsf{cons}(F\{a \mapsto X\}, \mathsf{map}([a]F, L)), \end{cases}$$

[4] Similarly, for disequations.

where $\_\{a \mapsto \_\}$ is a binary term-former representing (explicit) substitutions; see [14, Example 43] for more details. Since we are not imposing a type discipline on nominal terms it is possible to construct ill-typed terms, for instance $\mathsf{map}(a, [a]t)$. In what follows we ignore those expressions by noticing that a type discipline will not allow such constructions. Then sufficient completeness can be checked using the following NEP:

$$\forall Y_1 Y_2 Y_3 L' : \mathsf{map}([a]F, L) \not\approx_\alpha \mathsf{map}([b]Y_1, \mathsf{nil}) \wedge \\ \mathsf{map}([a]F, L) \not\approx_\alpha \mathsf{map}([b]Y_2, \mathsf{cons}(Y_3, L')),$$

If the problem has a solution then $\mathcal{R}_{\mathsf{map}}$ is not complete, and the solution indicates the missing pattern cases in the definition.

Solutions of Nominal Equational Problems. We are interested in solutions for NEPs in the ground nominal algebra. From now on, $\mathcal{A}$ denotes an infinite subalgebra of CORE with domain A. Below we define solutions using idempotent substitutions, which can be seen as a representation for valuations that map variables to elements of the ground term algebra.

We first extend the interpretation function under a valuation ς $[\![\cdot]\!]_\varsigma$ (see Section 2) to the negated form of freshness and α-equality constraints.

Definition 6. *Let ς be a (fixed but arbitrarily given) valuation. A negative constraint $a \#\!\!\!\!\setminus t$ (resp. $s \not\approx_\alpha t$) is valid under ς when:*

- *it is not the case that $a \#_{\mathsf{sem}} [\![t]\!]_\varsigma$, this is written $[\![a \#\!\!\!\!\setminus t]\!]_\varsigma$; and, respectively,*
- *it is not the case that $[\![s]\!]_\varsigma = [\![t]\!]_\varsigma$, this is written $[\![s \not\approx_\alpha t]\!]_\varsigma$.*

In standard unification algorithms, idempotent substitutions are used as a compact representation of a set of valuations in the ground term algebra. Similarly, given a valuation in the ground term algebra, one can build a ground substitution representing it. In the case of the ground nominal algebra, where elements are α-equivalence classes of terms, the representative is generally not unique, but any representative can be used.

Definition 7. *Given a substitution $\sigma = [X_1/t_1, \ldots, X_n/t_n]$, for any valuation ς, we denote by ς^σ the valuation such that $\varsigma^\sigma(X) = \varsigma(X)$ if $X \notin \mathtt{dom}(\sigma)$, and $\varsigma^\sigma(X) = [\![X\sigma]\!]_\varsigma$ otherwise.*

Given a valuation $\varsigma = [X_i \mapsto g_i \mid X_i \in \mathbb{X}, g_i \in A]$, and a finite set $\mathcal{X}$ of variables, we denote by $\sigma_{\mathcal{X}}^{\varsigma}$ any ground substitution such that for each $X_i \in \mathcal{X}$, $\sigma(X_i) = t_i$, if $g_i = [\![t_i]\!]_\varsigma$. We say that $\sigma_{\mathcal{X}}^{\varsigma}$ is a grounding substitution for $\mathcal{X}$.

The next lemma states that under mild conditions we can extend substitutions to valuations preserving semantic equality.

Lemma 1. *Given an idempotent substitution $\sigma = [X_1/t_1, \ldots, X_n/t_n]$ and a valuation ς we have: $[\![s\sigma]\!]_\varsigma = [\![s]\!]_{\varsigma^\sigma}$.*

The next definition allows us to use idempotent substitutions to represent solutions of constraints.

Definition 8 (Constraint $\mathcal{A}$-validation). *Let σ be an idempotent substitution whose domain includes all the variables occurring in a constraint C. Then σ $\mathcal{A}$-validates C iff $[\![C]\!]_{\varsigma^\sigma}$ is valid in $\mathcal{A}$ for any valuation ς.*

We now extend semantic validity to the syntax of systems. The interpretation for the logical connectives is defined as expected.

Definition 9 ($\mathcal{A}$-validation). *For an idempotent substitution σ whose domain includes all variables occurring in a system P, we say that σ $\mathcal{A}$-*validates *P iff*

1. *$P = \top$; or*
2. *$P = C$ and σ $\mathcal{A}$-validates C; or*
3. *$P = P_1 \wedge \ldots \wedge P_n$ and σ $\mathcal{A}$-validates each P_i, $1 \leq i \leq n$; or*
4. *$P = P_1 \vee \ldots \vee P_m$ and σ $\mathcal{A}$-validates at least one P_i, $1 \leq i \leq m$.*

Solutions of equational problems instantiate free variables and satisfy existential and universal requirements for auxiliary variables and parameters, respectively. To define this notion, we extend the domain of the substitution to include also existential and universally quantified variables as follows.

Definition 10 ($\mathcal{A}$-Solution). *Let $\mathcal{P} = \exists\overline{W}\forall\overline{Y} : P$ be a* NEP*. Let σ be an idempotent substitution such that $\mathtt{dom}(\sigma) = Fv(\mathcal{P})$. Then σ is an $\mathcal{A}$-*solution *of $\mathcal{P}$ iff there is a ground substitution δ, where $\mathtt{dom}(\delta) = \overline{W}$, such that for all ground substitution λ, where $\mathtt{dom}(\lambda) = \overline{Y}$, $\sigma\delta\lambda$ $\mathcal{A}$-validates P. The set of $\mathcal{A}$-solutions of $\mathcal{P}$ is denoted $\mathcal{S}_{\mathcal{A}}(\mathcal{P})$, or simply $\mathcal{S}(\mathcal{P})$ if $\mathcal{A}$ is clear from the context.*

Example 4. Consider the signature $\Sigma_{\mathsf{nat}} := \{\mathsf{zero} : 0, \mathsf{suc} : 1\}$ for natural numbers, and the nominal initial algebra $\mathcal{A}_{\mathsf{nat}}$ with zero and suc interpreted as expected. The problem $\mathcal{P} := \exists W\forall Y : W \not\approx_\alpha \mathsf{suc}(Y)$ has id as solution. Indeed, taking for example $\delta = [W/\mathsf{zero}]$ or $\delta = [W/a]$ and any choice of λ ($\mathtt{dom}(\lambda) = \{Y\}$), the composition $\mathtt{id}\delta\lambda$ $\mathcal{A}$-validates $W \not\approx_\alpha \mathsf{suc}(Y)$.

In Definition 10, δ is the substitution that instantiates auxiliary variables, so there can be many (possibly infinite) number of such δ's.

Lemma 2 (Equivariance of Solutions). *If σ is an $\mathcal{A}$-*solution *of the* NEP *$\mathcal{P}$ then for any permutation π, $\pi \cdot \sigma$ (defined by $[X_i/\pi \cdot t_i]$, as expected) is an $\mathcal{A}$-*solution *of $\pi \cdot \mathcal{P}$. In particular, if an $\mathcal{A}$-*solution *contains an atom not occurring in $\mathcal{P}$, that atom can be swapped for any other atom not occurring in $\mathcal{P}$.*

Lemma 2 is a direct consequence of the fact that interpretations are equivariant, and shows that solutions are closed by permutation. It allows us to use permutations to represent infinite choices for atoms in solutions.

Example 5. Consider the problem $\forall Y : X \not\approx_\alpha \lambda[a]Y$, built over the signature of Example 1. The set of solutions contains $\sigma = [X/a]$ as well as $(a\ b) \cdot [X/a] = [X/b]$; for any other atom b.

Lemma 3 (Closure by Instantiation). *If σ is an $\mathcal{A}$-solution of the* NEP $\mathcal{P} = \exists\overline{W}\forall\overline{Y} : P$ *then any idempotent substitution σ' obtained as an instance of σ such that* $\mathtt{dom}(\sigma') = \mathtt{dom}(\sigma)$ *is also an $\mathcal{A}$-solution of $\mathcal{P}$. In particular, for any such ground instance σ' of σ there is a ground substitution δ, where* $\mathtt{dom}(\delta) = \overline{W}$, *such that for all ground substitution λ, where* $\mathtt{dom}(\lambda) = \overline{Y}$, *$\sigma'\delta\lambda$ $\mathcal{A}$-validates P.*

Proof. By definition of $\mathcal{A}$-solution, to show that σ' is an $\mathcal{A}$-solution of $\mathcal{P}$ we need to consider all the valuations of the form $\varsigma^{\sigma'\delta\lambda}$ as indicated in Definitions 8, 9, 10. The result follows from the fact that for any valuation $\varsigma^{\sigma'\delta\lambda}$ there exists an equivalent valuation $\varsigma'^{\sigma\delta\lambda}$ by Lemma 1.

4 A rule-based procedure

In this section we present a set of simplification rules to solve NEPs. A simplification step, denoted $\mathcal{P} \Longrightarrow \mathcal{P}'$, transforms $\mathcal{P}$ into an equivalent problem $\mathcal{P}'$ from which solutions are easier to extract.

4.1 Simplification Rules

Rules may have application conditions (rule controls) that define a strategy of simplification. Our strategy gives priority to rules according to their role. We split the rules into groups $\mathcal{R}_i$ as shown in Figures 2, 3 and 4: $\mathcal{R}_1$ eliminates trivial constraints, $\mathcal{R}_2$ deals with clash and occurs check, $\mathcal{R}_3$ eliminates unneeded quantifiers, $\mathcal{R}_4$ and $\mathcal{R}_5$ decompose positive and negative constraints, respectively, $\mathcal{R}_6$ eliminates parameters and $\mathcal{R}_7$ instantiates variables. The Explosion and Elimination of Disjunction rules in $\mathcal{R}_8$ search for solutions as explained below. Finally, $\mathcal{R}_9$ eliminates the remaining universal quantifiers. A rule $R \in \mathcal{R}_i$ can only be applied if no rules from $\mathcal{R}_j$, where $j < i$, can be applied.

Since we are dealing with formulas that contain disjunction and conjunction connectives, we need to take into account the standard Boolean axioms. To simplify, instead of working modulo the Boolean axioms we apply a Boolean normalisation step before a rule is applied. Following Comon and Lescanne [10], we choose to take *conjunctive normal form*: Before the application of each rule $\mathcal{P}$ is reduced to a conjunction of disjunctions.

The explosion rule creates new branches by instantiating variables considering all possible ways of constructing terms (i.e., each $f \in \Sigma$, abstractions and atoms). Note that $\Sigma \cup \mathtt{Atoms}(P) \cup \{a'\}$ is a finite set (we can represent all possible constructions with a finite number of cases), so the rule is finitely branching.

The rule Elimination of Disjunctions also builds a finite number of branches. Therefore, our procedure builds a finitely branching tree of problems to be solved.

Rules $\mathcal{R}_1$-$\mathcal{R}_8$ are not sufficient to eliminate all parameters from a NEP (see Example 6) in contrast with the syntactic case [7], where similar rules produce parameterless normal forms. This is because we are dealing with both freshness and α-equality. Indeed, normal forms for rules $\mathcal{R}_1$-$\mathcal{R}_8$ may contain parameters, but only in disjunctions involving both freshness and equality constraints for the same parameter as the following lemma states. The rules in $\mathcal{R}_9$ (Figure 4) are introduced to deal with this problem.

$\mathcal{R}_1$: Trivial Rules

(T_1) $t \approx_\alpha t \Longrightarrow \top$ (T_2) $t \not\approx_\alpha t \Longrightarrow \bot$ (T_3) $a \approx_\alpha b \Longrightarrow \bot$
(T_4) $a\#b \Longrightarrow \top$ (T_5) $a\#a \Longrightarrow \bot$ (T_6) $a\not\#a \Longrightarrow \top$
(T_7) $a\not\#b \Longrightarrow \bot$ (T_8) $a\#t \wedge a\not\#t \Longrightarrow \bot$ (T_9) $a\#t \vee a\not\#t \Longrightarrow \top$

$\mathcal{R}_2$: Clash and Occurrence Check Rules

(CL_1) $s \not\approx_\alpha t \Longrightarrow \top$ (CL_2) $s \approx_\alpha t \Longrightarrow \bot$
Conditions for (CL_1) **and** (CL_2): $s(\epsilon) \neq t(\epsilon)$ and neither is a moderated variable.
(O_1) $\pi \cdot Z \approx_\alpha t \Longrightarrow \bot$ (O_2) $\pi \cdot Z \not\approx_\alpha t \Longrightarrow \top$
Conditions for (O_1) **and** (O_2): $Z \in \mathtt{vars}(t)$ and $t \not\equiv \pi' \cdot Z$

$\mathcal{R}_3$: Elimination of parameters and auxiliary unknowns.

(C_1) $\forall \overline{Y}, Y : P \Longrightarrow \forall \overline{Y} : P,$ $Y \notin \mathtt{vars}(P)$
(C_2) $\exists \overline{W}, W : P \Longrightarrow \exists \overline{W} : P,$ $W \notin \mathtt{vars}(P)$
(C_3) $\exists \overline{W}, W : \pi \cdot W \approx_\alpha t \wedge P \Longrightarrow \exists \overline{W} : P,$ $W \notin \mathtt{vars}(P, t)$

$\mathcal{R}_4$: Equality and freshness simplification

(E_1) $\pi \cdot X \approx_\alpha \gamma \cdot X \Longrightarrow \wedge \mathtt{ds}(\pi, \gamma)\#X$ (F_1) $a\#\pi \cdot X \Longrightarrow \pi^{-1}(a)\#X, \pi \neq \mathtt{id}$
(E_2) $[a]t \approx_\alpha [a]u \Longrightarrow t \approx_\alpha u$ (F_2) $a\#[a]t \Longrightarrow \top$
(E_3) $[a]t \approx_\alpha [b]u \Longrightarrow (b\ a) \cdot t \approx_\alpha u \wedge b\#t$ (F_3) $a\#[b]t \Longrightarrow a\#t$
(E_4) $f(\tilde{t}) \approx_\alpha f(\tilde{u}) \Longrightarrow \wedge_i t_i \approx_\alpha u_i$ (F_4) $a\#f(t_1, \ldots, t_n) \Longrightarrow \wedge_i a\#t_i$

$\mathcal{R}_5$: Disunification

(DC) $f(\tilde{t}) \not\approx_\alpha f(\tilde{u}) \Longrightarrow \vee_i t_i \not\approx_\alpha u_i$ (NF_1) $a\not\#\pi \cdot X \Longrightarrow \pi^{-1}(a)\not\#X, \pi \neq \mathtt{id}$
(D_1) $\pi \cdot X \not\approx_\alpha \gamma \cdot X \Longrightarrow \vee_i \mathtt{ds}(\pi, \gamma)\not\#X$ (NF_2) $a\not\#[a]t \Longrightarrow \bot$
(D_2) $[a]t \not\approx_\alpha [a]u \Longrightarrow t \not\approx_\alpha u$ (NF_3) $a\not\#[b]t \Longrightarrow a\not\#t$
(D_3) $[a]t \not\approx_\alpha [b]u \Longrightarrow (b\ a) \cdot t \not\approx_\alpha u \vee b\not\#t$ (NF_4) $a\not\#f(\tilde{t}) \Longrightarrow \vee_i a\not\#t_i$

$\mathcal{R}_6$: Simplification of Parameters

(U_1) $\forall \overline{Y}, Y : P \wedge \pi \cdot Y \not\approx_\alpha t \Longrightarrow \bot$ if $Y \notin \mathtt{vars}(t)$
(U_2) $\forall \overline{Y} : P \wedge (\pi \cdot Y \not\approx_\alpha t \vee Q) \Longrightarrow \forall \overline{Y} : P \wedge Q[Y/\pi^{-1} \cdot t]$, if $Y \notin \mathtt{vars}(t)$, $Y \in \overline{Y}$
(U_3) $\forall \overline{Y}, Y : P \wedge \pi \cdot Y \approx_\alpha t \Longrightarrow \bot$, if $\pi \cdot Y \not\equiv t$
(U_4) $\forall \overline{Y} : P \wedge (\pi_1 \cdot Z_1 \approx_\alpha t_1 \vee \cdots \vee \pi_n \cdot Z_n \approx_\alpha t_n \vee Q) \Longrightarrow \forall \overline{Y} : P \wedge Q$
(U_5) $\forall \overline{Y}, Y : P \wedge a\#Y \Longrightarrow \bot$
(U_6) $\forall \overline{Y}, Y : P \wedge a\not\#Y \Longrightarrow \bot$
Conditions for (U_4):

- Each equation in the disjunction contains at least one occurrence of a parameter and $\pi_i \cdot Z_i \not\equiv t_i$ for each $i = 1, \ldots, n$.
- Q does not contain any parameter.

$\mathcal{R}_7$: Instantiation Rules

(I_1) $\pi \cdot Z \approx_\alpha t \wedge P \Longrightarrow Z \approx_\alpha \pi^{-1} \cdot t \wedge P[Z/\pi^{-1} \cdot t]$

- If $\pi = \mathtt{id}$ then Z is not a parameter and Z occurs in P and if t is a variable then t occurs in P.
- If $\pi \neq \mathtt{id}$, then t is not of the form $\mathtt{id} \cdot Z'$.

(I_2) $\pi \cdot Z \not\approx_\alpha t \vee P \Longrightarrow Z \not\approx_\alpha \pi^{-1} \cdot t \vee P[Z/\pi^{-1} \cdot t]$

- If $\pi = \mathtt{id}$ then Z occurs in P and if t is a variable then t occurs in P.
- If $\pi \neq \mathtt{id}$ then t is not of the form $\mathtt{id} \cdot Z'$.

Fig. 2. Preserving Rules

$\mathcal{R}_8$: Explosion and Elimination of Disjunction

$(ED_1) \forall \overline{Y} : P \wedge (P_1 \vee P_2) \Longrightarrow \forall \overline{Y} : P \wedge P_1$, if $\mathtt{vars}(P_1) \cap \overline{Y} = \emptyset$ or $\mathtt{vars}(P_2) \cap \overline{Y} = \emptyset$.

(ED_2) $\forall \overline{Y_1}, \overline{Y_2} : P \wedge (P_1 \vee P_2) \Longrightarrow \forall \overline{Y_1}, \overline{Y_2} : P \wedge P_1$, if $\mathtt{vars}(P_1) \cap \overline{Y_2} = \emptyset$ and $\mathtt{vars}(P_2) \cap \overline{Y_1} = \emptyset$

(Exp) $\exists \overline{W} \forall \overline{Y} : P \Longrightarrow \exists \overline{W'} \exists \overline{W} \forall \overline{Y} : P \wedge X \approx_\alpha t$, for $t = f(\overline{W'})$ or $t = [a]W'$ or $t = a$

Conditions for (Exp)**:**

1. X is a free or existential variable occurring in P, $\overline{W'}$ are newly chosen auxiliary variables not occurring anywhere in the problem;
2. $f \in \Sigma$ and $a \in \mathtt{Atoms}(P) \cup \{a'\}$, where a' is a new atom;
3. there exists an equation $X \approx_\alpha u$ (or disequation $X \not\approx_\alpha u$) in P such that u is not a variable and contains at least one parameter; and
4. no other rule can be applied.

Fig. 3. Globally Preserving Rules

Example 6. Both $\mathcal{P} = a\#Y_1 \vee Y \approx_\alpha f(Y_1)$ and $\mathcal{P} = a\#Y_1 \vee a\not\#Y \vee Y_1 \approx_\alpha f(Y)$ are irreducible: neither (U_4) nor (ED_1) apply since all the disjuncts contain parameters; (ED_2) does not apply since each constraint has a parameter that occurs in another constraint; (Exp) does not apply because there is no equation or disequation with a free or existentially quantified variable in one side.

The following lemma characterises the irreducible disjunctions with respect to rules $\mathcal{R}_1$-$\mathcal{R}_8$ where parameters may remain.

Lemma 4. *Let P be a disjunction of constraints irreducible w.r.t. $\mathcal{R}_1$-$\mathcal{R}_8$. For each parameter Y such that $P = a\#Y \vee Q$ (resp. $P = a\not\#Y \vee Q$), for some atom a, the following holds:*

1. *$a\not\#Y$ (resp. $a\#Y$) cannot occur in Q;*
2. *Y has to occur in Q;*
3. *if Q contains an equational constraint then it has the form $Y \approx_\alpha t$, where $Y \notin \mathtt{vars}(t)$, or $Y' \approx_\alpha t$, with $Y \in \mathtt{vars}(t)$;*
4. *Q does not contain disequations or primitive freshness constraints for free or existentially quantified variables.*

Proof. In an irreducible disjunction of constraints at least one of the sides of equations (or disequations) is a variable, otherwise we could simplify the equation/disequation.

Condition 1. It holds, otherwise we could apply (T_9). **Condition 2.** It holds, otherwise we could apply (ED_2).

Condition 3. If Q had an equation of the form $X \approx_\alpha t$, for some free or existentially quantified variable, then t could not contain a parameter, otherwise we could apply rule (Exp). Therefore, $t = t[Z_1, \dots, Z_n]$, for $n \geq 0$ where each Z_i

$\mathcal{R}_9$: Simplification of parameters in freshness constraints
(U_7) $\forall\overline{Y}, Y : P \wedge (a\#Y \vee Q) \Longrightarrow \bot$ if $\mathcal{R}_1$-$\mathcal{R}_8$ do not apply (so Q does not contain $a\not\#Y$) and $Y \in \mathbf{vars}(Q)$.
(U_8) $\forall\overline{Y}, Y : P \wedge (a\not\#Y \vee Q) \Longrightarrow \bot$ if $\mathcal{R}_1$-$\mathcal{R}_8$ do not apply (so Q does not contain $a\#Y$) and $Y \in \mathbf{vars}(Q)$.

Fig. 4. Preserving Rules for (non)freshness constraints with parameters.

is either a free or existentially quantified variable, and one could apply rule ED_1. Thus, if an equation exists, one of the sides has to be a parameter, say $Y \approx_\alpha t$, and Y cannot occur in t otherwise rule O_2 applies.

Condition 4. If Q were to contain a disequation, say $X \not\approx_\alpha t$ then t could not contain a parameter, otherwise we could apply (Exp) as above, but then we could apply rule (ED_1). Therefore, if Q were to contain a disequation, it would be of the form $Y \not\approx_\alpha t$, then it would either reduce with (O_2) or with (U_2). Thus, Q does not contain disequations. Similary, if Q contained a primitive freshness constraint for a free or existentially quantified variable then (ED_1) would apply.

The remaining disjunctions with parameters can be simplified using the rules in $\mathcal{R}_9$, since they will not produce solutions (as shown in Theorem 1).

We end this section with an example of application of the simplification rules.

Example 7. Let $\mathcal{P}$ be a NEP, using the signature from Example 1, as follows:

$$\mathcal{P} = \forall Y : \lambda[a]X \not\approx_\alpha \lambda[a]\lambda[a]Y \overset{DC}{\Longrightarrow} \forall Y : [a]X \not\approx_\alpha [a]\lambda[a]Y \overset{D_2}{\Longrightarrow} \forall Y : X \not\approx_\alpha \lambda[a]Y$$

Rules in $\mathcal{R}_1$-$\mathcal{R}_7$ cannot be applied and the explosion rule produces six problems:

$$\begin{array}{ll}
\mathcal{P}_1 = \exists W_1 \forall Y : X \not\approx_\alpha \lambda[a]Y \wedge X \approx_\alpha \lambda W_1 & \mathcal{P}_4 = \exists W \forall Y : X \not\approx_\alpha [a]Y \wedge X = [b]W \\
\mathcal{P}_2 = \exists W_1, W_2 \forall Y : X \not\approx_\alpha [a]Y \wedge X = W_1 W_2 & \mathcal{P}_5 = \exists W \forall Y : X \not\approx_\alpha [a]Y \wedge X = a \\
\mathcal{P}_3 = \exists W \forall Y : X \not\approx_\alpha [a]Y \wedge X = [a]W & \mathcal{P}_6 = \forall Y : X \not\approx_\alpha [a]Y \wedge X = b
\end{array}$$

Reducing the first problem we get:

$$\begin{aligned}
\mathcal{P}_1 &\overset{I_1}{\Longrightarrow} \exists W_1 \forall Y : \lambda W_1 \not\approx_\alpha \lambda[a]Y \wedge X \approx_\alpha \lambda W_1 \\
&\overset{DC}{\Longrightarrow} \exists W_1 \forall Y : W_1 \not\approx_\alpha [a]Y \wedge X \approx_\alpha \lambda W_1 \\
&\overset{Exp}{\Longrightarrow} \exists W_1 W_2 \forall Y : W_1 \not\approx_\alpha [a]Y \wedge X \approx_\alpha \lambda W_1 \wedge W_1 \approx_\alpha \lambda W_2 \\
&\overset{I_1}{\Longrightarrow} \exists W_1 W_2 \forall Y : \lambda W_2 \not\approx_\alpha [a]Y \wedge X \approx_\alpha \lambda W_1 \wedge W_1 \approx_\alpha \lambda W_2 \\
&\overset{CL_1}{\Longrightarrow} \exists W_1 W_2 \forall Y : X \approx_\alpha \lambda W_1 \wedge W_1 \approx_\alpha \lambda W_2 \\
&\overset{I_1}{\Longrightarrow} \exists W_1 W_2 : X \approx_\alpha \lambda\lambda W_2 \wedge W_1 \approx_\alpha \lambda W_2.
\end{aligned}$$

At this point $\mathcal{P}_1$ has reached a normal form without any parameter. Solutions of $\mathcal{P}_1$ can be easily obtained by taking any instance of X of the form $\lambda\lambda t$. It is easy to check that this choice indeed generates solutions of $\mathcal{P}$. Similar reductions apply to $\mathcal{P}_i$, $2 \leq i \leq 6$.

As we will see in the next section, application of such simplification rules is *well-behaved* in the sense that we do not loose any solution along the way.

4.2 Soundness and Preservation of Solutions

The next step is to ensure that the application of rules does not change the set of solutions of an equational problem.

Definition 11 (Soundness and preservation of solution). *Let $\mathcal{A}$ be any infinite subalgebra of* CORE.

1. *A rule R is* $\mathcal{A}$-sound *if, $\mathcal{P} \Longrightarrow_R \mathcal{P}'$ implies $\mathcal{S}(\mathcal{P}') \subseteq \mathcal{S}(\mathcal{P})$.*
2. *A rule R is* $\mathcal{A}$-preserving *if, $\mathcal{P} \Longrightarrow_{\mathcal{R}} \mathcal{P}'$ implies $\mathcal{S}(\mathcal{P}) \subseteq \mathcal{S}(\mathcal{P}')$.*
3. *A rule R is* $\mathcal{A}$-globally preserving *if given any problem $\mathcal{P}$,*

$$\mathcal{S}(\mathcal{P}) \subseteq \bigcup_{\substack{\mathcal{P} \to_{\mathcal{R}} \pi \cdot \mathcal{P}_i \\ \mathtt{supp}(\pi) \cap \mathtt{Atoms}(\mathcal{P}) = \emptyset}} \mathcal{S}(\mathcal{P}_i).$$

All our rules, except those in $\mathcal{R}_8$, are sound and preserving (Theorem 1). The rules in $\mathcal{R}_8$ create branches in the derivation tree; they are sound and only globally preserving (Theorem 2).

Theorem 1. *The rules in $\mathcal{R}_1$ to $\mathcal{R}_7$ and the rules in $\mathcal{R}_9$ are $\mathcal{A}$-sound and $\mathcal{A}$-preserving for any infinite subalgebra $\mathcal{A}$ of* CORE.

Proof. **Rules in $\mathcal{R}_1$, $\mathcal{R}_2$, and $\mathcal{R}_3$** : soundness and preservation of solutions are easy to deduce. For instance, for clash rules, (CL_1) and (CL_2), it follows by inspection of deduction rules that the judgement $\vdash s\gamma \approx_\alpha t\gamma$ is not derivable for any valuation ς and corresponding grounding substitution $\gamma = \sigma^{\varsigma}_{\mathtt{vars}(s,t)}$ (see Definition 7) if the root constructors of s and t are different (hence every γ is a solution for the disequation). For (C_3) observe that we can take $[W/t]$ as a witness for W on a validation for $\exists \overline{W} : P$, if $W \notin \mathtt{vars}(P,t)$.
Rules in $\mathcal{R}_4$ and $\mathcal{R}_5$. It follows from soundness and preservation of simplification rules in [14]. We use the fact that nominal equality and freshness rules from Fig. 1 are reversible; for instance, let γ be a grounding substitution, a judgement $\vdash f(\tilde{s})\gamma \approx_\alpha f(\tilde{u})\gamma$ fails, which makes $f(\tilde{s})\gamma \not\approx_\alpha f(\tilde{u})\gamma$ valid, iff one of the premises $\vdash s_i\gamma \approx_\alpha u_i\gamma$ does not hold.
Rules in $\mathcal{R}_6$: The result is straightforward for rules U_1 and U_3.

U_2. To prove soundness for U_2 notice that the solution set of a conjunction is the intersection of the solution set of each of its members. We have to show that every solution of $Q[Y/\pi^{-1} \cdot t]$ is a solution of $(\pi \cdot Y \not\approx_\alpha t \vee Q)$. Let γ be a solution of $Q[Y/\pi^{-1} \cdot t]$ and take any substitution λ satisfying the conditions of Definition 10. So $(Q[Y/\pi^{-1} \cdot t])\gamma\lambda$ is valid and we need to show the validity of

$$(\pi \cdot Y \not\approx_\alpha t)\gamma\lambda \vee Q\gamma\lambda. \tag{1}$$

For each such λ there are two possible cases: First, $\vdash \pi \cdot Y\lambda \approx_\alpha t\gamma\lambda$ (note that λ is a ground substitution so both sides of this equation are ground); then we have that $\gamma\lambda = \gamma\lambda'[Y/\pi^{-1} \cdot t\gamma\lambda]$. By hypothesis, $\gamma\lambda$ validates $Q[Y/\pi^{-1} \cdot t]$

so $\gamma\lambda'[Y/\pi^{-1} \cdot t\gamma\lambda]$ validates Q. Second; $\not\vdash \pi \cdot Y\lambda \approx_\alpha t\gamma\lambda$, then $\gamma\lambda$ validates $\pi \cdot Y \not\approx_\alpha t$. Hence γ a solution of (1).

To prove preservation for $\boldsymbol{U}_2$, take γ a solution of $\forall\overline{Y}, Y : \pi \cdot Y \not\approx_\alpha t \vee Q$, we need to show that γ is also a solution of $\forall\overline{Y}, Y : Q[Y/\pi^{-1} \cdot t]$. Notice that γ is a solution of $\forall\overline{Y}, Y : \pi \cdot Y \not\approx_\alpha t$ or $\forall\overline{Y}, Y : Q$ but it clearly cannot solve the first problem. Hence, γ solves $\forall\overline{Y}, Y : Q$. By Definition 10, for all substitutions λ with domain $\overline{Y} \cup \{Y\}$ we have that $\lambda\gamma$ validates Q. In particular, the substitution $[Y/\pi^{-1} \cdot t\gamma]\lambda\gamma$ which is equivalent to $[Y/\pi^{-1} \cdot t]\lambda\gamma$ (since γ is away from λ) must also validate Q. Consequently, $\lambda\gamma$ validates $(Q[Y/\pi^{-1} \cdot t])$.

$\boldsymbol{U}_4$. Soundness for this rule follows trivially. For preservation of solutions, we show that any solution of $\forall\overline{Y} : \bigvee_i Z_i \approx_\alpha t_i \vee Q$ is a solution of $\forall\overline{Y} : Q$. The shape of the first problem induces a requirement that the disjunction $\bigvee_i Z_i \approx_\alpha t_i$ does not have a solution. To show this we prove that the negated form $\bigwedge_i Z_i \not\approx_\alpha t_i$ has at least one solution. Notice that such a solution is a witness for the failure of $\bigvee_i Z_i \approx_\alpha t_i$, since all of those equations have at least one parameter. Lemma 5 shows that this is true.

$\boldsymbol{U}_5$ and $\boldsymbol{U}_6$. We need to show that every solution of $\forall\overline{Y}, Y : P \wedge a\#Y$ is also a solution of $\bot$, i.e., no such solution exists for the lhs of the rule. In fact, the existence of such γ would imply that (taking $\lambda = [Y/a]$) $a\#a$ which is impossible. For U_6 we do the same reasoning with $\lambda = [Y/[a]a]$.

Rules in $\mathcal{R}_7$. Soundness and preservation of (I_1) has been proved in previous works, since rule (I_1) is used in standard nominal unification algorithms [23]. Rule (I_2) is a direct adaptation of the rule used in the standard (syntactic) case, proved sound and preserving in [10]. Indeed, $\gamma \in \mathcal{S}(\pi \cdot Z \not\approx_\alpha t \vee P)$ if, and only if, for any grounding instance γ' of γ, $\gamma' \in \mathcal{S}(Z \not\approx_\alpha \pi^{-1} \cdot t)$ or $\gamma' \in \mathcal{S}(P)$ (by Lemma 3). Finally, notice that $\gamma \in \mathcal{S}(P) \setminus \mathcal{S}(Z \not\approx_\alpha \pi^{-1} \cdot t)$ if and only if $\gamma \in \mathcal{S}(P[Z/\pi^{-1} \cdot t])$.

Rules in $\mathcal{R}_9$. Soundness follows trivially, since $\bot$ has no solution. We show below that U_7 is $\mathcal{A}$-preserving; the proof is analogous for rule (U_8).

Let $\mathcal{P} = \exists\overline{W}\forall\overline{Y}, Y : P \wedge (a\#Y \vee Q)$ where Q is fully reduced by $\mathcal{R}_1$-$\mathcal{R}_8$, $Y \in \mathtt{vars}(Q)$ and Q does not contain $a\#Y$. We prove that $\mathcal{P}$ does not have solutions by induction on the number of freshness constraints in $a\#Y \vee Q$.

Base case: Q contains just equational constraints, each containing at least one occurrence of the parameter Y, as specified in Lemma 4. Suppose by contradiction that there exists an $\mathcal{A}$-solution γ. Thus, γ is away from $\overline{Y} \cup \{Y\}$, $\mathtt{dom}(\gamma) = \overline{X} = \mathtt{Fv}(\mathcal{P})$, there is a ground substitution δ with $\mathtt{dom}(\delta) = \overline{W}$ and for all λ away from $\overline{X}, \overline{W}$, with $\mathtt{dom}(\lambda) = \overline{Y} \cup \{Y\}$, $\gamma\delta\lambda$ $\mathcal{A}$-validates $P \wedge (a\#Y \vee Q)$. Then, it $\mathcal{A}$-validates both P and $(a\#Y \vee Q)$. The latter implies that $\gamma\delta\lambda$ $\mathcal{A}$-validates Q for every λ (but then Q has a solution, which is impossible due to the form of the equational constraints) or Q implies $a\not\#Y$ (since there is at least one $f \in \Sigma$ such that $f : n$ and $n > 0$, and therefore $a\#Y$ is false for an infinite number of ground terms $Y\lambda$). The latter is impossible since $a\#Y$ is defined as $a \notin \mathtt{supp}(Y)$, which is defined as $(a\,a') \cdot Y = Y$ for a new a', and reduced problems cannot contain fixed point equations or their negations (these are simplified using rules (E_1) and (D_1), respectively).

The inductive step is proved similarly, using Lemma 4 as in the base case to deduce that the constraints in Q cannot entail $a\#Y$.

Theorem 2. *Let $\mathcal{A}$ be any infinite subalgebra of* CORE.

1. *Rule (Exp) is $\mathcal{A}$-sound and $\mathcal{A}$-globally preserving.*
2. *Rules (ED_1) and (ED_2) are $\mathcal{A}$-sound and $\mathcal{A}$-globally preserving.*

Lemma 5 guarantees the existence of a solution for a conjunction of non-trivial disequations as long as the algebra considered has sufficient ground terms.

Lemma 5. *Let $\mathcal{P}$ be a conjunction of non-trivial disequations. Let $\mathcal{A}$ be any infinite subalgebra of* CORE. *Then $\mathcal{P}$ has at least one solution in $\mathcal{A}$.*

Proof. The proof proceeds by induction on the number of distinct variables occurring in $\mathcal{P}$. For the base case $\mathcal{P}$ has no variables. Then every substitution solves $\mathcal{P}$, since by hypothesis $\mathcal{P}$ does not contain any trivial disequation $t \not\approx_\alpha t$.

Assume the result holds for problems with $m-1$ variables. Let $\mathcal{P}$ be a conjunction of non-trivial disequations such that $|\texttt{vars}(P)| = m$ and $X \in \texttt{vars}(P)$. For each disequation $s \not\approx_\alpha t \in \mathcal{P}$, the equation $s \approx_\alpha t$ has at most one solution (modulo α-renaming) when the variables distinct from X are considered as constants. Let $\mathcal{S}$ the set of such solutions for all these equations. Since A (the domain of $\mathcal{A}$) is infinite, there exists $a \in \mathcal{A}$ such that $[X/a] \notin \mathcal{S}$. Therefore, $[X/a]$ is a solution for $\mathcal{P}$. Now, consider the problem $\mathcal{P}' = \mathcal{P}[X/a]$ which has $m-1$ variables. The result follows by induction hypothesis.

4.3 Termination

To prove termination we define a measure function for NEPs that strictly decreases with each application of a rule. The measure uses the following auxiliary functions:

Definition 12 (Auxiliary Functions). *The function* $\texttt{sizePar}(t)$ *denotes the sum of the sizes of the parameter positions in t:*

$$\texttt{sizePar}(t) := \sum_{p_j \in \texttt{PosPar}(t)} |p_j|$$

where $\texttt{PosPar}(t) = \{p_j \mid t|_{p_j} = Y_i$ *for some parameter* $Y_i\}$.

Given a disjunction of equations, disequations, freshness, and negated freshness constraints $d = C_1 \vee \ldots \vee C_n$ we define auxiliary functions ϕ_1 and ϕ_2 over d.

1. *$\phi_1(d)$ is the number of distinct parameters in d.*
2. *$\phi_2(d)$ is the multiset* $\{\texttt{MSP}(C_1), \ldots, \texttt{MSP}(C_n)\}$ *where* $\texttt{MSP}(C)$ *is defined by:*
   (a) $\texttt{MSP}(C) = 0$ *if C is an equation or disequation and a member of C is a solved parameter (a parameter Y is* solved *in d if there exists a disequation $Y \not\approx_\alpha u$ in d and Y occurs only once in d); or if C is a primitive freshness or a primitive negated freshness constraint;*

(b) *otherwise*, $\mathtt{MSP}(s \approx_\alpha t) = \mathtt{MSP}(s \not\approx_\alpha t) = max(\mathtt{sizePar}(s), \mathtt{sizePar}(t))$ *and* $\mathtt{MSP}(a\#t) = \mathtt{MSP}(a\not\#t) = \mathtt{sizePar}(t)$.

Definition 13 (Measure). *Let* $\mathcal{P} = \exists\overline{W}\forall\overline{Y}d_1 \wedge \ldots \wedge d_n$ *be a nominal equational problem in conjunctive normal form.* $\mathcal{P}$ *is measured using the tuple:*

$$\Phi(\mathcal{P}) = (N_u, N_d, \psi_1(\mathcal{P}), M, \psi_2(\mathcal{P})), \text{ where}$$

1. N_u *is the number of free variables that are unsolved in* $\mathcal{P}$. *A variable* X *is* solved *if there is an equation* $X \approx_\alpha t$ *and* X *occurs only once in* $\mathcal{P}$.
2. N_d *is a multiset that contains for each disjunction* d_i *in* $\mathcal{P}$ *the number of variables that are not d-solved in* d_i.
   A variable X *is d-solved in* d_i *if* $d_i = X \not\approx_\alpha t \vee Q$ *and* X *does not occur in* Q.
3. $\Psi_1(\mathcal{P})$ *is the multiset* $\{(\phi_1(d_1), \phi_2(d_1)), \ldots, (\phi_1(d_n), \phi_2(d_n))\}$
4. M *is the multiset* $\{M(d_1), \ldots, M(d_n)\}$ *where* $M(d)$ *is the multiset of sizes of the constraints in* d. *The size of a constraint is the size of its largest member, or 0 if it has a solved variable or it is a primitive (negated) freshness.*
5. $\Psi_2(\mathcal{P})$ *is the total size of* $\mathcal{P}$ *(that is, the number of function symbols, atoms, variables, quantifiers, conjunctions, disjunctions,* $\top$, $\bot$ *in* $\mathcal{P}$.

Using this measure we can prove the termination of the simplification process.

Theorem 3. *The procedure defined in Section 4 for application of rules, expressed as* $\mathfrak{R} := \mathcal{R}_1\mathcal{R}_2 \ldots \mathcal{R}_9$, *terminates.*

5 Nominal Equational Solved Forms

We have shown that the simplification process terminates and each application of the transformation rules preserves solutions. We now characterise the normal forms, called *solved forms*. Intuitively, solved forms are simple enough that one can easily extract solutions from it. A first example of well-known solved form is that of *unification solved form*: a conjunction of equations $X_i = t_i$ such that each X_i occurs only once. It directly represents a solution mapping $X_i \mapsto t_i$.

We show in Theorem 4 existence of solutions for certain solved forms, and in Theorem 5 we prove that our procedure is complete with respect to solved forms.

Definition 14 (Solved Forms).

1. *A* NEP $\mathcal{P}$ *is in* parameterless solved form *if it contains no universal quantifiers.*
2. *A* NEP *is a* definition with constraints *if it is* $\top, \bot$ *or a conjunction of the form*

$$\mathcal{P} = \exists\overline{W} : \left(\bigwedge_{i=1}^{n} Z_i \approx_\alpha t_i\right) \wedge \left(\bigwedge_{j=1}^{m} Z'_j \not\approx_\alpha v_j\right) \wedge \left(\bigwedge_{l=1}^{p} C_l\right),$$

   such that:
   - *each* Z_i *occurs only once in* $\mathcal{P}$;
   - *each* Z'_j *is syntactically different from* v_j; *and*

– *each C_l is either a positive, $a\#X$, or negative, $a\not\#X$, freshness constraint such that each pair a, X occurs at most once in $\mathcal{P}$.*

3. *A* NEP *is in* unification solved form *if it is a definition with constraints which does not contain negative constraints.*

Theorem 4 below shows that a problem reduced to definition with constraints solved form has at least one solution.

Theorem 4. *Let $\mathcal{A}$ be any infinite subalgebra of* CORE*. If $\mathcal{P} \not\equiv \bot$ is in definition with constraints solved form, then it has at least one solution.*

Proof. First assume $\mathcal{P}$ is in unification solved form (see Definition 14). Let ∇ be the context containing all constraints C_l occurring in $\mathcal{P}$. Furthermore, define the substitution σ that assigns to each free variable X_i the term t_i, and the substitution δ mapping each existential variable W_k to t_k. Then $[\![\nabla\sigma\delta]\!]_\varsigma$, which is equivalent to $[\![\nabla]\!]_{\varsigma^{\sigma\delta}}$ by Lemma 1, is valid in $\mathcal{A}$. Consequently,

$$[\![\nabla \vdash X_i\sigma \approx_\alpha t_i\sigma\delta]\!]_\varsigma \text{ and } [\![\nabla \vdash W_k\delta \approx_\alpha t_k\delta]\!]_\varsigma$$

are valid judgements. So, σ is an $\mathcal{A}$-solution of $\mathcal{P}$ with existential witnesses given by δ. In the general case, when $\mathcal{P}$ is in *definition with constraints* solved form containing also negative constraints, the construction is similar. We can guarantee a solution for the disunification part of the problem, $\bigwedge_{j=1}^{m} Z'_j \not\approx_\alpha v_j$, by Lemma 5.

Definition 15. *A set $\mathfrak{R}$ of rules for solving nominal equational problems is* complete *w.r.t. a kind of solved forms S if for each $\mathcal{P}$ there exists a family of* NEP*s $\mathcal{Q}_i$ in S-solved form such that $\mathcal{P} \overset{*}{\Longrightarrow}_{\mathfrak{R}} \mathcal{Q}_i$ and $\mathcal{S}(\mathcal{P}) = \bigcup_i \mathcal{S}(\mathcal{Q}_i)$.*

The next result states that a NEP's normal form with respect to the simplification rules given in the previous section is a definition with constraints. In particular, all parameters are removed from the problem. The proof is by case analysis, considering all possible occurrences of parameters in a problem.

Theorem 5 (Completeness). *Let $\mathcal{A}$ be any infinite subalgebra of* CORE*. Then the rules in Figures 2, 3, and 4 are complete for parameterless solved forms and definition with constraints solved forms.*

6 Conclusion

In this paper, we introduced *nominal equational problems* (NEPs) as an extension of standard first-order equational problems to nominal terms which, besides equations and disequations, includes freshness and non-freshness constraints. We proposed a sound and preserving rule-based algorithm to solve NEPs in the nominal ground algebra CORE, and showed that this algorithm is complete for two main types of solved forms: parameterless and definition with constraints. As future work, we aim to investigate the purely equational approach to nominal syntax via the formulation of freshness constraints using fixed-point equations with the И-quantifier [21], as well as the solvability of nominal equational problems in more complex algebras.

References

1. Ayala-Rincón, M., Fernández, M., Nantes-Sobrinho, D., Vale, D.: On Solving Nominal Disunification Constraints. ENTCS **348**, 3 – 22 (2020). https://doi.org/10.1016/j.entcs.2020.02.002, proc. 14th Int. Workshop on Logical and Semantic Frameworks, with Applications LSFA 2019
2. Baader, F., Schulz, K.U.: Combination techniques and decision problems for disunification. Theor. Comput. Sci. **142**(2), 229–255 (1995). https://doi.org/10.1016/0304-3975(94)00277-0
3. Buntine, W.L., Bürckert, H.J.: On solving equations and disequations. J. ACM **41**(4), 591–629 (Jul 1994). https://doi.org/10.1145/179812.179813
4. Byrd, W.E., Friedman, D.P.: αKanren A Fresh Name in Nominal Logic Programming (2008), http://webyrd.net/alphamk/alphamk.pdf, earlier version available in the Proc. 2007 Workshop on Scheme and Functional Programming, Université Laval Technical Report DIUL-RT-0701
5. Calvès, C., Fernández, M.: Matching and alpha-equivalence check for nominal terms. Journal of Computer and System Sciences **76**(5), 283 – 301 (2010). https://doi.org/10.1016/j.jcss.2009.10.003
6. Cheney, J., Urban, C.: αProlog: A Logic Programming Language with Names, Binding and α-Equivalence. In: Proc. 20th International Conference on Logic Programming ICLP. LNCS, vol. 3132, pp. 269–283. Springer (2004). https://doi.org/10.1007/978-3-540-27775-0_19
7. Comon, H.: Unification et disunification: Théorie et applications. Ph.D. thesis, Institut National Polytechnique de Grenoble, France (1988)
8. Comon, H.: Disunification: a Survey. In: Lassez, J.L., Plotkin, G. (eds.) Computational Logic: Essays in Honor of Alan Robinson, pp. 322–359. MIT Press (1991)
9. Comon, H., Fernández, M.: Negation elimination in equational formulae. In: Proc. 17th International Symposium on Mathematical Foundations of Computer Science MFCS. LNCS, vol. 629, pp. 191–199. Springer (1992). https://doi.org/10.1007/3-540-55808-X_17
10. Comon, H., Lescanne, P.: Equational problems and disunification. J. Symb. Comput. **7**(3/4), 371–425 (1989). https://doi.org/10.1016/S0747-7171(89)80017-3
11. Fernández, M.: Narrowing based procedures for equational disunification. Appl. Algebra Eng. Commun. Comput. **3**, 1–26 (1992). https://doi.org/10.1007/BF01189020
12. Fernández, M.: AC complement problems: Satisfiability and negation elimination. Journal of Symbolic Computation **22**(1), 49–82 (1996). https://doi.org/10.1006/jsco.1996.0041
13. Fernández, M.: Negation elimination in empty or permutative theories. Journal of Symbolic Computation **26**(1), 97–133 (1998). https://doi.org/10.1006/jsco.1998.0203
14. Fernández, M., Gabbay, M.J.: Nominal rewriting. Information and Computation **205**(6), 917 – 965 (2007). https://doi.org/10.1016/j.ic.2006.12.002
15. Gabbay, M.J., Mathijssen, A.: Nominal (Universal) algebra: equational logic with names and binding. J. of Logic and Computation **19**(6), 1455–1508 (2009). https://doi.org/10.1093/logcom/exp033
16. Kounalis, E., Lugiez, D., Pottier, L.: A Solution of the Complement Problem in Associative-Commutative Theories. In: Proc. 16th International Symposium on Mathematical Foundations of Computer Science MFCS. LNCS, vol. 520, pp. 287–297. Springer (1991). https://doi.org/10.1007/3-540-54345-7_72

17. Levy, J., Villaret, M.: An efficient nominal unification algorithm. In: Proc. of the 21st International Conference on Rewriting Techniques and Applications, RTA. LIPIcs, vol. 6, pp. 209–226 (2010). https://doi.org/10.4230/LIPIcs.RTA.2010.209
18. Lugiez, D.: Higher order disunification: Some decidable cases. In: First International Conference on Constraints in Computational Logics, CCL. LNCS, vol. 845, pp. 121–135. Springer (1994). https://doi.org/10.1007/BFb0016848
19. Lugiez, D., Moysset, J.L.: Complement problems and tree automata in AC-like theories (extended abstract). In: Proc. 10th Annual Symposium on Theoretical Aspects of Computer Science STACS. LNCS, vol. 665, pp. 515–524. Springer (1993). https://doi.org/10.1007/3-540-56503-5_51
20. Near, J.P., Byrd, W.E., Friedman, D.P.: αleanTAP: A Declarative Theorem Prover for First-Order Classical Logic. In: Proc. 24th International Conference on Logic Programming ICLP. LNCS, vol. 5366, pp. 238–252. Springer (2008). https://doi.org/10.1007/978-3-540-89982-2_26
21. Pitts, A.M.: Nominal Sets: Names and Symmetry in Computer Science. Cambridge University Press (2013). https://doi.org/10.1017/CBO9781139084673
22. Ravishankar, V., Cornell, K.A., Narendran, P.: Asymmetric Unification and Disunification. In: Description Logic, Theory Combination, and All That. vol. 11560 (2019). https://doi.org/10.1007/978-3-030-22102-7_23
23. Urban, C., Pitts, A.M., Gabbay, M.: Nominal unification. Theor. Comput. Sci. **323**(1-3), 473–497 (2004). https://doi.org/10.1016/j.tcs.2004.06.016

Finding Cut-Offs in Leaderless Rendez-Vous Protocols is Easy *

A. R. Balasubramanian(✉)[1], Javier Esparza[1], Mikhail Raskin[1]

Technische Universität München, Munich, Germany
bala.ayikudi@tum.de, esparza@in.tum.de, raskin@in.tum.de

Abstract. In rendez-vous protocols an arbitrarily large number of indistinguishable finite-state agents interact in pairs. The cut-off problem asks if there exists a number B such that all initial configurations of the protocol with at least B agents in a given initial state can reach a final configuration with all agents in a given final state. In a recent paper [17], Horn and Sangnier prove that the cut-off problem is equivalent to the Petri net reachability problem for protocols with a leader, and in EXPSPACE for leaderless protocols. Further, for the special class of symmetric protocols they reduce these bounds to PSPACE and NP, respectively. The problem of lowering these upper bounds or finding matching lower bounds is left open. We show that the cut-off problem is P-complete for leaderless protocols, NP-complete for symmetric protocols with a leader, and in NC for leaderless symmetric protocols, thereby solving all the problems left open in [17].

Keywords: rendez-vous protocols · cut-off problem · Petri nets

1 Introduction

Distributed systems are often designed for an unbounded number of participant agents. Therefore, they are not just one system, but an infinite family of systems, one for each number of agents. Parameterized verification addresses the problem of checking that all systems in the family satisfy a given specification.

In many application areas, agents are indistinguishable. This is the case in computational biology, where cells or molecules have no identities; in some security applications, where the agents' identities should stay private; or in applications where the identities can be abstracted away, like certain classes of multithreaded programs [15,2,31,3,18,25]. Following [3,18], we use the term *replicated systems* for distributed systems with indistinguishable agents. Replicated systems include population protocols, broadcast protocols, threshold automata, and many other models [15,2,11,7,16]. They also arise after applying a *counter abstraction* [28,3]. In finite-state replicated systems the global state of the system is determined by the function (usually called a *configuration*) that assigns

* This project has received funding from the European Research Council (ERC) under the European Union's Horizon 2020 research and innovation programme under grant agreement No 787367 (PaVeS).

S. Kiefer and C. Tasson (Eds.): FOSSACS 2021, LNCS 12650, pp. 42–61, 2021.
https://doi.org/10.1007/978-3-030-71995-1_3

to each state the number of agents that currently occupy it. This feature makes many verification problems decidable [4,10].

Surprisingly, there is no a priori relation between the complexity of a parameterized verification question (i.e., whether a given property holds for all initial configurations, or, equivalently, whether its negation holds for some configuration), and the complexity of its corresponding single-instance question (whether the property holds for a fixed initial configuration). Consider replicated systems where agents interact in pairs [15,17,2]. The complexity of single-instance questions is very robust. Indeed, checking most properties, including all properties expressible in LTL and CTL, is PSPACE-complete [9]. On the contrary, the complexity of parameterized questions is very fragile, as exemplified by the following example. While the existence of a reachable configuration that populates a given state with *at least* one agent is in P, and so well below PSPACE, the existence of a reachable configuration that populates a given state with *exactly* one agent is as hard as the reachability problem for Petri nets, and so non-elementary [6]. This fragility makes the analysis of parameterized questions very interesting, but also much harder.

Work on parameterized verification has concentrated on whether every initial configuration satisfies a given property (see e.g. [15,11,3,18,7]). However, applications often lead to questions of the form "do all initial configurations *in a given set* satisfy the property?", "do infinitely many initial configurations satisfy the property?", or "do all but finitely many initial configurations satisfy the property?". An example of the first kind is proving correctness of population protocols, where the specification requires that for a given partition $\mathcal{I}_0, \mathcal{I}_1$ of the set of initial configurations, and a partition Q_0, Q_1 of the set of states, runs starting from $\mathcal{I}_0$ eventually trap all agents within Q_0, and similarly for $\mathcal{I}_1$ and Q_1 [12]. An example of the third kind is the existence of *cut-offs*; cut-off properties state the existence of an initial configuration such that for all larger initial configurations some given property holds [8,4]. A systematic study of the complexity of these questions is still out of reach, but first results are appearing. In particular, Horn and Sangnier have recently studied the complexity of the *cut-off problem* for parameterized rendez-vous networks [17]. The problem takes as input a network with one single initial state *init* and one single final state *fin*, and asks whether there exists a cut-off B such that for every number of agents $n \geq B$, the final configuration in which all agents are in state *fin* is reachable from the initial configuration in which all agents are in state *init*.

Horn and Sangnier study two versions of the cut-off problem, for leaderless networks and networks with a leader. Intuitively, a leader is a distinguished agent with its own set of states. They show that in the presence of a leader the cut-off problem and the reachability problem for Petri nets problems are inter-reducible, which shows that the cut-off problem is in the Ackermannian complexity class $\mathcal{F}_\omega$ [22], and non-elementary [6]. For the leaderless case, they show that the problem is in EXPSPACE. Further, they also consider the special case of symmetric networks, for which they obtain better upper bounds: PSPACE for the case of a

Horn and Sangnier	Asymmetric rendez-vous	Symmetric rendez-vous
Presence of a leader	Decidable, non-elementary	PSPACE
Absence of a leader	EXPSPACE	NP

This paper	Asymmetric rendez-vous	Symmetric rendez-vous
Presence of a leader	Decidable, non-elementary	NP-complete
Absence of a leader	P-complete	NC

Table 1. Summary of the results by Horn and Sangnier and the results of this paper.

leader, and NP in the leaderless case. These results are summarized at the top of Table 1.

In [17] the question of improving the upper bounds or finding matching lower bounds is left open. In this paper we close it with a surprising answer: All elementary upper bounds of [17] can be dramatically improved. In particular, our main result shows that the EXPSPACE bound for the leaderless case can be brought down to P. Further, the PSPACE and NP bounds of the symmetric case can be lowered to NP and NC, respectively, as shown at the bottom of Table 1. We also obtain matching lower bounds. Finally, we provide almost tight upper bounds for the size of the cut-off B; more precisely, we show that if B exists, then $B \in 2^{n^{O(1)}}$ for a protocol of size n.

Our results follow from two lemmas, called the Scaling and Insertion Lemmas, that connect the *continuous semantics* for Petri nets to their standard semantics. In the continuous semantics of Petri nets transition firings can be scaled by a positive rational factor; for example, a transition can fire with factor 1/3, taking "1/3 of a token" from its input places. The continuous semantics is a relaxation of the standard one, and its associated reachability problem is much simpler (polynomial instead of non-elementary [14,6,5]). The Scaling Lemma[1] states that given two markings M, M' of a Petri net, if M' is reachable from M in the continuous semantics, then nM' is reachable from nM in the standard semantics for some $n \in 2^{m^{O(1)}}$, where m is the total size of the net and the markings. The Insertion Lemma states that, given four markings M, M', L, L', if M' is reachable from M in the continuous semantics and the *marking equation* $L' = L + \mathcal{A}\mathbf{x}$ has a solution $\mathbf{x} \in \mathbb{Z}^T$ (observe that $\mathbf{x}$ can have negative components), then $nM' + L'$ is reachable from $nM + L$ in the standard semantics for some $n \in 2^{m^{O(1)}}$. We think that these lemmas can be of independent interest.

The paper is organized as follows. Section 2 contains preliminaries; in particular, it defines the cut-off problem for rendez-vous networks and reduces it to the cut-off problem for Petri nets. Section 3 gives a polynomial time algorithm for the leaderless cut-off problem for acyclic Petri nets. Section 4 introduces the Scaling and Insertion Lemmas, and Section 5 presents the novel polynomial

[1] Heavily based on previous results by Fraca and Haddad [14].

time algorithm for the cut-off problem. Sections 6 and 7 present the results for symmetric networks, for the cases with and without leaders, respectively.

Due to lack of space, full proofs of some of the lemmas can be found in the appendix.

2 Preliminaries

Multisets Let E be a finite set. For a semi-ring S, a vector from E to S is a function $v : E \to S$. The set of all vectors from E to S will be denoted by S^E. In this paper, the semi-rings we will be concerned with are the natural numbers $\mathbb{N}$, the integers $\mathbb{Z}$ and the non-negative rationals $\mathbb{Q}_{\geq 0}$ (under the usual addition and multiplication operators). The *support* of a vector v is the set $[\![v]\!] := \{e : v(e) \neq 0\}$ and its *size* is the number $\|v\| = \sum_{e \in [\![v]\!]} abs(v(e))$ where $abs(x)$ denotes the absolute value of x. Vectors from E to $\mathbb{N}$ are also called discrete multisets (or just multisets) and vectors from E to $\mathbb{Q}_{\geq 0}$ are called continuous multisets.

Given a multiset M and a number α we let $\alpha \cdot M$ be the multiset given by $(\alpha \cdot M)(e) = M(e) \cdot \alpha$ for all $e \in E$. Given two multisets M and M' we say that $M \leq M'$ if $M(e) \leq M'(e)$ for all $e \in E$ and we let $M + M'$ be the multiset given by $(M + M')(e) = M(e) + M'(e)$ and if $M' \leq M$, we let $M - M'$ be the multiset given by $(M - M')(e) = M(e) - M'(e)$. The empty multiset is denoted by $\mathbf{0}$. We sometimes denote multisets using a set-like notation, e.g. $\wr a, 2 \cdot b, c \int$ denotes the multiset given by $M(a) = 1, M(b) = 2, M(c) = 1$ and $M(e) = 0$ for all $e \notin \{a, b, c\}$.

Given an $I \times J$ matrix A with I and J sets of indices, $I' \subseteq I$ and $J' \subseteq J$, we let $A_{I' \times J'}$ denote the restriction of M to rows indexed by I' and columns indexed by J'.

Rendez-vous protocols and the cut-off problem. Let Σ be a fixed finite set which we will call the communication alphabet and we let $RV(\Sigma) = \{!a, ?a : a \in \Sigma\}$. The symbol $!a$ denotes that the message a is sent and $?a$ denotes that the message a is received.

Definition 1. *A rendez-vous protocol $\mathcal{P}$ is a tuple $(Q, \Sigma, init, fin, R)$ where Q is a finite set of states, Σ is the communication alphabet, $init, fin \in Q$ are the initial and final states respectively and $R \subseteq Q \times RV(\Sigma) \times Q$ is the set of rules.*

The size $|\mathcal{P}|$ of a protocol is defined as the number of bits needed to encode $\mathcal{P}$ in $\{0, 1\}^*$ using some standard encoding. A configuration C of $\mathcal{P}$ is a multiset of states, where $C(q)$ should be interpreted as the number of agents in state q. We use $\mathcal{C}(\mathcal{P})$ to denote the set of all configurations of $\mathcal{P}$. An initial (final) configuration C is a configuration such that $C(q) = 0$ if $q \neq init$ (resp. $C(q) = 0$ if $q \neq fin$). We use C^n_{init} (C^n_{fin}) to denote the initial (resp. final) configuration such that $C^n_{init}(init) = n$ (resp. $C^n_{fin}(fin) = n$).

The operational semantics of a rendez-vous protocol $\mathcal{P}$ is given by means of a transition system between the configurations of $\mathcal{P}$. We say that there is a transition between C and C', denoted by $C \Rightarrow C'$ iff there exists $a \in \Sigma$, $p, q, p', q' \in Q$ such that $(p, !a, p'), (q, ?a, q') \in R$, $C \geq \wr p, q \wr$ and $C' = C - \wr p, q \wr + \wr p', q' \wr$. As usual, $\stackrel{*}{\Rightarrow}$ denotes the reflexive and transitive closure of $\Rightarrow$.

The cut-off problem for rendez-vous protocols, as defined in [17], is:

Given: A rendez-vous protocol $\mathcal{P}$
Decide: Is there $B \in \mathbb{N}$ such that $C^n_{init} \stackrel{*}{\Rightarrow} C^n_{fin}$ for every $n \geq B$?

If such a B exists then we say that $\mathcal{P}$ admits a cut-off and that B is a cut-off for $\mathcal{P}$.

Petri nets. Rendez-vous protocols can be seen as a special class of Petri nets.

Definition 2. *A Petri net is a tuple $\mathcal{N} = (P, T, Pre, Post)$ where P is a finite set of places, T is a finite set of transitions, Pre and Post are matrices whose rows and columns are indexed by P and T respectively and whose entries belong to $\mathbb{N}$. The incidence matrix $\mathcal{A}$ of $\mathcal{N}$ is defined to be the $P \times T$ matrix given by $\mathcal{A} = Post - Pre$. Further by the* weight *of $\mathcal{N}$, we mean the largest absolute value appearing in the matrices Pre and Post.*

The size $|\mathcal{N}|$ of $\mathcal{N}$ is defined as the number of bits needed to encode $\mathcal{N}$ in $\{0,1\}^*$ using some suitable encoding. For a transition $t \in T$ we let ${}^\bullet t = \{p : Pre[p,t] > 0\}$ and $t^\bullet = \{p : Post[p,t] > 0\}$. We extend this notation to set of transitions in the obvious way. Given a Petri net $\mathcal{N}$, we can associate with it a graph where the vertices are $P \cup T$ and the edges are $\{(p,t) : p \in {}^\bullet t\} \cup \{(t,p) : p \in t^\bullet\}$. A Petri net $\mathcal{N}$ is called acyclic if its associated graph is acyclic.

A *marking* of a Petri net is a multiset $M \in \mathbb{N}^P$, which intuitively denotes the number of *tokens* that are present in every place of the net. For $t \in T$ and markings M and M', we say that M' is reached from M by firing t, denoted $M \xrightarrow{t} M'$, if for every place p, $M(p) \geq Pre[p,t]$ and $M'(p) = M(p) + \mathcal{A}[p,t]$.

A *firing sequence* is any sequence of transitions $\sigma = t_1, t_2, \dots, t_k \in T^*$. The support of σ, denoted by $[\![\sigma]\!]$, is the set of all transitions which appear in σ. We let $\sigma\sigma'$ denote the concatenation of two sequences σ, σ'.

Given a firing sequence $\sigma = t_1, t_2, \dots, t_k \in T^*$, we let $M \xrightarrow{\sigma} M'$ denote that there exist $M_1, \dots, M_{k-1}$ such that $M \xrightarrow{t_1} M_1 \xrightarrow{t_2} M_2 \dots M_{k-1} \xrightarrow{t_k} M'$. Further, $M \rightarrow M'$ denotes that there exists $t \in T$ such that $M \xrightarrow{t} M'$, and $M \xrightarrow{*} M'$ denotes that there exists $\sigma \in T^*$ such that $M \xrightarrow{\sigma} M'$.

Marking equation of a Petri net system. In the following, a *Petri net system* is a triple $(\mathcal{N}, M, M')$ where $\mathcal{N}$ is a Petri net and $M \neq M'$ are markings. The *marking equation* for $(\mathcal{N}, M, M')$ is the equation

$$M' = M + \mathcal{A}\mathbf{v}$$

over the variables $\mathbf{v}$. It is well known that $M \xrightarrow{\sigma} M'$ implies $M' = M + \mathcal{A}\overrightarrow{\sigma}$, where $\overrightarrow{\sigma} \in \mathbb{N}^T$ is the the *Parikh image* of σ, defined as the vector whose component $\overrightarrow{\sigma}[t]$ for transition t is equal to the number of times t appears in σ. Therefore, if $M \xrightarrow{\sigma} M'$ then $\overrightarrow{\sigma}$ is a nonnegative integer solution of the marking equation. The converse does not hold.

From rendez-vous protocols to Petri nets. Let $\mathcal{P} = (Q, \Sigma, init, fin, R)$ be a rendez-vous protocol. Create a Petri net $\mathcal{N}_\mathcal{P} = (P, T, Pre, Post)$ as follows. The set of places is Q. For each letter $a \in \Sigma$ and for each pair of rules $r = (q, !a, s), r' = (q', ?a, s') \in R$, add a transition $t_{r,r'}$ to $\mathcal{N}_\mathcal{P}$ and set

- $Pre[p,t] = 0$ for every $p \notin \{q, q'\}$, $Post[p,t] = 0$ for every $p \notin \{s, s'\}$
- If $q = q'$ then $Pre[q,t] = -2$, otherwise $Pre[q,t] = Pre[q',t] = -1$
- If $s = s'$ then $Post[s,t] = 2$, otherwise $Post[s,t] = Post[s',t] = 1$.

It is clear that any configuration of a protocol $\mathcal{P}$ is also a marking of $\mathcal{N}_\mathcal{P}$, and vice versa. Further, the following proposition is obvious.

Proposition 1. *For any two configurations C and C' we have that $C \xRightarrow{*} C'$ over the protocol $\mathcal{P}$ iff $C \xrightarrow{*} C'$ over the Petri net $\mathcal{N}_\mathcal{P}$.*

Consequently, the cut-off problem for Petri nets, defined by

Given : A Petri net system $(\mathcal{N}, M, M')$
Decide: Is there $B \in \mathbb{N}$ such that $n \cdot M \xrightarrow{*} n \cdot M'$ for every $n \geq B$?

generalizes the problem for rendez-vous protocols.

3 The cut-off problem for acyclic Petri nets

We show that the cut-off problem for acyclic Petri nets can be solved in polynomial time. The reason for considering this special case first is that it illustrates one of the main ideas of the general case in a very pure form.

Let us fix a Petri net system $(\mathcal{N}, M, M')$ for the rest of this section, where $\mathcal{N} = (P, T, Pre, Post)$ is acyclic and $\mathcal{A}$ is its incidence matrix. It is well-known that in acyclic Petri nets the reachability relation is characterized by the marking equation (see e.g. [24]):

Proposition 2 ([24]). *Let $(\mathcal{N}, M, M')$ be an acyclic Petri net system. For every sequence $\sigma \in T^*$, we have $M \xrightarrow{\sigma} M'$ iff $\overrightarrow{\sigma}$ is a solution of the marking equation. Consequently, $M \xrightarrow{*} M'$ iff the marking equation has a nonnegative integer solution.*

This proposition shows that the reachability problem for acyclic Petri nets reduces to the feasibilty problem (i.e., existence of solutions) of systems of linear diophantine equations over the nonnegative integers. So the reachability problem for acyclic Petri nets is in NP, and in fact both the reachability and the feasibility problems are NP-complete [13].

There are two ways to relax the conditions on the solution so as to make the feasibility problem polynomial. Feasibility over the nonnegative *rationals* and feasibility over all integers are both in P. The first is due to the polynomiality of linear programming. For the second, feasibility can be decided in polynomial time after computing the Smith or Hermite normal forms (see e.g. [29]), which can themselves be computed in polynomial time [19]. We show that the cut-off problem can be reduced to these two relaxed problems.

3.1 Characterizing acyclic systems with cut-offs

Horn and Sangnier proved in [17] a very useful charaterization of the rendez-vous protocols with a cut-off: A rendez-vous protocol $\mathcal{P}$ admits a cut-off iff there exists $n \in \mathbb{N}$ such that $C^n_{init} \xrightarrow{*} C^n_{fin}$ and $C^{n+1}_{init} \xrightarrow{*} C^{n+1}_{fin}$. The proof immediately generalizes to the case of Petri nets:

Lemma 1 ([17]). *A Petri net system $(\mathcal{N}, M, M')$ (acyclic or not) admits a cut-off iff there exists $n \in \mathbb{N}$ such that $n \cdot M \xrightarrow{*} n \cdot M'$ and $(n+1) \cdot M \xrightarrow{*} (n+1) \cdot M'$. Moreover if $n \cdot M \xrightarrow{*} n \cdot M'$ and $(n+1) \cdot M \xrightarrow{*} (n+1) \cdot M'$, then n^2 is a cut-off for the system.*

Using this lemma, we characterize those acyclic Petri net systems which admit a cut-off.

Theorem 1. *An acyclic Petri net system $(\mathcal{N}, M, M')$ admits a cut-off iff the marking equation has solutions $\mathbf{x} \in \mathbb{Q}^T_{\geq 0}$ and $\mathbf{y} \in \mathbb{Z}^T$ such that $[\![\mathbf{y}]\!] \subseteq [\![\mathbf{x}]\!]$.*

Proof. ($\Rightarrow$): Suppose $(\mathcal{N}, M, M')$ admits a cut-off. Hence there exists $b \in \mathbb{N}$ such that for all $n \geq b$ we have $nM \xrightarrow{*} nM'$. Let $bM \xrightarrow{\sigma'} bM'$ and $(b+1)M \xrightarrow{\tau'} (b+1)M'$. Then, notice that $(2b+1)M \xrightarrow{\sigma'\tau'} (2b+1)M'$ and $(2b+2)M \xrightarrow{\tau'\tau'} (2b+2)M'$. Hence, if we let $n = 2b+1$, $\sigma = \sigma'\tau'$ and $\tau = \tau'\tau'$ we have, $nM \xrightarrow{\sigma} nM'$, $(n+1)M \xrightarrow{\tau} (n+1)M'$ and $[\![\tau]\!] \subseteq [\![\sigma]\!]$. By Proposition 2, there exist $\mathbf{x}', \mathbf{y}' \in \mathbb{N}^T$ such that $[\![\mathbf{y}']\!] \subseteq [\![\mathbf{x}']\!]$, $nM' = nM + \mathcal{A}\mathbf{x}'$ and $(n+1)M' = (n+1)M + \mathcal{A}\mathbf{y}'$. Letting $\mathbf{x} = \mathbf{x}'/n$ and $\mathbf{y} = \mathbf{y}' - \mathbf{x}'$, we get our required vectors.

($\Leftarrow$): Suppose $\mathbf{x} \in \mathbb{Q}^T_{\geq 0}$ and $\mathbf{y} \in \mathbb{Z}^T$ are solutions of the marking equation such that $[\![\mathbf{y}]\!] \subseteq [\![\mathbf{x}]\!]$. Let μ be the least common multiple of the denominators of the components of $\mathbf{x}$, and let α be the largest absolute value of the numbers in the vector $\mathbf{y}$. By definition of μ we have $\alpha(\mu\mathbf{x}) \in \mathbb{N}^T$. Also, since $[\![\mathbf{y}]\!] \subseteq [\![\mathbf{x}]\!]$ it follows by definition of α that $\alpha(\mu\mathbf{x}) + \mathbf{y} \geq \mathbf{0}$ and hence $\alpha(\mu\mathbf{x}) + \mathbf{y} \in \mathbb{N}^T$. Since $M' = M + \mathcal{A}\mathbf{x}$ and $M' = M + \mathcal{A}\mathbf{y}$ we get

$$\alpha\mu M' = \alpha\mu M + \mathcal{A}(\alpha\mu\mathbf{x}) \quad \text{and} \quad (\alpha\mu + 1)M' = (\alpha\mu + 1)M + \mathcal{A}(\alpha\mu\mathbf{x} + \mathbf{y})$$

Taking $\alpha\mu = n$, by Proposition 2 we get that $nM \xrightarrow{*} nM'$ and $(n+1)M \xrightarrow{*} (n+1)M'$. By Lemma 1, $(\mathcal{N}, M, M')$ admits a cut-off.

Intuitively, the existence of the rational solution $\mathbf{x} \in \mathbb{Q}^T_{\geq 0}$ guarantees $nM \xrightarrow{*} nM'$ for infinitely many n, and the existence of the integer solution $\mathbf{y} \in \mathbb{Z}^T$ guarantees that for one of those n we have $(n+1)M \xrightarrow{*} (n+1)M'$ as well.

Example 1. The net system given by the net on Figure 1 along with the markings $M = \wr i \int$ and $M' = \wr f \int$ admits a cut-off. The conditions of the theorem are satisfied by $\mathbf{x} = (\frac{1}{5}, \frac{1}{5}, \frac{1}{5}, \frac{1}{5})$ and $\mathbf{y} = (-1, 1, 1, 1)$.

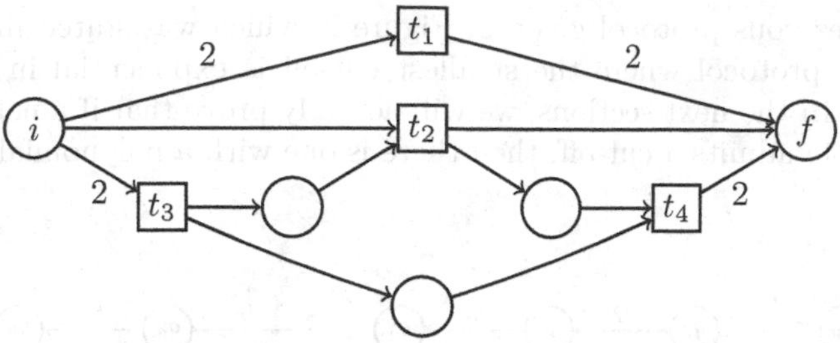

Fig. 1. A net with cut-off 2.

3.2 Polynomial time algorithm

We derive a polynomial time algorithm for the cut-off problem from the characterization of Theorem 1. The first step is the following lemma. A very similar lemma is proved in [14], but since the proof is short we give it for the sake of completeness:

Lemma 2. *If the marking equation is feasible over $\mathbb{Q}_{\geq 0}$, then it has a solution with maximum support. Moreover, such a solution can be found in polynomial time.*

Proof. If $\mathbf{y}, \mathbf{z} \in \mathbb{Q}^T_{\geq 0}$ are solutions of the marking equation, then we have $M' = M + \mathcal{A}((\mathbf{y} + \mathbf{z})/2)$ and $[\![\mathbf{y}]\!] \cup [\![\mathbf{z}]\!] \subseteq [\![(\mathbf{y} + \mathbf{z})/2]\!]$. Hence if the marking equation if feasible over $\mathbb{Q}_{\geq 0}$, then it has a solution with maximum support.

To find such a solution in polynomial time we proceed as follows. For every transition t we solve the linear program $M' = M + \mathcal{A}\mathbf{v}, \mathbf{v} \geq \mathbf{0}, \mathbf{v}(t) > 0$. (Recall that solving linear programs over the rationals can be done in polynomial time). Let $\{t_1, \ldots, t_n\}$ be the set of transitions whose associated linear programs are feasible over $\mathbb{Q}^T_{\geq 0}$, and let $\{\mathbf{u}_1, \ldots, \mathbf{u}_n\}$ be solutions to these programs. Then $1/n \cdot \sum_{i=1}^{n} \mathbf{u}_i$ is a solution of the marking equation with maximum support.

We now have all the ingredients to give a polynomial time algorithm.

Theorem 2. *The cut-off problem for acyclic net systems can be solved in polynomial time.*

Proof. First, we check that the marking equation has a solution over the non-negative rationals. If such a solution does not exist, by Theorem 1 the given net system does not admit a cut-off.

Suppose such a solution exists. By Lemma 2 we can find a non-negative rational solution $\mathbf{x}$ with maximum support in polynomial time. Let U contain all the transitions t such that $\mathbf{x}_t = 0$. We now check in polynomial time if the marking equation has a solution $\mathbf{y}$ over $\mathbb{Z}^T$ such that $\mathbf{y}_t = 0$ for every $t \in U$. By Theorem 1 such a solution exists iff the net system admits a cut-off.

The rendez-vous protocol given in Figure 2, which was stated in [17], is an example of a protocol where the smallest cut-off is exponential in the size of the protocol. In the next sections, we will actually prove that if a net system $\mathcal{N}$ (acyclic or not) admits a cut-off, then there is one with a polynomial number of bits in $|\mathcal{N}|$.

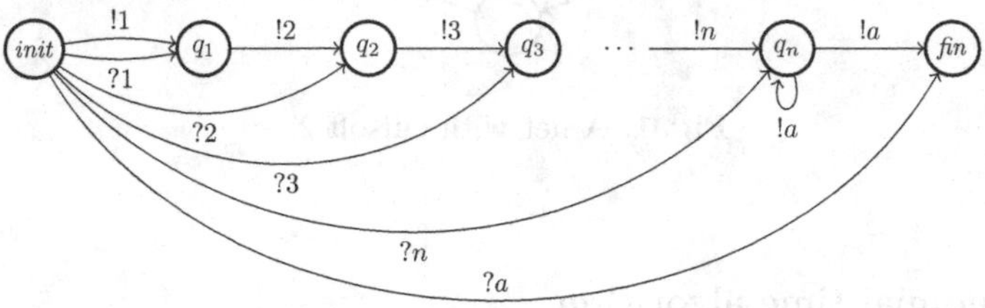

Fig. 2. Example of a protocol with an exponential cut-off

4 The Scaling and Insertion lemmas

Similar to the case of acyclic net systems, we would like to provide a characterization of net systems admitting a cut-off and then use this characterization to derive a polynomial time algorithm. Unfortunately, in general net systems there is no characterization of reachability akin to Proposition 2 for acyclic systems. To this end, we prove two intermediate lemmas to help us come up with a characterization for cut-off admissible net systems in the general case. We believe that these two lemmas could be of independent interest in their own right. Further, the proofs of both lemmas are provided so that it will enable us later on to derive a bound on the cut-off for net systems.

4.1 The Scaling Lemma

The Scaling Lemma shows that, given a Petri net system $(\mathcal{N}, M, M')$, whether $nM \xrightarrow{*} nM'$ holds for some $n \geq 1$ can be decided in polynomial time; moreover, if $nM \xrightarrow{*} nM'$ holds for some n, then it holds for some n with at most $(|\mathcal{N}|(\log \|M\| + \log \|M'\|))^{O(1)}$ bits. The name of the lemma is due to the fact that the firing sequence leading from nM to nM' is obtained by *scaling up* a *continuous firing sequence* from M to M'; the existence of such a continuous sequence can be decided in polynomial time [14].

In the rest of the section we first recall continuous Petri nets and the charaterization of [14], and then present the Scaling Lemma[2].

[2] The lemma is implicitly proved in [14], but the bound on the size of n is hidden in the details of the proof, and we make it explicit.

Reachability in continuous Petri nets. Petri nets can be given a *continuous semantics* (see e.g. [1,30,14]), in which markings are continuous multisets; we call them *continuous markings.* A continuous marking M enables a transition t *with factor* $\lambda \in \mathbb{Q}_{\geq 0}$ if $M(p) \geq \lambda \cdot Pre[p,t]$ for every place p; we also say that M enables λt. If M enables λt, then λt can fire or occur, leading to a new marking M' given by $M'(p) = M(p) + \lambda \cdot \mathcal{A}[p,t]$ for every $p \in P$. We denote this by $M \xrightarrow[\mathbb{Q}]{\lambda t} M'$, and say that M' is reached from M by firing λt. A *continuous firing sequence* is any sequence of transitions $\sigma = \lambda_1 t_1, \lambda_2 t_2, \ldots, \lambda_k t_k \in (\mathbb{Q}_{\geq 0} \times T)^*$. We let $M \xrightarrow[\mathbb{Q}]{\sigma} M'$ denote that there exist continuous markings $M_1, \ldots, M_{k-1}$ such that $M \xrightarrow[\mathbb{Q}]{\lambda_1 t_1} M_1 \xrightarrow[\mathbb{Q}]{\lambda_2 t_2} M_2 \cdots M_{k-1} \xrightarrow[\mathbb{Q}]{\lambda_k t_k} M'$. Further, $M \xrightarrow[\mathbb{Q}]{*} M'$ denotes that $M \xrightarrow[\mathbb{Q}]{\sigma} M'$ holds for some continuous firing sequence σ.

The *Parikh image* of $\sigma = \lambda_1 t_1, \lambda_2 t_2, \ldots, \lambda_k t_k \in (\mathbb{Q}_{\geq 0} \times T)^*$ is the vector $\overrightarrow{\sigma} \in \mathbb{Q}^T_{\geq 0}$ where $\overrightarrow{\sigma}[t] = \sum_{i=1}^{k} \delta_{i,t}\lambda_i$, where $\delta_{i,t} = 1$ if $t_i = t$ and 0 otherwise. The support of σ is the support of its Parikh image $\overrightarrow{\sigma}$. If $M \xrightarrow[\mathbb{Q}]{\sigma} M'$ then $\overrightarrow{\sigma}$ is a solution of the marking equation over $\mathbb{Q}^T_{\geq 0}$, but the converse does not hold. In [14], Fraca and Haddad strengthen this necessary condition to make it also sufficient, and use the resulting characterization to derive a polynomial algorithm.

Theorem 3 ([14]). *Let $(\mathcal{N}, M, M')$ be a Petri net system.*

- *$M \xrightarrow[\mathbb{Q}]{\sigma} M'$ iff $\overrightarrow{\sigma}$ is a solution of the marking equation over $\mathbb{Q}^T_{\geq 0}$, and there exist continuous firing sequences τ, τ' and continuous markings L and L' such that $[\![\tau]\!] = [\![\sigma]\!] = [\![\tau']\!]$, $M \xrightarrow[\mathbb{Q}]{\tau} L$, and $L' \xrightarrow[\mathbb{Q}]{\tau'} M'$.*
- *It can be decided in polynomial time if $M \xrightarrow[\mathbb{Q}]{*} M'$ holds.*

Scaling. It follows easily from the definitions that $nM \xrightarrow{*} nM'$ holds for some $n \geq 1$ iff $M \xrightarrow[\mathbb{Q}]{*} M'$. Indeed, if $M \xrightarrow[\mathbb{Q}]{\sigma} M'$ for some $\sigma = \lambda_1 t_1, \lambda_2 t_2, \ldots, \lambda_k t_k \in (\mathbb{Q}_{\geq 0} \times T)^*$, then we can scale this continuous firing sequence to a discrete sequence $nM \xrightarrow[\mathbb{Q}]{n\sigma} nM'$ where n is the smallest number such that $n\lambda_1, \ldots, n\lambda_k \in \mathbb{N}$, and $n\sigma = t_1^{n\lambda_1} t_2^{n\lambda_2} \ldots t_k^{n\lambda_k}$. So Theorem 3 immediately implies that the existence of $n \geq 1$ such that $nM \xrightarrow{*} nM'$ can be decided in polynomial time. The following lemma also gives a bound on n.

Lemma 3. *Let $(\mathcal{N}, M, M')$ be a Petri net system with weight w such that $M \xrightarrow[\mathbb{Q}]{\sigma} M'$ for some continuous firing sequence $\sigma \in (\mathbb{Q}_{\geq 0} \times T)^*$. Let m be the number of transitions in $[\![\sigma]\!]$ and let ℓ be $\|\overrightarrow{\sigma}\|$. Let k be the smallest natural number such that $k\overrightarrow{\sigma} \in \mathbb{N}^T$. Then, there exists a firing sequence $\tau \in T^*$ such that $[\![\tau]\!] = [\![\sigma]\!]$ and*

$$\left(16w(w+1)^{2m}k\ell \cdot M\right) \xrightarrow{\tau} \left(16w(w+1)^{2m}k\ell \cdot M'\right)$$

Lemma 4. (Scaling Lemma). *Let $(\mathcal{N}, M, M')$ be a Petri net system such that $M \xrightarrow[\mathbb{Q}]{\sigma} M'$. There exists a number n with a polynomial number of bits in $|\mathcal{N}|(\log \|M\| + \log \|M'\|)$ such that $nM \xrightarrow{\tau} nM'$ for some τ with $[\![\tau]\!] = [\![\sigma]\!]$.*

4.2 The Insertion Lemma

In the acyclic case, the existence of a cut-off is characterized by the existence of solutions to the marking equation $\mathbb{Q}^T_{\geq 0}$ and $\mathbb{Z}^T$. Intuitively, in the general case we replace the existence of solutions over $\mathbb{Q}^T_{\geq 0}$ by the conditions of the Scaling Lemma, and the existence of solutions over $\mathbb{Z}^T$ by the Insertion Lemma:

Lemma 5 (Insertion Lemma). *Let M, M', L, L' be markings of $\mathcal{N}$ satisfying $M \xrightarrow{\sigma} M'$ for some $\sigma \in T^*$ and $L' = L + \mathcal{A}\mathbf{y}$ for some $\mathbf{y} \in \mathbb{Z}^T$ such that $[\![\mathbf{y}]\!] \subseteq [\![\sigma]\!]$. Then $\mu M + L \xrightarrow{*} \mu M' + L'$ for $\mu = \|\mathbf{y}\|(\|\overrightarrow{\sigma}\|nw + nw + 1)$, where w is the weight of $\mathcal{N}$, and n is the number of places in ${}^\bullet[\![\sigma]\!]$.*

The idea of the proof is a follows: In a first stage, we asynchronously execute multiple "copies" of the firing sequence σ from multiple "copies" of the marking M, until we reach a marking at which all places of ${}^\bullet[\![\sigma]\!]$ contain a sufficiently large number of tokens. At this point we temporarily interrupt the executions of the copies of σ to *insert* a firing sequence with Parikh mapping $\|\mathbf{y}\|\overrightarrow{\sigma} + \mathbf{y}$. The net effect of this sequence is to transfer some copies of M to M', leaving the other copies untouched, and exactly one copy of L to L'. In the third stage, we resume the interrupted executions of the copies of σ, which completes the transfer of the remaining copies of M to M' .

Proof. Let $\mathbf{x}$ be the Parikh image of σ, i.e., $\mathbf{x} = \overrightarrow{\sigma}$. Since $M \xrightarrow{\sigma} M'$, by the marking equation we have $M' = M + \mathcal{A}\mathbf{x}$

First stage: Let $\lambda_x = \|x\|$, $\lambda_y = \|y\|$ and $\mu = \lambda_y(\lambda_x nw + nw + 1)$. Let $\sigma := r_1, r_2, \ldots, r_k$ and let $M =: M_0 \xrightarrow{r_1} M_1 \xrightarrow{r_2} M_2 \ldots M_{k-1} \xrightarrow{r_k} M_k := M$. Notice that for each place $p \in {}^\bullet[\![\sigma]\!]$, there exists a marking $M_{i_p} \in \{M_0, \ldots, M_{k-1}\}$ such that $M_{i_p}(p) > 0$.

Since each of the markings in $\{M_{i_p}\}_{p \in {}^\bullet[\![\sigma]\!]}$ can be obtained from M by firing a (suitable) prefix of σ, it is easy to see that from the marking $\mu M + L = \lambda_y M + L + (\lambda_x \lambda_y nw + \lambda_y nw)M$ we can reach the marking $\texttt{First} := \lambda_y M + L + \sum_{p \in {}^\bullet[\![\sigma]\!]} (\lambda_x \lambda_y w + \lambda_y w) M_{i_p}$. This completes our first stage.

Second stage - Insert: Since $[\![\mathbf{y}]\!] \subseteq [\![\sigma]\!]$, if $\mathbf{y}(t) \neq 0$ then $\mathbf{x}(t) \neq 0$. Since $\mathbf{x}(t) \geq 0$ for every transition, it now follows that $(\lambda_y \mathbf{x} + \mathbf{y})(t) \geq 0$ for every transition t and $(\lambda_y \mathbf{x} + \mathbf{y})(t) > 0$ precisely for those transitions in $[\![\sigma]\!]$.

Let ξ be any firing sequence such that $\overrightarrow{\xi} = \lambda_y \mathbf{x} + \mathbf{y}$. Notice that for every place $p \in {}^\bullet[\![\sigma]\!]$, $\texttt{First}(p) \geq \lambda_x \lambda_y w + \lambda_y w \geq \|(\lambda_y \mathbf{x} + \mathbf{y})\| \cdot w$. By an easy induction on $\|\xi\|$, it follows that that $\texttt{First} \xrightarrow{\xi} \texttt{Second}$ for some marking $\texttt{Second}$. By the marking equation, it follows that $\texttt{Second} = \lambda_y M' + L' + \sum_{p \in {}^\bullet[\![\sigma]\!]} (\lambda_x \lambda_y w + \lambda_y w) M_{i_p}$. This completes our second stage.

Third stage: Notice that for each place $p \in {}^\bullet[\![\sigma]\!]$, by construction of M_{i_p}, there is a firing sequence which takes the marking M_{i_p} to the marking M'. It then follows that there is a firing sequence which takes the marking $\texttt{Second}$ to the marking $\lambda_y M' + L' + \sum_{p \in {}^\bullet[\![\sigma]\!]} (\lambda_x \lambda_y w + \lambda_y w) M' = \mu M' + L'$. This completes our third stage and also completes the desired firing sequence from $\mu M + L$ to $\mu M' + L'$.

5 Polynomial time algorithm for the general case

Let $(\mathcal{N}, M, M')$ be a net system with $\mathcal{N} = (P, T, Pre, Post)$, such that $\mathcal{A}$ is its incidence matrix. As in Section 3, we first characterize the Petri net systems that admit a cut-off, and then provide a polynomial time algorithm.

5.1 Characterizing systems with cut-offs

We generalize the characterization of Theorem 1 for acyclic Petri net systems to general systems.

Theorem 4. *A Petri net system $(\mathcal{N}, M, M')$ admits a cut-off iff there exists some rational firing sequence σ such that $M \xrightarrow[\mathbb{Q}]{\sigma} M'$ and the marking equation has a solution $\mathbf{y} \in \mathbb{Z}^T$ such that $[\![\mathbf{y}]\!] \subseteq [\![\sigma]\!]$.*

Proof. ($\Rightarrow$): Assume $(\mathcal{N}, M, M')$ admits a cut-off. Hence there exists $B \in \mathbb{N}$ such that for all $n \geq B$ we have $nM \xrightarrow{*} nM'$. Similar to the proof of theorem 1, we can show that there exist $n \in \mathbb{N}$ and firing sequences τ, τ' such that $nM \xrightarrow{\tau} nM'$, $(n+1)M \xrightarrow{\tau'} (n+1)M'$ and $[\![\tau']\!] \subseteq [\![\tau]\!]$.

Let $\tau = t_1 t_2 \cdots t_k$. Construct the rational firing sequence $\sigma := t_1/n\, t_2/n \cdots t_k/n$. From the fact that $nM \xrightarrow{\tau} nM'$, we can easily conclude by induction on k that $M \xrightarrow[\mathbb{Q}]{\sigma} M'$. Further, by the marking equation we have $nM' = nM + \mathcal{A}\overrightarrow{\tau}$ and $(n+1)M' = (n+1)M + \mathcal{A}\overrightarrow{\tau'}$. Let $\mathbf{y} = \overrightarrow{\tau'} - \overrightarrow{\tau}$. Then $\mathbf{y} \in \mathbb{Z}^T$ and $M' = M + \mathcal{A}\mathbf{y}$. Further, since $[\![\tau']\!] \subseteq [\![\tau]\!] = [\![\sigma]\!]$, we have $[\![\mathbf{y}]\!] \subseteq [\![\sigma]\!]$.

($\Leftarrow$): Assume there exists a rational firing sequence σ and a vector $\mathbf{y} \in \mathbb{Z}^T$ such that $[\![\mathbf{y}]\!] \subseteq [\![\sigma]\!]$, $M \xrightarrow[\mathbb{Q}]{\sigma} M'$ and $M' = M + \mathcal{A}\mathbf{y}$. Let $s = |\mathcal{N}|(\log \|M\| + \log \|M'\|)$. It is well known that if a system of linear equations over the integers is feasible, then there is a solution which can be described using a number of bits which is polynomial in the size of the input (see e.g. [20]). Hence, we can assume that $\|\mathbf{y}\|$ can be described using $s^{O(1)}$ bits.

By Lemma 4 there exists n (which can be described using $s^{O(1)}$ bits) and a firing sequence τ with $[\![\tau]\!] = [\![\sigma]\!]$ such that $nM \xrightarrow{\tau} nM'$. Hence $knM \xrightarrow{*} knM'$ is also possible for any $k \in \mathbb{N}$. By Lemma 5, there exists μ (which can once again be described using $s^{O(1)}$ bits) such that $\mu nM + M \xrightarrow{*} \mu nM' + M'$ is possible. By Lemma 1 the system $(\mathcal{N}, M, M')$ admits a cut-off with a polynomial number of bits in s.

Notice that we have actually proved that if a net system admits a cut-off then it admits a cut-off with a polynomial number of bits in its size. Since the cut-off problem for a rendez-vous protocol $\mathcal{P}$ can be reduced to a cut-off problem for the Petri net system $(\mathcal{N}_{\mathcal{P}}, \llbracket init \rrbracket, \llbracket fin \rrbracket)$, it follows that,

Corollary 1. *If the system $(\mathcal{N}, M, M')$ admits a cut-off then it admits a cut-off with a polynomial number of bits in $|\mathcal{N}|(\log \|M\| + \log \|M'\|)$. Hence, if a rendez-vous protocol $\mathcal{P}$ admits a cut-off then it admits a cut-off with a polynomial number of bits in $|\mathcal{P}|$.*

5.2 Polynomial time algorithm

We use the characterization given in the previous section to provide a polynomial time algorithm for the cut-off problem. The following lemma, which was proved in [14] and whose proof is given in the appendix, enables us to find a firing sequence between two markings with maximum support.

Lemma 6. *[14] Among all the rational firing sequences σ such that $M \xrightarrow[\mathbb{Q}]{\sigma} M'$, there is one with maximum support. Moreover, the support of such a firing sequence can be found in polynomial time.*

We now have all the ingredients to prove the existence of a polynomial time algorithm.

Theorem 5. *The cut-off problem for net systems can be solved in polynomial time.*

Proof. First, we check that there is a rational firing sequence σ with $M \xrightarrow[\mathbb{Q}]{\sigma} M'$, which can be done in polynomial time by ([14], Proposition 27). If such a sequence does not exist, by Theorem 4 the given net system does not admit a cut-off.

Suppose such a sequence exists. By Lemma 6 we can find in polynomial time, the maximum support S of all the firing sequences τ such that $M \xrightarrow[\mathbb{Q}]{\tau} M'$. We now check in polynomial time if the marking equation has a solution $\mathbf{y}$ over $\mathbb{Z}^T$ such that $\mathbf{y}(t) = 0$ for every $t \notin S$. By Theorem 4 such a solution exists iff the net system admits a cut-off.

This immediately proves that the cut-off problem for rendez-vous protocols is also in polynomial time. By an easy logspace reduction from the Circuit Value Problem [21], we prove that

Lemma 7. *The cut-off problem for rendez-vous protocols is P-hard.*

Clearly, this also proves that the cut-off problem for Petri nets is P-hard.

6 Symmetric rendez-vous protocols

In [17] Horn and Sangnier introduce symmetric rendez-vous protocols, where sending and receiving a message at each state has the same effect, and show that the cut-off problem is in NP. We improve on their result and shown that it is in NC.

Recall that NC is the set of problems in P that can be solved in polylogarithmic *parallel* time, i.e., problems which can be solved by a uniform family of circuits with polylogarithmic depth and polynomial number of gates. Two well-known problems which lie in NC are graph reachability and feasibility of linear equations over the finite field $\mathbb{F}_2$ of size 2 [27,23]. We proceed to formally define symmetric protocols and state our results.

Definition 3. *A rendez-vous protocol* $\mathcal{P} = (Q, \Sigma, init, fin, R)$ *is* symmetric, *iff its set of rules is symmetric under swapping* $!a$ *and* $?a$ *for each* $a \in \Sigma$, *i.e., for each* $a \in \Sigma$, *we have* $(q, !a, q') \in R$ *iff* $(q, ?a, q') \in R$.

Horn and Sangnier show that, because of their symmetric nature, there is a very easy characterization for cut-off admitting symmetric protocols.

Proposition 3. *([17], Lemma 18) A symmetric protocol* $\mathcal{P}$ *admits a cut-off iff there exists an even number e and an odd number o such that* $C^e_{init} \xrightarrow{*} C^e_{fin}$ *and* $C^o_{init} \xrightarrow{*} C^o_{fin}$.

From a symmetric protocol $\mathcal{P}$, we can derive a graph $G(\mathcal{P})$ where the vertices are the states and there is an edge between q and q' iff there exists $a \in \Sigma$ such that $(q, a, q') \in R$. The following proposition is immediate from the definition of symmetric protocols:

Proposition 4. *Let* $\mathcal{P}$ *be a symmetric protocol. There exists an even number e such that* $C^e_{init} \xrightarrow{*} C^e_{fin}$ *iff there is a path from init to fin in the graph* $G(\mathcal{P})$.

Proof. The left to right implication is obvious. For the other side, suppose there is a path $init, q_1, q_2, \ldots, q_{m-1}, fin$ in the graph $G(\mathcal{P})$. Then notice that $⦃2 \cdot init⦄ \rightarrow ⦃2 \cdot q_1⦄ \rightarrow ⦃2 \cdot q_2⦄ \cdots \rightarrow ⦃2 \cdot q_{m-1}⦄ \rightarrow ⦃2 \cdot q_f⦄$ is a valid run of the protocol.

Since graph reachability is in NC , this takes care of the "even" case from Proposition 3. Hence, we only need to take care of the "odd" case from Proposition 3.

Fix a symmetric protocol $\mathcal{P}$ for the rest of the section. As a first step, for each state $q \in Q$, we compute if there is a path from *init* to q and if there is a path from q to *fin* in the graph $\mathcal{G}(\mathcal{P})$. Since graph reachability is in NC this computation can be carried out in NC by parallely running graph reachability for each $q \in Q$. If such paths exist for a state q then we call q a good state, and otherwise a bad state. The following proposition easily follows from the symmetric nature of $\mathcal{P}$:

Proposition 5. *If* $q \in Q$ *is a good state, then* $⦃2 \cdot init⦄ \xrightarrow{*} ⦃2 \cdot q⦄$ *and* $⦃2 \cdot q⦄ \xrightarrow{*} ⦃2 \cdot fin⦄$.

Similar to the general case of rendez-vous protocols, given a symmetric protocol $\mathcal{P}$ we can construct a Petri net $\mathcal{N}_{\mathcal{P}}$ whose places are the states of $\mathcal{P}$ and which faithfully represents the reachability relation of configurations of $\mathcal{P}$. Observe that this construction can be carried out in parallel over all the states in Q and over all pairs of rules in R. Let $\mathcal{N} = (P, T, Pre, Post)$ be the Petri net that we construct out of the symmetric protocol $\mathcal{P}$ and let $\mathcal{A}$ be its incidence matrix. We now write the marking equation for $\mathcal{N}$ as follows: We introduce a variable $\mathbf{v}[t]$ for each transition $t \in T$ and we construct an equation system Eq enforcing the following three conditions:

- $\mathbf{v}[t] = 0$ for every $t \in T$ such that ${}^{\bullet}t \cup t^{\bullet}$ contains a bad state.
  By definition of a bad state, such transitions will never be fired on any run from an initial to a final configuration and so our requirement is safe.

- $\sum_{t \in T} \mathcal{A}[q,t] \cdot \mathbf{v}[t] = 0$ for each $q \notin \{init, fin\}$.
  Notice that the net-effect of any run from an initial to a final configuration on any state not in $\{init, fin\}$ is 0 and hence this condition is valid as well.
- $\sum_{t \in T} \mathcal{A}[init,t] \cdot \mathbf{v}[t] = -1$ and $\sum_{t \in T} \mathcal{A}[fin,t] \cdot \mathbf{v}[t] = 1$.

It is clear that the construction of Eq can be carried out in parallel over each $q \in Q$ and each $t \in T$. Finally, we solve Eq over arithmetic modulo 2, i.e., we solve Eq over the field $\mathbb{F}_2$ which as mentioned before can be done in NC. We have:

Lemma 8. *There exists an odd number o such that $C^o_{init} \xrightarrow{*} C^o_{fin}$ iff the equation system Eq has a solution over $\mathbb{F}_2$.*

Proof. (Sketch.) The left to right implication is true because of taking modulo 2 on both sides of the marking equation. For the other side, we use an idea similar to Lemma 5. Let $\mathbf{x}$ be a solution to Eq over $\mathbb{F}_2$. Using Proposition 5 we first populate all the good states of Q with enough processes such that all the good states except *init* have an even number of processes. Then, we fire exactly once, all the transitions t such that $\mathbf{x}[t] = 1$. Since $\mathbf{x}$ satisfies Eq, we can now argue that in the resulting configuration, the number of processes at each bad state is 0 and the number of processes in each good state except *fin* is even. Hence, we can once again use Proposition 5 to conclude that we can move all the processes which are not at *fin* to the final state *fin*.

Theorem 6. *The problem of deciding whether a symmetric protocol admits a cut-off is in* NC.

Proof. By Proposition 3 it suffices to find an even number e and an odd number o such that $C^e_{init} \xrightarrow{*} C^e_{fin}$ and $C^o_{init} \xrightarrow{*} C^o_{fin}$. By Proposition 4 the former can be done in NC. By Lemma 8 and by the fact that the equation system Eq can be constructed and solved in NC, it follows that the latter can also be done in NC.

7 Symmetric protocols with leaders

In this section, we extend symmetric rendez-vous protocols by adding a special process called leader. We state the cut-off problem for such protocols and prove that it is NP-complete.

Definition 4. *A symmetric leader protocol is a pair of symmetric protocols $\mathcal{P} = (\mathcal{P}^L, \mathcal{P}^F)$ where $\mathcal{P}^L = (Q^L, \Sigma, init^L, fin^L, R^L)$ is the leader protocol and $\mathcal{P}^F = (Q^F, \Sigma, init^F, fin^F, R^F)$ is the follower protocol where $Q^L \cap Q^F = \emptyset$.*

A configuration of a symmetric leader protocol $\mathcal{P}$ is a multiset over $Q^L \cup Q^F$ such that $\sum_{q \in Q^L} C(q) = 1$. This corresponds to the intuition that exactly one process can execute the leader protocol. For each $n \in \mathbb{N}$, let C^n_{init} (resp. C^n_{fin}) denote the initial (resp. final) configuration of $\mathcal{P}$ given by $C^n_{init}(init^L) = 1$ (resp. $C^n_{fin}(fin^L) = 1$) and $C^n_{init}(init^F) = n$ (resp. $C^n_{fin}(fin^F) = n$). We say that $C \Rightarrow C'$

if there exists $(p, !a, p'), (q, ?a, q') \in R^L \cup R^F$, $C \geq \llbracket p, q \rrbracket$ and $C' = C - \llbracket p, q \rrbracket + \llbracket p', q' \rrbracket$. Since we allow at most one process to execute the leader protocol, given a configuration C, we can let $lead(C)$ denote the unique state $q \in Q^L$ such that $C(q) > 0$.

Definition 5. *The cut-off problem for symmetric leader protocols is the following.*

Input: A symmetric leader protocol $\mathcal{P} = (\mathcal{P}^L, \mathcal{P}^F)$.
Output: Is there $B \in \mathbb{N}$ such that for all $n \geq B$, $C^n_{init} \overset{*}{\Rightarrow} C^n_{fin}$.

We know the following fact regarding symmetric leader protocols.

Proposition 6. *([17], Lemma 18) A symmetric leader protocol admits a cut-off iff there exists an even number e and an odd number o such that $C^e_{init} \overset{*}{\Rightarrow} C^e_{fin}$ and $C^o_{init} \overset{*}{\Rightarrow} C^o_{fin}$.*

The main theorem of this section is

Theorem 7. *The cut-off problem for symmetric leader protocols is* NP*-complete*

7.1 A non-deterministic polynomial time algorithm

Let $\mathcal{P} = (\mathcal{P}^L, \mathcal{P}^F)$ be a symmetric leader protocol with $\mathcal{P}^L = (Q^L, \Sigma, init^L, fin^L, R^L)$ and $\mathcal{P}^F = (Q^F, \Sigma, init^F, fin^F, R^F)$. Similar to the previous section, from $\mathcal{P}^F$ we can construct a graph $\mathcal{G}(\mathcal{P}^F)$ where the vertices are given by the states Q^F and the edges are given by the rules in R^F. In $\mathcal{G}(\mathcal{P}^F)$, we can clearly remove all vertices which are not reachable from the state $init^F$ and which do not have a path to fin^F. In the sequel, we will assume that such vertices do not exist in $\mathcal{G}(\mathcal{P}^F)$.

Similar to the general case, we will construct a Petri net $\mathcal{N}_\mathcal{P}$ from the given symmetric leader protocol $\mathcal{P}$. However, the construction is made slightly complicated due to the presence of a leader.

From $\mathcal{P} = (\mathcal{P}^L, \mathcal{P}^F)$, we construct a Petri net $\mathcal{N} = (P, T, Pre, Post)$ as follows: Let P be $Q^L \cup Q^F$. For each $a \in \Sigma$ and $r = (q, !a, s), r' = (q', ?a, s') \in R^L \cup R^F$ such that *at most one of r and r' belongs to R^L*, we will have a transition $t_{r,r'} \in T$ in $\mathcal{N}$ such that

- $Pre[p, t] = 0$ for every $p \notin \{q, q'\}$, $Post[p, t] = 0$ for every $p \notin \{s, s'\}$
- If $q = q'$ then $Pre[q, t] = -2$, otherwise $Pre[q, t] = Pre[q', t] = -1$
- If $s = s'$ then $Post[s, t] = 2$, otherwise $Post[s, t] = Post[s', t] = 1$.

Transitions $t_{r,r'}$ in which exactly one of r, r' is in R^L will be called *leader transitions* and transitions in which both of r, r' are in R^F will be called *follower-only transitions*. Notice that if t is a leader transition, then there is a unique place $p \in {}^\bullet t \cap Q^L$ and a unique place $p \in t^\bullet \cap Q^L$. These places will be denoted by $t.from$ and $t.to$ respectively.

As usual, we let $\mathcal{A}$ denote the incidence matrix of the constructed net $\mathcal{N}$. The following proposition is obvious from the construction of the net $\mathcal{N}$

Proposition 7. *For two configurations C and C', we have that $C \xRightarrow{*} C'$ in the protocol $\mathcal{P}$ iff $C \xrightarrow{*} C$ in the net $\mathcal{N}$.*

Because $\mathcal{P}$ is symmetric we have the following fact, which is easy to verify.

Proposition 8. *If $q \in Q^F$, then $\wr 2 \cdot init^F \int \xrightarrow{*} \wr 2 \cdot q \int \xrightarrow{*} \wr 2 \cdot fin^F \int$*

For any vector $\mathbf{x} \in \mathbb{N}^T$, we define $lead(\mathbf{x})$ to be the set of all leader transitions such that $\mathbf{x}[t] > 0$. The graph of the vector $\mathbf{x}$, denoted by $\mathcal{G}(\mathbf{x})$ is defined as follows: The set of vertices is the set $\{t.from : t \in lead(\mathbf{x})\} \cup \{t.to : t \in lead(\mathbf{x})\}$. The set of edges is the set $\{(t.from, t.to) : t \in lead(\mathbf{x})\}$. Further, for any two vectors $\mathbf{x}, \mathbf{y} \in \mathbb{N}^T$ and a transition $t \in T$, we say that $\mathbf{x} = \mathbf{y}[t\text{--}]$ iff $\mathbf{x}[t] = \mathbf{y}[t] - 1$ and $\mathbf{x}[t'] = \mathbf{y}[t']$ for all $t' \neq t$.

Definition 6. *Let C be a configuration and let $\mathbf{x} \in \mathbb{N}^T$. We say that the pair $(C, \mathbf{x})$ is compatible if $C + \mathcal{A}\mathbf{x} \geq \mathbf{0}$ and every vertex in $\mathcal{G}(\mathbf{x})$ is reachable from $lead(C)$.*

The following lemma states that *as long as there are enough followers in every state*, it is possible for the leader to come up with a firing sequence from a compatible pair.

Lemma 9. *Suppose $(C, \mathbf{x})$ is a compatible pair such that $C(q) \geq 2\|\mathbf{x}\|$ for every $q \in Q^F$. Then there is a configuration D and a firing sequence ξ such that $C \xrightarrow{\xi} D$ and $\overrightarrow{\xi} = \mathbf{x}$.*

Proof. (Sketch.) We prove by induction on $\|\mathbf{x}\|$. If $\mathbf{x}[t] > 0$ for some follower-only transition, then it is easy to verify that if we let C' be such that $C \xrightarrow{t} C'$ and $\mathbf{x}'$ be $\mathbf{x}[t\text{--}]$, then $(C', \mathbf{x}')$ is compatible and $C(q) \geq 2\|\mathbf{x}'\|$ for every $q \in Q^F$.

Suppose $\mathbf{x}[t] > 0$ for some leader transition. Let $p = lead(C)$. If p belongs to some cycle $S = p, r_1, p_1, r_2, p_2, \ldots, p_k, r_{k+1}, p$ in the graph $\mathcal{G}(\mathbf{x})$, then we let $C \xrightarrow{r_1} C'$ and $\mathbf{x}' = \mathbf{x}[t\text{--}]$. It is easy to verify that $C' + \mathcal{A}\mathbf{x}' \geq \mathbf{0}$, $C'(q) \geq 2\|\mathbf{x}'\|$ for every $q \in Q^F$ and $lead(C') = p_1$. Any path P in $\mathcal{G}(\mathbf{x})$ from p to some vertex s either goes through p_1 or we can use the cycle S to traverse from p_1 to p first and then use P to reach s. This gives a path from p_1 to every vertex s in $\mathcal{G}(\mathbf{x}')$.

If p does not belong to any cycle in $\mathcal{G}(\mathbf{x})$, then using the fact that $C + \mathcal{A}\mathbf{x} \geq 0$, we can show that there is exactly one out-going edge t from p in $\mathcal{G}(\mathbf{x})$. We then let $C \xrightarrow{t} C'$ and $\mathbf{x}' = \mathbf{x}[t\text{--}]$. Since any path in $\mathcal{G}(\mathbf{x})$ from p has to necessarily use this edge t, it follows that in $\mathcal{G}(\mathbf{x}')$ there is a path from $t.to = lead(C')$ to every vertex.

Lemma 10. *Let $par \in \{0, 1\}$. There exists $k \in \mathbb{N}$ such that $C^k_{init} \xrightarrow{*} C^k_{fin}$ and $k \equiv par \pmod 2$ iff there exists $n \in \mathbb{N}$, $\mathbf{x} \in \mathbb{N}^T$ such that $n \equiv par \pmod 2$, $(C^n_{init}, \mathbf{x})$ is compatible and $C^n_{fin} = C^n_{init} + \mathcal{A}\mathbf{x}$.*

Proof. (Sketch.) The left to right implication is easy and follows from the marking equation along with induction on the number of leader transitions in the run. For the other side, we use an idea similar to Lemma 5. Let $(C^n_{init}, \mathbf{x})$ be the given compatible pair. We first use Proposition 8 to populate all the states of Q^F with enough processes such that all the states of Q^F except $init^F$ have an even number of processes. Then we use Lemma 9 to construct a firing sequence ξ which can be fired from C^n_{init} and such that $\overrightarrow{\xi} = \mathbf{x}$. By means of the marking equation, we then argue that in the resulting configuration, the leader is in the final state, n followers are in the state fin^F and every other follower state has an even number of followers. Once again, using Proposition 8 we can now move all the processes which are not at fin^F to the final state fin^F.

Lemma 11. *Given a symmetric leader protocol, checking whether a cut-off exists can be done in NP.*

Proof. By Proposition 6 it suffices to find an even number e and an odd number o such that $C^e_{init} \xrightarrow{*} C^e_{fin}$ and $C^o_{init} \xrightarrow{*} C^o_{fin}$. Suppose we want to check that there exists $2k \in \mathbb{N}$ such that $C^{2k}_{init} \xrightarrow{*} C^{2k}_{fin}$. We first non-deterministically guess a set of leader transitions $S = \{t_1, \ldots, t_k\}$ and check that for each $t \in S$, we can reach $t.from$ and $t.to$ from $init^L$ using only the transitions in S.

Once we have guessed all this, we write a polynomially sized integer linear program as follows: We let $\mathbf{v}$ denote $|T|$ variables, one for each transition in T and we let n be another variable, with all these variables ranging over $\mathbb{N}$. We then enforce the following conditions: $C^{2n}_{fin} = C^{2n}_{init} + \mathcal{A}\mathbf{v}$ and $\mathbf{v}[t] = 0 \iff t \notin S$ and solve the resulting linear program, which we can do in non-deterministic polynomial time [26]. If there exists a solution, then we accept. Otherwise, we reject.

By Lemma 10 and by the definition of compatibility, it follows that at least one of our guesses gets accepted iff there exists $2k \in \mathbb{N}$ such that $C^{2k}_{init} \xrightarrow{*} C^{2k}_{fin}$. Similarly we can check if exists $2l+1 \in \mathbb{N}$ such that $C^{2l+1}_{init} \xrightarrow{*} C^{2l+1}_{fin}$.

By a reduction from 3-SAT, we prove that

Lemma 12. *The cut-off problem for symmetric leader protocols is NP-hard.*

References

1. Alla, H., David, R.: Continuous and hybrid Petri nets. J. Circuits Syst. Comput. **8**(1), 159–188 (1998)
2. Angluin, D., Aspnes, J., Diamadi, Z., Fischer, M.J., Peralta, R.: Computation in networks of passively mobile finite-state sensors. Distributed Computing **18**(4), 235–253 (2006). https://doi.org/10.1007/s00446-005-0138-3
3. Basler, G., Mazzucchi, M., Wahl, T., Kroening, D.: Symbolic counter abstraction for concurrent software. In: Bouajjani, A., Maler, O. (eds.) 21st International Conference on Computer Aided Verification, CAV 2009, Grenoble, France, June 26 - July 2, 2009, Proceedings. Lecture Notes in Computer Science, vol. 5643, pp. 64–78. Springer (2009). https://doi.org/10.1007/978-3-642-02658-4_9

4. Bloem, R., Jacobs, S., Khalimov, A., Konnov, I., Rubin, S., Veith, H., Widder, J.: Decidability of Parameterized Verification. Synthesis Lectures on Distributed Computing Theory, Morgan & Claypool Publishers (2015). https://doi.org/10.2200/S00658ED1V01Y201508DCT013
5. Blondin, M.: The abc of Petri net reachability relaxations. ACM SIGLOG News **7**(3) (2020)
6. Czerwinski, W., Lasota, S., Lazic, R., Leroux, J., Mazowiecki, F.: The reachability problem for Petri nets is not elementary. In: Charikar, M., Cohen, E. (eds.) 51st Annual ACM SIGACT Symposium on Theory of Computing, STOC 2019, Phoenix, AZ, USA, June 23-26, 2019, Proceedings. pp. 24–33. ACM (2019). https://doi.org/10.1145/3313276.3316369
7. Delzanno, G., Sangnier, A., Traverso, R., Zavattaro, G.: On the complexity of parameterized reachability in reconfigurable broadcast networks. In: FSTTCS. LIPIcs, vol. 18, pp. 289–300. Schloss Dagstuhl - Leibniz-Zentrum für Informatik (2012)
8. Emerson, E.A., Kahlon, V.: Model checking large-scale and parameterized resource allocation systems. In: TACAS. Lecture Notes in Computer Science, vol. 2280, pp. 251–265. Springer (2002)
9. Esparza, J.: Decidability and complexity of Petri net problems - an introduction. In: Petri Nets. Lecture Notes in Computer Science, vol. 1491, pp. 374–428. Springer (1996)
10. Esparza, J.: Parameterized verification of crowds of anonymous processes. In: Dependable Software Systems Engineering, NATO Science for Peace and Security Series - D: Information and Communication Security, vol. 45, pp. 59–71. IOS Press (2016)
11. Esparza, J., Finkel, A., Mayr, R.: On the verification of broadcast protocols. In: LICS. pp. 352–359. IEEE Computer Society (1999)
12. Esparza, J., Ganty, P., Leroux, J., Majumdar, R.: Verification of population protocols. Acta Informatica **54**(2), 191–215 (2017). https://doi.org/10.1007/s00236-016-0272-3
13. Esparza, J., Nielsen, M.: Decidability issues for Petri nets - a survey. J. Inf. Process. Cybern. **30**(3), 143–160 (1994)
14. Fraca, E., Haddad, S.: Complexity analysis of continuous Petri nets. Fundam. Informaticae **137**(1), 1–28 (2015)
15. German, S.M., Sistla, A.P.: Reasoning about systems with many processes. Journal of the ACM **39**(3), 675–735 (1992). https://doi.org/10.1145/146637.146681
16. Gmeiner, A., Konnov, I., Schmid, U., Veith, H., Widder, J.: Tutorial on parameterized model checking of fault-tolerant distributed algorithms. In: SFM. Lecture Notes in Computer Science, vol. 8483, pp. 122–171. Springer (2014)
17. Horn, F., Sangnier, A.: Deciding the existence of cut-off in parameterized rendezvous networks. In: CONCUR. LIPIcs, vol. 171, pp. 46:1–46:16. Schloss Dagstuhl - Leibniz-Zentrum für Informatik (2020)
18. Kaiser, A., Kroening, D., Wahl, T.: Dynamic cutoff detection in parameterized concurrent programs. In: Touili, T., Cook, B., Jackson, P.B. (eds.) 22nd International Conference on Computer Aided Verification, CAV 2010, Edinburgh, UK, July 15-19, 2010, Proceedings. Lecture Notes in Computer Science, vol. 6174, pp. 645–659. Springer (2010). https://doi.org/10.1007/978-3-642-14295-6_55
19. Kannan, R., Bachem, A.: Polynomial algorithms for computing the Smith and Hermite normal forms of an integer matrix. SIAM J. Comput. **8**(4), 499–507 (1979)

20. Kannan, R., Monma, C.L.: On the computational complexity of integer programming problems. In: Henn, R., Korte, B., Oettli, W. (eds.) Optimization and Operations Research. pp. 161–172. Springer Berlin Heidelberg, Berlin, Heidelberg (1978)
21. Ladner, R.E.: The circuit value problem is Log Space complete for P. SIGACT News **7**(1), 18–20 (1975)
22. Leroux, J., Schmitz, S.: Reachability in vector addition systems is primitive-recursive in fixed dimension. In: LICS. pp. 1–13. IEEE (2019)
23. Mulmuley, K.: A fast parallel algorithm to compute the rank of a matrix over an arbitrary field. Comb. **7**(1), 101–104 (1987). https://doi.org/10.1007/BF02579205
24. Murata, T.: Petri nets: Properties, analysis and applications. Proceedings of the IEEE **77**(4), 541–580 (1989)
25. Navlakha, S., Bar-Joseph, Z.: Distributed information processing in biological and computational systems. Communications of the ACM **58**(1), 94–102 (2015). https://doi.org/10.1145/2678280
26. Papadimitriou, C.H.: On the complexity of integer programming. J. ACM **28**(4), 765–768 (1981). https://doi.org/10.1145/322276.322287
27. Papadimitriou, C.H.: Computational complexity. Academic Internet Publ. (2007)
28. Pnueli, A., Xu, J., Zuck, L.D.: Liveness with (0, 1, infty)-counter abstraction. In: CAV. Lecture Notes in Computer Science, vol. 2404, pp. 107–122. Springer (2002)
29. Pohst, M.E., Zassenhaus, H.: Algorithmic algebraic number theory, Encyclopedia of mathematics and its applications, vol. 30. Cambridge University Press (1989)
30. Recalde, L., Haddad, S., Suárez, M.S.: Continuous Petri nets: Expressive power and decidability issues. Int. J. Found. Comput. Sci. **21**(2), 235–256 (2010)
31. Soloveichik, D., Cook, M., Winfree, E., Bruck, J.: Computation with finite stochastic chemical reaction networks. Nat. Comput. **7**(4), 615–633 (2008)

Fixpoint Theory – Upside Down

Paolo Baldan[1], Richard Eggert[2(✉)], Barbara König[2], and Tommaso Padoan[1]

[1] Università di Padova, Padova, Italy
[2] Universität Duisburg-Essen, Duisburg, Germany
✉ richard.eggert@uni-due.de

Abstract. Knaster-Tarski's theorem, characterising the greatest fixpoint of a monotone function over a complete lattice as the largest post-fixpoint, naturally leads to the so-called coinduction proof principle for showing that some element is below the greatest fixpoint (e.g., for providing bisimilarity witnesses). The dual principle, used for showing that an element is above the least fixpoint, is related to inductive invariants. In this paper we provide proof rules which are similar in spirit but for showing that an element is above the greatest fixpoint or, dually, below the least fixpoint. The theory is developed for non-expansive monotone functions on suitable lattices of the form $\mathbb{M}^Y$, where Y is a finite set and $\mathbb{M}$ an MV-algebra, and it is based on the construction of (finitary) approximations of the original functions. We show that our theory applies to a wide range of examples, including termination probabilities, behavioural distances for probabilistic automata and bisimilarity. Moreover it allows us to determine original algorithms for solving simple stochastic games.

1 Introduction

Fixpoints are ubiquitous in computer science as they allow to provide a meaning to inductive and coinductive definitions (see, e.g., [26,23]). A monotone function $f : L \to L$ over a complete lattice $(L, \sqsubseteq)$, by Knaster-Tarski's theorem [28], admits a least fixpoint μf and greatest fixpoint νf which are characterised as the least pre-fixpoint and the greatest post-fixpoint, respectively. This immediately gives well-known proof principles for showing that a lattice element $l \in L$ is *below* νf or *above* μf

$$\frac{l \sqsubseteq f(l)}{l \sqsubseteq \nu f} \qquad\qquad \frac{f(l) \sqsubseteq l}{\mu f \sqsubseteq l}$$

On the other hand, showing that a given element l is *above* νf or *below* μf is more difficult. One can think of using the characterisation of least and largest fixpoints via Kleene's iteration. E.g., the largest fixpoint is the least element of the (possibly transfinite) descending chain obtained by iterating f from $\top$. Then showing that $f^i(\top) \sqsubseteq l$ for some i, one concludes that $\nu f \sqsubseteq l$. This proof principle is related to the notion of ranking functions. However, this is a less satisfying notion of witness since f has to be applied i times, and this can be inefficient or unfeasible when i is an infinite ordinal.

S. Kiefer and C. Tasson (Eds.): FOSSACS 2021, LNCS 12650, pp. 62–81, 2021.
https://doi.org/10.1007/978-3-030-71995-1_4

The aim of this paper is to present an alternative proof rule for this purpose for functions over lattices of the form $L = \mathbb{M}^Y$ where Y is a finite set and $\mathbb{M}$ is an MV-chain, i.e., a totally ordered complete lattice endowed with suitable operations of sum and complement. This allows us to capture several examples, ranging from ordinary relations, for dealing with bisimilarity, behavioural metrics, termination probabilities and simple stochastic games.

Assume $f : \mathbb{M}^Y \to \mathbb{M}^Y$ monotone and consider the question of proving that some fixpoint $a : Y \to \mathbb{M}$ is the largest fixpoint νf. The idea is to show that there is no "slack" or "wiggle room" in the fixpoint a that would allow us to further increase it. This is done by associating with every $a : Y \to \mathbb{M}$ a function $f_a^\#$ on $\mathbf{2}^Y$ whose greatest fixpoint gives us the elements of Y where we have a potential for increasing a by adding a constant. If no such potential exists, i.e. $\nu f_a^\#$ is empty, we conclude that a is νf. A similar function $f_\#^a$ (specifying decrease instead of increase) exists for the case of least fixpoints. Note that the premise is $\nu f_\#^a = \emptyset$, i.e. the witness remains coinductive. The proof rules are:

$$\frac{f(a) = a \qquad \nu f_a^\# = \emptyset}{\nu f = a} \qquad\qquad \frac{f(a) = a \qquad \nu f_\#^a = \emptyset}{\mu f = a}$$

For applying the rule we compute a greatest fixpoint on $\mathbf{2}^Y$, which is finite, instead of working on the potentially infinite $\mathbb{M}^Y$. The rule does not work for all monotone functions $f : \mathbb{M}^Y \to \mathbb{M}^Y$, but we show that whenever f is non-expansive the rule is valid. Actually, it is not only sound, but also reversible, i.e., if $a = \nu f$ then $\nu f_a^\# = \emptyset$, providing an if-and-only-if characterisation.

Quite interestingly, under the same assumptions on f, using a restricted function f_a^*, the rule can be used, more generally, when a is just a *pre-fixpoint* ($f(a) \sqsubseteq a$) and it allows to conclude that $\nu f \sqsubseteq a$. A dual result holds for *post-fixpoints* in the case of least fixpoints.

$$\frac{f(a) \sqsubseteq a \qquad \nu f_a^* = \emptyset}{\nu f \sqsubseteq a} \qquad\qquad \frac{a \sqsubseteq f(a) \qquad \nu f_*^a = \emptyset}{a \sqsubseteq \mu f}$$

As already mentioned, the theory above applies to many interesting scenarios: witnesses for non-bisimilarity, algorithms for simple stochastic games [11] and lower bounds for termination probabilities and behavioural metrics in the setting of probabilistic systems [1] and probabilistic automata [2]. In particular we were inspired by, and generalise, the self-closed relations of Fu [16], also used in [2].

Motivating Example. Consider a Markov chain (S, T, η) with a finite set of states S, where $T \subseteq S$ are the terminal states and every state $s \in S \backslash T$ is associated with a probability distribution $\eta(s) \in \mathcal{D}(S)$.[3] Intuitively, $\eta(s)(s')$ denotes the probability of state s choosing s' as its successor. Assume that, given a fixed state $s \in S$, we want to determine the termination probability of s, i.e. the probability of reaching any terminal state from s. As a concrete example, take the Markov chain given in Fig. 1, where u is the only terminal state.

[3] $\mathcal{D}(S)$ is the set of all maps $p : S \to [0, 1]$ such that $\sum_{s \in S} p(s) = 1$.

$$\mathcal{T} : [0,1]^S \to [0,1]^S$$
$$\mathcal{T}(t)(s) = \begin{cases} 1 & \text{if } v \in T \\ \sum_{s' \in S} \eta(s)(s') \cdot t(s') & \text{otherwise} \end{cases}$$

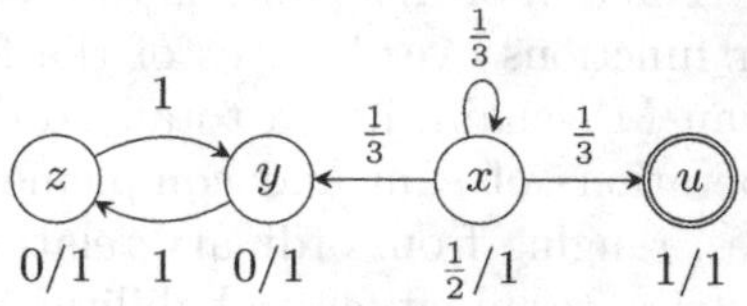

Fig. 1: Function $\mathcal{T}$ (left) and a Markov chain with two fixpoints of $\mathcal{T}$ (right)

The termination probability arises as the least fixpoint of a function $\mathcal{T}$ defined as in Fig. 1. The values of $\mu\mathcal{T}$ are indicated in green (left value).

Now consider the function t assigning to each state the termination probability written in red (right value). It is not difficult to see that t is another fixpoint of $\mathcal{T}$, in which states y and z convince each other incorrectly that they terminate with probability 1, resulting in a vicious cycle that gives "wrong" results. We want to show that $\mu\mathcal{T} \neq t$ without knowing $\mu\mathcal{T}$. Our idea is to compute the set of states that still has some "wiggle room", i.e., those states which could reduce their termination probability by δ if all their successors did the same. This definition has a coinductive flavour and it can be computed as a greatest fixpoint on the finite powerset $\mathbf{2}^S$ of states, instead of on the infinite lattice $S^{[0,1]}$.

We hence consider a function $\mathcal{T}_\#^t : \mathbf{2}^{[S]^t} \to \mathbf{2}^{[S]^t}$, dependent on t, defined as follows. Let $[S]^t$ be the set of all states s where $t(s) > 0$, i.e., a reduction is in principle possible. Then a state $s \in [S]^t$ is in $\mathcal{T}_\#^t(S')$ iff $s \notin T$ and for all s' for which $\eta(s)(s') > 0$ it holds that $s' \in S'$, i.e. all successors of s are in S'.

The greatest fixpoint of $\mathcal{T}_\#^t$ is $\{y, z\}$. The fact that it is not empty means that there is some "wiggle room", i.e., the value of t can be reduced on the elements $\{y, z\}$ and thus t cannot be the least fixpoint of f. Moreover, the intuition that t can be improved on $\{y, z\}$ can be made precise, leading to the possibility of performing the improvement and search for the least fixpoint from there.

Contributions. In the paper we formalise the theory outlined above, showing that the proof rules work for non-expansive monotone functions f on lattices of the form $\mathbb{M}^Y$, where Y is a finite set and $\mathbb{M}$ an MV-algebra (§3 and §4). Additionally, given a decomposition of f we show how to obtain the corresponding approximation compositionally (§5). Then, in order to show that our approach covers a wide range of examples and allows us to derive original algorithms, we discuss various applications: termination probability, behavioural distances for probabilistic automata and bisimilarity (§6) and simple stochastic games (§7).

Proofs and further material can be found in the full version of the paper [5].

2 Lattices and MV-Algebras

In this section, we review some basic notions used in the paper.

A preordered or partially ordered set $(P, \sqsubseteq)$ is often denoted simply as P, omitting the order relation. Given $x, y \in P$, with $x \sqsubseteq y$, we denote by $[x, y]$ the interval $\{z \in P \mid x \sqsubseteq z \sqsubseteq y\}$. The *join* and the *meet* of a subset $X \subseteq P$ (if they exist) are denoted $\bigsqcup X$ and $\bigsqcap X$, respectively.

A *complete lattice* is a partially ordered set $(L, \sqsubseteq)$ such that each subset $X \subseteq L$ admits a join $\bigsqcup X$ and a meet $\bigsqcap X$. A complete lattice $(L, \sqsubseteq)$ always has a least element $\bot = \bigsqcup \emptyset$ and a greatest element $\top = \bigsqcap \emptyset$.

A function $f : L \to L$ is *monotone* if for all $l, l' \in L$, if $l \sqsubseteq l'$ then $f(l) \sqsubseteq f(l')$. By Knaster-Tarski's theorem [28, Thm. 1], any monotone function on a complete lattice has a least and a greatest fixpoint, denoted respectively μf and νf, characterised as the meet of all pre-fixpoints respectively the join of all post-fixpoints: $\mu f = \bigsqcap\{l \mid f(l) \sqsubseteq l\}$ and $\nu f = \bigsqcup\{l \mid l \sqsubseteq f(l)\}$.

Let $(C, \sqsubseteq)$, $(A, \leq)$ be complete lattices. A *Galois connection* is a pair of monotone functions $\langle\alpha, \gamma\rangle$ such that $\alpha : C \to A$, $\gamma : A \to C$ and for all $a \in A$ and $c \in C$: $\alpha(c) \leq a \iff c \sqsubseteq \gamma(a)$. Equivalently, for all $a \in A$ and $c \in C$, (i) $c \sqsubseteq \gamma(\alpha(c))$ and (ii) $\alpha(\gamma(a)) \leq a$. In this case we will write $\langle\alpha, \gamma\rangle : C \to A$. For a Galois connection $\langle\alpha, \gamma\rangle : C \to A$, the function α is called the left (or lower) adjoint and γ the right (or upper) adjoint.

Galois connections are at the heart of abstract interpretation [13,14]. In particular, when $\langle\alpha, \gamma\rangle$ is a Galois connection, given $f^C : C \to C$ and $f^A : A \to A$, monotone functions, if $f^C \circ \gamma \sqsubseteq \gamma \circ f^A$, then $\nu f^C \sqsubseteq \gamma(\nu f^A)$. If equality holds, i.e., $f^C \circ \gamma = \gamma \circ f^A$, then greatest fixpoints are preserved along the connection, i.e., $\nu f^C = \gamma(\nu f^A)$.

Given a set Y and a complete lattice L, the set of functions $L^Y = \{f \mid f : Y \to L\}$, endowed with pointwise order, i.e., for $a, b \in L^Y$, $a \sqsubseteq b$ if $a(y) \sqsubseteq b(y)$ for all $y \in Y$, is a complete lattice.

In the paper we will mostly work with lattices of the kind $\mathbb{M}^Y$ where $\mathbb{M}$ is a special kind of lattice with a rich algebraic structure, i.e. an MV-algebra [21].

Definition 1 (MV-algebra). *An* MV-algebra *is a tuple* $\mathbb{M} = (M, \oplus, 0, \overline{(\cdot)})$ *where* $(M, \oplus, 0)$ *is a commutative monoid and* $\overline{(\cdot)} : M \to M$ *maps each element to its* complement, *such that for all* $x, y \in M$ *(1)* $\overline{\overline{x}} = x$; *(2)* $x \oplus \overline{0} = \overline{0}$; *(3)* $\overline{(\overline{x} \oplus y)} \oplus y = \overline{(\overline{y} \oplus x)} \oplus x$.

We denote $1 = \overline{0}$, *multiplication* $x \otimes y = \overline{\overline{x} \oplus \overline{y}}$ *and subtraction* $x \ominus y = x \otimes \overline{y}$.

Definition 2 (natural order). *Let* $\mathbb{M} = (M, \oplus, 0, \overline{(\cdot)})$ *be an MV-algebra. The* natural order *on* $\mathbb{M}$ *is defined, for* $x, y \in M$, *by* $x \sqsubseteq y$ *if* $x \oplus z = y$ *for some* $z \in M$. *When* $\sqsubseteq$ *is total* $\mathbb{M}$ *is called an* MV-chain.

The natural order gives an MV-algebra a lattice structure where $\bot = 0$, $\top = 1$, $x \sqcup y = (x \ominus y) \oplus y$ and $x \sqcap y = \overline{\overline{x} \sqcup \overline{y}} = x \otimes (\overline{x} \oplus y)$. We call the MV-algebra *complete*, if it is a complete lattice, which is not true in general, e.g., $([0,1] \cap \mathbb{Q}, \leq)$.

Example 3. A prototypical example of an MV-algebra is $([0,1], \oplus, 0, \overline{(\cdot)})$ where $x \oplus y = \min\{x + y, 1\}$ and $\overline{x} = 1 - x$ for $x, y \in [0,1]$. This means that $x \otimes y = \max\{x + y - 1, 0\}$ and $x \ominus y = \max\{0, x - y\}$ (truncated subtraction). The operators $\oplus$ and $\otimes$ are also known as strong disjunction and conjunction in Łukasiewicz logic [22]. The natural order is $\leq$ (less or equal) on the reals.

Another example is $(\{0,\ldots,k\},\oplus,0,\overline{(\cdot)})$ where $n \oplus m = \min\{n+m,k\}$ and $\overline{n} = k-n$ for $n,m \in \{0,\ldots,k\}$. Both MV-algebras are complete and MV-chains.

Boolean algebras (with disjunction and complement) also form MV-algebras that are complete, but in general not MV-chains.

MV-algebras are the algebraic semantics of Łukasiewicz logic. They can be shown to correspond to intervals of the kind $[0,u]$ in suitable groups, i.e., abelian lattice-ordered groups with a strong unit u [21].

3 Non-expansive Functions and Their Approximations

As mentioned in the introduction, our interest is for fixpoints of monotone functions $f : \mathbb{M}^Y \to \mathbb{M}^Y$, where $\mathbb{M}$ is an MV-chain and Y is a finite set. We will see that for non-expansive functions we can over-approximate the sets of points in which a given $a \in \mathbb{M}^Y$ can be increased in a way that is preserved by the application of f. This will be the core of the proof rules outlined earlier.

Non-expansive Functions on MV-Algebras. For defining non-expansiveness it is convenient to introduce a norm.

Definition 4 (norm). *Let $\mathbb{M}$ be an MV-chain and let Y be a finite set. Given $a \in \mathbb{M}^Y$ we define its* norm *as $\|a\| = \max\{a(y) \mid y \in Y\}$.*

Given a finite set Y we extend $\oplus$ and $\otimes$ to $\mathbb{M}^Y$ pointwise. Given $Y' \subseteq Y$ and $\delta \in \mathbb{M}$, we write $\delta_{Y'}$ for the function defined by $\delta_{Y'}(y) = \delta$ if $y \in Y'$ and $\delta_{Y'}(y) = 0$, otherwise. Whenever this does not generate confusion, we write δ instead of δ_Y. It can be seen that $\|\cdot\|$ has the properties of a norm, i.e., for all $a, b \in \mathbb{M}^Y$ and $\delta \in \mathbb{M}$, it holds that (1) $\|a \oplus b\| \sqsubseteq \|a\| \oplus \|b\|$, (2) $\|\delta \otimes a\| = \delta \otimes \|a\|$ and and $\|a\| = 0$ implies that a is the constant 0. Moreover, it is clearly monotonic, i.e., if $a \sqsubseteq b$ then $\|a\| \sqsubseteq \|b\|$.

We next introduce non-expansiveness. Despite the fact that we will finally be interested in endo-functions $f : \mathbb{M}^Y \to \mathbb{M}^Y$, in order to allow for a compositional reasoning we work with functions where domain and codomain can be different.

Definition 5 (non-expansiveness). *Let $f : \mathbb{M}^Y \to \mathbb{M}^Z$ be a function, where $\mathbb{M}$ is an MV-chain and Y, Z are finite sets. We say that it is* non-expansive *if for all $a, b \in \mathbb{M}^Y$ it holds $\|f(b) \ominus f(a)\| \sqsubseteq \|b \ominus a\|$.*

Note that $(a, b) \mapsto \|a \ominus b\|$ is the supremum lifting of a directed version of Chang's distance [21]. It is easy to see that all non-expansive functions on MV-chains are monotone.

Approximating the Propagation of Increases. Let $f : \mathbb{M}^Y \to \mathbb{M}^Z$ be a monotone function and take $a, b \in \mathbb{M}^Y$ with $a \sqsubseteq b$. We are interested in the difference $b(y) \ominus a(y)$ for some $y \in Y$ and on how the application of f "propagates" this increase. The reason is that, understanding that no increase can be propagated will be crucial to establish when a fixpoint of a non-expansive function f is actually the largest one, and, more generally, when a (pre-)fixpoint of f is above the largest fixpoint.

In order to formalise the above intuition, we rely on tools from abstract interpretation. In particular, the following pair of functions, which, under a suitable condition, form a Galois connection, will play a major role. The left adjoint $\alpha_{a,\delta}$ takes as input a set Y' and, for $y \in Y'$, it increases the values $a(y)$ by δ, while the right adjoint $\gamma_{a,\delta}$ takes as input a function $b \in \mathbb{M}^Y$, $b \in [a, a \oplus \delta]$ and checks for which parameters $y \in Y$ the value $b(y)$ exceeds $a(y)$ by δ.

We also define $[Y]_a$, the subset of elements in Y where $a(y)$ is not 1 and thus there is a potential to increase, and δ_a, which gives us the minimal such increase.

Definition 6 (functions to sets, and vice versa). *Let* $\mathbb{M}$ *be an MV-algebra and let* Y *be a finite set. Define the set* $[Y]_a = \{y \in Y \mid a(y) \neq 1\}$ *and* $\delta_a = \min\{\overline{a(y)} \mid y \in [Y]_a\}$ *with* $\min \emptyset = 1$.

For $0 \sqsubset \delta \in \mathbb{M}$ *we consider the functions* $\alpha_{a,\delta} : \mathbf{2}^{[Y]_a} \to [a, a \oplus \delta]$ *and* $\gamma_{a,\delta} : [a, a \oplus \delta] \to \mathbf{2}^{[Y]_a}$*, defined, for* $Y' \in \mathbf{2}^{[Y]_a}$ *and* $b \in [a, a \oplus \delta]$*, by*

$$\alpha_{a,\delta}(Y') = a \oplus \delta_{Y'} \qquad \gamma_{a,\delta}(b) = \{y \in [Y]_a \mid b(y) \ominus a(y) \sqsupseteq \delta\}.$$

When δ is sufficiently small, the pair $\langle \alpha_{a,\delta}, \gamma_{a,\delta} \rangle$ is a Galois connection.

Lemma 7 (Galois connection). *Let* $\mathbb{M}$ *be an MV-algebra and* Y *be a finite set. For* $0 \neq \delta \sqsubseteq \delta_a$*, the pair* $\langle \alpha_{a,\delta}, \gamma_{a,\delta} \rangle : \mathbf{2}^{[Y]_a} \to [a, a \oplus \delta]$ *is a Galois connection.*

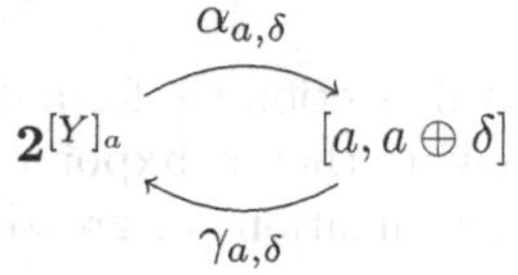

Whenever f is non-expansive, it is easy to see that it restricts to a function $f : [a, a \oplus \delta] \to [f(a), f(a) \oplus \delta]$ for all $\delta \in \mathbb{M}$.

As mentioned before, a crucial result shows that for all non-expansive functions, under the assumption that Y, Z are finite and the order on $\mathbb{M}$ is total, we can suitably approximate the propagation of increases. In order to state this result, a useful tool is a notion of approximation of a function.

Definition 8 ((δ, a)-approximation). *Let* $\mathbb{M}$ *be an MV-chain, let* Y*,* Z *be finite sets and let* $f : \mathbb{M}^Y \to \mathbb{M}^Z$ *be a non-expansive function. For* $a \in \mathbb{M}^Y$ *and any* $\delta \in \mathbb{M}$ *we define* $f^{\#}_{a,\delta} : \mathbf{2}^{[Y]_a} \to \mathbf{2}^{[Z]_{f(a)}}$ *as* $f^{\#}_{a,\delta} = \gamma_{f(a),\delta} \circ f \circ \alpha_{a,\delta}$.

Given $Y' \subseteq [Y]_a$, its image $f^{\#}_{a,\delta}(Y') \subseteq [Z]_{f(a)}$ is the set of points $z \in [Z]_{f(a)}$ such that $\delta \sqsubseteq f(a \oplus \delta_{Y'})(z) \ominus f(a)(z)$, i.e., the points to which f propagates an increase of the function a with value δ on the subset Y'.

We first show that $f^{\#}_{a,\delta}$ is antitone in the parameter δ, a non-trivial result.

Lemma 9 (anti-monotonicity). *Let* $\mathbb{M}$ *be an MV-chain, let* Y*,* Z *be finite sets, let* $f : \mathbb{M}^Y \to \mathbb{M}^Z$ *be a non-expansive function and let* $a \in \mathbb{M}^Y$*. For* $\theta, \delta \in \mathbb{M}$*, if* $\theta \sqsubseteq \delta$ *then* $f^{\#}_{a,\delta} \subseteq f^{\#}_{a,\theta}$.

Since $f^{\#}_{a,\delta}$ increases when δ decreases and there are finitely many such functions, there must be a value ι^f_a such that all functions $f^{\#}_{a,\delta}$ for $0 \sqsubset \delta \sqsubseteq \iota^f_a$ are equal. This function is denoted by $f^{\#}_a$ and is called the *a-approximation of* f.

We next show that indeed, for all non-expansive functions, the a-approximation properly approximates the propagation of increases.

Theorem 10 (approximation of non-expansive functions). *Let* $\mathbb{M}$ *be a complete MV-chain, let* Y, Z *be finite sets and let* $f : \mathbb{M}^Y \to \mathbb{M}^Z$ *be a non-expansive function. Then there exists* $\iota_a^f \in \mathbb{M}$, *the largest value below or equal to* δ_a *such that* $f_{a,\delta}^{\#} = f_{a,\delta'}^{\#}$ *for all* $0 \sqsubset \delta, \delta' \sqsubseteq \iota_a^f$.

We denote this function by $f_a^{\#}$ *and call it the* a-approximation *of* f. *Then for all* $0 \sqsubset \delta \in \mathbb{M}$:

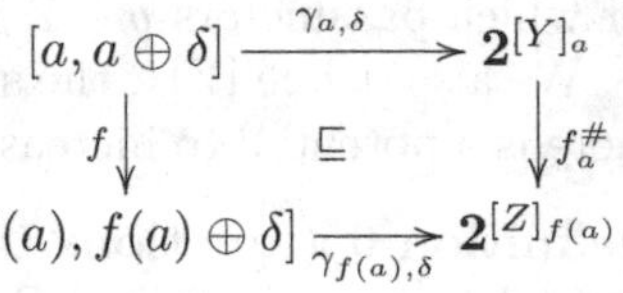

a. $\gamma_{f(a),\delta} \circ f \sqsubseteq f_a^{\#} \circ \gamma_{a,\delta}$
b. *for* $\delta \sqsubseteq \delta_a$: $\delta \sqsubseteq \iota_a^f$ *iff* $\gamma_{f(a),\delta} \circ f = f_a^{\#} \circ \gamma_{a,\delta}$

$$
\begin{array}{ccc}
[a, a \oplus \delta] & \xrightarrow{\gamma_{a,\delta}} & 2^{[Y]_a} \\
f \downarrow & \sqsubseteq & \downarrow f_a^{\#} \\
[f(a), f(a) \oplus \delta] & \xrightarrow{\gamma_{f(a),\delta}} & 2^{[Z]_{f(a)}}
\end{array}
$$

Note that if $Y = Z$ and a is a fixpoint of f, i.e., $a = f(a)$, condition (a) above corresponds exactly to soundness in the sense of abstract interpretation [13], while condition (b) corresponds to (γ-)completeness (see also §2).

4   Proof Rules

In this section we formalise the proof technique outlined in the introduction for showing that a fixpoint is the largest and, more generally, for checking over-approximations of greatest fixpoints of non-expansive functions.

Consider a monotone function $f : \mathbb{M}^Y \to \mathbb{M}^Y$ for some finite set Y. We first focus on the problem of establishing whether some given fixpoint a of f coincides with νf (without explicitly knowing νf), and, in case it does not, finding an "improvement", i.e., a post-fixpoint of f, larger than a. Observe that when a is a fixpoint, $[Y]_a = [Y]_{f(a)}$ and thus the a-approximation of f (Thm. 10) is an endofunction $f_a^{\#} : [Y]_a \to [Y]_a$. We have the following result, which relies on the fact that due to Thm. 10 $\gamma_{a,\delta}$ preserves fixpoints (of f and $f_a^{\#}$).

Theorem 11 (soundness and completeness for fixpoints). *Let* $\mathbb{M}$ *be a complete MV-chain,* Y *a finite set and* $f : \mathbb{M}^Y \to \mathbb{M}^Y$ *be a non-expansive function. Let* $a \in \mathbb{M}^Y$ *be a fixpoint of* f. *Then* $\nu f_a^{\#} = \emptyset$ *if and only if* $a = \nu f$.

Whenever a is a fixpoint, but not yet the largest fixpoint of f, we can increase it and obtain a post-fixpoint.

Lemma 12. *Let* $\mathbb{M}$ *be a complete MV-chain,* $f : \mathbb{M}^Y \to \mathbb{M}^Y$ *a non-expansive function,* $a \in \mathbb{M}$ *a fixpoint of* f, *and let* $f_a^{\#}$ *be the corresponding* a-approximation *and* ι_a^f *as in Thm. 10. Then* $\alpha_{a,\iota_a^f}(\nu f_a^{\#}) = a \oplus (\iota_a^f)_{\nu f_a^{\#}}$ *is a post-fixpoint of* f.

Using these results one can perform an alternative fixpoint iteration where we iterate to the largest fixpoint from below: start with a post-fixpoint $a_0 \sqsubseteq f(a_0)$ (which is clearly below νf) and obtain, by (possibly transfinite) iteration, an ascending chain that converges to a, the least fixpoint above a_0. Now check with Thm. 11 whether $Y' = \nu f_a^{\#} = \emptyset$. If yes, we have reached $\nu f = a$. If not,

$\alpha_{a,\iota_a^f}(Y') = a \oplus (\iota_a^f)_{Y'}$ is again a post-fixpoint (cf. Lem. 12) and we continue this procedure until – for some ordinal – we reach the largest fixpoint νf, for which we have $\nu f_{\nu f}^{\#} = \emptyset$.

Interestingly, the soundness result in Thm. 11 can be generalised to the case in which a is a pre-fixpoint instead of a fixpoint. In this case, the a-approximation for a function $f : \mathbb{M}^Y \to \mathbb{M}^Y$ is a function $f_a^{\#} : [Y]_a \to [Y]_{f(a)}$ where domain and codomain are different, hence it would not be meaningful to look for fixpoints. However, as explained below, it can be restricted to an endofunction.

Theorem 13 (soundness for pre-fixpoints). *Let $\mathbb{M}$ be a complete MV-chain, Y a finite set and $f : \mathbb{M}^Y \to \mathbb{M}^Y$ be a non-expansive function. Given a pre-fixpoint $a \in \mathbb{M}^Y$ of f, let $[Y]_{a=f(a)} = \{y \in [Y]_a \mid a(y) = f(a)(y)\}$. Let us define $f_a^* : [Y]_{a=f(a)} \to [Y]_{a=f(a)}$ as $f_a^*(Y') = f_a^{\#}(Y') \cap [Y]_{a=f(a)}$, where $f_a^{\#} : \mathbf{2}^{[Y]_a} \to \mathbf{2}^{[Y]_{f(a)}}$ is the a-approximation of f. If $\nu f_a^* = \emptyset$ then $\nu f \sqsubseteq a$.*

Roughly, the intuition for the above result is the following: the value of $f(a)$ on some y might or might not depend "circularly" on the value of a on y itself. In a purely inductive setting, without such circular dependencies, $\mu f = \nu f$ and hence a being a pre-fixpoint means that we over-approximate νf. However, we might have vicious cycles, as explained in the introduction, that destroy the over-approximation since the values are too low. Now, since we restrict to non-expansive functions, it must be the case that there is a cycle, such that all elements on this cycle are points where a and $f(a)$ coincide. It is hence sufficient to check whether a given pre-fixpoint could be increased on its subpart which corresponds to a fixpoint, i.e., the idea is to restrict to $[Y]_{a=f(a)}$. We detect such situations by looking for "wiggle room" as for fixpoints.

Completeness does not generalise to pre-fixpoints, i.e., it is not true that if a is a pre-fixpoint of f and $\nu f \sqsubseteq a$ then $\nu f_a^* = \emptyset$. A pre-fixpoint might contain slack even though it is above the greatest fixpoint. A counterexample is in Ex. 25.

The Dual View for Least Fixpoints. The theory developed so far can be easily dualised to check under-approximations of least fixpoints. Given a complete MV-algebra $\mathbb{M} = (M, \oplus, 0, \overline{(\cdot)})$ and a monotone function $f : \mathbb{M}^Y \to \mathbb{M}^Y$, in order to show that a post-fixpoint $a \in \mathbb{M}^Y$ satisfies $a \sqsubseteq \mu f$, we can in fact simply work in the dual MV-algebra, $\mathbb{M}^{op} = (M, \sqsupseteq, \otimes, \overline{(\cdot)}, 1)$. It is convenient to formulate the conditions using $\ominus$ and the original order.

We next outline the dualised setting. The notation for the dual case is obtained from that of the original (primal) case, exchanging subscripts and superscripts.

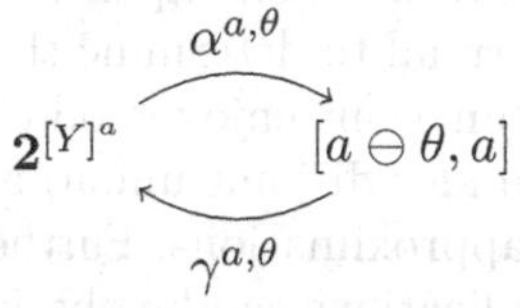

Given $a \in \mathbb{M}^Y$, define $[Y]^a = \{y \in Y \mid a(y) \neq 0\}$ and $\delta^a = \min\{a(y) \mid y \in [Y]^a\}$. For $\theta \in \mathbb{M}$, we consider the pair of functions $\langle \alpha^{a,\theta}, \gamma^{a,\theta} \rangle : \mathbf{2}^{[Y]^a} \to [a \ominus \theta, a]$ where, for $Y' \in \mathbf{2}^{[Y]^a}$, we let $\alpha^{a,\theta}(Y') = a \ominus \theta_{Y'}$ and, for $b \in [a \ominus \theta, a]$, $\gamma^{a,\theta}(b) = \{y \in Y \mid a(y) \ominus b(y) \sqsupseteq \theta\}$.

A function $f : \mathbb{M}^Y \to \mathbb{M}^Z$ is non-expansive in the dual MV-algebra when it is in the primal one. Its approximation in the sense of Thm. 10 is denoted $f_{\#}^a$.

Table 1: Basic functions $f\colon \mathbb{M}^Y \to \mathbb{M}^Z$ (constant, reindexing, minimum, maximum, average), function composition, disjoint union and the corresponding approximations $f_a^\#\colon \mathbf{2}^{[Y]_a} \to \mathbf{2}^{[Z]_{f(a)}}$, $f_\#^a\colon \mathbf{2}^{[Y]^a} \to \mathbf{2}^{[Z]^{f(a)}}$.

Notation: $\mathcal{R}^{-1}(z) = \{y \in Y \mid y\mathcal{R}z\}$, $supp(p) = \{y \in Y \mid p(y) > 0\}$ for $p \in \mathcal{D}(Y)$, $Min_a = \{y \in Y \mid a(y) \text{ minimal}\}$, $Max_a = \{y \in Y \mid a(y) \text{ maximal}\}$, $a\colon Y \to \mathbb{M}$

function f	definition of f	$f_a^\#(Y')$ (above), $f_\#^a(Y')$ (below)
c_k $(k \in \mathbb{M}^Z)$	$f(a) = k$	$\emptyset$ $\emptyset$
u^* $(u\colon Z \to Y)$	$f(a) = a \circ u$	$u^{-1}(Y')$ $u^{-1}(Y')$
$\min_\mathcal{R}$ $(\mathcal{R} \subseteq Y \times Z)$	$f(a)(z) = \min_{y\mathcal{R}z} a(y)$	$\{z \in [Z]_{f(a)} \mid Min_{a\|_{\mathcal{R}^{-1}(z)}} \subseteq Y'\}$ $\{z \in [Z]^{f(a)} \mid Min_{a\|_{\mathcal{R}^{-1}(z)}} \cap Y' \neq \emptyset\}$
$\max_\mathcal{R}$ $(\mathcal{R} \subseteq Y \times Z)$	$f(a)(z) = \max_{y\mathcal{R}z} a(y)$	$\{z \in [Z]_{f(a)} \mid Max_{a\|_{\mathcal{R}^{-1}(z)}} \cap Y' \neq \emptyset\}$ $\{z \in [Z]^{f(a)} \mid Max_{a\|_{\mathcal{R}^{-1}(z)}} \subseteq Y'\}$
av_D $(\mathbb{M} = [0,1]$, $Z = D \subseteq \mathcal{D}(Y))$	$f(a)(p) = \sum_{y\in Y} p(y) \cdot a(y)$	$\{p \in [D]_{f(a)} \mid supp(p) \subseteq Y'\}$ $\{p \in [D]^{f(a)} \mid supp(p) \subseteq Y'\}$
$h \circ g$ $(g\colon \mathbb{M}^Y \to \mathbb{M}^W$, $h\colon \mathbb{M}^W \to \mathbb{M}^Z)$	$f(a) = h(g(a))$	$h^\#_{g(a)} \circ g^\#_a(Y')$ $h^{g(a)}_\# \circ g^a_\#(Y')$
$\biguplus_{i\in I} f_i$ I finite $(f_i\colon \mathbb{M}^{Y_i} \to \mathbb{M}^{Z_i}$, $Y = \bigcup_{i\in I} Y_i$, $Z = \biguplus_{i\in I} Z_i)$	$f(a)(z) = f_i(a\|_{Y_i})(z)$ $(z \in Z_i)$	$\biguplus_{i\in I} (f_i)^\#_{a\|_{Y_i}}(Y' \cap Y_i)$ $\biguplus_{i\in I} (f_i)^{a\|_{Y_i}}_\#(Y' \cap Y_i)$

Then the dualisations of Thm. 11 and 13 hold, i.e., if a is a fixpoint of f, then $\nu f_\#^a = \emptyset$ iff $\mu f = a$, and whenever a is a post-fixpoint, $\nu f_*^a = \emptyset$ implies $a \sqsubseteq \mu f$.

5 (De)Composing Functions and Approximations

Given a non-expansive function f and a (pre/post-)fixpoint a, it is often non-trivial to determine the corresponding approximations. However, non-expansive functions enjoy good closure properties (closure under composition, and closure under disjoint union) and we will see that the same holds for the corresponding approximations. Furthermore it turns out that the functions needed in the applications can be obtained from just a few templates. This gives us a toolbox for assembling approximations with relative ease.

Theorem 14. *All basic functions listed in Table 1 are non-expansive. Furthermore non-expansive functions are closed under composition and disjoint union. The approximations are the ones listed in the third column of the table.*

6 Applications

6.1 Termination Probability

We start by making the example from the introduction (§1) more formal. Consider a Markov chain (S, T, η), as defined in the introduction (Fig. 1), where we restrict the codomain of $\eta\colon S\backslash T \to \mathcal{D}(S)$ to $D \subseteq \mathcal{D}(S)$, where D is finite (to ensure that all involved sets are finite). Furthermore let $\mathcal{T}\colon [0,1]^S \to [0,1]^S$ be the function from the introduction whose least fixpoint $\mu\mathcal{T}$ assigns to each state its termination probability.

Lemma 15. *The function $\mathcal{T}$ can be written as $\mathcal{T} = (\eta^* \circ \mathrm{av}_D) \uplus c_k$ where $k\colon T \to [0,1]$ is the constant function 1 defined only on terminal states.*

From this representation and Thm. 14 it is obvious that $\mathcal{T}$ is non-expansive.

Lemma 16. *Let $t\colon S \to [0,1]$. The approximation for $\mathcal{T}$ in the dual sense is $\mathcal{T}^t_\#\colon \mathbf{2}^{[S]^t} \to \mathbf{2}^{[S]^{\mathcal{T}(t)}}$ with*

$$\mathcal{T}^t_\#(S') = \{s \in [S]^{\mathcal{T}(t)} \mid s \notin T \wedge supp(\eta(s)) \subseteq S'\}.$$

It is well-known that the function $\mathcal{T}$ can be tweaked in such a way that it has a unique fixpoint, coinciding with $\mu\mathcal{T}$, by determining all states which cannot reach a terminal state and setting their value to zero [3]. Hence fixpoint iteration from above does not bring us any added value here. It does however make sense to use the proof rule in order to guarantee lower bounds via post-fixpoints.

Furthermore, termination probability is a special case of the considerably more complex stochastic games that will be studied in §7, where the trick of modifying the function is not applicable.

6.2 Behavioural Metrics for Probabilistic Automata

Before we start discussing probabilistic automata, we first consider the Hausdorff and the Kantorovich lifting and the corresponding approximations.

Hausdorff Lifting. Given a metric on a set X, the Hausdorff metric is obtained by lifting the original metric to $\mathbf{2}^X$. Here we define this for general distance functions on $\mathbb{M}$, not restricting to metrics. In particular the Hausdorff lifting is given by a function $\mathcal{H} : \mathbb{M}^{X\times X} \to \mathbb{M}^{\mathbf{2}^X \times \mathbf{2}^X}$ where

$$\mathcal{H}(d)(X_1, X_2) = \max\{\max_{x_1\in X_1} \min_{x_2\in X_2} d(x_1, x_2), \max_{x_2\in X_2} \min_{x_1\in X_1} d(x_1, x_2)\}.$$

An alternative characterisation due to Mémoli [20], also in [4], is more convenient for our purposes. If we let $u : \mathbf{2}^{X\times X} \to \mathbf{2}^X \times \mathbf{2}^X$ with $u(C) = (\pi_1[C], \pi_2[C])$, where π_1, π_2 are the projections $\pi_i : X \times X \to X$ and $\pi_i[C] = \{\pi_i(c) \mid c \in C\}$. Then $\mathcal{H}(d)(X_1, X_2) = \min\{\max_{(x_1,x_2)\in C} d(x_1, x_2) \mid C \subseteq X \times X \;\wedge\; u(C) = (X_1, X_2)\}$. Relying on this, we can obtain the result below, from which we deduce that $\mathcal{H}$ is non-expansive and construct its approximation as the composition of the corresponding functions from Table 1.

Lemma 17. *$\mathcal{H} = \min_u \circ \max_\in$ where $\max_\in : \mathbb{M}^{X\times X} \to \mathbb{M}^{\mathbf{2}^{X\times X}}$ ($\in \subseteq (X\times X)\times \mathbf{2}^{X\times X}$ is the "is-element-of"-relation on $X \times X$), $\min_u : \mathbb{M}^{\mathbf{2}^{X\times X}} \to \mathbb{M}^{\mathbf{2}^X \times \mathbf{2}^X}$.*

Kantorovich Lifting. The Kantorovich (also known as Wasserstein) lifting converts a metric on X to a metric on probability distributions over X. As for the Hausdorff lifting, we lift distance functions that are not necessarily metrics.

Furthermore, in order to ensure finiteness of all the sets involved, we restrict to $D \subseteq \mathcal{D}(X)$, some finite set of probability distributions over X. A *coupling* of $p, q \in D$ is a probability distribution $c \in \mathcal{D}(X \times X)$ whose left and right marginals are p, q, i.e., $p(x_1) = m_c^L(x_1) := \sum_{x_2\in X} c(x_1, x_2)$ and $q(x_2) = m_c^R(x_2) := \sum_{x_1\in X} c(x_1, x_2)$. The set of all couplings of p, q, denoted by $\Omega(p, q)$, forms a polytope with finitely many vertices [24]. The set of all polytope vertices that are obtained by coupling any $p, q \in D$ is also finite and is denoted by $VP_D \subseteq \mathcal{D}(X \times X)$.

The Kantorovich lifting is given by $\mathcal{K} : [0, 1]^{X\times X} \to [0, 1]^{D\times D}$ where

$$\mathcal{K}(d)(p, q) = \min_{c\in\Omega(p,q)} \sum_{(x_1,x_2)\in X\times X} c(x_1, x_2) \cdot d(x_1, x_2).$$

The coupling c can be interpreted as the optimal transport plan to move goods from suppliers to customers [30]. Again there is an alternative characterisation, which shows non-expansiveness of $\mathcal{K}$:

Lemma 18. *Let $u : VP_D \to D\times D$, $u(c) = (m_c^L, m_c^R)$. Then $\mathcal{K} = \min_u \circ \mathrm{av}_{VP_D}$, where $\mathrm{av}_{VP_D} : [0, 1]^{X\times X} \to [0, 1]^{VP_D}$, $\min_u : [0, 1]^{VP_D} \to [0, 1]^{D\times D}$.*

Probabilistic Automata. We now compare our approach with [2], which describes the first method for computing behavioural distances for probabilistic automata. Although the behavioural distance arises as a least fixpoint, it is in fact better, even the only known method, to iterate from above, in order to reach this least fixpoint. This is done by guessing and improving couplings, similar to strategy iteration discussed later in §7. A major complication, faced in [2], is that the procedure can get stuck at a fixpoint which is not the least and one has to determine that this is the case and decrease the current candidate. In fact this paper was our inspiration to generalise this technique to a more general setting.

A *probabilistic automaton* is a tuple $\mathcal{A} = (S, L, \eta, \ell)$, where S is a non-empty finite set of states, L is a finite set of labels, $\eta : S \to \mathbf{2}^{\mathcal{D}(S)}$ assigns finite sets of probability distributions to states and $\ell : S \to L$ is a labelling function. (In the following we again replace $\mathcal{D}(S)$ by a finite subset D.)

The *probabilistic bisimilarity pseudometrics* is the least fixpoint of the function $\mathcal{M} : [0, 1]^{S\times S} \to [0, 1]^{S\times S}$ where for $d : S \times S \to [0, 1]$, $s, t \in S$:

$$\mathcal{M}(d)(s, t) = \begin{cases} 1 & \text{if } \ell(s) \neq \ell(t) \\ \mathcal{H}(\mathcal{K}(d))(\eta(s), \eta(t)) & \text{otherwise} \end{cases}$$

where $\mathcal{H}$ is the Hausdorff lifting (for $\mathbb{M} = [0, 1]$) and $\mathcal{K}$ is the Kantorovich lifting defined earlier. Now assume that d is a fixpoint of $\mathcal{M}$, i.e., $d = \mathcal{M}(d)$. In order to check whether $d = \mu f$, [2] adapts the notion of a self-closed relation from [16].

Definition 19 ([2]). *A relation $M \subseteq S \times S$ is* self-closed *wrt.* $d = \mathcal{M}(d)$ *if, whenever* $s\,M\,t$*, then*

- $\ell(s) = \ell(t)$ *and* $d(s,t) > 0$,
- *if* $p \in \eta(s)$ *and* $d(s,t) = \min_{q' \in \eta(t)} \mathcal{K}(d)(p,q')$*, then there exists* $q \in \eta(t)$ *and* $c \in \Omega(p,q)$ *such that* $d(s,t) = \sum_{u,v \in S} d(u,v) \cdot c(u,v)$ *and* $supp(c) \subseteq M$,
- *if* $q \in \eta(t)$ *and* $d(s,t) = \min_{p' \in \eta(s)} \mathcal{K}(d)(p',q)$*, then there exists* $p \in \eta(s)$ *and* $c \in \Omega(p,q)$ *such that* $d(s,t) = \sum_{u,v \in S} d(u,v) \cdot c(u,v)$ *and* $supp(c) \subseteq M$.

The largest self-closed relation, denoted by $\approx_d$ is empty if and only if $d = \mu f$ [2]. We now investigate the relation between self-closed relations and post-fixpoints of approximations. For this we will first show that $\mathcal{M}$ can be composed from non-expansive functions, which proves that it is indeed non-expansive. Furthermore, this decomposition will help in the comparison.

Lemma 20. *The fixpoint function $\mathcal{M}$ characterizing probabilistic bisimilarity pseudometrics can be written as:*

$$\mathcal{M} = \max_\rho \circ (((\eta \times \eta)^* \circ \mathcal{H} \circ \mathcal{K}) \uplus c_l)$$

where $\rho\colon (S \times S) \uplus (S \times S) \to (S \times S)$ *with* $\rho((s,t),i) = (s,t)$.[4] *Furthermore* $l\colon S \times S \to [0,1]$ *is defined as* $l(s,t) = 0$ *if* $\ell(s) = \ell(t)$ *and* $l(s,t) = 1$ *if* $\ell(s) \neq \ell(t)$.

Hence $\mathcal{M}$ is a composition of non-expansive functions and thus non-expansive itself. We do not spell out $\mathcal{M}_\#^d$ explicitly, but instead show how it is related to self-closed relations.

Proposition 21. *Let* $d\colon S \times S \to [0,1]$ *where* $d = \mathcal{M}(d)$*. Then* $\mathcal{M}_\#^d\colon \mathbf{2}^{[S\times S]^d} \to \mathbf{2}^{[S\times S]^d}$*, where* $[S \times S]^d = \{(s,t) \in S \times S \mid d(s,t) > 0\}$.

Then M is a self-closed relation wrt. d if and only if $M \subseteq [S \times S]^d$ and M is a post-fixpoint of $\mathcal{M}_\#^d$.

6.3 Bisimilarity

In order to define standard bisimilarity we use a variant $\mathcal{G}$ of the Hausdorff lifting $\mathcal{H}$ from §6.2 where max and min are swapped and which we denote by $\mathcal{G}$.

Now we can define the fixpoint function for bisimilarity and its corresponding approximation. For simplicity we consider unlabelled transition systems, but it would be straightforward to handle labelled transitions.

Let X be a finite set of states and $\eta : X \to \mathbf{2}^X$ a function that assigns a set of successors $\eta(x)$ to a state $x \in X$. For the fixpoint function for bisimilarity $\mathcal{B} : \{0,1\}^{X\times X} \to \{0,1\}^{X\times X}$ we use the Hausdorff lifting $\mathcal{G}$ with $\mathbb{M} = \{0,1\}$.

Lemma 22. *Bisimilarity on η is the greatest fixpoint of* $\mathcal{B} = (\eta \times \eta)^* \circ \mathcal{G}$.

[4] Here we use $i \in \{0,1\}$ as indices to distinguish the elements in the disjoint union.

Since we are interested in the greatest fixpoint, we are working in the primal sense. Bisimulation relations are represented by their characteristic functions $d\colon X \times X \to \{0,1\}$, in fact the corresponding relation can be obtained by taking the complement of $[X \times X]_d = \{(x_1, x_2) \in X_1 \times X_2 \mid d(x_1, x_2) = 0\}$.

Lemma 23. *Let $d\colon X \times X \to \{0,1\}$. The approximation for the bisimilarity function $\mathcal{B}$ in the primal sense is $\mathcal{B}^{\#}_d : \mathbf{2}^{[X\times X]_d} \to \mathbf{2}^{[X\times X]_{\mathcal{B}(d)}}$ with*

$$\begin{aligned}\mathcal{B}^{\#}_d(R) = \{(x_1,x_2) \in [X \times X]_{\mathcal{B}(d)} \mid \\ &\forall y_1 \in \eta(x_1)\exists y_2 \in \eta(x_2)\big((y_1,y_2) \notin [X \times X]_d \vee (y_1,y_2) \in R)\big) \\ &\wedge \forall y_2 \in \eta(x_2)\exists y_1 \in \eta(x_1)\big((y_1,y_2) \notin [X \times X]_d \vee (y_1,y_2) \in R)\}\end{aligned}$$

We conclude this section by discussing how this view on bisimilarity can be useful: first, it again opens up the possibility to compute bisimilarity – a greatest fixpoint – by iterating from below, through smaller fixpoints. This could potentially be useful if it is easy to compute the least fixpoint of $\mathcal{B}$ inductively and continue from there.

Furthermore, we obtain a technique for witnessing non-bisimilarity of states. While this can also be done by exhibiting a distinguishing modal formula [17,9] or by a winning strategy for the spoiler in the bisimulation game [27], to our knowledge there is no known method that does this directly, based on the definition of bisimilarity.

With our technique however, we can witness non-bisimilarity of two states $x_1, x_2 \in X$ by presenting a pre-fixpoint d (i.e., $\mathcal{B}(d) \leq d$) such that $d(x_1, x_2) = 0$ (equivalent to $(x_1, x_2) \in [X \times X]_d$) and $\nu\mathcal{B}^{\#}_d = \emptyset$, since this implies $\nu\mathcal{B}(x_1, x_2) \leq d(x_1, x_2) = 0$ by our proof rule.

There are two issues to discuss: first, how can we characterise a pre-fixpoint of $\mathcal{B}$ (which is quite unusual, since bisimulations are post-fixpoints)? In fact, the condition $\mathcal{B}(d) \leq d$ can be rewritten to: for all $(x_1, x_2) \in [X \times X]_d$ there exists $y_1 \in \eta(x_1)$ such that for all $y_2 \in \eta(x_2)$ we have $(y_1, y_2) \in [X \times X]_d$ (*or* vice versa). Second, at first sight it does not seem as if we gained anything since we still have to do a fixpoint computation on relations. However, the carrier set is $[X \times X]_d$, i.e., a set of non-bisimilarity witnesses and this set can be small even though X might be large.

Example 24. We consider the transition system depicted below.

Our aim is to construct a witness showing that x, u are not bisimilar. This witness is a function $d\colon X \times X \to \{0,1\}$ with $d(x,u) = 0 = d(y,u)$ and for all other pairs the value is 1.

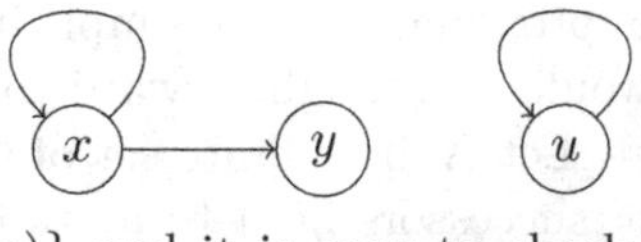

Hence $[X \times X]_{d=\mathcal{B}(d)} = [X \times X]_d = \{(x,u),(y,u)\}$ and it is easy to check that d is a pre-fixpoint of $\mathcal{B}$ and that $\nu\mathcal{B}^*_d = \emptyset$: we iterate over $\{(x,u),(y,u)\}$ and first remove (y,u) (since y has no successors) and then (x,u). This implies that $\nu\mathcal{B} \leq d$ and hence $\nu\mathcal{B}(x,u) = 0$, which means that x, u are not bisimilar.

Example 25. We modify Ex. 24 and consider a function d where $d(x, u) = 0$ and all other values are 1. Again d is a pre-fixpoint of $\mathcal{B}$ and $\nu\mathcal{B} \leq d$ (since only reflexive pairs are in the bisimilarity). However $\nu\mathcal{B}_d^* \neq \emptyset$, since $\{(x, u)\}$ is a post-fixpoint. This is a counterexample to completeness discussed after Thm. 13.

Intuively speaking, the states y, u over-approximate and claim that they are bisimilar, although they are not. (This is permissible for a pre-fixpoint.) This tricks x, u into thinking that there is some wiggle room and that one can increase the value of (x, u). This is true, but only because of the limited, local view, since the "true" value of (y, u) is 0.

7 Simple Stochastic Games

Introduction to Simple Stochastic Games. In this section we show how our techniques can be applied to simple stochastic games [11,10]. A simple stochastic game is a state-based two-player game where the two players, Min and Max, each own a subset of states they control, for which they can choose the successor. The system also contains sink states with an assigned payoff and averaging states which randomly choose their successor based on a given probability distribution. The goal of Min is to minimise and the goal of Max to maximise the payoff.

Simple stochastic games are an important type of games that subsume parity games and the computation of behavioural distances for probabilistic automata (cf. §6.2, [2]). The associated decision problem is known to lie in $\mathsf{NP} \cap \mathsf{coNP}$, but it is an open question whether it is contained in P. There are known randomised subexponential algorithms [7].

It has been shown that it is sufficient to consider positional strategies, i.e., strategies where the choice of the player is only dependent on the current state. The expected payoffs for each state form a so-called value vector and can be obtained as the least solution of a fixpoint equation (see below).

A *simple stochastic game* is given by a finite set V of nodes, partitioned into $\mathit{MIN}, \mathit{MAX}, \mathit{AV}$ (average) and SINK, and the following data: $\eta_{\min} : \mathit{MIN} \to \mathbf{2}^V$, $\eta_{\max} : \mathit{MAX} \to \mathbf{2}^V$ (successor functions for Min and Max nodes), $\eta_{\mathrm{av}} : \mathit{AV} \to D$ (probability distributions, where $D \subseteq \mathcal{D}(V)$ finite) and $w : \mathit{SINK} \to [0,1]$ (weights of sink nodes).

The fixpoint function $\mathcal{V}\colon [0,1]^V \to [0,1]^V$ is defined below for $a\colon V \to [0,1]$ and $v \in V$:

$$\mathcal{V}(a)(v) = \begin{cases} \min_{v' \in \eta_{\min}(v)} a(v') & v \in \mathit{MIN} \\ \max_{v' \in \eta_{\max}(v)} a(v') & v \in \mathit{MAX} \\ \sum_{v' \in V} \eta_{\mathrm{av}}(v)(v') \cdot a(v') & v \in \mathit{AV} \\ w(v) & v \in \mathit{SINK} \end{cases}$$

The *least* fixpoint of $\mathcal{V}$ specifies the average payoff for all nodes when Min and Max play optimally. In an infinite game the payoff is 0. In order to avoid infinite games and guarantee uniqueness of the fixpoint, many authors [18,10,29] restrict to stopping games, which are guaranteed to terminate for every pair of Min/Max-strategies. Here we deal with general games where more than one fixpoint may exist. Such a scenario has been studied in [19], which considers value iteration to under- and over-approximate the value vector. The over-approximation faces challenges with cyclic dependencies, similar to the vicious cycles described earlier. Here we focus on strategy iteration, which is usually less efficient than value iteration, but yields a precise result instead of approximating it.

Example 26. We consider the game depicted below. Here min is a Min node with $\eta_{\min}(\mathrm{min}) = \{\mathbf{1}, \mathrm{av}\}$, max is a Max node with $\eta_{\max}(\mathrm{max}) = \{\boldsymbol{\varepsilon}, \mathrm{av}\}$, $\mathbf{1}$ is a sink node with payoff 1, $\boldsymbol{\varepsilon}$ is a sink node with some small payoff $\varepsilon \in (0,1)$ and av is an average node which transitions to both min and max with probability $\frac{1}{2}$.

Min should choose av as successor since a payoff of 1 is bad for Min. Given this choice of Min, Max should not declare av as successor since this would create an infinite play and hence the payoff is 0. Therefore Max has to choose $\boldsymbol{\varepsilon}$ and be content with a payoff of ε, which is achieved from all nodes different from $\mathbf{1}$.

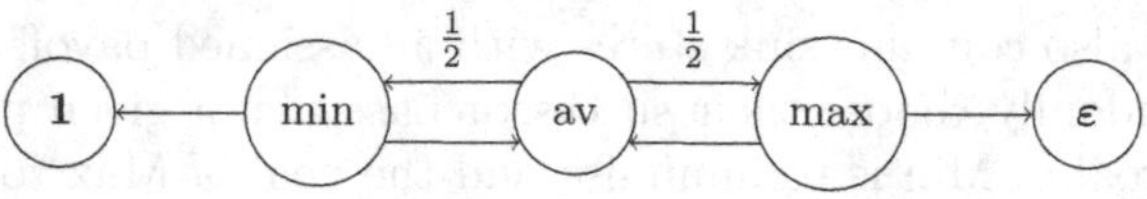

In order to be able to determine the approximation of $\mathcal{V}$ and to apply our techniques, we consider the following equivalent definition.

Lemma 27. *$\mathcal{V} = (\eta^*_{\min} \circ \min_{\in}) \uplus (\eta^*_{\max} \circ \max_{\in}) \uplus (\eta^*_{\mathrm{av}} \circ \mathrm{av}_D) \uplus c_w$, where $\in \subseteq V \times \mathbf{2}^V$ is the "is-element-of"-relation on V.*

As a composition of non-expansive functions, $\mathcal{V}$ is non-expansive as well. Since we are interested in the least fixpoint we work in the dual sense and obtain the following approximation, which intuitively says: we can decrease a value at node v by a constant only if, in the case of a Min node, we decrease the value of one successor where the minimum is reached, in the case of a Max node, we decrease the values of all successors where the maximum is reached, and in the case of an average node, we decrease the values of all successors.

Lemma 28. *Let $a\colon V \to [0,1]$. The approximation for the value iteration function $\mathcal{V}$ in the dual sense is $\mathcal{V}^a_\#\colon \mathbf{2}^{[V]^a} \to \mathbf{2}^{[V]^{\mathcal{V}(a)}}$ with*

$$\mathcal{V}^a_\#(V') = \{v \in [V]^{\mathcal{V}(a)} \mid \big(v \in \mathit{MIN} \wedge \mathit{Min}_{a|_{\eta_{\min}(v)}} \cap V' \neq \emptyset\big) \vee \\ \big(v \in \mathit{MAX} \wedge \mathit{Max}_{a|_{\eta_{\max}(v)}} \subseteq V'\big) \vee \big(v \in \mathit{AV} \wedge \mathit{supp}(\eta_{\mathrm{av}}(v)) \subseteq V'\big)\}$$

Strategy Iteration from Above and Below. We describe two algorithms based on strategy iteration, first introduced by Hoffman and Karp in [18], that are novel, as far as we know. The first iterates to the least fixpoint from above and uses the techniques described in §4. The second iterates from below: the role of our results is not directly visible in the code of the algorithm, but its non-trivial correctness proof is based on the proof rule introduced earlier.

Determine $\mu\mathcal{V}$ (from above)

1. Guess a Min-strategy $\tau^{(0)}$, $i := 0$
2. $a^{(i)} := \mu\mathcal{V}_{\tau^{(i)}}$
3. $\tau^{(i+1)} := sw_{\min}(\tau^{(i)}, a^{(i)})$
4. If $\tau^{(i+1)} \neq \tau^{(i)}$ $i := i + 1$ then goto 2.
5. Compute $V' = \nu\mathcal{V}_{\#}^{a}$, where $a = a^{(i)}$.
6. If $V' = \emptyset$ then stop and return $a^{(i)}$. Otherwise set $a^{(i+1)} := a - (\iota_{\mathcal{V}}^{a})_{V'}$, $\tau^{(i+2)} := sw_{\min}(\tau^{(i)}, a^{(i+1)})$, $i := i+2$, goto 2.

(a) Strategy iteration from above

Determine $\mu\mathcal{V}$ (from below)

1. Guess a Max-strategy $\sigma^{(0)}$, $i := 0$
2. $a^{(i)} := \mu\mathcal{V}_{\sigma^{(i)}}$
3. $\sigma^{(i+1)} := sw_{\max}(\sigma^{(i)}, a^{(i)})$
4. If $\sigma^{(i+1)} \neq \sigma^{(i)}$ set $i := i+1$ and goto 2. Otherwise stop and return $a^{(i)}$.

(b) Strategy iteration from below

Fig. 2: Strategy iteration from above and below

We first recap the underlying notions: a Min-strategy is a mapping $\tau\colon MIN \to V$ such that $\tau(v) \in \eta_{\min}(v)$ for every $v \in MIN$. With such a strategy, Min decides to always leave a node v via $\tau(v)$. Analogously $\sigma\colon MAX \to V$ fixes a Max-strategy. Fixing a strategy for either player induces a modified value function. If τ is a Min-strategy, we obtain $\mathcal{V}_\tau$ which is defined exactly as $\mathcal{V}$ but for $v \in MIN$ where we set $\mathcal{V}_\tau(a)(v) = a(\tau(v))$. Analogously, for σ a Max-strategy, $\mathcal{V}_\sigma$ is obtained by setting $\mathcal{V}_\sigma(a)(v) = a(\sigma(v))$ when $v \in MAX$. If both players fix their strategies, the game reduces to a Markov chain.

In order to describe our algorithms we also need the notion of a *switch*. Assume that τ is a Min-strategy and let a be a (pre-)fixpoint of $\mathcal{V}_\tau$. Min can now potentially improve her strategy for nodes $v \in MIN$ where $\min_{v' \in \eta_{\min}(v)} a(v') < a(\tau(v))$, called *switch nodes*. This results in a Min-strategy $\tau' = sw_{\min}(\tau, a)$, where[5] $\tau'(v) = \arg\min_{v' \in \eta_{\min}(v)} a^{(i)}(v')$ for a switch node v and τ', τ agree otherwise. Also, $sw_{\max}(\sigma, a)$ is defined analogously for Max strategies.

Now strategy iteration from above works as described in Figure 2a. The computation of $\mu\mathcal{V}_{\tau^{(i)}}$ in the second step intuitively means that Max chooses his best answering strategy and we compute the least fixpoint based on this answering strategy. At some point no further switches are possible and we have reached a fixpoint a, which need not yet be the least fixpoint. Hence we use the techniques from §4 to decrease a and obtain a new pre-fixpoint $a^{(i+1)}$, from which we can continue. The correctness of this procedure partially follows from Thm. 11 and Lem. 12, however we also need to show the following: first, we can compute $a^{(i)} = \mu\mathcal{V}_{\tau^{(i)}}$ efficiently by solving a linear program (cf. Lem. 29) by adapting [11]. Second, the chain of the $a^{(i)}$ decreases, which means that the algorithm will eventually terminate (cf. Thm. 30).

[5] If the minimum is achieved in several nodes, Min simply chooses one of them. However, she will only switch if this strictly improves the value.

Strategy iteration from below is given in Figure 2b. At first sight, the algorithm looks simpler than strategy iteration from above, since we do not have to check whether we have already reached $\nu\mathcal{V}$, reduce and continue from there. However, in this case the computation of $\mu\mathcal{V}_{\sigma^{(i)}}$ via a linear program is more involved (cf. Lem. 29), since we have to pre-compute (via greatest fixpoint iteration over $\mathbf{2}^V$) the nodes where Min can force a cycle based on the current strategy of Max, thus obtaining payoff 0.

This algorithm does not directly use our technique but we can use our proof rules to prove the correctness of the algorithm (Thm. 30). In particular, the proof that the sequence $a^{(i)}$ increases is quite involved: we have to show that $a^{(i)} = \mu\mathcal{V}_{\sigma^{(i)}} \leq \mu\mathcal{V}_{\sigma^{(i+1)}} = a^{(i+1)}$. We prove this, using our proof rules, by showing that $a^{(i)}$ is below the least fixpoint of $\mathcal{V}_{\sigma^{(i+1)}}$.

The algorithm generalises strategy iteration by Hoffman and Karp [18]. Note that we cannot simply adapt their proof, since we do not assume that the game is stopping, which is a crucial ingredient.

Lemma 29. *The least fixpoints of $\mathcal{V}_\tau$ and $\mathcal{V}_\sigma$ can be determined by solving linear programs.*

Theorem 30. *Strategy iteration from above and below both terminate and compute the least fixpoint of $\mathcal{V}$.*

Example 31. Ex. 26 is well suited to explain our two algorithms.

Starting with strategy iteration from above, we may guess $\tau^{(0)}(\min) = \mathbf{1}$. In this case, Max would choose av as successor and we would reach a fixpoint, where each node except for ε is associated with a payoff of 1. Next, our algorithm would detect the vicious cycle formed by min, av and max. We can reduce the values in this vicious cycle and reach the correct payoff values for each node.

For strategy iteration from below assume that $\sigma^{(0)}(\max) = \text{av}$. Given this strategy of Max, Min can force the play to stay in a cycle formed by min, av and max. Thus, the payoff achieved by the Max strategy $\sigma^{(0)}$ and an optimal play by Min would be 0 for each of these nodes. In the next iteration Max switches and chooses ε as successor, i.e. $\sigma^{(1)}(\max) = \varepsilon$, which results in the correct values.

We implemented strategy iteration from above and below and classical Kleene iteration in MATLAB. In Kleene iteration we terminate with a tolerance of 10^{-14}, i.e., we stop if the change from one iteration to the next is below this bound. We tested the algorithms on random stochastic games and found that Kleene iteration is always the fastest, but only converges and it is known that the rate of convergence can be exponentially slow [10]. Strategy iteration from below is usually slightly faster than strategy iteration from above. More details can be found in the full version [5].

8 Conclusion

It is well-known that several computations in the context of system verification can be performed by various forms of fixpoint iteration and it is worthwhile to study such methods at a high level of abstraction, typically in the setting of complete lattices and monotone functions. Going beyond the classical results by Tarski [28], combination of fixpoint iteration with approximations [14,6] and with up-to techniques [25] has proven to be successful. Here we treated a more specific setting, where the carrier set consists of functions from a finite set into an MV-chain and the fixpoint functions are non-expansive (and hence monotone), and introduced a novel technique to obtain upper bounds for greatest and lower bounds for least fixpoints, including associated algorithms. Such techniques are widely applicable to a wide range of examples and so far they have been studied only in quite specific scenarios, such as in [2,16,19].

In the future we plan to lift some of the restrictions of our approach. First, an extension to an infinite domain Y would of course be desirable, but since several of our results currently depend on finiteness, such a generalisation does not seem to be easy. Another restriction, to total orders, seems easier to lift: in particular, if the partially ordered MV-algebra $\tilde{\mathbb{M}}$ is of the form $\mathbb{M}^I$ where I is a finite index set and $\mathbb{M}$ an MV-chain. (E.g., finite Boolean algebras are of this type.) Then our function space is $\tilde{\mathbb{M}}^Y = (\mathbb{M}^I)^Y \cong \mathbb{M}^{Y\times I}$ and we have reduced to the setting presented in this paper. This will allow us to handle featured transition systems [12] where transitions are equipped with boolean formulas. We also plan to determine the largest possible increase that can be added to a fixpoint that is not yet the greatest fixpoint in order to maximally speed up fixpoint iteration from below (this might be larger than ι_a^f).

There are several other application examples that did not fit into this paper, but that can also be handled by our approach: for instance behavioural distances for metric transition systems [15] and other types of systems [4]. We also plan to investigate other types of games, such as energy games [8]. While here we introduced strategy iteration techniques for simple stochastic games, we also want to check whether we can provide an improvement to value iteration techniques, combining our approach with [19].

We also plan to study whether some examples can be handled with other types of Galois connections: here we used an additive variant, but looking at multiplicative variants (multiplication by a constant factor) might also be fruitful.

Acknowledgements: We are grateful to Ichiro Hasuo for making us aware of stochastic games as application domain. Furthermore we would like to thank Matthias Kuntz and Timo Matt for their help with experiments.

References

1. Bacci, G., Bacci, G., Larsen, K.G., Mardare, R.: On-the-fly exact computation of bisimilarity distances. Logical Methods in Computer Science **13**(2:13), 1–25 (2017)
2. Bacci, G., Bacci, G., Larsen, K.G., Mardare, R., Tang, Q., van Breugel, F.: Computing probabilistic bisimilarity distances for probabilistic automata. In: Proc. of CONCUR '19. LIPIcs, vol. 140, pp. 9:1–9:17. Schloss Dagstuhl – Leibniz Center for Informatics (2019)

3. Baier, C., Katoen, J.P.: Principles of Model Checking. MIT Press (2008)
4. Baldan, P., Bonchi, F., Kerstan, H., König, B.: Coalgebraic behavioral metrics. Logical Methods in Computer Science **14**(3) (2018), selected Papers of the 6th Conference on Algebra and Coalgebra in Computer Science (CALCO 2015)
5. Baldan, P., Eggert, R., König, B., Padoan, T.: Fixpoint theory – upside down (2021), `https://arxiv.org/abs/2101.08184`, arXiv:2101.08184
6. Baldan, P., König, B., Padoan, T.: Abstraction, up-to techniques and games for systems of fixpoint equations. In: Proc. of CONCUR '20. LIPIcs, vol. 171, pp. 25:1–25:20. Schloss Dagstuhl – Leibniz Center for Informatics (2020), `https://doi.org/10.4230/LIPIcs.CONCUR.2020.25`
7. Björklund, H., Vorobyov, S.: Combinatorial structure and randomized subexponential algorithms for infinite games. Theoretical Computer Science **349**(3), 347–360 (2005)
8. Brim, L., Chaloupka, J., Doyen, L., Gentilini, R., Raskin, J.F.: Faster algorithms for mean-payoff games. Formal Methods in System Design **38**(2), 97–118 (2011)
9. Cleaveland, R.: On automatically explaining bisimulation inequivalence. In: Proc. of CAV '90. pp. 364–372. Springer (1990), LNCS 531
10. Condon, A.: On algorithms for simple stochastic games. In: Advances In Computational Complexity Theory. DIMACS Series in Discrete Mathematics and Theoretical Computer Science, vol. 13, pp. 51–71 (1990)
11. Condon, A.: The complexity of stochastic games. Information and Computation **96**(2), 203–224 (1992). https://doi.org/10.1016/0890-5401(92)90048-K, `https://doi.org/10.1016/0890-5401(92)90048-K`
12. Cordy, M., Classen, A., Perrouin, G., Schobbens, P.Y., Heymans, P., Legay, A.: Simulation-based abstractions for software product-line model checking. In: Proc. of ICSE '12 (International Conference on Software Engineering). pp. 672–682. IEEE (2012)
13. Cousot, P., Cousot, R.: Abstract interpretation: A unified lattice model for static analysis of programs by construction or approximation of fixpoints. In: Proc. of POPL '77 (Los Angeles, California). pp. 238–252. ACM (1977)
14. Cousot, P., Cousot, R.: Temporal abstract interpretation. In: Wegman, M.N., Reps, T.W. (eds.) Proc. of POPL '00. pp. 12–25. ACM (2000)
15. de Alfaro, L., Faella, M., Stoelinga, M.: Linear and branching system metrics. IEEE Transactions on Software Engineering **35**(2), 258–273 (2009)
16. Fu, H.: Computing game metrics on Markov decision processes. In: Proc. of ICALP '12, Part II. pp. 227–238. Springer (2012), LNCS 7392
17. Hennessy, M., Milner, R.: Algebraic laws for nondeterminism and concurrency. Journal of the ACM **32**, 137–161 (1985)
18. Karp, R.M., Hoffman, A.J.: On nonterminating stochastic games. Management Science **12**(5), 359–370 (1966)
19. Kelmendi, E., Krämer, J., Křetínský, J., Weininger, M.: Value iteration for simple stochastic games: Stopping criterion and learning algorithm. In: Proc. of CAV '18. pp. 623–642. Springer (2018), LNCS 10981
20. Mémoli, F.: Gromov-Wasserstein distances and the metric approach to object matching. Foundations of Computational Mathematics **11**(4), 417–487 (2011)
21. Mundici, D.: MV-algebras. A short tutorial, available at `http://www.matematica.uns.edu.ar/IXCongresoMonteiro/Comunicaciones/Mundici_tutorial.pdf`
22. Mundici, D.: Advanced Łukasiewicz calculus and MV-algebras, Trends in Logic, vol. 35. Springer (2011)
23. Nielson, F., Nielson, H.R., Hankin, C.: Principles of Program Analysis. Springer (2010)

24. Peyré, G., Cuturi, M.: Computational optimal transport (2020), `https://arxiv.org/abs/2009.14817`, arXiv:1803.00567
25. Pous, D.: Complete lattices and up-to techniques. In: Proc. of APLAS '07. pp. 351–366. Springer (2007), LNCS 4807
26. Sangiorgi, D.: Introduction to Bisimulation and Coinduction. Cambridge University Press (2011)
27. Stirling, C.: Bisimulation, model checking and other games. Notes for Mathfit instructional meeting on games and computation, Edinburgh (June 1997), `http://homepages.inf.ed.ac.uk/cps/mathfit.pdf`
28. Tarski, A.: A lattice-theoretical theorem and its applications. Pacific Journal of Mathematics **5**, 285–309 (1955)
29. Tripathi, R., Valkanova, E., Kumar, V.A.: On strategy improvement algorithms for simple stochastic games. Journal of Discrete Algorithms **9**, 263–278 (2011)
30. Villani, C.: Optimal Transport – Old and New, A Series of Comprehensive Studies in Mathematics, vol. 338. Springer (2009)

“Most of” leads to undecidability: Failure of adding frequencies to LTL

Bartosz Bednarczyk✉[1,2] and Jakub Michaliszyn[2]

[1] Computational Logic Group, Technische Universität Dresden, Dresden, Germany
[2] Institute of Computer Science, University of Wrocław, Wrocław, Poland
{bartosz.bednarczyk, jakub.michaliszyn}@cs.uni.wroc.pl

Abstract. Linear Temporal Logic (LTL) interpreted on finite traces is a robust specification framework popular in formal verification. However, despite the high interest in the logic in recent years, the topic of their quantitative extensions is not yet fully explored. The main goal of this work is to study the effect of adding weak forms of percentage constraints (*e.g.* that *most of* the positions in the past satisfy a given condition, or that σ is the *most-frequent* letter occurring in the past) to fragments of LTL. Such extensions could potentially be used for the verification of influence networks or statistical reasoning. Unfortunately, as we prove in the paper, it turns out that percentage extensions of even tiny fragments of LTL have undecidable satisfiability and model-checking problems. Our undecidability proofs not only sharpen most of the undecidability results on logics with arithmetics interpreted on words known from the literature, but also are fairly simple. We also show that the undecidability can be avoided by restricting the allowed usage of the negation, and discuss how the undecidability results transfer to first-order logic on words.

1 Introduction

Linear Temporal Logic [29] (LTL) interpreted on finite traces is a robust logical framework used in formal verification [1,18,19]. However, LTL is not perfect: it can express whether some event happens or not, but it cannot provide any insight on how frequently such an event occurs or for how long such an event took place. In many practical applications, such *quantitative* information is important: think of optimising a server based on how frequently it receives messages or optimising energy consumption knowing for how long a system is usually used in rush hours. Nevertheless, there is a solution: one can achieve such goals by adding quantitative features to LTL.

It is known that adding quantitative operators to LTL often leads to undecidability. The proofs, however, typically involve operators such as “next” or “until”, and are often quite complicated (see the discussion on the related work below). In this work, we study the logic $\mathrm{LTL}_{\mathbf{F}}$, a fragment of LTL where the only allowed temporal operator is “sometimes in the future” $\mathbf{F}$. We extend its language with two types of operators, sharing a similar “percentage” flavour: with the *Past-Majority* $\mathbf{PM}\,\varphi$ operator (stating that most of the past positions satisfy

S. Kiefer and C. Tasson (Eds.): FOSSACS 2021, LNCS 12650, pp. 82–101, 2021.
https://doi.org/10.1007/978-3-030-71995-1_5

a formula φ), and with the *Most-Frequent-Letter* **MFL** σ predicates (meaning that the letter σ is among the most frequent letters appearing in the past). These operators can be used to express a number of interesting properties, such as *if a process failed to enter the critical section, then the other process was in the critical section the majority of time.* Of course, for practical applications, we could also consider richer languages, such as parametrised versions of these operators, *e.g.* stating that *at least a fraction p of positions in the past satisfies a formula.* However, we show, as our main result, that even these very simple percentage operators raise undecidability when combined with **F**.

To make the undecidability proof for both operators similar, we define an intermediate operator, **Half**, which is satisfied when exactly half of the past positions satisfy a given formula. The **Half** operator can be expressed easily with **PM**, but not with **MFL** — we show, however, that we can simulate it to an extent enough to show the undecidability. Our proof method relies on enforcing a model to be in the language $(\{wht\}\{shdw\})^+$, for some letters *wht* and *shdw*, which a priori seems to be impossible without the "next" operator. Then, thanks to the specific shape of the models, we show that one can "transfer" the truth of certain formulae from positions into their successors, hence the "next" operator can be partially expressed. With a combination of these two ideas, we show that it is possible to write equicardinality statements in the logic. Finally, we perform a reduction from the reachability problem of Two-counter Machines [26]. In the reduction, the equicardinality statements will be responsible for handling zero-tests. The idea of transferring predicates from each position into its successor will be used for switching the machine into its next configuration.

The presented undecidability proof of LTL with percentage operators can be adjusted to extensions of fragments of first-order logic on finite words. We show that $\mathrm{FO}^2_{\mathsf{M}}[<]$, *i.e.* the two-variable fragment of first-order logic admitting the majority quantifier M and linear order predicate $<$ has an undecidable satisfiability problem. Here the meaning of a formula $\mathsf{M}x.\varphi(x,y)$ is that at least a half of possible interpretations of x satisfies $\varphi(x,y)$. Our result sharpens an existing undecidability proof for (full) FO with Majority from [23] (since in our case the number of variables is limited) but also $\mathrm{FO}^2[<, succ]$ with arithmetics from [25] (since our counting mechanism is weaker and the successor relation *succ* is disallowed). On the positive side, we show that the undecidability heavily depends on the presence of the negation in front of the percentage operators. To do so, we introduce a logic, extending the full LTL, in which the usage of percentage operators is possible, but suitably restricted. For this logic, we show that the satisfiability problem is decidable.

All the above-mentioned results can be easily extended to the model checking problem, where the question is whether a given Kripke structure satisfies a given formula. The full version of the paper is available on arXiv [4].

1.1 Related work

The first paper studying the addition of quantitative features to logic was [21], where the authors proved undecidability of Weak MSO with Cardinalities. They also developed a model of so-called Parikh Automaton, a finite automaton imposing a semi-linear constraint on the set of its final configurations. Such an automaton was successfully used to decide logics with counting as well as logics on data words [27,17]. Its expressiveness was studied in [11].

Another idea in the realm of quantitative features is availability languages [20], which extend regular expressions by numerical occurrence constraints on the letters. However, their high expressivity leads to undecidable emptiness problems. Weak forms of arithmetics have also attracted interest from researchers working on temporal logics. Several extensions of LTL were studied, including extensions with counting [24], periodicity constraints [14], accumulative values [7], discounting [2], averaging [9] and frequency constraints [8]. A lot of work was done to understand LTL with timed constraints, *e.g.* a metric LTL was considered in [28]. However, its complexity is high and its extensions are undecidable [3].

Arithmetical constraints can also be added to the First-Order logic (FO) on words via so-called counting quantifiers. It is known that weak MSO on words is decidable with threshold counting and modulo-counting (thanks to the famous Büchi theorem [10]), while even FO on words with percentage quantifiers becomes undecidable [23]. Extensions of fragments of FO on words are often decidable, *e.g.* the two-variable fragment FO^2 with counting [12] or FO^2 with modulo-counting [25]. The investigation of decidable extensions of FO^2 is limited by the undecidability of FO^2 on words with Presburger constraints [25].

Among the above-mentioned logics, the formalisms of this paper are most similar to Frequency LTL [8]. The satisfiability problem for Frequency LTL was claimed to be undecidable, but the undecidability proof as presented in [8] is bugged (see [9, Sec. 8] for discussion). It was mentioned in [9] that the undecidability proof from [8] can be patched, but no correction was published so far. Our paper not only provides a valid proof but also sharpens the result, as we use a way less expressive language (*e.g.* we are allowed to use neither the "until" operator nor the "next" operator). We also believe that our proof is simpler. The second-closest formalism to ours is average-LTL [9]. The main difference is that the averages of average-LTL are computed based on the future, while in our paper, the averages are based on the past. The second difference, as in the previous case, is that their undecidability proof uses more expressive operators, such as the "until" operator.

2 Preliminaries

We recall definitions concerning logics on words and temporal logics (*cf.* [15]).

Words and logics. Let AP be a countably-infinite set of *atomic propositions*, called here also *letters*. A finite *word* $\mathfrak{w} \in (2^{\mathsf{AP}})^*$ is a non-empty finite sequence of *positions* labelled with sets of letters from AP. A set of words is called a *language*. Given a word $\mathfrak{w}$, we denote its i-th position with $\mathfrak{w}_i$ (where the first position is $\mathfrak{w}_0$) and its prefix up to the i-th position with $\mathfrak{w}_{\leq i}$. We usually use the letters p, q, i, j to denote positions. With $|\mathfrak{w}|$ we denote the length of $\mathfrak{w}$.

The syntax of $\mathrm{LTL}_{\mathbf{F}}$, a fragment of LTL with only the *finally* operator $\mathbf{F}$, is defined with the grammar: $\varphi, \varphi' ::= a$ (with $a \in \mathsf{AP}$) $\mid \neg\varphi \mid \varphi \wedge \varphi' \mid \mathbf{F}\varphi$.

The satisfaction relation $\models$ is defined for words as follows:

$$\begin{array}{ll} \mathfrak{w}, i \models a & \text{if } a \in \mathfrak{w}_i \\ \mathfrak{w}, i \models \neg\varphi & \text{if not } \mathfrak{w}, i \models \varphi \\ \mathfrak{w}, i \models \varphi_1 \land \varphi_2 & \text{if } \mathfrak{w}, i \models \varphi_1 \text{ and } \mathfrak{w}, i \models \varphi_2 \\ \mathfrak{w}, i \models \mathbf{F}\,\varphi & \text{if } \exists j \text{ such that } |\mathfrak{w}| > j \geq i \text{ and } \mathfrak{w}, j \models \varphi. \end{array}$$

We write $\mathfrak{w} \models \varphi$ if $\mathfrak{w}, 0 \models \varphi$. The usual Boolean connectives: $\top, \bot, \lor, \rightarrow, \leftrightarrow$ can be defined, hence we will use them as abbreviations. Additionally, we use the *globally* operator $\mathbf{G}\,\varphi := \neg\mathbf{F}\,\neg\varphi$ to speak about events happening globally in the future.

Percentage extension. In our investigation, *percentage operators* **PM**, **MFL** and **Half** are added to $\mathrm{LTL}_{\mathbf{F}}$.

The operator $\mathbf{PM}\,\varphi$ (read as: *majority in the past*) is satisfied if at least half of the positions in the past satisfy φ:

$$\mathfrak{w}, i \models \mathbf{PM}\,\varphi \text{ if } |\{j < i \colon \mathfrak{w}, j \models \varphi\}| \geq \tfrac{i}{2}$$

For example, the formula $\mathbf{G}\,(r \leftrightarrow \neg g) \land \mathbf{G}\,\mathbf{PM}\,r \land \mathbf{G}\,\mathbf{F}\,(g \land \mathbf{PM}\,g)$ is true over words where each *request* r is eventually fulfilled by a *grant* g, and where each grant corresponds to at least one request. This can be also seen as the language of balanced parentheses, showing that with the operator **PM** one can define properties that are not regular.

The operator $\mathbf{MFL}\,\sigma$ (read as: *most-frequent letter in the past*), for $\sigma \in \mathsf{AP}$, is satisfied if σ is among the letters with the highest number of appearances in the past, *i.e.*

$$\mathfrak{w}, i \models \mathbf{MFL}\,\sigma \text{ if } \forall \tau \in \mathsf{AP}.\ |\{j < i \colon \mathfrak{w}, j \models \sigma\}| \geq |\{j < i \colon \mathfrak{w}, j \models \tau\}|$$

For example, the formula $\mathbf{G}\,\neg(r \land g) \land \mathbf{G}\ \mathbf{MFL}\,r \land \mathbf{G}\,\mathbf{F}\,(g \land \mathbf{MFL}\,g)$ again defines words where each request is eventually fulfilled, but this time the formula allows for states where nothing happens (*i.e.* when both r and g are false).

The last operator, **Half** is used to simplify the forthcoming undecidability proofs. This operator can be satisfied only at even positions, and its intended meaning is *exactly half of the past positions satisfy a given formula.*

$$\mathfrak{w}, i \models \mathbf{Half}\,\varphi \text{ if } |\{j < i \colon \mathfrak{w}, j \models \varphi\}| = \tfrac{i}{2}$$

It is not difficult to see that the operator $\mathbf{Half}\,\varphi$ can be defined in terms of the past-majority operator as $\mathbf{PM}\,(\varphi) \land \mathbf{PM}\,(\neg\varphi)$ and that $\mathbf{Half}\,\varphi$ can be satisfied only at even positions.

In the next sections, we distinguish different logics by enumerating the allowed operators in the subscripts, *e.g.* $\mathrm{LTL}_{\mathbf{F},\mathbf{PM}}$ or $\mathrm{LTL}_{\mathbf{F},\mathbf{MFL}}$.

Computational problems Kripke structures are commonly used in verification to formalise abstract models. A Kripke structure is composed of a finite set S of *states*, a set of *initial* states $I \subseteq S$, a total *transition* relation $R \subseteq S \times S$, and a finite *labelling function* $\ell : S \rightarrow 2^{\mathsf{AP}}$. A *trace* of a Kripke structure is a finite word

$\ell(s_0), \ell(s_1), \ldots, \ell(s_k)$ for any $s_0, s_1, \ldots, s_k$ satisfying $s_0 \in I$ and $(s_i, s_{i+1}) \in R$ for all $i < k$.

The *model-checking problem* amounts to checking whether *some* trace of a given Kripke structure satisfies a given formula φ. In the *satisfiability problem*, or simply in *SAT*, we check whether an input formula φ has a *model*, *i.e.* a finite word $\mathfrak{w}$ witnessing $\mathfrak{w} \models \varphi$.

3 Playing with Half Operator

Before we jump into the encoding of Minsky machines, we present some exercises to help the reader understand the expressive power of the logic $\mathrm{LTL}_{\mathbf{F},\mathbf{Half}}$. The tools established in the exercises play a vital role in the undecidability proofs provided in the following section.

We start from the definition of shadowy words.

Definition 1. *Let wht and shdw be fixed distinct atomic propositions from* AP. *A word* $\mathfrak{w}$ *is* shadowy *if its length is even, all even positions of* $\mathfrak{w}$ *are labelled with wht, all odd positions of* $\mathfrak{w}$ *are labelled with shdw, and no position is labelled with both letters.*

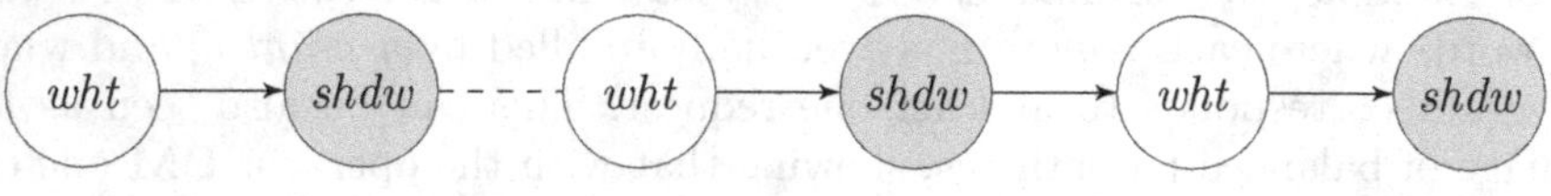

We will call the positions satisfying *wht* simply *white* and their successors satisfying *shdw* simply their *shadows*.

The following exercise is simple in LTL, but becomes much more challenging without the $\mathbf{X}$ operator.

Exercise 1. There is an $\mathrm{LTL}_{\mathbf{F},\mathbf{Half}}$ formula $\psi_{shadowy}$ defining shadowy words.

Solution. We start with the "base" formula $\varphi_{init}^{ex1} := wht \wedge \mathbf{G}\,(wht \leftrightarrow \neg shdw) \wedge \mathbf{G}\,(wht \rightarrow \mathbf{F}\,shdw)$, which states that the position 0 is labelled with *wht*, each position is labelled with exactly one letter among *wht*, *shdw* and that every white eventually sees a shadow in the future. What remains to be done is to ensure that only odd positions are shadows and that only even positions are white.

In order to do that, we employ the formula $\varphi_{odd}^{ex1} := \mathbf{G}\,((\mathbf{Half}\,wht) \leftrightarrow wht)$. Since **Half** is never satisfied at odd positions, the formula φ_{odd}^{ex1} stipulates that odd positions are labelled with *shdw*. An inductive argument shows that all the even positions are labelled with *wht*: for the position 0, it follows from φ_{init}^{ex1}. For an even position $p > 0$, assuming (inductively) that all even positions are labelled with *wht*, the formula φ_{odd}^{ex1} ensures that p is labelled with *wht*.

Putting it all together, the formula $\psi_{shadowy} := \varphi_{init}^{ex1} \wedge \varphi_{odd}^{ex1}$ is as required. □

In the next exercise, we show that it is possible to transfer the presence of certain letters from white positions into their shadows. It justifies the usage of "shadows" in the paper.

We introduce the so-called *counting terms*. For a formula φ, word $\mathfrak{w}$ and a position p, by $\#_{\varphi}^{<}(\mathfrak{w}, p)$ we denote the total number of positions among $0, \ldots, p-1$

satisfying φ, *i.e.* the size of $\{p' < p \mid \mathfrak{w}, p' \models \varphi\}$. We omit $\mathfrak{w}$ in counting terms if it is known from the context.

Exercise 2. Let σ and $\tilde{\sigma}$ be distinct letters from $\mathsf{AP} \setminus \{wht, shdw\}$. There is an $\mathrm{LTL}_{\mathbf{F},\mathbf{Half}}$ formula $\varphi^{trans}_{\sigma \rightsquigarrow \tilde{\sigma}}$, such that $\mathfrak{w} \models \varphi^{trans}_{\sigma \rightsquigarrow \tilde{\sigma}}$ iff:

1. $\mathfrak{w}$ is shadowy,
2. only white (resp., shadow) positions of $\mathfrak{w}$ can be labelled σ (resp., $\tilde{\sigma}$) and
3. for any even position p we have: $\mathfrak{w}, p \models \sigma \Leftrightarrow \mathfrak{w}, p{+}1 \models \tilde{\sigma}$.

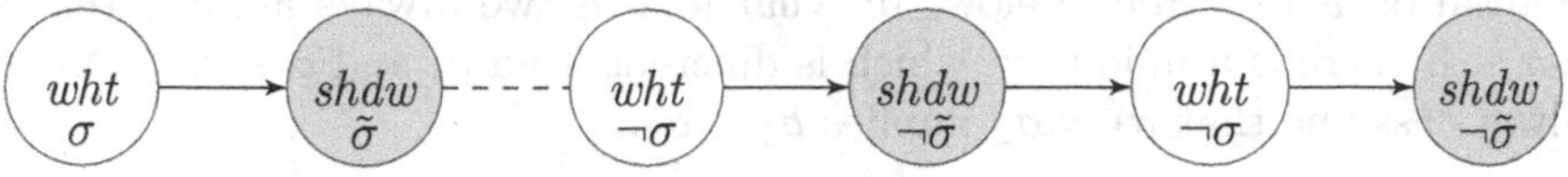

Solution. Note that the first two conditions can be expressed with the conjunction of $\psi_{shadowy}$, $\mathbf{G}\,(\sigma \to wht)$ and $\mathbf{G}\,(\tilde{\sigma} \to shdw)$. The last condition is more involving. Assuming that the words under consideration satisfy conditions 1–2, it is easy to see that the third condition is equivalent to expressing that all white positions p satisfy the equation (♡):

$$(\heartsuit): \quad \#^{<}_{wht \wedge \sigma}(\mathfrak{w}, p) = \#^{<}_{shdw \wedge \tilde{\sigma}}(\mathfrak{w}, p)$$

supplemented with the condition (♢), ensuring that the last white position satisfies the condition 3, *i.e.*

$$(\diamondsuit): \quad \text{for the last white position } p \text{ we have: } \mathfrak{w}, p \models \sigma \Leftrightarrow \mathfrak{w}, p{+}1 \models \tilde{\sigma}.$$

The proof of the following lemma can be found in the appendix.

Lemma 1. *Let $\mathfrak{w}$ be a word satisfying the conditions 1–2. Then $\mathfrak{w}$ satisfies the condition 3 iff $\mathfrak{w}$ satisfies (♢) and for all white positions p the equation (♡) holds.*

Going back to Exercise 2, we show how to define (♡) and (♢) in $\mathrm{LTL}_{\mathbf{F},\mathbf{Half}}$, taking advantage of shadowness of the intended models. Take an arbitrary white position p of $\mathfrak{w}$. The equation (♡) for p is clearly equivalent to:

$$(\heartsuit'): \#^{<}_{wht \wedge \sigma}(\mathfrak{w}, p) + \left(\frac{p}{2} - \#^{<}_{shdw \wedge \tilde{\sigma}}(\mathfrak{w}, p)\right) = \frac{p}{2}$$

Since p is even, we infer that $\frac{p}{2} \in \mathbb{N}$. From the shadowness of $\mathfrak{w}$, we know that there are exactly $\frac{p}{2}$ shadows in the past of p. Moreover, each shadow satisfies either $\tilde{\sigma}$ or $\neg\tilde{\sigma}$. Hence, the expression $\frac{p}{2} - \#^{<}_{shdw \wedge \tilde{\sigma}}(\mathfrak{w}, p)$ from (♡′), can be replaced with $\#^{<}_{shdw \wedge \neg\tilde{\sigma}}(\mathfrak{w}, p)$. Finally, since *wht* and *shdw* label disjoint positions, the property that every white position p satisfies (♡) can be written as an $\mathrm{LTL}_{\mathbf{F},\mathbf{Half}}$ formula $\varphi_{(\heartsuit)} := \mathbf{G}\,(wht \to \mathbf{Half}\,([wht \wedge \sigma] \vee [shdw \wedge \neg\tilde{\sigma}]))$. Its correctness follows from the correctness of each arithmetic transformation and the semantics of $\mathrm{LTL}_{\mathbf{F},\mathbf{Half}}$.

For the property (♢), we first need to define formulae detecting the last and the second to last positions of the model. Detecting the last position is easy: since the last position of $\mathfrak{w}$ is shadow, it is sufficient to express that it sees only shadows in its future, *i.e.* $\varphi^{ex2}_{last} := \mathbf{G}\,(shdw)$. Similarly, a position is second to last if it is white and it sees only white or last positions in the future, which results in a formula $\varphi^{ex2}_{stl} := wht \wedge \mathbf{G}\,(wht \vee \varphi^{ex2}_{last})$. Note that the correctness of φ^{ex2}_{last} and φ^{ex2}_{stl} follows immediately from shadowness. Hence, we can define the formula $\varphi_{(\diamondsuit)}$ as $\mathbf{F}\,(\varphi^{ex2}_{stl} \wedge \sigma) \leftrightarrow \mathbf{F}\,(\varphi^{ex2}_{last} \wedge \tilde{\sigma})$. The conjunction of $\varphi_{(\heartsuit)}$ and $\varphi_{(\diamondsuit)}$ formulae gives us to $\varphi^{trans}_{\sigma \rightsquigarrow \tilde{\sigma}}$. □

We consider a generalisation of shadowy models, where each shadow mimics all letters from a finite set $\Sigma \subseteq \mathsf{AP}$ rather than just a single letter σ. Such a generalisation is described below. In what follows, we always assume that for each $\sigma \in \Sigma$ there is a unique $\tilde{\sigma}$, which is different from σ, and $\tilde{\sigma} \notin \Sigma$. Moreover, we always assume that $\sigma_1 \neq \sigma_2$ implies $\tilde{\sigma}_1 \neq \tilde{\sigma}_2$.

Definition 2. *Let $\Sigma \subseteq \mathsf{AP} \setminus \{wht, shdw\}$ be a finite set. A shadowy word $\mathfrak{w}$ is called* truly Σ-shadowy, *if for every letter $\sigma \in \Sigma$ only the white (resp. shadow) positions of $\mathfrak{w}$ can be labelled with σ (resp. $\tilde{\sigma}$) and every white position p of $\mathfrak{w}$ satisfies $\mathfrak{w}, p \models \sigma \Leftrightarrow \mathfrak{w}, p{+}1 \models \tilde{\sigma}$.*

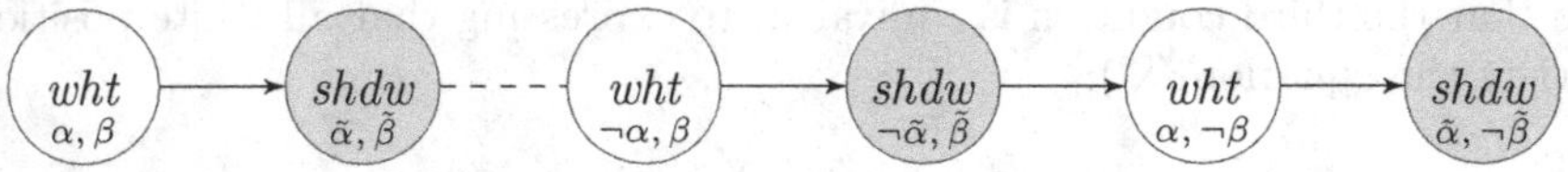

Knowing the solution for the previous exercise, it is easy to come up with a formula $\psi^{truly-\Sigma}_{shadowy}$ defining truly Σ-shadowy models: just take the conjunction of $\psi_{shadowy}$ and $\varphi^{trans}_{\sigma \rightsquigarrow \tilde{\sigma}}$ over all letters $\sigma \in \Sigma$. The correctness follows immediately from from Exercise 2.

Corollary 1. *The formula $\psi^{truly-\Sigma}_{shadowy}$ defines the language of truly Σ-shadowy words.*

The next exercise shows how to compare cardinalities in $\mathrm{LTL}_{\mathbf{F},\mathbf{Half}}$ over truly Σ-shadowy models. We are not going to introduce any novel techniques here, but the exercise is of great importance: it is used in the next section to encode zero tests of Minsky machines.

Exercise 3. Let Σ be a finite subset of $\mathsf{AP} \setminus \{wht, shdw\}$ and let $\alpha \neq \beta \in \Sigma$. There exists an $\mathrm{LTL}_{\mathbf{F},\mathbf{Half}}$ formula $\psi_{\#\alpha=\#\beta}$ such that for any truly Σ-shadowy word $\mathfrak{w}$ and any of its white positions p: the equivalence $\mathfrak{w}, p \models \psi_{\#\alpha=\#\beta} \Leftrightarrow \#^{<}_{wht \wedge \alpha}(\mathfrak{w}, p) = \#^{<}_{wht \wedge \beta}(\mathfrak{w}, p)$ holds.

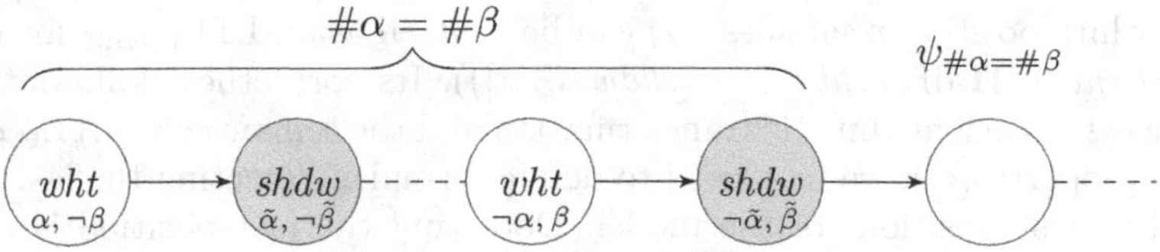

The solution is in the appendix, here we briefly discuss the main idea. Follow the previous exercise. The main difficulty is to express the equality of counting terms, written as $\mathsf{LHS} = \mathsf{RHS}$. Note that it is clearly equivalent to $\mathsf{LHS} + (\frac{p}{2} - \mathsf{RHS}) = \frac{p}{2}$. Unfold $\frac{p}{2}$ on the left hand side, *i.e.* replace it with the total number of shadows in the past. Use the fact that $\mathfrak{w}$ satisfies $\varphi^{trans}_{\sigma \rightsquigarrow \tilde{\sigma}}$, which implies the equality $\#^{<}_{wht \land \beta}(\mathfrak{w}, p) = \#^{<}_{shdw \land \tilde{\beta}}(\mathfrak{w}, p)$. Finally, get rid of subtraction and write an $\mathrm{LTL}_{\mathbf{F},\mathbf{Half}}$ formula by employing **Half**. The presented exercises show that the expressive power of $\mathrm{LTL}_{\mathbf{F},\mathbf{Half}}$ is so high that, under a mild assumption of truly-shadowness, it allows us to perform cardinality comparison. We are now only a step away from showing undecidability of the logic, which is tackled next.

4 Undecidability of LTL extensions

This section is dedicated to the main technical contribution of the paper, namely that $\mathrm{LTL}_{\mathbf{F},\mathbf{Half}}$, $\mathrm{LTL}_{\mathbf{F},\mathbf{PM}}$ and $\mathrm{LTL}_{\mathbf{F},\mathbf{MFL}}$ have undecidable satisfiability and model checking problems. We start from $\mathrm{LTL}_{\mathbf{F},\mathbf{Half}}$. Then, the undecidability of $\mathrm{LTL}_{\mathbf{F},\mathbf{PM}}$ will follow immediately from the fact that **Half** is definable by **PM**. Finally, we will show how the undecidability proof can be adjusted to $\mathrm{LTL}_{\mathbf{F},\mathbf{MFL}}$.

We start by recalling the basics on Minsky Machines.

Minsky machines A *deterministic Minsky machine* is, roughly speaking, a finite transition system equipped with two unbounded-size natural counters, where each counter can be incremented, decremented (only in the case it is positive), and tested for being zero. Formally, a Minsky machine $\mathcal{A}$ is composed of a finite set of *states* Q with a distinguished *initial* state q_0 and a transition function $\delta : (Q \times \{0,+\}^2) \to (\{-1,0,1\}^2 \times (Q \setminus \{q_0\})$ satisfying three additional requirements: whenever $\delta(q,f,s) = (\bar{f}, \bar{s}, q')$ holds, $\bar{f} = -1$ implies $f = +$, $\bar{s} = -1$ implies $s = +$ (*i.e.* it means that only the positive counters can be decremented) and $q \neq q'$ (the machine cannot enter the same state two times in a row). Intuitively, the first coordinate of δ describes the current state of the machine, the second and the third coordinates tell us whether the current value of the i-th counter is zero or positive, the next two coordinates denote the update on the counters and the last coordinate denotes the target state.

We define a *run* of a Minsky machine $\mathcal{A}$ as a sequence of consecutive transitions of $\mathcal{A}$. Formally, a run of $\mathcal{A}$ is a finite word $\mathfrak{w} \in (Q \times \{0,+\}^2 \times \{-1,0,1\}^2 \times Q \setminus \{q_0\})^+$ such that, when denoting $\mathfrak{w}_i$ as $(q^i, f^i, s^i, \bar{f}^i, \bar{s}^i, q^i_N)$, all the following conditions are satisfied:

1. $q^0 = q_0$ and $f^0 = s^0 = 0$,
2. for each i we have $\delta(q^i, f^i, s^i) = (\bar{f}^i, \bar{s}^i, q^i_N)$,
3. for each $i < |\mathfrak{w}|$ we have $q^i_N = q^{i+1}$,
4. for each i, f^i equals 0 iff $\bar{f}^0 + \cdots + \bar{f}^{i-1} = 0$, and $+$ otherwise; similarly s^i is 0 if $\bar{s}^0 + \cdots + \bar{s}^{i-1} = 0$ and $+$ otherwise.

It is not hard to see that this definition is equivalent to the classical one [26]. We say that a Minsky machine *reaches* a state $q \in Q$ if there is a run with a letter containing q on its last coordinate. It is well known that the problem of checking whether a given Minsky machine reaches a given state is undecidable [26].

4.1 "Half of" meets the halting problem

We start from presenting the overview of the claimed reduction. Until the end of Section 4, let us fix a Minsky machine $\mathcal{A} = (Q, q_0, \delta)$ and its state $\mathsf{q} \in Q$. Our ultimate goal is to define an $\mathrm{LTL}_{\mathbf{F},\mathbf{Half}}$ formula $\psi_{\mathcal{A}}^{\mathsf{q}}$ such that $\psi_{\mathcal{A}}^{\mathsf{q}}$ has a model iff $\mathcal{A}$ reaches q. To do so, we define a formula $\psi_{\mathcal{A}}$ such that there is a one-to-one correspondence between the models of $\psi_{\mathcal{A}}$ and runs of $\mathcal{A}$. Expressing the reachability of q, and thus $\psi_{\mathcal{A}}^{\mathsf{q}}$, based on $\psi_{\mathcal{A}}$ is easy.

Intuitively, the formula $\psi_{\mathcal{A}}$ describes a shadowy word $\mathfrak{w}$ encoding on its white positions the consecutive letters of a run of $\mathcal{A}$. In order to express it, we introduce a set $\Sigma_{\mathcal{A}}$, composed of the following distinguished atomic propositions:

- $from_q$ and to_q for all states $q \in Q$,
- $fVal_c$ and $sVal_c$ for counter values $c \in \{0, +\}$, and
- fOP_{op} and sOP_{op} for all operations $op \in \{-1, 0, 1\}$.

We formalise the one-to-one correspondence as the function *run*, which takes an appropriately defined shadowy model and returns a corresponding run of $\mathcal{A}$. More precisely, the function $run(\mathfrak{w})$ returns a run whose ith configuration is $(q, f, s, \bar{f}, \bar{s}, q_N)$ if and only if the ith white configuration of $\mathfrak{w}$ is labelled with $from_q, fVal_f, sVal_s, fOP_{\bar{f}}, sOP_{\bar{s}}$ and to_{q_N}.

The formula $\psi_{\mathcal{A}}$ ensures that its models are truly $\Sigma_{\mathcal{A}}$-shadowy words representing a run satisfying properties P1–P4. To construct it, we start from $\psi_{shadowy}^{truly-\Sigma_{\mathcal{A}}}$ and extending it with four conjuncts. The first two of them represent properties P1–P2 of runs. They can be written in $\mathrm{LTL}_{\mathbf{F}}$ in an obvious way.

To ensure the satisfaction of the property P3, we observe that in some sense the letters $from_q$ and to_q are paired in a model, *i.e.* always after reaching a state in $\mathcal{A}$ you need to get out of it (the initial state is an exception here, but we assumed that there are no transitions to the initial state). Thus, to identify for which q we should set the $from_q$ letter on the position p, it is sufficient to see for which state we do not have a corresponding pair, *i.e.* for which state q the number of white $from_q$ to the left of p is not equal to the number of white to_q to the left of p. We achieve this in the spirit of Exercise 3.

Finally, the satisfaction of the property P4 can be achieved by checking for each position p whether the number of white fOP_{+1} to the left of p is the same as the number of white fOP_{-1} to the left of p, and similarly for the second counter. This reduces to checking an equicardinality of certain sets, which can be done by employing shadows and Exercise 3.

The reduction Now we are ready to present the claimed reduction.
We first restrict the class of models under consideration to truly $\Sigma_{\mathcal{A}}$-shadowy words (for the feasibility of equicardinality encoding) with a formula $\psi_{shadowy}^{truly-\Sigma_{\mathcal{A}}}$. Then, we express that the models satisfy properties P1 and P2. The first property can be expressed with $\psi_{P1} := from_{q_0} \wedge fVal_0 \wedge sVal_0$.
The property P2 will be a conjunction of two formulae. The first one, namely ψ_{P2}^1, is an immediate implementation of P2. The second one, *i.e.* ψ_{P2}^2, is not necessary, but simplifies the proof; we require that no position is labelled by more

than six letters from $\Sigma_{\mathcal{A}}$.

$$\psi^1_{P2} := \mathbf{G}\,(\mathit{wht} \to \bigvee_{\delta(q,f,s)=(\bar{f},\bar{s},q_N)} \mathit{from}_q \land \mathit{fVal}_f \land \mathit{sVal}_s \land \mathit{fOP}_{\bar{f}} \land \mathit{sOP}_{\bar{s}} \land \mathit{to}_{q_N}),$$

$$\psi^2_{P2} := \mathbf{G} \bigwedge_{\substack{p_1,\ldots,p_7 \in \Sigma_{\mathcal{A}} \\ p_1,\ldots,p_7 \text{ are pairwise different}}} \neg(p_1 \land p_2 \land \cdots \land p_7).$$

We put $\psi_{P2} := \psi^1_{P2} \land \psi^2_{P2}$ and $\psi_{\mathit{enc\text{-}basics}} := \psi^{\mathit{truly}-\Sigma_{\mathcal{A}}}_{\mathit{shadowy}} \land \psi_{P1} \land \psi_{P2}$.

We now formalise the correspondence between intended models and runs. Let *run* be the function which takes a word $\mathfrak{w}$ satisfying $\psi_{\mathit{enc\text{-}basics}}$ and returns the word $\mathfrak{w}^{\mathcal{A}}$ such that $|\mathfrak{w}^{\mathcal{A}}| = |\mathfrak{w}|/2$ and for each position i we have:

$$(\leftrightsquigarrow) : \mathfrak{w}^{\mathcal{A}}_i = (q, f, s, \bar{f}, \bar{s}, q_N) \text{ iff}$$
$$\mathfrak{w}_{2i} \supseteq \{\, \mathit{wht}, \mathit{from}_q, \mathit{fVal}_f, \mathit{sVal}_s, \mathit{fOP}_{\bar{f}}, \mathit{sOP}_{\bar{s}}, \mathit{to}_{q_N} \,\}.$$

The definition of $\psi_{\mathit{enc\text{-}basics}}$ makes the function run correctly defined and unambiguous, and that the results of run satisfy properties P1 and P2.

Fact 5 *The function run is uniquely defined and returns words satisfying P1 and P2.*

What remains to be done is to ensure properties P3 and P4. Both formulas rely on the tools established in Exercise 3 and are defined as follows:

$$\psi_{P3} := \mathbf{G}\,(\mathit{wht} \to \bigwedge_{q \in Q\setminus\{q_0\}} (\mathit{from}_q \lor \psi_{\#\mathit{from}_q = \#\mathit{to}_q})).$$

$$\begin{aligned}\psi_{P4} := \;& \mathbf{G}\,(\mathit{fVal}_0 \to \psi_{\#\mathit{fOP}_{+1} = \#\mathit{fOP}_{-1}}) \\ & \land \mathbf{G}\,(\mathit{sVal}_0 \to \psi_{\#\mathit{sOP}_{+1} = \#\mathit{sOP}_{-1}}) \\ & \land \mathbf{G}\,(\mathit{wht} \to (\mathit{fVal}_0 \leftrightarrow \neg \mathit{fVal}_+)) \land \mathbf{G}\,(\mathit{wht} \to (\mathit{sVal}_0 \leftrightarrow \neg \mathit{sVal}_+))\end{aligned}$$

Lemma 2. *If $\mathfrak{w}$ satisfies $\psi_{\mathit{enc\text{-}basics}} \land \psi_{P3}$, then run($\mathfrak{w}$) satisfies P1–P3.*

Proof. The satisfaction of the properties P1 and P2 by *run*($\mathfrak{w}$) follows from Fact 5. Ad absurdum, assume that *run*($\mathfrak{w}$) does not satisfy P3. It implies the existence of a white position p in $\mathfrak{w}$ such that $\mathfrak{w}, p \models \mathit{to}_q$ but $\mathfrak{w}, p{+}2 \models \mathit{from}_{q'}$ for some $q \neq q'$. By our definition of Minsky machines, we conclude that $\mathfrak{w}, p \models \mathit{from}_{q''}$ for some $q'' \neq q$. Thus, $\mathfrak{w}, p \not\models \mathit{from}_q$.

From the satisfaction of ψ_{P3} by $\mathfrak{w}$ we know that $\mathfrak{w}, p \models \psi_{\#\mathit{from}_q = \#\mathit{to}_q}$. Let k be the total number of positions labelled with from_q before p. Since $\mathfrak{w}, p \models \psi_{\#\mathit{from}_q = \#\mathit{to}_q}$ holds, by Exercise 3 we infer that the number of positions satisfying to_q before p is also equal to k. Since $\mathfrak{w}, p{+}2 \not\models \mathit{from}_q$ and from the satisfaction of ψ_{P3} by $\mathfrak{w}$ we once more conclude $\mathfrak{w}, p{+}2 \models \psi_{\#\mathit{from}_q = \#\mathit{to}_q}$. But such a situation clearly cannot happen due to the fact that the number of to_q in the past is equal to $k+1$, while the number of from_q in the past is k. □

Finally, let us define $\psi_{\mathcal{A}}$ as $\psi_{enc\text{-}basics} \wedge \psi_{P3} \wedge \psi_{P4}$. The use of $\leftrightarrow$ in ψ_{P4} guarantees that $fVal_0$ labels exactly the white positions having the counter empty (and similarly for the second counter). The counters are never decreased from 0, thus the white positions not satisfying $fVal_0$ are exactly those having the first counter positive.

The proof of the forthcoming fact relies on the correctness of Exercise 3 and is quite similar to the proof of Lemma 2, and is presented in the appendix.

Lemma 3. *If* $\mathfrak{w}$ *satisfies* $\psi_{\mathcal{A}}$*, then* $run(\mathfrak{w})$ *is a run of* $\mathcal{A}$.

Lastly, to show that the encoding is correct, we need to show that each run has a corresponding model. It is again easy: it can be shown by constructing an appropriate $\mathfrak{w}$; the white positions are defined according to ($\leftrightsquigarrow$), and the shadows can be constructed accordingly.

Fact 6 *If* $\mathfrak{w}^{\mathcal{A}}$ *is a run of* $\mathcal{A}$*, then there is a word* $\mathfrak{w} \models \psi_{\mathcal{A}}$ *s.t.* $run(\mathfrak{w}) = \mathfrak{w}^{\mathcal{A}}$.

Let $\psi_{\mathcal{A}}^{\mathsf{q}} := \psi_{\mathcal{A}} \wedge \mathbf{F}\,(to_{\mathsf{q}})$. Observe that the formula $\psi_{\mathcal{A}}^{\mathsf{q}}$ is satisfiable if and only if $\mathcal{A}$ reaches q. The "if" part follows from Lemma 3 and the satisfaction of the conjunct $\mathbf{F}\,(to_{\mathsf{q}})$ from $\psi_{\mathcal{A}}$. The "only if" part follows from Fact 6. Hence, from undecidability of the reachability problem Minsky machines we infer our main theorem:

Theorem 1. *The satisfiability problem for* $\mathrm{LTL}_{\mathbf{F},\mathbf{Half}}$ *is undecidable.*

6.1 Undecidability of model-checking

For a given alphabet Σ, we can define a Kripke structure $\mathcal{K}_{\Sigma}$ whose set of traces is the language $(2^{\Sigma})^{+}$: the set of states S of $\mathcal{K}_{\Sigma}$ is composed of all subsets of Σ, all states are initial (*i.e.* $I = S$), the transition relation is the maximal relation ($R = S \times S$) and $\ell(X) = X$ for any subset $X \subseteq \Sigma$. It follows that a formula φ over an alphabet Σ is satisfiable if and only if there is a trace of $\mathcal{K}_{\Sigma}$ satisfying φ. From the undecidability of the satisfiability problem for $\mathrm{LTL}_{\mathbf{F},\mathbf{Half}}$ we get:

Theorem 2. *Model-checking of* $\mathrm{LTL}_{\mathbf{F},\mathbf{Half}}$ *formulae over Kripke structures is undecidable.*

The decidability can be regained if additional constraints on the shape of Kripke structures are imposed: model-checking of $\mathrm{LTL}_{\mathbf{F},\mathbf{Half}}$ formulae over *flat* structures is decidable [13].

As discussed earlier, the **Half** operator can be expressed in terms of the **PM** operator. Hence, we conclude:

Corollary 2. *Model-checking and satisfiability problems for* $\mathrm{LTL}_{\mathbf{F},\mathbf{PM}}$ *are undecidable.*

6.2 Most-Frequent Letter and Undecidability

We next turn our attention to the **MFL** operator, which turns out to be a little bit problematic. Typically, formulae depend only on the atomic propositions that they explicitly mentioned. Here, it is not the case. Consider a formula $\varphi = \mathbf{MFL}\, a$ and words $\mathfrak{w}_1 = \{a\}\{\}\{a\}$ and $\mathfrak{w}_2 = \{a,b\}\{b\}\{a,b\}$. Clearly, $\mathfrak{w}_1, 2 \models \varphi$ whereas $\mathfrak{w}_2, 2 \not\models \varphi$. This can be fixed in many ways – for example, by parametrising **MFL** with a domain, so that it expresses that "a is the most frequent letter among $b_1, \ldots, b_n$". We show, however, that even this very basic version of **MFL** is undecidable. The proof is an adaptation of our previous proofs with a little twist inside.

First, we adjust the definition of shadowy words. A word w is *strongly shadowy* if w is shadowy and for each even position of $\mathfrak{w}$ we have that *wht* and *shdw* are the most frequent letters among the other labelling $\mathfrak{w}$ while for odd positions *wht* is the most frequent. Note that the words constructed in the previous sections were strongly shadowy because each letter σ appeared only at whites or at shadows.

Exercise 4. There exists an $\mathrm{LTL}_{\mathbf{F},\mathbf{MFL}}$ formula $\psi^{MFL}_{shadowy}$ defining strongly shadowy words.

Proof. It suffices to revisit Exercise 1 and to modify the formula φ^{ex1}_{odd} stipulating that odd positions are exactly those labelled with *shdw* (since it is the only formulae employing **Half**). We claim that φ^{ex1}_{odd} can be expressed with

$$\varphi^{MFL}_{odd} := \mathbf{G}\left[\mathbf{MFL}\,(wht) \wedge (wht \leftrightarrow \mathbf{MFL}\,(shdw))\right]$$

Indeed, take any word $\mathfrak{w} \models \varphi^{ex1}_{init} \wedge \varphi^{MFL}_{odd}$. Of course we have $\mathfrak{w}, 0 \models wht$ (due to φ^{ex1}_{init}). Moreover, $\mathfrak{w}, 1 \models shdw$ holds: otherwise we would get contradiction with *shdw* not being the most frequent letter in the past of 1. Now assume $p > 1$ and assume that the word $\mathfrak{w}_0, \ldots, \mathfrak{w}_{p-1}$ is strongly shadowy. Consider two cases. If p is odd, then both *wht* and *shdw* are the most frequent letters in the past of $p-1$ and $p-1$ is labelled by *wht*. Then, *shdw* is not the most frequent letter in the past of p and thus p is labelled by *shdw* and *wht* is the most frequent letter in the past of p. If p is even, $p-2$ is labelled by *wht* and the most frequent letters in the past of $p-2$ are *wht* and *shdw*, and $p-1$ is labelled by *shdw*. Thus both *wht* and *shdw* are the most frequent letters in the past of p and therefore *wht* is labelled by *wht*. Thus, $\mathfrak{w}_0, \ldots, \mathfrak{w}_p$ is strongly shadowy. By induction, $\mathfrak{w}$ is strongly shadowy. It can be readily checked that every strongly shadowy word satisfies $\psi^{MFL}_{shadowy}$. □

We argue that over the strongly shadowy models, the formulae **Half** σ and **MFL** σ are equivalent.

Lemma 4. *For all strongly shadowy words* $\mathfrak{w} \models \psi^{MFL}_{shadowy}$, *all even positions* $2i$ *and all letters* σ *we have the equivalence* $\mathfrak{w}, 2i \models \mathbf{Half}\,\sigma$ *iff* $\mathfrak{w}, 2i \models \mathbf{MFL}\,\sigma$.

Proof. If $\mathfrak{w}, 2i \models \mathbf{MFL}\,\sigma$, then $\mathfrak{w}, 2i \models \mathbf{MFL}\, wht$ due to the strongly shadowness of $\mathfrak{w}$. Hence $\#^{<}_{\sigma}(\mathfrak{w}, 2i) = \#^{<}_{wht}(\mathfrak{w}, 2i) = \frac{2i}{2}$, implying $\mathfrak{w}, 2i \models \mathbf{Half}\,\sigma$.

Now, assume that $\mathfrak{w}, 2i \models \mathbf{Half}\,\sigma$ holds, so σ appears i times in the past. Since $\mathfrak{w}$ is strongly shadowy we know that *wht* is the most frequent letter. Moreover, *wht* appears $\frac{2i}{2} = i$ times in the past. Hence, $\mathfrak{w}, 2i \models \mathbf{MFL}\,\sigma$. □

We say that a letter σ is *importunate* in a word $\mathfrak{w}$ if σ labels more than half of the positions in some even prefix of $\mathfrak{w}$. Notice that strongly shadowy words cannot have importunate letters.

With the above lemma, it is tempting to finish the proof as follows: replace each $\mathbf{Half}\,(\varphi)$ in the formulae from Section 4.1 with $\mathbf{MFL}\,(p_\varphi)$ for some fresh atomic proposition p_φ and require that $\mathbf{G}\,(\varphi \leftrightarrow p_\varphi)$ holds. A formula obtained from φ in this way will be called a *dehalfication* of φ and will be denoted with $\mathsf{dehalf}(\varphi)$. The next lemma shows that $\mathsf{dehalf}(\cdot)$ preserves satisfaction of certain $\mathrm{LTL}_{\mathbf{F},\mathbf{Half}}$ formulae.

Lemma 5. *Let φ be an $\mathrm{LTL}_{\mathbf{F},\mathbf{Half}}$ formula without nested* **Half** *operators and without* **F** *modality, Λ be the set of all formulae λ such that $\mathbf{Half}\,\lambda$ appears in φ and let $\mathfrak{w}$ be a word such that $\mathfrak{w} \models \psi^{MFL}_{shadowy} \wedge \bigwedge_{\lambda\in\Lambda} \mathbf{G}\,(p_\lambda \leftrightarrow \lambda)$. Then for all even positions $2p$ of $\mathfrak{w}$ we have that $\mathfrak{w}, 2p \models \mathsf{dehalf}(\varphi)$ implies $\mathfrak{w}, 2p \models \varphi$. Moreover, $\mathfrak{w} \models \mathbf{G}\,(wht \rightarrow \mathsf{dehalf}(\varphi))$ implies $\mathfrak{w} \models \mathbf{G}\,(wht \rightarrow \varphi)$.*

Proof. The proof goes via structural induction over $\mathrm{LTL}_{\mathbf{F},\mathbf{Half}}$ formulae without nested **Half** operators and without **F** operators. The only interesting case is when $\varphi = \mathbf{Half}\,\lambda$, which follows from Lemma 4. □

Note, however, that the above lemma works only one way: it fails when the formula φ is satisfied in more than half of the positions of some prefix, as that would make p_φ importunate leading to unsatisfiablity of $\psi^{MFL}_{shadowy}$.

6.3 Most-Frequent Letter: the reduction

The next step is to construct a formula defining truly $\Sigma_{\mathcal{A}}$-shadowy words, which are the crucial part of $\psi^{MFL}_{enc\text{-}basics}$. To do it, we first need to rewrite a formula $\varphi^{trans}_{\sigma\rightsquigarrow\tilde{\sigma}}$, transferring the truth of a letter σ from whites into their shadows. The main ingredient of $\varphi^{trans}_{\sigma\rightsquigarrow\tilde{\sigma}}$ is the formula $\varphi_{(\heartsuit)} := \mathbf{G}\,(wht \rightarrow \mathbf{Half}\,([wht \wedge \sigma] \vee [shdw \wedge \neg\tilde{\sigma}]))$, which we replace with $\mathsf{dehalf}(\varphi_{(\heartsuit)})$. We call the obtained formula $(\varphi^{trans}_{\sigma\rightsquigarrow\tilde{\sigma}})^{MFL}$ and show its correctness below.

First, by Lemma 5 we know that every model of $(\varphi^{trans}_{\sigma\rightsquigarrow\tilde{\sigma}})^{MFL}$ is also a model of $\varphi^{trans}_{\sigma\rightsquigarrow\tilde{\sigma}}$. Then, the models of $\varphi^{trans}_{\sigma\rightsquigarrow\tilde{\sigma}}$ can be made strongly shadowy, so dehalfication of $\varphi^{trans}_{\sigma\rightsquigarrow\tilde{\sigma}}$ is satisfiability-preserving.

Lemma 6. *Let p_φ be a fresh letter for $\varphi := [wht \wedge \sigma] \vee [shdw \wedge \neg\tilde{\sigma}]$. Take $\mathfrak{w}$, a strongly shadowy word satisfying $\mathfrak{w} \models \varphi^{trans}_{\sigma\rightsquigarrow\tilde{\sigma}}$ without any occurrences of p_φ. Then $\mathfrak{w}'$, the word obtained by labelling with p_φ all the positions of $\mathfrak{w}$ satisfying φ, is strongly shadowy.*

Hence, we obtain the correctness of $(\varphi^{trans}_{\sigma\rightsquigarrow\tilde{\sigma}})^{MFL}$. By applying the same strategy to other conjuncts of $\psi_{enc\text{-}basics}$ and Fact 5, we obtain $\psi^{MFL}_{enc\text{-}basics}$ satisfying:

Corollary 3. *The function run (taking as input the words satisfying $\psi^{MFL}_{enc\text{-}basics}$) is uniquely defined and returns words satisfying P1 and P2. Moreover the formulae $\psi^{MFL}_{enc\text{-}basics}$ and $\psi_{enc\text{-}basics}$ are equi-satisfiable.*

Towards completing the undecidability proof we need to prepare the rewritings of the formulae ψ_{P3} and ψ_{P4}. For ψ_{P3} we proceed similarly to the previous case. We know that the models of $\psi^{MFL}_{enc\text{-}basics} \wedge \mathsf{dehalf}(\psi_{P3})$ satisfy P3 (due to Lemma 5 they satisfy ψ_{P3} and hence, by Lemma 2, also P3). To observe the existence of such models, we show again that the satisfiability of ψ_{P3} is preserved by dehalfication.

Lemma 7. *Let p_q be a fresh letter for $\varphi_q := [wht \wedge from_q] \vee [shdw \wedge \neg \widetilde{to_q}]$ indexed over $q \in Q \setminus \{q_0\}$. Take $\mathfrak{w}$, a strongly shadowy word satisfying $\mathfrak{w} \models \psi^{MFL}_{enc\text{-}basics} \wedge \psi_{P3}$ without any occurrences of p_q. Then $\mathfrak{w}'$, the word obtained by labelling with p_q all the positions of $\mathfrak{w}$ satisfying φ_q, is strongly shadowy.*

From Lemma 2, Lemma 7 and Lemma 5 we immediately conclude:

Corollary 4. *If $\mathfrak{w}$ satisfies $\psi^{MFL}_{enc\text{-}basics} \wedge \mathsf{dehalf}(\psi_{P3})$, then $run(\mathfrak{w})$ satisfies P1–P3. Moreover the formulae $\psi^{MFL}_{enc\text{-}basics} \wedge \mathsf{dehalf}(\psi_{P3})$ and $\psi_{enc\text{-}basics} \wedge \psi_{P3}$ are equi-satisfiable.*

The last formula to rewrite is ψ_{P4}. We focus only on its first part, speaking about the first counter, *i.e.*
$\mathbf{G}\,(fVal_0 \rightarrow \mathbf{Half}\,([wht \wedge fOP_{+1}] \vee [shdw \wedge \neg \widetilde{fOP_{-1}}]) \wedge \mathbf{G}\,(wht \rightarrow (fVal_0 \leftrightarrow \neg fVal_+))$
Note that this time we cannot simply dehalfise this formula: the letter responsible for the inner part of **Half** would necessarily be importunate – consider an initial fragment of a run of $\mathcal{A}$ in which $\mathcal{A}$ increments its first counter without decrementing it. Fortunately, we cannot say the same when the machine decrements the counter and hence, it suffices to express the equivalent (due to even length of shadowy models) statement ψ'_{P4} as follows: $\mathbf{G}\,(fVal_0 \rightarrow \mathbf{Half}\,\neg([wht \wedge fOP_{+1}] \vee [shdw \wedge \neg \widetilde{fOP_{-1}}]) \wedge \mathbf{G}\,(wht \rightarrow (fVal_0 \leftrightarrow \neg fVal_+))$.

As we did before, we show that dehalfication of ψ'_{P4} preserves satisfiability:

Lemma 8. *Let p_φ be a fresh letter for $\varphi := \neg([wht \wedge fOP_{+1}] \vee [shdw \wedge \neg \widetilde{fOP_{-1}}])$. Take $\mathfrak{w}$, a strongly shadowy word satisfying $\mathfrak{w} \models \psi^{MFL}_{enc\text{-}basics} \wedge \mathsf{dehalf}(\psi_{P3}) \wedge \psi'_{P4}$ without any occurrences of p_φ. Then $\mathfrak{w}'$, the word obtained by labelling with p_φ all the positions of $\mathfrak{w}$ satisfying φ, is strongly shadowy.*

Finally, let $(\psi^{\mathfrak{q}}_{\mathcal{A}})^{MFL} := \psi^{MFL}_{enc\text{-}basics} \wedge \mathsf{dehalf}(\psi_{P3}) \wedge \mathsf{dehalf}(\psi_{P4}) \wedge \mathbf{F}\,to_{\mathfrak{q}}$. From Lemma 3, Lemma 8 and Lemma 5 we immediately conclude:

Corollary 5. *If $\mathfrak{w}$ satisfies $(\psi^{\mathfrak{q}}_{\mathcal{A}})^{MFL}$ then it satisfies P1–P4. Moreover the formulae $(\psi^{\mathfrak{q}}_{\mathcal{A}})^{MFL}$ and $\psi^{\mathfrak{q}}_{\mathcal{A}}$ are equi-satisfiable.*

Thus, by Theorem 1 and the above corollary, we obtain the undecidability of $\mathrm{LTL}_{\mathbf{F},\mathbf{MFL}}$. Undecidability of the model-checking problem is concluded by virtually the same argument as in Section 6.1. Hence:

Theorem 3. *The model-checking and the satisfiability problems for* $\mathrm{LTL}_{\mathbf{F},\mathbf{MFL}}$ *are undecidable.*

7 Decidable variants

We have shown that $\mathrm{LTL}_{\mathbf{F}}$ with frequency operators lead to undecidability. Without the operators that can express $\mathbf{F}$ (*e.g.* $\mathbf{F}$, $\mathbf{G}$ or $\mathbf{U}$), the decision problems become NP-complete. Below we assume the standard semantics of LTL operator $\mathbf{X}$, *i.e.* $\mathfrak{w}, i \models \mathbf{X}\varphi$ iff $i{+}1 < |\mathfrak{w}|$ and $\mathfrak{w}, i{+}1 \models \varphi$.

Theorem 4. *Model-checking and satisfiability problems for* $\mathrm{LTL}_{\mathbf{X},\mathbf{MFL},\mathbf{PM}}$ *are NP-complete.*

The complexity of $\mathrm{LTL}_{\mathbf{X},\mathbf{MFL},\mathbf{PM}}$ is so low because the truth of the formula depends only on some initial fragment of a trace. This is a big restriction of the expressive power. Thus, we consider a different approach motivated by [7].

In the new setting, we allow to use arbitrary LTL formulae as well as percentage operators as long as the they are not mixed with $\mathbf{G}$. We introduce a logic $\mathrm{LTL}^{\%}$, which extends the classical LTL [29] with the percentage operators of the form $\mathbf{P}_{\bowtie k\%}\varphi$ for any $\bowtie \in \{\leq, <, =, >, \geq\}$, $k \in \mathbb{N}$ and $\varphi \in \mathrm{LTL}$. By way of example, the formula $\mathbf{P}_{<20\%}(a)$ is true at a position p if less then 20% of positions before p satisfy a. The past majority operator is a special case of the percentage operator: $\mathbf{PM} \equiv \mathbf{P}_{\geq 50\%}$. Formally:

$$\mathfrak{w}, i \models \mathbf{P}_{\bowtie k\%}\varphi \text{ if } |\{j < i \colon \mathfrak{w}, j \models \varphi\}| \bowtie \tfrac{k}{100} i$$

To avoid undecidability, the percentage operators cannot appear under negation or be nested. Therefore, the syntax of $\mathrm{LTL}^{\%}$ is defined with the grammar $\varphi, \varphi' ::= \psi_{\mathrm{LTL}} \mid \varphi \vee \varphi' \mid \varphi \wedge \varphi' \mid \mathbf{F}(\psi_{\mathrm{LTL}} \wedge \mathbf{P}_{\bowtie k\%}\psi'_{\mathrm{LTL}})$, where ψ_{LTL}, ψ'_{LTL} are (full) LTL formulae.

The main tool used in the decidability proof is the Parikh Automata [21]. A Parikh automaton $\mathcal{P} = (\mathcal{A}, \mathcal{E})$ over the alphabet Σ is composed of a finite-state automaton $\mathcal{A}$ accepting words from Σ^* and a semi-linear set $\mathcal{E}$ given as a system of linear inequalities with integer coefficients, where the variables are x_a for $a \in \Sigma$. We say that $\mathcal{P}$ accepts a word $\mathfrak{w}$ if $\mathcal{A}$ accepts $\mathfrak{w}$ and the mapping assigning to each variable x_a from $\mathcal{E}$ the total number of positions of $\mathfrak{w}$ carrying the letter a, is a solution to $\mathcal{E}$. Checking non-emptiness of the language of $\mathcal{P}$ can be done in NP [17]. Our main decidability results is obtained by constructing an appropriate Parikh automaton recognising the models of an input $\mathrm{LTL}^{\%}$ formula.

Theorem 5. *Model-checking and satisfiability problems for* $\mathrm{LTL}^{\%}$ *are decidable.*

Proof. Let $\varphi \in \mathrm{LTL}^{\%}$. By turning φ into a DNF, we can focus on checking satisfiability of some of its conjuncts. Hence, w.l.o.g. we assume that $\varphi = \varphi_0 \wedge \bigwedge_{i=1}^{n} \varphi_i$, where φ_0 is in LTL and all φ_i have the form $\mathbf{F}(\psi^{i,1}_{\mathrm{LTL}} \wedge \mathbf{P}_{\bowtie k_i\%}\psi^{i,2}_{\mathrm{LTL}})$ for some LTL formulae $\psi^{i,1}_{\mathrm{LTL}}$ and $\psi^{i,2}_{\mathrm{LTL}}$. Observe that a word $\mathfrak{w}$ is a model of φ iff it satisfies φ_0 and for each conjunct φ_i we can pick a witness position p_i from $\mathfrak{w}$ such that $\mathfrak{w}, p_i \models \psi^{i,1}_{\mathrm{LTL}} \wedge \mathbf{P}_{\bowtie k_i\%}\psi^{i,2}_{\mathrm{LTL}}$. Moreover, the percentage constraints inside such formulae speak only about the prefix $\mathfrak{w}_{<p_i}$. Thus, knowing the position p_i and the number of positions before p_i satisfying $\psi^{i,2}_{\mathrm{LTL}}$, the percentage constraint inside φ_i can be imposed globally rather than locally. It suggests the use of Parikh

automata: the LTL part of φ can be checked by the appropriate automaton $\mathcal{A}$ (due to the correspondence that for an LTL formula over finite words one can build a finite-state automaton recognising the models of such a formula [19]) and the global constraints, speaking about the satisfaction of percentage operators, can be ensured with a set of linear inequalities $\mathcal{E}$.

Our plan is as follows: we decorate the intended models $\mathfrak{w}$ with additional information on witnesses, such that the witness position p_i for φ_i will be labelled by w_i (and there will be a unique such position in a model), all positions before p_i will be labelled by b_i and, among them, we distinguish with a letter s_i some special positions, *i.e.* those satisfying $\psi^{i,2}_{\mathrm{LTL}}$. More formally, for each φ_i we produce an LTL formula φ'_i according to the following rules:

- there is a unique position p_i such that $\mathfrak{w}, p_i \models w_i$ (selecting a witness for φ_i),
- for all $j < p_i$ we have $\mathfrak{w}, j \models b_i$ (the positions before p_i are labelled with b_i),
- $\mathfrak{w} \models \mathbf{G}\,(s_i \to [b_i \wedge \psi^{i,2}_{\mathrm{LTL}}])$ (distribution of the special positions among b_i) and
- $\mathfrak{w}, p_i \models \psi^{i,1}_{\mathrm{LTL}}$ (a precondition for φ_i).

Let $\varphi' := \varphi_0 \wedge \bigwedge_{i=1}^{n} \varphi'_i \wedge \bigwedge_{i=1}^{n} \mathbf{F}\,(p_i \wedge \mathbf{P}_{\bowtie k_i\%} s_i)$. Note that $\mathfrak{w} \models \varphi'$ implies $\mathfrak{w} \models \varphi$. Moreover, any model $\mathfrak{w} \models \varphi$ can be labelled with letters b_i, s_i, w_i such that the decorated word satisfies φ'. Let $\varphi'' := \varphi_0 \wedge \bigwedge_{i=1}^{n} \varphi'_i$ and let $\mathcal{E}$ be the system of n inequalities with $\mathcal{E}_i = 100 \cdot x_{b_i} \bowtie k_i \cdot x_{s_i}$. Now observe that any model of φ' satisfies $\mathcal{E}$ (i.e. the value assigned to x_a is the total number of positions labelled with a), due to the satisfaction of counting operators, and vice versa: every word $\mathfrak{w} \models \varphi''$ satisfying $\mathcal{E}$ is a model of φ''. It gives us a sufficient characterisation of models of φ. Let $\mathcal{A}$ be a finite automaton recognising the models of φ'', then a Parikh automaton $\mathcal{P} = (\mathcal{A}, \mathcal{E})$, as we already discussed, is non-empty if and only if φ has a model. Since checking non-emptiness of $\mathcal{P}$ is decidable, we can conclude that LTL$^{\%}$ is decidable. □

A rough complexity analysis yields an NExpTime upper bound on the problem: the automaton $\mathcal{P}$ that we constructed is exponential in φ (translating φ to DNF does not increase the complexity since we only guess one conjunct, which is of polynomial size in φ). Moreover, checking non-emptiness can be done non-deterministically in time polynomial in the size of the automaton. The NExpTime bound is not optimal: we conjuncture that the problem is PSpace-complete. We believe that by employing techniques similar to [7], one can construct $\mathcal{P}$ and check its non-emptiness on the fly, which should result in the PSpace upper bound.

For the model-checking problem, we observe that determining whether some trace of a Kripke structure $\mathcal{K} = (S, I, R, l)$ satisfies φ is equivalent to checking the satisfiability of formula $\varphi_{\mathcal{K}} \wedge \varphi$, where $\varphi_{\mathcal{K}}$ is a formula describing all the traces of $\mathcal{K}$. Such a formula can be constructed in a standard manner. For simplicity, we treat S as a set of auxiliary letters, and consider the conjunction of (1) $\bigvee_{s \in I} s$, (2) $\mathbf{G}\,(\mathbf{X}\top \to \bigvee_{(s,s') \in R}(s \wedge \mathbf{X}\,s'))$ and (3) $\bigwedge_{s \in S} \mathbf{G}\,(s \to \bigwedge_{p \in \ell(s)} p)$, expressing that the trace starts with an initial state, consecutive positions describe consecutive states and that the trace is labelled by the appropriate letters. Thus, the model-checking problem can be reduced in polynomial time to the satisfiability problem.

8 Two-Variable First-Order Logic with Majority

The *Two-Variable First-Order Logic on words* ($\mathrm{FO}^2[<]$) is a robust fragment of First-Order Logic FO interpreted on finite words. It involves quantification over variables x and y (ranging over the words' positions) and it admits a linear order predicate $<$ (interpreted as a natural order on positions) and the equality predicate $=$. Henceforth we assume the usual semantics of $\mathrm{FO}^2[<]$ (*cf.* [16]).

In this section, we investigate the logic $\mathrm{FO}^2_{\mathsf{M}}[<]$, namely the extension of $\mathrm{FO}^2[<]$ with the so-called *Majority quantifier* M. Such quantifier was intensively studied due to its close connection with circuit complexity and algebra, see *e.g.* [22,5,6]. Intuitively, the formula $\mathsf{M}x.\varphi$ specifies that at least half of all the positions in a model, after substituting x with them, satisfy φ. Formally $\mathfrak{w} \models \mathsf{M}x.\varphi$ holds, if and only if $\frac{|\mathfrak{w}|}{2} \leq |\{p \mid \mathfrak{w}, p \models \varphi[x/p]\}|$. We stress that the formula $\mathsf{M}x.\varphi$ may contain free occurrences of the variable y.

Note that the Majority quantifier shares similarities to the **PM** operator, but in contrast to **PM**, the M quantifier counts *globally*. We take advantage of such similarities and by reusing the technique developed in the previous sections, we show that the satisfiability problem for $\mathrm{FO}^2_{\mathsf{M}}[<]$ is also undecidable. We stress that our result significantly sharpens an existing undecidability result for FO with Majority from [23] (since in our case the number of variables is limited) as well as for $\mathrm{FO}^2[<, \mathit{succ}]$ with Presburger Arithmetics from [25] (since our counting mechanism is limited and the successor relation *succ* is disallowed).

Proof plan There are three possible approaches to proving the undecidability of $\mathrm{FO}^2_{\mathsf{M}}[<]$. The first one is to reproduce all the results for $\mathrm{LTL}_{\mathbf{F},\mathbf{PM}}$, which is rather uninspiring. The second one is to define a translation from $\mathrm{LTL}_{\mathbf{F},\mathbf{PM}}$ to $\mathrm{FO}^2_{\mathsf{M}}[<]$ that produces an equisatisfiable formula. But because of models of odd length, this involves a lot of case study. Here we present a third approach, which, we believe, gives the best insight: we show a translation from $\mathrm{LTL}_{\mathbf{F},\mathbf{PM}}$ to $\mathrm{FO}^2_{\mathsf{M}}[<]$ that works for $\mathrm{LTL}_{\mathbf{F},\mathbf{PM}}$ formulae whose all models are shadowy. Since we only use such models in the undecidability proof of $\mathrm{LTL}_{\mathbf{F},\mathbf{PM}}$, this shows the undecidability of $\mathrm{FO}^2_{\mathsf{M}}[<]$.

Shadowy models We first focus on defining shadowy words in $\mathrm{FO}^2_{\mathsf{M}}[<]$. Before we start, let us introduce a bunch of useful macros in order to simplify the forthcoming formulae. Their names coincide with their intuitive meaning and their semantics.

- $\underline{\mathsf{Half}}x.\varphi := \mathsf{M}x.\varphi \land \mathsf{M}x.\neg\varphi$,
- $\underline{\mathit{first}}(x) := \neg\exists y\, y < x$, $\underline{\mathit{second}}(x) := \exists y\, y < x \land \forall y\, y < x \rightarrow \underline{\mathit{first}}(y)$,
- $\underline{\mathit{last}}(x) := \neg\exists y\, y > x$, $\underline{\mathit{sectolast}}(x) := \exists y\, y > x \land \forall y\, y > x \rightarrow \underline{\mathit{last}}(y)$

Lemma 9. *There is an* $\mathrm{FO}^2_{\mathsf{M}}[<]$ *formula* $\psi^{FO}_{shadowy}$ *defining shadowy words.*

Proof. Let φ^{lem9}_{base} be a formula defining the language of all (non-empty) words, where the letters *wht* and *shdw* label disjoint positions in the way that the first position satisfies *wht* and the total number of *shdw* and *wht* coincide. It can be written, *e.g.* with $\forall x(wht(x) \leftrightarrow \neg shdw(x)) \land \exists x(\underline{first}(x) \land wht(x)) \land \underline{\mathsf{Half}}x.wht(x) \land \underline{\mathsf{Half}}x.shdw(x)$. To define shadowy words, it would be sufficient to specify that no neighbouring positions carry the same letter among $\{wht, shdw\}$. This can be done with, rather complicated at the first glance, formulae:

$$\varphi^{forbid}_{wht\cdot wht}(x) := wht(x) \rightarrow \underline{\mathsf{Half}}y.\,([y < x \land wht(y)] \lor [x < y \land shdw(y)])\,,$$

$$\varphi^{forbid}_{shdw\cdot shdw}(x) := shdw(x) \rightarrow \underline{\mathsf{Half}}y.\,([(y{<}x \lor x{=}y) \land shdw(y)] \lor [x{<}y \land wht(y)])\,.$$

Finally, let $\psi^{FO}_{shadowy} := \varphi^{lem9}_{base} \land \forall x.\left(\varphi^{forbid}_{wht\cdot wht}(x) \land \varphi^{forbid}_{shdw\cdot shdw}(x)\right)$.

Showing that shadowness implies the satisfaction of $\psi^{FO}_{shadowy}$ can be done by routine induction. For the opposite direction, take $\mathfrak{w} \models \psi^{FO}_{shadowy}$. Since $\mathfrak{w} \models \varphi^{lem9}_{base}$ the only possibility for $\mathfrak{w}$ to not be shadowy is to have two consecutive positions $p, p{+}1$ carrying the same letter. W.l.o.g assume they are both white. Let w be the number of white positions to the left of p and let s be the number of shadows to the right of p. By applying $\varphi^{forbid}_{wht\cdot wht}$ to p we infer that $w + s = \frac{1}{2}|\mathfrak{w}|$. On the other hand, by applying $\varphi^{forbid}_{wht\cdot wht}$ to $p{+}1$ it follows that $(w{+}1){+}s = \frac{1}{2}|\mathfrak{w}|$, which contradicts the previous equation. Hence, $\mathfrak{w}$ is shadowy. □

Translation It is a classical result from [16] that $\mathrm{FO}^2[<]$ can express $\mathrm{LTL}_{\mathbf{F}}$. We define a translation $\mathfrak{tr}_v(\varphi)$ from $\mathrm{LTL}_{\mathbf{F},\mathbf{PM}}$ to $\mathrm{FO}^2_{\mathsf{M}}[<]$, parametrised by a variable v (where v is either x or y and $\bar{v}$ denotes the different variable from v), inductively. We write $v \leq \bar{v}$ rather than $v < \bar{v} \lor v = \bar{v}$ for simplicity. For $\mathrm{LTL}_{\mathbf{F}}$ cases, we follow [16]: $\mathfrak{tr}_v(a) := a(v)$, for a fresh unary predicate a for each $a \in \mathsf{AP}$, $\mathfrak{tr}_v(\neg\varphi) := \neg\mathfrak{tr}_v(\varphi)$, $\mathfrak{tr}_v(\varphi \land \varphi') := \mathfrak{tr}_v(\varphi) \land \mathfrak{tr}_v(\varphi')$, $\mathfrak{tr}_v(\mathbf{F}\,\varphi) := \exists\bar{v}\ (v \leq \bar{v}) \land \mathfrak{tr}_{\bar{v}}(\varphi)$. For $\mathbf{PM}$, we propose $\mathfrak{tr}_v(\mathbf{PM}\,\varphi) := \mathsf{M}\bar{v}((\bar{v} < v \land \mathfrak{tr}_{\bar{v}}(\varphi)) \lor (\bar{v} \geq v \land wht(\bar{v})))$. Finally, for a given $\mathrm{LTL}_{\mathbf{F},\mathbf{PM}}$ formula φ, let $\mathfrak{tr}(\varphi)$ stand for $\psi^{FO}_{shadowy} \land \exists x.(\underline{first}(x) \land \mathfrak{tr}_x(\varphi))$.

The following lemma shows the correctness of the presented translation.

Lemma 10. *An* $\mathrm{LTL}_{\mathbf{F},\mathbf{PM}}$ *formula* φ *has a shadowy model iff* $\mathfrak{tr}(\varphi)$ *has a model.*

Since the formulae used in our undecidability proof for $\mathrm{LTL}_{\mathbf{F},\mathbf{PM}}$ have only shadowy models, by Lemma 10 we conclude that $\mathrm{FO}^2_{\mathsf{M}}[<]$ is also undecidable.

Theorem 6. *The satisfiability problem for* $\mathrm{FO}^2_{\mathsf{M}}[<]$ *is undecidable.*

9 Conclusions

We have provided a simple proof showing that adding different percentage operators to $\mathrm{LTL}_{\mathbf{F}}$ yields undecidability. We showed that our technique can be applied to an extension of first-order logic on words, and we hope that our work will turn useful in showing undecidability for other extensions of temporal logics. Decidability results for logics with percentage operators in restricted contexts were also provided.

Acknowledgements

Bartosz Bednarczyk was supported by the Polish Ministry of Science and Higher Education program "Diamentowy Grant" no. DI2017 006447. Jakub Michaliszyn was supported by NCN grant no. 2017/27/B/ST6/00299.

References

1. Jorge A. Baier, Sheila A. McIlraith: Planning with First-Order Temporally Extended Goals using Heuristic Search. In: AAAI (2006)
2. Shaull Almagor, Udi Boker, Orna Kupferman: Discounting in LTL. In: TACAS (2014). https://doi.org/10.1007/978-3-642-54862-8_37
3. Rajeev Alur, Thomas A. Henzinger: Real-Time Logics: Complexity and Expressiveness. Inf. Comput. (1993)
4. Bednarczyk, B., Michaliszyn, J.: "most of" leads to undecidability: Failure of adding frequencies to LTL `https://arxiv.org/abs/2007.01233`
5. Christoph Behle, Andreas Krebs: Regular Languages in MAJ[>] with three variables. ECCC (2011)
6. Christoph Behle, Andreas Krebs, Stephanie Reifferscheid: Regular Languages Definable by Majority Quantifiers with Two Variables. In: DLT (2009). https://doi.org/10.1007/978-3-642-02737-6_7
7. Udi Boker, Krishnendu Chatterjee, Thomas A. Henzinger, Orna Kupferman: Temporal Specifications with Accumulative Values. ACM Trans. Comput. Log. (2014). https://doi.org/10.1145/2629686
8. Benedikt Bollig, Normann Decker, Martin Leucker: Frequency Linear-time Temporal Logic. In: TASE (2012). https://doi.org/10.1109/TASE.2012.43
9. Patricia Bouyer, Nicolas Markey, Raj Mohan Matteplackel: Averaging in LTL. In: CONCUR (2014). https://doi.org/10.1007/978-3-662-44584-6_19
10. Büchi, J.R.: Weak Second-Order Arithmetic and Finite Automata. Mathematical Logic Quarterly (1960). https://doi.org/10.1002/malq.19600060105
11. Cadilhac, M., Finkel, A., McKenzie, P.: Affine Parikh automata. RAIRO Theor. Informatics Appl. (2012). https://doi.org/10.1051/ita/2012013
12. Witold Charatonik, Piotr Witkowski: Two-variable Logic with Counting and a Linear Order. Logical Methods in Computer Science (2016). https://doi.org/10.2168/LMCS-12(2:8)2016
13. Normann Decker, Peter Habermehl, Martin Leucker, Arnaud Sangnier, Daniel Thoma: Model-Checking Counting Temporal Logics on Flat Structures. In: CONCUR (2017). https://doi.org/10.4230/LIPIcs.CONCUR.2017.29
14. Stéphane Demri: LTL over integer periodicity constraints. Theor. Comput. Sci. (2006). https://doi.org/10.1016/j.tcs.2006.02.019
15. Demri, S., Goranko, V., Lange, M.: Temporal Logics in Computer Science: Finite-State Systems. Cambridge Tracts in Theoretical Computer Science, Cambridge University Press (2016). https://doi.org/10.1017/CBO9781139236119
16. Kousha Etessami, Moshe Y. Vardi, Thomas Wilke: First-Order Logic with Two Variables and Unary Temporal Logic. Inf. Comput. (2002). https://doi.org/10.1006/inco.2001.2953
17. Diego Figueira, Leonid Libkin: Path Logics for Querying Graphs: Combining Expressiveness and Efficiency. In: LICS (2015). https://doi.org/10.1109/LICS.2015.39
18. Giuseppe De Giacomo, Moshe Y. Vardi: Linear Temporal Logic and Linear Dynamic Logic on Finite Traces. In: IJCAI (2013)
19. Giuseppe De Giacomo, Moshe Y. Vardi: Synthesis for LTL and LDL on Finite Traces. In: IJCAI (2015)
20. Jochen Hoenicke, Roland Meyer, Ernst-Rüdiger Olderog: Kleene, Rabin, and Scott Are Available. In: CONCUR (2010). https://doi.org/10.1007/978-3-642-15375-4_32
21. Felix Klaedtke, Harald Rueß: Monadic Second-Order Logics with Cardinalities. In: ICALP (2003). https://doi.org/10.1007/3-540-45061-0_54

22. Andreas Krebs: Typed semigroups, majority logic, and threshold circuits. Ph.D. thesis, University of Tübingen, Germany (2008)
23. Klaus-Jörn Lange: Some Results on Majority Quantifiers over Words. In: CCC (2004). https://doi.org/10.1109/CCC.2004.1313817
24. François Laroussinie, Antoine Meyer, Eudes Petonnet: Counting LTL. In: TIME (2010). https://doi.org/10.1109/TIME.2010.20
25. Kamal Lodaya, Sreejith, A.V.: Two-Variable First Order Logic with Counting Quantifiers: Complexity Results. In: DLT (2017). https://doi.org/10.1007/978-3-319-62809-7_19
26. Minsky, M.L.: Computation: Finite and Infinite Machines. Prentice-Hall Series in Automatic Computation, Prentice-Hall (1967)
27. Matthias Niewerth: Data definition languages for XML repository management systems. Ph.D. thesis, Technical University of Dortmund, Germany (2016)
28. Joël Ouaknine, James Worrell: On the decidability and complexity of Metric Temporal Logic over finite words. Logical Methods in Computer Science (2007)
29. Amir Pnueli: The Temporal Logic of Programs. In: FOCS (1977). https://doi.org/10.1109/SFCS.1977.32

Combining Semilattices and Semimodules*

Filippo Bonchi and Alessio Santamaria(✉)

Dipartimento di Informatica, Università degli Studi di Pisa, Pisa, Italy
filippo.bonchi@unipi.it, alessio.santamaria@di.unipi.it

Abstract. We describe the canonical weak distributive law $\delta\colon \mathcal{SP} \to \mathcal{PS}$ of the powerset monad $\mathcal{P}$ over the S-left-semimodule monad $\mathcal{S}$, for a class of semirings S. We show that the composition of $\mathcal{P}$ with $\mathcal{S}$ by means of such δ yields *almost* the monad of convex subsets previously introduced by Jacobs: the only difference consists in the absence in Jacobs's monad of the empty convex set. We provide a handy characterisation of the canonical weak lifting of $\mathcal{P}$ to $\mathbb{EM}(\mathcal{S})$ as well as an algebraic theory for the resulting composed monad. Finally, we restrict the composed monad to finitely generated convex subsets and we show that it is presented by an algebraic theory combining semimodules and semilattices with bottom, which are the algebras for the finite powerset monad $\mathcal{P}_f$.

Keywords: algebraic theories · monads · weak distributive laws.

1 Introduction

Monads play a fundamental role in different areas of computer science since they embody notions of computations [32], like nondeterminism, side effects and exceptions. Consider for instance automata theory: deterministic automata can be conveniently regarded as certain kind of coalgebras on $\mathbb{S}\mathrm{et}$ [33], nondeterministic automata as the same kind of coalgebras but on $\mathbb{EM}(\mathcal{P}_f)$ [35], and weighted automata on $\mathbb{EM}(\mathcal{S})$ [4]. Here, $\mathcal{P}_f$ is the finite powerset monad, modelling nondeterministic computations, while $\mathcal{S}$ is the monad of semimodules over a semiring S, modelling various sorts of quantitative aspects when varying the underlying semiring S. It is worth mentioning two facts: first, rather than taking coalgebras over $\mathbb{EM}(T)$, the category of algebras for the monad T, one can also consider coalgebras over $\mathbb{Kl}(T)$, the Kleisli category induced by T [20]; second, these two approaches based on monads have lead not only to a deeper understanding of the subject, but also to effective proof techniques [6,7,14], algorithms [1,8,22,36,39] and logics [19,21,27].

Since compositionality is often the key to master complex structures, computer scientists devoted quite some efforts to *compose monads* [40] or the equivalent notion of *algebraic theories* [24]. Indeed, the standard approach of composing monads by means of *distributive laws* [3] turned out to be somehow unsatisfactory. On the one hand, distributive laws do not exist in many relevant cases:

* Supported by the Ministero dell'Università e della Ricerca of Italy under Grant No. 201784YSZ5, PRIN2017 – ASPRA (*Analysis of Program Analyses*).

S. Kiefer and C. Tasson (Eds.): FOSSACS 2021, LNCS 12650, pp. 102–123, 2021.
https://doi.org/10.1007/978-3-030-71995-1_6

see [28,41] for some no-go theorems; on the other hand, proving their existence is error-prone: see [28] for a list of results that were mistakenly assuming the existence of a distributive law of the powerset monad over itself.

Nevertheless, some sort of weakening of the notion of distributive law–e.g., distributive laws of functors over monads [26]–proved to be ubiquitous in computer science: they are GSOS specifications [38], they are sound coinductive up-to techniques [7] and complete abstract domains [5]. In this paper we will exploit *weak distributive laws* in the sense of [15] that have been recently shown successful in composing the monads for nondeterminism and probability [17].

The goal of this paper is to somehow combine the monads $\mathcal{P}_f$ and $\mathcal{S}$ mentioned above. Our interest in $\mathcal{S}$ relies on the wide expressiveness provided by the possibility of varying S: for instance by taking S to be the Boolean semiring, one obtains the monad $\mathcal{P}_f$; by fixing S to be the field of reals, coalgebras over $\mathbb{EM}(\mathcal{S})$ turn out be linear dynamical systems [34].

We proceed as follows. Rather than composing $\mathcal{P}_f$, we found it convenient to compose the *full*, not necessarily finite, powerset monad $\mathcal{P}$ with $\mathcal{S}$. In this way we can reuse several results in [12] that provide necessary and sufficient conditions on the semiring S for the existence of a canonical weak [15] distributive law $\delta\colon \mathcal{SP} \to \mathcal{PS}$. Our first contribution (Theorem 21) consists in showing that such δ has a convenient alternative characterisation, whenever the underlying semiring is a *positive semifield*, a condition that is met, e.g., by the semirings of Booleans and non-negative reals.

Such characterisation allows us to give a handy definition of the *canonical weak lifting* of $\mathcal{P}$ over $\mathbb{EM}(\mathcal{S})$ (Theorem 24) and to observe that such lifting is *almost* the same as the monad $\mathcal{C}\colon \mathbb{EM}(\mathcal{S}) \to \mathbb{EM}(\mathcal{S})$ defined by Jacobs in [25] (Remark 25): the only difference is the absence in $\mathcal{C}$ of the empty subset. Such difference becomes crucial when considering the composed monads, named $\mathcal{CM}\colon \mathbb{S}\mathrm{et} \to \mathbb{S}\mathrm{et}$ in [25] and $\mathcal{P}_c\mathcal{S}\colon \mathbb{S}\mathrm{et} \to \mathbb{S}\mathrm{et}$ in this paper: the latter maps a set X into the set of convex subsets of $\mathcal{S}X$, while the former additionally requires the subsets to be non-empty. It turns out that while $\mathbb{K}\mathrm{l}(\mathcal{CM})$ is not $\mathbb{CPPO}$-enriched, a necessary condition for the coalgebraic framework in [20], $\mathbb{K}\mathrm{l}(\mathcal{P}_c\mathcal{S})$ indeed is (Theorem 30).

Composing monads by means of weak distributive laws is rewarding in many respects: here we exploit the fact that algebras for the composed monad $\mathcal{P}_c\mathcal{S}$ coincide with δ-algebras, namely algebras for both $\mathcal{P}$ and $\mathcal{S}$ satisfying a certain pentagonal law. One can extract from this law some distributivity axioms that, together with the axioms for semimodules (algebras for the monad $\mathcal{S}$) and those for complete semilattices (algebras for the monad $\mathcal{P}$), provide an algebraic theory presenting the monad $\mathcal{P}_c\mathcal{S}$ (Theorem 32).

We conclude by coming back to the finite powerset monad $\mathcal{P}_f$. By replacing, in the above theory, complete semilattices with semilattices with bottom (algebras for the monad $\mathcal{P}_f$) one obtains a theory presenting the monad $\mathcal{P}_{fc}\mathcal{S}$ of *finitely generated* convex subsets (Theorem 35), which is formally defined as a restriction of the canonical $\mathcal{P}_c\mathcal{S}$. The theory, displayed in Table 1, consists of the

Table 1. The sets of axioms $E_{S\mathcal{L}}$ for semilattices (left), $E_{\mathcal{LSM}}$ for S-semimodules (right) and $E_{\mathcal{D}'}$ for their distributivity (bottom).

$(x \sqcup y) \sqcup z = x \sqcup (y \sqcup z)$	$(x+y)+z = x+(y+z)$	$(\lambda +_S \mu) \cdot x = \lambda \cdot x + \mu \cdot x$
$x \sqcup y = y \sqcup x$	$x + y = y + x$	$0_S \cdot x = 0$
$x \sqcup \bot = x$	$x + 0 = x$	$(\lambda\mu) \cdot x = \lambda \cdot (\mu \cdot x)$
$x \sqcup x = x$		$\lambda \cdot (x+y) = \lambda \cdot x + \lambda \cdot y$
		$\lambda \cdot 0 = 0$
$\lambda \cdot \bot = \bot$ for $\lambda \neq 0_S$	$\lambda \cdot (x \sqcup y) = (\lambda \cdot x) \sqcup (\lambda \cdot y)$	
$x + \bot = \bot$	$x + (y \sqcup z) = (x+y) \sqcup (x+z)$	

theory presenting the monad $\mathcal{P}_f$ and the theory presenting the monad $\mathcal{S}$ with four distributivity axioms.

To save space we had to omit most of the proofs of the results in this article: the interested reader can find them in [9].

Notation. We assume the reader to be familiar with monads and their maps. Given a monad (M, η^M, μ^M) on $\mathbb{C}$, $\mathbb{EM}(M)$ and $\mathbb{Kl}(M)$ denote, respectively, the Eilenberg-Moore category and the Kleisli category of M. The latter is defined as the category whose objects are the same as $\mathbb{C}$ and a morphism $f\colon X \to Y$ in $\mathbb{Kl}(M)$ is a morphism $f\colon X \to M(Y)$ in $\mathbb{C}$. We write $U^M\colon \mathbb{EM}(M) \to \mathbb{C}$ and $U_M\colon \mathbb{Kl}(M) \to \mathbb{C}$ for the canonical forgetful functors, and $F^M\colon \mathbb{C} \to \mathbb{EM}(M)$, $F_M\colon \mathbb{C} \to \mathbb{Kl}(M)$ for their respective left adjoints. Recall, in particular, that $F^M(X) = (X, \mu^M_X)$ and, for $f\colon X \to Y$, $F^M(f) = M(f)$. Given n a natural number, we denote by $\underline{n}$ the set $\{1, \dots, n\}$.

2 (Weak) Distributive laws

Given two monads S and T on a category $\mathbb{C}$, is there a way to compose them to form a new monad ST on $\mathbb{C}$? This question was answered by Beck [3] and his theory of *distributive laws*, which are natural transformations $\delta\colon TS \to ST$ satisfying four axioms and that provide a canonical way to endow the composite functor ST with a monad structure. We begin by recalling the classic definition. In the following, let (T, η^T, μ^T) and (S, η^S, μ^S) be two monads on a category $\mathbb{C}$.

Definition 1. *A* distributive law *of the monad S over the monad T is a natural transformation $\delta\colon TS \to ST$ such that the following diagrams commute.*

$$\begin{array}{ccccc} TSS & \xrightarrow{\delta S} & STS & \xrightarrow{S\delta} & SST \\ {\scriptstyle T\mu^S}\downarrow & & & & \downarrow{\scriptstyle \mu^S T} \\ TS & & \xrightarrow{\delta} & & ST \end{array} \qquad \begin{array}{ccccc} TTS & \xrightarrow{T\delta} & TST & \xrightarrow{\delta T} & STT \\ {\scriptstyle \mu^T S}\downarrow & & & & \downarrow{\scriptstyle S\mu^T} \\ TS & & \xrightarrow{\delta} & & ST \end{array} \tag{1}$$

$$\begin{array}{ccc} & T & \\ {\scriptstyle T\eta^S}\swarrow & & \searrow{\scriptstyle \eta^S T} \\ TS & \xrightarrow{\delta} & ST \end{array} \qquad \begin{array}{ccc} & S & \\ {\scriptstyle \eta^T S}\swarrow & & \searrow{\scriptstyle S\eta^T} \\ TS & \xrightarrow{\delta} & ST \end{array}$$

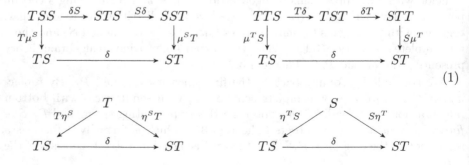

One important result of Beck's theory is the bijective correspondence between distributive laws, liftings to Eilenberg-Moore algebras and extensions to Kleisli categories, in the following sense.

Definition 2. *A* lifting *of the monad* S *to* $\mathbb{EM}(T)$ *is a monad* $(\tilde{S}, \eta^{\tilde{S}}, \mu^{\tilde{S}})$ *where*

$$\begin{array}{ccc} \mathbb{EM}(T) & \xrightarrow{\tilde{S}} & \mathbb{EM}(T) \\ {\scriptstyle F^T}\uparrow & & \uparrow{\scriptstyle F^T} \\ \mathbb{C} & \xrightarrow{S} & \mathbb{C} \end{array} \quad \textit{commutes,} \quad U^T\eta^{\tilde{S}} = \eta^S U^T, \quad U^T\mu^{\tilde{S}} = \mu^S U^T.$$

An extension *of the monad* T *to* $\mathbb{Kl}(S)$ *is a monad* $(\tilde{T}, \eta^{\tilde{T}}, \mu^{\tilde{T}})$ *such that*

$$\begin{array}{ccc} \mathbb{C} & \xrightarrow{T} & \mathbb{C} \\ {\scriptstyle F_S}\downarrow & & \downarrow{\scriptstyle F_S} \\ \mathbb{Kl}(S) & \xrightarrow{\tilde{T}} & \mathbb{Kl}(S) \end{array} \quad \textit{commutes,} \quad \eta^{\tilde{T}} F_S = F_S\eta^T, \quad \mu^{\tilde{T}} F_S = F_S\mu^T.$$

Böhm [11] and Street [37] have studied various weaker notions of distributive law; here we shall use the one that consists in dropping the axiom involving η^T in Definition 1, following the approach of Garner [15].

Definition 3. *A* weak distributive law of S over T *is a natural transformation* $\delta\colon TS \to ST$ *such that the diagrams in* (1) *regarding* μ^S, μ^T *and* η^S *commute.*

There are suitable weaker notions of liftings and extensions which also bijectively correspond to weak distributive laws as proved in [11,15].

Definition 4. *A* weak lifting *of* S *to* $\mathbb{EM}(T)$ *consists of a monad* $(\tilde{S}, \eta^{\tilde{S}}, \mu^{\tilde{S}})$ *on* $\mathbb{EM}(T)$ *and two natural transformations*

$$U^T\tilde{S} \xrightarrow{\iota} SU^T \xrightarrow{\pi} U^T\tilde{S}$$

such that $\pi\iota = id_{U^T\tilde{S}}$ *and such that the following diagrams commute:*

$$\begin{array}{ccccc} U^T\tilde{S}\tilde{S} & \xrightarrow{\iota\tilde{S}} & SU^T\tilde{S} & \xrightarrow{S\iota} & SSU^T \\ {\scriptstyle U^T\mu^{\tilde{S}}}\downarrow & & & & \downarrow{\scriptstyle \mu^S U^T} \\ U^T\tilde{S} & & \xrightarrow{\iota} & & SU^T \end{array} \qquad \begin{array}{ccc} & U^T & \\ {\scriptstyle U^T\eta^{\tilde{S}}}\swarrow & & \searrow{\scriptstyle \eta^S U^T} \\ U^T\tilde{S} & \xrightarrow{\iota} & SU^T \end{array} \tag{2}$$

$$\begin{array}{ccccc} SSU^T & \xrightarrow{S\pi} & SU^T\tilde{S} & \xrightarrow{\pi\tilde{S}} & U^T\tilde{S}\tilde{S} \\ {\scriptstyle \mu^S U^T}\downarrow & & & & \downarrow{\scriptstyle U^T\mu^{\tilde{S}}} \\ SU^T & & \xrightarrow{\pi} & & U^T\tilde{S} \end{array} \qquad \begin{array}{ccc} & U^T & \\ {\scriptstyle \eta^S U^T}\swarrow & & \searrow{\scriptstyle U^T\eta^{\tilde{S}}} \\ SU^T & \xrightarrow{\pi} & U^T\tilde{S} \end{array} \tag{3}$$

A weak extension *of* T *to* $\mathbb{Kl}(S)$ *is a functor* $\tilde{T}\colon \mathbb{Kl}(S) \to \mathbb{Kl}(S)$ *together with a natural transformation* $\mu^{\tilde{T}}: \tilde{T}\tilde{T} \to \tilde{T}$ *such that* $F_S T = \tilde{T}F_S$ *and* $\mu^{\tilde{T}} F_S = F_S\mu^T$.

Theorem 5 ([3,11,15]). *There is a bijective correspondence between (weak) distributive laws* $TS \to ST$*, (weak) liftings of* S *to* $\mathbb{EM}(T)$ *and (weak) extensions of* T *to* $\mathbb{Kl}(S)$.

3 The Powerset and Semimodule Monads

The Monad $\mathcal{P}$. Let us now consider, as S, the *powerset* monad $(\mathcal{P}, \eta^{\mathcal{P}}, \mu^{\mathcal{P}})$, where $\eta^{\mathcal{P}}_X(x) = \{x\}$ and $\mu^{\mathcal{P}}_X(\mathcal{U}) = \bigcup_{U \in \mathcal{U}} U$. Its algebras are precisely the complete semilattices and we have that $\mathbb{K}\mathrm{l}(\mathcal{P})$ is isomorphic to the category $\mathbb{R}\mathrm{el}$ of sets and relations. Hence, giving a distributive law $T\mathcal{P} \to \mathcal{P}T$ is the same as giving an extension of T to $\mathbb{R}\mathrm{el}$: for this to happen the notion of weak cartesian functor and natural transformation is crucial.

Definition 6. *A functor $T\colon \mathbb{S}\mathrm{et} \to \mathbb{S}\mathrm{et}$ is said to be* weakly cartesian *if and only if it preserves weak pullbacks. A natural transformation $\varphi\colon F \to G$ is said to be* weakly cartesian *if and only if its naturality squares are weak pullbacks.*

Kurz and Velebil [29] proved, using an original argument of Barr [2], that an endofunctor T on $\mathbb{S}\mathrm{et}$ has at most one extension to $\mathbb{R}\mathrm{el}$ and this happens precisely when it is weakly cartesian; similarly a natural transformation $\varphi\colon F \to G$, with F and G weakly cartesian, has at most one extension $\tilde{\varphi}\colon \tilde{F} \to \tilde{G}$, precisely when it is weakly cartesian. The following result is therefore immediate.

Proposition 7 ([15, Corollary 16]). *For any monad (T, η^T, μ^T) on $\mathbb{S}\mathrm{et}$:*

1. *There exists a unique distributive law of $\mathcal{P}$ over T if and only if T, η^T and μ^T are weakly Cartesian.*
2. *There exists a unique* weak *distributive law of $\mathcal{P}$ over T if and only if T and μ^T are weakly Cartesian.*

The Monad $\mathcal{S}$. Recall that a *semiring* is a tuple $(S, +, \cdot, 0, 1)$ such that $(S, +, 0)$ is a commutative monoid, $(S, \cdot, 1)$ is a monoid, $\cdot$ distributes over $+$ and 0 is an annihilating element for $\cdot$. In other words, a semiring is a ring where not every element has an additive inverse. Natural numbers $\mathbb{N}$ with the usual operations of addition and multiplication form a semiring. Similarly, integers, rationals and reals form semirings. Also the Booleans $\mathbb{B}\mathrm{ool} = \{0, 1\}$ with $\vee$ and $\wedge$ acting as $+$ and $\cdot$, respectively, form a semiring.

Every semiring S generates a *semimodule* monad $\mathcal{S}$ on $\mathbb{S}\mathrm{et}$ as follows. Given a set X, $\mathcal{S}(X) = \{\varphi\colon X \to S \mid \mathit{supp}\,\varphi \text{ finite}\}$, where $\mathit{supp}\,\varphi = \{x \in X \mid \varphi(x) \neq 0\}$. For $f\colon X \to Y$, define for all $\varphi \in \mathcal{S}(X)$

$$\mathcal{S}(f)(\varphi) = \Big(y \mapsto \sum_{x \in f^{-1}\{y\}} \varphi(x)\Big)\colon Y \to S.$$

This makes $\mathcal{S}$ a functor. The unit $\eta^{\mathcal{S}}_X\colon X \to \mathcal{S}(X)$ is given by $\eta^{\mathcal{S}}_X(x) = \Delta_x$, where Δ_x is the Dirac function centred in x, while the multiplication $\mu^{\mathcal{S}}_X\colon \mathcal{S}^2(X) \to \mathcal{S}(X)$ is defined for all $\Psi \in \mathcal{S}^2(X)$ as

$$\mu^{\mathcal{S}}_X(\Psi) = \Big(x \mapsto \sum_{\varphi \in \mathit{supp}\,\Psi} \Psi(\varphi) \cdot \varphi(x)\Big)\colon X \to S.$$

Table 2. Definition of some properties of a semiring S. Here $a, b, c, d \in S$.

Positive	$a+b=0 \implies a=0=b$
Semifield	$a \neq 0 \implies \exists x. a \cdot x = x \cdot a = 1$
Refinable	$a+b=c+d \implies \exists x,y,z,t. x+y=a,\ z+t=b,\ x+z=c,\ y+t=d$
(A)	$a+b=1 \implies a=0$ or $b=0$
(B)	$a \cdot b=0 \implies a=0$ or $b=0$
(C)	$a+c=b+c \implies a=b$
(D)	$\forall a,b.\ \exists x. a+x=b$ or $b+x=a$
(E)	$a+b=c \cdot d \implies \exists t\colon \{(x,y) \in S^2 \mid x+y=d\} \to S$ such that $\sum_{x+y=d} t(x,y)x = a,\quad \sum_{x+y=d} t(x,y)y = b,\quad \sum_{x+y=d} t(x,y) = c.$

An algebra for $\mathcal{S}$ is precisely a *left-S-semimodule*, namely a set X equipped with a binary operation $+$, an element 0 and a unary operation $\lambda\cdot$ for each $\lambda \in S$, satisfying the equations in Table 1. Indeed, if X carries a semimodule structure then one can define a map $a\colon \mathcal{S}X \to X$ as, for $\varphi \in \mathcal{S}X$,

$$a(\varphi) = \sum_{x \in X} \varphi(x) \cdot x \tag{4}$$

where the above sum is finite because so is $supp\, \varphi$. Vice versa, if (X, a) is an $\mathcal{S}$-algebra, then the corresponding left-semimodule structure on X is obtained by defining for all $\lambda \in S$ and $x, y \in X$

$$x +^a y = a(x \mapsto 1, y \mapsto 1), \qquad 0^a = a(\varepsilon), \qquad \lambda \cdot^a x = a(x \mapsto \lambda). \tag{5}$$

Above and in the remainder of the paper, we write the list $(x_1 \mapsto s_1, \ldots, x_n \mapsto s_n)$ for the only function $\varphi\colon X \to S$ with support $\{x_1, \ldots, x_n\}$ mapping x_i to s_i and we write the empty list ε for the function constant to 0. For instance, for $a = \mu^{\mathcal{S}}_X\colon \mathcal{S}\mathcal{S}X \to \mathcal{S}X$, the left-semimodule structure is defined for all $\varphi_1, \varphi_2 \in \mathcal{S}X$ and $x \in X$ as

$$(\varphi_1 +^{\mu^{\mathcal{S}}} \varphi_2)(x) = \varphi_1(x) + \varphi_2(x), \qquad 0^{\mu^{\mathcal{S}}}(x) = 0, \qquad (\lambda \cdot^{\mu^{\mathcal{S}}} \varphi_1)(x) = \lambda \cdot \varphi_1(x).$$

Proposition 7 tells us exactly when a (weak) distributive law of the form $T\mathcal{P} \to \mathcal{P}T$ exists for an arbitrary monad T on $\mathbb{S}$et. Take then $T = \mathcal{S}$: when are the functor $\mathcal{S}$ and the natural transformations $\eta^{\mathcal{S}}$ and $\mu^{\mathcal{S}}$ weakly cartesian? The answer has been given in [12] (see also [18]), where a complete characterisation in purely algebraic properties for S is provided. In Table 2 we recall such properties.

Theorem 8 ([12]). *Let S be a semiring.*

1. *The functor $\mathcal{S}$ is weakly cartesian if and only if S is positive and refinable.*
2. *$\eta^{\mathcal{S}}$ is weakly cartesian if and only if S enjoys* (A) *in Table 2.*
3. *If $\mathcal{S}$ is weakly cartesian, then $\mu^{\mathcal{S}}$ is weakly cartesian if and only if S enjoys* (B) *and* (E) *in Table 2.*

Remark 9. In [12, Proposition 9.1] it is proved that if S enjoys (C) and (D), then S is refinable; if S is a positive semifield, then it enjoys (B) and (E). In the next Proposition we prove that if S is a positive semifield then it is also refinable, hence $\mathcal{S}$ and $\mu^{\mathcal{S}}$ are weakly cartesian.

Proposition 10. *If S is a positive semifield, then it is refinable.*

Proof. Let a, b, c and d in S be such that $a+b=c+d$. If $a+b=0$, then take $x=y=z=t=0$, otherwise take

$$x=\frac{ac}{c+d},\quad y=\frac{ad}{c+d},\quad z=\frac{bc}{c+d},\quad t=\frac{bd}{c+d}.$$

Then $x+y=a$, $z+t=b$, $x+z=c$, $y+t=d$. □

Example 11. It is known that, for $\mathcal{S}=\mathbb{N}$, a distributive law $\delta\colon \mathcal{SP}\to\mathcal{PS}$ exists. Indeed one can check that all conditions of Theorem 8 are satisfied, therefore we can apply Proposition 7.1. In this case, the monad $\mathcal{S}X$ is naturally isomorphic to the commutative monoid monad, which given a set X returns the collection of all *multisets* of elements of X. The law δ is well known (see e.g. [15,23]): given a multiset $\langle A_1,\ldots,A_n\rangle$ of subsets of X in $\mathcal{SP}X$, where the A_i's need not be distinct, it returns the set of multisets $\{\langle a_1,\ldots,a_n\rangle \mid a_i\in A_i\}$.

Convex Subsets of Left-semimodules. Theorem 8 together with Proposition 7.1 tell us that whenever the element 1 of S can be decomposed as a non-trivial sum there is no distributive law $\delta\colon \mathcal{SP}\to\mathcal{PS}$. Semirings with this property abound, for example $\mathbb{Q}$, $\mathbb{R}$, $\mathbb{R}^+$ with the usual operations of sum an multiplication, as well as $\mathbb{B}$ool (since $1\vee 1=1$). Such semirings are precisely those for which the notion of *convex subset* of their left-semimodules is non-trivial. For the existence of a *weak* distributive law, however, this condition on 1_S is not required: convexity will indeed play a crucial role in the definition of the weak distributive law.

Definition 12. *Let S be a semiring, X an S-left-semimodule and $A\subseteq X$. The* convex closure *of A is the set*

$$\overline{A}=\left\{\sum_{i=1}^{n}\lambda_i\cdot a_i \mid n\in\mathbb{N},\, a_i\in A,\, \sum_{i=1}^{n}\lambda_i=1\right\}\subseteq X.$$

The set A is said to be convex *if and only if $A=\overline{A}$.*

Recalling that the category of S-left-semimodules is isomorphic to $\mathbb{EM}(\mathcal{S})$, we can use (4) to translate Definition 12 of convex subset of a semimodule into the following notion of convex subset of a $\mathcal{S}$-algebra $a\colon \mathcal{S}X\to X$.

Definition 13. *Let S be a semiring, $(X,a)\in\mathbb{EM}(\mathcal{S})$, $A\subseteq X$. The* convex closure *of A in (X,a) is the set*

$$\overline{A}^{a}=\left\{a(\varphi)\mid \varphi\in\mathcal{S}X,\, supp\,\varphi\subseteq A,\sum_{x\in X}\varphi(x)=1\right\}.$$

A is said to be convex *in (X, a) if and only if $A = \overline{A}^a$. We denote by $\mathcal{P}_c^a X$ the set of convex subsets of X with respect to a.*

Remark 14. Observe that $\emptyset$ is convex, because $\overline{\emptyset}^a = \emptyset$, since there is no $\varphi \in \mathcal{S}X$ with empty support such that $\sum_{x \in X} \varphi(x) = 1$.

Example 15. Suppose S is such that $\eta^{\mathcal{S}}$ is weakly cartesian (equivalently (A) holds: $x + y = 1 \implies x = 0$ or $y = 0$), for example $S = \mathbb{N}$, and let $(X, a) \in \mathbb{EM}(\mathcal{S})$. A $\varphi \in \mathcal{S}X$ such that $\sum_{x \in X} \varphi(x) = 1$ and $supp\, \varphi \subseteq A$ is a function that assigns 1 to *exactly one* element of A and 0 to all the other elements of X. These functions are precisely all the Δ_x for those elements $x \in A$. Since $a \colon \mathcal{S}X \to X$ is a structure map for an $\mathcal{S}$-algebra, it maps the function Δ_x into x. Therefore $\overline{A}^a = \{a(\Delta_x) \mid x \in A\} = \{x \mid x \in A\} = A$. Thus *all* $A \in \mathcal{PS}X$ are convex.

Example 16. When $S = \mathbb{B}\text{ool}$, we have that $\mathcal{S}$ is naturally isomorphic to $\mathcal{P}_f$, the finite powerset monad, whose algebras are idempotent commutative monoids or equivalently semilattices with a bottom element. So, for $(X, a) \in \mathbb{EM}(\mathcal{S})$, a $\varphi \in \mathcal{S}X$ such that $\sum_{x \in X} \varphi(x) = 1$ and $supp\, \varphi \subseteq A$ is any finitely supported function from X to $\mathbb{B}\text{ool}$ that assigns 1 to at least one element of A. Intuitively, such a φ selects a non-empty finite subset of A, then $a(\varphi)$ takes the join of all the selected elements. Thus, $\overline{A}^a$ adds to A all the possible joins of non-empty finite subsets of A: A is convex if and only if it is closed under binary joins.

4 The Weak Distributive Law $\delta \colon \mathcal{SP} \to \mathcal{PS}$

Weak extensions of $\mathcal{S}$ to $\mathbb{K}\text{l}(\mathcal{P}) = \mathbb{R}\text{el}$ only consist of extensions of the functor $\mathcal{S}$ and of the multiplication $\mu^{\mathcal{S}}$, for which necessary and sufficient conditions are listed in Theorem 8. Hence for semirings S satisfying those criteria a weak distributive law $\delta \colon \mathcal{SP} \to \mathcal{PS}$ does exist, and it is unique because there is only one extension of the functor $\mathcal{S}$ to $\mathbb{R}\text{el}$.

Theorem 17. *Let S be a positive, refinable semiring satisfying* (B) *and* (E) *in Table 2. Then there exists a unique weak distributive law $\delta \colon \mathcal{SP} \to \mathcal{PS}$ defined for all sets X and $\Phi \in \mathcal{SP}X$ as:*

$$\delta_X(\Phi) = \left\{ \varphi \in \mathcal{S}X \mid \exists \psi \in \mathcal{S}(\ni_X). \begin{cases} \forall A \in \mathcal{P}X.\, \Phi(A) = \sum\limits_{x \in A} \psi(A, x) & (a) \\ \forall x \in X.\, \varphi(x) = \sum\limits_{A \ni x} \psi(A, x) & (b) \end{cases} \right\} \tag{6}$$

where $\ni_X$ is the set $\{(A, x) \in \mathcal{P}X \times X \mid x \in A\}$.

The above δ, which is obtained by following the standard recipe of Proposition 7, is illustrated by the following example.

Example 18. Take $S = \mathbb{R}^+$ with the usual operations of sum and multiplication. Consider $X = \{x, y, z, a, b\}$, $A_1 = \{x, y\}$, $A_2 = \{y, z\}$ and $A_3 = \{a, b\}$. Let $\Phi \in \mathcal{S}(\mathcal{P}X)$ be defined as

$$\Phi = (A_1 \mapsto 5, \quad A_2 \mapsto 9, \quad A_3 \mapsto 13)$$

and $\Phi(A) = 0$ for all other sets $A \subseteq X$, so $supp\,\Phi = \{A_1, A_2, A_3\}$. In order to find an element $\varphi \in \delta_X(\Phi)$, we can first take a $\psi \in \mathcal{S}(\ni_X)$ satisfying condition (a) in (6) and then compute the $\varphi \in \mathcal{S}X$ using condition (b).

Among the $\psi \in \mathcal{S}(\ni_X)$, consider for instance the following:

$$\psi = \begin{pmatrix} (A_1,x) \mapsto 2 & (A_2,y) \mapsto 4 & (A_3,a) \mapsto 6 \\ (A_1,y) \mapsto 3 & (A_2,z) \mapsto 5 & (A_3,b) \mapsto 7 \end{pmatrix}.$$

Since $\Phi(A_1) = \psi(A_1,x) + \psi(A_1,y)$, $\Phi(A_2) = \psi(A_2,y) + \psi(A_2,z)$ and $\Phi(A_3) = \psi(A_3,a) + \psi(A_3,b)$, we have that ψ satisfies condition (a) in (6). Condition (b) forces φ to be the following:

$$\varphi = (x \mapsto 2, \quad y \mapsto 3+4, \quad z \mapsto 5, \quad a \mapsto 6, \quad b \mapsto 7).$$

Remark 19. If S enjoys (A) in Table 2, then the transformation δ given in (6) is actually a distributive law, and for $S = \mathbb{N}$ we recover the well-known δ of Example 11. Example 18 can be repeated with $S = \mathbb{N}$: then Φ is the multiset where the set A_1 occurs five times, A_2 nine times and A_3 thirteen times. The elements of $\delta_X(\Phi)$ are all those multisets containing one element per copy of A_1, A_2 and A_3 in $supp\,\Phi$. The φ provided indeed contains five elements of A_1 (two copies of x and three of y), nine elements of A_2 (four copies of y and five of z), thirteen elements of A_3 (six copies of a and seven of b).

As Example 18 shows, each element φ of $\delta_X(\Phi)$ is determined by a function ψ choosing for each set $A \in supp\,\Phi$ a finite number of elements $x_1^A, \ldots, x_m^A$ in A and $s_1^A, \ldots, s_m^A$ in S in such a way that $\sum_{j=1}^m s_j^A = \Phi(A)$. The function φ maps each x_j^A to s_j^A if the sets in $supp\,\Phi$ are *disjoint*; if however there are x_j^A and x_k^B such that $x_j^A = x_k^B$ (like y in Example 18), then x_j^A is mapped to $s_j^A + s_k^B$.

Among those ψ's, there are some special, *minimal* ones as it were, that choose for each A in $supp\,\Phi$ exactly *one* element of A, and assign to it $\Phi(A)$. The induced φ in $\delta_X(\Phi)$ can be described as $\sum_{A \in u^{-1}\{x\}} \Phi(A)$ (equivalently $\mathcal{S}(u)(\Phi)$[1]) where $u\colon supp\,\Phi \to X$ is a function selecting an element of A for each $A \in supp\,\Phi$ (that is $u(A) \in A$). We denote the set of such φ's by $\mathfrak{c}(\Phi)$.

$$\mathfrak{c}(\Phi) = \{\mathcal{S}(u)(\Phi) \mid u\colon supp\,\Phi \to X \text{ such that } \forall A \in supp\,\Phi.\, u(A) \in A\} \tag{7}$$

Example 20. Take X, A_1 and A_2 as in Example 18, but a different, smaller, $\Phi \in \mathcal{S}(\mathcal{P}X)$ defined as $\Phi = (A_1 \mapsto 1, \quad A_2 \mapsto 2)$. There are only four functions $u\colon supp\,\Phi \to X$ such that $u(A) \in A$ and thus only four functions φ in $\mathfrak{c}(\Phi)$:

$$\begin{array}{ll|l} u_1 = (A_1 \mapsto x, & A_2 \mapsto y) & \varphi_1 = (x \mapsto 1,\ y \mapsto 2) \\ u_2 = (A_1 \mapsto x, & A_2 \mapsto z) & \varphi_2 = (x \mapsto 1,\ z \mapsto 2) \\ u_3 = (A_1 \mapsto y, & A_2 \mapsto y) & \varphi_3 = (y \mapsto 3) \\ u_4 = (A_1 \mapsto y, & A_2 \mapsto z) & \varphi_4 = (y \mapsto 1,\ z \mapsto 2) \end{array}$$

Observe that the function $\varphi = (x \mapsto 1, y \mapsto 1, z \mapsto 1)$ belongs to $\delta_X(\Phi)$ but not to $\mathfrak{c}(\Phi)$. Nevertheless φ can be retrieved as the convex combination $\frac{1}{2} \cdot \varphi_1 + \frac{1}{2} \cdot \varphi_2$.

[1] More precisely, we should write $\mathcal{S}(u)(\Phi')$ where Φ' is the restriction of Φ to $supp\,\Phi$.

Our key result states that every $\varphi \in \delta_X(\Phi)$ can be written as a convex combination (performed in the $\mathcal{S}$-algebra $(\mathcal{S}X, \mu^{\mathcal{S}}_X)$) of functions in $\mathfrak{c}(\Phi)$, at least when S is a positive semifield, which by Remark 9 and Proposition 10 satisfies all the conditions that make (6) a weak distributive law. The proof is laborious and omitted here: we only remark that divisions in S play a crucial role in it.

Theorem 21. *Let S be a positive semifield. Then for all sets X and $\Phi \in \mathcal{SP}X$*

$$\delta_X(\Phi) = \left\{ \mu^{\mathcal{S}}_X(\Psi) \mid \Psi \in \mathcal{S}^2X.\ \sum_{\varphi \in \mathcal{S}X} \Psi(\varphi) = 1,\ supp\,\Psi \subseteq \mathfrak{c}(\Phi) \right\} = \overline{\mathfrak{c}(\Phi)}^{\mu^{\mathcal{S}}_X}. \quad (8)$$

Remark 22. If we drop the hypothesis of semifield and only have the minimal assumptions of Theorem 17, then (8) does not hold any more: $S = \mathbb{N}$ is a counterexample. Indeed, in this case every subset of $\mathcal{S}X$ is convex with respect to $\mu^{\mathcal{S}}_X$ (see Example 15), therefore we would have $\delta_X(\Phi) = \mathfrak{c}(\Phi)$, which is false: the function φ of Example 18 is an example of an element in $\delta_X(\Phi) \setminus \mathfrak{c}(\Phi)$.

Remark 23. When $S = \mathbb{B}\mathrm{ool}$ (which is a positive semifield), the monad $\mathcal{S}$ coincides with the monad $\mathcal{P}_f$. The function $\mathfrak{c}(\cdot)$ in (7) can then be described as

$$\mathfrak{c}(\mathcal{A}) = \{\mathcal{P}_f(u)(\mathcal{A}) \mid u \colon \mathcal{A} \to X \text{ such that } \forall A \in \mathcal{A}.\, u(A) \in A\}$$

for all $\mathcal{A} \in \mathcal{P}_f\mathcal{P}X$. It is worth remarking that this is the transformation χ appearing in Example 9 of [27] (which is in turn equivalent to the one in Example 2.4.7 of [31]). This transformation was erroneously supposed to be a distributive law, as it fails to be natural (see [28]). However, by taking its convex closure, as displayed in (8), one can turn it into a *weak* distributive law.

5 The Weak Lifting of $\mathcal{P}$ to $\mathbb{EM}(\mathcal{S})$

By exploiting the characterisation of the weak distributive law δ (Theorem 21), we can now describe the weak lifting of $\mathcal{P}$ to $\mathbb{EM}(\mathcal{S})$ generated by δ.

Recall from Definition 13 that $\mathcal{P}^a_c X$ is the set of convex subsets of X with respect to the $\mathcal{S}$-algebra $a \colon \mathcal{S}X \to X$. The functions $\iota_{(X,a)} \colon \mathcal{P}^a_c X \to \mathcal{P}X$ and $\pi_{(X,a)} \colon \mathcal{P}X \to \mathcal{P}^a_c X$ are defined for all $A \in \mathcal{P}^a_c X$ and $B \in \mathcal{P}X$ as

$$\iota_{(X,a)}(A) = A \qquad \text{and} \qquad \pi_{(X,a)}(B) = \overline{B}^a, \quad (9)$$

that is $\iota_{(X,a)}$ is just the obvious set inclusion and $\pi_{(X,a)}$ performs the convex closure in a. The function $\alpha_a \colon \mathcal{SP}^a_c X \to \mathcal{P}^a_c X$ is defined for all $\Phi \in \mathcal{SP}^a_c X$ as

$$\alpha_a(\Phi) = \{a(\varphi) \mid \varphi \in \mathfrak{c}(\Phi)\}. \quad (10)$$

To be completely formal, above we should have written $\mathfrak{c}(\mathcal{S}(\iota)(\Phi))$ in place of $\mathfrak{c}(\Phi)$, but it is immediate to see that the two sets coincide. Proving that $\alpha_a \colon \mathcal{SP}^a_c X \to \mathcal{P}^a_c X$ is well defined (namely, $\alpha_a(\Phi)$ is a convex set) and forms an $\mathcal{S}$-algebra requires some ingenuity and will be shown later in Section 5.1. The assignment $(X, a) \mapsto (\mathcal{P}_c^a X, \alpha_a)$ gives rise to a functor $\tilde{\mathcal{P}}\colon \mathbb{EM}(\mathcal{S}) \to \mathbb{EM}(\mathcal{S})$ defined on morphisms $f\colon (X,a) \to (X',a')$ as

$$\tilde{\mathcal{P}}(f)(A) = \mathcal{P}f(A) \tag{11}$$

for all $A \in \mathcal{P}_c^a X$. For all (X,a) in $\mathbb{EM}(\mathcal{S})$, $\eta^{\tilde{\mathcal{P}}}_{(X,a)}\colon (X,a) \to \tilde{\mathcal{P}}(X,a)$ and $\mu^{\tilde{\mathcal{P}}}_{(X,a)}\colon \tilde{\mathcal{P}}\tilde{\mathcal{P}}(X,a) \to \tilde{\mathcal{P}}(X,a)$ are defined for $x \in X$ and $\mathcal{A} \in \mathcal{P}_c^{\alpha_a}(\mathcal{P}_c^a X)$ as

$$\eta^{\tilde{\mathcal{P}}}_{(X,a)}(x) = \{x\} \qquad \text{and} \qquad \mu^{\tilde{\mathcal{P}}}_{(X,a)}(\mathcal{A}) = \bigcup_{A \in \mathcal{A}} A. \tag{12}$$

Theorem 24. *Let $\mathcal{S}$ be a positive semifield. Then the canonical weak lifting of the powerset monad $\mathcal{P}$ to $\mathbb{EM}(\mathcal{S})$, determined by (8), consists of the monad $(\tilde{\mathcal{P}}, \eta^{\tilde{\mathcal{P}}}, \mu^{\tilde{\mathcal{P}}})$ on $\mathbb{EM}(\mathcal{S})$ defined as in (10), (11), (12) and the natural transformations $\iota\colon U^{\mathcal{S}}\tilde{\mathcal{P}} \to \mathcal{P}U^{\mathcal{S}}$ and $\pi\colon \mathcal{P}U^{\mathcal{S}} \to U^{\mathcal{S}}\tilde{\mathcal{P}}$ defined as in (9).*

It is worth spelling out the left-semimodule structure on $\mathcal{P}_c^a X$ corresponding to the $\mathcal{S}$-algebra $\alpha_a\colon \mathcal{S}\mathcal{P}_c^a X \to \mathcal{P}_c^a X$. Let us start with $\lambda \cdot^{\alpha_a} A$ for some $A \in \mathcal{P}_c^a X$. By (5), $\lambda \cdot^{\alpha_a} A = \alpha_a(\Phi)$ where $\Phi = (A \mapsto \lambda)$. By (10), $\alpha_a(\Phi) = \{a(\varphi) \mid \varphi \in \mathfrak{c}(\Phi)\}$. Following the definition of $\mathfrak{c}(\Phi)$ given in (7), one has to consider functions $u\colon \mathit{supp}\,\Phi \to X$ such that $u(B) \in B$ for all $B \in \mathit{supp}\,\Phi$: if $\lambda \neq 0$, then $\mathit{supp}\,\Phi = \{A\}$ and thus, for each $x \in A$, there is exactly one function $u_x\colon \mathit{supp}\,\Phi \to X$ mapping A into x. It is immediate to see that $\mathcal{S}(u_x)(\Phi)$ is exactly the function $(x \mapsto \lambda)$ and thus $a(\mathcal{S}(u_x)(\Phi))$ is, by (5), $\lambda \cdot^a x$. Now if $\lambda = 0$, then $\mathit{supp}\,\Phi = \emptyset$, so there is *exactly one* function $u\colon \mathit{supp}\,\Phi \to X$ and $\mathcal{S}(u)(\Phi)$ is the function mapping all $x \in X$ into 0 and thus, by (5), $a(\mathcal{S}(u)(\Phi)) = 0^a$. Summarising,

$$\lambda \cdot^{\alpha_a} A = \begin{cases} \{\lambda \cdot^a x \mid x \in A\} & \text{if } \lambda \neq 0 \\ \{0^a\} & \text{if } \lambda = 0 \end{cases} \tag{13}$$

Following similar lines of thoughts, one can check that

$$A +^{\alpha_a} B = \{x +^a y \mid x \in A,\ y \in B\} \qquad \text{and} \qquad 0^{\alpha_a} = \{0^a\}. \tag{14}$$

Remark 25. By comparing (14) and (13) with (4) and (5) in [25], it is immediate to see that our monad $\tilde{\mathcal{P}}$ coincides with a slight variation of Jacobs's convex powerset monad $\mathcal{C}$, the only difference being that we do allow for $\emptyset$ to be in $\mathcal{P}_c^a X$. Jacobs insisted on the necessity of $\mathcal{C}(X)$ to be the set of *non-empty* convex subsets of X, because otherwise he was not able to define a semimodule structure on $\mathcal{C}(X)$ such that $0 \cdot \emptyset = \{0^a\}$. However, we do manage to do so, since by (13), $0 \cdot A = 0^a$ for all A and in particular for $A = \emptyset$. At first sight, this may look like an ad-hoc solution, but this is not the case: it is intrinsic in the definition of the unique weak lifting of $\mathcal{P}$ to $\mathbb{EM}(\mathcal{S})$, as stated by Theorem 24 and shown next.

5.1 Proof of Theorem 24

By Theorem 5, the weak distributive law (6) corresponds to a weak lifting $\tilde{\mathcal{P}}$ of $\mathcal{P}$ to $\mathbb{EM}(\mathcal{S})$, which we are going to show coincides with the data of (9)-(12). The image along $\tilde{\mathcal{P}}$ of a $\mathcal{S}$-algebra (X, a) will be a set Y together with a structure map α_a that makes it a $\mathcal{S}$-algebra in turn. Garner [15, Proposition 13] gives us the recipe to build Y and α_a appropriately. Y is obtained by splitting the following idempotent in $\mathbb{S}\text{et}$:

$$e_{(X,a)} = \mathcal{P}X \xrightarrow{\eta^{\mathcal{S}}_{\mathcal{P}X}} \mathcal{S}(\mathcal{P}X) \xrightarrow{\delta_X} \mathcal{P}(\mathcal{S}X) \xrightarrow{\mathcal{P}a} \mathcal{P}X \tag{15}$$

as a composite $e_{(X,a)} = \iota_{(X,a)} \circ \pi_{(X,a)}$, where $\pi_{(X,a)}$ is the corestriction of $e_{(X,a)}$ to its image and $\iota_{(X,a)}$ is the set-inclusion of the image of $e_{(X,a)}$ into $\mathcal{P}X$. In other words, Y is the set of fixed points of $e_{(X,a)}$. α_a is obtained as the composite

$$\alpha_a = \mathcal{S}Y \xrightarrow{\mathcal{S}\iota_{(X,a)}} \mathcal{S}\mathcal{P}X \xrightarrow{\delta_X} \mathcal{P}\mathcal{S}X \xrightarrow{\mathcal{P}a} \mathcal{P}X \xrightarrow{\pi_{(X,a)}} Y.$$

Let us, then, fix an $\mathcal{S}$-algebra (X, a). Given $A \in \mathcal{P}X$, we have $\eta^{\mathcal{S}}_{\mathcal{P}X}(A) = \Delta_A \colon \mathcal{P}X \to S$, the Dirac-function centred in A. The set $\delta_X(\eta^{\mathcal{S}}_{\mathcal{P}X}(A))$ has a simple description, shown in the next Lemma.

Lemma 26. *For all $A \in \mathcal{P}X$*

$$\delta_X(\eta^{\mathcal{S}}_{\mathcal{P}X}(A)) = \left\{ \varphi \in \mathcal{S}X \mid \mathit{supp}\, \varphi \subseteq A, \sum_{x \in X} \varphi(x) = 1 \right\}.$$

The image along A of the idempotent e is therefore

$$e(A) = \mathcal{P}a(\delta_X(\eta^{\mathcal{S}}_{\mathcal{P}X}(A))) = \left\{ a(\varphi) \mid \varphi \in \mathcal{S}X, \mathit{supp}\, \varphi \subseteq A, \sum_{x \in X} \varphi(x) = 1 \right\} = \overline{A}^a.$$

Hence the idempotent e computes the convex closure of elements of $\mathcal{P}X$ and its fixed points are precisely the convex subsets of X with respect to the structure map a. Therefore, the carrier set of $\tilde{\mathcal{P}}(X, a)$ is precisely $\mathcal{P}^a_c X$, the natural transformations π and ι are, respectively, the convex closure operator and the set-inclusion of $\mathcal{P}^a_c X$ into $\mathcal{P}X$ as in (9).

$\mathcal{P}^a_c X$ is then equipped with a structure map $\alpha_a \colon \mathcal{S}\mathcal{P}^a_c X \to \mathcal{P}^a_c X$ given by

$$\alpha_a = \mathcal{S}\mathcal{P}^a_c X \xrightarrow{\mathcal{S}\iota_{(X,a)}} \mathcal{S}\mathcal{P}X \xrightarrow{\delta_X} \mathcal{P}\mathcal{S}X \xrightarrow{\mathcal{P}a} \mathcal{P}X \xrightarrow{\pi_{(X,a)}} \mathcal{P}^a_c X.$$

Let us try to calculate α_a: given $\Phi \colon \mathcal{P}^a_c X \to S$ with finite support, we have that $\mathcal{S}(\iota_{(X,a)})(\Phi)$ is just the extension of Φ to $\mathcal{P}X$ which assigns 0 to each non-convex subset of X. If we write ι instead of $\iota_{(X,a)}$ for short, we have

$$\alpha_a(\Phi) = \overline{\mathcal{P}a(\delta_X(\mathcal{S}(\iota)(\Phi)))}^a. \tag{16}$$

Next, we can use the following technical result.

Proposition 27. *Let (X, a) be a $\mathcal{S}$-algebra. If $\mathcal{A}$ is a convex subset of $(\mathcal{S}X, \mu^{\mathcal{S}}_X)$, then $\mathcal{P}a(\mathcal{A})$ is convex in (X, a).*

Since $\delta_X(\Phi')$ is the convex closure of $\mathfrak{c}(\Phi')$ in $(\mathcal{S}X, \mu^{\mathcal{S}}_X)$ for every $\Phi' \in \mathcal{SP}X$, by Proposition 27 we can avoid to perform the a-convex closure in (16). Therefore

$$\alpha_a(\Phi) = \mathcal{P}a(\delta_X(\mathcal{S}(\iota)(\Phi))) = \mathcal{P}a\big(\overline{\mathfrak{c}(\mathcal{S}(\iota)(\Phi))}^{\mu^{\mathcal{S}}_X}\big).$$

In the next Proposition we show that also the $\mu^{\mathcal{S}}_X$-convex closure is superfluous, due to the fact that $\Phi \in \mathcal{SP}^a_c X$ (and not simply $\mathcal{SP}X$), thus obtaining (10).

Proposition 28. *Let S be a positive semifield, (X, a) a $\mathcal{S}$-algebra, $\Phi \in \mathcal{SP}^a_c X$. Then $\mathcal{P}a(\delta_X(\mathcal{S}(\iota)(\Phi))) = \mathcal{P}a(\mathfrak{c}(\mathcal{S}(\iota)(\Phi)))$.*

Proof. In this proof we shall simply write Φ instead of the more verbose $\mathcal{S}(\iota)(\Phi)$. We want to prove that

$$\mathcal{P}a\big(\delta_X(\Phi)\big) =$$
$$\Big\{a(\psi) \mid \psi \in \mathcal{S}X.\, \exists u\colon \mathit{supp}\,\Phi \to X.\, u(A) \in A,\, \forall x \in X.\, \psi(x) = \sum_{\substack{A \in \mathit{supp}\,\Phi \\ u(A)=x}} \Phi(A)\Big\} \quad (17)$$

where we have, by Theorem 21, that

$$\mathcal{P}a\big(\delta_X(\Phi)\big) = \{a(\mu^{\mathcal{S}}_X(\Psi)) \mid \Psi \in \mathcal{S}^2X,\, \sum_{\varphi \in \mathcal{S}X} \Psi(\varphi) = 1,\, \mathit{supp}\,\Psi \subseteq \mathfrak{c}(\Phi)\}.$$

First of all, $\emptyset$ is *not* a $\mathcal{S}$-algebra, because there is no map $\mathcal{S}(\emptyset) \to \emptyset$ given that $\mathcal{S}(\emptyset) = \{\emptyset\colon \emptyset \to S\}$, hence $X \neq \emptyset$. Next, if $\Phi = \varepsilon\colon \mathcal{P}X \to S$, namely the function constant to 0, then $\mathfrak{c}(\Phi) = \{\varepsilon\colon X \to S\}$ therefore one can easily see that the left-hand side of (17) is equal to $\{a(\varepsilon\colon X \to S)\}$. For the same reason, the right-hand side is also equal to $\{a(\varepsilon\colon X \to S)\}$. Moreover, if $\Phi(\emptyset) \neq 0$, then there is no $u\colon \mathit{supp}\,\Phi \to X$ such that $u(\emptyset) \in \emptyset$, so $\mathfrak{c}(\Phi) = \emptyset$ and so is the left-hand side of (17); for the same reason, also the right-hand side is empty.

Suppose then, for the rest of the proof, that $\Phi \neq 0$ and that $\Phi(\emptyset) = 0$.

For the right-to-left inclusion in (17): given $\psi \in \mathfrak{c}(\Phi)$, consider $\Psi = \eta^{\mathcal{S}}_{\mathcal{S}X}(\psi) = \Delta_\psi \in \mathcal{S}^2X$. Then Ψ clearly satisfies all the required properties and $\mu^{\mathcal{S}}_X(\Psi) = \psi$.

The left-to-right inclusion is more laborious. Let $\Psi \in \mathcal{S}^2X$ be such that $\sum_{\chi \in \mathcal{S}X} \Psi(\chi) = 1$ and such that $\mathit{supp}\,\Psi \subseteq \mathfrak{c}(\Phi)$, that is, for all $\varphi \in \mathit{supp}\,\Psi$ there is $u^\varphi\colon \mathit{supp}\,\Phi \to X$ such that $u^\varphi(A) \in A$ for all $A \in \mathit{supp}\,\Phi$ and $\varphi = \mathcal{S}(u^\varphi)(\Phi)$. We have to show that $a(\mu(\Psi)) = a(\psi)$ for some $\psi \in \mathcal{S}X$ of the form $\sum_{A \in \mathit{supp}\,\Phi} \Phi(A) \cdot u(A)$ for some choice function $u\colon \mathit{supp}\,\Phi \to X$. Notice that the given Ψ is a convex linear combination of functions φ's in $\mathcal{S}X$ like the one we have to produce: the trick will be to exploit the fact that each $A \in \mathit{supp}\,\Phi$ is convex. Here we shall only give a sketch of the proof. Suppose $\mathit{supp}\,\Phi = \{A_1, \dots, A_n\}$ and $\mathit{supp}\,\Psi = \{\varphi^1, \dots, \varphi^m\}$. Call u^j the choice function that generates φ^j. Then Ψ is of this form:

$$\Psi = \left(\underbrace{\begin{pmatrix} u^1(A_1) \mapsto \Phi(A_1) \\ \vdots \\ u^1(A_n) \mapsto \Phi(A_n) \end{pmatrix}}_{\varphi^1} \mapsto \Psi(\varphi^1), \dots, \underbrace{\begin{pmatrix} u^m(A_1) \mapsto \Phi(A_1) \\ \vdots \\ u^m(A_n) \mapsto \Phi(A_n) \end{pmatrix}}_{\varphi^m} \mapsto \Psi(\varphi^m) \right)$$

Define the following element of $\mathcal{S}^2 X$:

$$\Psi' = \Bigg(\underbrace{\begin{pmatrix} u^1(A_1) \mapsto \Psi(\varphi^1) \\ \vdots \\ u^m(A_1) \mapsto \Psi(\varphi^m) \end{pmatrix}}_{\chi^1} \mapsto \Phi(A_1), \ldots, \underbrace{\begin{pmatrix} u^1(A_n) \mapsto \Psi(\varphi^1) \\ \vdots \\ u^m(A_n) \mapsto \Psi(\varphi^m) \end{pmatrix}}_{\chi^n} \mapsto \Phi(A_n) \Bigg)$$

Observe that $u^1(A_i), \ldots, u^m(A_i) \in A_i$ by definition, and A_i is convex by assumption: since $\sum_{j=1}^m \Psi(\varphi^j) = 1$, we have that $a(\chi^i) \in A_i$. Set then $u(A_i) = a(\chi^i)$ and define $\psi = \mathcal{S}(a)(\Psi')$: we have $\psi \in \mathfrak{c}(\Phi)$ with u as the generating choice function. It is not difficult to see that $\mu_X^{\mathcal{S}}(\Psi) = \mu_X^{\mathcal{S}}(\Psi')$, therefore we have

$$a(\psi) = a\big(\mathcal{S}(a)(\Psi')\big) = a\big(\mu_X^{\mathcal{S}}(\Psi')\big) = a\big(\mu_X^{\mathcal{S}}(\Psi)\big)$$

as desired. □

The rest of the proof of Theorem 24, concerning the action of $\tilde{\mathcal{P}}$ on morphisms and the unit and multiplication of the monad $\tilde{\mathcal{P}}$, consists in following the recipe provided by Garner [15].

6 The Composite Monad: an Algebraic Presentation

We can now compose the two monads $\mathcal{P}$ and $\mathcal{S}$ by considering the monad arising from the composition of the following two adjunctions:

$$\mathbb{S}\mathrm{et} \underset{U^{\mathcal{S}}}{\overset{F^{\mathcal{S}}}{\rightleftarrows}} \mathbb{EM}(\mathcal{S}) \underset{U^{\tilde{\mathcal{P}}}}{\overset{F^{\tilde{\mathcal{P}}}}{\rightleftarrows}} \mathbb{EM}(\tilde{\mathcal{P}}) \qquad (\bot,\ \bot)$$

Direct calculations show that the resulting endofunctor on $\mathbb{S}\mathrm{et}$, which we call $\mathcal{P}_c\mathcal{S}$, maps a set X and a function $f\colon X \to Y$ into, respectively,

$$\mathcal{P}_c\mathcal{S}X = \mathcal{P}_c^{\mu_X^{\mathcal{S}}}(\mathcal{S}X) \quad \text{and} \quad \mathcal{P}_c\mathcal{S}(f)(\mathcal{A}) = \{\mathcal{S}(f)(\Phi) \mid \Phi \in \mathcal{A}\} \tag{18}$$

for all $\mathcal{A} \in \mathcal{P}_c\mathcal{S}X$. For all sets X, $\eta_X^{\mathcal{P}_c\mathcal{S}}\colon X \to \mathcal{P}_c\mathcal{S}X$ and $\mu_X^{\mathcal{P}_c\mathcal{S}}\colon \mathcal{P}_c\mathcal{S}\mathcal{P}_c\mathcal{S}X \to \mathcal{P}_c\mathcal{S}X$ are defined as

$$\eta_X^{\mathcal{P}_c\mathcal{S}}(x) = \{\Delta_x\} \quad \text{and} \quad \mu_X^{\mathcal{P}_c\mathcal{S}}(\mathscr{A}) = \bigcup_{\Omega \in \mathscr{A}} \alpha_{\mu_X^{\mathcal{S}}}(\Omega) \tag{19}$$

for all $x \in X$ and $\mathscr{A} \in \mathcal{P}_c\mathcal{S}\mathcal{P}_c\mathcal{S}X$.

Theorem 29. *Let $\mathcal{S}$ be a positive semifield. Then the canonical weak distributive law $\delta\colon \mathcal{S}\mathcal{P} \to \mathcal{P}\mathcal{S}$ given in Theorem 21 induces a monad $\mathcal{P}_c\mathcal{S}$ on $\mathbb{S}\mathrm{et}$ with endofunctor, unit and multiplication defined as in* (18) *and* (19)*.*

Recall from Remark 25 that the monad $\mathcal{C}\colon \mathbb{EM}(\mathcal{S}) \to \mathbb{EM}(\mathcal{S})$ from [25] coincides with our lifting $\tilde{\mathcal{P}}$ modulo the absence of the empty set. The same happens for the composite monad, which is named $\mathcal{CM}$ in [25]. The absence of $\emptyset$ in $\mathcal{CM}$ turns out to be rather problematic for Jacobs. Indeed, in order to use the standard framework of coalgebraic trace semantics [20], one would need the Kleisli category $\mathbb{Kl}(\mathcal{CM})$ to be enriched over $\mathbb{CPPO}$, the category of ω-complete partial orders with *bottom* and continuous functions. $\mathbb{Kl}(\mathcal{CM})$ is not $\mathbb{CPPO}$-enriched since there is no bottom element in $\mathcal{CM}(X)$. Instead, in $\mathcal{P}_c\mathcal{S}X$ the bottom is exactly the empty set; moreover, $\mathbb{Kl}(\mathcal{P}_c\mathcal{S})$ enjoys the properties required by [20].

Theorem 30. *The category* $\mathbb{Kl}(\mathcal{P}_c\mathcal{S})$ *is enriched over* $\mathbb{CPPO}$ *and satisfies the left-strictness condition: for all* $f\colon X \to \mathcal{P}_c\mathcal{S}Y$ *and* Z *set,* $\bot_{Y,Z} \circ f = \bot_{X,Z}$.

It is immediate that every homset in $\mathbb{Kl}(\mathcal{P}_c\mathcal{S})$ carries a complete partial order. Showing that composition of arrows in $\mathbb{Kl}(\mathcal{P}_c\mathcal{S})$ preserves joins (of ω-chains) requires more work: the proof, omitted here, crucially relies on the algebraic theory presenting the monad $\mathcal{P}_c\mathcal{S}$, illustrated next.

An Algebraic Presentation. Recall that an *algebraic theory* is a pair $\mathcal{T} = (\Sigma, E)$ where Σ is a *signature*, whose elements are called *operations*, to each of which is assigned a cardinal number called its *arity*, while E is a class of formal *equations* between Σ-terms. An *algebra* for the theory $\mathcal{T}$ is a set A together with, for each operation o of arity κ in Σ, a function $o_A\colon A^\kappa \to A$ satisfying the equations of E. A *homomorphism* of algebras is a function $f\colon A \to B$ respecting the operations of Σ in their realisations in A and B. Algebras and homomorphisms of an algebraic theory $\mathcal{T}$ form a category $\mathbb{Alg}(\mathcal{T})$.

Definition 31. *Let* M *be a monad on* $\mathbb{Set}$, *and* $\mathcal{T}$ *an algebraic theory. We say that* $\mathcal{T}$ presents M *if and only if* $\mathbb{EM}(M)$ *and* $\mathbb{Alg}(\mathcal{T})$ *are isomorphic.*

Left S-semimodules are algebras for the theory $\mathcal{LSM} = (\Sigma_{\mathcal{LSM}}, E_{\mathcal{LSM}})$ where $\Sigma_{S\mathcal{LSM}} = \{+, 0\} \cup \{\lambda \cdot \mid \lambda \in S\}$ and $E_{\mathcal{LSM}}$ is the set of axioms in Table 1. As already mentioned in Section 3, left S-semimodules are exactly $\mathcal{S}$-algebras and morphisms of S-semimodules coincide with those of $\mathcal{S}$-algebras. Thus, the theory $\mathcal{LSM}$ presents the monad $\mathcal{S}$.

Similarly, semilattices are algebras for the theory $\mathcal{SL} = (\Sigma_{\mathcal{SL}}, E_{\mathcal{SL}})$ where $\Sigma_{\mathcal{SL}} = \{\sqcup, \bot\}$ and $E_{\mathcal{SL}}$ is the set of axioms in Table 1. It is well known that semilattices are algebras for the *finite* powerset monad. Actually, this monad is presented by $\mathcal{SL}$. In order to present the full powerset monad $\mathcal{P}$ we need to take joins of arbitrary arity. A *complete semilattice* is a set X equipped with joins $\bigsqcup_{x\in A} x$ for all–not necessarily finite–$A \subseteq X$. Formally the (infinitary) theory of *complete semilattices* is given as $\mathcal{CSL} = (\Sigma_{\mathcal{CSL}}, E_{\mathcal{CSL}})$ where $\Sigma_{\mathcal{CSL}} = \{\bigsqcup_I \mid I \text{ set}\}$ and $E_{\mathcal{CSL}}$ is the set of axioms displayed in Table 3 (for a detailed treatment of infinitary algebraic theories see, for example, [30]).

We can now illustrate the theory (Σ, E) presenting the composed monad $\mathcal{P}_c\mathcal{S}$: the operations in Σ are exactly those of complete semilattices and S-

Table 3. The sets of axioms $E_{\mathcal{CSL}}$ for complete semilattices: the second axiom generalises the usual idempotency and commutativity properties of finitary $\sqcup$, while the third one generalises associativity and neutrality of $\bigsqcup_{\emptyset} = \bot$.

$$\bigsqcup_{i\in\{0\}} x_i = x_0$$
$$\bigsqcup_{j\in J} x_j = \bigsqcup_{i\in I} x_{f(i)} \text{ for all } f\colon I \to J \text{ surjective}$$
$$\bigsqcup_{i\in I} x_i = \bigsqcup_{j\in J} \bigsqcup_{i\in f^{-1}\{j\}} x_i \text{ for all } f\colon I \to J$$

semimodules, while the axioms are those of complete semilattices and S-semimodules together with the set $E_{\mathcal{D}}$ of *distributivity* axioms illustrated below.

$$\lambda \cdot \bigsqcup_{i\in I} x_i = \bigsqcup_{i\in I} \lambda \cdot x_i \quad \text{for } \lambda \neq 0, \qquad \bigsqcup_{i\in I} x_i + \bigsqcup_{j\in J} y_j = \bigsqcup_{(i,j)\in I\times J} x_i + y_j \tag{20}$$

In short, $\Sigma = \Sigma_{\mathcal{CSL}} \cup \Sigma_{\mathcal{LSM}}$ and $E = E_{\mathcal{CSL}} \cup E_{\mathcal{LSM}} \cup E_{\mathcal{D}}$.

Theorem 32. *The monad $\mathcal{P}_c\mathcal{S}$ is presented by the algebraic theory (Σ, E).*

The presentation crucially relies on the fact that $\mathcal{P}_c\mathcal{S}$ is obtained by composing $\mathcal{P}$ and $\mathcal{S}$ via δ. Indeed, we know from general results in [11,15] that $\mathcal{P}_c\mathcal{S}$-algebras are in one to one correspondence with δ-algebras [3], namely triples (X, a, b) such that $a\colon \mathcal{S}X \to X$ is a $\mathcal{S}$-algebra, $b\colon \mathcal{P}X \to X$ is a $\mathcal{P}$-algebra and the following diagram commutes.

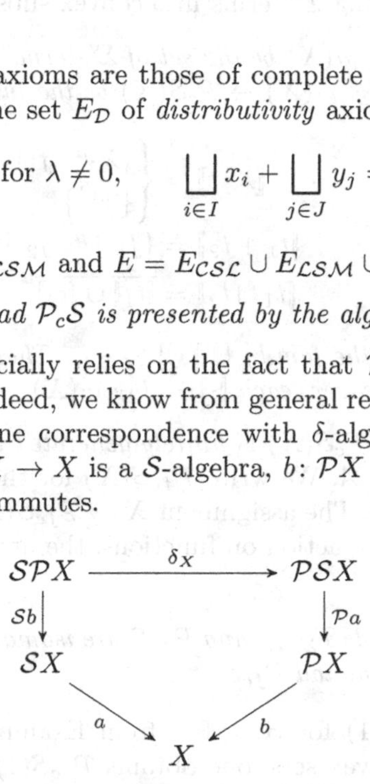

(21)

The $\mathcal{S}$-algebra a corresponds to a S-semimodule $(X, +, 0, \lambda\cdot)$, the $\mathcal{P}$-algebra b to a complete lattice $(X, \bigsqcup_I)$ and the commutativity of diagram (21) expresses exactly the distributivity axioms in (20).

Example 33. Let S be $\mathbb{R}^+$ and let $[a, b]$ with $a, b \in \mathbb{R}^+$ denote the set $\{x \in \mathbb{R}^+ \mid a \leq x \leq b\}$ and $[a, \infty)$ the set $\{x \in \mathbb{R}^+ \mid a \leq x\}$. For $1 = \{x\}$, $\mathcal{P}_c\mathcal{S}(1) = \{\emptyset\} \cup \{[a, b] \mid a, b \in \mathbb{R}^+\} \cup \{[a, +\infty) \mid a \in \mathbb{R}^+\}$. The $\mathcal{P}_c\mathcal{S}$-algebra $\mu_1^{\mathcal{P}_c\mathcal{S}}\colon \mathcal{P}_c\mathcal{S}\mathcal{P}_c\mathcal{S}1 \to \mathcal{P}_c\mathcal{S}1$ induces a δ-algebra where the structure of complete lattice is given as[2]

$$\bigsqcup_{i\in I} A_i = \begin{cases} [\inf_{i\in I}, a_i, \sup_{i\in I} b_i] & \text{if, for all } i \in I,\ A_i = [a_i, b_i] \wedge \sup_{i\in I} b_i \in \mathbb{R}^+ \\ [\inf_{i\in I} a_i, \infty) & \text{otherwise} \end{cases}$$

The $\mathbb{R}^+$-semimodule is as expected, e.g., $[a_1, b_1] + [a_2, b_2] = [a_1 + a_2, b_1 + b_2]$.

[2] For the sake of brevity, we are ignoring the case where some $A_i = \emptyset$.

Finite Joins and Finitely Generated Convex Sets. We now consider the algebraic theory (Σ', E') obtained by restricting (Σ, E) to finitary joins. More precisely, we fix

$$\Sigma' = \Sigma_{\mathcal{SL}} \cup \Sigma_{\mathcal{LSM}} \qquad E' = E_{\mathcal{SL}} \cup E_{\mathcal{LSM}} \cup E_{\mathcal{D}'}$$

where $(\Sigma_{\mathcal{SL}}, E_{\mathcal{SL}})$ is the algebraic theory for semilatices, $(\Sigma_{\mathcal{LSM}}, E_{\mathcal{LSM}})$ is the one for S-semimodules, and $E_{\mathcal{D}'}$ is the set of distributivity axioms illustrated in Table 1. Thanks to the characterisation provided by Theorem 32, we easily obtain a function translating Σ'-terms into convex subsets.

Proposition 34. *Let $T_{\Sigma',E'}(X)$ be the set of Σ'-terms with variables in X quotiented by E'. Let $[\![\cdot]\!]_X : T_{\Sigma',E'}(X) \to \mathcal{P}_c\mathcal{S}(X)$ be the function defined as*

$$[\![x]\!] = \{\Delta_x\} \text{ for } x \in X \qquad [\![\lambda \cdot t]\!] = \begin{cases} \{\lambda \cdot^{\mu^S} f \mid f \in [\![t]\!]\} & \text{if } \lambda \neq 0 \\ \{0^{\mu^S}\} & \text{otherwise} \end{cases}$$
$$[\![0]\!] = \{0^{\mu^S}\} \qquad [\![t_1 + t_2]\!] = \{f_1 +^{\mu^S} f_2 \mid f_1 \in [\![t_1]\!],\ f_2 \in [\![t_2]\!]\}$$
$$[\![\bot]\!] = \emptyset \qquad [\![t_1 \sqcup t_2]\!] = \overline{[\![t_1]\!] \cup [\![t_2]\!]}^{\mu^S}$$

Let $[\![\cdot]\!] : T_{\Sigma',E'} \to \mathcal{P}_c\mathcal{S}$ be the family $\{[\![\cdot]\!]_X\}_{X \in |\mathbb{S}\mathrm{et}|}$. Then $[\![\cdot]\!] : T_{\Sigma',E'} \to \mathcal{P}_c\mathcal{S}$ is a map of monads and, moreover, each $[\![\cdot]\!]_X : T_{\Sigma',E'}(X) \to \mathcal{P}_c\mathcal{S}(X)$ is injective.

We say that a set $\mathcal{A} \in \mathcal{P}_c\mathcal{S}(X)$ is *finitely generated* if there exists a finite set $\mathcal{B} \subseteq \mathcal{S}(X)$ such that $\overline{\mathcal{B}} = \mathcal{A}$. We write $\mathcal{P}_{fc}\mathcal{S}(X)$ for the set of all $\mathcal{A} \in \mathcal{P}_c\mathcal{S}(X)$ that are finitely generated. The assignment $X \mapsto \mathcal{P}_{fc}\mathcal{S}(X)$ gives rise to a monad $\mathcal{P}_{fc}\mathcal{S} : \mathbb{S}\mathrm{et} \to \mathbb{S}\mathrm{et}$ where the action on functions, the unit and the multiplication are defined as for $\mathcal{P}_c\mathcal{S}$.

Theorem 35. *The monads $T_{\Sigma',E'}$ and $\mathcal{P}_{fc}\mathcal{S}$ are isomorphic. Therefore (Σ', E') is a presentation for the monad $\mathcal{P}_{fc}\mathcal{S}$.*

Example 36. Recall $\mathcal{P}_c\mathcal{S}(1)$ for $S = \mathbb{R}^+$ from Example 33. By restricting to the finitely generated convex sets, one obtains $\mathcal{P}_{fc}\mathcal{S}(1) = \{\emptyset\} \cup \{[a, b] \mid a, b \in \mathbb{R}^+\}$, that is the sets of the form $[a, \infty)$ are not finitely generated. Table 4 illustrates the isomorphism $[\![\cdot]\!] : T_{\Sigma',E'}(1) \to \mathcal{P}_c\mathcal{S}(1)$. It is worth observing that every closed interval $[a, b]$ is denoted by a term in $T_{\Sigma',E'}(1)$ for $1 = \{x\}$: indeed, $[\![(a \cdot x) \sqcup (b \cdot x)]\!] = [a, b]$. For $2 = \{x, y\}$, $\mathcal{P}_{fc}\mathcal{S}(2)$ is the set containing all convex polygons: for instance the term $(r_1 \cdot x + s_1 \cdot y) \sqcup (r_2 \cdot x + s_2 \cdot y) \sqcup (r_3 \cdot x + s_3 \cdot y)$ denote a triangle with vertexes (r_i, s_i). For $n = \{x_0, \ldots x_{n-1}\}$, it is easy to see that $\mathcal{P}_{fc}\mathcal{S}(n)$ contains all convex n-polytopes.

7 Conclusions: Related and Future Work

Our work was inspired by [17] where Goy and Petrisan compose the monads of powerset and probability distributions by means of a weak distributive law in the sense of Garner [15]. Our results also heavily rely on the work of Clementino

Table 4. The inductive definition of the function $[\![\cdot]\!]_1 \colon T_{\Sigma',E'}(1) \to \mathcal{P}_c\mathcal{S}(1)$ for $1 = \{x\}$.

$$
\begin{array}{ll}
[\![x]\!] = [1,1] & [\![\lambda \cdot t]\!] = \begin{cases} [\lambda \cdot a, \lambda \cdot b] & \text{if } \lambda \neq 0,\ [\![t]\!] = [a,b] \\ \emptyset & \text{if } \lambda \neq 0,\ [\![t]\!] = \emptyset \\ [0,0] & \text{otherwise} \end{cases} \\
[\![0]\!] = [0,0] & [\![t_1 + t_2]\!] = \begin{cases} [a_1 + a_2, b_1 + b_2] & \text{if } [\![t_i]\!] = [a_i, b_i] \\ \emptyset & \text{otherwise} \end{cases} \\
[\![\bot]\!] = \emptyset & [\![t_1 \sqcup t_2]\!] = \begin{cases} [\mathit{min}\ a_i, \mathit{max}\ b_i] & \text{if } [\![t_i]\!] = [a_i, b_i] \\ [a_1, b_1] & \text{if } [\![t_1]\!] = [a_1, b_1],\ [\![t_2]\!] = \emptyset \\ [a_2, b_2] & \text{if } [\![t_2]\!] = [a_2, b_2],\ [\![t_1]\!] = \emptyset \\ \emptyset & \text{otherwise} \end{cases}
\end{array}
$$

et al. [12] that illustrates necessary and sufficient conditions on a semiring S for the existence of a weak distributive law $\delta \colon \mathcal{SP} \to \mathcal{PS}$. However, to the best of our knowledge, the alternative characterisation of δ provided by Theorem 21 was never shown.

Such characterisation is essential for giving a handy description of the lifting $\tilde{\mathcal{P}} \colon \mathbb{EM}(\mathcal{S}) \to \mathbb{EM}(\mathcal{S})$ (Theorem 24) as well as to observe the strong relationships with the work of Jacobs (Remark 25) and the one of Klin and Rot (Remark 23). The weak distributive law δ also plays a key role in providing the algebraic theories presenting the composed monad $\mathcal{P}_c\mathcal{S}$ (Theorem 24) and its finitary restriction $\mathcal{P}_{fc}\mathcal{S}$ (Theorem 35). These two theories resemble those appearing in, respectively, [17] and [10] where the monad of probability distributions plays the role of the monad $\mathcal{S}$ in our work.

Theorem 30 allows to reuse the framework of coalgebraic trace semantics [20] for modelling over $\mathbb{Kl}(\mathcal{P}_c\mathcal{S})$ systems with both nondeterminism and quantitative features. The alternative framework based on coalgebras over $\mathbb{EM}(\mathcal{P}_c\mathcal{S})$ directly leads to *nondeterministic weighted automata*. A proper comparison with those in [13] is left as future work. Thanks to the abstract results in [7], language equivalence for such coalgebras could be checked by means of coinductive up-to techniques. It is worth remarking that, since δ is a weak distributive law, then thanks to the work in [16], up-to techniques are also sound for "convex-bisimilarity" (in coalgebraic terms, behavioural equivalence for the lifted functor $\tilde{\mathcal{P}} \colon \mathbb{EM}(\mathcal{S}) \to \mathbb{EM}(\mathcal{S})$).

We conclude by recalling that we have two main examples of positive semifields: $\mathbb{B}$ool and $\mathbb{R}^+$. Booleans could lead to a coalgebraic modal logic and trace semantics for *alternating automata* in the style of [27]. For $\mathbb{R}^+$, we hope that exploiting the ideas in [34] our monad could shed some lights on the behaviour of linear dynamical systems featuring some sort of nondeterminism.

References

1. Barlocco, S., Kupke, C., Rot, J.: Coalgebra learning via duality. In: International Conference on Foundations of Software Science and Computation Structures. pp. 62–79. Springer (2019). https://doi.org/10.1007/978-3-030-17127-8_4
2. Barr, M.: Relational algebras. In: MacLane, S., Applegate, H., Barr, M., Day, B., Dubuc, E., Phreilambud, Pultr, A., Street, R., Tierney, M., Swierczkowski, S. (eds.) Reports of the Midwest Category Seminar IV. pp. 39–55. Lecture Notes in Mathematics, Springer, Berlin, Heidelberg (1970). https://doi.org/10.1007/BFb0060439
3. Beck, J.: Distributive laws. In: Appelgate, H., Barr, M., Beck, J., Lawvere, F.W., Linton, F.E.J., Manes, E., Tierney, M., Ulmer, F., Eckmann, B. (eds.) Seminar on Triples and Categorical Homology Theory. pp. 119–140. Lecture Notes in Mathematics, Springer, Berlin, Heidelberg (1969). https://doi.org/10.1007/BFb0083084
4. Bonchi, F., Bonsangue, M.M., Boreale, M., Rutten, J.J.M.M., Silva, A.: A coalgebraic perspective on linear weighted automata. Inf. Comput. **211**, 77–105 (2012). https://doi.org/10.1016/j.ic.2011.12.002
5. Bonchi, F., Ganty, P., Giacobazzi, R., Pavlovic, D.: Sound up-to techniques and complete abstract domains. In: Dawar, A., Grädel, E. (eds.) Proceedings of the 33rd Annual ACM/IEEE Symposium on Logic in Computer Science, LICS 2018, Oxford, UK, July 09-12, 2018. pp. 175–184. ACM (2018). https://doi.org/10.1145/3209108.3209169
6. Bonchi, F., König, B., Petrisan, D.: Up-to techniques for behavioural metrics via fibrations. In: Schewe, S., Zhang, L. (eds.) 29th International Conference on Concurrency Theory, CONCUR 2018, September 4-7, 2018, Beijing, China. LIPIcs, vol. 118, pp. 17:1–17:17. Schloss Dagstuhl - Leibniz-Zentrum für Informatik (2018). https://doi.org/10.4230/LIPIcs.CONCUR.2018.17
7. Bonchi, F., Petrisan, D., Pous, D., Rot, J.: Coinduction up-to in a fibrational setting. In: Henzinger, T.A., Miller, D. (eds.) Joint Meeting of the Twenty-Third EACSL Annual Conference on Computer Science Logic (CSL) and the Twenty-Ninth Annual ACM/IEEE Symposium on Logic in Computer Science (LICS), CSL-LICS '14, Vienna, Austria, July 14 - 18, 2014. pp. 20:1–20:9. ACM (2014). https://doi.org/10.1145/2603088.2603149
8. Bonchi, F., Pous, D.: Hacking nondeterminism with induction and coinduction. Commun. ACM **58**(2), 87–95 (2015). https://doi.org/10.1145/2713167
9. Bonchi, F., Santamaria, A.: Combining Semilattices and Semimodules (Dec 2020), http://arxiv.org/abs/2012.14778
10. Bonchi, F., Sokolova, A., Vignudelli, V.: The theory of traces for systems with nondeterminism and probability. In: 2019 34th Annual ACM/IEEE Symposium on Logic in Computer Science (LICS). pp. 1–14. IEEE (2019). https://doi.org/10.1109/LICS.2019.8785673
11. Böhm, G.: The weak theory of monads. Advances in Mathematics **225**(1), 1–32 (Sep 2010). https://doi.org/10.1016/j.aim.2010.02.015
12. Clementino, M.M., Hofmann, D., Janelidze, G.: The monads of classical algebra are seldom weakly cartesian. Journal of Homotopy and Related Structures **9**(1), 175–197 (Apr 2014). https://doi.org/10.1007/s40062-013-0063-2
13. Droste, M., Kuich, W., Vogler, H.: Handbook of weighted automata. Springer Science & Business Media (2009). https://doi.org/10.1007/978-3-642-01492-5
14. Erkens, R., Rot, J., Luttik, B.: Up-to techniques for branching bisimilarity. In: International Conference on Current Trends in Theory and Practice of Informatics. pp. 285–297. Springer (2020). https://doi.org/10.1007/978-3-030-38919-2_24

15. Garner, R.: The Vietoris Monad and Weak Distributive Laws. Applied Categorical Structures **28**(2), 339–354 (Apr 2020). https://doi.org/10.1007/s10485-019-09582-w
16. Goy, A., Petrisan, D.: Combining weak distributive laws: Application to up-to techniques (2020), https://arxiv.org/abs/2010.00811
17. Goy, A., Petrişan, D.: Combining probabilistic and non-deterministic choice via weak distributive laws. In: Proceedings of the 35th Annual ACM/IEEE Symposium on Logic in Computer Science. pp. 454–464. LICS '20, Association for Computing Machinery, New York, NY, USA (Jul 2020). https://doi.org/10.1145/3373718.3394795
18. Gumm, H.P., Schröder, T.: Monoid-labeled transition systems. Electronic Notes in Theoretical Computer Science **44**(1), 185–204 (May 2001). https://doi.org/10.1016/S1571-0661(04)80908-3
19. Hasuo, I.: Generic weakest precondition semantics from monads enriched with order. Theoretical Computer Science **604**, 2–29 (2015). https://doi.org/10.1016/j.tcs.2015.03.047
20. Hasuo, I., Jacobs, B., Sokolova, A.: Generic trace semantics via coinduction. Log. Methods Comput. Sci. **3**(4) (2007). https://doi.org/10.2168/LMCS-3(4:11)2007
21. Hasuo, I., Shimizu, S., Cîrstea, C.: Lattice-theoretic progress measures and coalgebraic model checking. In: Bodík, R., Majumdar, R. (eds.) Proceedings of the 43rd Annual ACM SIGPLAN-SIGACT Symposium on Principles of Programming Languages, POPL 2016, St. Petersburg, FL, USA, January 20 - 22, 2016. pp. 718–732. ACM (2016). https://doi.org/10.1145/2837614.2837673
22. van Heerdt, G., Kupke, C., Rot, J., Silva, A.: Learning weighted automata over principal ideal domains. In: International Conference on Foundations of Software Science and Computation Structures. pp. 602–621. Springer, Cham (2020). https://doi.org/10.1007/978-3-030-45231-5_31
23. Hyland, M., Nagayama, M., Power, J., Rosolini, G.: A category theoretic formulation for engeler-style models of the untyped lambda. Electron. Notes Theor. Comput. Sci. **161**, 43–57 (2006). https://doi.org/10.1016/j.entcs.2006.04.024
24. Hyland, M., Plotkin, G.D., Power, J.: Combining effects: Sum and tensor. Theor. Comput. Sci. **357**(1-3), 70–99 (2006). https://doi.org/10.1016/j.tcs.2006.03.013
25. Jacobs, B.: Coalgebraic Trace Semantics for Combined Possibilitistic and Probabilistic Systems. Electronic Notes in Theoretical Computer Science **203**(5), 131–152 (Jun 2008). https://doi.org/10.1016/j.entcs.2008.05.023
26. Klin, B.: Bialgebras for structural operational semantics: An introduction. Theor. Comput. Sci. **412**(38), 5043–5069 (2011). https://doi.org/10.1016/j.tcs.2011.03.023
27. Klin, B., Rot, J.: Coalgebraic trace semantics via forgetful logics. In: Pitts, A.M. (ed.) Foundations of Software Science and Computation Structures - 18th International Conference, FoSSaCS 2015, Held as Part of the European Joint Conferences on Theory and Practice of Software, ETAPS 2015, London, UK, April 11-18, 2015. Proceedings. Lecture Notes in Computer Science, vol. 9034, pp. 151–166. Springer (2015). https://doi.org/10.1007/978-3-662-46678-0_10
28. Klin, B., Salamanca, J.: Iterated covariant powerset is not a monad. In: Staton, S. (ed.) Proceedings of the Thirty-Fourth Conference on the Mathematical Foundations of Programming Semantics, MFPS 2018, Dalhousie University, Halifax, Canada, June 6-9, 2018. Electronic Notes in Theoretical Computer Science, vol. 341, pp. 261–276. Elsevier (2018). https://doi.org/10.1016/j.entcs.2018.11.013

29. Kurz, A., Velebil, J.: Relation lifting, a survey. Journal of Logical and Algebraic Methods in Programming **85**(4), 475–499 (Jun 2016). https://doi.org/10.1016/j.jlamp.2015.08.002
30. Manes, E.G.: Algebraic Theories, Graduate Texts in Mathematics, vol. 26. Springer, New York, NY (1976). https://doi.org/10.1007/978-1-4612-9860-1
31. Manes, E., Mulry, P.: Monad compositions I: General constructions and recursive distributive laws. Theory and Applications of Categories [electronic only] **18**, 172–208 (2007), http://www.tac.mta.ca/tac/volumes/18/7/18-07abs.html
32. Moggi, E.: Notions of computation and monads. Inf. Comput. **93**(1), 55–92 (1991). https://doi.org/10.1016/0890-5401(91)90052-4
33. Rutten, J.J.M.M.: Automata and coinduction (an exercise in coalgebra). In: Sangiorgi, D., de Simone, R. (eds.) CONCUR '98: Concurrency Theory, 9th International Conference, Nice, France, September 8-11, 1998, Proceedings. Lecture Notes in Computer Science, vol. 1466, pp. 194–218. Springer (1998). https://doi.org/10.1007/BFb0055624
34. Rutten, J.J.M.M.: A tutorial on coinductive stream calculus and signal flow graphs. Theor. Comput. Sci. **343**(3), 443–481 (2005). https://doi.org/10.1016/j.tcs.2005.06.019
35. Silva, A., Bonchi, F., Bonsangue, M.M., Rutten, J.J.M.M.: Generalizing determinization from automata to coalgebras. Log. Methods Comput. Sci. **9**(1) (2013). https://doi.org/10.2168/LMCS-9(1:9)2013
36. Smolka, S., Foster, N., Hsu, J., Kappé, T., Kozen, D., Silva, A.: Guarded kleene algebra with tests: verification of uninterpreted programs in nearly linear time. Proceedings of the ACM on Programming Languages **4**(POPL), 1–28 (2019). https://doi.org/10.1145/3371129
37. Street, R.: Weak Distributive Laws. Theory and Applications of Categories **22**(12), 313–320 (2009), http://www.tac.mta.ca/tac/volumes/22/12/22-12abs.html
38. Turi, D., Plotkin, G.: Towards a mathematical operational semantics. In: Proceedings of Twelfth Annual IEEE Symposium on Logic in Computer Science. pp. 280–291. IEEE (1997). https://doi.org/10.1109/LICS.1997.614955
39. Urbat, H., Schröder, L.: Automata learning: An algebraic approach. In: Proceedings of the 35th Annual ACM/IEEE Symposium on Logic in Computer Science. pp. 900–914 (2020). https://doi.org/10.1145/3373718.3394775
40. Varacca, D., Winskel, G.: Distributing probability over non-determinism. Math. Struct. Comput. Sci. **16**(1), 87–113 (2006). https://doi.org/10.1017/S0960129505005074
41. Zwart, M., Marsden, D.: No-Go Theorems for Distributive Laws. In: 2019 34th Annual ACM/IEEE Symposium on Logic in Computer Science (LICS). pp. 1–13 (Jun 2019). https://doi.org/10.1109/LICS.2019.8785707

One-way Resynchronizability of Word Transducers*

Sougata Bose[1], S.N. Krishna[2], Anca Muscholl[1], and ✉ Gabriele Puppis[3]

[1] LaBRI, University of Bordeaux, Bordeaux, France
[2] Dept. of Computer Science & Engineering IIT Bombay, Bombay, India
[3] Dept. of Mathematics, Computer Science, and Physics, Univ. of Udine, Udine, Italy
gabriele.puppis@uniud.it

Abstract. The origin semantics for transducers was proposed in 2014, and it led to various characterizations and decidability results that are in contrast with the classical semantics. In this paper we add a further decidability result for characterizing transducers that are close to one-way transducers in the origin semantics. We show that it is decidable whether a non-deterministic two-way word transducer can be resynchronized by a bounded, regular resynchronizer into an origin-equivalent one-way transducer. The result is in contrast with the usual semantics, where it is undecidable to know if a non-deterministic two-way transducer is equivalent to some one-way transducer.

Keywords: String transducers · Resynchronizers · One-way transducers

1 Introduction

Regular word-to-word functions form a robust and expressive class of transformations, as they correspond to deterministic two-way transducers, to deterministic streaming string transducers [1], and to monadic second-order logical transductions [11]. However, the transition from word languages to functions over words is often quite tricky. One of the challenges is to come up with effective characterizations of restricted transformations. A first example is the characterization of functions computed by one-way transducers (known as *rational functions*). It turns out that it is decidable whether a regular function is rational [14], but the algorithm is quite involved [3]. In addition, non-determinism makes the problem intractable: it is undecidable whether the relation computed by a non-deterministic two-way transducer can be also computed by a one-way transducer, [2]. A second example is the problem of knowing whether a regular word function can be described by a first-order logical transduction. This question is still open in general [16], and it is only known how to decide if a *rational* function is definable in first-order logic [13].

Word transducers with origin semantics were introduced by Bojańczyk [4] and shown to provide a machine-independent characterization of regular word-

* Work supported by ANR DeLTA (ANR-16-CE40-0007) and ReLaX.

S. Kiefer and C. Tasson (Eds.): FOSSACS 2021, LNCS 12650, pp. 124–143, 2021.
https://doi.org/10.1007/978-3-030-71995-1_7

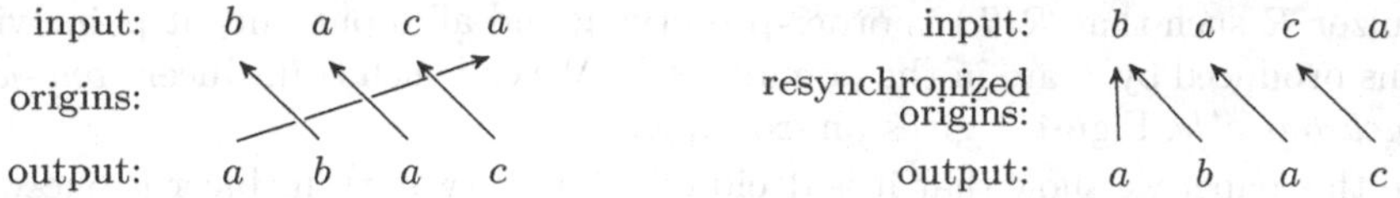

Fig. 1: On the left, an input-output pair for a transducer T that reads wd and outputs dw, $d \in \Sigma$, $w \in \Sigma^*$, the arrows denoting origins. On the right, the same input-output pair, but with origins modified by a resynchronizer $\mathcal{R}$. The resynchronized relation $\mathcal{R}(T)$ is order-preserving, and T is one-way resynchronizable.

to-word functions. The origin semantics, as the name suggests, means tagging the output by the positions of the input that generated that output.

A nice phenomenon is that origins can restore decidability for some interesting problems. For example, the equivalence of word relations computed by one-way transducers, which is undecidable in the classical semantics [18,19], is PSPACE-complete for two-way non-deterministic transducers in the origin semantics [7]. Another, deeper, observation is that the origin semantics provides an algebraic approach that can be used to decide fragments. For example, [4] provides an effective characterization of first-order definable word functions under the origin semantics. As for the problem of knowing whether a regular word function is rational, it becomes almost trivial in the origin semantics.

A possible objection against the origin semantics is that the comparison of two transducers in the origin semantics is too strict. Resynchronizations were proposed in order to overcome this issue. A resynchronization is a binary relation between input-output pairs with origins, that preserves the input and the output, changing only the origins. Resynchronizations were introduced for one-way transducers [15], and later for two-way transducers [7]. For one-way transducers *rational* resynchronizations are transducers acting on the synchronization languages, whereas for two-way transducers, *regular* resynchronizations are described by regular properties over the input that restrict the change of origins. The class of bounded[4] regular resynchronizations was shown to behave very nicely, preserving the class of transductions defined by non-deterministic, two-way transducers: for any bounded regular resynchronization $\mathcal{R}$ and any two-way transducer T, the resynchronized relation $\mathcal{R}(T)$ can be computed by another two-way transducer [7]. In particular, non-deterministic, two-way transducers can be effectively compared modulo bounded regular resynchronizations.

As mentioned above, it is easy to know if a two-way transducer is equivalent under the origin semantics to some one-way transducer [4], since this is equivalent to being order-preserving. But what happens if this is not the case? Still, the given transducer T can be "close" to some order-preserving transducer. What we mean here by "close" is that there exists some bounded regular resyn-

[4] "Bounded" refers here to the number of source positions that are mapped to the same target position. It rules out resynchronizations such as the universal one.

chronizer $\mathcal{R}$ such that $\mathcal{R}(T)$ is order-preserving and all input-output pairs with origins produced by T are in the domain of $\mathcal{R}$. We call such transducers *one-way resynchronizable*. Figure 1 gives an example.

In this paper we show that it is decidable if a two-way transducer is one-way resynchronizable. We first solve the problem for bounded-visit two-way transducers. A bounded-visit transducer is one for which there is a uniform bound for the number of visits of any input position. Then, we use the previous result to show that one-way resynchronizability is decidable for arbitrary two-way transducers, so without the bounded-visit restriction. This is done by constructing, if possible, a bounded, regular resynchronization from the given transducer to a bounded-visit transducer with regular language outputs. Finally, we show that bounded regular resynchronizations are closed under composition, and this allows to combine the previous construction with our decidability result for bounded-visit transducers.

Related work and paper overview. The synthesis problem for resynchronizers asks to compute a resynchronizer from one transducer to another one, when the two transducers are given as input. The problem was studied in [6] and shown to be decidable for unambiguous two-way transducers (it is open for unrestricted transducers). The paper [21] shows that the containment version of the above problem is undecidable for unrestricted one-way transducers.

The origin semantics for streaming string transducers (SST) [1] has been studied in [5], providing a machine-independent characterization of the sets of origin graphs generated by SSTs. An open problem here is to characterize origin graphs generated by aperiodic streaming string transducers [10,16]. Going beyond words, [17] investigates decision problems of tree transducers with origin, and regains the decidability of the equivalence problem for non-deterministic top-down and MSO transducers by considering the origin semantics. An open problem for tree transducers with origin is that of synthesizing resynchronizers as in the word case.

We will recall regular resynchronizations in Section 3. Section 4 provides the proof ingredients for the bounded-visit case, and the proof of decidability of one-way resynchronizability in the bounded-visit case can be found in Section 5. Finally, in Section 6 we sketch the proof in the general case. A full version of the paper is available at `https://arxiv.org/abs/2101.08011`.

2 Preliminaries

Let Σ be a finite input alphabet. Given a word $w \in \Sigma^*$ of length $|w| = n$, a *position* is an element of its domain $\mathsf{dom}(w) = \{1, \ldots, n\}$. For every position i, $w(i)$ denotes the letter at that position. A *cut* of w is any number from 1 to $|w| + 1$, so a cut identifies a position *between* two consecutive letters of the input. The cut $i = 1$ represents the position just before the first input letter, and $i = |w| + 1$ the position just after the last letter of w.

Two-way transducers. We use two-way transducers as defined in [3,6], with a slightly different presentation than in classical papers such as [22]. As usual for two-way machines, for any input $w \in \Sigma^*$, $w(0) = \vdash$ and $w(|w|+1) = \dashv$, where $\vdash, \dashv \notin \Sigma$ are special markers used as delimiters.

A *two-way transducer* (or just *transducer* from now on) is a tuple $T = (Q, \Sigma, \Gamma, \Delta, I, F)$, where Σ, Γ are respectively the input and output alphabets, $Q = Q_{\prec} \uplus Q_{\succ}$ is the set of states, partitioned into left-reading states from $Q_{\prec}$ and right-reading states from $Q_{\succ}$, $I \subseteq Q_{\succ}$ is the set of initial states, $F \subseteq Q$ is the set of final states, and $\Delta \subseteq Q \times (\Sigma \uplus \{\vdash, \dashv\}) \times \Gamma^* \times Q$ is the finite transition relation. Left-reading states read the letter to the left, whereas right-reading states read the letter to the right. This partitioning will also determine the head movement during a transition, as explained below.

As usual, to define runs of transducers we first define configurations. Given a transducer T and a word $w \in \Sigma^*$, a *configuration* of T on w is a state-cut pair (q, i), with $q \in Q$ and $1 \leq i \leq |w|+1$. A configuration (q, i), $1 \leq i \leq |w|+1$ means that the automaton is in state q and its head is *between* the $(i-1)$-th and the i-th letter of w. The transitions that depart from a configuration (q, i) and read a are denoted $(q, i) \xrightarrow{a} (q', i')$, and must satisfy one of the following: (1) $q \in Q_{\succ}$, $q' \in Q_{\succ}$, $a = w(i)$, $(q, a, v, q') \in \Delta$, and $i' = i + 1$, (2) $q \in Q_{\succ}$, $q' \in Q_{\prec}$, $a = w(i)$, $(q, a, v, q') \in \Delta$, and $i' = i$, (3) $q \in Q_{\prec}$, $q' \in Q_{\succ}$, $a = w(i-1)$, $(q, a, v, q') \in \Delta$, and $i' = i$, (4) $q \in Q_{\prec}$, $q' \in Q_{\prec}$, $a = w(i-1)$, $(q, a, v, q') \in \Delta$, and $i' = i - 1$. When T has only right-reading states (i.e. $Q_{\prec} = \emptyset$), its head can only move rightward. In this case we call T a *one-way transducer*.

A *run* of T on w is a sequence $\rho = (q_1, i_1) \xrightarrow{a_{j_1}|v_1} (q_2, i_2) \xrightarrow{a_{j_2}|v_2} \cdots \xrightarrow{a_{j_m}|v_m} (q_{m+1}, i_{m+1})$ of configurations connected by transitions. Note that the positions $j_1, j_2, \ldots, j_m$ of letters do not need to be ordered from smaller to bigger, and can differ slightly (by $+1$ or -1) from the cuts $i_1, i_2, \ldots, i_{m+1}$, since cuts take values in between consecutive letters.

A configuration (q, i) on w is *initial* (resp. *final*) if $q \in I$ and $i = 1$ (resp. $q \in F$ and $i = |w|+1$). A run is *successful* if it starts with an initial configuration and ends with a final configuration. The *output* associated with a successful run ρ as above is the word $v_1 v_2 \cdots v_m \in \Gamma^*$. A transducer T defines a relation $[\![T]\!] \subseteq \Sigma^* \times \Gamma^*$ consisting of all the pairs (u, v) such that v is the output of some successful run ρ of T on u.

Origin semantics. In the origin semantics for transducers [4] the output is tagged with information about the position of the input where it was produced. If reading the i-th letter of the input we output v, then all letters of v are tagged with i, and we say they have *origin* i. We use the notation (v, i) for $v \in \Gamma^*$ to denote that all positions in the output word v have origin i, and we view (v, i) as word over the alphabet $\Gamma \times \mathbb{N}$. The outputs associated with a successful run $\rho = (q_1, i_1) \xrightarrow{b_1|v_1} (q_2, i_2) \xrightarrow{b_2|v_2} (q_3, i_3) \cdots \xrightarrow{b_m|v_m} (q_{m+1}, i_{m+1})$ in the origin semantics are the words of the form $\nu = (v_1, j_1)(v_2, j_2) \cdots (v_m, j_m)$ over $\Gamma \times \mathbb{N}$ where, for all $1 \leq k \leq m$, $j_k = i_k$ if $q_k \in Q_{\succ}$, and $j_k = i_k - 1$ if $q_k \in Q_{\prec}$. Under the origin semantics, the relation defined by T, denoted $[\![T]\!]_o$, is the set of pairs $\sigma = (u, \nu)$ —called *synchronized pairs*— such that $u \in \Sigma^*$ and $\nu \in (\Gamma \times \mathbb{N})^*$ is the output of some successful run on u.

Equivalently, a synchronized pair (u, ν) can be described as a triple (u, v, orig), where v is the projection of ν on Γ, and $\mathit{orig} : \mathsf{dom}(v) \to \mathsf{dom}(u)$ associates with each position of v its origin in u. So for $\nu = (v_1, j_1)(v_2, j_2) \cdots (v_m, j_m)$ as above, $v = v_1 \dots v_m$, and, for all positions i s.t. $|v_1 \dots v_{k-1}| < i \leq |v_1 \dots v_k|$, we have $\mathit{orig}(i) = j_k$. Given two transducers T_1, T_2, we say they are *origin-equivalent* if $[\![T_1]\!]_o = [\![T_2]\!]_o$. Note that two transducers T_1, T_2 can be equivalent in the classical semantics, $[\![T_1]\!] = [\![T_2]\!]$, while they can have different origin semantics, so $[\![T_1]\!]_o \neq [\![T_2]\!]_o$.

Bounded-visit transducers. Let $k > 0$ be some integer, and ρ some run of a two-way transducer T. We say that ρ is *k-visit* if for every $i \geq 0$, it has at most k occurrences of configurations from $Q \times \{i\}$. We call a transducer T *k-visit* if for every $\sigma \in [\![T]\!]_o$ there is some successful, k-visit run ρ of T with output σ (actually we should call the transducer k-visit *in the origin semantics*, but for simplicity we omit this). For example, the relation $\{(w, \overline{w}) \mid w \in \Sigma^*\}$, where $\overline{w}$ denotes the reverse of w, can be computed by a 3-visit transducer. A transducer is called *bounded-visit* if it is k-visit for some k.

Common guess. It is often useful to work with a variant of two-way transducers that can guess beforehand some annotation on the input and inspect it consistently when visiting portions of the input multiple times. This feature is called *common guess* [5], and strictly increases the expressive power of two-way transducers, including bounded-visit ones.

3 One-way resynchronizability

3.1 Regular resynchronizers

Resynchronizations are used to compare transductions in the origin semantics. A *resynchronization* is a binary relation $\mathcal{R} \subseteq (\Sigma^* \times (\Gamma \times \mathbb{N})^*)^2$ over synchronized pairs such that $(\sigma, \sigma') \in \mathcal{R}$ implies that $\sigma = (u, v, \mathit{orig})$ and $\sigma' = (u, v, \mathit{orig}')$ for some origin mappings $\mathit{orig}, \mathit{orig}' : \mathsf{dom}(v) \to \mathsf{dom}(u)$. In other words, a resynchronization will only change the origin mapping, but neither the input, nor the output. Given a relation $S \subseteq \Sigma^* \times (\Gamma \times \mathbb{N})^*$ with origins, the *resynchronized relation* $\mathcal{R}(S)$ is defined as $\mathcal{R}(S) = \{\sigma' \mid (\sigma, \sigma') \in \mathcal{R},\ \sigma \in S\}$. For a transducer T we abbreviate $\mathcal{R}([\![T]\!]_o)$ by $\mathcal{R}(T)$. The typical use of a resynchronization $\mathcal{R}$ is to ask, given two transducers T, T', whether $\mathcal{R}(T)$ and T' are origin-equivalent.

Regular resynchronizers (originally called MSO resynchronizers) were introduced in [7] as a resynchronization mechanism that preserves definability by two-way transducers. They were inspired by MSO (monadic second-order) transductions [9,12] and they are formally defined as follows. A *regular resynchronizer* is a tuple $\mathcal{R} = (\overline{I}, \overline{O}, \mathsf{ipar}, \mathsf{opar}, (\mathsf{move}_\tau)_\tau, (\mathsf{next}_{\tau,\tau'})_{\tau,\tau'})$ consisting of

- some monadic parameters (colors) $\overline{I} = (I_1, \ldots, I_m)$ and $\overline{O} = (O_1, \ldots, O_n)$,
- MSO sentences ipar, opar, defining languages over expanded input and output alphabets, i.e. over $\Sigma' = \Sigma \times 2^{\{1,\ldots,m\}}$ and $\Gamma' = \Gamma \times 2^{\{1,\ldots,n\}}$, respectively,
- MSO formulas $\mathsf{move}_\tau(y, z)$, $\mathsf{next}_{\tau,\tau'}(z, z')$ with two free first-order variables and parametrized by expanded output letters τ, τ' (called types, see below).

To apply a regular resynchronizer as above, one first guesses the valuation of all the predicates I_j, O_k, and uses it to interpret the parameters $\overline{I}$ and $\overline{O}$. Based on the chosen valuation of the parameters $\overline{O}$, each position x of the output v gets an associated *type* $\tau_x = (v(x), b_1, \ldots, b_n) \in \Gamma \times \{0,1\}^n$, where b_j is 1 or 0 depending on whether $x \in O_j$ or not. We refer to the output word together with the valuation of the output parameters as *annotated output*, so a word over $\Gamma \times \{0,1\}^n$. Similarly, the *annotated input* is a word over $\Sigma \times \{0,1\}^m$. The annotated input and output word must satisfy the formulas ipar and opar, respectively.

The origins of output positions are constrained using the formulas move_τ and $\mathsf{next}_{\tau,\tau'}$, which are *parametrized by output types and evaluated over the annotated input.* Intuitively, the formula $\mathsf{move}_\tau(y, z)$ states how the origin of every output position of type τ changes from y to z. We refer to y and z as *source* and *target* origin, respectively. The formula $\mathsf{next}_{\tau,\tau'}(z, z')$ instead constrains the target origins z, z' of any two consecutive output positions with types τ and τ', respectively.

Formally, $\mathcal{R} = (\overline{I}, \overline{O}, \mathsf{ipar}, \mathsf{opar}, (\mathsf{move}_\tau), (\mathsf{next}_{\tau,\tau'}))$ defines the resynchronization consisting of all pairs (σ, σ'), with $\sigma = (u, v, \mathit{orig})$, $\sigma' = (u, v, \mathit{orig}')$, $u \in \Sigma^*$, and $v \in \Gamma^*$, for which there exist $u' \in \Sigma'^*$ and $v' \in \Gamma'^*$ such that

- $\pi_\Sigma(u') = u$ and $\pi_\Gamma(v') = v$
- u' satisfies ipar and v' satisfies opar,
- $(u', \mathit{orig}(x), \mathit{orig}'(x))$ satisfies move_τ for all τ-labeled output positions $x \in \mathsf{dom}(v')$, and
- $(u', \mathit{orig}'(x), \mathit{orig}'(x+1))$ satisfies $\mathsf{next}_{\tau,\tau'}$ for all $x, x+1 \in \mathsf{dom}(v')$ such that x and $x+1$ have label τ and τ', respectively.

Example 1. Consider the following resynchronization $\mathcal{R}$. A pair (σ, σ') belongs to $\mathcal{R}$ if $\sigma = (uv, uwv, \mathit{orig})$, $\sigma' = (uv, uwv, \mathit{orig}')$, with $u, v, w \in \Sigma^+$. The origins orig and orig' are both the identity over u and v. The origin of every position of w in σ (hence a source origin) is either the first or the last position of v. The origin of every position of w in σ' (a target origin) is the first position of v.

This resynchronization is described by a regular resynchronizer that uses two input parameters I_1, I_2 to mark the last and the first positions of v in the input, and one output parameter O to mark the factor w in the output. The formula $\mathsf{move}_\tau(y, z)$ is either $(I_1(y) \vee I_2(y)) \wedge I_2(z)$ or $(y = z)$, depending on whether the type τ describes a position inside w or a position outside w.

We now turn to describing some important restrictions on (regular) resynchronizers. Let $\mathcal{R} = (\overline{I}, \overline{O}, \mathsf{ipar}, \mathsf{opar}, (\mathsf{move}_\tau), (\mathsf{next}_{\tau,\tau'}))$ be a resynchronizer.

- $\mathcal{R}$ is *k-bounded* (or just *bounded*) if for every annotated input $u' \in \Sigma'^*$, every output type $\tau \in \Gamma'$, and every position z, there are at most k positions y such that (u', y, z) satisfies move_τ. Recall that y, z are input positions.
- $\mathcal{R}$ is *T-preserving* for a given transducer T, if every $\sigma \in [\![T]\!]_o$ belongs to the domain of $\mathcal{R}$.
- $\mathcal{R}$ is *partially bijective* if each move_τ formula defines a partial, bijective function from source origins to target origins. Observe that this property implies that $\mathcal{R}$ is 1-bounded.

The boundedness restriction rules out resynchronizations such as the universal one, that imposes no restriction on the change of origins. It is a decidable restriction [7], and it guarantees that definability by two-way transducers is effectively preserved under regular resynchronizations, modulo common guess. More precisely, Theorem 16 in [7] shows that, given a bounded regular resynchronizer $\mathcal{R}$ and a transducer T, one can construct a transducer T' with common guess that is origin-equivalent to $\mathcal{R}(T)$.

Example 1 (continued). Consider again the regular resynchronizer $\mathcal{R}$ described in the previous example. Note that $\mathcal{R}$ is 2-bounded, since at most two source origins are redirected to the same target origin. If we used an additional output parameter to distinguish, among the positions of w, those that have source origin in the first position of v and those that have source origin in the last position of v, we would get a 1-bounded, regular resynchronizer.

We state below two crucial properties of regular resynchronizers (the second lemma is reminiscent of Lemma 11 from [21], which proves closure of bounded resynchronizers with vacuous $\mathsf{next}_{\tau,\tau'}$ relations).

Lemma 1. *Every bounded, regular resynchronizer is effectively equivalent to some* 1*-bounded, regular resynchronizer.*

Lemma 2. *The class of bounded, regular resynchronizers is effectively closed under composition.*

3.2 Main result

Given a two-way transducer T one can ask if it is origin-equivalent to some one-way transducer. It was observed in [4] that this property holds if and only if all synchronized pairs defined by T are *order-preserving*, namely, for all $\sigma = (u, v, orig) \in [\![T]\!]_o$ and all $y, y' \in \mathsf{dom}(v)$, with $y < y'$, we have $orig(y) \leq orig(y')$. The decidability of the above question should be contrasted to the analogous question in the classical semantics: "is a given two-way transducer classically equivalent to some one-way transducer?" The latter problem turns out to be decidable for functional transducers [14,3], but is undecidable for arbitrary two-way transducers [2].

Here we are interested in a different, more relaxed notion:

Definition 1. *A transducer T is called* one-way resynchronizable *if there exists a bounded, regular resynchronizer $\mathcal{R}$ that is T-preserving and such that $\mathcal{R}(T)$ is order-preserving.*

Note that if T' is an order-preserving transducer, then one can construct rather easily a one-way transducer T'' such that $T' =_o T''$, by eliminating non-productive U-turns from accepting runs.

Moreover, note that without the condition of being T-preserving every transducer T would be one-way resynchronizable, using the empty resynchronization.

Example 2. Consider the transducer T_1 that moves the last letter of the input wa to the front by a first left-to-right pass that outputs the last letter a, followed by a right-to-left pass without output, and finally by a left-to-right pass that produces the remaining w. Let $\mathcal{R}$ be the bounded regular resynchronizer that redirects the origin of the last a to the first position. Assuming an output parameter O with an interpretation constrained by opar that marks the last position of the output, the formula $\mathsf{move}_{(a,1)}(y, z)$ says that target origin z (source origin y, resp.) of the last a is the first (last, resp.) position of the input. It is easy to see that $\mathcal{R}(T_1)$ is origin-equivalent to the one-way transducer that on input wa, guesses a and outputs aw. Thus, T_1 is one-way resynchronizable. See also Figure 1.

Example 3. Consider the transducer T_2 that reads inputs of the form $u\#v$ and outputs vu in the obvious way, by a first left-to-right pass that outputs v, followed by a right-to-left pass, and a finally a left-to-right pass that outputs u. Using the characterization with the notion of cross-width that we introduce below, it can be shown that T_2 is not one-way resynchronizable.

In order to give a flavor of our results, we anticipate here the two main theorems, before introducing the key technical concepts of cross-width and inversion (these will be defined further below).

Theorem 1. *For every bounded-visit transducer T, the following are equivalent:*

(1) T is one-way resynchronizable,
(2) the cross-width of T is finite,
(3) no successful run of T has inversions,
(4) there is a partially bijective, regular resynchronizer $\mathcal{R}$ that is T-preserving and such that $\mathcal{R}(T)$ is order-preserving.

Moreover, condition (3) is decidable.

We will use Theorem 1 to show that one-way resynchronizability is decidable for arbitrary two-way transducers (not just bounded-visit ones).

Theorem 2. *It is decidable whether a given two-way transducer T is one-way resynchronizable.*

Let us now introduce the first key concept, that of cross-width:

Definition 2 (cross-width). *Let* $\sigma = (u, v, orig)$ *be a synchronized pair and let* $X_1, X_2 \subseteq \mathsf{dom}(v)$ *be sets of output positions such that, for all* $x_1 \in X_1$ *and* $x_2 \in X_2$, $x_1 < x_2$ *and* $orig(x_1) > orig(x_2)$. *We call such a pair* (X_1, X_2) *a* cross *and define its* width *as* $\min(|orig(X_1)|, |orig(X_2)|)$, *where* $orig(X) = \{orig(x) \mid x \in X\}$ *is the set of origins corresponding to a set* X *of output positions. The* cross-width *of a synchronized pair* σ *is the maximal width of the crosses in* σ. *A transducer has* bounded cross-width *if for some integer* k, *all synchronized pairs associated with successful runs of* T *have cross-width at most* k.

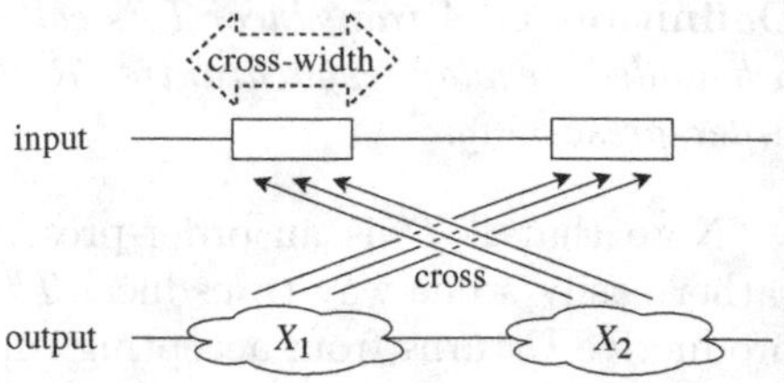

For instance, the transducer T_2 in Example 3 has unbounded cross-width. In contrast, the transducer T_1 in Example 2 has cross-width one.

The other key notion of *inversion* will be introduced formally in the next section (page 135), as it requires a few technical definitions. The notion however is very similar in spirit to that of cross, with the difference that a single inversion is sufficient for witnessing a family of crosses with arbitrarily large cross-width.

4 Proof overview for Theorem 1

This section provides an overview of the proof of Theorem 1, and introduces the main ingredients.

We will use flows (a concept inspired from crossing sequences [22,3] and revised in Section 4.1) in order to derive the key notion of inversion. Roughly speaking, an inversion in a run involves two loops that produce outputs in an order that is reversed compared to the order on origins. Inversions were also used in the characterization of one-way definability of two-way transducers under the classical semantics [3]. There, they were used for deriving some combinatorial properties of outputs. Here we are only interested in detecting inversions, and this is a simple task.

Flows will also be used to associate factorization trees with runs (the existence of factorization trees of bounded height was established by the celebrated Simon's factorization theorem [23]). We will use a structural induction on these factorization trees and the assumption that there is no inversion in every run to construct a regular resynchronization witnessing one-way resynchronizability of the transducer at hand.

Another important ingredient underlying the main characterization is given by the notion of dominant output interval (Section 4.2), which is used to formalize the invariant of our inductive construction.

4.1 Flows and inversions

Intervals. An *interval* of a word is a set of consecutive positions in it. An interval is often denoted by $I = [i, i')$, with $i = \min(I)$ and $i' = \max(I) + 1$. Given two intervals $I = [i, i')$ and $J = [j, j')$, we write $I < J$ if $i' \leq j$, and we say that I, J are adjacent if $i' = j$. The union of two adjacent intervals $I = [i, i')$, $J = [j, j')$, denoted $I \cdot J$, is the interval $[i, j')$ (if I, J are not adjacent, then $I \cdot J$ is undefined).

Subruns. Given a run ρ of a transducer, a *subrun* is a factor of ρ. Note that a subrun of a two-way transducer may visit a position of the input several times. For an input interval $I = [i, j)$ and a run ρ, we say that a subrun ρ' of ρ *spans over* I if i (resp. j) is the smallest (resp. greatest) input position labeling some transition of ρ'. The left hand-side of the figure at page 134 gives an example of an interval I of an input word together with the subruns $\alpha_1, \alpha_2, \alpha_3, \beta_1, \beta_2, \beta_3, \gamma_1$ that span over it. Subruns spanning over an interval can be left-to-right, left-to-left, right-to-left, or right-to-right depending on where the starting and ending positions are w.r.t. the endpoints of the interval.

Flows. Flows are used to summarize subruns of a two-way transducer that span over a given interval. The definition below is essentially taken from [3], except for replacing "functional" by "K-visit". Formally, a *flow* of a transducer T is a graph with vertices divided into two groups, L-vertices and R-vertices, labeled by states of T, and with directed edges also divided into two groups, productive and non-productive edges. The graph satisfies the following requirements. Edge sources are either an L-vertex labeled by a right-reading state, or an R-vertex labeled by a left-reading state, and symmetrically for edge destinations; moreover, edges are of one of the following types: LL, LR, RL, RR. Second, each node is the endpoint of exactly one edge. Finally, L (R, resp.) vertices are totally ordered, in such a way that for every LL (RR, resp.) edge (v, v'), we have $v < v'$. We will only consider flows of K-visiting transducers, so flows with at most $2K$ vertices. For example, the flow in the left-hand side of the figure at page 134 has six L-vertices on the left, and six R-vertices on the right. The edges α_1, α_2, α_3 are LL, LR, and RR, respectively.

Given a run ρ of T and an interval $I = [i, i')$ on the input, *the flow of ρ on I*, denoted $flow_\rho(I)$, is obtained by identifying every configuration at position i (resp. i') with an L (resp. R) vertex, labeled by the state of the configuration, and every subrun spanning over I with an edge connecting the appropriate vertices (this subrun is called the *witnessing subrun* of the edge of the flow). An edge is said to be *productive* if its witnessing subrun produces non-empty output.

Flow monoid. The composition of two flows F and G is defined when the R-vertices of F induce the same sequence of labels as the L-vertices of G. In this case, the composition results in the flow $F \cdot G$ that has as vertices the L-vertices of F and the R-vertices of G, and for edges the directed paths in the graph obtained by glueing the R-vertices of F with the L-vertices of G so that states are matched. Productiveness of edges is inherited by paths, implying that an edge of $F \cdot G$ is productive if and only if the corresponding path contains at least one edge (from F or G) that is productive. When the composition is undefined, we simply write $F \cdot G = \bot$. The above definitions naturally give rise to a *flow monoid* associated with the transducer T, where elements are the flows of T, extended with a dummy element $\bot$, and the product operation is given by the composition of flows, with the convention that $\bot$ is absorbing. It is easy to verify that for any two adjacent intervals $I < J$ of a run ρ, $\mathit{flow}_\rho(I) \cdot \mathit{flow}_\rho(J) = \mathit{flow}_\rho(I \cdot J)$. We denote by M_T the *flow monoid* of a K-visiting transducer T.

Let us estimate the size of M_T. If Q is the set of states of T, there are at most $|Q|^{2K}$ possible sequences of L and R-vertices; and the number of edges (marked as productive or not) is bounded by $\binom{2K}{K} \cdot (2K)^K \cdot 2^K \leq (2K+1)^{2K}$. Including the dummy element $\bot$ in the flow monoid, we get $|M_T| \leq (|Q| \cdot (2K+1))^{2K} + 1 =: \mathbf{M}$.

Loops. A loop of a run ρ over input w is an interval $I = [i, j)$ with a flow $F = \mathit{flow}_\rho(I)$ such that $F \cdot F = F$ (call F *idempotent*). The run ρ can be pumped on a loop $I = [i, j)$ as expected: given $n > 0$, we let $\mathit{pump}^n_I(\rho)$ be the run obtained from ρ by glueing the subruns that span over the intervals $[1, i)$ and $[j, |w| + 1)$ with n copies of the subruns spanning over I (see figure to the right).

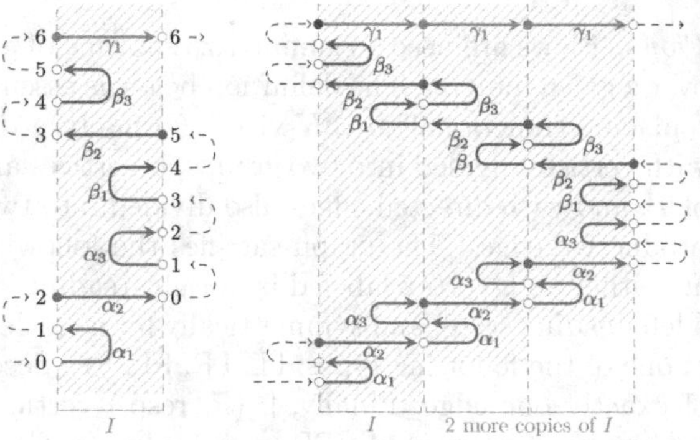

The lemma below shows that the occurrence order relative to subruns witnessing LR or RL edges of a loop (called *straight edges*, for short) is preserved when pumping the loop. This seemingly straightforward lemma is needed for detecting inversions and its proof is surprisingly non-trivial. For example, the external edge connecting the two L-vertices 1, 2 in the figure above appears before edge α_2, and also before every copy of α_2 in the run where loop I is pumped.

Lemma 3. *Let ρ be a run of T on u, let $J < I < K$ be a partition of the domain of u into intervals, with I loop of ρ, and let $F = \mathit{flow}_\rho(J)$, $E = \mathit{flow}_\rho(I)$, and $G = \mathit{flow}_\rho(K)$ be the corresponding flows. Consider an arbitrary edge f of either F or G, and a straight edge e of the idempotent flow E. Let ρ_f and ρ_e be the witnessing subruns of f and e, respectively. Then the occurrence order of ρ_f and ρ_e in ρ is the same as the occurrence order of ρ_f and any copy of ρ_e in $\mathit{pump}^n_I(\rho)$.*

We can now recall the key notion of inversion:

Definition 3 (inversion). *An* inversion *of ρ is a tuple (I, e, I', e') such that*

- *I, I' are loops of ρ and $I < I'$,*
- *e, e' are productive straight edges in $flow_\rho(I)$ and $flow_\rho(I')$ respectively,*
- *the subrun witnessing e' precedes the subrun witnessing e in the run order*

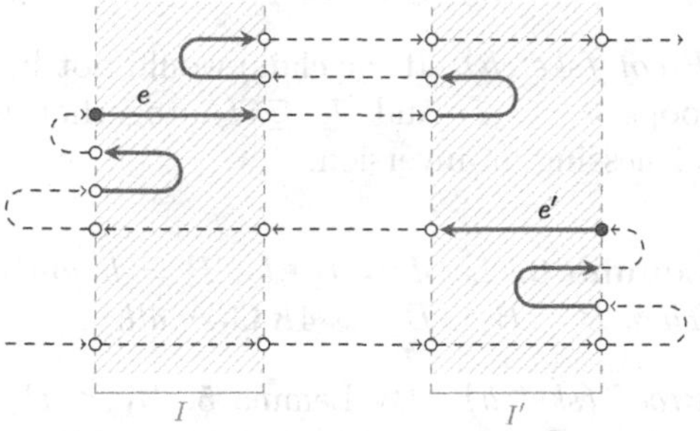

(see the figure to the right).

4.2 Dominant output intervals

In this section we identify some particular intervals of the output that play an important role in the inductive construction of the resynchronizer for a one-way resynchronizable transducer.

Given $n \in \mathbb{N}$, we say that a set B of output positions is *n-large* if $|orig(B)| > n$; otherwise, we say that B is *n-small.* Recall that here we work with a K-visiting transducer T, for some constant K, and that $\mathbf{M} = (|Q| \cdot (2K+1))^{2K} + 1$ is an upper bound to the size of the flow monoid M_T. We will extensively use the derived constant $\mathbf{C} = \mathbf{M}^{2K}$ to distinguish between large and small sets of output positions. The intuition behind this constant is that any set of output positions that is $\mathbf{C}$-large must traverse a loop of ρ. This is captured by the lemma below. The proof uses algebraic properties of the flow monoid M_T [20] (see also Theorem 7.2 in [3], which proves a similar result, but with a larger constant derived from Simon's factorization theorem):

Lemma 4. *Let I be an input interval and B a set of output positions with origins inside I. If B is $\mathbf{C}$-large, then there is a loop $J \subseteq I$ of ρ such that $flow_\rho(J)$ contains a productive straight edge witnessed by a subrun that intersects B (in particular, $out(J) \cap B \neq \emptyset$).*

We need some more notations for outputs. Given an input interval I we denote by $out_\rho(I)$ the set of output positions whose origins belong to I (note that this might not be an output interval). An *output block* of I is a maximal interval contained in $out_\rho(I)$.

The *dominant output interval* of I, denoted $bigout_\rho(I)$, is the smallest output interval that contains all $\mathbf{C}$-large output blocks of I. In particular, $bigout_\rho(I)$ either is empty or begins with the first $\mathbf{C}$-large output block of I and ends with the last $\mathbf{C}$-large outblock block of I. We will often omit the subscript ρ from the notations $flow_\rho(I)$, $out_\rho(I)$, $bigout_\rho(I)$, etc., when no confusion arises.

We now fix a successful run ρ of the K-visiting transducer T. The rest of the section presents some technical lemmas that will be used in the inductive constructions for the proof of the main theorem. *In the lemmas below, we assume that all successful runs of T (in particular, ρ) avoid inversions.*

Lemma 5. *Let $I_1 < I_2$ be two input intervals and B_1, B_2 output blocks of I_1, I_2, respectively. If both B_1, B_2 are* **C***-large, then $B_1 < B_2$.*

Proof (sketch). If the claim would not hold, then Lemma 4 would provide some loops $J_1 \subseteq I_1$ and $J_2 \subseteq I_2$, together with some productive edges in them, witnessing an inversion. □

Lemma 6. *Let $I = I_1 \cdot I_2$, $B = bigout(I)$, and $B_i = bigout(I_i)$ for $i = 1, 2$. Then $B \setminus (B_1 \cup B_2)$ is $4K\mathbf{C}$-small.*

Proof (sketch). By Lemma 5, $B_1 < B_2$. Moreover, all **C**-large output blocks of I_1 or I_2 are also **C**-large output blocks of I, so B contains both B_1 and B_2. Suppose, by way of contradiction, that $B \setminus (B_1 \cup B_2)$ is $4K\mathbf{C}$-large. This means that there is a $2K\mathbf{C}$-large set $S \subseteq B \setminus (B_1 \cup B_2)$ with origins entirely to the left of I_2, or entirely to the right of I_1. Suppose, w.l.o.g., that the former case holds, and decompose S as a union of maximal output blocks $B'_1, B'_2, \ldots, B'_n$ with origins either entirely inside I_1, or entirely outside. Since $S \cap B_1 = \emptyset$, every block B'_i with origins inside I_1 is **C**-small. Similarly, one can prove that every block B'_i with origins outside I_1 is **C**-small too. Moreover, since ρ is K-visiting, we get $n \leq 2K$. Altogether, this contradicts the assumption that S is $2K\mathbf{C}$-large. □

Lemma 7. *Let $I = I_1 \cdot I_2 \cdots I_n$, such that I is a loop and $flow(I) = flow(I_k)$ for all k. Then $bigout(I)$ can be decomposed as $B_1 \cdot J_1 \cdot B_2 \cdot J_2 \cdot \ldots \cdot J_{n-1} \cdot B_n$, where*

1. *for all $1 \leq k \leq n$, $B_k = bigout(I_k)$ (with B_k possibly empty);*
2. *for all $1 \leq k < n$, the positions in J_k have origins inside $I_k \cup I_{k+1}$ and J_k is $2K\mathbf{C}$-small.*

Proof (sketch). The proof idea is similar to the previous lemma. First, using properties of idempotent flows, one shows that all output positions strictly between B_k and B_{k+1}, for any $k = 1, \ldots, n-1$, have origin in $I_k \cup I_{k+1}$. Then, one observes that every output block of I_k disjoint from B_k is **C**-small, and since T is K-visiting there are at most K such blocks. This shows that every output interval J_k between B_k and B_{k+1} is $2K\mathbf{C}$-small. For an illustration see the figure to the right. The **C**-large blocks in I_1 are shown in red; in blue those for I_2, in purple those for I_3. So $bigout(I_1)$ is the entire output between the two red dots, $bigout(I_2)$ between the two blue dots, and $bigout(I_3)$ between the purple dots. All three blocks are non-empty, and $bigout(I_1 \cdot I_2 \cdot I_3)$ goes from the first red to the second purple dot. Black non-dashed arrows stand for **C**-small blocks. □

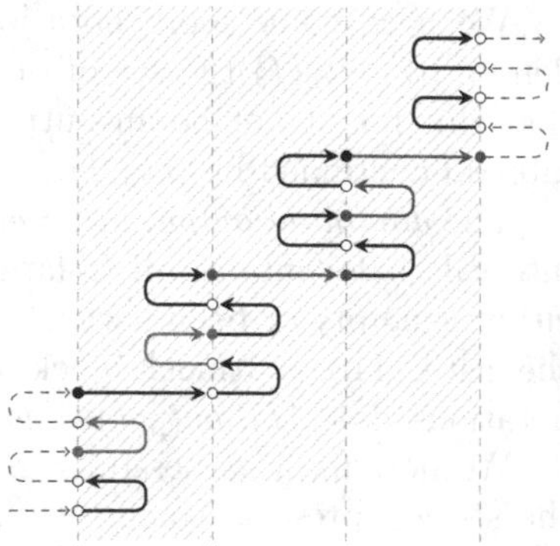

5 Proof of Theorem 1

This section is devoted to proving the characterization of one-way resynchronizability in the bounded-visit case. We will use the notion of *bounded-traversal* from [21], that was shown to characterize the class of bounded regular resynchronizers, in as much as bounded-delay characterizes rational resynchronizers [15].

Definition 4 (traversal [21]). *Let $\sigma = (u, v, orig)$ and $\sigma' = (u, v, orig')$ be two synchronized pairs with the same input and output words.*

Given two input positions $y, y' \in \mathsf{dom}(u)$, we say that y traverses *y' if there is a pair (y, z) of source and target origins associated with the same output position such that y' is between y and z, with $y' \neq z$ and possibly $y' = y$. More precisely:*

- *(y, y') is a* left-to-right traversal *if $y \leq y'$ and for some output position x, $orig(x) = y$ and $z = orig'(x) > y'$;*
- *(y, y') is a* right-to-left traversal *if $y \geq y'$ and for some output position x, $orig(x) = y$ and $z = orig'(x) < y'$.*

A pair (σ, σ') of synchronized pairs with input u and output v is said to have k-bounded traversal, *with $k \in \mathbb{N}$, if every $y' \in \mathsf{dom}(u)$ is traversed by at most k distinct positions of $\mathsf{dom}(u)$.*

A resynchronizer $\mathcal{R}$ has bounded traversal *if there is some $k \in \mathbb{N}$ such that every $(\sigma, \sigma') \in \mathcal{R}$ has k-bounded traversal.*

Lemma 8 ([21]). *A regular resynchronizer is bounded if and only if it has bounded traversal.*

Proof (of Theorem 1). First of all, observe that the implication $4 \rightarrow 1$ is straightforward. To prove the implication $1 \rightarrow 2$, assume that there is a k-bounded, regular resynchronizer $\mathcal{R}$ that is T-preserving and such that $\mathcal{R}(T)$ is order-preserving. Lemma 8 implies that $\mathcal{R}$ has t-bounded traversal, for some constant t. We head towards proving that T has cross-width bounded by $t + k$. Consider two synchronized pairs $\sigma = (u, v, orig)$ and $\sigma' = (u, v, orig')$ such that $\sigma \in [\![T]\!]_o$ and $(\sigma, \sigma') \in \mathcal{R}$, and consider a cross (X_1, X_2) of σ. We claim that $|orig(X_1)|$ or $|orig(X_2)|$ is at most $t + k$. Let $x_1 = \min(orig(X_1))$, $x'_1 = \max(orig'(X_1))$, $x_2 = \max(orig(X_1))$, and $x'_2 = \min(orig'(X_2))$. Since (X_1, X_2) is a cross, we have $x_1 > x_2$, and since σ' is order-preserving, we have $x'_1 \leq x'_2$. Now, if $x'_1 > x_2$, then at least $|orig(X_2)| - k$ input positions from X_2 traverse x'_1 to the right (the $-k$ term is due to the fact that at most k input positions can be resynchronized to x'_1). Symmetrically, if $x'_1 \leq x_2$, then at least $|orig(X_1)| - k$ input positions from X_1 traverse x_2 to the left (the $-k$ term accounts for the case where some positions are resynchronized to x'_1 and $x'_1 = x_2$). This implies $\min(|orig(X_1)|, |orig(X_2)|) \leq t + k$, as claimed.

The remaining implications rely on the assumption that T is bounded-visit.

The implication $2 \rightarrow 3$ is shown by contraposition: one considers a successful run ρ with an inversion, and shows that crosses of arbitrary width emerge after pumping the loops of the inversion (here Lemma 3 is crucial).

The proof of 3 → 4 is more involved, we only sketch it here. Assuming that no successful run of T has inversions we build a partially bijective, regular resynchronizer $\mathcal{R}$ that is T-preserving and $\mathcal{R}(T)$ is order-preserving. The resynchronizer $\mathcal{R}$ uses some parameters to guess a successful run ρ of T on u and a factorization tree of bounded height for ρ. Formally, a *factorization tree* for a sequence α of monoid elements (e.g. the flows $\mathit{flow}_\rho([y, y])$ for all input positions y) is an ordered, unranked tree whose yield is the sequence α. The leaves of the factorization tree are labeled with the elements of α. All other nodes have at least two children and are labeled by the monoid product of the child labels (in our case by the flows of ρ induced by the covered factors in the input). In addition, if a node has more than two children, then all its children must have the same label, representing an idempotent element of the monoid. By Simon's factorization theorem [23], every sequence of monoid elements has some factorization tree of height at most linear in the size of the monoid (in our case, at most $3|M_T|$, see e.g. [8]).

Parameters. We use input parameters to encode the successful run ρ and a factorization tree for ρ of height at most $H = 3|M_T|$. These parameters specify, for each input interval corresponding to a subtree, the start and end positions of the interval and the label of the root of the subtree. Correctness of these annotations can be enforced by an MSO sentence ipar. The run and the factorization tree also need to be encoded over the output, using output parameters. More precisely, given a level in the tree and an output position, we need to be able to determine the flow and the productive edge that generated that position. We omit the technical details for checking correctness of the output annotation using the formulas opar, move_τ and $\mathsf{next}_{\tau,\tau'}$.

Moving origins. For each level ℓ of the factorization tree, a partial resynchronization relation $\mathcal{R}_\ell$ is defined. The relation is partial in the sense that some output positions may not have a source-target origin pair defined at a given level. But once a source-target pair is defined for some output position at a given level, it remains defined for all higher levels.

In the following we write $\mathit{bigout}(p)$ for the dominant output interval associated with the input interval $I(p)$ corresponding to a node p in the tree. For every level ℓ of the factorization tree, the resynchronizer $\mathcal{R}_\ell$ will be a partial function from source origins to target origins, and will satisfy the following:

- the set of output positions for which $\mathcal{R}_\ell$ defines target origins is the union of the intervals $\mathit{bigout}(p)$ for all nodes p at level ℓ;
- $\mathcal{R}_\ell$ only moves origins within the same interval at level ℓ, that is, $\mathcal{R}_\ell$ defines only pairs (y, z) of source-target origins such that $y, z \in I(p)$ for some node p at level ℓ;
- the target origins defined by $\mathcal{R}_\ell$ are order-preserving within every interval at level ℓ, that is, for all output positions $x < x'$, if $\mathcal{R}_\ell$ defines the target origins of x, x' to be z, z', respectively, and if $z, z' \in I(p)$ for some node p at level ℓ, then $z \leq z'$;
- $\mathcal{R}_\ell$ is $\ell \cdot 4K\mathbf{C}$-bounded, namely, there are at most $\ell \cdot 4K\mathbf{C}$ distinct source origins that are moved by $\mathcal{R}_\ell$ to the same target origin.

The construction of $\mathcal{R}_\ell$ is by induction on ℓ. For a binary node p at level ℓ with children p_1, p_2, the resynchronizer $\mathcal{R}_\ell$ inherits the source-origin pairs from level $\ell - 1$ for output positions that belong to $bigout(p_1) \cup bigout(p_2)$. Note that $bigout(p_1) < bigout(p_2)$ by Lemma 5, so $\mathcal{R}_\ell$ is order-preserving inside $bigout(p_1) \cup bigout(p_2)$. Output positions inside $bigout(p) \setminus (bigout(p_1) \cup bigout(p_2))$ are moved in an order-preserving manner to one of the extremities of $I(p)$, or to the last position of $I(p_1)$. Boundedness of $\mathcal{R}_\ell$ is guaranteed by Lemma 6.

The case where p is an idempotent node at level ℓ with children $p_1, p_2, \ldots, p_n$ follows a similar approach. For brevity, let $I_i = I(p_i)$ and $B_i = bigout(p_i)$, and observe that, by Lemma 5, $B_1 < B_2 < \cdots < B_n$. Lemma 7 provides a decomposition of $bigout(p)$ as $B_1 \cdot J_1 \cdot B_2 \cdot J_2 \cdot \ldots \cdot J_{n-1} \cdot B_n$, for some $2K\mathbf{C}$-small output intervals J_k with origins inside $I_k \cup I_{k+1}$, for $k = 1, \ldots, n-1$. As before, the resynchronizer $\mathcal{R}_\ell$ behaves exactly as $\mathcal{R}_{\ell-1}$ for the output positions inside the B_k's. For any other output position, say $x \in J_k$, the resynchronizer $\mathcal{R}_\ell$ will move the origin either to the last position of I_k or to the first position of I_{k+1}, depending on whether the source origin of x belongs to I_k or I_{k+1}. □

6 Proof overview of Theorem 2

The main obstacle towards dropping the bounded-visit restriction from Theorem 1, while maintaining the effectiveness of the characterization, is the lack of a bound on the number of flows. Indeed, for a transducer T that is not necessarily bounded-visit, there is no bound on the number of flows that encode successful runs of T, and thus the proofs of the implications $2 \to 3 \to 4$ are not applicable anymore. However, the proofs of the implications $1 \to 2$ and $4 \to 1$ remain valid, even for a transducer T that is not bounded-visit.

The idea for proving Theorem 2 is to transform T into an equivalent bounded-visit transducer $low(T)$, so that the property of one-way resynchronizability is preserved. More precisely, given a two-way transducer T, we construct:

1. a bounded-visit transducer $low(T)$ that is classically equivalent to T,
2. a 1-bounded, regular resynchronizer $\mathcal{R}$ that is T-preserving and such that $\mathcal{R}(T) =_o low(T)$.

We can apply our characterization of one-way resynchronizability in the bounded-visit case to the transducer $low(T)$. If $low(T)$ is one-way resynchronizable, then by Theorem 1 we obtain another partially bijective, regular resynchronizer $\mathcal{R}'$ that is $low(T)$-preserving and such that $\mathcal{R}'(low(T)))$ is order-preserving. Thanks to Lemma 2, the resynchronizers $\mathcal{R}$ and $\mathcal{R}'$ can be composed, so we conclude that the original transducer T is one-way resynchronizable. Otherwise, if $low(T)$ is not one-way resynchronizable, we show that neither is T. This is precisely shown in the lemma below.

Lemma 9. *For all transducers T, T', with T' bounded-visit, and for every partially bijective, regular resynchronizer $\mathcal{R}$ that is T-preserving and such that*

$\mathcal{R}(T) =_o T'$, *T is one-way resynchronizable if and only if T' is one-way resynchronizable.*

There are however some challenges in the approach described above. First, as T may output arbitrarily many symbols with origin in the same input position, and $low(T)$ is bounded-visit, we need $low(T)$ to be able to produce arbitrarily long outputs within a single transition. For this reason, we allow $low(T)$ to be a transducer with *regular outputs*. The transition relation of such a transducer consists of finitely many tuples of the form (q, a, L, q'), with $q, q' \in Q$, $a \in \Sigma$, and $L \subseteq \Gamma^*$ a regular language over the output alphabet. The semantics of a transition rule (q, a, L, q') is that, upon reading a, the transducer can switch from state q to state q', and move its head accordingly, while outputting any word from L. We also need to use transducers with common guess. Both extensions, regular outputs and common guess, already appeared in prior works (cf. [5,7]), and the proof of Theorem 1 in the bounded-visit case can be easily adapted to these features.

There is still another problem: we cannot always expect that there exists a bounded-visit transducer $low(T)$ classically equivalent to T. Consider, for instance, the transducer that performs several passes on the input, and on each left-to-right pass, at an arbitrary input position, it copies as output the letter under its head. It is easy to see that the Parikh image of the output is an exact multiple of the Parikh image of the input, and standard pumping arguments show that no bounded-visit transducer can realize such a relation.

A solution to this second problem is as follows. Before trying to construct $low(T)$, we test whether T satisfies the following condition on vertical loops (these are runs starting and ending at the same position and at the same state). There should exist some K such that T is K*-sparse*, meaning that the number of different origins of outputs generated inside some vertical loop is at most K. If this condition is not met, then we show that T has unbounded cross-width, and hence, by the implication $1 \to 2$ of Theorem 1, T is not one-way resynchronizable. Otherwise, if the condition holds, then we show that a bounded-visit transducer $low(T)$ equivalent to T can indeed be constructed.

7 Complexity

We discuss the effectiveness and complexity of our characterization. For a k-visit transducer T, the effectiveness of the characterization relies on detecting inversions in successful runs of T. It is not difficult to see that this can be decided in space that is polynomial in the size of T and the bound k. We can also show that one-way resynchronizability is PSPACE-hard. For this we recall that the emptiness problem for two-way finite automata is PSPACE-complete. Let A be a two-way automaton accepting some language L, and let Σ be a binary alphabet disjoint from that of L. The function $\{(w \cdot a_1 \dots a_n, a_n \dots a_1) \mid w \in L, a_1 \dots a_n \in \Sigma^*, n \geq 0\}$ can be realized by a two-way transducer T of size polynomial in $|A|$, and T is one-way resynchronizable if and only if L is empty.

In the unrestricted case, we showed that one-way resynchronizability is decidable (Theorem 2). We briefly outline the complexity of the decision procedure:

1. First one checks that T is K-sparse for some K. To do this, we construct from T the regular language L of all inputs with some positions marked that correspond to origins produced within the same vertical loop. Bounded sparsity is equivalent to having a uniform bound on the number of marked positions in every input from L. Standard techniques for two-way automata allow to decide this in space that is polynomial in the size of T. Moreover, this also gives us a computable exponential bound to the largest constant K for which T can be K-sparse.
2. Next, we construct from the K-sparse transducer T a bounded-visit transducer T' that is classically equivalent to T and has exponential size.
3. Finally, we decide one-way resynchronizability of T' by detecting inversions in successful runs of T' (Theorem 1).

Summing up, one can decide one-way resynchronizability of unrestricted two-way transducers in exponential space. It is open if this bound is optimal. We also do not have any interesting bound on the size of the resynchronizer that witnesses one-way resynchronizability, both in the bounded-visit case and in the unrestricted case. Similarly, we lack upper and lower bounds on the size of the resynchronized one-way transducers, when these exist.

8 Conclusions

As the main contribution of this paper, we provided a characterization for the subclass of two-way transducers that are one-way resynchronizable, namely, that can be transformed by some bounded, regular resynchronizer, into an origin-equivalent one-way transducer.

There are similar definability problems that emerge in the origin semantics. For instance, one could ask whether a given two-way transducer can be resynchronized, through some bounded, regular resynchronization, to a relation that is origin-equivalent to a first-order transduction. This can be seen as a relaxation of the first-order definability problem in the origin semantics, namely, the problem of telling whether a two-way transducer is origin-equivalent to some first-order transduction, shown decidable in [4]. It is worth contrasting the latter problem with the challenging open problem whether a given transduction is equivalent to a first-order transduction in the classical setting.

Acknowledgments. We thank the FoSSaCS reviewers for their constructive and useful comments.

References

1. Rajeev Alur and Pavel Cerný. Expressiveness of streaming string transducer. In *IARCS Annual Conference on Foundation of Software Technology and Theoretical*

Computer Science (FSTTCS'10), volume 8 of *LIPIcs*, pages 1–12. Schloss Dagstuhl - Leibniz-Zentrum für Informatik, 2010.

2. Félix Baschenis, Olivier Gauwin, Anca Muscholl, and Gabriele Puppis. One-way definability of sweeping transducers. In *IARCS Annual Conference on Foundation of Software Technology and Theoretical Computer Science (FSTTCS'15)*, volume 45 of *LIPIcs*, pages 178–191. Schloss Dagstuhl - Leibniz-Zentrum für Informatik, 2015.
3. Félix Baschenis, Olivier Gauwin, Anca Muscholl, and Gabriele Puppis. One-way definability of two-way word transducers. *Logical Methods in Computer Science*, 14(4):1–54, 2018.
4. Mikolaj Bojańczyk. Transducers with origin information. In *International Colloquium on Automata, Languages and Programming (ICALP'14)*, number 8572 in LNCS, pages 26–37. Springer, 2014.
5. Mikolaj Bojańczyk, Laure Daviaud, Bruno Guillon, and Vincent Penelle. Which classes of origin graphs are generated by transducers? In *International Colloquium on Automata, Languages and Programming (ICALP'17)*, volume 80 of *LIPIcs*, pages 114:1–114:13. Schloss Dagstuhl - Leibniz-Zentrum für Informatik, 2017.
6. Sougata Bose, Shankara Narayanan Krishna, Anca Muscholl, Vincent Penelle, and Gabriele Puppis. On synthesis of resynchronizers for transducers. In *International Symposium on Mathematical Foundations of Computer Science (MFCS'19)*, volume 138 of *LIPIcs*, pages 69:1–69:14. Schloss Dagstuhl - Leibniz-Zentrum für Informatik, 2019.
7. Sougata Bose, Anca Muscholl, Vincent Penelle, and Gabriele Puppis. Origin-equivalence of two-way word transducers is in PSPACE. In *IARCS Annual Conference on Foundations of Software Technology and Theoretical Computer Science (FSTTCS'18)*, volume 122 of *LIPIcs*, pages 1–18. Schloss Dagstuhl - Leibniz-Zentrum für Informatik, 2018.
8. Thomas Colcombet. Factorisation forests for infinite words. In *Fundamentals of Computation Theory (FCT)*, volume 4639 of *LNCS*, pages 226–237. Springer, 2007.
9. Bruno Courcelle and Joost Engelfriet. *Graph Structure and Monadic Second-Order Logic - A Language-Theoretic Approach*, volume 138 of *Encyclopedia of mathematics and its applications*. Cambridge University Press, 2012.
10. Luc Dartois, Ismaël Jecker, and Pierre-Alain Reynier. Aperiodic string transducers. *Int. J. Found. Comput. Sci.*, 29(5):801–824, 2018.
11. Joost Engelfriet and Hendrik Jan Hoogeboom. MSO definable string transductions and two-way finite-state transducers. *ACM Trans. Comput. Log.*, 2(2):216–254, 2001.
12. Joost Engelfriet and Hendrik Jan Hoogeboom. Finitary compositions of two-way finite-state transductions. *Fundamenta Informaticae*, 80:111–123, 2007.
13. Emmanuel Filiot, Olivier Gauwin, and Nathan Lhote. Logical and algebraic characterizations of rational transductions. *Logical Methods in Computer Science*, 15(4), 2019.
14. Emmanuel Filiot, Olivier Gauwin, Pierre-Alain Reynier, and Frédéric Servais. From two-way to one-way finite state transducers. In *ACM/IEEE Symposium on Logic in Computer Science (LICS'13)*, pages 468–477, 2013.
15. Emmanuel Filiot, Ismaël Jecker, Christof Löding, and Sarah Winter. On equivalence and uniformisation problems for finite transducers. In *Proc. of nternational Colloquium on Automata, Languages, and Programming (ICALP'16)*, number 125 in LIPIcs, pages 1–14. Schloss Dagstuhl - Leibniz-Zentrum für Informatik, 2016.

16. Emmanuel Filiot, Shankara Narayanan Krishna, and Ashutosh Trivedi. First-order definable string transformations. In *IARCS Annual Conference on Foundations of Software Technology and Theoretical Computer Science (FSTTCS'14)*, LIPIcs, pages 147–159. Schloss Dagstuhl - Leibniz-Zentrum für Informatik, 2014.
17. Emmanuel Filiot, Sebastian Maneth, Pierre-Alain Reynier, and Jean-Marc Talbot. Decision problems of tree transducers with origin. *Inf. Comput.*, 261(Part):311–335, 2018.
18. T. V. Griffiths. The unsolvability of the equivalence problem for lambda-free nondeterministic generalized machines. *J. ACM*, 15(3):409–413, 1968.
19. Oscar H. Ibarra. The unsolvability of the equivalence problem for e-free NGSM's with unary input (output) alphabet and applications. *SIAM J. of Comput.*, 7(4):524–532, 1978.
20. Ismael Jecker. Personal communication.
21. Denis Kuperberg and Jan Martens. Regular resynchronizability of origin transducers is undecidable. In *International Symposium on Mathematical Foundations of Computer Science (MFCS'20)*, volume 170 of *LIPIcs*, pages 1–14. Schloss Dagstuhl - Leibniz-Zentrum für Informatik, 2020.
22. John C. Shepherdson. The reduction of two-way automata to one-way automata. *IBM Journal of Research and Development*, 3(2):198–200, 1959.
23. Imre Simon. Factorization forests of finite height. *Theoretical Computer Science*, 72(1):65–94, 1990.

Fair Refinement for Asynchronous Session Types*

Mario Bravetti[1], Julien Lange[2] (✉), and Gianluigi Zavattaro[1]

[1] University of Bologna / INRIA FoCUS Team, Bologna, Italy
{mario.bravetti,gianluigi.zavattaro}@unibo.it
[2] Royal Holloway, University of London, Egham, UK
julien.lange@rhul.ac.uk

Abstract. Session types are widely used as abstractions of asynchronous message passing systems. Refinement for such abstractions is crucial as it allows improvements of a given component without compromising its compatibility with the rest of the system. In the context of session types, the most general notion of refinement is the asynchronous session subtyping, which allows to anticipate message emissions but only under certain conditions. In particular, asynchronous session subtyping rules out candidates subtypes that occur naturally in communication protocols where, e.g., two parties simultaneously send each other a finite but unspecified amount of messages before removing them from their respective buffers. To address this shortcoming, we study fair compliance over asynchronous session types and fair refinement as the relation that preserves it. This allows us to propose a novel variant of session subtyping that leverages the notion of controllability from service contract theory and that is a sound characterisation of fair refinement. In addition, we show that both fair refinement and our novel subtyping are undecidable. We also present a sound algorithm, and its implementation, which deals with examples that feature potentially unbounded buffering.

Keywords: Session types · Asynchronous communication · Subtyping.

1 Introduction

The coordination of software components via message-passing techniques is becoming increasingly popular in modern programming languages and development methodologies based on actors and microservices, e.g., Rust, Go, and the Twelve-Factor App methodology [1]. Often the communication between two concurrent or distributed components takes place over point-to-point FIFO channels.

Abstract models such as communicating finite-state machines [5] and asynchronous session types [21] are essential to reason about the correctness of such systems in a rigorous way. In particular these models are important to reason about mathematically grounded techniques to improve concurrent and distributed systems in a compositional way. The key question is whether a component can be *refined* independently of the others, without compromising the

* Research partly supported by the H2020-MSCA-RISE project ID 778233 "Behavioural Application Program Interfaces (BEHAPI)".

S. Kiefer and C. Tasson (Eds.): FOSSACS 2021, LNCS 12650, pp. 144–163, 2021.
https://doi.org/10.1007/978-3-030-71995-1_8

correctness of the whole system. In the theory of session types, the most general notion of refinement is the asynchronous session subtyping [14, 15, 26], which leverages asynchrony by allowing the refined component to anticipate message emissions, but only under certain conditions. Notably asynchronous session subtyping rules out candidate subtypes that occur naturally in communication protocols where, e.g., two parties simultaneously send each other a finite but unspecified amount of messages before removing them from their buffers.

We illustrate this key limitation of asynchronous session subtyping with Figure 1, which depicts possible communication protocols between a spacecraft and a ground station. For convenience, the protocols are represented as session types (bottom) and equivalent communicating finite-state machines (top). Consider T_S and T_G first. Session type T_S is the abstraction of the spacecraft. It may send a finite but unspecified number of telemetries (*tm*), followed by a message *over* — this phase of the protocol typically models a `for` loop and its exit. In the second phase, the spacecraft receives a number of telecommands (*tc*), followed by a message *done*. Session type T_G is the abstraction of the ground station. It is the *dual* of T_S, written $\overline{T_S}$, as required in standard binary session types without subtyping. Since T_G and T_S are dual of each other, the theory of session types guarantees that they form a *correct composition*, namely both parties terminate successfully, with empty queues.

However, it is clear that this protocol is not efficient: the communication is half-duplex, i.e., it is never the case that more than one party is sending at any given time. Using full-duplex communication is crucial in distributed systems with intermittent connectivity, e.g., in this case ground stations are not always visible from low orbit satellites.

The abstraction of a more efficient ground station is given by type T'_G, which sends telecommands before receiving telemetries. It is clear that T'_G and T_S forms a correct composition. Unfortunately T'_G is not an asynchronous subtype of T_G according to earlier definitions of session subtyping [14,15,26]. Hence they cannot formally guarantee that T'_G is a safe replacement for T_G. Concretely, these subtyping relations allow for anticipation of emissions (output) only when they are preceded by a *bounded* number of receptions (input), but this does not hold between T'_G and T_G because the latter starts with a loop of inputs. Note that the composition of T'_G and T_S is not existentially bounded, hence it cannot be verified by related communicating finite-state machines techniques [4,19,20,24].

In this paper we address this limitation of previous asynchronous session subtyping relations. To do this, we move to an alternative notion of correct composition. In [14] the authors show that their subtyping relation is fully abstract w.r.t. the notion of *orphan-message-free* composition. More precisely, it captures exactly a notion of refinement that preserves the possibility for all sent messages to be consumed along *all* possible computations of the receiver. In the spacecraft example, given the initial loop of outputs in T'_G, there is an extreme case in which it performs infinitely many outputs without consuming any incoming messages. Nevertheless, this limit case cannot occur under the natural assumption that

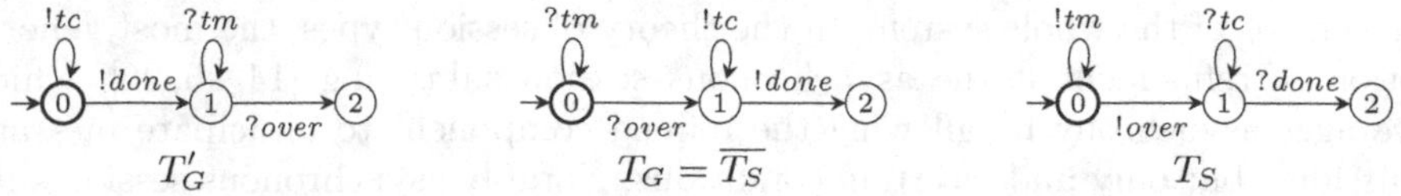

$$T'_G = \mu\mathbf{t}.\oplus\{tc : \mathbf{t}, done : \mu\mathbf{t}'.\&\{tm : \mathbf{t}', over : \mathbf{end}\}\}$$
$$T_G = \mu\mathbf{t}.\&\{tm : \mathbf{t}, over : \mu\mathbf{t}'.\oplus\{tc : \mathbf{t}', done : \mathbf{end}\}\}$$
$$T_S = \mu\mathbf{t}.\oplus\{tm : \mathbf{t}, over : \mu\mathbf{t}'.\&\{tc : \mathbf{t}', done : \mathbf{end}\}\}$$

Fig. 1. Satellite protocols. T'_G is the refined session type of the ground station, T_G is the session type of ground station, and T_S is the session type of the spacecraft.

the loop of outputs eventually terminates, i.e., only a finite (but unspecified) amount of messages can be emitted.

The notion of correct composition that we use is based on *fair* compliance, which requires each component to always be able to eventually reach a successful final state. This is a liveness property, holding under *full fairness* [32], used also in the theory of should testing [30] where "every reachable state is required to be on a path to success". This is a natural constraint since even programs that conceptually run indefinitely must account for graceful termination (e.g., to release acquired resources). Previously, fair compliance has been considered to reason formally about component/service composition with *synchronous* session types [29] and *synchronous* behavioural contracts [11]. A preliminary formalisation of fair compliance for *asynchronous* behavioural contracts was presented in [10], but considering an operational model very different from session types.

Given a notion of fair compliance defined on an operational model for asynchronous session types, we define *fair refinement* as the relation that preserves it. Then, we propose a novel variant of session subtyping called *fair asynchronous session subtyping*, that leverages the notion of controllability from service contract theory, and which is a sound characterisation of fair refinement. We show that both fair refinement and fair asynchronous session subtyping are undecidable, but give a sound algorithm for the latter. Our algorithm covers session types that exhibit complex behaviours (including the spacecraft example and variants). Our algorithm has been implemented in a tool available online [31].

Structure of the paper The rest of this paper is structured as follows. In § 2 we recall syntax and semantics of asynchronous session types, we define *fair compliance* and the corresponding *fair refinement*. In § 3 we introduce *fair asynchronous subtyping*, the first relation of its kind to deal with examples such as those in Figure 1. In § 4 we propose a sound algorithm for subtyping that supports examples with unbounded accumulations, including the ones discussed in this paper. In § 5 we discuss the implementation of this algorithm. Finally, in § 6 we discuss related works and future work. We give proofs for all our results and examples of output from our tool in [9].

2 Refinement for Asynchronous Session Types

In this section we first recall the syntax of two-party session types, their reduction semantics, and a notion of compliance centred on the successful termination of interactions. We define our notion of refinement based on this compliance and show that it is generally undecidable whether a type is a refinement of another.

2.1 Preliminaries: Asynchronous Session Types

Syntax The formal syntax of two-party session types is given below. We follow the simplified notation used in, e.g., [7,8], without dedicated constructs for sending an output/receiving an input. Additionally we abstract away from message payloads since they are orthogonal to the results of this paper.

Definition 1 (Session Types). *Given a set of labels $\mathcal{L}$, ranged over by l, the syntax of two-party session types is given by the following grammar:*

$$T ::= \oplus\{l_i : T_i\}_{i\in I} \quad | \quad \&\{l_i : T_i\}_{i\in I} \quad | \quad \mu\mathbf{t}.T \quad | \quad \mathbf{t} \quad | \quad \mathbf{end}$$

Output selection $\oplus\{l_i : T_i\}_{i\in I}$ represents a guarded internal choice, specifying that a label l_i is sent over a channel, then continuation T_i is executed. Input branching $\&\{l_i : T_i\}_{i\in I}$ represents a guarded external choice, specifying a protocol that waits for messages. If message l_i is received, continuation T_i takes place. In selections and branchings each branch is tagged by a label l_i, taken from a global set of labels $\mathcal{L}$. In each selection/branching, these labels are assumed to be pairwise distinct. In the sequel, we leave implicit the index set $i \in I$ in input branchings and output selections when it is clear from the context. Types $\mu\mathbf{t}.T$ and $\mathbf{t}$ denote standard recursion constructs. We assume recursion to be guarded in session types, i.e., in $\mu\mathbf{t}.T$, the recursion variable $\mathbf{t}$ occurs within the scope of a selection or branching. Session types are closed, i.e., all recursion variables $\mathbf{t}$ occur under the scope of a corresponding binder $\mu\mathbf{t}.T$. Terms of the session syntax that are not closed are dubbed (session) terms. Type **end** denotes the end of the interactions.

The dual of session type T, written $\overline{T}$, is inductively defined as follows: $\overline{\oplus\{l_i : T_i\}_{i\in I}} = \&\{l_i : \overline{T}_i\}_{i\in I}$, $\overline{\&\{l_i : T_i\}_{i\in I}} = \oplus\{l_i : \overline{T}_i\}_{i\in I}$, $\overline{\mathbf{end}} = \mathbf{end}$, $\overline{\mathbf{t}} = \mathbf{t}$, and $\overline{\mu\mathbf{t}.T} = \mu\mathbf{t}.\overline{T}$.

Operational characterisation Hereafter, we let ω range over words in $\mathcal{L}^*$, write ϵ for the empty word, and write $\omega_1 {\cdot} \omega_2$ for the concatenation of words ω_1 and ω_2, where each word may contain zero or more labels. Also, we write $T\{^{T'}/_{\mathbf{t}}\}$ for T where every free occurrence of $\mathbf{t}$ is replaced by T'.

We give an asynchronous semantics of session types via transition systems whose states are configurations of the form: $[T_1, \omega_1] \| [T_2, \omega_2]$ where T_1 and T_2 are session types equipped with two sequences ω_1 and ω_2 of incoming messages (representing unbounded buffers). We use s, s', etc. to range over configurations.

In this paper, we use explicit unfoldings of session types, as defined below.

Definition 2 (Unfolding). *Given session type T, we define* $\mathsf{unfold}(T)$*:*

$$\mathsf{unfold}(T) = \begin{cases} \mathsf{unfold}(T'\{T/\mathbf{t}\}) & \text{if } T = \mu\mathbf{t}.T' \\ T & \text{otherwise} \end{cases}$$

Definition 2 is standard, e.g., an equivalent function is used in the first session subtyping [18]. Notice that $\mathsf{unfold}(T)$ unfolds all the recursive definitions in front of T, and it is well defined for session types with guarded recursion.

Definition 3 (Transition Relation). *The transition relation $\rightarrow$ over configurations is the minimal relation satisfying the rules below (plus symmetric ones):*

1. *if* $j \in I$ *then* $[\oplus\{l_i : T_i\}_{i\in I}, \omega_1]||[T_2, \omega_2] \rightarrow [T_j, \omega_1]||[T_2, \omega_2{\cdot}l_j]$;
2. *if* $j \in I$ *then* $[\&\{l_i : T_i\}_{i\in I}, l_j{\cdot}\omega_1]||[T_2, \omega_2] \rightarrow [T_j, \omega_1]||[T_2, \omega_2]$;
3. *if* $[\mathsf{unfold}(T_1), \omega_1]||[T_2, \omega_2] \rightarrow s$ *then* $[T_1, \omega_1]||[T_2, \omega_2] \rightarrow s$.

We write $\rightarrow^*$ for the reflexive and transitive closure of the $\rightarrow$ relation.

Intuitively a configuration s reduces to configuration s' when either (1) a type outputs a message l_j, which is added at the end of its partner's queue; (2) a type consumes an expected message l_j from the head of its queue; or (3) the unfolding of a type can execute one of the transitions above.

Next, we define successful configurations as those configurations where both types have terminated (reaching **end**) and both queues are empty. We use this to give our definition of compliance which holds when it is possible to reach a successful configuration from all reachable configurations.

Definition 4 (Successful Configuration). *The notion of successful configuration is formalised by a predicate $s\surd$ defined as follows:*

$$[T, \omega_T]||[S, \omega_S]\surd \;\; \textit{iff} \;\; \mathsf{unfold}(T) = \mathsf{unfold}(S) = \mathbf{end} \;\; \textit{and} \;\; \omega_T = \omega_S = \epsilon$$

Definition 5 (Compliance). *Given a configuration s we say that it is a correct composition if, whenever $s \rightarrow^* s'$, there exists a configuration s'' such that $s' \rightarrow^* s''$ and $s''\surd$.*
Two session types T and S are compliant *if $[T, \epsilon]||[S, \epsilon]$ is a correct composition.*

Observe that our definition of compliance is stronger than what is generally considered in the literature on session types, e.g., [16, 23, 24], where two types are deemed compliant if all messages that are sent are eventually received, and each non-terminated type can always eventually make a move. Compliance is analogous to the notion of *correct session* in [29] but in an asynchronous setting.

A consequence of Definition 5 is that it is generally *not* the case that a session type T is compliant with its dual $\overline{T}$, as we show in the example below.

Example 1. The session type $T = \&\{l_1 : \mathbf{end},\ l_2 : \mu\mathbf{t}.\oplus\{l_3 : \mathbf{t}\}\}$ and its dual $\overline{T} = \oplus\{l_1 : \mathbf{end},\ l_2 : \mu\mathbf{t}.\&\{l_3 : \mathbf{t}\}\}$ are not compliant. Indeed, when $\overline{T}$ sends label l_2, the configuration $[\mathbf{end}, \epsilon]||[\mathbf{end}, \epsilon]$ is no longer reachable.

2.2 Fair Refinement for Asynchronous Session Types

We introduce a notion of refinement that preserves compliance. This follows previous work done in the context of behavioural contracts [11] and *synchronous* multi-party session types [29]. The key difference with these works is that we are considering asynchronous communication based on (unbounded) FIFO queues. Asynchrony makes fair refinement undecidable, as we show below.

Definition 6 (Refinement). *A session type T refines S, written $T \sqsubseteq S$, if for every S' s.t. S and S' are compliant then T and S' are also compliant.*

In contrast to traditional (synchronous and asynchronous) subtyping for session types [14,18,26], this refinement is not covariant on outputs, i.e., it does not always allow a refined type to have output selections with less labels.[3]

Example 2. Let $T = \mu\mathbf{t}. \oplus\{l_1 : \mathbf{t}\}$ and $S = \mu\mathbf{t}. \oplus\{l_1 : \mathbf{t},\ l_2 : \mathbf{end}\}$. We have that T is a synchronous (and asynchronous) subtype of S. However T is *not* a refinement of S. In particular, the type $\overline{S} = \mu\mathbf{t}. \&\{l_1 : \mathbf{t},\ l_2 : \mathbf{end}\}$ is compliant with S but not with T, since T does not terminate.

Next, we show that the refinement relation $\sqsubseteq$ is generally undecidable. The proof of undecidability exploits results from the tradition of computability theory, i.e., Turing completeness of queue machines. The crux of the proof is to reduce the problem of checking the reachability of a given state in a queue machine to the problem of checking the refinement between two session types.

Preliminaries Below we consider only state reachability in queue machines, and not the typical notion of the language recognised by a queue machine (see, e.g., [7] for a formalisation of queue machines). Hence, we use a simplified formalisation, where no input string is considered.

Definition 7 (Queue Machine). *A queue machine M is defined by a six-tuple $(Q, \Sigma, \Gamma, \$, s, \delta)$ where:*

- *Q is a finite set of states;*
- *$\Sigma \subset \Gamma$ is a finite set denoting the input alphabet;*
- *Γ is a finite set denoting the queue alphabet (ranged over by A, B, C, X);*
- *$\$ \in \Gamma - \Sigma$ is the initial queue symbol;*
- *$s \in Q$ is the start state;*
- *$\delta : Q \times \Gamma \to Q \times \Gamma^*$ is the transition function (Γ^* is the set of sequences of symbols in Γ).*

Considering a queue machine $M = (Q, \Sigma, \Gamma, \$, s, \delta)$, a *configuration* of M is an ordered pair (q, γ) where $q \in Q$ is its *current state* and $\gamma \in \Gamma^*$ is the *queue*. The starting configuration is $(s, \$)$, composed of the start state s and the initial queue symbol $\$$.

Next, we define the transition relation $(\to_M)$, leading a configuration to another, and the related notion of state reachability.

[3] The synchronous subtyping in [18] follows a channel-oriented approach; hence it has the opposite direction and is contravariant on outputs.

Definition 8 (State Reachability). *Given a machine $M = (Q, \Sigma, \Gamma, \$, s, \delta)$, the transition relation $\rightarrow_M$ over configurations $Q \times \Gamma^*$ is defined as follows. For $p, q \in Q$, $A \in \Gamma$, and $\alpha, \gamma \in \Gamma^*$, we have $(p, A\alpha) \rightarrow_M (q, \alpha\gamma)$ whenever $\delta(p, A) = (q, \gamma)$. Let $\rightarrow_M^*$ be the reflexive and transitive closure of $\rightarrow_M$. A target state $q_f \in Q$ is* reachable *in M if there is $\gamma \in \Gamma^*$ s.t. $(s, \$) \rightarrow_M^* (q_f, \gamma)$.*

Since queue machines can deterministically encode Turing machines (see, e.g., [7]), checking state reachability for queue machines is undecidable.

Theorem 1. *Given a queue machine M and a target state q_f it is possible to reduce the problem of checking the reachability of q_f in M to the problem of checking refinement between two session types.*

In the light of the undecidability of reachability in queue machines, we can conclude that refinement (Definition 6) is also undecidable.

2.3 Controllability for Asynchronous Session Types

Given a notion of compliance, controllability amounts to checking the existence of a compliant partner (see, e.g., [12, 25, 33]). In our setting, a session type is *controllable* if there exists another session type with which it is compliant.

Checking for controllability algorithmically is not trivial as it requires to consider infinitely many potential partners. For the synchronous case, an algorithmic characterisation was studied in [29]. In the asynchronous case, the problem is even harder because each of the infinitely many potential partners may generate an infinite state computation (due to unbounded buffers). The main contribution of this subsection is to give an algorithmic characterisation of controllability in the asynchronous setting. Doing this is important because controllability is an essential ingredient for defining fair asynchronous subtyping, see Section 3.

Definition 9 (Characterisation of Controllability, $T\,\mathsf{ctrl}$). *Given a session type T, we define the judgement $T\,\mathsf{ok}$ inductively as follows:*

$$\frac{}{\mathbf{end}\,\mathsf{ok}} \qquad \frac{\mathbf{end} \in T \qquad T\{\mathbf{end}/\mathbf{t}\}\,\mathsf{ok}}{\mu\mathbf{t}.T\,\mathsf{ok}} \qquad \frac{T\,\mathsf{ok}}{\&\{l : T\}\,\mathsf{ok}} \qquad \frac{\forall i \in I.\ T_i\,\mathsf{ok}}{\oplus\{l_i : T_i\}_{i \in I}\,\mathsf{ok}}$$

where **end** $\in T$ *holds if* **end** *occurs in* T.

We write $T\,\mathsf{ctrl}$ if there exists T' such that (i) T' is obtained from T by syntactically replacing every input prefix $\&\{l_i : T_i\}_{i \in I}$ occurring in T with a term $\&\{l_j : T_j\}$ (with $j \in I$) and (ii) $T'\,\mathsf{ok}$ holds.

Notice that a type T such that $T\,\mathsf{ctrl}$ is indeed controllable, in that $\overline{T'}$, the dual of type T' considered above, is compliant with T (the predicate **end** $\in T$ in the premise of the rule for recursion guarantees that a successful configuration is always reachable while looping). Moreover the above definition naturally yields a simple algorithm that decides whether or not $T\,\mathsf{ctrl}$ holds for a type T, i.e., we first pick a single branch for each input prefix syntactically occurring in T (there are finitely many of them) and then we inductively check if $T'\,\mathsf{ok}$ holds.

The following theorem shows that the judgement $T\,\mathsf{ctrl}$, as defined above, precisely characterises controllability (i.e., the existence of a compliant type).

Theorem 2. *T ctrl holds if and only if there exists a session type S such that T and S are compliant.*

Example 3. Consider the session type $T = \mu\mathbf{t}.\ \&\{l_1 : \&\{l_2 : \oplus\{l_4 : \mathbf{end},\ l_5 : \mu\mathbf{t}'.\oplus\{l_6 : \mathbf{t}'\}\},\ l_3 : \mathbf{t}\}\}$. T ctrl does *not* hold because it is not possible to construct a T' as specified in Definition 9 for which T' ok holds. By Theorem 2, there is no session type S that is compliant with T. Hence T is not controllable.

3 Fair Asynchronous Session Subtyping

In this section, we present our novel variant of asynchronous subtyping which we dub *fair asynchronous subtyping.*

We need to define a distinctive notion of unfolding. Function selUnfold(T) unfolds type T by replacing recursion variables with their corresponding definitions only if they are guarded by an output selection. In the definition, we use the predicate $\oplus g(\mathbf{t}, T)$ which holds if all instances of variable $\mathbf{t}$ are output selection guarded, i.e., $\mathbf{t}$ occurs free in T only inside subterms $\oplus\{l_i : T_i\}_{i \in I}$.

Definition 10 (Selective Unfolding). *Given a term T, define* selUnfold(T) =

$$\begin{cases} \oplus\{l_i : T_i\}_{i\in I} & \text{if } T = \oplus\{l_i : T_i\}_{i\in I} \\ \&\{l_i : \mathsf{selUnfold}(T_i)\}_{i\in I} & \text{if } T = \&\{l_i : T_i\}_{i\in I} \\ T'\{^{\mu\mathbf{t}.T'}/_{\mathbf{t}}\} & \text{if } T = \mu\mathbf{t}.T',\ \oplus g(\mathbf{t}, T') \\ \mu\mathbf{t}.\mathsf{selUnfold}(\mathsf{selRepl}(\mathbf{t}, \hat{\mathbf{t}}, T')\{^{\mu\mathbf{t}.T'}/_{\hat{\mathbf{t}}}\}) \text{ with } \hat{\mathbf{t}} \text{ fresh} & \text{if } T = \mu\mathbf{t}.T',\ \neg\oplus g(\mathbf{t}, T') \\ \mathbf{t} & \text{if } T = \mathbf{t} \\ \mathbf{end} & \text{if } T = \mathbf{end} \end{cases}$$

where, selRepl($\mathbf{t}, \hat{\mathbf{t}}, T'$) *is obtained from T' by replacing the free occurrences of $\mathbf{t}$ that are inside a subterm $\oplus\{l_i : S_i\}_{i\in I}$ of T' by $\hat{\mathbf{t}}$.*

Example 4. Consider the type $T = \mu\mathbf{t}.\&\{l_1 : \mathbf{t},\ l_2 : \oplus\{l_3 : \mathbf{t}\}\}$, then we have

$$\mathsf{selUnfold}(T) = \mu\mathbf{t}.\&\{l_1 : \mathbf{t},\ l_2 : \oplus\{l_3 : \mu\mathbf{t}.\ \&\{l_1 : \mathbf{t},\ l_2 : \oplus\{l_3 : \mathbf{t}\}\}\}\}$$

i.e., the type is only unfolded within output selection sub-terms. Note that $\hat{\mathbf{t}}$ is used to identify where unfolding must take place, e.g.,
$\mathsf{selRepl}(\mathbf{t}, \hat{\mathbf{t}}, \&\{l_1 : \mathbf{t},\ l_2 : \oplus\{l_3 : \mathbf{t}\}\}) = \&\{l_1 : \mathbf{t},\ l_2 : \oplus\{l_3 : \hat{\mathbf{t}}\}\}$.

The last auxiliary notation required to define our notion of subtyping is that of *input contexts*, which are used to record inputs that may be delayed in a candidate super-type.

Definition 11 (Input Context). *An input context $\mathcal{A}$ is a session type with several holes defined by the syntax:*

$$\mathcal{A} ::= \quad [\,]^k \quad | \quad \&\{l_i : \mathcal{A}_i\}_{i\in I} \quad | \quad \mu\mathbf{t}.\mathcal{A} \quad | \quad \mathbf{t}$$

where the holes $[\,]^k$, *with* $k \in K$, *of an input context* $\mathcal{A}$ *are assumed to be pairwise distinct. We assume that recursion is guarded, i.e., in an input context* $\mu\mathbf{t}.\mathcal{A}$, *the recursion variable* $\mathbf{t}$ *must occur within a subterm* $\&\{l_i : \mathcal{A}_i\}_{i\in I}$.

We write holes$(\mathcal{A})$ *for the set of hole indices in* $\mathcal{A}$. *Given a type* T_k *for each* $k \in K$, *we write* $\mathcal{A}[T_k]^{k\in K}$ *for the type obtained by filling each hole* k *in* $\mathcal{A}$ *with the corresponding* T_k.

In contrast to previous work [6,7,13–15,26], these input contexts may contain recursive constructs. This is crucial to deal with examples such as Figure 1.

We are now ready to define the *fair asynchronous subtyping* relation, written $\leq$. The rationale behind asynchronous session subtyping is that under asynchronous communication it is unobservable whether or not an output is anticipated before an input, as long as this output is executed along all branches of the candidate super-type. Besides the usage of our new recursive input contexts the definition of fair asynchronous subtyping differs from those in [6,7,13–15,26] in that controllability plays a fundamental role: the subtype is not required to mimic supertype inputs leading to uncontrollable behaviours.

Definition 12 (Fair Asynchronous Subtyping, $\leq$).
A relation $\mathcal{R}$ *on session types is a controllable subtyping relation whenever* $(T, S) \in \mathcal{R}$ *implies:*

1. *if* $T = \mathbf{end}$ *then* $\mathsf{unfold}(S) = \mathbf{end}$;
2. *if* $T = \mu\mathbf{t}.T'$ *then* $(T'\{^{T}/_{\mathbf{t}}\}, S) \in \mathcal{R}$;
3. *if* $T = \&\{l_i : T_i\}_{i\in I}$ *then* $\mathsf{unfold}(S) = \&\{l_j : S_j\}_{j\in J}$, $I \supseteq K$, *and* $\forall k \in K.\,(T_k, S_k) \in \mathcal{R}$, *where* $K = \{k \in J \mid S_k$ *is controllable*$\}$;
4. *if* $T = \oplus\{l_i : T_i\}_{i\in I}$ *then* $\mathsf{selUnfold}(S) = \mathcal{A}[\oplus\{l_i : S_{ki}\}_{i\in I}]^{k\in K}$ *and* $\forall i \in I.\,(T_i, \mathcal{A}[S_{ki}]^{k\in K}) \in \mathcal{R}$.

T is a controllable subtype of S if there is a controllable subtyping relation $\mathcal{R}$ s.t. $(T, S) \in \mathcal{R}$.
T is a fair asynchronous subtype *of S, written $T \leq S$, whenever: S controllable implies that T is a controllable subtype of S.*

Notice that the top-level check for controllability in the above definition is consistent with the inner controllability checks performed in Case (3).

Subtyping simulation game Session type T is a fair asynchronous subtype of S if S is not controllable or if T is a controllable subtype of S. Intuitively, the above co-inductive definition says that it is possible to play a simulation game between a subtype T and its supertype S as follows. Case (1) says that if T is the **end** type, then S must also be **end**. Case (2) says that if T is a recursive definition, then it simply unfolds this definition while S does not need to reply. Case (3) says that if T is an input branching, then the sub-terms in S that are controllable can reply by inputting at most some of the labels l_i in the branching (contravariance of inputs), and the simulation game continues (see Example 5). Case (4) says that if T is an output selection, then S can reply by outputting *all* the labels l_i in the selection, possibly after executing some inputs, after which the simulation game continues. We comment further on Case (4) with Example 6.

Example 5. Consider $T = \&\{l_1 : \mathbf{end},\ l_2 : \mathbf{end}\}$ and $S = \&\{l_1 : \mathbf{end},\ l_3 : \mu\mathbf{t}.\oplus\{l_4 : \mathbf{t}\}\}$. We have $T \leq S$. Once branch l_3, that is uncontrollable, is removed from S, we can apply contravariance for input branching. We have $I = \{1, 2\} \supseteq \{1\} = K$ in Definition 12.

Example 6. Consider T_G and T'_G from Figure 1. For the pair (T'_G, T_G), we apply Case (4) of Definition 12 for which we compute

$$\mathsf{selUnfold}(T_G) = \mathcal{A}[\oplus\{tc : \mu\mathbf{t}'. \oplus \{tc : \mathbf{t}', done : \mathbf{end}\}, done : \mathbf{end}\}]$$

with $\mathcal{A} = \mu\mathbf{t}.\&\{tm : \mathbf{t}, over : [\,]^1\}$. Observe that $\mathcal{A}$ contains a recursive sub-term, such contexts are not allowed in previous works [14,15,26].

The use of selective unfolding makes it possible to express T_G in terms of a *recursive* input context $\mathcal{A}$ with holes filled by types (i.e., closed terms) that start with an output prefix. Indeed selective unfolding does not unfold the recursion variable $\mathbf{t}$ (*not* guarded by an output selection), which becomes part of the input context $\mathcal{A}$. Instead it unfolds the recursion variable $\mathbf{t}'$ (which is guarded by an output selection) so that the term that fills the hole, which is required to start with an output prefix, is a closed term.

Case (4) of Definition 12 requires us to check that the following pairs are in the relation: (*i*) $(T'_G, \mathcal{A}[\mu\mathbf{t}'. \oplus \{tc : \mathbf{t}', done : \mathbf{end}\}])$ and (*ii*) $(\mu\mathbf{t}'. \&\{tm : \mathbf{t}', over : \mathbf{end}\}, \mathcal{A}[\mathbf{end}])$. Observe that $T_G = \mathcal{A}[\mu\mathbf{t}'. \oplus \{tc : \mathbf{t}', done : \mathbf{end}\}]$. Hence, we have $T'_G \leq T_G$ with

$$\mathcal{R} = \{(T'_G, T_G), (\mathbf{end}, \mathbf{end}), (\mu\mathbf{t}'.\&\{tm \colon \mathbf{t}', over \colon \mathbf{end}\}, \mu\mathbf{t}.\&\{tm \colon \mathbf{t}, over \colon \mathbf{end}\})\}$$

and $\mathcal{R}$ is a controllable subtyping relation.

We show that fair asynchronous subtyping is sound w.r.t. fair refinement. In fact, fair asynchronous subtyping can be seen as a sound coinductive characterisation of fair refinement. Namely this result gives an operational justification to the syntactical definition of fair asynchronous session subtyping. Note that $\leq$ is not complete w.r.t. $\sqsubseteq$, see Example 7.

Theorem 3. *Given two session types T and S, if $T \leq S$ then $T \sqsubseteq S$.*

Example 7. Let $T = \oplus\{l_1 : \&\{l_3 : \mathbf{end}\}\}$ and $S = \&\{l_3 : \oplus\{l_1 : \mathbf{end},\ l_2 : \mathbf{end}\}\}$. We have $T \sqsubseteq S$, but T is not a fair asynchronous subtype of S since $\{l_1\} \neq \{l_1, l_2\}$, i.e., covariance of outputs is not allowed.

Unfortunately, fair asynchronous session subtyping is also undecidable. The proof is similar to the one of undecidability of fair refinement, in particular we proceed by reduction from the termination problem in queue machines.

Theorem 4. *Given two session types T and S, it is in general undecidable to check whether $T \leq S$.*

4 A Sound Algorithm for Fair Asynchronous Subtyping

We propose an algorithm which soundly verifies whether a session type is a fair asynchronous subtype of another. The algorithm relies on building a tree whose nodes are labelled by configurations of the simulation game induced by Definition 12. The algorithm analyses the tree to identify *witness* subtrees which contain input contexts that are growing following a recognisable pattern.

Example 8. Recall the satellite communication example (Figure 1). The spacecraft with protocol T_S may be a replacement for an older generation of spacecraft which follows the more complicated protocol T'_S, see Figure 2. Type T'_S notably allows the reception of telecommands to be interleaved with the emission of telemetries. The new spacecraft may safely replace the old one because $T_S \leq T'_S$.

However, checking $T_S \leq T'_S$ leads to an infinite accumulation of input contexts, hence it requires to consider infinitely many pairs of session types. E.g., after T_S selects the output label *tm* twice, the subtyping simulation game considers the pair (T_S, T''_S), where also T''_S is in Figure 2. The pairs generated for this example illustrate a common recognisable pattern where some branches grow infinitely (the *tc*-branch), while others stay stable throughout the derivation (the *done*-branch). The crux of our algorithm is to use a finite parametric characterisation of the infinitely many pairs occurring in the check of $T_S \leq T'_S$.

The *simulation tree* for $T \leq S$, written *simtree*(T, S), is the labelled tree representing the simulation game for $T \leq S$, i.e., *simtree*(T, S) is a tuple $(N, n_0, \twoheadrightarrow, \lambda)$ where N is its set of nodes, $n_0 \in N$ is its root, $\twoheadrightarrow$ is its transition function, and λ is its labelling function, such that $\lambda(n_0) = (S, T)$. We omit the formal definition of $\twoheadrightarrow$, as it is straightforward from Definition 12 following the subtyping simulation game discussed after that definition. We give an example below.

Notice that the simulation tree *simtree*(T, S) is defined only when S is controllable, since $T \leq S$ holds without needing to play the subtyping simulation game if S is not controllable. We say that a branch of *simtree*(T, S) is *successful* if it is infinite or if it finishes in a leaf labelled by $(\mathbf{end}, \mathbf{end})$. All other branches are *unsuccessful*. Under the assumption that S is controllable, we have that all branches of *simtree*(T, S) are successful if and only if $T \leq S$. As a consequence checking whether all branches of *simtree*(T, S) are successful is generally undecidable. It is possible to identify a branch as successful if it visits finitely many pairs (or node labels), see Example 6; but in general a branch may generate infinitely many pairs, see Examples 8 and 12.

In order to support types that generate unbounded accumulation, we characterise finite subtrees — called witness subtrees, see Definition 13 — such that all the branches that traverse these finite subtrees are guaranteed to be successful.

Notation We give a few auxiliary definitions and notations. Hereafter $\mathcal{A}$ and $\mathcal{A}'$ range over *extended* input contexts, i.e., input contexts that may contain distinct holes with the same index. These are needed to deal with unfoldings of input contexts, see Example 9.

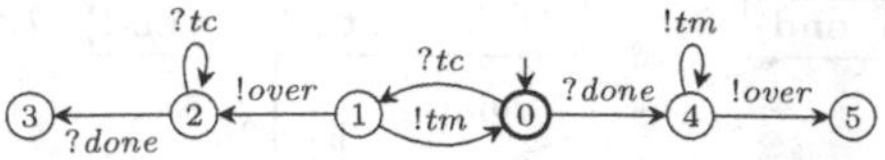

$$
\begin{array}{rl}
T'_S = & \mu\mathbf{t}\ .\&\{\ tc :\quad \oplus\{tm : \mathbf{t}, over : \mu\mathbf{t}'.\ \&\{tc : \mathbf{t}', done : \mathbf{end}\}\}, \\
& \qquad done : \mu\mathbf{t}''. \oplus \{tm : \mathbf{t}'', over : \mathbf{end}\}\} \\
T''_S = & \quad \&\{\ tc : \quad \&\{\ tc : \quad T'_S, \\
& \qquad\qquad done : \mu\mathbf{t}''. \oplus \{tm : \mathbf{t}'', over : \mathbf{end}\}\ \}, \\
& \qquad done : \mu\mathbf{t}''. \oplus \{tm : \mathbf{t}'', over : \mathbf{end}\} \qquad\qquad \}
\end{array}
$$

Fig. 2. T'_S is an alternative session type for T_S, see Example 8.

The set of *reductions* of an input context $\mathcal{A}$ is the minimal set $\mathcal{S}$ s.t. (*i*) $\mathcal{A} \in \mathcal{S}$; (*ii*) if $\&\{l_i : \mathcal{A}_i\}_{i \in I} \in \mathcal{S}$ then $\forall i \in I. \mathcal{A}_i \in \mathcal{S}$ and (*iii*) if $\mu\mathbf{t}.\mathcal{A}' \in \mathcal{S}$ then $\mathcal{A}'\{^{\mu\mathbf{t}.\mathcal{A}'}/_{\mathbf{t}}\} \in \mathcal{S}$. Notice that due to unfolding (item (*iii*)), the reductions of an input context may contain extended input contexts. Moreover, given a reduction $\mathcal{A}'$ of $\mathcal{A}$, we have that $holes(\mathcal{A}') \subseteq holes(\mathcal{A})$.

Example 9. Consider the following extended input contexts:

$$\mathcal{A}_1 = \mu\mathbf{t}.\ \&\{l_1 : [\,]^1,\ l_2 : \&\{l_3 : \mathbf{t}\}\} \qquad \mathcal{A}_2 = \&\{l_3 : \mu\mathbf{t}.\ \&\{l_1 : [\,]^1,\ l_2 : \&\{l_3 : \mathbf{t}\}\}\}$$

$$\mathsf{unfold}(\mathcal{A}_1) = \&\{l_1 : [\,]^1,\ l_2 : \&\{l_3 : \mu\mathbf{t}.\ \&\{l_1 : [\,]^1,\ l_2 : \&\{l_3 : \mathbf{t}\}\}\}\}$$

Context $\mathcal{A}_2$ is a reduction of $\mathcal{A}_1$, i.e., one can reach $\mathcal{A}_2$ from $\mathcal{A}_1$, by unfolding $\mathcal{A}_1$ and executing the input l_2. Context $\mathsf{unfold}(\mathcal{A}_1)$ is also a reduction of $\mathcal{A}_1$. Observe that $\mathsf{unfold}(\mathcal{A}_1)$ contains two distinct holes indexed by 1.

Given an extended context $\mathcal{A}$ and a set of hole indices K such that $K \subseteq holes(\mathcal{A})$, we use the following shorthands. Given a type T_k for each $k \in K$, we write $\mathcal{A}\lfloor T_k \rfloor^{k \in K}$ for the extended context obtained by replacing each hole $k \in K$ in $\mathcal{A}$ by T_k. Also, given an extended context $\mathcal{A}'$ we write $\mathcal{A}\langle \mathcal{A}' \rangle^K$ for the extended context obtained by replacing each hole $k \in K$ in $\mathcal{A}$ by $\mathcal{A}'$. When $K = \{k\}$, we often omit K and write, e.g., $\mathcal{A}\langle \mathcal{A}' \rangle^k$ and $\mathcal{A}\lfloor T_k \rfloor^k$.

Example 10. Using the above notation and posing $\mathcal{A} = \&\{tc : [\,]^1, done : [\,]^2\}$, we can rewrite T''_S (Figure 2) as $\mathcal{A}\langle \mathcal{A}\lfloor T'_S \rfloor^1 \rangle^1 \lfloor \mu\mathbf{t}''. \oplus \{tm : \mathbf{t}'', over : \mathbf{end}\} \rfloor^2$.

Example 11. Consider the session type below

$$S = \&\{l_1 : \&\{l_1 : T_1,\ l_2 : T_2,\ l_3 : T_3\},\ l_2 : \&\{l_1 : T_1,\ l_2 : T_2,\ l_3 : T_3\},\ l_3 : T_3\}.$$

Posing $\mathcal{A} = \&\{l_1 : [\,]^1, l_2 : [\,]^2, l_3 : [\,]^3\}$ we have $holes(\mathcal{A}) = \{1, 2, 3\}$. Assuming $J = \{1, 2\}$ and $K = \{3\}$, we can rewrite S as $\mathcal{A}\langle \mathcal{A}\lfloor T_j \rfloor^{j \in J} \rangle^J \lfloor T_k \rfloor^{k \in K}$.

Example 12. Figure 3 shows the partial simulation tree for $T_S \leq T'_S$, from Figures 1 and 2 (ignore the dashed edges for now). Notice how the branch leading to the top part of the tree visits only finitely many node labels (see dotted box), however the bottom part of the tree generates infinitely many labels, see the path along the $!tm$ transitions in the dashed box.

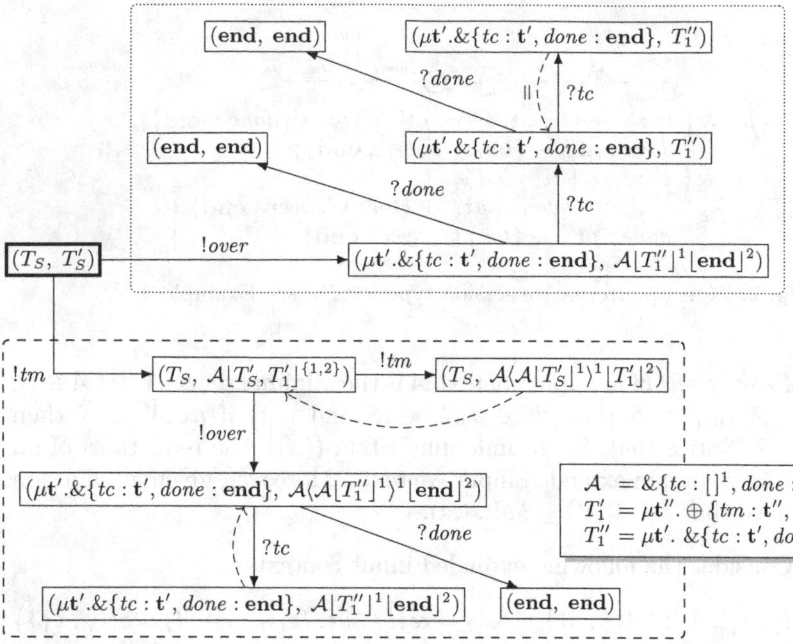

Fig. 3. Simulation tree for $T_S \leq T'_S$ (Figures 1 and 2), the root of the tree is in bold.

Witness subtrees Next, we define witness trees which are finite subtrees of a simulation tree which we prove to be successful. The role of the witness subtree is to identify branches that satisfy a certain accumulation pattern. It detects an input context $\mathcal{A}$ whose holes fall in two categories: (i) growing holes (indexed by indices in J below) which lead to an infinite growth and (ii) constant holes (indexed by indices in K below) which stay stable throughout the simulation game. The definition of witness trees relies on the notion of *ancestor* of a node n, which is a node n' (different from n) on the path from the root n_0 to n. We illustrate witness trees with Figure 3 and Example 13.

Definition 13 (Witness Tree). *A tree* $(N, n_0, \twoheadrightarrow, \lambda)$ *is a* witness tree *for* $\mathcal{A}$, *such that* $holes(\mathcal{A}) = I$, *with* $\emptyset \subseteq K \subset I$ *and* $J = I \setminus K$, *if all the following conditions are satisfied:*

1. *for all* $n \in N$ *either* $\lambda(n) = (T, \mathcal{A}'\langle \mathcal{A}\lfloor S_j \rfloor^{j \in J} \rangle^J \lfloor S_k \rfloor^{k \in K})$ *or* $\lambda(n) = (T, \mathcal{A}'\langle \mathcal{A}\langle \mathcal{A}\lfloor S_j \rfloor^{j \in J} \rangle^J \rangle^J \lfloor S_k \rfloor^{k \in K})$, *where* $\mathcal{A}'$ *is a reduction of* $\mathcal{A}$, *and it holds that*
   - $holes(\mathcal{A}') \subseteq K$ *implies that* n *is a leaf and*
   - *if* $\lambda(n) = (T, \mathcal{A}[S_i]^{i \in I})$ *and* n *is not a leaf then* unfold(T) *starts with an output selection;*
2. *each leaf* n *of the tree satisfies one of the following conditions:*
   (a) $\lambda(n) = (T, S)$ *and* n *has an ancestor* n' *s.t.* $\lambda(n') = (T, S)$

(b) $\lambda(n) = (T, \mathcal{A}\langle\mathcal{A}\lfloor S_j\rfloor^{j\in J}\rangle^J\lfloor S_k\rfloor^{k\in K})$ *and n has an ancestor* n' *s.t.* $\lambda(n') = (T, \mathcal{A}[S_i]^{i\in I})$

(c) $\lambda(n) = (T, \mathcal{A}[S_i]^{i\in I})$ *and*
n has an ancestor n' *s.t.* $\lambda(n') = (T, \mathcal{A}\langle\mathcal{A}\lfloor S_j\rfloor^{j\in J}\rangle^J\lfloor S_k\rfloor^{k\in K})$

(d) $\lambda(n) = (T, \mathcal{A}'[S_k]^{k\in K'})$ *where* $K' \subseteq K$

and for all leaves (T, S) *of type* (2c) *or* (2d) $T \leq S$ *holds.*

Intuitively Condition (1) says that a witness subtree consists of nodes that are labelled by pairs (T, S) where S contains a fixed context $\mathcal{A}$ (or a reduction/repetition thereof) whose holes are partitioned in growing holes (J) and constant holes (K). Whenever all growing holes have been removed from a pair (by reduction of the context) then this means that the pair is labelling a leaf of the tree. In addition, if the initial input is limited to only one instance of $\mathcal{A}$, the l.h.s. type starts with an output selection so that this input cannot be consumed in the subtyping simulation game.

Condition 2 says that all leaves of the tree must validate certain conditions from which we can infer that their continuations in the full simulation tree lead to successful branches. Leaves satisfying Condition (2a) straightforwardly lead to successful branches as the subtyping simulation game, starting from the corresponding pair, has been already checked starting from its ancestor having the same label. Leaves satisfying Condition (2b) lead to an infinite but regular "increase" of the types in J-indexed holes — following the same pattern of accumulation from their ancestor. The next two kinds of leaves must additionally satisfy the subtyping relation — using witness trees inductively or based on the fact they generate finitely many labels. Leaves satisfying Condition (2c) lead to regular "decrease" of the types in J-indexed holes — following the same pattern of reduction from their ancestor. Leaves satisfying Condition (2d) use only constant K-indexed holes because, by reduction of the context $\mathcal{A}'$, the growing holes containing the accumulation $\mathcal{A}$ have been removed.

Remark 1. Definition 13 is parameterised by an input context $\mathcal{A}$. We explain how such contexts can be identified while building a simulation tree in Section 5.

Example 13. In the tree of Figure 3 we highlight two subtrees. The subtree in the dotted box is not a witness subtree because it does not validate Condition (1) of Definition 13, i.e., there is an intermediary node with a label in which the r.h.s type does not contain $\mathcal{A}$.

The subtree in the dashed box is a witness subtree with 3 leaves, where the dashed edges represent the ancestor relation, $\mathcal{A} = \&\{tc : [\,]^1, done : [\,]^2\}$, $J = \{1\}$ and $K = \{2\}$. We comment on the leaves clockwise, starting from (**end**, **end**), which satisfies Condition (2d). The next leaf satisfies condition (2c), while the final leaf satisfies Condition (2b).

Algorithm Given two session types T and S we first check whether S is uncontrollable. If this is the case we immediately conclude that $T \leq S$. Otherwise, we proceed in four steps.

S1 We compute a finite fragment of *simtree*(T, S), stopping whenever (i) we encounter a leaf (successful or not), (ii) we encounter a node that has an ancestor as defined in Definition 13 (Conditions (2a), (2b), and (2c)), (iii) or the length of the path from the root of *simtree*(T, S) to the current node exceeds a bound set to two times the depth of the AST of S. This bound allows the algorithm to explore paths that will traverse the super-type at least twice. We have empirically confirmed that it is sufficient for all examples mentioned in Section 5.
S2 We remove subtrees from the tree produced in **S1** corresponding to successful branches of the simulation game which contain finitely many labels. Concretely, we remove each subtree whose each leaf n is either successful or has an ancestor n' such that n' is in the same subtree and $\lambda(n) = \lambda(n')$.
S3 We extract subtrees from the tree produced in **S2** that are potential *candidates* to be subsequently checked. The extraction of these finite candidate subtrees is done by identifying the forest of subtrees rooted in ancestor nodes which do not have ancestors themselves.
S4 We check that each of the candidate subtrees from **S3** is a witness tree.

If an unsuccessful leaf is found in **S1**, then the considered session types are not related. In **S1**, if the generation of the subtree reached the bound before reaching an ancestor or a leaf, then the algorithm is unable to give a decisive verdict, i.e., the result is *unknown*. Otherwise, if all checks in **S4** succeed then the session types are in the fair asynchronous subtyping relation. In all other cases, the result is *unknown* because a candidate subtree is not a witness.

Example 14. We illustrate the algorithm above with the tree in Figure 3. After **S1**, we obtain the whole tree in the figure (11 nodes). After **S2**, all nodes in the dotted boxed are removed. After **S3** we obtain the (unique) candidate subtree contained in the dashed box. This subtree is identified as a witness subtree in **S4**, hence we have $T_S \leq T'_S$.

We state the main theorem that establishes the soundness of our algorithm, where $\twoheadrightarrow^*$ is the reflexive and transitive closure of $\twoheadrightarrow$.

Theorem 5. *Let T and S be session types s.t. simtree$(T, S) = (N, n_0, \twoheadrightarrow, \lambda)$. If simtree$(T, S)$ contains a witness subtree with root n then for every node $n' \in N$ s.t. $n \twoheadrightarrow^* n'$, either n' is a successful leaf, or there exists n'' s.t. $n' \twoheadrightarrow n''$.*

We can conclude that if the candidate subtrees of *simtree*(T, S) identified with the strategy explained above are also witness subtrees, then we have $T \leq S$.

5 Implementation

To evaluate our algorithm, we have produced a Haskell implementation of it, which is available on GitHub [31]. Our tool takes two session types T and S as input then applies Steps **S1** to **S4** to check whether $T \leq S$. A user-provided bound can be given as an optional argument. We have run our tool on a dozen of examples handcrafted to test the limits of our algorithm (inc. the examples discussed in this paper), as well as on the 174 tests taken from [6]. All of these tests terminate under a second.

For debugging and illustration purposes, the tool can optionally generate graphical representations of the simulation and witness trees, and check whether the given types are controllable. We give examples of these in [9].

Our tool internally uses automata to represent session types and uses strong bisimilarity instead of syntactic equality between session types. Using automata internally helps us identify candidate input contexts as we can keep track of states that correspond to the input context computed when applying Case (4) of Definition 12. In particular, we augment each local state in the automata representation of the candidate supertype with two counters: the *c*-counter keeps track of how many times a state has been used in an input *c*ontext; the *h*-counter keeps track of how many times a state has occurred within a *h*ole of an input context. We illustrate this with Figure 4 which illustrates the internal data structures our tool manipulates when checking $T_S \leq T'_S$ from Figures 1 and 2. The state indices of the automata in Figure 4 correspond to the ones in Figure 1 (2nd column) and Figure 2 (3rd column).

The first row of Figure 4 represents the root of the simulation tree, where both session types are in their respective initial state and no transition has been executed. We use state labels of the form $n_{c,h}$ where n is the original identity of the state, c is the value of the *c*-counter, and h is the value of the *h*-counter. The second row depicts the configuration after firing transition $!tm$, via Case (4) of Definition 12. While the candidate subtype remains in state 0 (due to a self-loop) the candidate supertype is unfolded with $\mathsf{selUnfold}(T'_S)$ (Definition 10). The resulting automaton contains an additional state and two transitions. All previously existing states have their *h*-counter incremented, while the new state has its *c*-counter incremented. The third row of the figure shows the configuration after firing transition $!over$, using Case (4) of Definition 12 again. In this step, another copy of state 0 is added. Its *c*-counter is set to 2 since this state has been used in a context twice; and the *h*-counters of all other states are incremented.

Using this representation, we construct a candidate input context by building a tree whose root is a state $q_{c,h}$ such that $c > 1$. The nodes of the tree are taken from the states reachable from $q_{c,h}$, stopping when a state $q'_{c',h'}$ such that $c' < c$ is found. A leaf $q'_{c',h'}$ becomes a hole of the input context. The hole is a constant (K) hole when $h' = c$, and growing (J) otherwise. Given this strategy and the configurations in Figure 4, we successfully identify the context $\mathcal{A} = \&\{tc : [\,]^1, done : [\,]^2\}$ with $J = \{1\}$ and $K = \{2\}$.

6 Related and Future Work

Related work We first compare with previous work on refinement for asynchronous communication by some of the authors of this paper. The work in [10] also considers fair compliance, however here we consider binary (instead of multiparty) communication and we use a unique input queue for all incoming messages instead of distinct named input channels. Moreover, here we provide a

Last transition	State of T_S	Representation of T'_S
ϵ	0	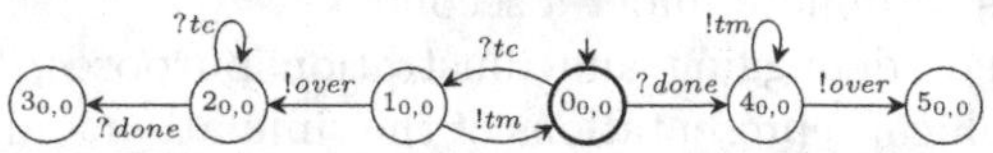
$!tm$	0	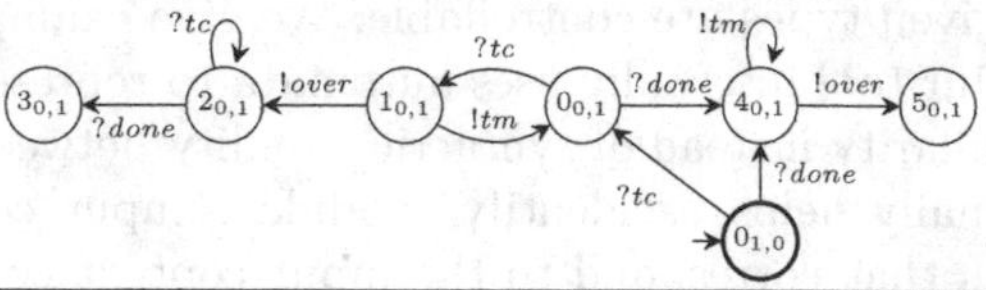
$!over$	1	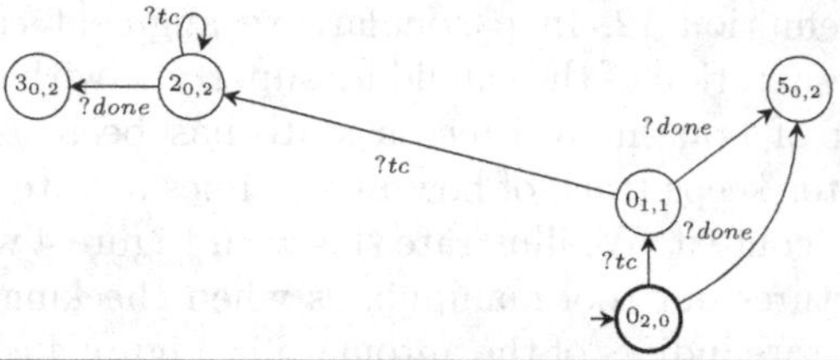

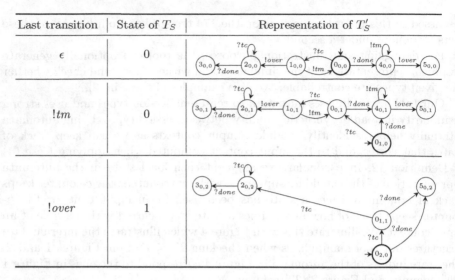

Fig. 4. Internal representation of the simulation tree for $T_S \leq T'_S$ (fragment).

sound characterisation of fair refinement using coinductive subtyping and provide a sound algorithm and its implementation. In [13] the asynchronous subtyping of [7, 14, 15, 26] is used to characterise refinement for a notion of correct composition based on the impossibility to reach a deadlock, instead of the possibility to reach a final successful configuration as done in the present paper. The refinement from [13] does not support examples such as those in Figure 1.

Concerning previous notions of synchronous subtyping, Gay and Hole [17,18] first introduced the notion of subtyping for *synchronous* session types, which is decidable in quadratic time [22]. This subtyping only supports covariance of outputs and contravariance of inputs, but does not address anticipation of outputs. Padovani studied a notion of fair subtyping for *synchronous* multi-party session types in [29]. This work notably considers the notion of *viability* which corresponds, in the synchronous multiparty setting, to our notion of controllability. We use the term controllability instead of viability following the tradition of service contract theories like those based on Petri nets [25, 33] or process calculi [12]. In contrast to [29], asynchronous communication makes it much more involved to characterise controllability in a decidable way, as we do in this paper. Fair refinement in [29] is characterised by defining a coinductive relation on normal form of types, obtained by removing inputs leading to uncontrollable continuations. Instead of using normal forms, we remove these inputs during the asynchronous subtyping check. A limited form of variance on output is also admitted in [29]. Covariance between the outputs of a subtype and those of a supertype is possible when the additional branches in the supertype are not needed to have compliance with potential partners. In [29] this check is made possible by exploiting a *difference* operation [29, Definition 3.15] on types, which synthesises a new type representing branches of one type that are absent in the other. We observe that the same approach cannot work to introduce variance on outputs in an asynchronous setting. Indeed the interplay between output anticipation and recursion could generate differences in the branches of a subtype and a supertype that cannot be statically represented by a (finite) session type.

Padovani also studied an alternative notion of fair *synchronous* subtyping in [28]. Although the contribution of that paper refers to session types, the formal framework therein seems to deviate from the usual session type approach. In particular, it considers shared channel communication instead of binary channels: when a partner emits a message, it is possible to have a race among several potential receivers for consuming it. As a consequence of this alternative semantics, the subtyping in [28] does not admit variance on input. Another difference with respect to session type literature is the notion of *success* among interacting sessions: a composition of session is successful if at least one participant reaches an internal successful state. This approach has commonalities with testing [27], where only the test composed with the system under test is expected to succeed, but differs from the typical notion of success considered for session types. In [2,3] (resp. [14]) it was proved that the Gay-Hole synchronous session subtyping (resp. orphan message free asynchronous subtyping) coincides with refinement induced by a successful termination notion requiring interacting processes to be *both* in the **end** state (with empty buffers, in the asynchronous case).

Several variants of asynchronous session subtyping have been proposed in [14, 15, 26] and further studied in our earlier work [6, 7, 13]. All these variants have been shown to be undecidable [7, 8, 23]. Moreover, all these subtyping relations are (implicitly) based on an unfair notion of compliance. Concretely, the definition of asynchronous subtyping introduced in this paper differs from the one in [14,15] since no additional constraint guaranteeing absence of orphan-messages is considered. Such a constraint requires the subtype not to have output loops whenever an output anticipation is performed, thus guaranteeing that at least one input is performed in all possible paths. In this paper, absence of orphan messages is guaranteed by enforcing types to (fairly) reach a successful termination. Moreover, our novel subtyping differs from those in [14, 15, 26] since we use recursive input contexts (and not just finite ones) for the first time— this is necessary to obtain $T'_G \leq T_G$ and $T_S \leq T'_S$ (see Figures 1 and 2). Notice that not imposing the above mentioned orphan-message-free constraint of [14, 15] is consistent with recursive input contexts that allows for input loops in the supertype whenever an output anticipation is performed. In [6], we proposed a sound algorithm for the asynchronous subtyping in [14]. The sound algorithm that we present in this paper substantially differs from that of [6]. Here we use witness trees that take under consideration both increasing and decreasing of accumulated input. In [6], instead, only regular growing accumulation is considered.

Future work In future work, we will investigate how to support output variance in fair asynchronous subtyping. We also plan to study fairness in the context of asynchronous multiparty session types, as fair compliance and refinement extend naturally to several partners. Finally, we will investigate a more refined termination condition for our algorithm using ideas from [6, Definition 11].

References

1. Adam Wiggins. The Twelve Factor methodology. https://12factor.net, 2017.
2. F. Barbanera and U. de'Liguoro. Two notions of sub-behaviour for session-based client/server systems. In *Proc. of the 12th International ACM SIGPLAN Conference on Principles and Practice of Declarative Programming, PPDP'10*, pages 155–164. ACM, 2010.
3. G. T. Bernardi and M. Hennessy. Modelling session types using contracts. *Mathematical Structures in Computer Science*, 26(3):510–560, 2016.
4. A. Bouajjani, C. Enea, K. Ji, and S. Qadeer. On the completeness of verifying message passing programs under bounded asynchrony. In *CAV (2)*, volume 10982 of *Lecture Notes in Computer Science*, pages 372–391. Springer, 2018.
5. D. Brand and P. Zafiropulo. On communicating finite-state machines. *J. ACM*, 30(2):323–342, 1983.
6. M. Bravetti, M. Carbone, J. Lange, N. Yoshida, and G. Zavattaro. A sound algorithm for asynchronous session subtyping. In *CONCUR*, volume 140 of *LIPIcs*, pages 38:1–38:16. Schloss Dagstuhl - Leibniz-Zentrum für Informatik, 2019.
7. M. Bravetti, M. Carbone, and G. Zavattaro. Undecidability of asynchronous session subtyping. *Inf. Comput.*, 256:300–320, 2017.
8. M. Bravetti, M. Carbone, and G. Zavattaro. On the boundary between decidability and undecidability of asynchronous session subtyping. *Theor. Comput. Sci.*, 722:19–51, 2018.
9. M. Bravetti, J. Lange, and G. Zavattaro. Fair refinement for asynchronous session types (extended version). CoRR abs/2101.08181, 2021
10. M. Bravetti and G. Zavattaro. Contract Compliance and Choreography Conformance in the Presence of Message Queues. In *WS-FM'08*, volume 5387 of *Lecture Notes in Computer Science*, pages 37–54. Springer, 2008.
11. M. Bravetti and G. Zavattaro. A foundational theory of contracts for multi-party service composition. *Fundam. Inform.*, 89(4):451–478, 2008.
12. M. Bravetti and G. Zavattaro. A theory of contracts for strong service compliance. *Math. Struct. Comput. Sci.*, 19(3):601–638, 2009.
13. M. Bravetti and G. Zavattaro. Relating session types and behavioural contracts: The asynchronous case. In *SEFM*, volume 11724 of *Lecture Notes in Computer Science*, pages 29–47. Springer, 2019.
14. T. Chen, M. Dezani-Ciancaglini, A. Scalas, and N. Yoshida. On the preciseness of subtyping in session types. *Logical Methods in Computer Science*, 13(2), 2017.
15. T.-C. Chen, M. Dezani-Ciancaglini, and N. Yoshida. On the preciseness of subtyping in session types. In *PPDP 2014*, pages 146–135. ACM Press, 2014.
16. P. Deniélou and N. Yoshida. Multiparty compatibility in communicating automata: Characterisation and synthesis of global session types. In *ICALP 2013*, pages 174–186, 2013.
17. S. J. Gay and M. Hole. Types and subtypes for client-server interactions. In *ESOP 1999*, pages 74–90, 1999.
18. S. J. Gay and M. Hole. Subtyping for session types in the pi calculus. *Acta Inf.*, 42(2-3):191–225, 2005.
19. B. Genest, D. Kuske, and A. Muscholl. A Kleene theorem and model checking algorithms for existentially bounded communicating automata. *Inf. Comput.*, 204(6):920–956, 2006.
20. B. Genest, D. Kuske, and A. Muscholl. On communicating automata with bounded channels. *Fundam. Inform.*, 80(1-3):147–167, 2007.

21. K. Honda, N. Yoshida, and M. Carbone. Multiparty asynchronous session types. *J. ACM*, 63(1):9, 2016.
22. J. Lange and N. Yoshida. Characteristic formulae for session types. In *TACAS*, volume 9636 of *Lecture Notes in Computer Science*, pages 833–850. Springer, 2016.
23. J. Lange and N. Yoshida. On the undecidability of asynchronous session subtyping. In *Proc. of 20th Int. Conference on Foundations of Software Science and Computation Structures, FOSSACS'17*, volume 10203 of *Lecture Notes in Computer Science*, pages 441–457, 2017.
24. J. Lange and N. Yoshida. Verifying asynchronous interactions via communicating session automata. In *CAV (1)*, volume 11561 of *Lecture Notes in Computer Science*, pages 97–117. Springer, 2019.
25. N. Lohmann. Why does my service have no partners? In *WS-FM*, volume 5387 of *Lecture Notes in Computer Science*, pages 191–206. Springer, 2008.
26. D. Mostrous, N. Yoshida, and K. Honda. Global principal typing in partially commutative asynchronous sessions. In *ESOP*, volume 5502 of *Lecture Notes in Computer Science*, pages 316–332. Springer, 2009.
27. R. D. Nicola and M. Hennessy. Testing Equivalences for Processes. *Theoretical Computer Science*, 34:83–133, 1984.
28. L. Padovani. Fair subtyping for open session types. In *ICALP*, volume 7966 of *Lecture Notes in Computer Science*, pages 373–384. Springer, 2013.
29. L. Padovani. Fair subtyping for multi-party session types. *Math. Struct. Comput. Sci.*, 26(3):424–464, 2016.
30. A. Rensink and W. Vogler. Fair testing. *Inf. Comput.*, 205(2):125–198, 2007.
31. M. Bravetti, J. Lange, and G. Zavattaro. Fair refinement for asynchronous session types. https://github.com/julien-lange/fair-asynchronous-subtyping, 2020
32. R. van Glabbeek and P. Höfner. Progress, justness, and fairness. *ACM Comput. Surv.*, 52(4):69:1–69:38, 2019.
33. D. Weinberg. Efficient controllability analysis of open nets. In *WS-FM*, volume 5387 of *Lecture Notes in Computer Science*, pages 224–239. Springer, 2008.

Running Time Analysis of Broadcast Consensus Protocols* **

Philipp Czerner[1] [✉] and Stefan Jaax[1]

Fakultät für Informatik, Technische Universität München, Garching bei München, Germany
{czerner,jaax}@in.tum.de

Abstract. Broadcast consensus protocols (BCPs) are a model of computation, in which anonymous, identical, finite-state agents compute by sending/receiving global broadcasts. BCPs are known to compute all number predicates in $\mathsf{NL} = \mathsf{NSPACE}(\log n)$ where n is the number of agents. They can be considered an extension of the well-established model of population protocols. This paper investigates execution time characteristics of BCPs. We show that every predicate computable by population protocols is computable by a BCP with expected $\mathcal{O}(n \log n)$ interactions, which is asymptotically optimal. We further show that every log-space, randomized Turing machine can be simulated by a BCP with $\mathcal{O}(n \log n \cdot T)$ interactions in expectation, where T is the expected runtime of the Turing machine. This allows us to characterise polynomial-time BCPs as computing exactly the number predicates in ZPL, i.e. predicates decidable by log-space, randomised Turing machine with zero-error in expected polynomial time where the input is encoded as unary.

Keywords: broadcast protocols · complexity theory · distributed computing

1 Introduction

In recent years, models of distributed computation following the *computation-by-consensus* paradigm attracted considerable interest in research (see for example [9,25,26,8,13]). In such models, network agents compute number predicates, i.e. Boolean-valued functions of the type $\mathbb{N}^k \to \{0,1\}$, by reaching a stable consensus whose value determines the outcome of the computation. Perhaps the most prominent model following this paradigm are *population protocols* [5,6], a model in which anonymous, identical, finite-state agents interact randomly in pairwise rendezvous to agree on a common Boolean output.

Due to anonymity and locality of interactions, it is an inherent property of population protocols that agents are generally unable to detect with absolute

* This work was supported by an ERC Advanced Grant (787367: PaVeS) and by the Research Training Network of the Deutsche Forschungsgemeinschaft (DFG) (378803395: ConVeY).

** The full version of this paper can be found at https://arxiv.org/abs/2101.03780 .

S. Kiefer and C. Tasson (Eds.): FOSSACS 2021, LNCS 12650, pp. 164–183, 2021.
https://doi.org/10.1007/978-3-030-71995-1_9

certainty when the computation has stabilized. This makes sequential composition of protocols difficult, and further complicates the implementation of control structures such as loops or branching statements. To overcome this drawback, two kinds of approaches have been suggested in the literature: 1.) Let agents guess when the computation has stabilized, leading to composable, but merely *approximately correct* protocols [7,24], or 2.) extend population protocols by global communication primitives that enable agents to query global properties of the agent population [13,8,26].

Approaches of the first kind are for the most part based on simulations of global broadcasts by means of *epidemics*. In epidemics-based approaches the spread of the broadcast signal is simulated by random pairwise rendezvous, akin to the spread of a viral epidemic in a population. When the broadcasting agent meets a certain fraction of "infected" agents, it may decide with reasonable certainty that the broadcast has propagated throughout the entire population, which then leads to the initiation of the next computation phase. Of course, the decision to start the next phase may be premature, in which case the rest of the execution may be faulty. However, epidemics can also be used to implement phase clocks that help keep the failure probability low (see e.g. [7]).

In [13], Blondin, Esparza, and one of the authors of this paper introduced *broadcast consensus protocols* (BCPs), an extension of population protocols by reliable, global, and atomic broadcasts. BCPs find their precursor in the broadcast protocol model introduced by Emerson and Namjoshi in [17] to describe bus-based hardware protocols. This model has been investigated intensely in the literature, see e.g. [18,19,15,28]. Broadcasts also arise naturally in biological systems. For example, Uhlendorf *et al.* analyse applications of broadcasts in the form of an external, global light source for controlling a population of yeasts [12].

The authors of [13] show that BCPs compute precisely the predicates in $\mathsf{NL} = \mathsf{NSPACE}(\log n)$, where n is the number of agents. For comparison, it is known that population protocols compute precisely the *Presburger predicates*, which are the predicates definable in the first-order theory of the integers with addition and the usual order; a class much less expressive than the former.

An epidemics-based approach was used in [7] to show that population protocols can simulate with high probability a step of a virtual register machine with expected $\mathcal{O}(n \log^5(n))$ interactions, where n is the number of agents. This result stimulated further research into time bounds for classical problems such as leader election (see e.g. [21,1,16,29,11]) and majority (see e.g. [4,2]). In their seminal paper [5], Angluin *et al.* already showed that population protocols can stably compute Presburger predicates with $\mathcal{O}(n^2 \log n)$ interactions in expectation. Belleville *et al.* further showed that leaderless protocols require a quadratic number of interactions in expectation to stabilize to the correct output for a wide class of predicates [10]. The aforementioned bounds apply to *stabilisation time*: the time it takes to go from an initial configuration to a stable consensus that cannot be destroyed by future interactions. In [24], Kosowski and Uznanski considered the weaker notion of *convergence time*: the time it takes on average to ultimately transition to the correct consensus (although this consensus could in principle be destroyed by future interactions), and they show that sublinear convergence time is achievable.

By contrast, to the best of our knowledge, time characteristics of BCPs have not been discussed in the literature. The NL-powerful result presented in [13] does not establish any time bounds. In fact, [13] only considers a non-probabilistic variant of BCPs with a global fairness assumption instead of probabilistic choices.

Contributions of the paper. This paper initiates the runtime analysis of BCPs in terms of expected number of interactions to reach a stable consensus. To simplify the definition of probabilistic execution semantics, we introduce a restricted, deterministic variant of BCPs without rendezvous transitions. In Section 2, we define probabilistic execution semantics for the restricted version of BCPs, and we provide an introductory example for a fast protocol computing majority in Section 3.

In Section 4, we show that these restrictions of our BCP model are inconsequential in terms of expected number of interactions: both rendezvous and nondeterministic choices can be simulated with a constant runtime overhead.

In Section 5, we show that every Presburger predicate can be computed by BCPs with $\mathcal{O}(n \log n)$ interactions and with constant space, where n denotes the number of agents in the population. This result is asymptotically optimal.

In more generality, in Section 6, we use BCPs to simulate Turing machines (TMs). In particular, we show that any randomised, logarithmically space-bound, polynomial-time TM can be simulated by a BCP with an overhead of $\mathcal{O}(n \log n)$ interactions per step. Conversely, any polynomial-time BCP can be simulated by such a TM. This result can be considered an improvement of the NL bound from [13], now in a probabilistic setting. We also give a corresponding upper bound, which yields the following succinct characterisation: polynomial-time BCPs compute exactly the number predicates in ZPL, which are the languages decidable by randomised log-space polynomial-time TMs with zero-error (the log-space analogue to ZPP).

Bounding the time requires a careful analysis of each step in the simulation of the Turing machine. Thus, our proof diverges in significant ways from the proof establishing the NL lower bound in [13]. Most notably, we now make use of epidemics in order to implement clocks that help reduce failure rates.

2 Preliminaries

Complexity classes. As is usual, we define NL as the class of languages decidable by a nondeterministic log-space TM. Additionally, by ZPL we denote the set of languages decided by a randomised log-space TM A, s.t. A only terminates with the correct result (zero-error) and that it terminates within $\mathcal{O}(\text{poly}\, n)$ steps in expectation, as defined by Nisan in [27].

Multisets. A *multiset* over a finite set E is a mapping $M \colon E \to \mathbb{N}$. The set of all multisets over E is denoted $\mathbb{N}^E$. For every $e \in E$, $M(e)$ denotes the number of occurrences of e in M. We sometimes denote multisets using a set-like notation, e.g. $\wr f, g, g \int$ is the multiset M such that $M(f) = 1$, $M(g) = 2$ and

$M(e) = 0$ for every $e \in E \setminus \{f, g\}$. Addition, comparison and scalar multiplication are extended to multisets componentwise, i.e. $(M + M')(e) \stackrel{\text{def}}{=} M(e) + M'(e)$, $(\lambda M)(e) \stackrel{\text{def}}{=} \lambda M(e)$ and $M \leq M' \stackrel{\text{def}}{\iff} M(e) \leq M'(e)$ for every $M, M' \in \mathbb{N}^Q$, $e \in E$, and $\lambda \in \mathbb{N}$. For $M' \leq M$ we also define componentwise subtraction, i.e. $(M - M')(e) \stackrel{\text{def}}{=} M(e) - M'(e)$ for every $e \in E$. For every $e \in E$, we write $\boldsymbol{e} \stackrel{\text{def}}{=} \wr e \int$. We lift functions $f \colon E \to E'$ to multisets by defining $f(M)(e') \stackrel{\text{def}}{=} \sum_{f(e)=e'} M(e)$ for $e' \in E'$. Finally, we define the *support* and *size* of $M \in \mathbb{N}^E$ respectively as $[\![M]\!] \stackrel{\text{def}}{=} \{e \in E : M(e) > 0\}$ and $|M| \stackrel{\text{def}}{=} \sum_{e \in E} M(e)$.

Broadcast Consensus Protocols. A *broadcast consensus protocol* [13] (BCP) is a tuple $\mathcal{P} = (Q, \Sigma, \delta, I, O)$ where

- Q is a non-empty, finite set of *states*,
- Σ is a non-empty, finite *input alphabet*,
- δ is the *transition function* (defined below),
- $I \colon \Sigma \to Q$ is the *input mapping*, and
- $O \subseteq Q$ is a set of *accepting states*.

The function δ maps every state $q \in Q$ to a pair (r, f) consisting of the *successor state* $r \in Q$ and the *response function* $f \colon Q \to Q$.

Configurations. A *configuration* is a multiset $C \in \mathbb{N}^Q$. Intuitively, a configuration C describes a collection of identical finite-state *agents* with Q as set of states, containing $C(q)$ agents in state q for every $q \in Q$. We say that $C \in \mathbb{N}^Q$ is a 1-*consensus* if $[\![C]\!] \subseteq O$, and a 0-*consensus* if $[\![C]\!] \subseteq Q \setminus O$.

Step relation. A broadcast $\delta(q) = (r, f)$ is executed in three steps: (1) an agent at state q broadcasts a signal and leaves q; (2) all other agents receive the signal and move to the states indicated by the function f, i.e. an agent in state s moves to $f(s)$; and (3) the broadcasting agent enters state r.

Formally, for two configurations C, C' we write $C \to C'$, whenever there exists a state $q \in Q$ s.t. $C(q) \geq 1$, $\delta(q) = (r, f)$, and $C' = f(C - \boldsymbol{q}) + \boldsymbol{r}$ is the configuration computed from C by the above three steps. By $\xrightarrow{*}$ we denote the reflexive-transitive closure of $\to$.

For example, consider a configuration $C \stackrel{\text{def}}{=} \wr a, a, b \int$ and a broadcast transition $a \mapsto b, \{a \mapsto c, b \mapsto d\}$. To execute this transition, we move an agent from state a to state b and apply the transition function to all other agents, so we end up in $C' \stackrel{\text{def}}{=} \wr b \int + \wr c, d \int$.

Broadcast transitions. We write broadcast transitions as $q \mapsto r, S$ with S a set of expressions $q' \mapsto r'$. This refers to $\delta(q) = (r, f)$, with $f(q') = r'$ for $(q' \mapsto r') \in S$. We usually omit identity mappings $q' \mapsto q'$ when specifying S.

For graphic representations of broadcast protocols we use a different notation, which separates sending and receiving broadcasts. There we identify a transition $\delta(q) = (r, f)$ with a name α and specify it by writing $q \xrightarrow{!\alpha} r$ and $q' \xrightarrow{?\alpha} r'$ for $f(q') = r'$. Intuitively, $q' \xrightarrow{?\alpha} r'$ can be understood as an agent transitioning from q' to r' upon receiving the signal α, and $q \xrightarrow{!\alpha} r$ means that an agent in state q may transmit the signal α and simultaneously transition to state r.

As defined, δ is a total function, so each state is associated with a unique broadcast. If we do not specify a transition $\delta(q) = (r, f)$ explicitly, we assume that it simply maps each state to itself, i.e. $q \mapsto q, \{r \mapsto r : r \in Q\}$. We refer to those transitions as *silent*.

Executions. An *execution* is an infinite sequence $\pi = C_0C_1C_2...$ of configurations with $C_i \to C_{i+1}$ for every i. It has some fixed number of agents $n \stackrel{\text{def}}{=} |C_0| = |C_1| = ...$. Given a BCP and an initial configuration $C_0 \in \mathbb{N}^Q$, we generate a random execution with the following Markov chain: to perform a step at configuration C_i, a state $q \in Q$ is picked at random with probability distribution $p(q) = C_i(q)/|C_i|$, and the (uniquely defined) transition $\delta(q)$ is executed, giving the successor configuration C_{i+1}. We refer to the random variable corresponding to the trace of this Markov chain as *random execution.*

Stable Computation. Let π denote an execution and $\mathtt{inf}(\pi)$ the configurations occurring infinitely often in π. If $\mathtt{inf}(\pi)$ contains only b-consensuses, we say that π *stabilises* to b. For a predicate $\varphi : \mathbb{N}^\Sigma \to \{0, 1\}$ we say that $\mathcal{P}$ *(stably) computes* φ, if for all inputs $X \in \mathbb{N}^\Sigma$, the random execution of $\mathcal{P}$ with initial configuration $C_0 = I(X)$ stabilises to $\varphi(X)$ with probability 1.

Finally, for an execution $\pi = C_0C_1C_2...$ we let T_π denote the smallest i s.t. all configurations in $C_iC_{i+1}...$ are $\varphi(X)$-consensuses, or ∞ if no such i exists. We say that a BCP $\mathcal{P}$ *computes* φ *within* $f(n)$ *interactions*, if for all initial configurations C_0 with n agents the random execution π starting at C_0 has $\mathbb{E}(T_\pi) \le f(n) < \infty$, i.e. $\mathcal{P}$ stabilises within $f(n)$ steps in expectation. If $f \in \mathcal{O}(\text{poly}(n))$, then we call $\mathcal{P}$ a *polynomial-time* BCP.

Global States. Often, it is convenient to have a shared global state between all agents. If, for a BCP $\mathcal{P} = (Q, \Sigma, \delta, I, O)$ we have $Q = S \times G$, $I(\Sigma) \subseteq Q \times \{j\}$ for some $j \in G$, and $f((s, j)) \in Q \times \{j'\}$ for each $\delta((q, j)) = ((r, j'), f)$, then we say that $\mathcal{P}$ has *global states* G. A configuration C has *global state* j, if $[\![C]\!] \subseteq Q \times \{j\}$ for $j \in G$. Note that, starting from a configuration with global state j, $\mathcal{P}$ can only reach configurations with a global state. Hence for $\mathcal{P}$ we will generally only consider configurations with a global state. To make our notation more concise, when specifying a transition $\delta(q) = (r, f)$ for $\mathcal{P}$, we will write f as a mapping from S to S, as q, r already determine the mapping of global states.

Population Protocols. A population protocol [5] replaces broadcasts by local rendezvous. It can be specified as a tuple $(Q, \Sigma, \delta, I, O)$ where Q, Σ, I, O are defined as in BCPs, and $\delta\colon Q^2 \to Q^2$ defines *rendezvous transitions.* A step of the protocol at C is made by picking two agents uniformly at random, and applying δ to their states: first $q_1 \in Q$ is picked with probability $C(q_1)/|C|$, then $q_2 \in Q$ is picked with probability $C'(q_2)/|C'|$, where $C' \stackrel{\text{def}}{=} C - \wr q_1 \wr$. The successor configuration then is $C - \wr q_1, q_2 \wr + \wr r_1, r_2 \wr$ where $\delta(q_1, q_2) = (r_1, r_2)$.

Broadcast Protocols. Later on we will construct BCPs out of smaller building blocks which we call *broadcast protocols (BPs).* A BP is a pair (Q, δ), where Q and δ are defined as for BCPs. We extend the applicable definitions from above to BPs, in particular the notions of configurations, executions, and global states.

3 Example: Majority

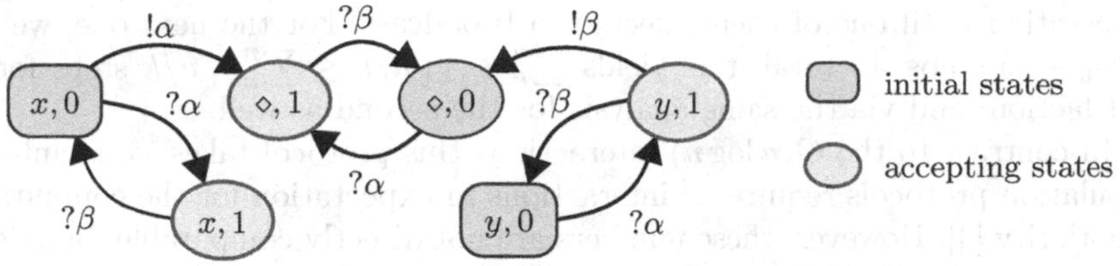

Fig. 1. A fast broadcast consensus protocol computing the majority predicate.

As an introductory example, we construct a broadcast consensus protocol for the *majority predicate* $\varphi(x, y) = x > y$. Figure 1 depicts the protocol graphically. We have the set of states $\{x, y, \diamond\} \times \{0, 1\}$, with global states $\{0, 1\}$, where the states $O \stackrel{\text{def}}{=} \{(x, 1), (y, 1), (\diamond, 1)\}$ are accepting, and $I(x) = (x, 0)$ and $I(y) = (y, 0)$. The transitions are

$$(x, 0) \mapsto (\diamond, 1), \emptyset \qquad (\alpha)$$
$$(y, 1) \mapsto (\diamond, 0), \emptyset \qquad (\beta)$$

Note that we use the more compact notation for transitions in the presence of global states, written in long form (α) would be

$$(x, 0) \mapsto (\diamond, 1), \{(x, 0) \mapsto (x, 1), (y, 0) \mapsto (y, 1), (\diamond, 0) \mapsto (\diamond, 1)\} \qquad (\alpha)$$

To make the presentation of the following sample execution more readable, we shorten the state (i, j) to i_j. For input $x = 3$ and $y = 2$, an execution could look like this:

$$\langle x_0, x_0, x_0, y_0, y_0 \rangle \xrightarrow{\alpha} \langle \diamond_1, x_1, x_1, y_1, y_1 \rangle \xrightarrow{\beta} \langle \diamond_0, x_0, x_0, \diamond_0, y_0 \rangle$$
$$\xrightarrow{\alpha} \langle \diamond_1, \diamond_1, x_1, \diamond_1, y_1 \rangle \xrightarrow{\beta} \langle \diamond_0, \diamond_0, x_0, \diamond_0, \diamond_0 \rangle \xrightarrow{\alpha} \langle \diamond_1, \diamond_1, \diamond_1, \diamond_1, \diamond_1 \rangle$$

Intuitively, there is a preliminary global consensus, which is stored in the global state. Initially, it is rejecting, as $x > y$ is false in the case $x = y = 0$. However, any x agent is enough to tip the balance, moving to an accepting global state. Now any y agent could speak up, flipping the consensus again.

The two factions initially belonging to x and y, respectively, alternate in this manner by sending signals α and β. Strict alternation is ensured as an agent will not broadcast to confirm the global consensus, only to change it.

After emitting the signal, the agent from the corresponding faction goes into state $\diamond$, where it can no longer influence the computation. In the end, the majority faction remains and determines the final consensus.

Considering these alternations with shrinking factions, the expected number of steps of the protocol until stabilization can be bounded by $2\sum_{k=1}^{n} n/k =$

$\mathcal{O}(n \log n)$. To see that this holds, we consider the factions separately: let n_0 denote the number of agents the first faction starts with (i.e. agents initially in state $(x, 0)$), and n_1 the number at the end. When we are waiting for the first transition of this faction all n_0 agents are enabled, so we wait n/n_0 steps in expectation until one of them executes a broadcast. For the next one, we wait $n/(n_0 - 1)$ steps. In total, this yields $\sum_{k=n_1+1}^{n_0} n/k \leq \sum_{k=1}^{n} n/k$ steps for the first faction, and via the same analysis for the second as well.

In contrast to the $\mathcal{O}(n \log n)$ interactions this protocol takes, constant-state population protocols require n^2 interactions in expectation for the computation of majority [4]. However, these numbers are not directly comparable: broadcasts may not be parallelizable, while it is uncontroversial to assume that n rendezvous occur in parallel time 1.

4 Comparison with other Models

To facilitate the definition of an execution model, we only consider deterministic BCPs, in the sense that for each state there is a unique transition to execute. Blondin, Esparza and Jaax [14] analysed a more general model, i.e. they allow multiple transitions for a single state, picking one of them uniformly at random when an agent in that state sends a broadcast. Additionally, as they consider BCPs as an extension of population protocols, they include rendezvous transitions. We now show that we can simulate both extensions within a constant-factor overhead.

4.1 Non-Deterministic Broadcast Protocols

The following construction allows for two broadcast transitions to be executed uniformly at random from a single state. This can easily be extended to any constant number of transitions using the usual construction of a binary tree with rejection sampling.

Now assume that we are given a BCP $(Q, \Sigma, \delta_0, I, F)$ with another set of broadcast transitions δ_1 and we want each agent to pick one transition uniformly at random from δ_0 or δ_1 whenever it executes a broadcast.

We implement this using a synthetic coin, i.e. we are utilising randomness provided by the scheduler to enable individual agents to make random choices. This idea has also been used for population protocols [1,3]. Compared to these implementations, broadcasts allow for a simpler approach.

The idea is that we partition the agents into types, so that half of the agents have type 0 and the other half have type 1. Additionally, there is a global coin shared across all agents. To flip the coin, a random agent announces its type (the coin is set to heads if the agent is type 0, tails if it is type 1) and a second random agent executes a broadcast transition from either δ_0 or δ_1, depending on the state of the global coin that has just been set. These two steps repeat, the former flipping the coin fairly and the latter then executing the actual transitions. Figure 2 sketches this procedure.

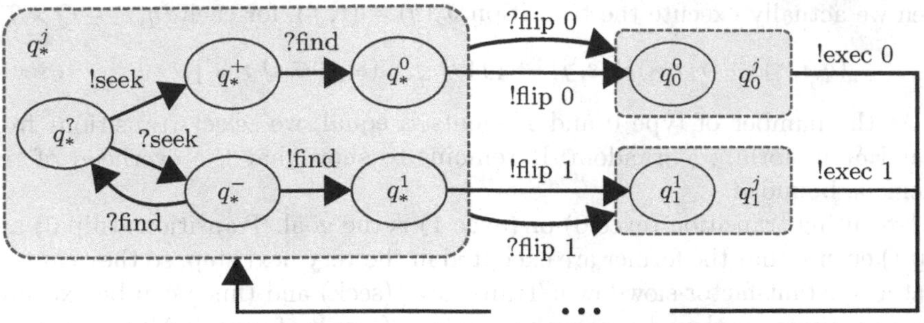

Fig. 2. Transition diagram for implementing multiple broadcasts per state, for $q \in Q$, with (q, i, j) written as q_j^i. Dashed nodes represent multiple states, with $j \in T$. Transitions resulting from executing the broadcasts in δ_0, δ_1 are not shown.

Intuitively, we start with no agents having either type 0 or 1. When such a typeless agent is picked by the scheduler to announce its type (to flip the global coin) it instead broadcasts that it is searching for a partner. Once this has happened twice, these two agents are matched, one is assigned type 0 and the other type 1. Thus we ensure that there is the exact same number of type 0 and type 1 agents at all times, meaning that we get a perfectly fair coin. Additionally we make progress regardless of whether an agent with or without a type is chosen.

To describe the construction formally, we introduce a set of types $T \overset{\text{def}}{=} \{?, +, -, 0, 1\}$, and choose the set of states $Q' \overset{\text{def}}{=} Q \times T \times \{*, 0, 1\}$, with global states $\{*, 0, 1\}$ used to represent the state of the synthetic coin. We use $(q, ?)$ as initial state instead of $q \in I$, and start with global state $*$. To pick types, we need transitions

$$(q, ?, *) \mapsto (q, +, *), \{(r, ?) \mapsto (r, -) : r \in Q\} \qquad \text{for } q \in Q \qquad \text{(seek)}$$

$$\begin{aligned}(q, -, *) \mapsto (q, 1, *), \{&(r, -) \mapsto (r, ?) : r \in Q\} \\ \cup\, \{&(r, +) \mapsto (r, 0) : r \in Q\}\end{aligned} \qquad \text{for } q \in Q \qquad \text{(find)}$$

So an agent of type ? announces that it seeks a partner, moving itself to type + and the others to type −. Then any type − agent may broadcast that a match has been found, moving itself to type 1 and the type + agent to type 0. The other type − agents revert to type ?. This ensures that the number of type 0 and 1 agents is always equal. Note that there may be an odd number of agents, in which case one agent of type + remains.

The following transitions effectively flip the global coin, by having an agent of type 0 or 1 announce that we now execute a broadcast transition from respectively δ_0 or δ_1. Here, we have $q \in Q, \circ \in \{0, 1\}$.

$$(q, \circ, *) \mapsto (q, \circ, \circ), \emptyset \qquad \text{(flip } \circ\text{)}$$

Then we actually execute the transition $\delta_\circ(q) = (r, f)$, for each $(q, i) \in Q \times T$.

$$(q, i, \circ) \mapsto (r, i, *), \{(s, j) \mapsto (f(s), j) : (s, j) \in Q \times T\} \qquad \text{(exec } \circ\text{)}$$

As the number of type 0 and 1 agents is equal, we select transitions from δ_0 and δ_1 uniformly at random. It remains to show that the overhead of this scheme is bounded.

Executing transition (exec 0) or (exec 1) is the goal. Transitions (flip 0) and (flip 1) ensure that the former are executed in the very next step, so they cause at most a constant-factor slowdown. Transitions (seek) and (find) can be executed at most n times, as they decrease the number of agent of type ?. All that remains is the implicit silent transition of states $(q, +, j)$, which occurs with probability at most $1/n$ in each step.

Hence, to execute $m \geq n$ steps of the simulated protocol our construction takes at most $(2m + 2n) \cdot n/(n-1) \leq 8m$ steps in expectation.

4.2 Population Protocols

Another extension to BCPs is the addition of rendez-vous transitions. Here we are given a map $R : Q^2 \to Q^2$. At each step, we flip a coin and either execute a broadcast transition as usual, or pick two distinct agents uniformly at random, in state q and r, respectively. These interact and move to the two states $R(q, r)$.

Again, we can simulate this extension with only a constant-factor increase in the expected number of steps. Given a BCP (Q, Σ, B, I, F), the idea is to add states $\{\tilde{q} : q \in Q\} \cup \{r_q : r, q \in Q\}$ and insert "activating" transitions $q \mapsto \tilde{q}, \{r \mapsto r_q : r \in Q\}$ for $q \in Q$ and "deactivating" transitions $r_q \mapsto s, \{\tilde{q} \mapsto t\} \cup \{u_q \mapsto u : u \in Q\}$ for each $R(q, r) = (s, t)$. So a state q first signals that it wants to start a rendez-vous transition. Then, any other state r answers, both executing the transition and signalling to all other states that it has occurred.

Each state in Q has exactly 2 broadcast transitions, so (using the scheme described above) the probability of executing any "activating" transition is exactly $\frac{1}{2}$, the same as doing one of the original broadcast transitions in B. After doing an activating transition we may do nothing for a few steps by executing the broadcast transition on $\tilde{q}$, but eventually we execute a "deactivating" transition and go back. The probability of executing a broadcast on $\tilde{q}$ is $1/n$, so simulating a single rendez-vous transition takes $1 + n/(n-1) \leq 3$ steps in expectation.

5 Protocols for Presburger Arithmetic

While Blondin, Esparza and Jaax [14] show that BCPs are more expressive than population protocols, they leave the question open whether BCPs provide a runtime speed-up for the class of Presburger predicates computable by population protocols. We already saw that Majority can be computed within $\mathcal{O}(n \log n)$ interactions in BCPs. This also holds in general for Presburger predicates:

Theorem 1. *Every Presburger predicate is computable by a BCP within at most* $\mathcal{O}(n \log n)$ *interactions.*

We remark that the $\mathcal{O}(n \log n)$ bound is asymptotically optimal: e.g. the stable consensus for the parity predicate ($x = 1 \bmod 2$) must alternate with configuration size, which clearly requires every agent to perform at least one broadcast in the computation, and thus yields a lower bound of $\sum_{k=1}^{n} \frac{n}{k} = \Omega(n \log n)$ steps like in the coupon collector's problem [20].

It is known [22] that every Presburger predicate can be expressed as Boolean combination of linear inequalities and linear congruence equations over the integers, i.e. as Boolean combination of predicates of the form $\sum_i \alpha_i x_i < c$, and $\sum_i \alpha_i x_i = c \bmod m$, where the α_i, c and m are integer constants. In Section 5.1 we construct BCPs that compute arbitrary linear inequalities, before we sketch the construction for congruences and Boolean combinations in Section 5.2.

5.1 Linear Inequalities

Proposition 1. *Let $\alpha_1, \ldots, \alpha_k, c \in \mathbb{Z}$ and let $\varphi(x_1, \ldots, x_k) \overset{def}{\iff} \sum_{i=1}^{k} \alpha_i x_i < c$ denote a linear inequality. There exists a broadcast consensus protocol that computes φ within $\mathcal{O}(n \log n)$ interactions in expectation.*

Proof. We assume wlog that $\alpha_i \neq 0$ for $i = 1, ..., k$ and that $\alpha_1, ..., \alpha_k$ are pairwise distinct. Let $A \overset{\text{def}}{=} \max\{|\alpha_1|, |\alpha_2| \ldots, |\alpha_k|, |c|\}$. We define a BCP $\mathcal{P} = (Q \times G, \Sigma, \delta, I, O)$ with global states G, where

$$Q \overset{\text{def}}{=} \{0, \alpha_1, ..., \alpha_k\} \qquad \Sigma \overset{\text{def}}{=} \{x_1, \ldots, x_k\}$$
$$G \overset{\text{def}}{=} [-2A, 2A] \qquad O \overset{\text{def}}{=} \{(q, v) : v < c\}$$

As inputs we get $I(x_i) \overset{\text{def}}{=} (\alpha_i, 0)$ for each $i = 1, ..., k$. The transitions δ are constructed as follows. For every $v \in [-2A, 2A]$ and every α_i satisfying $v + \alpha_i \in [-2A, 2A]$, we add the following transition to T:

$$(\alpha_i, v) \mapsto (0, v + \alpha_i), \emptyset \qquad (\alpha_i)$$

Intuitively, in the first component of its state an agent stores its contribution to $\sum_i \alpha_i x_i$, the left-hand side of the inequality. The global state is used to store a counter value, initially set to 0. Each agent adds its contribution to the counter, as long as it does not overflow. The counter goes from $-2A$ to $2A$, which allows it to store the threshold plus any single contribution. The final counter value then determines the outcome of the computation.

Correctness. Let $\mathsf{ctr}(C)$ denote the global state (and thus current counter value) of configuration C. Further, let

$$\mathsf{sum}(C) \overset{\text{def}}{=} \sum_{(\alpha,v) \in Q} C(\alpha, v) \cdot \alpha + \mathsf{ctr}(C)$$

denote the sum of all agents' contributions and the current value of the counter. Every initial configuration C_0 has $\mathsf{ctr}(C) = 0$ and thus $\mathsf{sum}(C) = \sum_i \alpha x_i$. Each transition α increases the counter by α but sets the agent's contribution to 0 (from α), so $\mathsf{sum}(C)$ is constant throughout the execution.

Recall that our output mapping depends only on the value of the counter, so our agents always form a consensus (though not necessarily a stable one). If this consensus and $\varphi(C_0)$ disagree, then, we claim, a non-silent transition is enabled.

To see this, note that the current consensus depends on whether $\mathsf{ctr}(C) < c$. If that is the case, but $\varphi(C_0) = 0$, then $\mathsf{sum}(C) \geq c$ and some agent with positive contribution $\alpha > 0$ exists. Due to $\mathsf{ctr}(C) < c$, transition α is enabled. Conversely, if $\mathsf{ctr}(C) \geq c$ and $\varphi(C_0) = 1$, some transition α with $\alpha < 0$ will be enabled.

Finally, note that each non-silent transition increases the number of agents with contribution 0 by one, so at most n can be executed in total. So the execution converges and reaches, by the above argument, a correct consensus.

Convergence time. Each agent executes at most one non-silent transition. To estimate the total number of steps, we partition the agents by their current contribution: for a configuration C let $C^+ \stackrel{\text{def}}{=} C \restriction \{(q, v) \in Q : q > 0\}$ denote the agents with positive contribution, and define C^- analogously. We have that either $\mathsf{ctr}(C) < 0$ and all transitions of agents in C^+ would be enabled, or $\mathsf{ctr}(C) \geq 0$ and the transitions of C^- could be executed.

If C^+ is enabled, then we have to wait at most $n/|C^+|$ steps in expectation until a transition is executed, which reduces $|C^+|$ by one. In total we get $n/|C_0^+| + n/(|C_0^+| - 1) + ... + n/1 \in \mathcal{O}(n \log n)$. The same holds for C^-, yielding our overall bound of $\mathcal{O}(n \log n)$.

5.2 Modulo Predicates and Boolean Combinations

Proposition 2. *Let* $\varphi(x_1, ..., x_k) \stackrel{def}{\iff} \sum_{i=1}^{k} \alpha_i x_i \equiv c \pmod{l} < c$ *denote a linear inequality, with* $\alpha_1, \ldots, \alpha_k, c, l \in \mathbb{Z}, l \geq 2$. *There exists a broadcast consensus protocol that computes* φ *within* $\mathcal{O}(n \log n)$ *interactions in expectation.*

Proof (sketch). The idea is the same as for Proposition 1, but instead of taking care not to overflow the counter we simply perform the additions modulo l.

Proposition 3 (Boolean combination of predicates). *Let* φ *be a Boolean combination of predicates* $\varphi_1, ..., \varphi_k$*, which are computed by BCPs* $\mathcal{P}_1, ..., \mathcal{P}_k$*, respectively, within* $\mathcal{O}(n \log n)$ *interactions. Then there is a protocol computing* φ *within* $\mathcal{O}(n \log n)$ *interactions.*

Proof (sketch). We do a simple parallel composition of the k BCPs, which is the same construction as used for ordinary population protocols (see for example [5, Lemma 6]). A detailed proof can be found in the full version of this paper.

6 Protocols for all Predicates in ZPL

BCPs compute precisely the predicates in NL with input encoded in unary, which corresponds to $\mathsf{NSPACE}(n)$ when encoded in binary. The proof of the NL

lower bound by Blondin, Esparza and Jaax [14] goes through multiple stages of reduction and thus does not reveal which predicates can be computed *efficiently*. We will now take a more direct approach, using a construction similar to the one by Angluin, Aspnes and Eisenstat [7]. A step of a randomised Turing machine (RTM) can be simulated using variants of the protocols for Presburger predicates from Section 5, which we combine with a clock to determine whether the step has finished, with high probability.

Instead of simulating RTMs directly, it is more convenient to first reduce them to counter machines. Here, we will use counter machines that are both randomised and capable of multiplying and dividing by two, with the latter also determining the remainder. This ensures that the reduction is performed efficiently, i.e. with overhead of $\mathcal{O}(n \log n)$ interactions per step.

We first show the other direction: simulating BCPs with RTMs.

Lemma 1. *Polynomial-time BCPs compute at most the predicates in* ZPL *with input encoded in unary.*

Proof. An RTM can store the number of agents in each state as binary counters. Picking an agent uniformly at random can be done in $\mathcal{O}(\log n)$ time by picking a random number between 1 and n and comparing it to the agents in the different states. Simulating a transition can also be done with logarithmic overhead. It can further be shown that stabilization of the execution is decidable in time $\mathcal{O}(\log n)$ (see the full version of this paper for details). As the BCP uses only $\mathcal{O}(\text{poly}\, n)$ interactions (in expectation) the RTM is also $\mathcal{O}(\text{poly}\, n)$ time-bounded.

Theorem 2. *Polynomial-time BCPs compute exactly the predicates in* ZPL *with input encoded in unary.*

The proof of Theorem 2 will take up the remainder of this section.

Counter machines. Let $\text{Cmd} \stackrel{\text{def}}{=} \{\mathsf{mul}_2, \mathsf{inc}, \mathsf{divmod}_2, \mathsf{iszero}\}$ denote a set of commands, and $\text{Ret} \stackrel{\text{def}}{=} \{\mathsf{done}_0, \mathsf{done}_1\}$ a set of completion statuses. A *multiplicative counter machine with k counters (k-CM)* $A = (S, \mathcal{T}_1, \mathcal{T}_2)$ consists of a finite set of states S with $\mathsf{init}, 0, 1 \in S$ and two transition functions $\mathcal{T}_1, \mathcal{T}_2$ mapping a state $q \in S$ to a tuple (i, j, q'_0, q'_1) where $i \in \{1, ..., k\}$ refers to a counter, $j \in \text{Cmd}$ is a command, and $q'_0, q'_1 \in S$ are successor states (q'_1 is not used for mul_2 and inc operations). Additionally, we require that $\mathcal{T}_1, \mathcal{T}_2$ map $q \in \{0, 1\}$ to $(1, \mathsf{iszero}, q, q)$, effectively executing no operation from those states.

The idea is that A, starting in state init, picks transitions uniformly at random from either $\mathcal{T}_1$ or $\mathcal{T}_2$. Apart from this randomness, the transitions are deterministic. Eventually, A ends up in either state 0 or 1, at which point it cannot perform further actions, thereby indicating whether the input is accepted or rejected.

Step-execution function. A *CM-configuration* is a tuple $K = (q, x_1, ..., x_k) \in Q \times \mathbb{N}^k$. We define the *step-execution function* step as follows, with $x \in \mathbb{N}$:

- $\mathsf{step}(\mathsf{mul}_2, x) \stackrel{\text{def}}{=} (\mathsf{done}_0, 2x)$,
- $\mathsf{step}(\mathsf{inc}, x) \stackrel{\text{def}}{=} (\mathsf{done}_0, x + 1)$,

– $\mathsf{step}(\mathsf{divmod}, 2x + b) \stackrel{\text{def}}{=} (\mathsf{done}_b, x)$, for $b \in \{0, 1\}$, and
– $\mathsf{step}(\mathsf{iszero}, x) \stackrel{\text{def}}{=} (\mathsf{done}_b, x)$, where b is 1 if $x > 0$ and 0 else.

For two CM-configurations $K = (q, x_1, ..., x_k)$ and $K' = (q', x'_1, ..., x'_k)$ where $\mathcal{T}_\circ(q) = (i, j, q'_0, q'_1)$ for $\circ \in \{1, 2\}$ we write $K \xrightarrow{\circ} K'$ if $\mathsf{step}(j, x_i) = (\mathsf{done}_b, x'_i)$, $q' = q'_b$ for some $b \in \{0, 1\}$, and $x_r = x'_r$ for $r \neq i$. Note that for each K and $\circ$ there is exactly one K' with $K \xrightarrow{\circ} K'$.

The reasoning for introducing the step-execution function is that we want to construct a broadcast protocol (BP) which simulates just one step of the CM. Later on we can use this BP as a building block in a more general protocol.

Computation. Let $\varphi : \mathbb{N}^l \to \{0, 1\}$ denote a predicate, for $l \leq k$, and $C \in \mathbb{N}^l$ an input to φ. We sample a *random (CM-)execution* $\pi = K_0K_1K_2...$ *for input* C, where $K_0, ...$ are CM-configurations, via a Markov chain. For the initial configuration we have $K_0 \stackrel{\text{def}}{=} (\mathsf{init}, C(1), ..., C(l), 0, ..., 0)$, and K_i is determined as the unique configuration with $K_{i-1} \xrightarrow{\circ} K_i$, where $\circ \in \{1, 2\}$ is chosen uniformly at random. (So π is the random variable defined as trace of the Markov Chain.)

We say that A *computes* φ *within* $f(n)$ *steps* if for each $C \in \mathbb{N}^l$ with $|C| = n$ the random execution for input C reaches a configuration in $\{\varphi(C)\} \times \mathbb{N}^k$ after at most $f(n)$ steps in expectation. Finally, A is n-*bounded* if the random executions for inputs C with $|C| = n$ can only reach configurations in $Q \times \mathbb{N}^k_{\leq n}$.

Theorem 3. *Let* φ *be a predicate decidable by a log-space bounded RTM within* $\mathcal{O}(f(n))$ *steps in expectation with unary input encoding. There exists an* n-*bounded CM that accepts* φ *within* $\mathcal{O}(f(n)\log(n))$ *steps in expectation.*

Proof (sketch). This can be shown by first representing the Turing machine by a stack machine with two stacks that contain the tape content to the left/right of the current machine head position. In this representation, head movements and tape updates amount to performing pop/push operations on the stack. Moreover, we can simulate an $c \cdot n$-bounded stack by c many n-bounded stacks. An n-bounded stack, in turn, can be represented in a counter machine with a constant number of 2^n-bounded counters. The stack content is represented as the base-2 number corresponding to the binary sequence stored in the stack. Popping then amounts to a divmod_2 operation, and pushing amounts to doubling the counter value, followed by adding 1 or 0, respectively.

A detailed proof can be found in the full version of this paper.

We formally define two types of BPs, ones that simulate a step of the CM, and ones behaving like a clock.

Definition 1. *Let BP* $\mathcal{P} = (Q \times G, \delta)$ *denote a BP with global states* G *where* $0, 1, \perp \in Q$ *and* $\mathrm{Cmd}, \mathrm{Ret} \subseteq G$. *We define the injection* $\varphi : G \times \mathbb{N}_{\leq n} \to \mathbb{N}^{Q \times G}$ *as* $\varphi(j, x) \stackrel{\text{def}}{=} x \cdot \wr(1, j)\wr + (n - x) \cdot \wr(0, j)\wr$. *The configurations in* $\varphi(\mathrm{Cmd} \times \mathbb{N})$ *are called* initial, *the ones in* $\varphi(\mathrm{Ret} \times \mathbb{N})$ final. *We call a configuration* C failing, *if* $C(\perp, i) > 0$ *for some* $i \in G$.

We say that $\mathcal{P}$ *is* CM-simulating *if the sets of final and failing configurations are closed under reachability, and from every initial configuration* $\varphi(j, w)$ *the only reachable final configuration is* $\varphi(\mathrm{step}(j, w))$, *if both are well-defined.*

Definition 2. *Let $\mathcal{P} = (Q, \delta)$ denote a BP with $0, 1 \in Q$ and* Time($\mathcal{P}$) *the number of steps until $\mathcal{P}$, starting in configuration $\wr 0, ..., 0 \int$, reaches $\wr 1, ..., 1 \int$, or ∞ if it does not. If* Time($\mathcal{P}$) *is almost surely finite and no agent is in state 1 before* Time($\mathcal{P}$)*, then we call $\mathcal{P}$ a* clock-BP.

Now we begin by constructing a CM-simulating BP. The value of a given counter is scattered across the population: each agent stores its contribution to this counter value in its state. The counter value is the sum of all contributions. Usually, an agent's contribution is either 1 or 0, thus n agents can maximally store a counter value equal to n, which is not problematic, since the counter machine is assumed to be n-bounded. The difficult part is multiplying and dividing the counter by two. Besides contributions 0 and 1, we will also allow intermediate contributions $\frac{1}{2}$ and 2. By executing a single broadcast, we can multiply (or divide) all the individual contributions by 2, by setting all contributions of value 1 to $\frac{1}{2}$, or 2, respectively. Then, over time, we "normalise" the agents to all have contribution 0 or 1 again in a manner which is specified below. This process takes some time, and we cannot determine with perfect reliability whether it is finished, so we only bound the time with high probability. Here and in the following, we say that some event (dependent on the population size n) happens *with high probability*, if for *all* $k > 0$ the event happens with probability $1 - \mathcal{O}(n^{-k})$.

In this and subsequent lemmata we use $\mathcal{G}(p)$, for $0 < p < 1$, to denote the geometric distribution, that is the number of *trials* until a coin flip with probability p succeeds, which has expectation $1/p$. We start with a statement about the tail distributions of sums of geometric variables.

Lemma 2. *Let $n \geq 3$ and $X_1, ..., X_n$ denote independent random variables with sum X and $X_i \sim \mathcal{G}(i/n)$. Then for any $k \geq 1$ there is an l s.t.*

$$\mathbb{P}(X \geq l \cdot n \ln n) \leq n^{-k}$$

Proof. See the full version of this paper.

Lemma 3. *There is a CM-simulating BP s.t. starting from an initial configuration it reaches a final configuration within $\mathcal{O}(n \log n)$ steps with high probability.*

Proof. Let $\mathcal{P} = (Q \times G, \delta)$ denote our BP, with $Q \stackrel{\text{def}}{=} \{0, \frac{1}{2}, 1, 2, *\}$ and $G \stackrel{\text{def}}{=}$ Cmd $\cup$ Ret $\cup$ {high}. The following transitions initialise the computation, with $b \in \{0, 1\}$:

$$(b, \mathsf{mul}_2) \mapsto (2b, \mathsf{done}_0),\ \{1 \mapsto 2, 0 \mapsto 0\} \qquad (\alpha_1)$$

$$(b, \mathsf{divmod}_2) \mapsto (\tfrac{b}{2}, \mathsf{done}_0),\ \{1 \mapsto \tfrac{1}{2}, 0 \mapsto 0\} \qquad (\alpha_2)$$

$$(b, \mathsf{inc}) \mapsto (b, \mathsf{high}),\ \emptyset \qquad (\alpha_3)$$

Additionally, we need transitions that move agents back into states 0 and 1.

$$(0, \mathsf{high}) \mapsto (1, \mathsf{done}_0),\ \emptyset \qquad (\beta_1)$$

$$(2, \mathsf{done}_0) \mapsto (1, \mathsf{high}),\ \emptyset \qquad (\beta_2)$$

$$(\tfrac{1}{2}, \mathsf{done}_0) \mapsto (0, \mathsf{done}_1),\ \emptyset \qquad (\beta_3)$$

$$(\tfrac{1}{2}, \mathsf{done}_1) \mapsto (1, \mathsf{done}_0),\ \emptyset \qquad (\beta_4)$$

This requires some explanation. Basically, we have the invariant that for a configuration C the current value of the counter is $b + \sum_{i \in Q, j \in G} i \cdot C((i,j))$, where b is 1 if the global state is high and 0 else. There is a "canonical" representation of each counter value, where $b = 0$ and the individual contributions $i \in Q$ are only 0 and 1. The transitions (α_1-α_3) update the represented counter value in a single step, but cause a "noncanonical" representation. The transitions (β_1-β_4) preserve the value of the counter and cause the representation to eventually become canonical.

This corresponds to final configurations from Definition 1: as long as the representation is noncanonical, i.e. an agent with value $\frac{1}{2}$, 2 or $*$ exists, the configuration is not final. Conversely, once we reach a final configuration our representation is canonical, and, as the value of the counter is preserved, we reach the correct final configuration.

$$(1, \mathsf{iszero}) \mapsto (1, \mathsf{done}_1),\ \emptyset \qquad (\alpha_4)$$
$$(0, \mathsf{iszero}) \mapsto (0, \mathsf{done}_0),\ \{1 \mapsto *\} \qquad (\alpha_5)$$
$$(*, \mathsf{done}_0) \mapsto (1, \mathsf{done}_1),\ \{* \mapsto 1\} \qquad (\beta_5)$$

For iszero we do something similar, but the value of the counter does not change. If the initial transition is executed by an agent with value 1, we can go to the global state done_1 directly. Otherwise, we replace 1 by $*$ and go to done_0, so if no agents with value 1 exist, we are finished. Else some agent with value $*$ executes (β_5) and we move to the correct final configuration.

Final configurations can only contain states $\{0,1\} \times \mathrm{Ret}$. As we have no outgoing transitions from those states, they are indeed closed under reachability.

It remains to be shown that starting from a configuration C_0 we reach a final configuration within $\mathcal{O}(n \log n)$ steps with high probability. Note that transitions (α_1-α_5) are executed at most once. Moreover, these are the only transitions enabled at C_0, so let C_1 denote the successor configuration after executing (α_1-α_5), i.e. $C_0 \rightarrow C_1$. From now on, we consider only transitions (β_1-β_5).

Let $M \overset{\text{def}}{=} \{\frac{1}{2}, 2, *\} \times G$ denote the set of "noncanonical" states, and, for a configuration C, let $\Phi(C) \overset{\text{def}}{=} 2\sum_{q \in M} C(q) + b$ denote a potential function, with b being 1 if the global state of C is high and 0 else. Now we can observe that executing a (β_1-β_5) transition strictly decreases Φ, and that $0 \leq \Phi(C) \leq 2n$ for any configuration C. So after at most $2n$ non-silent transitions, we have reached a final configuration.

Fix some transition (β_j), let $q \in Q \times G$ denote the state initiating (β_j), and let C, C', C'' denote configurations with $C \xrightarrow{\beta_j} C' \xrightarrow{*} C''$, meaning that C'' is a configuration reachable from C after executing (β_j). Then, we claim, $C(q) > C''(q)$.

To see that this holds for transitions (β_2-β_5), note that for $i \in \{\frac{1}{2}, 2, *\}$ the number of agents with value i can only decrease when executing transitions (β_1-β_5). For (β_1) this is slightly more complicated, as (β_3) increases the number of agents with value 0. However, (β_1) is reachable only after (α_1) or (α_3) has been executed, while (β_3) requires (α_2). Thus, our claim follows.

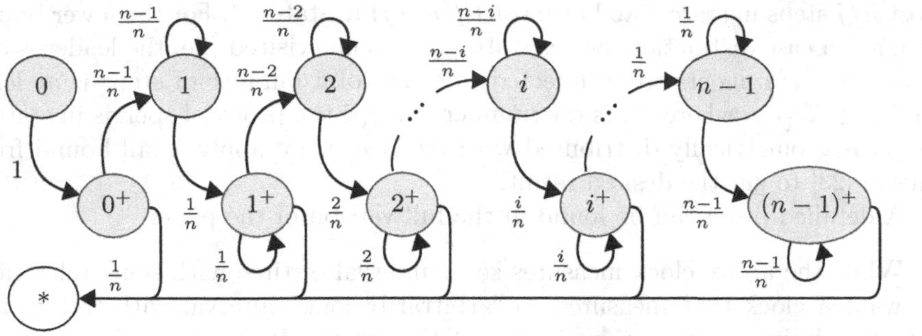

Fig. 3. State diagram of the clock implementation. Nodes with i agents in state c_3 are labelled i or i^+, the latter denoting that the other agents are in states c_1^+ and c_2^+. The final state $*$ has all agents in state 1. Arcs are labelled with transition probabilities.

Let X_k denote the number of silent transitions before executing (β_j) for the k-th time, $k = 1, ..., l$, and let r_k denote the number of agents in state q at that time. Then $n \geq r_1 > r_2 > ... > r_l \geq 1$ and X_k is distributed according to $\mathcal{G}(r_k/n)$. So we can use Lemma 2 to show that the sum of X_k is $\mathcal{O}(n \log n)$ with high probability. There are only 5 transitions (β_j), so the same holds for the total number of steps until reaching a final state.

Our next construction is the clock-BP, which indicates that some amount of time has passed (with high probability). Angluin, Aspnes and Eisenstat used epidemics for this purpose [7], as do we. The idea is that one agent initiates an epidemic and waits until it sees an infected agent. Similar to standard analysis of the coupon collector's problem, this is likely to take $\Theta(n \log n)$ time.

Lemma 4. *There is a clock-BP* $\mathcal{P} = (Q, \delta)$ *s.t.* $\mathbb{E}(\text{Time}(\mathcal{P})) \in \mathcal{O}(n \log n)$ *and* $\text{Time}(\mathcal{P}) \in \Omega(n \log n)$ *with probability* $1 - \mathcal{O}(n^{-1/2})$.

Proof (sketch). For a clock we use states $\{0, 1, c_1, c_2, c_3, c_1^+, c_2^+\}$ and transitions

$$0 \mapsto c_1^+, \{0 \mapsto c_2^+\} \qquad (\alpha)$$
$$c_2^+ \mapsto c_3, \{c_2^+ \mapsto c_2, c_1^+ \mapsto c_1\} \qquad (\beta)$$
$$c_3 \mapsto c_3, \{c_2 \mapsto c_2^+, c_1 \mapsto c_1^+\} \qquad (\gamma)$$
$$c_1^+ \mapsto 1, \{c_2^+ \mapsto 1, c_3 \mapsto 1\} \qquad (\omega)$$

State 0 is the initial state, 1 the final state. States c_1 and c_2 denote "uninfected" agents, state c_3 "infected" ones. The former can become activated (moving to c_1^+ and c_2^+), causing one of them to become infected. Transition (α) marks a leader c_1, once they are infected the clock ends (via (ω)). In (β), a single activated agent becomes infected, deactivating the other agents. They get activated again via transition (γ). The state diagram is shown in Figure 3.

It remains to show that this protocol fulfils the stated time bounds. We prove $\mathbb{E}(\text{Time}(\mathcal{P})) \in \mathcal{O}(n \log n)$ by using that, in expectation, the protocol spends at most n/j steps in state j and at most $n/(n-j)$ in state j^+. For the lower bound we make a case distinction: either state $\lfloor\sqrt{n}\rfloor$ is not visited (i.e. the leader is one of the first $\sqrt{n}$ agents to be infected), or the total number of steps is at least $X_1 + ... + X_{\lfloor\sqrt{n}\rfloor}$, where X_j is the number of steps the protocol spends in state i. As X_j is geometrically distributed with mean n/j, we apply a tail bound from Janson [23] to get the desired result.

A detailed proof can be found in the full version of the paper.

While the above clock measures some interval of time with some reliability, we want a clock that measures an "arbitrarily long" interval with "arbitrarily high" reliability. Constructions for population protocols use phase clocks for this purpose, but broadcasts allow us to synchronise the agents, so we can directly execute the clock multiple times in sequence instead.

Lemma 5. *Let $k \in \mathbb{N}$ denote some constant. Then there is a clock-BP $\mathcal{P}$ s.t. $\mathbb{E}(\mathrm{Time}(\mathcal{P})) \in \mathcal{O}(n \log n)$, and $\mathrm{Time}(\mathcal{P}) < kn \log n$ with probability $\mathcal{O}(n^{-k})$.*

Proof (sketch). The idea is that we run $28k^2$ clocks in sequence, in groups of $2k$. Then it is likely that at least one clock in each group works, yielding the overall minimum running time. A detailed proof can be found in the full version of this paper.

As mentioned earlier, we combine the clock with the construction in Lemma 3. While we cannot reliably determine whether the operation has finished, we can use a clock to measure an interval of time long enough for the protocol to terminate with high probability. The next construction does just that. In particular, in contrast to Lemma 3, it uses its global state to indicate that it is done.

Lemma 6. *There is a CM-simulating BP s.t. starting from an initial configuration it reaches either a final or a failing configuration C almost surely and within $\mathcal{O}(n \log n)$ steps in expectation, and C is final with high probability. Additionally, all reachable configurations with global state in* Ret *are final or failing.*

Proof. Fix some $k \in \mathbb{N}$ and let $\mathcal{P} = (Q \times G, \delta)$ denote the BP we want to construct. Further, let $\mathcal{P}_1 = (Q_1 \times G_1, \delta_1)$ denote the BP from Lemma 3 and choose some c s.t. $\mathcal{P}_1$ reaches a final configuration after at most $cn \log n$ steps with probability at least $1 - n^{-k}$.

Now we use Lemma 5 to get a clock $\mathcal{P}_2 = (Q_2, \delta_2)$ that runs for at least $cn \log n$ steps with probability at least $1 - n^{-k}$.

We do a parallel composition of $\mathcal{P}_1$ and $\mathcal{P}_2$ to get $\mathcal{P}$. In particular, $Q \stackrel{\text{def}}{=} Q_1 \times Q_2$, $G \stackrel{\text{def}}{=} \{j_\circ : j \in G_1\} \cup \mathrm{Ret}$, where for Q we identify $(i, 0)$ with i for $i \in \{0, 1\ \bot\}$, and for G we identify j with $j_\circ$ for $j \in \mathrm{Cmd}$.

Intuitively, we use $\circ$ to rename the global states of $\mathcal{P}_1$, meaning that the global state $j \in G_1$ of $\mathcal{P}_1$ is now called $j_\circ$ in our protocol. We want $\mathcal{P}_1$ to start with the same initial state we have, which is why we identified j with $j_\circ$ for $j \in \mathrm{Cmd}$. However, we only want to enter a final configurations once the clock has run out, so the completion statuses of $\mathcal{P}_1$ are renamed into $j_\circ$ for $j \in \mathrm{Ret}$ and we enter a final configuration by setting to global state to a $j \in \mathrm{Ret}$.

For each $(q_1, j) \in Q_1 \times G_1$ and $q_2 \in Q_2$ with $\delta_1(q_1, j) = ((r_1, j'), f_1)$ and $\delta_2(q_2) = (r_2, f_2)$ we get the transition

$$(q_1, q_2, j_\circ) \mapsto (r_1, r_2, j'_\circ),\ \{(t_1, t_2) \mapsto (f_1(t_1), f(t_2)) : t_1 \in Q_1, t_2 \in Q_2\} \quad (\alpha)$$

These transitions, together with the way we identified states, ensure that $\mathcal{P}_1$ and $\mathcal{P}_2$ run normally, with the input being passed through to $\mathcal{P}_1$ transparently. However, note that the final configurations of $\mathcal{P}_1$ are not final for $\mathcal{P}$, meaning that the protocol never ends. Hence, for $q_1 \in Q_1, j \in \text{Ret}$ we add the transition

$$\begin{aligned}(q_1, 1, j_\circ) \mapsto (q_1, 0, j), \{(b, 1) \mapsto (b, 0) : b \in \{0, 1\}\} \\ \cup \{(i, 1) \mapsto (\bot, 0) : i \in Q_1 \setminus \{0, 1\}\}\end{aligned} \quad (\beta)$$

This terminates the protocol once the clock has run out. If $\mathcal{P}_1$ was in a final state, we will now enter a final state as well, else we move into a failing state.

Finally, we use the above BP to simulate the full l-CM.

Lemma 7. *Fix some predicate $\varphi : \mathbb{N}^k \to \{0, 1\}$ computable by an n-bounded l-CM within $\mathcal{O}(f(n)) \subseteq \mathcal{O}(\text{poly}\, n)$ steps. Then there is a BCP computing φ in $\mathcal{O}(f(n)\, n \log n)$ steps.*

Proof (sketch). For each counter we need n agents, so ln in total, but we can simply have each agent simulate a constant number of agents. To execute a step of the CM, we use the BP from Lemma 6. It succeeds only with high probability, but in the case of failure at least one agent will have local state $\bot$, from which that agent initiates a restart of the whole computation.

As the CM takes only a polynomial number of steps, we can fix a k s.t. a computation of our BCP without failures (i.e. one that succeeds on the first try) takes $\mathcal{O}(n^k)$ steps. A single step succeeds with high probability, so we can require it to fail with probability at most $\mathcal{O}(n^{-k-1})$. In total, the restarts increase the running time by a factor of $1/(1 - \mathcal{O}(n^{-1}))$, which is only a constant overhead.

A detailed proof can be found in the full version of this paper.

This completes the proof of Theorem 2. By Theorem 3, each predicate in ZPL (with input encoded in unary) is computable by a bounded l-CM. Lemma 7 then yields a polynomial-time BCP for that predicate.

We remark that our reductions also enable us to construct efficient BPPs for specific predicates. The predicate PowerOfTwo for example, as described in [14, Proposition 3], can trivially be decided by an $\mathcal{O}(\log n)$-time bounded RTM with input encoded as binary, so there is also a BCP computing that predicate within $\mathcal{O}(n \log^2 n)$ interactions.

References

1. Alistarh, D., Aspnes, J., Eisenstat, D., Gelashvili, R., Rivest, R.L.: Time-space trade-offs in population protocols. In: Proceedings of the twenty-eighth annual ACM-SIAM symposium on discrete algorithms. pp. 2560–2579. SIAM (2017)

2. Alistarh, D., Aspnes, J., Gelashvili, R.: Space-optimal majority in population protocols. In: Proceedings of the Twenty-Ninth Annual ACM-SIAM Symposium on Discrete Algorithms. pp. 2221–2239. SIAM (2018)
3. Alistarh, D., Gelashvili, R.: Recent algorithmic advances in population protocols. ACM SIGACT News **49**(3), 63–73 (2018)
4. Alistarh, D., Gelashvili, R., Vojnović, M.: Fast and exact majority in population protocols. In: Proceedings of the 2015 ACM Symposium on Principles of Distributed Computing. pp. 47–56 (2015)
5. Angluin, D., Aspnes, J., Diamadi, Z., Fischer, M.J., Peralta, R.: Computation in networks of passively mobile finite-state sensors. Distributed computing **18**(4), 235–253 (2006), https://www.cs.yale.edu/homes/aspnes/papers/podc04passive-dc.pdf
6. Angluin, D., Aspnes, J., Eisenstat, D.: Stably computable predicates are semilinear. In: Proceedings of the Twenty-fifth Annual ACM Symposium on Principles of Distributed Computing. pp. 292–299. PODC '06, ACM, New York, NY, USA (2006). https://doi.org/10.1145/1146381.1146425
7. Angluin, D., Aspnes, J., Eisenstat, D.: Fast computation by population protocols with a leader. Distributed Computing **21**(3), 183–199 (2008), https://www.cs.yale.edu/homes/aspnes/papers/disc2006-journal.pdf
8. Aspnes, J.: Clocked population protocols. In: Proceedings of the 2017 ACM Symposium on Principles of Distributed Computing (PODC). pp. 431–440 (2017). https://doi.org/10.1145/3087801.3087836
9. Aspnes, J., Ruppert, E.: An introduction to population protocols. In: Middleware for Network Eccentric and Mobile Applications, pp. 97–120. Springer (2009)
10. Belleville, A., Doty, D., Soloveichik, D.: Hardness of computing and approximating predicates and functions with leaderless population protocols. In: 44th International Colloquium on Automata, Languages, and Programming (ICALP 2017). Schloss Dagstuhl-Leibniz-Zentrum fuer Informatik (2017)
11. Berenbrink, P., Giakkoupis, G., Kling, P.: Optimal time and space leader election in population protocols. In: Proceedings of the 52nd Annual ACM SIGACT Symposium on Theory of Computing. pp. 119–129 (2020)
12. Bertrand, N., Dewaskar, M., Genest, B., Gimbert, H.: Controlling a population. In: Proc. 28th International Conference on Concurrency Theory (CONCUR). vol. 85, pp. 12:1–12:16 (2017). https://doi.org/10.4230/LIPIcs.CONCUR.2017.12
13. Blondin, M., Esparza, J., Jaax, S.: Expressive Power of Broadcast Consensus Protocols. In: Proceedings of the 30th International Conference on Concurrency Theory (CONCUR). pp. 31:1–31:16 (2019). https://doi.org/10.4230/LIPIcs.CONCUR.2019.31, http://drops.dagstuhl.de/opus/volltexte/2019/10933
14. Blondin, M., Esparza, J., Jaax, S.: Expressive power of broadcast consensus protocols (2019), https://arxiv.org/abs/1902.01668.pdf
15. Delzanno, G., Raskin, J., Begin, L.V.: Towards the automated verification of multithreaded Java programs. In: Proc. 8th International Conference on Tools and Algorithms for the Construction and Analysis of Systems (TACAS). pp. 173–187 (2002)
16. Doty, D., Soloveichik, D.: Stable leader election in population protocols requires linear time. Distributed Computing **31**(4), 257–271 (2018)
17. Emerson, E.A., Namjoshi, K.S.: On model checking for non-deterministic infinite-state systems. In: Proc. Thirteenth Annual IEEE Symposium on Logic in Computer Science (LICS). pp. 70–80 (1998). https://doi.org/10.1109/LICS.1998.705644

18. Esparza, J., Finkel, A., Mayr, R.: On the verification of broadcast protocols. In: Proc. 14th Annual IEEE Symposium on Logic in Computer Science (LICS). pp. 352–359 (1999). https://doi.org/10.1109/LICS.1999.782630
19. Finkel, A., Leroux, J.: How to compose Presburger-accelerations: Applications to broadcast protocols. In: Proc. 22nd Conference on Foundations of Software Technology and Theoretical Computer Science (FSTTCS). pp. 145–156 (2002)
20. Flajolet, P., Gardy, D., Thimonier, L.: Birthday paradox, coupon collectors, caching algorithms and self-organizing search. Discrete Applied Mathematics **39**(3), 207–229 (1992)
21. Gąsieniec, L., Staehowiak, G.: Fast space optimal leader election in population protocols. In: Proceedings of the Twenty-Ninth Annual ACM-SIAM Symposium on Discrete Algorithms. pp. 2653–2667. SIAM (2018)
22. Ginsburg, S., Spanier, E.H.: Semigroups, presburger formulas, and languages. Pacific J. Math. **16**(2), 285–296 (1966), https://projecteuclid.org:443/euclid.pjm/1102994974
23. Janson, S.: Tail bounds for sums of geometric and exponential variables. Statistics & Probability Letters **135**, 1–6 (2018), https://arxiv.org/pdf/1709.08157.pdf
24. Kosowski, A., Uznanski, P.: Brief announcement: Population protocols are fast. In: Newport, C., Keidar, I. (eds.) Proceedings of the 2018 ACM Symposium on Principles of Distributed Computing, PODC 2018, Egham, United Kingdom, July 23-27, 2018. pp. 475–477. ACM (2018), https://dl.acm.org/citation.cfm?id=3212788
25. Michail, O., Chatzigiannakis, I., Spirakis, P.G.: Mediated population protocols. Theoretical Computer Science **412**(22), 2434–2450 (2011)
26. Michail, O., Spirakis, P.G.: Terminating population protocols via some minimal global knowledge assumptions. Journal of Parallel and Distributed Computing pp. 1–10 (2015). https://doi.org/10.1016/j.jpdc.2015.02.005
27. Nisan, N.: On read once vs. multiple access to randomness in logspace. Theoretical Computer Science **107**(1), 135–144 (1993)
28. Schmitz, S., Schnoebelen, P.: The power of well-structured systems. In: Proc. 24th International Conference on Concurrency Theory (CONCUR). pp. 5–24 (2013)
29. Sudo, Y., Masuzawa, T.: Leader election requires logarithmic time in population protocols. Parallel Processing Letters **30**(01), 2050005 (2020)

Leafy automata for higher-order concurrency

Alex Dixon[1] (✉), Ranko Lazić[2], Andrzej S. Murawski[3], and Igor Walukiewicz[4]

[1] University of Warwick, Coventry, UK, alexander.dixon@warwick.ac.uk
[2] University of Warwick, Coventry, UK
[3] University of Oxford, Oxford, UK
[4] CNRS, Université de Bordeaux, Talence, France

Abstract. Finitary Idealized Concurrent Algol (FICA) is a prototypical programming language combining functional, imperative, and concurrent computation. There exists a fully abstract game model of FICA, which in principle can be used to prove equivalence and safety of FICA programs. Unfortunately, the problems are undecidable for the whole language, and only very rudimentary decidable sub-languages are known.

We propose leafy automata as a dedicated automata-theoretic formalism for representing the game semantics of FICA. The automata use an infinite alphabet with a tree structure. We show that the game semantics of any FICA term can be represented by traces of a leafy automaton. Conversely, the traces of any leafy automaton can be represented by a FICA term. Because of the close match with FICA, we view leafy automata as a promising starting point for finding decidable subclasses of the language and, more generally, to provide a new perspective on models of higher-order concurrent computation.

Moreover, we identify a fragment of FICA that is amenable to verification by translation into a particular class of leafy automata. Using a locality property of the latter class, where communication between levels is restricted and every other level is bounded, we show that their emptiness problem is decidable by reduction to Petri net reachability.

Keywords: Finitary Idealized Concurrent Algol, Higher-Order Concurrency, Automata over Infinite Alphabets, Game Semantics

1 Introduction

Game semantics is a versatile paradigm for giving semantics to a wide spectrum of programming languages [3,35]. It is well-suited for studying the observational equivalence of programs and, more generally, the behaviour of a program in an arbitrary context. About 20 years ago, it was discovered that the game semantics of a program can sometimes be expressed by a finite automaton or another simple computational model [20]. This led to algorithmic uses of game semantics for program analysis and verification [1,15,21,5,27,26,28,34,16,17]. Thus far, these advances concerned mostly languages without concurrency.

S. Kiefer and C. Tasson (Eds.): FOSSACS 2021, LNCS 12650, pp. 184–204, 2021.
https://doi.org/10.1007/978-3-030-71995-1_10

In this work, we consider Finitary Idealized Concurrent Algol (FICA) and its fully abstract game semantics [22]. It is a call-by-name language with higher-order features, side-effects, and concurrency implemented by a parallel composition operator and semaphores. It is finitary since, as it is common in this context, base types are restricted to finite domains. Quite surprisingly, the game semantics of this language is arguably simpler than that for the language without concurrency. The challenge comes from algorithmic considerations.

Following the successful approach from the sequential case [20,37,33,36,11], the first step is to find an automaton model abstracting the phenomena appearing in the semantics. The second step is to obtain program fragments from structural restrictions on the automaton model. In this paper we take both steps.

We propose *leafy automata*: an automaton model working on nested data. Data are used to represent pointers in plays, while the nesting of data reflects structural dependencies in the use of pointers. Interestingly, the structural dependencies in plays boil down to imposing a tree structure on the data. We show a close correspondence between the automaton model and the game semantics of FICA. For every program, there is a leafy automaton whose traces (data words) represent precisely the plays in the semantics of the program (Theorem 3). Conversely, for every leafy automaton, there is a program whose semantics consists of plays representing the traces of the automaton (Theorem 5). (The latter result holds modulo a saturation condition we explain later.) This equivalence shows that leafy automata are a suitable model for studying decidability questions for FICA.

Not surprisingly, due to their close connection to FICA, leafy automata turn out to have an undecidable emptiness problem. We use the undecidability argument to identify the source, namely communication across several unbounded levels, i.e., levels in which nodes can produce an unbounded number of children during the lifetime of the automaton. To eliminate the problem, we introduce a restricted variant of leafy automata, called *local*, in which every other level is bounded and communication is allowed to cross only one unbounded node. Emptiness for such automata can be decided via reduction to a number of instances of Petri net reachability problem.

We also identify a fragment of FICA, dubbed *local* FICA (LFICA), which maps onto local leafy automata. It is based on restricting the distance between semaphore and variable declarations and their uses inside the term. This is a first non-rudimentary fragment of FICA for which some verification tasks are decidable. Overall, this makes it possible to use local leafy automata to analyse LFICA terms and decide associated verification tasks.

Related work Concurrency, even with only first-order recursion, leads to undecidability [39]. Intuitively, one can encode the intersection of languages of two pushdown automata. From the automata side, much research on decidable cases has concentrated on bounding interactions between stacks representing different threads of the program [38,30,4]. From the game semantics side, the only known decidable fragment of FICA is Syntactic Control of Concurrency (SCC) [23], which imposes bounds on the number of threads in which arguments can be used.

This restriction makes it possible to represent the game semantics of programs by finite automata. In our work, we propose automata models that correspond to unbounded interactions with arbitrary FICA contexts, and importantly that remains true also when we restrict the terms to LFICA. Leafy automata are a model of computation over an infinite alphabet. This area has been explored extensively, partly motivated by applications to database theory, notably XML [41]. In this context, nested data first appeared in [7], where the authors considered shuffle expressions as the defining formalism. Later on, data automata [9] and class memory automata [8] have been adapted to nested data in [14,12]. They are similar to leafy automata in that the automaton is allowed to access states related to previous uses of data values at various depths. What distinguishes leafy automata is that the lifetime of a data value is precisely defined and follows a question and answer discipline in correspondence with game semantics. Leafy automata also feature run-time "zero-tests", activated when reading answers.

For most models over nested data, the emptiness problem is undecidable. To achieve decidability, the authors in [14,12] relax the acceptance conditions so that the emptiness problem can eventually be recast as a coverability problem for a well-structured transition system. In [10], this result was used to show decidability of equivalence for a first-order (sequential) fragment of Reduced ML. On the other hand, in [7] the authors relax the order of letters in words, which leads to an analysis based on semi-linear sets. Both of these restrictions are too strong to permit the semantics of FICA, because of the game-semantic WAIT condition, which corresponds to waiting until all sub-processes terminate.

Another orthogonal strand of work on concurrent higher-order programs is based on higher-order recursion schemes [24,29]. Unlike FICA, they feature recursion but the computation is purely functional over a single atomic type o.

Structure of the paper: In the next two sections we recall FICA and its game semantics from [22]. The following sections introduce leafy automata (LA) and their local variant (LLA), where we also analyse the associated decision problems and, in particular, show that the non-emptiness problem for LLA is decidable. Subsequently, we give a translation from FICA to LA (and back) and define a fragment LFICA of FICA which can be translated into LLA. We will occasionally refer the reader to the full paper [18] which includes appendices with proof details and worked examples.

2 Finitary Idealized Concurrent Algol (FICA)

Idealized Concurrent Algol [22] is a paradigmatic language combining higher-order with imperative computation in the style of Reynolds [40], extended to concurrency with parallel composition ($||$) and binary semaphores. We consider its finitary variant FICA over the finite datatype $\{0, \ldots, max\}$ ($max \geq 0$) with loops but no recursion. Its types θ are generated by the grammar

$$\theta ::= \beta \mid \theta \to \theta \qquad \beta ::= \mathbf{com} \mid \mathbf{exp} \mid \mathbf{var} \mid \mathbf{sem}$$

$$\frac{}{\Gamma \vdash \mathbf{skip} : \mathbf{com}} \qquad \frac{}{\Gamma \vdash \mathbf{div}_\theta : \theta} \qquad \frac{}{\Gamma \vdash i : \mathbf{exp}} \qquad \frac{\Gamma \vdash M : \mathbf{exp}}{\Gamma \vdash \mathbf{op}(M) : \mathbf{exp}}$$

$$\frac{\Gamma \vdash M : \mathbf{com} \qquad \Gamma \vdash N : \beta}{\Gamma \vdash M; N : \beta} \qquad \frac{\Gamma \vdash M : \mathbf{com} \qquad \Gamma \vdash N : \mathbf{com}}{\Gamma \vdash M || N : \mathbf{com}}$$

$$\frac{\Gamma \vdash M : \mathbf{exp} \qquad \Gamma \vdash N_1, N_2 : \beta}{\Gamma \vdash \mathbf{if}\, M\, \mathbf{then}\, N_1\, \mathbf{else}\, N_2 : \beta} \qquad \frac{\Gamma \vdash M : \mathbf{exp} \qquad \Gamma \vdash N : \mathbf{com}}{\Gamma \vdash \mathbf{while}\, M\, \mathbf{do}\, N : \mathbf{com}}$$

$$\frac{}{\Gamma, x : \theta \vdash x : \theta} \qquad \frac{\Gamma, x : \theta \vdash M : \theta'}{\Gamma \vdash \lambda x.M : \theta \to \theta'} \qquad \frac{\Gamma \vdash M : \theta \to \theta' \qquad \Gamma \vdash N : \theta}{\Gamma \vdash MN : \theta'}$$

$$\frac{\Gamma \vdash M : \mathbf{var} \qquad \Gamma \vdash N : \mathbf{exp}}{\Gamma \vdash M := N : \mathbf{com}} \qquad \frac{\Gamma \vdash M : \mathbf{var}}{\Gamma \vdash !M : \mathbf{exp}}$$

$$\frac{\Gamma \vdash M : \mathbf{sem}}{\Gamma \vdash \mathbf{release}(M) : \mathbf{com}} \qquad \frac{\Gamma \vdash M : \mathbf{sem}}{\Gamma \vdash \mathbf{grab}(M) : \mathbf{com}}$$

$$\frac{\Gamma, x : \mathbf{var} \vdash M : \mathbf{com}, \mathbf{exp}}{\Gamma \vdash \mathbf{newvar}\, x := i\, \mathbf{in}\, M : \mathbf{com}, \mathbf{exp}} \qquad \frac{\Gamma, x : \mathbf{sem} \vdash M : \mathbf{com}, \mathbf{exp}}{\Gamma \vdash \mathbf{newsem}\, x := i\, \mathbf{in}\, M : \mathbf{com}, \mathbf{exp}}$$

Fig. 1: FICA typing rules

where **com** is the type of commands; **exp** that of $\{0, \ldots, max\}$-valued expressions; **var** that of assignable variables; and **sem** that of semaphores. The typing judgments are displayed in Figure 1. **skip** and $\mathbf{div}_\theta$ are constants representing termination and divergence respectively, i ranges over $\{0, \cdots, max\}$, and **op** represents unary arithmetic operations, such as successor or predecessor (since we work over a finite datatype, operations of bigger arity can be defined using conditionals). Variables and semaphores can be declared locally via **newvar** and **newsem**. Variables are dereferenced using $!M$, and semaphores are manipulated using two (blocking) primitives, **grab**(s) and **release**(s), which grab and release the semaphore respectively. The small-step operational semantics of FICA is reproduced in the full paper [18, Appendix A]. We shall write **div** for $\mathbf{div}_{\mathbf{com}}$.

We are interested in *contextual equivalence* of terms. Two terms are contextually equivalent if there is no context that can distinguish them with respect to may-termination. More formally, a term $\vdash M : \mathbf{com}$ is said to terminate, written $M \Downarrow$, if there exists a terminating evaluation sequence from M to **skip**. Then *contextual (may-)equivalence* ($\Gamma \vdash M_1 \cong M_2$) is defined by: for all contexts $\mathcal{C}$ such that $\vdash \mathcal{C}[M] : \mathbf{com}$, $\mathcal{C}[M_1] \Downarrow$ if and only if $\mathcal{C}[M_2] \Downarrow$. The force of this notion is quantification over all contexts.

Since contextual equivalence becomes undecidable for FICA very quickly [23], we will look at the special case of testing equivalence with terms that always diverge, e.g. given $\Gamma \vdash M : \theta$, is it the case that $\Gamma \vdash M \cong \mathbf{div}_\theta$? Intuitively, equivalence with an always-divergent term means that $\mathcal{C}[M]$ will never converge (must diverge) if $\mathcal{C}$ uses M. At the level of automata, this will turn out to correspond to the emptiness problem.

In verification tasks, with the above equivalence test, we can check whether uses of M can ever lead to undesirable states. For example, for a given term $x : \mathbf{var} \vdash M : \theta$, the term

$$f : \theta \to \mathbf{com} \vdash \mathbf{newvar}\, x := 0\, \mathbf{in}\, (f(M)\, ||\, \mathbf{if}\, !x = 13\, \mathbf{then\ skip\ else\ div})$$

will be equivalent to **div** only when x is never set to 13 during a terminating execution. Note that, because of quantification over all contexts, f may use M an arbitrary number of times, also concurrently or in nested fashion, which is a very expressive form of quantification.

3 Game semantics

Game semantics for programming languages involves two players, called Opponent (O) and Proponent (P), and the sequences of moves made by them can be viewed as interactions between a program (P) and a surrounding context (O). In this section, we briefly present the fully abstract game model for FICA from [22], which we rely on in the paper. The games are defined using an auxiliary concept of an arena.

Definition 1. *An* arena *A is a triple $\langle M_A, \lambda_A, \vdash_A \rangle$ where:*

- *M_A is a set of* moves*;*
- *$\lambda_A : M_A \to \{O, P\} \times \{Q, A\}$ is a function determining for each $m \in M_A$ whether it is an* Opponent *or a* Proponent *move, and a* question *or an* answer*; we write $\lambda_A^{OP}, \lambda_A^{QA}$ for the composite of λ_A with respectively the first and second projections;*
- *$\vdash_A$ is a binary relation on M_A, called* enabling*, satisfying: if $m \vdash_A n$ for no m then $\lambda_A(n) = (O, Q)$, if $m \vdash_A n$ then $\lambda_A^{OP}(m) \neq \lambda_A^{OP}(n)$, and if $m \vdash_A n$ then $\lambda_A^{QA}(m) = Q$.*

We shall write I_A for the set of all moves of A which have no enabler; such moves are called *initial*. Note that an initial move must be an Opponent question. In arenas used to interpret base types all questions are initial and P-moves answering them are detailed in the table below, where $i \in \{0, \cdots, max\}$.

Arena	O-question	P-answers	Arena	O-question	P-answers
$[\![\mathbf{com}]\!]$	run	done	$[\![\mathbf{exp}]\!]$	q	i
$[\![\mathbf{var}]\!]$	read write(i)	i ok	$[\![\mathbf{sem}]\!]$	grb rls	ok ok

More complicated types are interpreted inductively using the *product* $(A \times B)$ and *arrow* $(A \Rightarrow B)$ constructions, given below.

$$\begin{array}{ll} M_{A\times B} = M_A + M_B & M_{A\Rightarrow B} = M_A + M_B \\ \lambda_{A\times B} = [\lambda_A, \lambda_B] & \lambda_{A\Rightarrow B} = [\langle \lambda_A^{PO}, \lambda_A^{QA}\rangle, \lambda_B] \\ \vdash_{A\times B} = \vdash_A + \vdash_B & \vdash_{A\Rightarrow B} = \vdash_A + \vdash_B + \{\, (b,a) \mid b \in I_B \text{ and } a \in I_A\} \end{array}$$

where $\lambda_A^{PO}(m) = O$ iff $\lambda_A^{QP}(m) = P$. We write $[\![\theta]\!]$ for the arena corresponding to type θ. Below we draw (the enabling relations of) $A_1 = [\![\mathbf{com} \to \mathbf{com} \to \mathbf{com}]\!]$ and $A_2 = [\![(\mathbf{var} \to \mathbf{com}) \to \mathbf{com}]\!]$ respectively, using superscripts to distinguish copies of the same move (the use of superscripts is consistent with our future use of tags in Definition 9).

$$
\begin{array}{lllll}
O & & & & \text{run} \\
P & \text{run}^2 & \text{run}^1 & & \text{done} \\
O & \text{done}^2 & \text{done}^1 & &
\end{array}
\qquad
\begin{array}{lllll}
O & & & & \text{run} \\
P & & & \text{run}^1 & \text{done} \\
O & \text{read}^{11} & \text{write}(i)^{11} & \text{done}^1 & \\
P & i^{11} & \text{ok}^{11} & &
\end{array}
$$

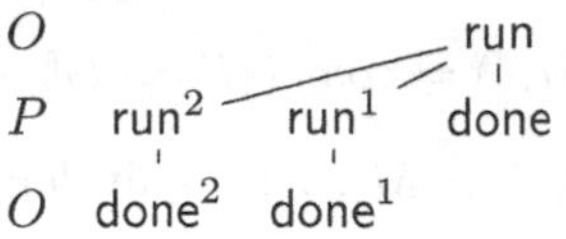

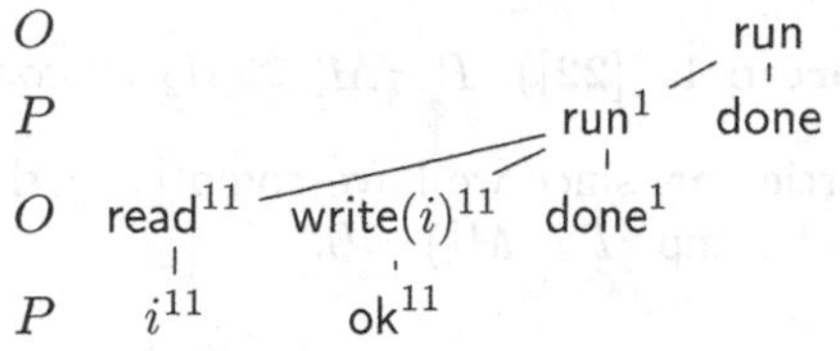

Given an arena A, we specify next what it means to be a legal play in A. For a start, the moves that players exchange will have to form a *justified sequence*, which is a finite sequence of moves of A equipped with pointers. Its first move is always initial and has no pointer, but each subsequent move n must have a unique pointer to an earlier occurrence of a move m such that $m \vdash_A n$. We say that n is (explicitly) justified by m or, when n is an answer, that n answers m. If a question does not have an answer in a justified sequence, we say that it is *pending* in that sequence. Below we give two justified sequences from A_1 and A_2 respectively.

$$
\text{run } \text{run}^1 \text{ run}^2 \text{ done}^1 \text{ done}^2 \text{ done} \qquad \text{run } \text{run}^1 \text{ read}^{11} \ 0^{11} \text{ write}(1)^{11} \text{ ok}^{11} \text{ read}^{11} \ 1^{11}
$$

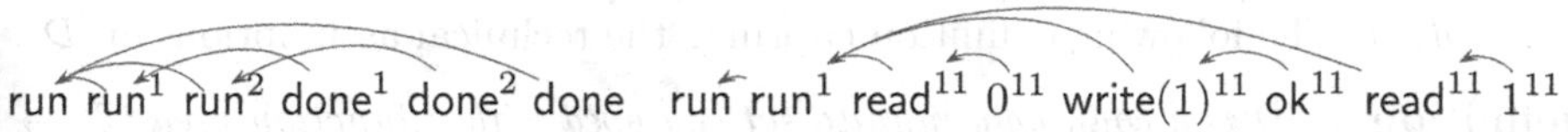

Not all justified sequences are valid. In order to constitute a legal play, a justified sequence must satisfy a well-formedness condition that reflects the "static" style of concurrency of our programming language: any started sub-processes must end before the parent process terminates. This is formalised as follows, where the letters q and a to refer to question- and answer-moves respectively, while m denotes arbitrary moves.

Definition 2. *The set P_A of plays over A consists of the justified sequences s over A that satisfy the two conditions below.*

FORK : *In any prefix $s' = \cdots q \cdots m$ of s, the question q must be pending when m is played.*

WAIT : *In any prefix $s' = \cdots q \cdots a$ of s, all questions justified by q must be answered.*

It is easy to check that the justified sequences given above are plays. A subset σ of P_A is *O-complete* if $s \in \sigma$ and $so \in P_A$ imply $so \in \sigma$, when o is an O-move.

Definition 3. *A* strategy *on A, written $\sigma : A$, is a prefix-closed O-complete subset of P_A.*

Suppose $\Gamma = \{x_1 : \theta_1, \cdots, x_l : \theta_l\}$ and $\Gamma \vdash M : \theta$ is a FICA-term. Let us write $[\![\Gamma \vdash \theta]\!]$ for the arena $[\![\theta_1]\!] \times \cdots \times [\![\theta_l]\!] \Rightarrow [\![\theta]\!]$. In [22] it is shown how to assign a strategy on $[\![\Gamma \vdash \theta]\!]$ to any FICA-term $\Gamma \vdash M : \theta$. We write $[\![\Gamma \vdash M]\!]$ to refer to that strategy. For example, $[\![\Gamma \vdash \mathbf{div}]\!] = \{\epsilon, \mathsf{run}\}$ and $[\![\Gamma \vdash \mathbf{skip}]\!] = \{\epsilon, \mathsf{run}, \mathsf{run}\,\mathsf{done}\}$. Given a strategy σ, we denote by $\mathsf{comp}(\sigma)$ the set of non-empty *complete* plays of σ, i.e. those in which all questions have been answered. The game-semantic interpretation $[\![\cdots]\!]$ turns out to provide a fully abstract model in the following sense.

Theorem 1 ([22]). *$\Gamma \vdash M_1 \cong M_2$ iff $\mathsf{comp}([\![\Gamma \vdash M_1]\!]) = \mathsf{comp}([\![\Gamma \vdash M_2]\!])$.*

In particular, since we have $\mathsf{comp}([\![\Gamma \vdash \mathbf{div}_\theta]\!]) = \emptyset$, $\Gamma \vdash M : \theta$ is equivalent to $\mathbf{div}_\theta$ iff $\mathsf{comp}([\![\Gamma \vdash M]\!]) = \emptyset$.

4 Leafy automata

We would like to be able to represent the game semantics of FICA using automata. To that end, we introduce *leafy automata* (LA). They are a variant of automata over nested data, i.e. a type of automata that read finite sequences of letters of the form $(t, d_0 d_1 \cdots d_j)$ $(j \in \mathbb{N})$, where t is a *tag* from a finite set Σ and each d_i $(0 \le i \le j)$ is a *data value* from an infinite set $\mathcal{D}$.

In our case, $\mathcal{D}$ will have the structure of a countably infinite forest and the sequences $d_0 \cdots d_j$ will correspond to branches of a tree. Thus, instead of $d_0 \cdots d_j$, we can simply write d_j, because d_j uniquely determines its ancestors: $d_0, \ldots, d_{j-1}$. The following definition captures the technical assumptions on $\mathcal{D}$.

Definition 4. *$\mathcal{D}$ is a countably infinite set equipped with a function $pred : \mathcal{D} \to \mathcal{D} \cup \{\bot\}$ (the* parent *function) such that the following conditions hold.*

- *Infinite branching: $pred^{-1}(\{d_\bot\})$ is infinite for any $d_\bot \in \mathcal{D} \cup \{\bot\}$.*
- *Well-foundedness: for any $d \in \mathcal{D}$, there exists $i \in \mathbb{N}$, called the* level *of d, such that $pred^{i+1}(d) = \bot$. Level-0 data values will be called* roots.

In order to define configurations of leafy automata, we will rely on finite subtrees of $\mathcal{D}$, whose nodes will be labelled with states. We say that $T \subseteq \mathcal{D}$ is a subtree of $\mathcal{D}$ iff T is closed ($\forall x \in T$: $pred(x) \in T \cup \{\bot\}$) and rooted ($\exists! x \in T$: $pred(x) = \bot$).

Next we give the formal definition of a level-k leafy automaton. Its set of states Q will be divided into layers, written $Q^{(i)}$ $(0 \le i \le k)$, which will be used to label level-i nodes. We will write $Q^{(i_1, \cdots, i_k)}$ to abbreviate $Q^{(i_1)} \times \cdots \times Q^{(i_k)}$, excluding any components $Q^{(i_j)}$ where $i_j < 0$. We distinguish $Q^{(0,-1)} = \{\dagger\}$.

Definition 5. *A level-k leafy automaton (k-LA) is a tuple $\mathcal{A} = \langle \Sigma, k, Q, \delta \rangle$, where*

- *$\Sigma = \Sigma_{\mathsf{Q}} + \Sigma_{\mathsf{A}}$ is a finite alphabet, partitioned into questions and answers;*
- *$k \ge 0$ is the level parameter;*
- *$Q = \sum_{i=0}^{k} Q^{(i)}$ is a finite set of states, partitioned into sets $Q^{(i)}$ of level-i states;*
- *$\delta = \delta_{\mathsf{Q}} + \delta_{\mathsf{A}}$ is a finite transition function, partitioned into question- and answer-related transitions;*

- $\delta_{\mathsf{Q}} = \sum_{i=0}^{k} \delta_{\mathsf{Q}}^{(i)}$, *where* $\delta_{\mathsf{Q}}^{(i)} \subseteq Q^{(0,1,\cdots,i-1)} \times \Sigma_{\mathsf{Q}} \times Q^{(0,1,\cdots,i)}$ *for* $0 \le i \le k$;
- $\delta_{\mathsf{A}} = \sum_{i=0}^{k} \delta_{\mathsf{A}}^{(i)}$, *where* $\delta_{\mathsf{A}}^{(i)} \subseteq Q^{(0,1,\cdots,i)} \times \Sigma_{\mathsf{A}} \times Q^{(0,1,\cdots,i-1)}$ *for* $0 \le i \le k$.

Configurations of LA are of the form (D, E, f), where D is a finite subset of $\mathcal{D}$ (consisting of data values that have been encountered so far), E is a finite subtree of $\mathcal{D}$, and $f : E \to Q$ is a level-preserving function, i.e. if d is a level-i data value then $f(d) \in Q^{(i)}$. A leafy automaton starts from the empty configuration $\kappa_0 = (\emptyset, \emptyset, \emptyset)$ and proceeds according to δ, making two kinds of transitions. Each kind manipulates a single leaf: for questions one new leaf is added, for answers one leaf is removed. Let the current configuration be $\kappa = (D, E, f)$.

- On reading a letter (t, d) with $t \in \Sigma_{\mathsf{Q}}$ and $d \notin D$ a fresh level-i data, the automaton adds a new leaf d in a configuration and updates the states on the branch to d. So it changes its configuration to $\kappa' = (D \cup \{d\}, E \cup \{d\}, f')$ provided that $pred(d) \in E$ and f' satisfies:

$$(f(pred^i(d)), \cdots, f(pred(d)), t, f'(pred^i(d)), \cdots, f'(pred(d)), f'(d)) \in \delta_{\mathsf{Q}}^{(i)},$$

  $\mathsf{dom}(f') = \mathsf{dom}(f) \cup \{d\}$, and $f'(x) = f(x)$ for all $x \notin \{pred(d), \cdots, pred^i(d)\}$.
- On reading a letter (t, d) with $t \in \Sigma_{\mathsf{A}}$ and $d \in E$ a level-i data which is a leaf, the automaton deletes d and updates the states on the branch to d. So it changes its configuration to $\kappa' = (D, E \setminus \{d\}, f')$ where f' satisfies:

$$(f(pred^i(d)), \cdots, f(pred(d)), f(d), t, f'(pred^i(d)), \cdots, f'(pred(d))) \in \delta_{\mathsf{A}}^{(i)},$$

  $\mathsf{dom}(f') = \mathsf{dom}(f) \setminus \{d\}$ and $f'(x) = f(x)$ for all $x \notin \{pred(d), \cdots, pred^i(d)\}$.
- Initially D,E, and f are empty; we proceed to $\kappa' = (\{d\}, \{d\}, \{d \mapsto q^{(0)}\})$ if (t, d) is read where $\dagger \xrightarrow{t} q^{(0)} \in \delta_{\mathsf{Q}}^{(0)}$. The last move is treated symmetrically.

In all cases, we write $\kappa \xrightarrow{(t,d)} \kappa'$. Note that a single transition can only change states on the branch ending in d. Other parts of the tree remain unchanged.

Example 1. Below we illustrate the effect of LA transitions. Let $D_1 = \{d_0, d_1, d_1'\}$ and $d_2 \notin D_1$. Let $\kappa_1 = (D_1, E_1, f_1)$, $\kappa_2 = (D_1 \cup \{d_2\}, E_2, f_2)$, $\kappa_3 = (D_1 \cup \{d_2\}, E_1, f_1)$, where the trees E_1, E_2 are displayed below and node annotations of the form (q) correspond to values of f_1, f_2, e.g. $f_1(d_0) = q^{(0)}$.

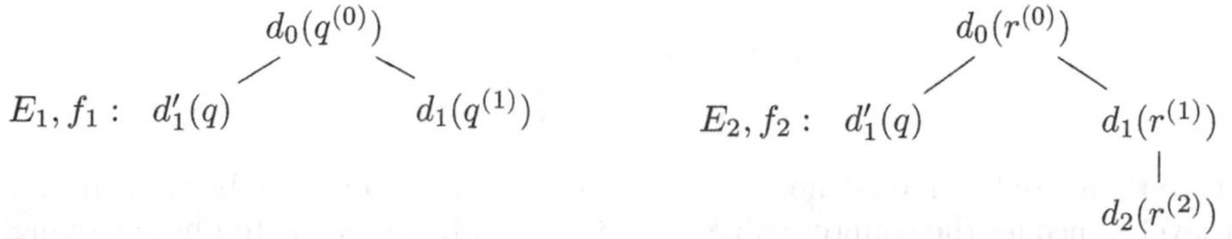

For κ_1 to evolve into κ_2 (on (t, d_2)), we need $(q^{(0)}, q^{(1)}, t, r^{(0)}, r^{(1)}, r^{(2)}) \in \delta_{\mathsf{Q}}^{(2)}$. On the other hand, to go from κ_2 to κ_3 (on (t, d_2)), we want $(r^{(0)}, r^{(1)}, r^{(2)}, t, q^{(0)}, q^{(1)}) \in \delta_{\mathsf{A}}^{(2)}$.

Definition 6. *A* trace *of a leafy automaton $\mathcal{A}$ is a sequence $w = l_1 \cdots l_h \in (\Sigma \times \mathcal{D})^*$ such that $\kappa_0 \xrightarrow{l_1} \kappa_1 \ldots \kappa_{h-1} \xrightarrow{l_h} \kappa_h$ where $\kappa_0 = (\emptyset, \emptyset, \emptyset)$. A configuration $\kappa = (D, E, f)$ is* accepting *if E and f are empty. A trace w is accepted by $\mathcal{A}$ if there is a non-empty sequence of transitions as above with κ_h accepting. The set of traces (resp. accepted traces) of $\mathcal{A}$ is denoted by $Tr(\mathcal{A})$ (resp. $L(\mathcal{A})$).*

Remark 1. When writing states, we will often use superscripts (i) to indicate the intended level. So $(q^{(0)}, \cdots, q^{(i-1)}) \xrightarrow{t} (r^{(0)}, \cdots, r^{(i)})$ refers to $(q^{(0)}, \cdots, q^{(i-1)}, t, r^{(0)}, \cdots, r^{(i)}) \in \delta_{\mathsf{Q}}^{(i)}$; similarly for $\delta_{\mathsf{A}}^{(i)}$ transitions. For $i = 0$, this degenerates to $\dagger \xrightarrow{t} r^{(0)}$ and $r^{(0)} \xrightarrow{t} \dagger$.

Example 2. Consider the 1-LA over $\Sigma_{\mathsf{Q}} = \{\mathsf{start}, \mathsf{inc}\}, \Sigma_{\mathsf{A}} = \{\mathsf{dec}, \mathsf{end}\}$. Let $Q^{(0)} = \{0\}$, $Q^{(1)} = \{0\}$ and define δ by: $\dagger \xrightarrow{\mathsf{start}} 0$, $0 \xrightarrow{\mathsf{inc}} (0, 0)$, $(0, 0) \xrightarrow{\mathsf{dec}} 0$, $0 \xrightarrow{\mathsf{end}} \dagger$. The accepted traces of this 1-LA have the form $(\mathsf{start}, d_0)\ (||_{i=0}^{n}(\mathsf{inc}, d_1^i)\ (\mathsf{dec}, d_1^i))\ (\mathsf{end}, d_0)$, i.e. they are valid histories of a single non-negative counter (histories such that the counter starts and ends at 0). In this case, all traces are simply prefixes of such words.

Remark 2. Note that, whenever a leafy automaton reads (t, d) $(t \in \Sigma_{\mathsf{Q}})$ and the level of d is greater than 0, then it must have read a unique question $(t', \mathit{pred}(d))$ earlier. Also, observe that an LA trace contains at most two occurrences of the same data value, such that the first is paired with a question and the second is paired with an answer. Because the question and the answer share the same data value, we can think of the answer as answering the question, like in game semantics. Indeed, justification pointers from answers to questions will be represented in this way in Theorem 3. Finally, we note that LA traces are invariant under tree automorphisms of $\mathcal{D}$.

Lemma 1. *The emptiness problem for* 2-LA *is undecidable. For* 1-LA, *it is reducible to the reachability problem for VASS in polynomial time and there is a reverse reduction in exponential time, so it is decidable in Ackermannian time [32] but not elementary [13].*

Proof. For 2-LA we reduce from the halting problem on two-counter-machines. Two counters can be simulated using configurations of the form

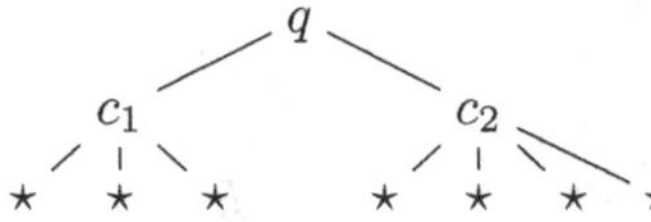

where there are two level-1 nodes, one for each counter. The number of children at level 2 encodes the counter value. Zero tests can be implemented by removing the corresponding level-1 node and creating a new one. This is possible only when the node is a leaf, i.e., it does not have children at level 2. The state of the 2-counter machine can be maintained at level 0, the states at level 1 indicate the name of the counter, and the level-2 states are irrelevant.

The translation from 1-LA to VASS is straightforward and based on representing 1-LA configurations by the state at level 0 and, for each state at level 1, the count of its occurrences. The reverse translation is based on the same idea and extends the encoding of a non-negative counter in Example 2, where the exponential blow up is simply due to the fact that vector updates in VASS are given in binary whereas 1-LA transitions operate on single branches. □

Lemma 2. 1-LA *equivalence is undecidable.*

Proof. We provide a direct reduction from the halting problem for 2-counter machines, where both counters are required to be zero initially as well as finally. The main obstacle is that implementing zero tests as in the proof of the first part of Lemma 1 is not available because we are restricted to leafy automata with levels 0 and 1 only. To overcome it, we exploit the power of the equivalence problem where one of the 1-LA will have the task not of correctly simulating zero tests but recognising zero tests that are incorrect. The complete argument can be found in the full paper [18, Appendix B]. □

5 Local leafy automata (LLA)

Here we identify a restricted variant of LA for which the emptiness problem is decidable. We start with a technical definition.

Definition 7. *A* k-LA *is* bounded *at level* i *(*$0 \le i \le k$*) if there is a bound* b *such that each node at level* i *can create at most* b *children during a run. We refer to* b *as the* branching bound.

Note that we are defining a "global" bound on the number of children that a node at level i may create across a whole run, rather than a "local" bound on the number of children a node may have in a given configuration.

To motivate the design of LLA, we observe that the undecidability argument (for the emptiness problem) for 2-LA used two consecutive levels (0 and 1) that are not bounded. For the node at level 0, this corresponded to the number of zero tests, while an unbounded counter is simulated at level 1. In the following we will eliminate consecutive unbounded levels by introducing an alternating pattern of bounded and unbounded levels. Even-numbered layers ($i = 0, 2, ...$) will be bounded, while odd-numbered layers will be unbounded. Observe in particular that the root (layer 0) is bounded. As we will see later, this alternation reflects the term/context distinction in game semantics: the levels corresponding to terms are bounded, and the levels coresponding to contexts are unbounded.

With this restriction alone, it is possible to reconstruct the undecidability argument for 4-LA, as two unbounded levels may still communicate. Thus we introduce a restriction on how many levels a transition can read and modify.

– when adding or removing a leaf at an odd level $2i + 1$, the automaton will be able to access levels $2i$, $2i - 1$ and $2i - 2$; while

– when adding or removing a leaf at an even level $2i$, the automaton will be able to access levels $2i-1$ and $2i-2$.

In particular, when an odd level produces a leaf, it will not be able to see the previous odd level. The above constraints mean that the transition functions $\delta_{\mathsf{Q}}^{(i)}, \delta_{\mathsf{Q}}^{(i)}$ can be presented in a more concise form, given below.

$$\delta_{\mathsf{Q}}^{(i)} \subseteq \begin{cases} Q^{(i-2,i-1)} \times \Sigma_{\mathsf{Q}} \times Q^{(i-2,i-1,i)} & \text{if } i \text{ is even} \\ Q^{(i-3,i-2,i-1)} \times \Sigma_{\mathsf{Q}} \times Q^{(i-3,i-2,i-1,i)} & \text{if } i \text{ is odd} \end{cases}$$

$$\delta_{\mathsf{A}}^{(i)} \subseteq \begin{cases} Q^{(i-2,i-1,i)} \times \Sigma_{\mathsf{A}} \times Q^{(i-2,i-1)} & \text{if } i \text{ is even} \\ Q^{(i-3,i-2,i-1,i)} \times \Sigma_{\mathsf{A}} \times Q^{(i-3,i-2,i-1)} & \text{if } i \text{ is odd} \end{cases}$$

In terms of the previous notation used for LA, $(q^{(i-2)}, q^{(i-1)}, x, r^{(i-2)}, r^{(i-1)}, r^{(i)}) \in \delta_{\mathsf{Q}}^{(i)}$ denotes all tuples of the form $(\vec{q}, q^{(i-2)}, q^{(i-1)}, x, \vec{q}, r^{(i-2)}, r^{(i-1)}, r^{(i)})$, where $\vec{q}$ ranges over $Q^{(0,\cdots,i-3)}$.

Definition 8. *A level-k* local leafy automaton *(k-*LLA*) is a k-*LA *whose transition function admits the above-mentioned presentation and which is bounded at all even levels.*

Theorem 2. *The emptiness problem for* LLA *is decidable.*

Proof (Sketch). Let b be a bound on the number of children created by each even node during a run.

The critical observation is that, once a node d at even level $2i$ has been created, all subsequent actions of descendants of d access (read and/or write) the states at levels $2i-1$ and $2i-2$ at most $2b$ times. The shape of the transition function dictates that this can happen only when child nodes at level $2i+1$ are added or removed. In addition, the locality property ensures that the automaton will never access levels $< 2i-2$ at the same time as node d or its descendants.

We will make use of these facts to construct *summaries* for nodes on even levels which completely describe such a node's lifetime, from its creation as a leaf until its removal, and in between performing at most $2b$ reads-writes of the parent and grandparent states. A summary is a sequence quadruples of states: two pairs of states of levels $2i-2$ and $2i-1$. The first pair are the states we expect to find on these levels, while the second are the states to which we update these levels. Hence a summary at level $2i$ is a complete record of a valid sequence of read-writes and stateful changes during the lifetime of a node on level $2i$.

We proceed by induction and show how to calculate the complete set of summaries at level $2i$ given the complete set of summaries at level $2i+2$. We construct a program for deciding whether a given sequence is a summary at level $2i$. This program can be evaluated via Vector Addition Systems with States (VASS). Since we can finitely enumerate all candidate summaries at level $2i$, this gives us a way to compute summaries at level $2i$. Proceeding this way, we finally calculate summaries at level 2. At this stage, we can reduce the emptiness problem for the given LLA to a reachability test on a VASS.

The complete argument is given in the full paper [18, Appendix C]. □

Let us remark also that the problem becomes undecidable if we remove either boundedness restriction, or allow transitions to look one level further.

6 From FICA to LA

Recall from Section 3 that, to interpret base types, game semantics uses moves from the set

$$\begin{aligned}\mathcal{M} &= M_{[\![\mathbf{com}]\!]} \cup M_{[\![\mathbf{exp}]\!]} \cup M_{[\![\mathbf{var}]\!]} \cup M_{[\![\mathbf{sem}]\!]} \\ &= \{\, \mathsf{run}, \mathsf{done}, \mathsf{q}, \mathsf{read}, \mathsf{grb}, \mathsf{rls}, \mathsf{ok} \,\} \cup \{\, i, \mathsf{write}(i) \mid 0 \le i \le \max \,\}.\end{aligned}$$

The game semantic interpretation of a term-in-context $\Gamma \vdash M : \theta$ is a strategy over the arena $[\![\Gamma \vdash \theta]\!]$, which is obtained through product and arrow constructions, starting from arenas corresponding to base types. As both constructions rely on the disjoint sum, the moves from $[\![\Gamma \vdash \theta]\!]$ are derived from the base types present in types inside Γ and θ. To indicate the exact occurrence of a base type from which each move originates, we will annotate elements of $\mathcal{M}$ with a specially crafted scheme of superscripts. Suppose $\Gamma = \{x_1 : \theta_1, \cdots, x_l : \theta_l\}$. The superscripts will have one of the two forms, where $\vec{i} \in \mathbb{N}^*$ and $\rho \in \mathbb{N}$:

- $(\vec{i}, \rho)$ will be used to represent moves from θ;
- $(x_v\vec{i}, \rho)$ will be used to represent moves from θ_v $(1 \le v \le l)$.

The annotated moves will be written as $m^{(\vec{i},\rho)}$ or $m^{(x_v\vec{i},\rho)}$, where $m \in \mathcal{M}$. We will sometimes omit ρ on the understanding that this represents $\rho = 0$. Similarly, when $\vec{i}$ is omitted, the intended value is ϵ. Thus, m stands for $m^{(\epsilon,0)}$.

The next definition explains how the $\vec{i}$ superscripts are linked to moves from $[\![\theta]\!]$. Given $X \subseteq \{m^{(\vec{i},\rho)} \mid \vec{i} \in \mathbb{N}^*, \rho \in \mathbb{N}\}$ and $y \in \mathbb{N} \cup \{x_1, \cdots, x_l\}$, we let $yX = \{m^{(y\vec{i},\rho)} \mid m^{(\vec{i},\rho)} \in X\}$.

Definition 9. *Given a type θ, the corresponding alphabet $\mathcal{T}_\theta$ is defined as follows*

$$\begin{aligned}\mathcal{T}_\beta &= \{\, m^{(\epsilon,\rho)} \mid m \in M_{[\![\beta]\!]}, \rho \in \mathbb{N} \,\} \qquad \beta = \mathbf{com}, \mathbf{exp}, \mathbf{var}, \mathbf{sem} \\ \mathcal{T}_{\theta_h \to \ldots \to \theta_1 \to \beta} &= \textstyle\bigcup_{u=1}^{h}(u\mathcal{T}_{\theta_u}) \cup \mathcal{T}_\beta\end{aligned}$$

For $\Gamma = \{x_1 : \theta_1, \cdots, x_l : \theta_l\}$, the alphabet $\mathcal{T}_{\Gamma\vdash\theta}$ is defined to be $\mathcal{T}_{\Gamma\vdash\theta} = \bigcup_{v=1}^{l}(x_v\mathcal{T}_{\theta_v}) \cup \mathcal{T}_\theta$.

Example 3. The alphabet $\mathcal{T}_{f:\mathbf{com}\to\mathbf{com}, x:\mathbf{com}\vdash\mathbf{com}}$ is $\{\mathsf{run}^{(f1,\rho)}, \mathsf{done}^{(f1,\rho)}, \mathsf{run}^{(f,\rho)}, \mathsf{done}^{(f,\rho)}, \mathsf{run}^{(x,\rho)}, \mathsf{done}^{(x,\rho)}, \mathsf{run}^{(\epsilon,\rho)}, \mathsf{done}^{(\epsilon,\rho)} \mid \rho \in \mathbb{N}\}$.

To represent the game semantics of terms-in-context, of the form $\Gamma \vdash M : \theta$, we are going to use *finite subsets* of $\mathcal{T}_{\Gamma\vdash\theta}$ as alphabets in leafy automata. The subsets will be finite, because ρ will be bounded. Note that $\mathcal{T}_\theta$ admits a natural partitioning into questions and answers, depending on whether the underlying move is a question or answer.

We will represent plays using data words in which the underpinning sequence of tags will come from an alphabet as defined above. Superscripts and data are used to represent justification pointers. Intuitively, we represent occurrences of questions with data values. Pointers from answers to questions just refer to these values. Pointers from questions use bounded indexing with the help of ρ.

Initial question-moves do not have a pointer and to represent such questions we simply use $\rho = 0$. For non-initial questions, we rely on the tree structure of $\mathcal{D}$ and use ρ to indicate the ancestor of the currently read data value that we mean to point at. Consider a trace $w(t_i, d_i)$ ending in a non-initial question, where d_i is a level-i data value and $i > 0$. In our case, we will have $t_i \in \mathcal{T}_{\Gamma \vdash \theta}$, i.e. $t_i = m^{(\cdots, \rho)}$. By Remark 2, trace w contains unique occurrences of questions $(t_0, d_0), \cdots, (t_{i-1}, d_{i-1})$ such that $pred(d_j) = d_{j-1}$ for $j = 1, \cdots, i$. The pointer from (t_i, d_i) goes to one of these questions, and we use ρ to represent the scenario in which the pointer goes to $(t_{i-(1+\rho)}, d_{i-(1+\rho)})$.

Pointers from answer-moves to question-moves are represented simply by using the same data value in both moves (in this case we use $\rho = 0$).

We will also use ϵ-tags ϵ_{Q} (question) and ϵ_{A} (answer), which do not contribute moves to the represented play. Each ϵ_{Q} will always be answered with ϵ_{A}. Note that the use of $\rho, \epsilon_{\mathsf{Q}}, \epsilon_{\mathsf{A}}$ means that several data words may represent the same play (see Examples 4, 6).

Example 4. Suppose $d_0 = pred(d_1), d_1 = pred(d_2) = pred(d_2'), d_2 = pred(d_3)$, and $d_2' = pred(d_3')$. Then the data word (run, d_0) (run^f, d_1) (run^{f1}, d_2) $(\mathsf{run}^{f1}, d_2')$ $(\mathsf{run}^{(x,2)}, d_3)$ $(\mathsf{run}^{(x,2)}, d_3')$ (done^x, d_3), which is short for $(\mathsf{run}^{(\epsilon,0)}, d_0)$ $(\mathsf{run}^{(f,0)}, d_1)$ $(\mathsf{run}^{(f1,0)}, d_2)$ $(\mathsf{run}^{(f1,0)}, d_2')$ $(\mathsf{run}^{(x,2)}, d_3)$ $(\mathsf{run}^{(x,2)}, d_3')$ $(\mathsf{done}^{(x,0)}, d_3)$, represents the play

$$\begin{array}{ccccccc} \mathsf{run} & \mathsf{run}^f & \mathsf{run}^{f1} & \mathsf{run}^{f1} & \mathsf{run}^x & \mathsf{run}^x & \mathsf{done}^x \\ O & P & O & O & P & P & O. \end{array}$$

Example 5. Consider the LA $\mathcal{A} = \langle Q, 3, \Sigma, \delta \rangle$, where $Q^{(0)} = \{0, 1, 2\}$, $Q^{(1)} = \{0\}$, $Q^{(2)} = \{0, 1, 2\}$, $Q^{(3)} = \{0\}$, $\Sigma_{\mathsf{Q}} = \{\mathsf{run}, \mathsf{run}^f, \mathsf{run}^{f1}, \mathsf{run}^{(x,2)}\}$, $\Sigma_{\mathsf{A}} = \{\mathsf{done}, \mathsf{done}^f, \mathsf{done}^{f1}, \mathsf{done}^x\}$, and δ is given by

$$\dagger \xrightarrow{\mathsf{run}} 0 \quad 0 \xrightarrow{\mathsf{run}^f} (1,0) \quad (1,0) \xrightarrow{\mathsf{done}^f} 2 \quad 2 \xrightarrow{\mathsf{done}} \dagger \quad (1,0) \xrightarrow{\mathsf{run}^{f1}} (1,0,0)$$
$$(1,0,0) \xrightarrow{\mathsf{run}^{(x,2)}} (1,0,1,0) \quad (1,0,1,0) \xrightarrow{\mathsf{done}^{(x,0)}} (1,0,2) \quad (1,0,2) \xrightarrow{\mathsf{done}^{f1}} (1,0)$$

Then traces from $Tr(\mathcal{A})$ represent all plays from $\sigma = [\![f : \mathbf{com} \to \mathbf{com}, x : \mathbf{com} \vdash fx]\!]$, including the play from Example 4, and $L(\mathcal{A})$ represents $\mathsf{comp}(\sigma)$.

Example 6. One might wish to represent plays of σ from the previous Example using data values $d_0, d_1, d_1', d_1'', d_2, d_2'$ such that $d_0 = pred(d_1) = pred(d_1') = pred(d_1''), d_1 = pred(d_2) = pred(d_2')$, so that the play from Example 4 is represented by $(\mathsf{run}^{(\epsilon,0)}, d_0)$ $(\mathsf{run}^{(f,0)}, d_1)$ $(\mathsf{run}^{(f1,0)}, d_2)$ $(\mathsf{run}^{(f1,0)}, d_2')$ $(\mathsf{run}^{(x,0)}, d_1')$ $(\mathsf{run}^{(x,0)}, d_1'')$ $(\mathsf{done}^{(x,0)}, d_1')$. Unfortunately, it is impossible to construct a 2-LA that would accept all representations of such plays. To achieve this, the automaton would have to make sure that the number of run^{f1}s is the same as that of run^xs. Because the former are labelled with level-2 values and the latter with incomparable level-1 values, the only point of communication (that could be used for comparison) is the root. However, the root cannot accommodate unbounded information, while plays of σ can feature an unbounded number of run^{f1}s, which could well be consecutive.

Before we state the main result linking FICA with leafy automata, we note some structural properties of the automata. Questions will create a leaf, and answers will remove a leaf. P-moves add leaves at odd levels (questions) and remove leaves at even levels (answers), while O-moves have the opposite effect at each level. Finally, when removing nodes at even levels we will not need to check if a node is a leaf. We call the last property *even-readiness.*

Even-readiness is a consequence of the WAIT condition in the game semantics. The condition captures well-nestedness of concurrent interactions – a term can terminate only after subterms terminate. In the leafy automata setting, this is captured by the requirement that only leaf nodes can be removed, i.e. a node can be removed only if all of its children have been removed beforehand. It turns out that, for *P-answers* only, this property will come for free. Formally, whenever the automaton arrives at a configuration $\kappa = (D, E, f)$, where $d \in E$ and there is a transition

$$(f(\mathit{pred}^{(2i)}(d)), \cdots, f(\mathit{pred}(d)), f(d), t, f'(\mathit{pred}^{(2i)}(d)), \cdots, f'(\mathit{pred}(d))) \in \delta_{\mathsf{A}}^{(2i)},$$

then d is a leaf. In contrast, our automata will not satisfy the same property for O-answers (the environment) and for such transitions it is crucial that the automaton actually checks that only leaves can be removed.

Theorem 3. *For any* FICA*-term* $\Gamma \vdash M : \theta$*, there exists an even-ready leafy automaton* $\mathcal{A}_M$ *over a finite subset of* $\mathcal{T}_{\Gamma\vdash\theta} + \{\epsilon_{\mathsf{Q}}, \epsilon_{\mathsf{A}}\}$ *such that the set of plays represented by data words from* $Tr(\mathcal{A}_M)$ *is exactly* $[\![\Gamma \vdash M : \theta]\!]$*. Moreover,* $L(\mathcal{A}_M)$ *represents* $\mathsf{comp}([\![\Gamma \vdash M : \theta]\!])$ *in the same sense.*

Proof (Sketch). Because every FICA-term can be converted to $\beta\eta$-normal form, we use induction on the structure of such normal forms. The base cases are: $\Gamma \vdash \mathbf{skip} : \mathbf{com}$ ($Q^{(0)} = \{0\}$; $\dagger \xrightarrow{\mathsf{run}} 0$, $0 \xrightarrow{\mathsf{done}} \dagger$), $\Gamma \vdash \mathbf{div} : \mathbf{com}$ ($Q^{(0)} = \{0\}$; $\dagger \xrightarrow{\mathsf{run}} 0$), and $\Gamma \vdash i : \mathbf{exp}$ ($Q^{(0)} = \{0\}$; $\dagger \xrightarrow{\mathsf{q}} 0$, $0 \xrightarrow{i} \dagger$).

The remaining cases are inductive. When referring to the inductive hypothesis for a subterm M_i, we shall use subscripts i to refer to the automata components, e.g. $Q_i^{(j)}$, $\xrightarrow{\mathsf{m}}_i$ etc. In contrast, $Q^{(j)}$, $\xrightarrow{\mathsf{m}}$ will refer to the automaton that is being constructed. Inference lines —— will indicate that the transitions listed under the line should be added to the new automaton provided the transitions listed above the line are present in the automaton obtained via induction hypothesis. We discuss a selection of technical cases below.

$\Gamma \vdash M_1 || M_2$ In this case we need to run the automata for M_1 and M_2 concurrently. To this end, their level-0 states will be combined ($Q^{(0)} = Q_1^{(0)} \times Q_2^{(0)}$), but not deeper states ($Q^{(j)} = Q_1^{(j)} + Q_2^{(j)}, 1 \le j \le k$). The first group of transitions activate and terminate the two components respectively: $\dfrac{\dagger \xrightarrow{\mathsf{run}}_1 q_1^{(0)} \quad \dagger \xrightarrow{\mathsf{run}}_2 q_2^{(0)}}{\dagger \xrightarrow{\mathsf{run}} (q_1^{(0)}, q_2^{(0)})}$,

$\frac{q_1^{(0)} \xrightarrow{\mathsf{done}}_1 \dagger \qquad q_2^{(0)} \xrightarrow{\mathsf{done}}_2 \dagger}{(q_1^{(0)}, q_2^{(0)}) \xrightarrow{\mathsf{done}} \dagger}$. The remaining transitions advance each component:

$$\frac{(q_1^{(0)},\cdots,q_1^{(j)}) \xrightarrow{\mathsf{m}}_1 (r_1^{(0)},\cdots,r_1^{(j')}) \qquad q_2^{(0)} \in Q_2^{(0)}}{((q_1^{(0)},q_2^{(0)}),\cdots,q_1^{(j)}) \xrightarrow{\mathsf{m}} ((r_1^{(0)},q_2^{(0)}),\cdots,r_1^{(j')})} \quad \frac{q_1^{(0)} \in Q_1^{(0)} \qquad (q_2^{(0)},\cdots,q_2^{(j)}) \xrightarrow{\mathsf{m}}_2 (r_2^{(0)},\cdots,r_2^{(j')})}{((q_1^{(0)},q_2^{(0)}),\cdots,q_2^{(j)}) \xrightarrow{\mathsf{m}} ((q_1^{(0)},r_2^{(0)}),\cdots,r_2^{(j')})}$$

where $\mathsf{m} \neq \mathsf{run}, \mathsf{done}$.

$\Gamma \vdash \mathbf{newvar}\, x := i\, \mathbf{in}\, M_1$ By [22], the semantics of this term is obtained from the semantics of $[\![\Gamma, x \vdash M_1]\!]$ by

1. restricting to plays in which the moves read^x, $\mathsf{write}(n)^x$ are followed immediately by answers,
2. selecting those plays in which each answer to a read^x-move is consistent with the preceding $\mathsf{write}(n)^x$-move (or equal to i, if no $\mathsf{write}(n)^x$ was made),
3. erasing all moves related to x, e.g. those of the form $m^{(x,\rho)}$.

To implement 1., we will lock the automaton after each read^x- or $\mathsf{write}(n)^x$-move, so that only an answer to that move can be played next. Technically, this will be done by adding an extra bit (lock) to the level-0 state. To deal with 2., we keep track of the current value of x, also at level 0. This makes it possible to ensure that answers to read^x are consistent with the stored value and that $\mathsf{write}(n)^x$ transitions cause the right change. Erasing from condition 3 is implemented by replacing all moves with the x subscript with $\epsilon_{\mathsf{Q}}, \epsilon_{\mathsf{A}}$-tags.

Accordingly, we have $Q^{(0)} = (Q_1^{(0)} + (Q_1^{(0)} \times \{lock\})) \times \{0, \cdots, max\}$ and $Q^{(j)} = Q_1^{(j)}$ $(1 \leq j \leq k)$. As an example of a transition, we give the transition related to writing: $\frac{(q_1^{(0)},\cdots,q_1^{(j)}) \xrightarrow{\mathsf{write}(z)^{(x,\rho)}}_1 (r_1^{(0)},\cdots,r_1^{(j')}) \qquad 0 \leq n, z \leq max}{((q_1^{(0)},n),\cdots,q_1^{(j)}) \xrightarrow{\epsilon_{\mathsf{Q}}} ((r_1^{(0)},lock,z),\cdots,r_1^{(j')})}$.

$\Gamma \vdash f M_h \cdots M_1 : \mathbf{com}$ *with* $(f : \theta_h \to \cdots \to \theta_1 \to \mathbf{com})$ Here we will need $Q^{(0)} = \{0,1,2\}$, $Q^{(1)} = \{0\}$, $Q^{(j+2)} = \sum_{u=1}^{h} Q_u^{(j)}$ $(0 \leq j \leq k)$. The first group of transitions corresponding to calling and returning from f: $\dagger \xrightarrow{\mathsf{run}} 0$, $\quad 0 \xrightarrow{\mathsf{run}^f} (1,0)$, $(1,0) \xrightarrow{\mathsf{done}^f} 2$, $\quad 2 \xrightarrow{\mathsf{done}} \dagger$. Additionally, in state $(1,0)$ we want to enable the environment to spawn an unbounded number of copies of each of $\Gamma \vdash M_u : \theta_u$ $(1 \leq u \leq h)$. This is done through rules that embed the actions of the automata for M_u while (possibly) relabelling the moves in line with our convention for representing moves from game semantics. Such transitions have the general form $\frac{(q_u^{(0)},\cdots,q_u^{(j)}) \xrightarrow{m^{(t,\rho)}}_u (q_u^{(0)},\cdots,q_u^{(j')})}{(1,0,q_u^{(0)},\cdots,q_u^{(j)}) \xrightarrow{m^{(t',\rho')}} (1,0,q_u^{(0)},\cdots,q_u^{(j')})}$. Note that this case also covers $f : \mathbf{com}$ $(h = 0)$.

More details and the remaining cases are covered in the full paper [18, Appendix D], along with an example of a term and the corresponding LA. □

7 Local FICA

In this section we identify a family of FICA terms that can be translated into LLA rather than LA. To achieve boundedness at even levels, we remove while[5]. To achieve restricted communication, we will constrain the distance between a variable declaration and its use. Note that in the translation, the application of function-type variables increases LA depth. So in LFICA we will allow the link between the binder **newvar/newsem** x and each use of x to "cross" at most one occurrence of a free variable. For example, the following terms

- $\mathbf{newvar}\, x := 0\, \mathbf{in}\, x := 1 \,||\, f(x := 2)$,
- $\mathbf{newvar}\, x := 0\, \mathbf{in}\, f(\mathbf{newvar}\, y\, \mathbf{in}\, f(y := 1) \,||\, x := !y)$

will be allowed, but not $\mathbf{newvar}\, x := 0\, \mathbf{in}\, f(f(x := 1))$.

To define the fragment formally, given a term Q in $\beta\eta$-normal form, we use a notion of the *applicative depth of a variable* $x : \beta$ ($\beta = \mathbf{var}, \mathbf{sem}$) *inside* Q, written $ad_x(Q)$ and defined inductively by the table below. The applicative depth is increased whenever a functional identifier is applied to a term containing x.

shape of Q	$ad_x(Q)$
x	1
$y\,(y \neq x)$, **skip**, **div**, i	0
$\mathbf{op}(M)$, $!M$, $\mathbf{release}(M)$, $\mathbf{grab}(M)$	$ad_x(M)$
$M; N$, $M \| N$, $M := N$, $\mathbf{while}\, M\, \mathbf{do}\, N$	$\max(ad_x(M), ad_x(N))$
$\mathbf{if}\, M\, \mathbf{then}\, N_1\, \mathbf{else}\, N_2$	$\max(ad_x(M), ad_x(N_1), ad_x(N_2))$
$\lambda y.M$, $\mathbf{newvar}\,/\mathbf{newsem}\, y := i\, \mathbf{in}\, M$	$ad_x(M[z/y])$, where z is fresh
$f M_1 \cdots M_k$	$1 + \max(ad_x(M_1), \cdots, ad_x(M_k))$

Note that in our examples above, in the first two cases the applicative depth of x is 2; and in the third case it is 3.

Definition 10 (Local FICA). *A FICA-term $\Gamma \vdash M : \theta$ is* local *if its $\beta\eta$-normal form does not contain any occurrences of* **while** *and, for every subterm of the normal form of the shape* $\mathbf{newvar}\,/\mathbf{newsem}\, x := i\, \mathbf{in}\, N$, *we have* $ad_x(N) \leq 2$. *We write* LFICA *for the set of local* FICA *terms.*

Theorem 4. *For any* LFICA*-term $\Gamma \vdash M : \theta$, the automaton $\mathcal{A}_M$ obtained from the translation in Theorem 3 can be presented as a* LLA.

Proof (Sketch). We argue by induction that the constructions from Theorem 3 preserve presentability as a LLA.

The case of parallel composition involves running copies of M_1 and M_2 in parallel without communication, with their root states stored as a pair at level 0. Note, though, that each of the automata transitions independently of the state of the other automaton. In consequence, if the automata M_1 and M_2 are LLA, so

[5] The automaton for **while** M **do** N may repeatedly visit the automata for M and N, generating an unbounded number of children at level 0 in the process.

will be the automaton for $M_1||M_2$. The branching bound after the construction is the sum of the two bounds for M_1 and M_2.

For $\Gamma \vdash \mathbf{newvar}\, x := i\, \mathbf{in}\, M$, because the term is in LFICA, so is $\Gamma, x : \mathbf{var} \vdash M$ and we have $ad_x(M) \leq 2$. Then we observe that in the translation of Theorem 3 ($\Gamma, x : \mathbf{var} \vdash M : \theta$) the questions related to x, (namely $\mathsf{write}(i)^{(x,\rho)}$ and $\mathsf{read}^{(x,\rho)}$) correspond to creating leaves at levels 1 or 3, while the corresponding answers ($\mathsf{ok}^{(x,\rho)}$ and $i^{(x,\rho)}$ respectively) correspond to removing such leaves. In the construction for $\Gamma \vdash \mathbf{newvar}\, x\, \mathbf{in}\, M$, such transitions need access to the root (to read/update the current state) and the root is indeed within the allowable range: in an LLA transitions creating/destroying leaves at level 3 can read/write at level 0. All other transitions (not labelled by x) proceed as in M and need not consult the root for additional information about the current state, as it is propagated. Consequently, if M is represented by a LLA then the interpretation of $\mathbf{newvar}\, x := i\, \mathbf{in}\, M$ is also a LLA. The construction does not affect the branching bound, because the resultant runs can be viewed as a subset of runs of the automaton for M, i.e. those in which reads and writes are related.

For $f M_h \cdots M_1$, we observe that the construction first creates two nodes at levels 0 and 1, and the node at level 1 is used to run an unbounded number of copies of (the automaton for) M_i. The copies do not need access to the states stored at levels 0 and 1, because they are never modified when the copies are running. Consequently, if each M_i can be translated into a LLA, the outcome of the construction in Theorem 3 is also a LLA. The new branching bound is the maximum over bounds from $M_1, \cdots, M_h$, because at even levels children are produced as in M_i and level 0 produces only 1 child. □

Corollary 1. *For any* LFICA*-term* $\Gamma \vdash M : \theta$*, the problem of determining whether* $\mathsf{comp}(\llbracket \Gamma \vdash M \rrbracket)$ *is empty is decidable.*

Theorems 1 and 2 imply the above. Thanks to Theorem 1, it is decidable if a LFICA term is equivalent to a term that always diverges (cf. example on page 187). In case of inequivalence, our results could also be applied to extract the distinguishing context, first by extracting the witnessing trace from the argument underpinning Theorem 2 and then feeding it to the Definability Theorem (Theorem 41 [22]). This is a valuable property given that in the concurrent setting bugs are difficult to replicate.

8 From LA to FICA

In this section, we show how to represent leafy automata in FICA. Let $\mathcal{A} = \langle \Sigma, k, Q, \delta \rangle$ be a leafy automaton. We shall assume that $\Sigma, Q \subseteq \{0, \cdots, max\}$ so that we can encode the alphabet and states using type **exp**. We will represent a trace w generated by $\mathcal{A}$ by a play $\mathsf{play}(w)$, which simulates each transition with two moves, by O and P respectively. The child-parent links in $\mathcal{D}$ will be represented by justification pointers. We refer the reader to [18, Appendix F] for details. Below we just state the lemma that identifies the types that correspond to our encoding, where we write $\theta^{max+1} \to \beta$ for $\underbrace{\theta \to \cdots \to \theta}_{max+1} \to \beta$.

Lemma 3. *Let $\mathcal{A}$ be a k-LA and $w \in Tr(\mathcal{A})$. Then* $\mathsf{play}(w)$ *is a play in* $[\![\theta_k]\!]$, *where* $\theta_0 = \mathbf{com}^{max+1} \to \mathbf{exp}$ *and* $\theta_{i+1} = (\theta_i \to \mathbf{com})^{max+1} \to \mathbf{exp}$ $(i \geq 0)$.

Before we state the main result, we recall from [22] that strategies corresponding to FICA terms satisfy a closure condition known as *saturation*: swapping two adjacent moves in a play belonging to such a strategy yields another play from the same strategy, as long as the swap yields a play and it is not the case that the first move is by O and the second one by P. Thus, saturated strategies express causal dependencies of P-moves on O-moves. Consequently, one cannot expect to find a FICA-term such that the corresponding strategy is the smallest strategy containing $\{\, \mathsf{play}(w) \mid w \in Tr(\mathcal{A}) \,\}$. Instead, the best one can aim for is the following result.

Theorem 5. *Given a k-LA $\mathcal{A}$, there exists a FICA term* $\vdash M_{\mathcal{A}} : \theta_k$ *such that* $[\![\vdash M_{\mathcal{A}} : \theta_k]\!]$ *is the smallest saturated strategy containing* $\{\, \mathsf{play}(w) \mid w \in Tr(\mathcal{A}) \,\}$.

Proof (Sketch). Our assumption $Q \subseteq \{0, \cdots, max\}$ allows us to maintain $\mathcal{A}$-states in the memory of FICA-terms. To achieve k-fold nesting, we use the higher-order structure of the term: $\lambda f^{(0)}.f^{(0)}(\lambda f^{(1)}.f^{(1)}(\lambda f^{(2)}.f^{(2)}(\cdots \lambda f^{(k)}.f^{(k)})))$. In fact, instead of the single variables $f^{(i)}$, we shall use sequences $f_0^{(i)} \cdots f_{max}^{(i)}$, so that a question $t_{\mathsf{Q}}^{(i)}$ read by $\mathcal{A}$ at level i can be simulated by using variable $f_{t_{\mathsf{Q}}^{(i)}}^{(i)}$ (using our assumption $\Sigma \subseteq \{0, \cdots, max\}$). Additionally, the term contains state-manipulating code that enables moves only if they are consistent with the transition function of $\mathcal{A}$. □

9 Conclusion and further work

We have introduced leafy automata, LA, and shown that they correspond to the game semantics of Finitary Idealized Concurrent Algol (FICA). The automata formulation makes combinatorial challenges posed by the equivalence problem explicit. This is exemplified by a very transparent undecidability proof of the emptiness problem for LA. Our hope is that LA will allow to discover interesting fragments of FICA for which some variant of the equivalence problem is decidable. We have identified one such instance, namely local leafy automata (LLA), and a fragment of FICA that can be translated to them. The decidability of the emptiness problem for LLA implies decidability of a simple instance of the equivalence problem. This in turn allows to decide some verification questions as in the example on page 187. Since these types of questions involve quantification over all contexts, the use of a fully-abstract semantics appears essential to solve them.

The obvious line of future work is to find some other subclasses of LA with decidable emptiness problem. Another interesting target is to find an automaton model for the call-by-value setting, where answers enable questions [2,25]. It would also be worth comparing our results with abstract machines [19], the Geometry of Interaction [31], and the π-calculus [6].

References

1. Abramsky, S., Ghica, D.R., Murawski, A.S., Ong, C.H.L.: Applying game semantics to compositional software modelling and verification. In: Proceedings of TACAS, Lecture Notes in Computer Science, vol. 2988, pp. 421–435. Springer-Verlag (2004)
2. Abramsky, S., McCusker, G.: Call-by-value games. In: Proceedings of CSL. Lecture Notes in Computer Science, vol. 1414, pp. 1–17. Springer-Verlag (1997)
3. Abramsky, S., McCusker, G.: Game semantics. In: Schwichtenberg, H., Berger, U. (eds.) Logic and Computation. Springer-Verlag (1998), proceedings of the NATO Advanced Study Institute, Marktoberdorf
4. Aiswarya, C., Gastin, P., Kumar, K.N.: Verifying communicating multi-pushdown systems via split-width. In: Automated Technology for Verification and Analysis - 12th International Symposium, ATVA 2014. Lecture Notes in Computer Science, vol. 8837, pp. 1–17. Springer (2014)
5. Bakewell, A., Ghica, D.R.: On-the-fly techniques for games-based software model checking. In: Proceedings of TACAS, Lecture Notes in Computer Science, vol. 4963, pp. 78–92. Springer (2008)
6. Berger, M., Honda, K., Yoshida, N.: Sequentiality and the pi-calculus. In: Proceedings of TLCA, Lecture Notes in Computer Science, vol. 2044, pp. 29–45. Springer-Verlag (2001)
7. Björklund, H., Bojańczyk, M.: Shuffle expressions and words with nested data. In: Proceedings of MFCS. Lecture Notes in Computer Science, vol. 4708, pp. 750–761 (2007)
8. Björklund, H., Schwentick, T.: On notions of regularity for data languages. Theor. Comput. Sci. **411**(4-5), 702–715 (2010)
9. Bojańczyk, M., David, C., Muscholl, A., Schwentick, T., Segoufin, L.: Two-variable logic on data words. ACM Trans. Comput. Log. **12**(4), 27:1–27:26 (2011)
10. Cotton-Barratt, C., Hopkins, D., Murawski, A.S., Ong, C.L.: Fragments of ML decidable by nested data class memory automata. In: Proceedings of FOSSACS. Lecture Notes in Computer Science, vol. 9034, pp. 249–263. Springer (2015)
11. Cotton-Barratt, C., Murawski, A.S., Ong, C.L.: ML, visibly pushdown class memory automata, and extended branching vector addition systems with states. ACM Trans. Program. Lang. Syst. **41**(2), 11:1–11:38 (2019)
12. Cotton-Barratt, C., Murawski, A.S., Ong, C.L.: Weak and nested class memory automata. In: Proceedings of LATA. LNCS, vol. 8977, pp. 188–199. Springer (2015)
13. Czerwiński, W., Lasota, S., Lazic, R., Leroux, J., Mazowiecki, F.: The reachability problem for Petri nets is not elementary. In: Proceedings of STOC. pp. 24–33. ACM (2019)
14. Decker, N., Habermehl, P., Leucker, M., Thoma, D.: Ordered navigation on multi-attributed data words. In: Proceedings of CONCUR. LNCS, vol. 8704, pp. 497–511. Springer (2014)
15. Dimovski, A., Ghica, D.R., Lazic, R.: A counterexample-guided refinement tool for open procedural programs. In: Proceedings of SPIN. Lecture Notes in Computer Science, vol. 3925, pp. 288–292. Springer-Verlag (2006)
16. Dimovski, A.S.: Symbolic game semantics for model checking program families. In: Proceedings of SPIN. Lecture Notes in Computer Science, vol. 9641, pp. 19–37. Springer (2016)
17. Dimovski, A.S.: Probabilistic analysis based on symbolic game semantics and model counting. In: Proceedings of GandALF. EPTCS, vol. 256, pp. 1–15 (2017)

18. Dixon, A., Lazic, R., Murawski, A.S., Walukiewicz, I.: Leafy automata for higher-order concurrency. CoRR **abs/2101.08720** (2021), https://arxiv.org/abs/2101.08720
19. Fredriksson, O., Ghica, D.R.: Abstract machines for game semantics, revisited. In: Proceedings of LICS. pp. 560–569 (2013)
20. Ghica, D.R., McCusker, G.: Reasoning about Idealized Algol using regular expressions. In: Proceedings of ICALP, Lecture Notes in Computer Science, vol. 1853, pp. 103–115. Springer-Verlag (2000)
21. Ghica, D.R., Murawski, A.S.: Compositional model extraction for higher-order concurrent programs. In: Proceedings of TACAS, Lecture Notes in Computer Science, vol. 3920, pp. 303–317. Springer (2006)
22. Ghica, D.R., Murawski, A.S.: Angelic semantics of fine-grained concurrency. Annals of Pure and Applied Logic **151(2-3)**, 89–114 (2008)
23. Ghica, D.R., Murawski, A.S., Ong, C.H.L.: Syntactic control of concurrency. Theoretical Computer Science pp. 234–251 (2006)
24. Hague, M.: Saturation of concurrent collapsible pushdown systems. In: Proceedings of FSTTCS. LIPIcs, vol. 24, pp. 313–325. Schloss Dagstuhl - Leibniz-Zentrum für Informatik (2013)
25. Honda, K., Yoshida, N.: Game-theoretic analysis of call-by-value computation. Theoretical Computer Science **221**(1–2), 393–456 (1999)
26. Hopkins, D., Murawski, A.S., Ong, C.H.L.: Hector: An Equivalence Checker for a Higher-Order Fragment of ML. In: Proceedings of CAV, Lecture Notes in Computer Science, vol. 7358, pp. 774–780. Springer (2012)
27. Hopkins, D., Ong, C.H.L.: Homer: A Higher-order Observational equivalence Model checkER. In: Proceedings of CAV, Lecture Notes in Computer Science, vol. 5643, pp. 654–660. Springer (2009)
28. Kiefer, S., Murawski, A.S., Ouaknine, J., Wachter, B., Worrell, J.: APEX: An Analyzer for Open Probabilistic Programs. In: Proceedings of CAV, Lecture Notes in Computer Science, vol. 7358, pp. 693–698. Springer (2012)
29. Kobayashi, N., Igarashi, A.: Model-checking higher-order programs with recursive types. In: Proceedings of ESOP. Lecture Notes in Computer Science, vol. 7792, pp. 431–450. Springer (2013)
30. La Torre, S., Madhusudan, P., Parlato, G.: Reducing context-bounded concurrent reachability to sequential reachability. In: Proceedings of CAV. Lecture Notes in Computer Science, vol. 5643, pp. 477–492. Springer (2009)
31. Lago, U.D., Tanaka, R., Yoshimizu, A.: The geometry of concurrent interaction: handling multiple ports by way of multiple tokens. In: Proceedings of LICS. pp. 1–12 (2017)
32. Leroux, J., Schmitz, S.: Reachability in vector addition systems is primitive-recursive in fixed dimension. In: Proceedings of LICS. pp. 1–13. IEEE (2019)
33. Murawski, A.S.: Games for complexity of second-order call-by-name programs. Theoretical Computer Science **343(1/2)**, 207–236 (2005)
34. Murawski, A.S., Ramsay, S.J., Tzevelekos, N.: Game semantic analysis of equivalence in IMJ. In: Proceedings of ATVA. Lecture Notes in Computer Science, vol. 9364, pp. 411–428. Springer (2015)
35. Murawski, A.S., Tzevelekos, N.: An invitation to game semantics. SIGLOG News **3**(2), 56–67 (2016)
36. Murawski, A.S., Walukiewicz, I.: Third-order Idealized Algol with iteration is decidable. Theoretical Computer Science **390**(2-3), 214–229 (2008)

37. Ong, C.H.L.: Observational equivalence of 3rd-order Idealized Algol is decidable. In: Proceedings of IEEE Symposium on Logic in Computer Science. pp. 245–256. Computer Society Press (2002)
38. Qadeer, S., Rehof, J.: Context-bounded model checking of concurrent software. In: Proceedings of TACAS. Lecture Notes in Computer Science, vol. 3440, pp. 93–107. Springer (2005)
39. Ramalingam, G.: Context-sensitive synchronization-sensitive analysis is undecidable. ACM Trans. Program. Lang. Syst. **22**(2), 416–430 (2000)
40. Reynolds, J.C.: The essence of Algol. In: de Bakker, J.W., van Vliet, J. (eds.) Algorithmic Languages, pp. 345–372. North Holland (1978)
41. Schwentick, T.: Automata for XML - A survey. J. Comput. Syst. Sci. **73**(3), 289–315 (2007)

Factorization in Call-by-Name and Call-by-Value Calculi via Linear Logic

Claudia Faggian[1] and Giulio Guerrieri[2](✉)

[1] Université de Paris, IRIF, CNRS, F-75013 Paris, France
[2] University of Bath, Department of Computer Science, Bath, UK
g.guerrieri@bath.ac.uk

Abstract. In each variant of the λ-calculus, factorization and normalization are two key properties that show how results are computed. Instead of proving factorization/normalization for the call-by-name (CbN) and call-by-value (CbV) variants separately, we prove them only once, for the bang calculus (an extension of the λ-calculus inspired by linear logic and subsuming CbN and CbV), and then we transfer the result via translations, obtaining factorization/normalization for CbN and CbV. The approach is robust: it still holds when extending the calculi with operators and extra rules to model some additional computational features.

1 Introduction

The λ-calculus is the model of computation underlying functional programming languages and proof assistants. Actually there are many λ-calculi, depending on the *evaluation mechanism* (for instance, call-by-name and call-by-value—CbN and CbV for short) and *computational features* that the calculus aims to model.

In λ-calculi, a rewriting relation formalizes computational steps in program execution, and normal forms are the results of computations. In each calculus, a key question is to define a *normalizing strategy*: How to compute a result? Is there a reduction strategy which is guaranteed to output a result, if any exists?

Proving that a calculus admits a normalizing strategy is complex, and many techniques have been developed. A well-known method first proves *factorization* [4,32,19,2]. Given a calculus with a rewriting relation $\to$, a strategy $\xrightarrow[\mathrm{L}]{} \subseteq \to$ *factorizes* if $\to^* \subseteq \xrightarrow[\mathrm{L}]{}^* \cdot \xrightarrow[\neg\mathrm{L}]{}^*$ ($\xrightarrow[\neg\mathrm{L}]{}$ is the dual of $\xrightarrow[\mathrm{L}]{}$), *i.e.* any reduction sequence can be rearranged so as to perform $\xrightarrow[\mathrm{L}]{}$-steps first and then the other steps. If, moreover, the strategy satisfies some "good properties", we can conclude that the strategy is normalizing. Factorization is important also because it is commonly used as a building block in the proof of other properties of the *how-to-compute* kind. For instance, *standardization*, which generalizes factorization: every reduction sequences can be rearranged according to a predefined order between redexes.

Two for One. CbN and CbV λ-calculi are two distinct rewriting systems. Quoting from Levy [20]: *the existence of two separate paradigms* (CbN and CbV) is troubling because to prove a certain property—such as factorization or normalization—for both systems *we always need to do it twice*.

S. Kiefer and C. Tasson (Eds.): FOSSACS 2021, LNCS 12650, pp. 205–225, 2021.
https://doi.org/10.1007/978-3-030-71995-1_11

The *first aim* of our paper is to develop a technique for deriving factorization for both the CbN [4] and CbV [27] λ-calculi as corollaries of a *single* factorization theorem, and similarly for normalization. A key tool in our study is the *bang calculus* [11,15], a calculus inspired by linear logic in which CbN and CbV embed.

The Bang Calculus. The bang calculus is a variant of the λ-calculus where an operator ! plays the role of a marker for non-linear management: duplicability and discardability of resources. The bang calculus is nothing but Simpson's linear λ-calculus [31] without linear abstraction, or the untyped version of the implicative fragment of Levy's Call-by-Push-Value [20], as first observed by Ehrhard [10].

The motivation to study the bang calculus is to have a general framework where both CbN and CbV λ-calculi can be simulated, via two distinct *translations* inspired by Girard's embeddings [14] of the intuitionistic arrow into linear logic. So, a certain property can be studied in the bang calculus and then automatically transferred to the CbN and CbV settings by translating back.

This approach has so far mainly be exploited semantically [21,10,11,15,9,7], but can be used it also to study operational properties [15,30,13]. In this paper, we push forward this operational direction.

The Least-Level Strategy. We study a strategy from the literature of linear logic [8], namely *least-level reduction* $\xrightarrow[L]{}$, which fires a redex at minimal level—the *level* of a redex is the number of ! under which the redex occurs.

We prove that the least-level reduction factorizes and normalizes in the bang calculus, and then we transfer the same results to CbN and CbV λ-calculi (for suitable definitions of least-level in CbN and CbV), by exploiting properties of their translations into the bang calculus. A single proof suffices. It is two-for-one! Or even better, three-for-one.

The rewriting study of the least level strategy in the bang calculus is based on simple techniques for factorization and normalization we developed recently with Accattoli [2], which simplify and generalize Takahashi's method [32].

Subtleties of the Embeddings. Transferring factorization and normalization results via translation is highly non-trivial, *e.g.* in CPS translations [27]. This applies also to transferring least-level factorization from the bang calculus to the CbN and CbV λ-calculi. To transfer the property smoothly, the translations should preserve levels and normal forms, which is delicate, in particular for CbV. For instance, the embedding of CbV into the bang calculus defined in [15,30] does not preserve levels and normal forms. As a consequence, the CbV translation studied in [15,30] cannot be used to derive least-level factorization or *any* normalization result in a CbV setting from the corresponding result in the bang calculus.

Here we adopt the refined CbV embedding of Bucciarelli *et al.* [7], which does preserve levels and normal forms. While the preservation of normal forms is already stressed in [7], the preservation of levels is proved here for the first time, and it is based on non-trivial properties of the embedding.

Beyond pure. Our *second aim* is to show that the developed technique for the joined factorization and normalization of CbN and CbV via the bang calculus is *robust.* We do so, by studying extensions of all three calculi with operators (or, in general, with extra rules) which model some additional computational features, such as non-deterministic or probabilistic choice. We then show that the technique scales up smoothly, under mild assumptions on the extension.

A Motivating Example. Let us illustrate our approach on a simple case, which we will use as a running example. De' Liguoro and Piperno's CbN non-deterministic λ-calculus $\Lambda_{\oplus}^{\mathtt{cbn}}$ [23] extends the CbN λ-calculus with an operator $\oplus$ whose reduction $\to_{\oplus}$ models *non-deterministic choice*: $t \oplus s$ rewrites to either t or s. It admits a standardization result, from which it follows that the leftmost-outermost reduction strategy (noted $\underset{\mathrm{LO}}{\to}_{\beta\oplus}$) is *complete*: if t has a normal form u then $t \underset{\mathrm{LO}}{\to}_{\beta\oplus}^{*} u$. In [22], de' Liguoro considers also a CbV variant $\Lambda_{\oplus}^{\mathtt{cbv}}$, extending Plotkin CbV λ-calculus [27] with an operator $\oplus$. One may prove standardization and completeness—again—from scratch, even though the proofs are similar.

The approach we propose here is to work in the bang calculus enriched with the operator $\oplus$. We show that the calculus satisfies *least-level factorization*, from which it follows that the least-level strategy (noted $\underset{\mathrm{L}}{\to}_{\beta_!\oplus}$) is *complete*, *i.e.* if t has a normal form u, then $t \underset{\mathrm{L}}{\to}_{\beta_!\oplus}^{*} u$. The translation then guarantees that analogous results hold also in $\Lambda_{\oplus}^{\mathtt{cbn}}$ and $\Lambda_{\oplus}^{\mathtt{cbv}}$, without proving them again.

The Importance of Being Modular. The bang calculus with operators is actually a general formalism for several calculi, one calculus for each kind of computational feature modeled by operators. Concretely, the reduction $\to$ consists of $\to_{\beta_!}$ (which subsumes CbN $\to_{\beta}$ and CbV $\to_{\beta_v}$) and other reduction rules $\to_{\rho}$.

We decompose the proof of factorization of $\to$ in modules, by using the *modular approach* we recently introduced together with Accattoli [3].

The key module is the least-level factorization of $\to_{\beta_!}$, because it is where the higher-order comes into play—this is done, once and for all. Then, we consider a generic reduction rule $\to_{\rho}$ to add to $\to_{\beta_!}$. Our general result is that if $\to_{\rho}$ has "good properties" and interacts well with $\to_{\beta_!}$ (which amounts to an easy test, combinatorial in nature), then we have least-level factorization for $\to_{\beta_!} \cup \to_{\rho}$.

Putting all together, when $\to_{\rho}$ is instantiated to a concrete reduction (such as $\to_{\oplus}$), the user of our method only has to verify a simple test (namely Proposition 34), to conclude that $\to_{\beta_!} \cup \to_{\rho}$ has least-level factorization. In particular, factorization for $\to_{\beta_!}$ is a ready-to-use black box the user need not to worry about—our proof is robust enough to hold whatever the other rules are. Finally, the embeddings automatically give least-level factorization for the corresponding CbV and CbN calculi. Section 7 illustrates our method in the case $\to_{\rho} = \to_{\oplus}$.

Subtleties of the Modular Extensions. To adopt the modular approach for factorization presented in [3], we have to face an important difficulty that arises when dealing with normalizing strategies, and which is not studied in [3].

A *normalizing* strategy cannot overlook redexes and it usually selects the redex r to fire through a property that r minimizes with respect to the redexes in the whole term, such as being a *least level* redex or being the *leftmost-outermost* (shortened to LO) redex—normalizing strategies are *positional.* The problem is that, in general, if $\rightarrow = \rightarrow_\beta \cup \rightarrow_\rho$, then $\underset{\text{LO}}{\rightarrow}$ reduction is not the union of $\underset{\text{LO}}{\rightarrow}_\beta$ and $\underset{\text{LO}}{\rightarrow}_\rho$: the normalizing strategy of the compound system is not obtained putting together the normalizing strategies of the components. Let us explain the issue on our running example $\rightarrow_{\beta\oplus}$, in the familiar case of leftmost-outermost reduction.

Example 1. Consider head reductions for $\rightarrow_\beta$ and for $\rightarrow_{\beta\oplus} = \rightarrow_\beta \cup \rightarrow_\oplus$, noted $\underset{\mathsf{h}}{\rightarrow}_\beta$ and $\underset{\mathsf{h}}{\rightarrow}_{\beta\oplus}$, respectively. In the term $s = (\mathsf{II})(x \oplus y)$ where $\mathsf{I} = \lambda x.x$, the subterm II (a β-redex) is in head position for both the reduction $\rightarrow_\beta$ and its extension $\rightarrow_{\beta\oplus}$. So, $s \underset{\mathsf{h}}{\rightarrow}_\beta \mathsf{I}(x \oplus y)$ and $s \underset{\mathsf{h}}{\rightarrow}_{\beta\oplus} \mathsf{I}(x \oplus y)$. And in the term $t = (x \oplus y)(\mathsf{II})$, the head position is occupied by $x \oplus y$, which is a $\oplus$-redex. Therefore, II is not the head redex in t, neither for β nor for $\beta\oplus$. In general, $\underset{\mathsf{h}}{\rightarrow}_{\beta\oplus} = \underset{\mathsf{h}}{\rightarrow}_\beta \cup \underset{\mathsf{h}}{\rightarrow}_\oplus$.

In contrast, for leftmost-outermost reduction $\underset{\text{LO}}{\rightarrow}_{\beta\oplus}$, which reduces the LO-redex, we have $\underset{\text{LO}}{\rightarrow}_{\beta\oplus} \neq \underset{\text{LO}}{\rightarrow}_\beta \cup \underset{\text{LO}}{\rightarrow}_\oplus$. Consider again the term $t = (x \oplus y)(\mathsf{II})$. Since $x \oplus y$ is not a β-redex, II is the LO-redex for $\rightarrow_\beta$. Instead, II is not the LO-redex for $\rightarrow_{\beta\oplus}$ (here the LO-redex is $x \oplus y$). So, $t \underset{\text{LO}}{\rightarrow}_\beta (x \oplus y)\mathsf{I}$ but $t \underset{\text{LO}}{\not\rightarrow}_{\beta\oplus} (x \oplus y)\mathsf{I}$.

The least-level factorization for $\rightarrow_{\beta_!}$, $\rightarrow_\beta$, and $\rightarrow_{\beta_v}$ we prove here is robust enough to make it ready to be used as a module in a larger proof, where it may combine with operators and other rules. The key point is to define the least-level reduction from the very beginning as a reduction firing a redex at minimal level with respect to a general set of redexes (including $\beta_!$, β or β_v, respectively), so that it is "ready" to be extended with other reduction rules (see Section 4).

Proofs. All proofs are available in [12], the long version of this paper.

2 Background in Abstract Rewriting

An (*abstract*) *rewriting system*, [33, Ch. 1] is a pair $(A, \rightarrow)$ consisting of a set A and a binary relation $\rightarrow \subseteq A \times A$ (called *reduction*) whose pairs are written $t \rightarrow s$ and called *steps*. A $\rightarrow$*-sequence* from t is a sequence of $\rightarrow$-steps. As usual, $\rightarrow^*$ (resp. $\rightarrow^=$) denotes the transitive-reflexive (resp. reflexive) closure of $\rightarrow$. We say that u is $\rightarrow$*-normal* (or a $\rightarrow$-normal form) if there is no t such that $u \rightarrow t$.

In general, a term may or may not reduce to a normal form. If it does, not all reduction sequences necessarily lead to normal form. A term is *weakly* or *strongly* normalizing, depending on if it may or must reduce to normal form. More precisely, a term t is *strongly* $\rightarrow$*-normalizing* if *every* maximal $\rightarrow$-sequence from t ends in a $\rightarrow$-normal form: any choice of $\rightarrow$-steps will eventually lead to a normal form. A term t is *weakly* $\rightarrow$*-normalizing* if $t \rightarrow^* u$ for some u $\rightarrow$-normal. If t is weakly but not strongly normalizing, how do we compute a normal form? This is the problem tackled by *normalization*: by repeatedly performing *only specific steps*, a normal form is eventually reached, provided that t can $\rightarrow$-reduce to any.

Definition 2 (Normalizing and complete strategy). *A reduction* $\underset{e}{\rightarrow} \subseteq \rightarrow$ *is a* strategy for $\rightarrow$ *if it has the same normal forms as* $\rightarrow$. *A strategy* $\underset{e}{\rightarrow}$ *for* $\rightarrow$ *is:*

- complete *if* $t \underset{e}{\rightarrow}^* u$ *whenever* $t \rightarrow^* u$ *with* u $\rightarrow$*-normal;*
- normalizing *if every weakly* $\rightarrow$*-normalizing term is strongly* $\underset{e}{\rightarrow}$*-normalizing.*

Note that if the strategy $\underset{e}{\rightarrow}$ is complete and *deterministic* (*i.e.* for every $t \in A$, $t \underset{e}{\rightarrow} s$ for at most one $s \in A$), then $\underset{e}{\rightarrow}$ is a normalizing strategy for $\rightarrow$.

Informally, a *strategy* for $\rightarrow$ is a way to control the fact that in a term there are different possible choices of a $\rightarrow$-step. A *normalizing strategy* for $\rightarrow$ is a strategy that is guaranteed to reach a $\rightarrow$-normal form, if it exists, from any term. This provides a useful tool to show that a term is not weakly $\rightarrow$-normalizing.

Proving Normalization. Factorization means that any $\rightarrow$-sequence from a term to another can be rearranged by performing a certain kind of steps first. It provides a simple technique to establish that a strategy is normalizing.

Definition 3 (Factorization). *Let* $(A, \rightarrow)$ *be a rewriting system with* $\rightarrow = \underset{e}{\rightarrow} \cup \underset{i}{\rightarrow}$. *The relation* $\rightarrow$ *satisfies* e-factorization, *written* $\mathtt{Fact}(\underset{e}{\rightarrow}, \underset{i}{\rightarrow})$, *if*

$$\mathtt{Fact}(\underset{e}{\rightarrow}, \underset{i}{\rightarrow}): \quad (\underset{e}{\rightarrow} \cup \underset{i}{\rightarrow})^* \subseteq \underset{e}{\rightarrow}^* \cdot \underset{i}{\rightarrow}^* \qquad \textbf{(Factorization)}$$

Lemma 4 (Normalization [2]). *Let* $\rightarrow = \underset{e}{\rightarrow} \cup \underset{\neg e}{\rightarrow}$, *and* $\underset{e}{\rightarrow}$ *be a strategy for* $\rightarrow$. *The strategy* $\underset{e}{\rightarrow}$ *is* complete *for* $\rightarrow$ *if the following conditions hold:*

1. *(*persistence*) if* $t \underset{\neg e}{\rightarrow} t'$ *then* t' *is not* $\rightarrow$*-normal;*
2. *(*factorization*)* $t \rightarrow^* u$ *implies* $t \underset{e}{\rightarrow}^* \cdot \underset{\neg e}{\rightarrow}^* u$.

The strategy $\underset{e}{\rightarrow}$ *is* normalizing *for* $\rightarrow$ *if it is complete and the following holds:*

3. *(*uniformity*) every weakly* $\underset{e}{\rightarrow}$*-normalizing term is strongly* $\underset{e}{\rightarrow}$*-normalizing.*

A sufficient condition for uniformity (and confluence) is the quasi-diamond.

Property 5 (Newman [25]) *If a reduction* $\rightarrow$ *is* quasi-diamond *(i.e.* $s \leftarrow t \rightarrow r$ *implies* $s = r$ *or* $s \rightarrow u \leftarrow r$ *for some* u*), then* $\rightarrow$ *is uniform and confluent (i.e.* $s\ {}^*\!\!\leftarrow r \rightarrow^* t$ *implies* $s \rightarrow^* u\ {}^*\!\!\leftarrow t$ *for some* u*).*

Proving Factorization. Hindley [17] first noted that a local property implies factorization. Let $\rightarrow = \underset{e}{\rightarrow} \cup \underset{i}{\rightarrow}$. We say that $\underset{i}{\rightarrow}$ *strongly postpones* after $\underset{e}{\rightarrow}$ if

$$\mathtt{SP}(\underset{e}{\rightarrow}, \underset{i}{\rightarrow}): \qquad \underset{i}{\rightarrow} \cdot \underset{e}{\rightarrow} \subseteq \underset{e}{\rightarrow}^* \cdot \underset{i}{\rightarrow}^= \qquad \textbf{(Strong Postponement)}$$

Lemma 6 (Hindley [17]). $\mathtt{SP}(\underset{e}{\rightarrow}, \underset{i}{\rightarrow})$ *implies* $\mathtt{Fact}(\underset{e}{\rightarrow}, \underset{i}{\rightarrow})$.

Strong postponement can rarely be used *directly*, because several interesting reductions—including β-reduction—do not satisfy it. However, it is at the heart of Takahashi's method [32] to prove head factorization of $\to_\beta$, via the following immediate property that can also be used to prove other factorizations (see [2]).

Property 7 (Characterization of factorization) *We have* $\mathtt{Fact}(\underset{e}{\to}, \underset{i}{\to})$ *if and only if there is a reduction* $\underset{i}{\Leftrightarrow}$ *such that* $\underset{i}{\Leftrightarrow}^* = \underset{i}{\to}^*$ *and* $\mathtt{SP}(\underset{e}{\to}, \underset{i}{\Leftrightarrow})$.

The core of Takahashi's method [32] to prove head factorization in the λ-calculus is to introduce a relation $\underset{i}{\Rightarrow}$, called *internal parallel reduction*, which verifies the conditions of Property 7. We will follow a similar path in Section 6.1, to prove *least-level* factorization in the bang calculus.

Compound systems: proving factorization in a modular way. In this paper, we will consider compound rewriting systems that are obtained by extending the λ-calculus with extra rules to model advanced computational features.

In an abstract setting, let us consider a rewrite system $(A, \to)$ where $\to = \to_\xi \cup \to_\rho$. Under which condition $\to$ admits factorization, assuming that both $\to_\xi$ and $\to_\rho$ do? To deal with this question, a technique for proving factorization for *compound systems* in a *modular* way has been introduced in [3]. The approach can be seen as an analogous for factorization of the classical technique for confluence based on Hindley-Rosen lemma [4]: if $\to_\xi, \to_\rho$ are e-factorizing reductions, their union $\to_\xi \cup \to_\rho$ also is, provided that two *local* conditions of commutation hold.

Lemma 8 (Modular factorization [3]). *Let* $\to_\xi = \underset{e}{\to}_\xi \cup \underset{i}{\to}_\xi$ *and* $\to_\rho = \underset{e}{\to}_\rho \cup \underset{i}{\to}_\rho$ *be* e*-factorizing relations. Let* $\underset{e}{\to} := \underset{e}{\to}_\xi \cup \underset{e}{\to}_\rho$ *and* $\underset{i}{\to} := \underset{i}{\to}_\xi \cup \underset{i}{\to}_\rho$. *The reduction* $\to_\xi \cup \to_\rho$ *fulfills factorization* $\mathtt{Fact}(\underset{e}{\to}, \underset{i}{\to})$ *if the following swaps hold:*

$$\underset{i}{\to}_\xi \cdot \underset{e}{\to}_\rho \subseteq \underset{e}{\to}_\rho \cdot \to_\xi^* \quad \textit{and} \quad \underset{i}{\to}_\rho \cdot \underset{e}{\to}_\xi \subseteq \underset{e}{\to}_\xi \cdot \to_\rho^* \qquad \textbf{(Linear Swaps)}$$

The subtlety here is to set $\underset{e}{\to}_\xi$ and $\underset{e}{\to}_\rho$ so that $\underset{e}{\to} = \underset{e}{\to}_\xi \cup \underset{e}{\to}_\rho$. As already shown in Example 1, when dealing with normalizing strategies one needs extra care.

3 λ-calculi: CbN, CbV, and bang

We present here a generic syntax for λ-calculi, possibly containing operators. All the variants of the λ-calculus we shall study use this language. We assume some familiarity with the λ-calculus, and refer to [4,18] for details.

Given a countable set Var of variables, denoted by $x, y, z, \ldots$, *terms* and *values* (whose sets are denoted by $\Lambda_\mathcal{O}$ and Val, respectively) are defined as follows:

$$t, s, r ::= v \mid ts \mid \mathbf{o}(t_1, \ldots, t_k) \quad \textit{Terms}: \Lambda_\mathcal{O} \qquad\qquad v ::= x \mid \lambda x.t \quad \textit{Values}: \mathsf{Val}$$

where $\mathbf{o}$ ranges over a set $\mathcal{O}$ of function symbols called *operators*, each one with its own arity $k \in \mathbb{N}$. If the operators are $\mathbf{o}_1, \ldots, \mathbf{o}_n$, the set of terms is indicated as $\Lambda_{\mathbf{o}_1 \ldots \mathbf{o}_n}$. When the set $\mathcal{O}$ of operators is empty, the calculus is called *pure*, and the sets of terms is denoted by Λ; otherwise, the calculus is *applied*.

Terms are identified up to renaming of bound variables, where abstraction is the only binder. We denote by $t\{s/x\}$ the capture-avoiding substitution of s for the free occurrences of x in t. *Contexts* (with exactly one hole $\langle\cdot\rangle$) are generated by the grammar below, and $\mathbf{c}\langle t\rangle$ stands for the term obtained from the context $\mathbf{c}$ by replacing the hole with the term t (possibly capturing free variables).

$$\mathbf{c} ::= \langle\cdot\rangle \mid t\mathbf{c} \mid \mathbf{c}t \mid \lambda x.\mathbf{c} \mid \mathbf{o}(t_1, \ldots, \mathbf{c}, \ldots, t_k) \qquad \textit{Contexts: } \mathcal{C}$$

A *rule* ρ is a binary relation on $\Lambda_{\mathcal{O}}$; we also call it ρ-*rule* and denote it by $\mapsto_\rho$, writing $t \mapsto_\rho t'$ rather than $(t, t') \in \rho$. The ρ-*reduction* $\to_\rho$ is the contextual closure of ρ. Explicitly, $t \to_\rho t'$ holds if $t = \mathbf{c}\langle r\rangle$ and $t' = \mathbf{c}\langle r'\rangle$ for some context $\mathbf{c}$ with $r \mapsto_\rho r'$; the term r is called a ρ-*redex*. The set of ρ-*redexes* is denoted by $\mathcal{R}_\rho$.

Given a set of rules Rules, the relation $\to\, = \bigcup_\rho \to_\rho$ (for $\rho \in \mathsf{Rules}$) can equivalently be defined as the contextual closure of $\mapsto\, = \bigcup_\rho \mapsto_\rho$.

3.1 Call-by-Name and Call-by-Value λ-calculi

Pure CbN and Pure CbV λ-calculi. The *pure call-by-name* (CbN for short) λ-calculus [4,18] is $(\Lambda, \to_\beta)$, the set of terms Λ together with the β-reduction $\to_\beta$, defined as the contextual closure of the usual β-rule, which we recall in (1) below.

The *pure call-by-value* (CbV for short) λ-calculus [27] is the set Λ endowed with the reduction $\to_{\beta_v}$, defined as the contextual closure of the β_v-rule in (2).

$$\text{CbN: } (\lambda x.t)s \mapsto_\beta t\{s/x\} \ \ (1) \qquad \text{CbV: } (\lambda x.t)v \mapsto_{\beta_v} t\{v/x\} \ \text{ with } v \in \mathsf{Val} \ \ (2)$$

CbN and CbV λ-calculi. A *CbN* (resp. *CbV*) *λ-calculus* is the set of terms endowed with a reduction $\to$ which extends $\to_\beta$ (resp. $\to_{\beta_v}$).

In particular, the *applied* setting with operators (when $\mathcal{O} \neq \emptyset$) models in the λ-calculus richer computational features, allowing $\mathbf{o}$-reductions as the contextual closure of $\mathbf{o}$-rules of the form $\mathbf{o}(t_1, \ldots, t_k) \mapsto_{\mathbf{o}} s$.

Example 9 (Non-deterministic λ-calculi). Let $\mathcal{O} = \{\oplus\}$ where $\oplus$ is a binary operator; let $\to_\oplus$ be the contextual closure of the (non-deterministic) rule below:

$$\oplus(t_1, t_2) \mapsto_\oplus t_1 \quad \text{and} \quad \oplus\,(t_1, t_2) \mapsto_\oplus t_2.$$

The *non-deterministic CbN λ-calculus* $\Lambda^{\mathtt{cbn}}_{\oplus} = (\Lambda_\oplus, \to_{\beta\oplus})$ is the set $\Lambda_\oplus$ with the reduction $\to_{\beta\oplus} = \to_\beta \cup \to_\oplus$. The *non-deterministic CbV λ-calculus* $\Lambda^{\mathsf{cbv}}_{\oplus} = (\Lambda_\oplus, \to_{\beta_v\oplus})$ is the set $\Lambda_\oplus$ with the reduction $\to_{\beta_v\oplus} = \to_{\beta_v} \cup \to_\oplus$.

3.2 Bang calculi

The bang calculus [11,15] is a variant of the λ-calculus inspired by linear logic. An operator ! plays the role of a marker for duplicability and discardability. Here we allow also the presence of operators other than !, ranging over a set $\mathcal{O}$. So, terms and contexts of the bang calculus (denoted by capital letters) are:

$$T, S, R ::= x \mid \lambda x.T \mid TS \mid !T \mid \mathbf{o}(T_1, \ldots, T_k) \qquad \textit{Terms: } \Lambda_{!\mathcal{O}}$$
$$\mathbf{C} ::= \langle \cdot \rangle \mid \lambda x.\mathbf{C} \mid T\mathbf{C} \mid \mathbf{C}T \mid !\mathbf{C} \mid \mathbf{o}(T_1, \ldots, \mathbf{C}, \ldots, T_k) \qquad \textit{Contexts: } \mathcal{C}_!$$

Terms of the form $!T$ are called *boxes* and their set is denoted by $!\Lambda_{!\mathcal{O}}$. When there are no operators other than ! (*i.e.* $\mathcal{O} = \emptyset$), the set of terms and the set of boxes are denoted by $\Lambda_!$ and $!\Lambda_!$, respectively. This syntax can be expressed in the one at the beginning of Section 3, where ! is an unary operator called *bang.*

The pure bang calculus. The *pure* bang calculus $(\Lambda_!, \to_{\beta_!})$ is the set of terms $\Lambda_!$ endowed with reduction $\to_{\beta_!}$, the closure under contexts in $\mathcal{C}_!$ of the $\beta_!$*-rule*:

$$(\lambda x.T)\,!S \mapsto_{\beta_!} T\{S/x\} \qquad (3)$$

Intuitively, in the bang calculus the bang-operator ! marks the only terms that can be erased and duplicated. Indeed, a *β-like redex* $(\lambda x.T)S$ can be fired by $\mapsto_{\beta_!}$ only when its argument S is a box, *i.e.* $S = !R$: if it is so, the content R of the box S (and not S itself) replaces any free occurrence of x in T.[3]

A proof of confluence of $\beta_!$-reduction $\to_{\beta_!}$ is in [15].

Notation 10 *We use the following notations to denote some notable terms.*

$$\iota := \lambda x.x \qquad \delta := \lambda x.xx \qquad I := \lambda x.!x \qquad \Delta := \lambda x.x\,!x.$$

Remark 11 (Notable terms). The term $I = \lambda x.!x$ plays the role of the identity in the bang calculus: $I\,!T \to_{\beta_!} !(x\{T/x\}) = !T$ for any term T. Instead, the term $\iota = \lambda x.x$, when applied to a box $!T$, opens the box, *i.e.* returns its content T: $\iota !T \to_{\beta_!} x\{T/x\} = T$. Finally, $\Delta\,!\Delta \to_{\beta_!} \Delta\,!\Delta \to_{\beta_!} \ldots$ is a diverging term.

A bang calculus. A *bang calculus* $(\Lambda_{!\mathcal{O}}, \to)$ is the set $\Lambda_{!\mathcal{O}}$ of terms endowed with a reduction $\to$ which extends $\to_{\beta_!}$. In this paper we shall consider calculi where $\to$ contains $\to_{\beta_!}$ and **o**-reductions $\to_{\mathbf{o}}$ ($\mathbf{o} \in \mathcal{O}$) defined from **o**-rules of the form $\mathbf{o}(T_1, \ldots, T_k) \mapsto_{\mathbf{o}} S$, and possibly other rules. So, $\to = \bigcup_\rho \to_\rho$ (for $\rho \in \mathsf{Rules}$), with $\mathsf{Rules} \supseteq \{!\beta, \mathbf{o} \mid \mathbf{o} \in \mathcal{O}\}$. We set $\to_{\mathcal{O}} = \bigcup_{\mathbf{o} \in \mathcal{O}} \to_{\mathbf{o}}$.

3.3 CbN and CbV translations into the bang calculus

Our motivation to study the bang calculus is to have a general framework where both CbN [4] and CbV [27] λ-calculi can be embedded, via two distinct translations. Here we show how these translations work. We extend the simulation results in [15,30,7] for the pure case to the case with operators (Proposition 13).

Following [7], the CbV translation defined here differs from [15,30] in the application case. Section 5 will show why this optimization is crucial.

CbN and *CbV translations* are two maps $(\cdot)^{\mathsf{n}} \colon \Lambda_{\mathcal{O}} \to \Lambda_{!\mathcal{O}}$ and $(\cdot)^{\mathsf{v}} \colon \Lambda_{\mathcal{O}} \to \Lambda_{!\mathcal{O}}$, respectively, translating terms of the λ-calculus into terms of the bang calculus:

[3] Syntax and reduction rule of the bang calculus follow [15], which is slightly different from [11]. Unlike [15] (but akin to [30,16]), here we do not use ι (aka der) as a primitive, since ι and its associated rule $\mapsto_{\mathsf{d}}$ can be simulated, see Remark 11 and (4).

$$x^{\mathsf{n}} = x \quad (\lambda x.t)^{\mathsf{n}} = \lambda x.t^{\mathsf{n}} \quad (\mathbf{o}(t_1,\ldots,t_k))^{\mathsf{n}} = \mathbf{o}(t_1^{\mathsf{n}},\ldots,t_k^{\mathsf{n}}) \quad (ts)^{\mathsf{n}} = t^{\mathsf{n}}\,!s^{\mathsf{n}}\,;$$

$$x^{\mathsf{v}} = !x \quad (\lambda x.t)^{\mathsf{v}} = !(\lambda x.t^{\mathsf{v}}) \quad (\mathbf{o}(t_1,\ldots,t_k))^{\mathsf{v}} = \mathbf{o}(t_1^{\mathsf{v}},\ldots,t_k^{\mathsf{v}}) \quad (ts)^{\mathsf{v}} = \begin{cases} T\,s^{\mathsf{v}} & \text{if } t^{\mathsf{v}} = !T \\ (\iota\, t^{\mathsf{v}})s^{\mathsf{v}} & \text{otherwise.} \end{cases}$$

Example 12. Consider the λ-term $\omega := \delta\delta$: then, $\delta^{\mathsf{n}} = \Delta$, $\delta^{\mathsf{v}} = !\Delta$ and $\omega^{\mathsf{n}} = \Delta\,!\Delta = \omega^{\mathsf{v}}$ (δ and Δ are defined in Notation 10). The λ-term ω is diverging in CbN and CbV λ-calculi, and so is $\omega^{\mathsf{n}} = \omega^{\mathsf{v}}$ in the bang calculus, see Remark 11.

For any term $t \in \Lambda_{\mathcal{O}}$, t^{n} and t^{v} are just different decorations of t by means of the bang-operator ! (recall that $\iota = \lambda x.x$). The translation $(\cdot)^{\mathsf{n}}$ puts the argument of any application into a box: in CbN any term is duplicable or discardable. On the other hand, only *values* (*i.e.* abstractions and variables) are translated by $(\cdot)^{\mathsf{v}}$ into boxes, as they are the only terms duplicable or discardable in CbV.

As in [15,30], we prove that the CbN translation $(\cdot)^{\mathsf{n}}$ (resp. CbV translation $(\cdot)^{\mathsf{v}}$) from the pure CbN (resp. CbV) λ-calculus into the bang calculus is *sound* and *complete*: it maps β-reductions (resp. β_v-reductions) of the λ-calculus into $\beta_!$-reductions of the bang calculus, and conversely $\beta_!$-reductions — when restricted to the image of the translation — into β-reductions (resp. β_v-reductions). The same holds if we consider any $\mathbf{o}$-reduction for operators, where we assume that the $\mathbf{o}$-rule commutes with the translations: if $\mathbf{o}(t_1,\ldots,t_k) \mapsto_{\mathbf{o}} s$ then $\mathbf{o}(t_1^{\mathsf{n}},\ldots,t_k^{\mathsf{n}}) \mapsto_{\mathbf{o}} s^{\mathsf{n}}$, and if $\mathbf{o}(t_1^{\mathsf{n}},\ldots,t_k^{\mathsf{n}}) \mapsto_{\mathbf{o}} S$ then $\mathbf{o}(t_1,\ldots,t_k) \mapsto_{\mathbf{o}} s$ with $s^{\mathsf{n}} = S$; similarly for $(\cdot)^{\mathsf{v}}$.

In the simulation, $\to_{\mathsf{d}}$ denotes the contextual closure of the rule:

$$\iota\,!T \mapsto_{\mathsf{d}} T \quad \text{(this is nothing but } (\lambda x.x)!T \mapsto_{\beta_!} T) \tag{4}$$

Clearly, $\to_{\mathsf{d}} \subseteq \to_{\beta_!}$ (Remark 11). We write $T \twoheadrightarrow_{\mathsf{d}} S$ if $T \to_{\mathsf{d}}^* S$ and S is d-normal.

Proposition 13 (Simulation of CbN and CbV). *Let $t \in \Lambda_{\mathcal{O}}$ and $\mathbf{o} \in \mathcal{O}$.*

1. CbN soundness: *If $t \to_\beta t'$ then $t^{\mathsf{n}} \to_{\beta_!} t'^{\mathsf{n}}$. If $t \to_{\mathbf{o}} t'$ then $t^{\mathsf{n}} \to_{\mathbf{o}} t'^{\mathsf{n}}$.*
   CbN completeness: *If $t^{\mathsf{n}} \to_{\beta_!} S$ then $S = t'^{\mathsf{n}}$ and $t \to_\beta t'$, for some $t' \in \Lambda_{\mathcal{O}}$. If $t^{\mathsf{n}} \to_{\mathbf{o}} S$ then $S = t'^{\mathsf{n}}$ and $t \to_{\mathbf{o}} t'$, for some $t' \in \Lambda_{\mathcal{O}}$.*
2. CbV soundness: *If $t \to_{\beta_v} t'$ then $t^{\mathsf{v}} \to_{\beta_!}\to_{\mathsf{d}}^{=} t'^{\mathsf{v}}$ with t'^{v} d-normal. If $t \to_{\mathbf{o}} t'$ then $t^{\mathsf{v}} \to_{\mathbf{o}}\to_{\mathsf{d}}^{=} t'^{\mathsf{v}}$ with t'^{v} d-normal.*
   CbV completeness: *If $t^{\mathsf{v}} \to_{\beta_!}\twoheadrightarrow_{\mathsf{d}} S$ then $t^{\mathsf{v}} \to_{\beta_!}\to_{\mathsf{d}}^{=} S$ with $S = t'^{\mathsf{v}}$ and $t \to_{\beta_v} t'$, for some $t' \in \Lambda_{\mathcal{O}}$. If $t^{\mathsf{v}} \to_{\mathbf{o}}\twoheadrightarrow_{\mathsf{d}} S$ then $t^{\mathsf{v}} \to_{\mathbf{o}}\to_{\mathsf{d}}^{=} S$ with $S = t'^{\mathsf{v}}$ and $t \to_{\mathbf{o}} t'$, for some $t' \in \Lambda_{\mathcal{O}}$.*

Example 14. Let $t = (\lambda z.z)x\,y$ and $t' = xy$. So $t \to_\beta t'$ with $t^{\mathsf{n}} = (\lambda z.z)!x\,!y \to_{\beta_!} x\,!y = t'^{\mathsf{n}}$; and $t \to_{\beta_v} t'$ with $t^{\mathsf{v}} = (\iota((\lambda z.!z)!x))!y \to_{\beta_!} (\iota\,!x)!y \to_{\mathsf{d}} x\,!y = t'^{\mathsf{v}}$.

4 The least-level strategy

The bang calculus $\Lambda_!$ has a natural normalizing strategy, derived from linear logic [8], namely the *least-level reduction*. It reduces only redexes at *least level*, where the *level* of a redex R in a term T is the number of bangs ! in which R is nested.

Least-level reduction is easily extended to a general bang calculus $(\Lambda_{!\mathcal{O}}, \rightarrow)$. The level of a redex R is then the number of bangs ! and operators $\mathbf{o}$ in which R is nested; intuitively, least-level reduction fires a redex which is *minimally nested.*

Below, we formalize the reduction in a way that is independent of the specific shape of the redexes, and even of specific definition of level one chooses. The interest of least-level reduction is in the properties it satisfies. All our developments will rely on such properties, rather than the specific definition of least level.

In this section, $\rightarrow = \bigcup_\rho \rightarrow_\rho$ for $\rho \in \mathsf{Rules}$ (for a generic set of rules Rules). We write $\mathcal{R} = \bigcup_\rho \mathcal{R}_\rho$ (again, with $\rho \in \mathsf{Rules}$) for the set of *all* redexes.

4.1 Least-level reduction in bang calculi

The *level* of a redex occurrence R in a term T is a measure of its depth. Formally, we indicate the *occurrence of a subterm* R in T with the context $\mathbf{C}$ such that $\mathbf{C}\langle R\rangle = T$. Its level is then the *level* $\ell(\mathbf{C}) \in \mathbb{N}$ of the hole in $\mathbf{C}$. The definition of *level* for contexts in a bang calculus $\Lambda_{!\mathcal{O}}$ is formalized as follows.

$$\begin{array}{cccc} \ell(\langle\cdot\rangle) = 0 & \ell(\lambda x.\mathbf{C}) = \ell(\mathbf{C}) & \ell(\mathbf{C}T) = \ell(\mathbf{C}) & \ell(T\mathbf{C}) = \ell(\mathbf{C}) \\ & \ell(!\mathbf{C}) = \ell(\mathbf{C}) + 1 & \ell(\mathbf{o}(\ldots, \mathbf{C}, \ldots)) = \ell(\mathbf{C}) + 1 & \end{array} \tag{5}$$

Note that the level increases by 1 in the scope of !, and of any operator $\mathbf{o} \in \mathcal{O}$.

A reduction step $T \rightarrow_\rho S$ is *at level* k if it fires a ρ-redex at level $k \in \mathbb{N}$; it is *least-level* if it reduces a redex whose level is minimal.

The *least level* $\ell\ell(T)$ of a term T expresses the minimal level of any redex occurrences in T; if no redex is in T, we set $\ell\ell(T) = \infty$. Formally:

Definition 15 (Least-level reduction). *Let* $\rightarrow = \bigcup_\rho \rightarrow_\rho$ *(for* $\rho \in \mathsf{Rules}$*) and* $\mathcal{R} = \bigcup_\rho \mathcal{R}_\rho$ *the set of redexes. Given a function* $\ell(\cdot)$ *from contexts to* $\mathbb{N}$*:*

– *The* least level *of a term* T *is defined as*[4]

$$\ell\ell(T) := \inf\{\ell(\mathbf{C}) \mid T = \mathbf{C}\langle R\rangle \text{ for some } R \in \mathcal{R}\} \in (\mathbb{N} \cup \{\infty\}). \tag{6}$$

– *A* ρ*-reduction step* $T \rightarrow_\rho S$ *is:*
  1. at level k, *noted* $T \rightarrow_{\rho:k} S$, *if* $T = \mathbf{C}\langle R\rangle$, $S = \mathbf{C}\langle R'\rangle$, $R \mapsto_\rho R'$, $\ell(\mathbf{C}) = k$;
  2. least-level, *noted* $T \underset{\mathrm{L}}{\rightarrow}_\rho S$, *if* $T \rightarrow_{\rho:k} S$ *and* $k = \ell\ell(T)$;
  3. internal, *noted* $T \underset{\neg\mathrm{L}}{\rightarrow}_\rho S$, *if* $T \rightarrow_{\rho:k} S$ *and* $k > \ell\ell(T)$.

– Least-level reduction *is* $\underset{\mathrm{L}}{\rightarrow} = \bigcup_\rho \underset{\mathrm{L}}{\rightarrow}_\rho$ *(for* $\rho \in \mathsf{Rules}$*).*

– Internal reduction *is* $\underset{\neg\mathrm{L}}{\rightarrow} = \bigcup_\rho \underset{\neg\mathrm{L}}{\rightarrow}_\rho$ *(for* $\rho \in \mathsf{Rules}$*).*

Note that $\rightarrow = \underset{\mathrm{L}}{\rightarrow} \cup \underset{\neg\mathrm{L}}{\rightarrow}$ and that our definitions solve the issue of Example 1. Indeed, the definition of least level $\ell\ell(T)$ of a term, and hence the definition of $\underset{\mathrm{L}}{\rightarrow}_\rho$, depend on the *whole* set $\mathcal{R} = \bigcup_\rho \mathcal{R}_\rho$ of redexes associated with $\rightarrow$.[5]

[4] Recall that $\inf \emptyset = \infty$, when $\emptyset$ is seen as the empty subset of $\mathbb{N}$ with the usual order.

[5] We should write $\ell\ell_{\mathcal{R}}(T)$, $\mathrm{L}_{\mathcal{R}}$ and $\underset{\mathrm{L}_{\mathcal{R}}}{\rightarrow}_\rho$, but we avoid it for the sake of readability.

Normal Forms. It is immediate that $\underset{L}{\rightarrow} \subsetneq \rightarrow$ is a *strategy* for $\rightarrow$. Indeed, $\underset{L}{\rightarrow}$ and $\rightarrow$ have the *same normal forms* because $\underset{L}{\rightarrow} \subseteq \rightarrow$ and if a term has a $\rightarrow$-redex, it has a redex at least-level, *i.e.* it has a $\underset{L}{\rightarrow}$-redex.

Remark 16 (Least level of normal forms). Note that $\ell\ell(T) = \infty$ if and only if T is $\rightarrow$-normal, because $\ell(\mathbf{C}) \in \mathbb{N}$ for all contexts $\mathbf{C}$.

A good least-level reduction. The beauty of least-level reduction for the bang calculus, is that it satisfies some elegant properties, which allow for neat proofs, in particular monotonicity and internal invariance (in Definition 17). The developments in the rest of the paper rely on such properties, and in fact will apply to any calculus whose reduction $\rightarrow$ has the properties described below.

Definition 17 (Good least-level). *A reduction $\rightarrow$ has a* good least-level *if:*

1. (monotonicity) *$T \rightarrow S$ implies $\ell\ell(T) \leq \ell\ell(S)$; and*
2. (internal invariance) *$T \underset{\neg L}{\rightarrow} S$ implies $\ell\ell(T) = \ell\ell(S)$.*

Point 1 states that no step can decrease the least level of a term. Point 2 says that internal steps cannot change the least level of a term. Therefore, only least-level steps may increase the least level. Together, they imply persistence: only least-level steps can approach normal forms.

Property 18 (Persistence) *If $\rightarrow$ has a good least-level, then $T \underset{\neg L}{\rightarrow} S$ implies that S is not $\rightarrow$-normal.*

Reduction $\rightarrow_{\beta_!}$ in the pure bang calculus $(\Lambda_!, \rightarrow_{\beta_!})$ has a good least-level. More in general, the same holds when extending the reduction with operators.

Proposition 19 (Good least-level of bang calculi). *Given $\Lambda_{!\mathcal{O}}$, let $\rightarrow = \rightarrow_{\beta_!} \cup \rightarrow_{\mathcal{O}}$, where each $\mathbf{o} \in \mathcal{O}$ has a redex of shape $\mathbf{o}(P_1, \ldots, P_k)$. The reduction $\rightarrow$ has a good least-level.*

4.2 Least-level for a bang calculus: examples.

Let us see more closely the least-level reduction for a bang calculus $(\Lambda_{!\mathcal{O}}, \rightarrow)$. For concreteness, we consider $\mathsf{Rules} = \{\beta_!, \mathbf{o} \mid \mathbf{o} \in \mathcal{O}\}$, hence the set of redexes is $\mathcal{R} = \mathcal{R}_{\beta_!} \cup \mathcal{R}_{\mathcal{O}}$, where $\mathcal{R}_{\mathcal{O}}$ is the set of terms $\mathbf{o}(T_1, \ldots, T_k)$ for any $\mathbf{o} \in \mathcal{O}$.

We observe that the least level $\ell\ell(T)$ of a term $T \in \Lambda_{!\mathcal{O}}$ can be easily defined in a direct way, by induction on T:

- $\ell\ell(T) = 0$ if $T \in \mathcal{R} = \mathcal{R}_{\beta_!} \cup \mathcal{R}_{\mathcal{O}}$,
- otherwise, $\ell\ell(x) = \infty$ and

$$\ell\ell(\lambda x.T) = \ell\ell(T) \qquad \ell\ell(!T) = \ell\ell(T) + 1 \qquad \ell\ell(TS) = \min\{\ell\ell(T), \ell\ell(S)\}.$$

Example 20 (Least level of a term). Let $R \in \mathcal{R}_{\beta_!}$. If $T_0 := R\,!R$, then $\ell\ell(T_0) = 0$. If $T_1 := x\,!R$ then $\ell\ell(T_1) = 1$. If $T_2 := \mathbf{o}(x, y)!R$ then $\ell\ell(T_2) = 0$, as $\mathbf{o}(x, y) \in \mathcal{R}_{\mathcal{O}}$.

Intuitively, least-level reduction fires a redex that is *minimally nested*, where a redex is any subterm whose form is in $\mathcal{R} = \mathcal{R}_{\beta_!} \cup \mathcal{R}_{\mathbf{O}}$. Note that least-level reduction can choose to fire one among possibly *several* redexes at minimal level.

Example 21. Let us revisit Example 20 with $R = \iota\,!z \in \mathcal{R}_{\beta_!}$ (so $R \mapsto_{\beta_!} z$, see Remark 11). Then $T_1 := x\,!R \underset{\mathrm{L}}{\rightarrow}_{\beta_!} x\,!z$ but $T_0 := R!R \underset{\mathrm{L}}{\not\rightarrow}_{\beta_!} R!z$ and $T_2 := \mathbf{o}(x,y)\,!R \underset{\mathrm{L}}{\not\rightarrow}_{\beta_!} \mathbf{o}(x,y)!z$. Also, $\mathbf{o}(x,R) \underset{\mathrm{L}}{\not\rightarrow}_{\beta_!} \mathbf{o}(x,z)$ although $\mathbf{o}(x,R) \rightarrow_{\beta_!} \mathbf{o}(x,z)$.

Let $S = \iota\,!(z\,!z)$ (so $S \mapsto_{\beta_!} z\,!z$). In $(\lambda z.S)!S$, two least-level steps are possible (the fired $\beta_!$-redex is underlined): $\underline{(\lambda z.S)!S} \underset{\mathrm{L}}{\rightarrow}_{\beta_!} \iota\,!(S\,!S)$, and $(\lambda z.\underline{S})!S \underset{\mathrm{L}}{\rightarrow}_{\beta_!} (\lambda z.z\,!z)!S$. But $(\lambda z.S)!S \underset{\mathrm{L}}{\not\rightarrow}_{\beta_!} (\lambda z.S)!(z\,!z)$ although $(\lambda z.S)!S \rightarrow_{\beta_!} (\lambda z.S)!(z\,!z)$.

4.3 Least-level for CbN and CbV λ-calculi

The definition of least-level reduction in Section 4.1 is independent of the specific notion of level chosen, and of the specific calculus. The idea is that the reduction strategy persistently fires a redex at minimal level, once such a notion is set.

Least-level reduction can indeed be defined also for the CbN and CbV λ-calculi, given an opportune definition of level. In CbN, we count the number of nested arguments and operators containing the redex occurrence. In CbV, we count the number of nested operators and *unapplied* abstractions containing the redex occurrence, where an abstraction is unapplied if it is not the right-hand side of an application. Formally, a redex occurrence is identified by a context (as explained in Section 4.1), and we define the *level* $\ell^{\mathrm{CbN}}(\mathbf{c}) \in \mathbb{N}$ and $\ell^{\mathrm{CbV}}(\mathbf{c}) \in \mathbb{N}$ of a context $\mathbf{c}$ in CbN and CbV λ-calculi, respectively, as follows.

$$\begin{aligned}
\ell^{\mathrm{CbN}}(\langle\cdot\rangle) &= 0 & \ell^{\mathrm{CbV}}(\langle\cdot\rangle) &= 0\\
\ell^{\mathrm{CbN}}(\lambda x.\mathbf{c}) &= \ell^{\mathrm{CbN}}(\mathbf{c}) & \ell^{\mathrm{CbV}}(\lambda x.\mathbf{c}) &= \ell^{\mathrm{CbV}}(\mathbf{c}) + 1\\
\ell^{\mathrm{CbN}}(\mathbf{c}t) &= \ell^{\mathrm{CbN}}(\mathbf{c}) & \ell^{\mathrm{CbV}}(\mathbf{c}t) &= \begin{cases}\ell^{\mathrm{CbV}}(\mathbf{c}') & \text{if } \mathbf{c} = \lambda x.\mathbf{c}'\\ \ell^{\mathrm{CbV}}(\mathbf{c}) & \text{otherwise}\end{cases}\\
\ell^{\mathrm{CbN}}(t\mathbf{c}) &= \ell^{\mathrm{CbN}}(\mathbf{c}) + 1 & \ell^{\mathrm{CbV}}(t\mathbf{c}) &= \ell^{\mathrm{CbV}}(\mathbf{c})\\
\ell^{\mathrm{CbN}}(\mathbf{o}(\dots,\mathbf{c},\dots)) &= \ell^{\mathrm{CbN}}(\mathbf{c}) + 1 & \ell^{\mathrm{CbV}}(\mathbf{o}(\dots,\mathbf{c},\dots)) &= \ell^{\mathrm{CbV}}(\mathbf{c}) + 1.
\end{aligned}$$

In both CbN and CbV λ-calculi, the *least level* of a term (denoted by $\ell\ell^{\mathrm{CbN}}(\cdot)$ and $\ell\ell^{\mathrm{CbV}}(\cdot)$) and *least-level* and *internal* reductions are given by Definition 15 (replace $\ell(\cdot)$ with $\ell^{\mathrm{CbN}}(\cdot)$ for CbN, and with $\ell^{\mathrm{CbV}}(\cdot)$ for CbV).

In Section 5 we will see that the definitions of CbN and CbV least level are not arbitrary, but induced by the CbN and CbV translations defined in Section 3.3.

5 Embedding of CbN and CbV by level

Here we refine the analysis of the CbN and CbV translations given in Section 3.3, by showing two new results: translations preserve normal forms (Proposition 22) and least-level (Proposition 25), back and forth. This way, to obtain least-level *factorization* or least-level *normalization* results, it suffices to prove them in the bang calculus. The translation transfers the results into the CbN and CbV λ-calculi (Theorem 26). We use here the expression "translate" in a strong sense: the results for CbN and CbV λ-calculi are obtained from the corresponding results in the bang calculus almost for free, just via CbN and CbV translations.

Preservation of normal forms. The targets of the CbN translation $(\cdot)^{\mathsf{n}}$ and CbV translation $(\cdot)^{\mathsf{v}}$ into the bang calculus can be *characterized syntactically.* A fine analysis of these fragments of the bang calculus (see [12] for details) proves that both CbN and CbV translations preserve normal forms, back and forth.

Proposition 22 (Preservation of normal forms). *Let* $t, s \in \Lambda_{\mathcal{O}}$ *and* $\mathbf{o} \in \mathcal{O}$.

1. CbN: t *is* β*-normal iff* t^{n} *is* $\beta_!$*-normal;* t *is* $\mathbf{o}$*-normal iff* t^{n} *is* $\mathbf{o}$*-normal.*
2. CbV: t *is* β_v*-normal iff* t^{v} *is* $\beta_!$*-normal;* t *is* $\mathbf{o}$*-normal iff* t^{v} *is* $\mathbf{o}$*-normal.*

By Remark 16, Proposition 22 can be seen as the fact that CbN and CbV translations preserve the least-level of a term, back and forth, when the least-level is infinite. Actually, this holds more in general for any value of the least-level.

Preservation of levels. We aim to show that least-level steps in CbN and CbV λ-calculi correspond to least-level steps in the bang calculus—back and forth—via CbN and CbV translations, respectively (Proposition 25). This result is subtle, one of the main technical contributions of this paper.

First, we extend the definition of translations to contexts. The *CbN and CbV translations for contexts* are two functions $(\cdot)^{\mathsf{n}} \colon \mathcal{C} \to \mathcal{C}_!$ and $(\cdot)^{\mathsf{v}} \colon \mathcal{C} \to \mathcal{C}_!$, respectively, mapping contexts of the λ-calculus into contexts of the bang calculus:

$$\langle\cdot\rangle^{\mathsf{n}} = \langle\cdot\rangle \qquad\qquad \langle\cdot\rangle^{\mathsf{v}} = \langle\cdot\rangle$$

$$(\lambda x.\mathbf{c})^{\mathsf{n}} = \lambda x.\mathbf{c}^{\mathsf{n}} \qquad\qquad (\lambda x.\mathbf{c})^{\mathsf{v}} = !(\lambda x.\mathbf{c}^{\mathsf{v}})$$

$$(\mathbf{o}(t_1, \dots, \mathbf{c}, \dots, t_k))^{\mathsf{n}} = \mathbf{o}(t_1^{\mathsf{n}}, \dots, \mathbf{c}^{\mathsf{n}}, \dots, t_k^{\mathsf{n}}) \qquad (\mathbf{o}(t_1, \dots, \mathbf{c}, \dots, t_k))^{\mathsf{v}} = \mathbf{o}(t_1^{\mathsf{v}}, \dots, \mathbf{c}^{\mathsf{v}}, \dots, t_k^{\mathsf{v}})$$

$$(\mathbf{c}t)^{\mathsf{n}} = \mathbf{c}^{\mathsf{n}}\,!(t^{\mathsf{n}}) \qquad\qquad (\mathbf{c}t)^{\mathsf{v}} = \begin{cases} \mathbf{C}\,t^{\mathsf{v}} & \text{if } \mathbf{c}^{\mathsf{v}} = !\mathbf{C} \\ (\iota\,\mathbf{c}^{\mathsf{v}})t^{\mathsf{v}} & \text{otherwise} \end{cases}$$

$$(t\mathbf{c})^{\mathsf{n}} = t^{\mathsf{n}}\,!(\mathbf{c}^{\mathsf{n}})\,; \qquad\qquad (t\mathbf{c})^{\mathsf{v}} = \begin{cases} T\,\mathbf{c}^{\mathsf{v}} & \text{if } t^{\mathsf{v}} = !T \\ (\iota\,t^{\mathsf{v}})\mathbf{c}^{\mathsf{v}} & \text{otherwise.} \end{cases}$$

Note that CbN (resp. CbV) level of a context defined in Section 4.3 increases by 1 whenever the CbN (resp. CbV) translation for contexts adds a !. Thus, CbN and CbV translations preserve, back and forth, the level of a redex and the least-level of a term. Said differently, the level for CbN and CbV is defined in Section 4.3 so as to enable the preservation of level via CbN and CbV translations.

Lemma 23 (Preservation of level via CbN translation).

1. For contexts: *For any context* $\mathbf{c} \in \mathcal{C}$*, one has* $\ell^{\mathrm{CbN}}(\mathbf{c}) = \ell(\mathbf{c}^{\mathsf{n}})$.
2. For reduction: *For any term* $t \in \Lambda_{\mathcal{O}}$: $t \to_{\beta:k} s$ *if and only if* $t^{\mathsf{n}} \to_{\beta_!:k} s^{\mathsf{n}}$*; and* $t \to_{\mathbf{o}:k} s$ *if and only if* $t^{\mathsf{n}} \to_{\mathbf{o}:k} s^{\mathsf{n}}$*, for any* $\mathbf{o} \in \mathcal{O}$.

3. For least-level of a term: *For any term* $t \in \Lambda_{\mathcal{O}}$*, one has* $\ell\ell^{\mathrm{CbN}}(t) = \ell\ell(t^{\mathsf{n}})$*.*

Lemma 24 (Preservation of level via CbV translation).

1. For contexts: *For any context* $\mathbf{c} \in \mathcal{C}$*, one has* $\ell^{\mathrm{CbV}}(\mathbf{c}) = \ell(\mathbf{c}^{\mathsf{v}})$*.*
2. For reduction: *For any term* $t \in \Lambda_{\mathcal{O}}$*:* $t \to_{\beta_v:k} s$ *if and only if* $t^{\mathsf{v}} \to_{\beta_!:k} \to^{=}_{\mathsf{d}:k} s^{\mathsf{v}}$*; and* $t \to_{\mathbf{o}:k} s$ *if and only if* $t^{\mathsf{v}} \to_{\mathbf{o}:k} \to^{=}_{\mathsf{d}:k} s^{\mathsf{v}}$*, for any* $\mathbf{o} \in \mathcal{O}$*.*
3. For least-level of a term: *For any term* $t \in \Lambda_{\mathcal{O}}$*, one has* $\ell\ell^{\mathrm{CbV}}(t) = \ell\ell(t^{\mathsf{v}})$*.*

From the two lemmas above it follows that CbN and CbV translations preserve least-level and internal reductions, back and forth.

Proposition 25 (Preservation of least-level and internal reductions). *Let* $t \in \Lambda_{\mathcal{O}}$ *and* $\mathbf{o} \in \mathcal{O}$*.*

1. CbN least-level: $t \underset{\mathrm{L}}{\to}_{\beta} s$ *iff* $t^{\mathsf{n}} \underset{\mathrm{L}}{\to}_{\beta_!} s^{\mathsf{n}}$*; and* $t \underset{\mathrm{L}}{\to}_{\mathbf{o}} s$ *iff* $t^{\mathsf{n}} \underset{\mathrm{L}}{\to}_{\mathbf{o}} s^{\mathsf{n}}$*.*
2. CbN internal: $t \underset{\neg\mathrm{L}}{\to}_{\beta} s$ *iff* $t^{\mathsf{n}} \underset{\neg\mathrm{L}}{\to}_{\beta_!} s^{\mathsf{n}}$*; and* $t \underset{\neg\mathrm{L}}{\to}_{\mathbf{o}} s$ *iff* $t^{\mathsf{n}} \underset{\neg\mathrm{L}}{\to}_{\mathbf{o}} s^{\mathsf{n}}$*.*
3. CbV least-level: $t \underset{\mathrm{L}}{\to}_{\beta_v} s$ *iff* $t^{\mathsf{v}} \underset{\mathrm{L}}{\to}_{\beta_!} \underset{\mathrm{L}}{\to}_{\mathsf{d}}^{=} s^{\mathsf{v}}$*; and* $t \underset{\mathrm{L}}{\to}_{\mathbf{o}} s$ *iff* $t^{\mathsf{v}} \underset{\mathrm{L}}{\to}_{\mathbf{o}} \underset{\mathrm{L}}{\to}_{\mathsf{d}}^{=} s^{\mathsf{v}}$*.*
4. CbV internal: $t \underset{\mathrm{L}}{\to}_{\beta_v} s$ *iff* $t^{\mathsf{v}} \underset{\neg\mathrm{L}}{\to}_{\beta_!} \underset{\neg\mathrm{L}}{\to}_{\mathsf{d}}^{=} s^{\mathsf{v}}$*; and* $t \underset{\mathrm{L}}{\to}_{\mathbf{o}} s$ *iff* $t^{\mathsf{v}} \underset{\neg\mathrm{L}}{\to}_{\mathbf{o}} \underset{\neg\mathrm{L}}{\to}_{\mathsf{d}}^{=} s^{\mathsf{v}}$*.*

As a consequence, least-level reduction induces factorization in CbN and CbV λ-calculi as soon as it does in the bang calculus. And, by Proposition 22, it is a normalizing strategy in CbN and CbV as soon as it is so in the bang calculus.

Theorem 26 (Factorization and normalization by translation). *Let* $\Lambda^{\mathsf{cbn}}_{\mathcal{O}} = (\Lambda_{\mathcal{O}}, \to_{\beta} \cup \to_{\mathcal{O}})$ *and* $\Lambda^{\mathsf{cbv}}_{\mathcal{O}} = (\Lambda_{\mathcal{O}}, \to_{\beta_v} \cup \to_{\mathcal{O}})$*.*

1. If $\Lambda_{!\mathcal{O}}$ *admits least-level factorization* $\mathtt{Fact}(\underset{\mathrm{L}}{\to}, \underset{\neg\mathrm{L}}{\to})$*, then so do* $\Lambda^{\mathsf{cbn}}_{\mathcal{O}}$ *and* $\Lambda^{\mathsf{cbv}}_{\mathcal{O}}$*.*
2. If $\Lambda_{!\mathcal{O}}$ *admits least-level normalization, then so do* $\Lambda^{\mathsf{cbn}}_{\mathcal{O}}$ *and* $\Lambda^{\mathsf{cbv}}_{\mathcal{O}}$*.*

A similar result will hold also when extending the pure calculi with a rule $\mapsto_{\rho}$ other than $\mapsto_{\mathbf{o}}$, as long as the translation preserves ρ-redexes, back and forth.

Remark 27 (Preservation of least-level and of normal forms). Preservation of normal form and least-level is delicate. For instance, it does not hold with the definition CbV translation $(\cdot)^{\mathsf{v}}$ in [15,30]. There, the translation $t = rs \in \Lambda$ would be $t^{\mathsf{v}} = (\iota\,!(r^{\mathsf{v}}))s^{\mathsf{v}}$ and then Proposition 22 and Proposition 25 would not hold: $\iota\,!(r^{\mathsf{v}})$ is a $\beta_!$-redex in t^{v} (see Remark 11) and hence t^{v} would not be normal even though so is t, and $\ell\ell(t^{\mathsf{v}}) = 0$ even though $\ell\ell^{\mathrm{CbV}}(t) \neq 0$. This is why we defined two distinct case when defining $(\cdot)^{\mathsf{v}}$ for applications, akin to Bucciarelli *et al.* [7].

6 Least-level factorization via bang calculus

We have shown that least-level factorization in a bang calculus $\Lambda_{!\mathcal{O}}$ implies least-level factorization in the corresponding CbN and CbV calculi, via forth-and-back translation. The central question now is *how to prove least-level factorization* for a bang calculus: this section is devoted to that, in the pure and applied cases.

Overview. Let us overview our approach by considering $\mathcal{O} = \{\mathbf{o}\}$, and $\to = \to_{\beta_!} \cup \to_{\mathbf{o}}$. Since by definition $\underset{\mathrm{L}}{\to} = \underset{\mathrm{L}}{\to}_{\beta_!} \cup \underset{\mathrm{L}}{\to}_{\mathbf{o}}$ (and $\underset{\neg\mathrm{L}}{\to} = \underset{\neg\mathrm{L}}{\to}_{\beta_!} \cup \underset{\neg\mathrm{L}}{\to}_{\mathbf{o}}$), Lemma 8 states that we can *decompose* least-level factorization of $\to$ in three modules:

1. prove least-level factorization of $\to_{\beta_!}$, *i.e.* $\to_{\beta_!}^* \subseteq \underset{\mathrm{L}}{\to}_{\beta_!}^* \cdot \underset{\neg\mathrm{L}}{\to}_{\beta_!}$;
2. prove least-level factorization of $\to_{\mathbf{o}}$, *i.e.* $\to_{\mathbf{o}}^* \subseteq \underset{\mathrm{L}}{\to}_{\mathbf{o}}^* \cdot \underset{\neg\mathrm{L}}{\to}_{\mathbf{o}}$;
3. prove the two linear swaps of Lemma 8.

Note that, for each of $\underset{\mathrm{L}}{\to}_{\beta_!}$ and $\underset{\mathrm{L}}{\to}_{\mathbf{o}}$, the least level is defined with respect to the set of *all* redexes $\mathcal{R} = \mathcal{R}_{\beta_!} \cup \mathcal{R}_{\mathbf{o}}$, so as to have $\underset{\mathrm{L}}{\to} = \underset{\mathrm{L}}{\to}_{\beta_!} \cup \underset{\mathrm{L}}{\to}_{\mathbf{o}}$. This approach solves the issue we mentioned in Example 1.

Clearly, Points 2 and 3 depend on the specific rule $\mapsto_{\mathbf{o}}$. However, the beauty of a modular approach is that Point 1 can be established in general: we do not need to know $\mapsto_{\mathbf{o}}$, only the shape of its redexes given by $\mathcal{R}_{\mathbf{o}}$. In Section 6.1 we provide a general result of least-level factorization for $\to_{\beta_!}$ (Theorem 28). In fact, we shall show a bit more: the way of decomposing the study of factorization that we have sketched, can be applied to study least-level factorization of any reduction $\to = \to_{\beta_!} \cup \to_{\rho}$, as long as $\to$ has a good least-level.

Once (1) is established (once and for all), to prove factorization of a reduction $\to_{\beta_!} \cup \to_{\mathbf{o}}$ we are only left with (2) and (3). In Section 6.3 we show that the proof of the two linear swaps can be reduced to a single, simple test, involving only the $\mapsto_{\mathbf{o}}$ step (Proposition 34). In Section 7, we will illustrate how all elements play together on a concrete case, applying them to non-deterministic λ-calculi.

6.1 Factorization of $\to_{\beta_!}$ in a bang calculus

We show that $\to_{\beta_!}$ *factorizes* via least-level reduction (Theorem 28). This holds for a definition of $\underset{\mathrm{L}}{\to}_{\beta_!}$ (as in Section 4) where the set of redexes $\mathcal{R}$ contains $\mathcal{R}_{\beta_!} \cup \mathcal{R}_{\mathcal{O}}$—this generalization has essentially no cost, and allows us to use Theorem 28 as a module in the factorization of larger reductions containing $\to_{\beta_!}$.

We prove factorization via Takahashi's parallel reduction method [32]. We define a reflexive reduction $\underset{\neg\mathrm{L}}{\Rightarrow}_{\beta_!}$ (called parallel internal $\beta_!$-reduction) which fulfills the conditions of Property 7, *i.e.* $\underset{\neg\mathrm{L}}{\Rightarrow}_{\beta_!}^* = \underset{\neg\mathrm{L}}{\to}_{\beta_!}^*$ and $\underset{\neg\mathrm{L}}{\Rightarrow}_{\beta_!} \cdot \underset{\mathrm{L}}{\to}_{\beta_!} \subseteq \underset{\mathrm{L}}{\to}_{\beta_!}^* \cdot \underset{\neg\mathrm{L}}{\Rightarrow}_{\beta_!}$.

The tricky point is to prove that $\underset{\neg\mathrm{L}}{\Rightarrow}_{\beta_!} \cdot \underset{\mathrm{L}}{\to}_{\beta_!} \subseteq \underset{\mathrm{L}}{\to}_{\beta_!}^* \cdot \underset{\neg\mathrm{L}}{\Rightarrow}_{\beta_!}$ We adapt the proof technique in [2]. All details are in [12]. Here we just give the definition of $\underset{\neg\mathrm{L}}{\Rightarrow}_{\beta_!}$.

We first introduce $\Rightarrow_{\beta_!:n}$ with $n \in \mathbb{N} \cup \{\infty\}$ (the parallel version of $\to_{\beta_!:n}$), which fires simultaneously a number of $\beta_!$-redexes at level at least $n \in \mathbb{N}$, and $\Rightarrow_{\beta_!:\infty}$ does not reduce any $\beta_!$-redex: $T \Rightarrow_{\beta_!:\infty} S$ implies $T = S$.

$$\frac{}{x \Rightarrow_{\beta_!:\infty} x} \qquad \frac{T \Rightarrow_{\beta_!:n} T'}{\lambda x.T \Rightarrow_{\beta_!:n} \lambda x.T'} \qquad \frac{T \Rightarrow_{\beta_!:m} T' \quad S \Rightarrow_{\beta_!:n} S'}{TS \Rightarrow_{\beta_!:\min\{m,n\}} T'S'} \qquad \frac{T \Rightarrow_{\beta_!:n} T'}{!T \Rightarrow_{\beta_!:n+1} !T'}$$

$$\frac{T \Rightarrow_{\beta_!:n} T' \quad S \Rightarrow_{\beta_!:m} S'}{(\lambda x.T)!S \Rightarrow_{\beta_!:0} T'\{S'/x\}}$$

The *parallel internal $\beta_!$-reduction* $\underset{\neg\mathrm{L}}{\Rrightarrow}_{\beta_!}$ is the parallel version of $\underset{\neg\mathrm{L}}{\rightarrow}_{\beta_!}$, which fires simultaneously a number of $\beta_!$-redexes that are not at minimal level. Formally,

$$T \underset{\neg\mathrm{L}}{\Rrightarrow}_{\beta_!} S \quad \text{if } T \Rightarrow_{\beta_!:n} S \text{ with } n = \infty \text{ or } n > \ell\ell(T).$$

Theorem 28 (Least-level factorization of $\rightarrow_{\beta_!}$). *Let $\rightarrow_\rho$ be the contextual closure of a rule $\mapsto_\rho$, and assume that $\rightarrow = \rightarrow_{\beta_!} \cup \rightarrow_\rho$ has good least-level in $\Lambda_{!\mathcal{O}}$. Then, $T \rightarrow^*_{\beta_!} S$ implies $T \underset{\mathrm{L}}{\rightarrow}_{\beta_!}{}^* \cdot \underset{\neg\mathrm{L}}{\rightarrow}_{\beta_!}{}^* S$.*

In particular, as $\rightarrow_{\beta_!}$ has a good least-level (Proposition 19) in $\Lambda_!$, we have:

Corollary 29 (Least-level factorization in the pure bang calculus). *In the pure bang calculus $(\Lambda_!, \rightarrow_{\beta_!})$, if $T \rightarrow^*_{\beta_!} S$ then $T \underset{\mathrm{L}}{\rightarrow}_{\beta_!}{}^* \cdot \underset{\neg\mathrm{L}}{\rightarrow}_{\beta_!}{}^* S$.*

Surface Digression. According to Definition 15, $\beta_!$-reduction $\rightarrow_{\beta_!:0}$ at level 0 (called *surface reduction* in Simpson [31]) can only fire redexes at level 0, *i.e.*, redexes that are not inside boxes or other operators. It can be equivalently defined as the closure of $\mapsto_{\beta_!}$ under contexts $\mathbf{S}$ defined by $\mathbf{S} ::= \langle\cdot\rangle \mid \lambda x.\mathbf{S} \mid \mathbf{S}T \mid T\mathbf{S}$. Since $\rightarrow_{\beta_!:0} \subseteq \underset{\mathrm{L}}{\rightarrow}_{\beta_!}$, from least-level factorization (Corollary 29) and monotonicity (Proposition 19), a new proof of a result already proven by Simpson [31] follows.

Corollary 30 (Surface factorization in the pure bang calculus). *In the pure bang calculus $(\Lambda_!, \rightarrow_{\beta_!})$, if $T \rightarrow^*_{\beta_!} S$ then $T \rightarrow^*_{\beta_!:0} \cdot \rightarrow^*_{\beta_!:k} S$ with $k > 0$.*

6.2 Pure calculi and least-level normalization

Least-level factorization of $\rightarrow_{\beta_!}$ implies in particular least-level factorization for $\rightarrow_\beta$ and $\rightarrow_{\beta_v}$. As a consequence, least-level reduction is a normalizing strategy for all three pure calculi: the bang calculus, the CbN, and the CbV λ-calculi.

The pure bang calculus. The least-level reduction $\underset{\mathrm{L}}{\rightarrow}_{\beta_!}$ is a *normalizing strategy* for $\rightarrow_{\beta_!}$. Indeed, it satisfies all ingredients in Lemma 4. Since we have least-level factorization (Corollary 29), same normal forms, and *persistence* (Proposition 19), $\underset{\mathrm{L}}{\rightarrow}_{\beta_!}$ is a *complete strategy* for $\rightarrow_{\beta_!}$: if $T \rightarrow^*_{\beta_!} S$ and S is $\beta_!$-normal, then $T \underset{\mathrm{L}}{\rightarrow}_{\beta_!}{}^* S$.

We already observed (Example 21) that the least-level reduction $\underset{\mathrm{L}}{\rightarrow}_{\beta_!}$ is non-deterministic, because several redexes at least level may be available. Such non-determinism is however harmless and inessential, because $\underset{\mathrm{L}}{\rightarrow}_{\beta_!}$ is *uniform*.

Lemma 31 (Quasi-Diamond). *In the pure bang calculus $(\Lambda_!, \rightarrow_{\beta_!})$, the reduction $\underset{\mathrm{L}}{\rightarrow}_{\beta_!}$ is quasi-diamond (Property 5), and therefore uniform.*

Putting all the ingredients together, we have (by Lemma 4):

Theorem 32 (Least-level normalization). *In the pure bang calculus $(\Lambda_!, \rightarrow_{\beta_!})$, the least-level reduction $\underset{\mathrm{L}}{\rightarrow}_{\beta_!}$ is a normalizing strategy for $\rightarrow_{\beta_!}$.*

Theorem 32 means not only that if T is weakly $\beta_!$-normalizing then T can reach its normal form by just performing least-level steps, but also that performing *whatever* least-level steps eventually leads to the normal form, if any.

Pure CbV and CbN λ-calculi. By forth-and-back translation (Theorem 26) the least-level factorization and normalization results for the pure bang calculus immediately transfers to the (pure) CbN and CbV settings.

Theorem 33 (CbV and CbN least-level normalization).
- CbN: *In* $(\Lambda, \to_\beta)$, $\underset{\text{L}}{\to}_\beta$ *is a normalizing strategy for* $\to_\beta$.
- CbV: *In* $(\Lambda, \to_{\beta_v})$, $\underset{\text{L}}{\to}_{\beta_v}$ *is a normalizing strategy for* $\to_{\beta_v}$.

6.3 Least-level Factorization, Modularly

Least-level factorization of $\to_{\beta_!}$ (Theorem 28) can be used to prove factorization for a more complex calculus. Indeed, a simple and modular *test* establishes least-level factorization of a reduction $\to_{\beta_!} \cup \to_\rho$ ($\to_\rho$ is a reduction added to $\to_{\beta_!}$), by adapting a similar result in [3]. The test relies on the fact that we have already proved Theorem 28, and it *simplifies* Lemma 8: the proof of the two linear swaps of Lemma 8 is reduced to a single, easier check, which only involves the rule $\mapsto_\rho$. As usual, the least level in $\underset{\text{L}}{\to}_{\beta_!}$ and $\underset{\text{L}}{\to}_\rho$ is defined with respect to the set $\mathcal{R} = \mathcal{R}_{\beta_!} \cup \mathcal{R}_\rho$ of redexes. An example of the use of this test is in Section 7.

Proposition 34 (Modular test for least-level factorization). *Let* $\to_\rho$ *be the contextual closure of a rule* $\mapsto_\rho$*, and assume that* $\to = \to_{\beta_!} \cup \to_\rho$ *has a good least-level in* $\Lambda_{!\mathcal{O}}$*. Then* $\to$ *factorizes via* $\underset{\text{L}}{\to} = \underset{\text{L}}{\to}_{\beta_!} \cup \underset{\text{L}}{\to}_\rho$ *if the following hold:*

1. *(*least-level factorization of $\to_\rho$*)* $\to_\rho^* \subseteq \underset{\text{L}}{\to}_\rho{}^* \cdot \underset{\neg\text{L}}{\to}_\rho{}^*$;
2. *(*substitutivity *of* $\mapsto_\rho$*)* $R \mapsto_\rho R'$ *implies* $R\{T/x\} \mapsto_\rho R'\{T/x\}$;
3. *(*root linear swap*)* $\underset{\neg\text{L}}{\to}_{\beta_!} \cdot \mapsto_\rho \subseteq \mapsto_\rho \cdot \to_{\beta_!}^*$.

7 Case study: non-deterministic λ-calculi

To show how to use our framework, we apply the tools we have developed on our running example (see Examples 1 and 9). We extend the bang calculus with a *non-deterministic* binary operator $\oplus$, that is, $(\Lambda_{!\oplus}, \to_{\beta_!\oplus})$ where $\to_{\beta_!\oplus} = \to_{\beta_!} \cup \to_\oplus$, and $\to_\oplus$ is the contextual closure of the (non-deterministic) rules:

$$\oplus(T, S) \mapsto_\oplus T \qquad\qquad \oplus(T, S) \mapsto_\oplus S.$$

First step: non-deterministic bang calculus. We analyze $\Lambda_{!\oplus}$. We use our modular test to prove least-level factorization for $\Lambda_{!\oplus}$: if $T \to_{\beta_!\oplus}^* U$ then $T \underset{\text{L}}{\to}_{\beta_!\oplus}{}^* \cdot \underset{\neg\text{L}}{\to}_{\beta_!\oplus}{}^* U$. By Lemma 4, an immediate consequence of the factorization result is that the least-level strategy is *complete*: if U is normal, $T \to_{\beta_!\oplus}^* U$ implies $T \underset{\text{L}}{\to}_{\beta\oplus}{}^* U$.

Second step: CbN and CbV non-deterministic calculi. By translation, we have *for free*, that the analogous results hold in $\Lambda_\oplus^{\mathtt{cbn}}$ and $\Lambda_\oplus^{\mathtt{cbv}}$, as defined in Example 9. So, least-level factorization holds for both calculi, and moreover
- *CbN completeness*: in $\Lambda_\oplus^{\mathtt{cbn}}$, if u is normal, $t \to_{\beta\oplus}^* u$ implies $t \underset{\text{L}}{\to}_{\beta\oplus}{}^* u$.
- *CbV completeness*: in $\Lambda_\oplus^{\mathtt{cbv}}$, if u is normal, $t \to_{\beta_v\oplus}^* u$ implies $t \underset{\text{L}}{\to}_{\beta_v\oplus}{}^* u$.

What do we really need to prove? The only result we need to prove is least-level factorization of $\to_{\beta_!\oplus}$. Completeness then follows by Lemma 4 and the translations will automatically take care of transferring the results.

To prove factorization of $\to_{\beta_!\oplus}$, most of the work is done, since least-level factorization of $\to_{\beta_!}$ is already established; we then use our test (Proposition 34) to extend $\to_{\beta_!}$ with $\to_{\oplus}$. The only ingredients we need are substitutivity of $\mapsto_{\oplus}$ (which is an obvious property), and the following easy lemma.

Lemma 35 (Roots). *Let* $\rho \in \{\beta_!, \oplus\}$. *If* $T \underset{\neg L}{\to}_\rho R \mapsto_\oplus S$ *then* $T \mapsto_\oplus \cdot \to_\rho^= S$.

Theorem 36 (Least-level factorization in non-deterministic calculi).

1. *In* $(\Lambda_{!\oplus}, \to)$, $\mathtt{Fact}(\underset{L}{\to}, \underset{\neg L}{\to})$ *holds for* $\to = \to_\oplus \cup \to_{\beta_!}$.
2. *Least-level factorization holds in* $(\Lambda_\oplus^{\mathtt{cbn}}, \to_\oplus \cup \to_\beta)$, *and in* $(\Lambda_\oplus^{\mathtt{cbv}}, \to_\oplus \cup \to_{\beta_v})$.

Proof. 1. It is enough to verify the hypotheses of Proposition 34, via Lemma 35. 2. It follows from Theorem 26 and Theorem 36.1. □

Completeness is the best that can be achieved in these calculi, because of the true non-determinism of $\to_\oplus$ and hence of least-level reduction and of any other complete strategy for $\to$. For instance, in $\Lambda_\oplus^{\mathtt{cbn}}$ there is no normalizing strategy for $\oplus(x, \delta\delta)$ in the sense of Definition 2, since $x \underset{L}{\leftarrow}_\oplus \oplus(x, \delta\delta) \underset{L}{\to}_\oplus \delta\delta \underset{L}{\to}_\beta \dots$.

8 Conclusions and Related Work

Combining translations (Theorem 26), least-level factorization for $\to_{\beta_!}$ (Theorem 28), and modularity (Proposition 34), gives us a powerful method to analyze factorization in various λ-calculi that *extend* the pure CbN and CbV calculi. The main novelty is transferring the results from a calculus to another via translations.

Related Work. Many calculi inspired by linear logic subsume CbN and CbV, such as [5,6,29,24] (other than the ones already cited). We chose the bang calculus for its simplicity, which eases the analysis of the CbN and CbV translations.

To study CbN and CbV in a uniform way, an approach orthogonal to ours is given by Ronchi della Rocca and Paolini's parametric λ-calculus [28]. It is a *meta-calculus*, where the reduction rule is *parametric* with respect to a subset of terms (called values) with suitable properties. Different choices for the set of values define different calculi—that is, different reductions. This allows for a uniform presentation of proof arguments, such as the proof of standardization, which is actually a *meta-proof* that can be instantiated in both CbN and CbV.

Least-level reduction is studied for calculi based on linear-logic in [34,1] and for linear logic proof-nets in [8,26]. It is studied for pure CbN λ-calculus in [2].

Acknowledgments. The authors thank Beniamino Accattoli for insightful comments and discussions. This work was partially supported by EPSRC Project EP/R029121/1 *Typed Lambda-Calculi with Sharing and Unsharing.*

References

1. Accattoli, B.: An Abstract Factorization Theorem for Explicit Substitutions. In: 23rd International Conference on Rewriting Techniques and Applications (RTA'12). Leibniz International Proceedings in Informatics (LIPIcs), vol. 15, pp. 6–21. Schloss Dagstuhl (2012). https://doi.org/10.4230/LIPIcs.RTA.2012.6
2. Accattoli, B., Faggian, C., Guerrieri, G.: Factorization and normalization, essentially. In: Programming Languages and Systems - 17th Asian Symposium, APLAS 2019. Lecture Notes in Computer Science, vol. 11893, pp. 159–180. Springer (2019). https://doi.org/10.1007/978-3-030-34175-6_9
3. Accattoli, B., Faggian, C., Guerrieri, G.: Factorize factorization. In: 29th EACSL Annual Conference on Computer Science Logic, CSL 2021. LIPIcs, vol. 183, pp. 6:1–6:25. Schloss-Dagstuhl (2021). https://doi.org/10.4230/LIPIcs.CSL.2021.6
4. Barendregt, H.P.: The Lambda Calculus: Its Syntax and Semantics, Studies in Logic and the Foundations of Mathematics, vol. 103. North Holland (1984)
5. Benton, P.N., Bierman, G.M., de Paiva, V., Hyland, M.: A term calculus for intuitionistic linear logic. In: International Conference on Typed Lambda Calculi and Applications, TLCA '93. Lecture Notes in Computer Science, vol. 664, pp. 75–90. Springer (1993). https://doi.org/10.1007/BFb0037099
6. Benton, P.N., Wadler, P.: Linear logic, monads and the lambda calculus. In: Proceedings, 11th Annual IEEE Symposium on Logic in Computer Science, LICS 1996. pp. 420–431. IEEE Computer Society (1996). https://doi.org/10.1109/LICS.1996.561458
7. Bucciarelli, A., Kesner, D., Ríos, A., Viso, A.: The bang calculus revisited. In: Functional and Logic Programming - 15th International Symposium, FLOPS 2020. Lecture Notes in Computer Science, vol. 12073, pp. 13–32. Springer (2020). https://doi.org/10.1007/978-3-030-59025-3_2
8. de Carvalho, D., Pagani, M., Tortora de Falco, L.: A semantic measure of the execution time in linear logic. Theor. Comput. Sci. **412**(20), 1884–1902 (2011). https://doi.org/10.1016/j.tcs.2010.12.017
9. Chouquet, J., Tasson, C.: Taylor expansion for Call-By-Push-Value. In: 28th EACSL Annual Conference on Computer Science Logic (CSL 2020). Leibniz International Proceedings in Informatics (LIPIcs), vol. 152, pp. 16:1–16:16. Schloss Dagstuhl (2020). https://doi.org/10.4230/LIPIcs.CSL.2020.16
10. Ehrhard, T.: Call-by-push-value from a linear logic point of view. In: Programming Languages and Systems - 25th European Symposium on Programming (ESOP 2016). Lecture Notes in Computer Science, vol. 9632, pp. 202–228 (2016). https://doi.org/10.1007/978-3-662-49498-1_9
11. Ehrhard, T., Guerrieri, G.: The bang calculus: an untyped lambda-calculus generalizing call-by-name and call-by-value. In: Proceedings of the 18th International Symposium on Principles and Practice of Declarative Programming (PPDP 2016). pp. 174–187. ACM (2016). https://doi.org/10.1145/2967973.2968608
12. Faggian, C., Guerrieri, G.: Factorization in call-by-name and call-by-value calculi via linear logic (long version). CoRR **abs/2101.08364** (2021), https://arxiv.org/abs/2101.08364
13. Faggian, C., Ronchi Della Rocca, S.: Lambda calculus and probabilistic computation. In: 34th Annual ACM/IEEE Symposium on Logic in Computer Science, LICS 2019. pp. 1–13. IEEE (2019). https://doi.org/10.1109/LICS.2019.8785699
14. Girard, J.: Linear logic. Theor. Comput. Sci. **50**, 1–102 (1987). https://doi.org/10.1016/0304-3975(87)90045-4

15. Guerrieri, G., Manzonetto, G.: The bang calculus and the two Girard's translations. In: Proceedings Joint International Workshop on Linearity & Trends in Linear Logic and Applications (Linearity-TLLA 2018). EPTCS, vol. 292, pp. 15–30 (2019). https://doi.org/10.4204/EPTCS.292.2
16. Guerrieri, G., Olimpieri, F.: Categorifying non-idempotent intersection types. In: 29th EACSL Annual Conference on Computer Science Logic, CSL 2021. LIPIcs, vol. 183, pp. 25:1–25:24. Schloss Dagstuhl (2021). https://doi.org/10.4230/LIPIcs.CSL.2021.25
17. Hindley, J.R.: The Church-Rosser Property and a Result in Combinatory Logic. Ph.D. thesis, University of Newcastle-upon-Tyne (1964)
18. Hindley, J.R., Seldin, J.P.: Introduction to Combinators and Lambda-Calculus. Cambridge University Press (1986)
19. Hirokawa, N., Middeldorp, A., Moser, G.: Leftmost Outermost Revisited. In: 26th International Conference on Rewriting Techniques and Applications (RTA 2015). Leibniz International Proceedings in Informatics (LIPIcs), vol. 36, pp. 209–222. Schloss Dagstuhl (2015). https://doi.org/10.4230/LIPIcs.RTA.2015.209
20. Levy, P.B.: Call-by-push-value: A subsuming paradigm. In: Typed Lambda Calculi and Applications, 4th International Conference (TLCA'99). Lecture Notes in Computer Science, vol. 1581, pp. 228–242 (1999). https://doi.org/10.1007/3-540-48959-2_17
21. Levy, P.B.: Call-by-push-value: Decomposing call-by-value and call-by-name. High. Order Symb. Comput. **19**(4), 377–414 (2006). https://doi.org/10.1007/s10990-006-0480-6
22. de' Liguoro, U.: Non-deterministic untyped λ-calculus. A study about explicit non determinism in higher-order functional calculi. Ph.D. thesis, Università di Roma La Sapienza (1991), http://www.di.unito.it/~deligu/papers/UdLTesi.pdf
23. de' Liguoro, U., Piperno, A.: Non deterministic extensions of untyped lambda-calculus. Inf. Comput. **122**(2), 149–177 (1995). https://doi.org/10.1006/inco.1995.1145
24. Maraist, J., Odersky, M., Turner, D.N., Wadler, P.: Call-by-name, call-by-value, call-by-need and the linear lambda calculus. Theor. Comput. Sci. **228**(1-2), 175–210 (1999). https://doi.org/10.1016/S0304-3975(98)00358-2
25. Newman, M.: On theories with a combinatorial definition of equivalence. Annals of Mathematics **43(2)** (1942)
26. Pagani, M., Tranquilli, P.: The conservation theorem for differential nets. Math. Struct. Comput. Sci. **27**(6), 939–992 (2017). https://doi.org/10.1017/S0960129515000456
27. Plotkin, G.D.: Call-by-name, call-by-value and the lambda-calculus. Theor. Comput. Sci. **1**(2), 125–159 (1975). https://doi.org/10.1016/0304-3975(75)90017-1
28. Ronchi Della Rocca, S., Paolini, L.: The Parametric Lambda Calculus - A Metamodel for Computation. Texts in Theoretical Computer Science. An EATCS Series, Springer (2004)
29. Ronchi Della Rocca, S., Roversi, L.: Lambda calculus and intuitionistic linear logic. Stud Logica **59**(3), 417–448 (1997). https://doi.org/10.1023/A:1005092630115
30. Santo, J.E., Pinto, L., Uustalu, T.: Modal embeddings and calling paradigms. In: 4th International Conference on Formal Structures for Computation and Deduction, FSCD 2019. LIPIcs, vol. 131, pp. 18:1–18:20. Schloss Dagstuhl (2019). https://doi.org/10.4230/LIPIcs.FSCD.2019.18
31. Simpson, A.K.: Reduction in a linear lambda-calculus with applications to operational semantics. In: Term Rewriting and Applications, 16th International

Conference (RTA 2005). Lecture Notes in Computer Science, vol. 3467, pp. 219–234 (2005). https://doi.org/10.1007/978-3-540-32033-3_17
32. Takahashi, M.: Parallel reductions in lambda-calculus. Inf. Comput. **118**(1), 120–127 (1995). https://doi.org/10.1006/inco.1995.1057
33. Terese: Term Rewriting Systems, Cambridge Tracts in Theoretical Computer Science, vol. 55. Cambridge University Press (2003)
34. Terui, K.: Light affine lambda calculus and polynomial time strong normalization. Archive for Mathematical Logic **46**(3-4), 253–280 (2007). https://doi.org/10.1007/s00153-007-0042-6

Generalized Bounded Linear Logic and its Categorical Semantics

Yōji Fukihara[1] (✉) and Shin-ya Katsumata[2]

[1] Kyoto University, Kyoto, Japan fukihara@kurims.kyoto-u.ac.jp
[2] National Institute of Informatics, Tokyo, Japan s-katsumata@nii.ac.jp

Abstract. We introduce a generalization of Girard et al.'s BLL called GBLL (and its affine variant GBAL). It is designed to capture the core mechanism of dependency in BLL, while it is also able to separate complexity aspects of BLL. The main feature of GBLL is to adopt a multi-object pseudo-semiring as a grading system of the !-modality. We analyze the complexity of cut-elimination in GBLL, and give a translation from BLL with constraints to GBAL with positivity axiom. We then introduce *indexed linear exponential comonads* (*ILEC* for short) as a categorical structure for interpreting the !-modality of GBLL. We give an elementary example of ILEC using folding product, and a technique to modify ILECs with symmetric monoidal comonads. We then consider a semantics of BLL using the folding product on the category of assemblies of a BCI-algebra, and relate the semantics with the realizability category studied by Hofmann, Scott and Dal Lago.

Keywords: Linear Logic · Categorical Semantics · Linear Exponential Comonad · Graded Comonad

1 Introduction

Girard's *linear logic* is a refinement of propositional logic by restricting weakening and contraction in proofs [15]. Linear logic also has an *of-course modality* !, which restores these structural rules to formulas of the form $!A$.

Later, Girard et al. extended the !-modality with quantitative information so that usage of !-modal formulas in proofs can be quantitatively controlled [16]. This extension, called *bounded linear logic* (BLL for short), is successfully applied to a logical characterization of P-time computations.

Their extension takes two steps. First, the !-modality is extended to the form $!_r A$, where the index r is an element of a semiring [16, Section 2.4]. The index r is called *grade* in modern terminology [11,13]. This extension and its variants have been employed in various logics and programming languages [7,30,14,26,28]. The categorical structure corresponding to $!_r A$ is identified as *graded linear exponential comonad* [7,13,22].

Second, the $!_r$-modality is further extended to the form $!_{x<p} A$, where p is a polynomial (called *resource polynomial*) giving the upper bound of x [16, Section 3]. The formula $!_{x<p} A$ also binds free occurrences of the resource variable

S. Kiefer and C. Tasson (Eds.): FOSSACS 2021, LNCS 12650, pp. 226–246, 2021.
https://doi.org/10.1007/978-3-030-71995-1_12

x in resource polynomials in A. Therefore, in BLL, both formulas and resource polynomials depend on the values stored in free resource variables. This dependency mechanism significantly increases the expressiveness of BLL, leading to a characterization of P-time complexity.

This characterization result was later revisited through a *realizability semantics* of BLL [16,19,10]. Inside this semantics, however, mechanisms for controlling complexity of program execution are hard-coded, and it is not very clear which semantics structure realizes the dependency mechanism of BLL. This leads us to seek a logical and categorical understanding of BLL's dependency mechanism hidden underneath the complexity-related features, such as resource polynomials and computability constraints.

As a result of the quest, we propose a generalization of BLL called GBLL, and study its categorical semantics. The central idea of the generalization is to replace the grading semiring of the $!_r$-modality with a particular *multi-object pseudo-semiring* realized as a 2-category. Let us see how this replacement works. In GBLL, each formula is formed by deriving a judgment of the form $\Delta \vdash A$, where Δ is a set (called *index set*) and A is a raw formula. We may think that such a well-formed formula $\Delta \vdash A$ denotes a Δ-indexed family $\{[\![A]\!]_i\}_{i\in\Delta}$ of denotations. The formation rule for !-modal formula in GBLL is the following:

$$\frac{\Delta' \vdash A \quad f \in \mathbf{Set}(\Delta, (\Delta')^*)}{\Delta \vdash !_f A} \qquad ((\_)^*\text{: Kleene closure})$$

where the function f abstractly represents dependency. This modality is enough to express the $!_{x<p}$-modality of BLL: we express the bindig $x < p$ under a resource variable context $\vec{y}$ as the *function* $f_p(\vec{y}) = (\vec{y}, 0) \cdots (\vec{y}, p(\vec{y}) - 1)$ that returns the list of environments extended with values less than $p(\vec{y})$. Then the denotation of the $!_f A$-modality is given by a variable-arity operator D. For each index $i \in \Delta$, the denotation is given by applying D to the denotations obtained by mapping A to list $f(i)$:

$$[\![!_f A]\!]_i = D([\![A]\!]_{j_1}, \cdots, [\![A]\!]_{j_n}) \text{ where } j_1 \cdots j_n = f(i).$$

A simple example of a variable-arity modal operator is the *folding product* $D(X_1, \cdots, X_n) = X_1 \otimes \cdots \otimes X_n$.

The pseudo-semiring structure on the class of functions of the form $\Delta \to (\Delta')^*$ is given as follows. For the multiplication $g \bullet f$, we adopt the *Kleisli composition* of the free monoid monad $(\_)^*$, while for the addition $f+g$, the pointwise concatenation $(f + g)(x) = f(x)g(x)$. However, these operations fail to satisfy one of the semiring axioms: $(f + g) \bullet h = f \bullet h + g \bullet h$. To fix this, we introduce (pointwise) list permutations as 2-cells between functions of type $\Delta \to (\Delta')^*$. These data form a 2-category **Idx**, which may be seen as a multi-object pseudo-semiring. Weakening, contraction, digging and dereliction in GBLL interact with these operations, much like the $!_r$-modality in [7].

We first study syntactic properties of GBLL. We introduce cut-elimination to GBLL and study its complexity property. It turns out that the proof technique used in BLL naturally extends to GBLL — as done in [16], we classify cuts into *reducible* and *irreducible* ones, introduce *proof weight*, and show that the reduction steps of reducible cuts will terminate in cubic time of proof weights. We also examine the expressive power of GBLL by giving a translation from an extension of BLL with *constraints* that are seen in Dal Lago et al.'s QBAL [10].

We next give a categorical semantics of GBLL. We introduce the concept of *indexed linear exponential comonad* (*ILEC*); it is an **Idx**-graded linear exponential comonad satisfying a commutativity condition with respect to an underlying indexed SMCCs. Then, we present a construction of ILEC from a symmetric monoidal closed category $\mathbb{C}$ with a symmetric monoidal comonad on it. We apply this construction to the case where $\mathbb{C}$ is the category of assemblies over a BCI algebra [2,20], and relate the semantics of GBLL with the constructed ILEC and the realizability category studied in [19,10].

Acknowledgment The first author was supported by JST ERATO HASUO Metamathematics for Systems Design Project (No. JPMJER1603). The authors are grateful to anonymous reviewers for comments, and Masahito Hasegawa, Naohiko Hoshino, Clovis Eberhart and Jérémy Dubut for fruitful discussions.

Preliminaries For a set Δ, by Δ^* we mean the set of finite sequences of Δ. The empty sequence is denoted by $()$. Juxtaposition of Δ^*-elements denotes the concatenation of sequences. For $x \in \Delta^*$, by $|x|$ we mean the length of x. We identify a natural number n and the set $\{0, \cdots, n-1\}$; note that $0 = \emptyset$. We also identify a sequence $x \in \Delta^*$ and the function "$\lambda i \in |x|$. the i-th element of x".

2 Generalized Bounded Linear Logic

2.1 Indexing 2-Category

We first introduce a 2-category **Idx** (and its variant $\mathbf{Idx}_a$), which may be seen as a multi-object pseudo-semiring. It consists of the following data[3]: 0-cells are sets (called index sets), and the hom-category $\mathbf{Idx}(\Delta, \Delta')$, which is actually a groupoid, is defined by:

- An object (1-cell) is a function $f : \Delta \to (\Delta')^*$.
- A morphism (2-cell) from f to g in $\mathbf{Idx}(\Delta, \Delta')$ is a Δ-indexed family of bijections $\{\sigma_x : |g(x)| \to |f(x)|\}_{x \in \Delta}$ such that $f(x)(\sigma_x(i)) = g(x)(i)$.

The identity 1-cell and the composition of 1-cells in **Idx** are denoted by i_Δ and $(\bullet)$, respectively. The composition is defined by $(g \bullet f)(x) \stackrel{\text{def}}{=} g(y_1) \cdots g(y_n)$ where $y_1 \cdots y_n = f(x)$. The hom-category $\mathbf{Idx}(\Delta, \Delta')$ has a symmetric strict monoidal structure:

- the monoidal unit is the constant empty-sequence function $0(x) = ()$,

[3] This is a full sub-2-category of the Kleisli 2-category $\mathbf{CAT}_{\mathcal{S}}$, where $\mathcal{S}$ is the 2-monad of symmetric strict monoidal category [21].

– the tensor product of f, g, denoted by $f + g$, is defined by the index-wise concatenation $(f + g)(x) \stackrel{\text{def}}{=} f(x)g(x)$.

We write $J : \mathbf{Set} \to \mathbf{Idx}$ for the inclusion, namely $J\Delta = \Delta$ and $(Jf)(x) = f(x)$ (the singleton sequence).

Proposition 2.1. *The composition $\bullet$ is symmetric strong monoidal in each argument. Especially, we have*

$$f \bullet 0 = 0 \quad 0 \bullet f = 0 \quad f \bullet (g + h) = f \bullet g + f \bullet h \quad (f + g) \bullet h \cong f \bullet h + g \bullet h.$$

We also define $\mathbf{Idx}_a$ by replacing "bijection" in the definition of 2-cell of $\mathbf{Idx}$ with "injection". The hom-category $\mathbf{Idx}_a(\Delta, \Delta')$ has the 1-cell 0 as the terminal object, hence is a symmetric *affine* monoidal category.

2.2 Formulas and Proofs

Definition of GBLL Formulas We first fix a set-indexed sets $\{\mathcal{A}(\Delta)\}_{\Delta \in \mathbf{Set}}$ of atomic propositions. Formulas are defined by the following BNF:

$$A ::= a \star r \mid A \otimes A \mid A \multimap A \mid !_f A$$

where $a \in \mathcal{A}(\Delta)$ for some set Δ, r is a function (called *reindexing function*) and f is a 1-cell in $\mathbf{Idx}$. Formula formation rules are introduced to derive the pair $\Delta \vdash A$ of an index set Δ and a formula A. They are defined as follows:

$$\frac{a \in \mathcal{A}(\Delta') \quad r \in \mathbf{Set}(\Delta, \Delta')}{\Delta \vdash a \star r} \qquad \frac{\Delta \vdash A \quad \Delta \vdash B}{\Delta \vdash A \otimes B} \qquad \frac{\Delta \vdash A \quad \Delta \vdash B}{\Delta \vdash A \multimap B}$$

$$\frac{\Delta' \vdash A \quad f \in \mathbf{Idx}(\Delta, \Delta')}{\Delta \vdash !_f A}$$

The formula $a \star r$ represents the atomic formula a precomposed with a reindexing function r. We write $\mathbf{Fml}(\Delta) = \{A \mid \Delta \vdash A\}$.

We next introduce the reindexing operation on formulas.

Definition 2.1. *For a reindexing function $r \in \mathbf{Set}(\Delta, \Delta')$, we define the* reindexing operator $(\_)|_r : \mathbf{Fml}(\Delta') \to \mathbf{Fml}(\Delta)$ *along r by*

$$a \star r|_{r'} \stackrel{\text{def}}{=} a \star (r \circ r'), \qquad (A \otimes B)|_r \stackrel{\text{def}}{=} A|_r \otimes B|_r,$$
$$(A \multimap B)|_r \stackrel{\text{def}}{=} A|_r \multimap B|_r, \qquad (!_f A)|_r \stackrel{\text{def}}{=} !_{f \bullet Jr} A.$$

We routinely extend reindexing operators to sequences of formulas well-formed under a common index set.

We quotient the set of well-formed formulas by the least congruent equivalence relation generated from the following binary relation:

$$\{(!_{Jr \bullet f} A, !_f(A|_r)) \mid r \in \mathbf{Set}(\Delta', \Delta''), f \in \mathbf{Idx}(\Delta, \Delta'), \Delta'' \vdash A\} \tag{2.1}$$

We see some formations of formulas in GBLL.

Example 2.1. Let us illustrate how a formula $!_{y<x^2}!_{z<x+y}A$ in BLL is represented in GBLL; here we assume that x, y, z are the only resource variables used in this formula. We first introduce a notation. Let E be a mathematical expression using variables $\mathbf{x}_1 \cdots \mathbf{x}_n$. Then by $[E]_n : \mathbb{N}^n \to (\mathbb{N}^{n+1})^*$ we mean the function

$$[E]_n(\vec{x}) = (\vec{x}, 0)(\vec{x}, 1) \cdots (\vec{x}, E[x_1/\mathbf{x}_1, \cdots, x_n/\mathbf{x}_n] - 1) \quad (\vec{x} \triangleq (x_1, \cdots, x_n) \in \mathbb{N}^n)$$

For instance, $[\mathbf{x}_1^2]_1(x) = (x, 0), \cdots, (x, x^2 - 1)$. Then from a well-formed formula $\mathbb{N}^3 \vdash A$, we obtain $\mathbb{N} \vdash !_{[\mathbf{x}_1^2]_1}!_{[\mathbf{x}_1+\mathbf{x}_2]_2}A$. Generalizing this, a BLL formula $!_{x<E}A$ containing resource variables $\mathbf{x}_1, \cdots, \mathbf{x}_n$ corresponds to the GBLL formula $!_{[E]_n}A$.

Example 2.2. We look at how we express the substitution of a resource polynomial $A[x := p(x_1, ..., x_n)]$. We define a function $\langle p \rangle_n : \mathbb{N}^n \to \mathbb{N}^{n+1}$ by

$$\langle p \rangle_n(x_1, ..., x_n) \stackrel{\text{def}}{=} (x_1, ..., x_n, p(x_1, ..., x_n)).$$

Then the reindexed formula $\mathbb{N}^n \vdash A|_{\langle p \rangle_n}$ corresponds to $A[x := p(x_1, \cdots, x_n)]$.

Example 2.3. We illustrate the equality between well-formed formulas. Consider a formula $\mathbb{N} \vdash A$ and a function $r \in \mathbf{Set}(\mathbb{N}^3, \mathbb{N})$. Then we equate formulas $\mathbb{N}^2 \vdash !_{[\mathbf{x}_1+\mathbf{x}_2]_2}(A|_r)$ and $\mathbb{N}^2 \vdash !_h A$, where $h \in \mathbf{Idx}(\mathbb{N}^2, \mathbb{N})$ is given by

$$h \stackrel{\text{def}}{=} Jr \bullet [\mathbf{x}_1 + \mathbf{x}_2]_2(x, y) = r(x, y, 0), \cdots, r(x, y, x + y - 1).$$

Definition of GBLL Proofs A *judgment* of GBLL is the form $\Delta \mid \Gamma \vdash A$, where Δ is an index set, Γ is a sequence of formulas well-formed under Δ, and A is a well-formed formula under Δ, respectively. The inference rules of GBLL are presented in Fig. 1. Similarly, we define GBAL to be the system obtained by replacing **Idx** in Fig. 1 with $\mathbf{Idx}_a$.

Example 2.4. We mimic a special case of the contraction rule in BLL

$$\frac{\Gamma, !_{x<x_i}A, !_{y<x_j}A\{^{x_i+y}/_x\} \vdash B}{\Gamma, !_{x<x_i+x_j}A \vdash B}$$

See also (!C)-rule of CBLL in Section 3.2. We use the *shift function* $s_{n,i} \in \mathbf{Set}(\mathbb{N}^{n+1}, \mathbb{N}^{n+1})$ defined by $s_{n,i}(x_1, \cdots, x_n, y) \stackrel{\text{def}}{=} (x_1, \cdots, x_n, x_i + y)$. Then we easily see $[\mathbf{x}_i]_n + Js_{n,i} \bullet [\mathbf{x}_j]_n = [\mathbf{x}_i + \mathbf{x}_j]_n$. By contraction rule of GBLL, we obtain the following derivation for well-formed formulas $\mathbb{N}^{n+1} \vdash A$ and $\mathbb{N}^n \vdash B$, mimicking the contraction of BLL:

$$\frac{!_{[\mathbf{x}_i]_n}A, !_{[\mathbf{x}_j]_n}(A|s_{n,i}) \vdash B}{!_{[\mathbf{x}_i+\mathbf{x}_j]_n}A = !_{[\mathbf{x}_i]_n+Js_{n,i}\bullet[\mathbf{x}_j]_n}A \vdash B}$$

Here, we use the formula equality $!_{Js_{n,i}\bullet[\mathbf{x}_j]_n}A = !_{[\mathbf{x}_j]_n}(A|s_{n,j})$.

$$\frac{\Delta \vdash A}{\Delta \mid A \vdash A}\ \text{(Ax) Axiom} \qquad \frac{\Delta \mid \Gamma, X, Y, \Gamma' \vdash A}{\Delta \mid \Gamma, Y, X, \Gamma' \vdash A}\ \text{(Exch) Exchange}$$

$$\frac{\Delta \mid \Gamma_1 \vdash A \quad \Delta \mid \Gamma_2, A \vdash B}{\Delta \mid \Gamma_1, \Gamma_2 \vdash B}\ \text{(Cut)}$$

$$\frac{\Delta \mid \Gamma, X, Y \vdash A}{\Delta \mid \Gamma, X \otimes Y \vdash A}\ (\otimes\mathrm{L}) \qquad \frac{\Delta \mid \Gamma_1 \vdash X \quad \Delta \mid \Gamma_2 \vdash Y}{\Delta \mid \Gamma_1, \Gamma_2 \vdash X \otimes Y}\ (\otimes\mathrm{R})$$

$$\frac{\Delta \mid \Gamma_1 \vdash X \quad \Delta \mid \Gamma_2, Y \vdash B}{\Delta \mid \Gamma_1, \Gamma_2, X \multimap Y \vdash B}\ (\multimap\mathrm{L}) \qquad \frac{\Delta \mid \Gamma, X \vdash Y}{\Delta \mid \Gamma \vdash X \multimap Y}\ (\multimap\mathrm{R})$$

$$\frac{\Delta \mid \Gamma \vdash B}{\Delta \mid \Gamma, !_0 A \vdash B}\ \text{(!W) Weakening} \qquad \frac{\Delta \mid \Gamma, A \vdash B}{\Delta \mid \Gamma, !_{\mathrm{id}} A \vdash B}\ \text{(!D) Dereliction}$$

$$\frac{\Delta \mid \Gamma, !_g A \vdash B \quad \sigma \in \mathbf{Idx}(\Delta, \Delta')(f, g)}{\Delta \mid \Gamma, !_f A \vdash B}\ \text{(!F) !-Functor}$$

$$\frac{\Delta \mid \Gamma, !_{f_1} A, !_{f_2} A \vdash B}{\Delta \mid \Gamma, !_{f_1 + f_2} A \vdash B}\ \text{(!C)Contraction}$$

$$\frac{\Delta' \mid !_{g_1} A_1, \cdots, !_{g_k} A_k \vdash B \quad f \in \mathbf{Idx}(\Delta, \Delta')}{\Delta \mid !_{g_1 \bullet f} A_1, \cdots, !_{g_k \bullet f} A_k \vdash !_f B}\ \text{(P!) Composition}$$

Fig. 1. GBLL Proof Rules

Example 2.5. The reindexing operator can be extended to proofs. Let r be a reindexing function in $\mathbf{Set}(\Delta, \Delta')$. Reindexing of the axiom rule $\Delta' \mid A \vdash A$, by r is the axiom rule $\Delta \mid A|_r \vdash A|_r$. Reindexing of other rules except (P!) can be easily defined—the judgment $\Delta' \mid \Gamma \vdash A$ in each rule is replaced with $\Delta \mid \Gamma|_r \vdash A|_r$ by reindexing. For (P!) rule, reindexing by r is given as follows:

$$\frac{\Delta'' \mid !_{g_1} A_1, \cdots, !_{g_k} A_k \vdash B \qquad f \bullet Jr \in \mathbf{Idx}(\Delta, \Delta'')}{\Delta \mid (!_{g_1 \bullet f} A_1)|_r, \cdots, (!_{g_k \bullet f} A_k)|_r \vdash (!_f B)|_r}$$

Remark 2.1. In this paper, indexing 2-category is either **Idx** or $\mathbf{Idx}_a$. Allowing more general indexing 2-categories in GBLL is a future work. In his PhD thesis, Breuvart designed a linear logic similar to GBLL upon an abstract indexing mechanism called *dependent semirings* [5, Definition 3.2.4.5]. It consists of categories $(\mathcal{S}, \mathcal{U})$ such that 1) each hom-set in $\mathcal{S}$ carries a (not necessarily commutative) ordered monoid structure $(0, +)$ and the composition of $\mathcal{S}$ distributes over $0, +$, and 2) $\mathcal{U}$ acts on $\mathcal{S}$ from both sides. Roughly speaking, $\mathcal{S}$ and $\mathcal{U}$ corresponds to our $\mathbf{Idx}^{\mathrm{op}}$ and $\mathbf{Set}^{\mathrm{op}}$, respectively. We expect that a unification of dependent semirings and 2-categories $\mathbf{Idx}, \mathbf{Idx}_a$ would yield a suitable generalization of indexing categories for GBLL. This generalization will subsume the non-graded linear logic, and allow us to compare GBLLs over different idexing categories.

2.3 Complexity of Cut-Elimination in GBLL

By a similar discussion to BLL [16], instances of Cut inference are divided in two classes: *reducible* cuts and *irreducible* cuts. We define the *weight* of proof $|\pi|$ for each proof $\pi \rhd \Delta \mid \Gamma \vdash A$ and *reduction steps* of proofs, such that every reduction steps will terminate, for each index $\delta \in \Delta$, in polynomial steps of $|\pi|(\delta)$.

Definition 2.2. *[16, Appedix A] In GBLL (resp. GBAL) proofs, an instance of the Cut inference is* irreducible *if there are at least one Composition rule below it or if its left premise is obtained by a Composition rule with nonempty context and the other premise is obtained by a Weakening, !-Functor, Dereliction, Contraction or Composition inference. A* reducible *cut is Cut inferences that is not irreducible.*

The definition of (ir)reducibility and weight is diverted from Girard's paper. Therefore, our system inherits from BLL the conditions under which cuts can be reduced. See also Section 2.4 in [16].

Definition 2.3. *A GBLL or GBAL proof is* irreducible *if it contains only irreducible cut inferences.*

Following [16], we introduce the concept of *weight* of a proof. It is a function $|\pi| : \Delta \to \mathbb{N}$ assigning a weight number $|\pi|(\delta)$ to a proof π at an index $\delta \in \Delta$. The weight number never increases at any reduction step of Cut in π. In the original BLL, weights are expressed by resource polynomials, while here, they are generalized to arbitrary functions. We remark that weights of the proofs involving Composition rules, which introduce $!_f$ modality, use the length of the lists constructed by f.

Definition 2.4. *For a given proof $\pi \rhd \Delta \mid \Gamma \vdash A$ of GBLL or GBAL, the* weight *of π is a function $|\pi| : \Delta \to \mathbb{N}$ inductively defined as follows. A) When $\Delta = \emptyset$, $|\pi|$ is the evident function. B) When $\Delta \neq \emptyset$, $|\pi|$ is defined by the following rules:*

1. *For an Axiom rule $\pi \rhd \Delta \mid A \vdash A$, $|\pi|(\delta) \stackrel{\text{def}}{=} 1$.*
2. *If π is obtained from π' by a unary rule except Contraction and Composition, $|\pi|(\delta) \stackrel{\text{def}}{=} |\pi'|(\delta) + 1$.*
3. *If π is obtained from π_1 and π_2 by a binary rule except Cut, $|\pi|(\delta) \stackrel{\text{def}}{=} |\pi_1|(\delta) + |\pi_2|(\delta) + 1$.*
4. *If π is obtained from π_1 and π_2 by a Cut rule, $|\pi|(\delta) \stackrel{\text{def}}{=} |\pi_1|(\delta) + |\pi_2|(\delta)$.*
5. *If π is obtained from π' by a Contraction rule, $|\pi|(\delta) \stackrel{\text{def}}{=} |\pi'|(\delta) + 2$*
6. *If π is obtained from π' by a Composition rule, such as*

$$\pi \rhd \frac{\begin{array}{c} \vdots \pi' \\ \Delta' \mid !_{\alpha_1} A_1, \cdots, !_{\alpha_k} A_k \vdash B \end{array}}{\Delta \mid !_{\alpha_1 \bullet f} A_1, \cdots, !_{\alpha_k \bullet f} A_k \vdash !_f B}$$

then $|\pi|(\delta) \stackrel{\text{def}}{=} \sum_{\gamma \in f(\delta)} (|\pi'|(\gamma) + 2k + 1) + k + 1$. *Note that the summation* $\sum_{\gamma \in f(\delta)}$ *scans all elements in the list* $f(\delta)$*, hence the weight depends on the length of* $f(\delta)$*.*

Theorem 2.1. *For every proof* $\pi \rhd \Delta \mid \Gamma \vdash A$ *and every* $\delta \in \Delta$*, reduction steps of reducible cuts will terminate in at most* $(|\pi|(\delta))^3$ *steps.*

Proof (sketch). The proof is almost the same as Section 2.2 and Appendix A of [16], except for the definition of the weight. Suppose that π one-step reduces into π'. From the definition of the weight, either 1) for all index $\delta \in \Delta$, the weight decreases (that is, $|\pi|(\delta) > |\pi'|(\delta)$), or 2) for all index $\delta \in \Delta$, the weight keeps (that is, $|\pi|(\delta) = |\pi'|(\delta)$). The reduction of the former type is called symmetric or axiom reduction [16, Section 2.2.1 and 2.2.2], while the latter commutative reduction [16, Section 2.2.3].

In the case where the weight keeps, we introduce another measure called the *cut size* $\|\pi\| : \Delta \to \mathbb{N}$ of a proof π. Its definition is the same as the definition of weight except for Cut rule. For a proof π obtained by Cut rule from π_1 and π_2, the cut size $\|\pi\|(\delta)$ is defined to be $\|\pi_1\|(\delta) + \|\pi_2\|(\delta) + |\pi_1|(\delta) + |\pi_2|(\delta)$.

In each commutative reduction from π to π' the cut size decrease at all index (that is, for all $\delta \in \Delta$, $\|\pi\|(\delta) > \|\pi'\|(\delta)$), and the cut size is at most the square of the weight (that is, for all $\delta \in \Delta, \|\pi\|(\delta) \leq (|\pi|(\delta))^2$). Therefore, the total number of steps is at most the cube of the weight. □

The number of reduction steps of a proof π and its weight depend on the length of lists computed by the **Idx**-morphisms occurring in π. However, to discuss the actual time complexity of cut-elimination, we further need to take into account the time complexity of the computation of **Idx**-morphisms. This would be achieved by looking at a subcategory of **Idx** computable within a certain time complexity. We leave this argument of analyzing the actual time complexity of cut-elimination as a future work.

3 Translation from Constrained BLL

We show that GBLL can express BLL via a translation. This translation is actually given to variants of these calculi, namely from *BLL with constraints* (called CBLL) to *GBAL with positivity axioms* (called GBAL$^+$).

CBLL is an extension of BLL with *constraints*, which are one of the features of Dal Lago and Hofmann's QBAL [10]. Constraints explicitly specify conditions imposed on resource variables, and it is natural to explicitly maintain these conditions throughout proofs. We also remark that in CBLL, weakening of !-formulas $!_{x<p+q}A \multimap !_{x<p}A$ is allowed, and atomic formulas are assumed to satisfy the positivity property (3.1).

GBAL$^+$ is designed for a sound translation from CBLL. Recall that GBAL is an extension of GBLL with weakening $!_{f+g}A \multimap !_f A$ on !-formulas. Then GBAL$^+$ is a further extension of GBAL with the following positivity axioms of atomic formulas: for every n-ary atomic formula $a \in \mathcal{A}$ in CBLL, we introduce an atomic formula $[a] \in \mathcal{A}(\mathbb{N}^n)$ to GBAL together with the axiom:

$$V_{\mathscr{C}}(F) \mid \emptyset \vdash [a] \star \langle p_1, \cdots, p_n \rangle \multimap [a] \star \langle q_1, \cdots, q_n \rangle \quad (\forall i. p_i \sqsubseteq_{\mathscr{C}} q_i).$$

Here the definition of each notation is given in Section 3.1 and 3.3. Positivity axiom induces proofs $V_{\mathscr{C}}(F) \mid A' \vdash A$ for every two formulas A, A' such that $A' \sqsubseteq_{\mathscr{C}} A$ (the relation $\sqsubseteq_{\mathscr{C}}$ for formulas is defined in Section 3.2).

3.1 Resource Polynomials and Constraints

We introduce basic concepts around CBLL, referring to its super-logic QBAL [10]. We put a reference in the beginning of each paragraph when the contents come from QBAL in [10].

[10, Definition 2.1] Given a countably infinite set $\mathcal{RV}$ of *resource variables*, a *resource monomial* over $\mathcal{RV}$ is a finite product of binomial coefficients $\prod_{i=1}^{m} \binom{x_i}{n_i}$, where the resource variables $x_1, \cdots, x_m$ are distinct and $n_1, \cdots, n_m \in \mathbb{N}$ are natural numbers. A *resource polynomial* over $\mathcal{RV}$ is a finite sum of resource monomials. We write 1 as $\binom{x}{0}$ and x as $\binom{x}{1}$ for short. Each positive natural number n denotes a resource polynomial $1 + 1 + \cdots + 1$. Resource polynomials are closed under sum, product, bounded sum and composition [10, Lemma 2.2].

[10, Definition 2.3] A *constraint* is an inequality $p \leq q$, where p and q are resource polynomials. We abbreviate $p+1 \leq q$ as $p < q$. A constraint $p \leq q$ *holds* (written $\models p \leq q$) if it is true in the standard model. A *constraint set* (denoted with $\mathscr{C}$, $\mathscr{D}$) is a finite set of constraints. A constraint $p \leq q$ *is a consequence* of a constraint set $\mathscr{C}$ (written $\mathscr{C} \models p \leq q$) if $p \leq q$ is a logical consequence of $\mathscr{C}$. For every constraint sets $\mathscr{C}$ and $\mathscr{D}$, we write $\mathscr{C} \models \mathscr{D}$ iff $\mathscr{C} \models p \leq q$ for every constraint $p \leq q$ in $\mathscr{D}$. For each constraint set $\mathscr{C}$, we define an order $\sqsubseteq_{\mathscr{C}}$ on resource polynomials by $p \sqsubseteq_{\mathscr{C}} q$ iff $\mathscr{C} \models p \leq q$.

[10, Definition 2.3] We define the polarity of occurrences of free resource variables. For a constraint $p \leq q$, we say that an occurrence of a resource variable x in p is called *negative*, while the one in q is called *positive*.

3.2 Formulas and Inference Rules of CBLL

Let $\mathcal{A}$ be a set of atomic formulas and assume that each atomic formula $a \in \mathcal{A}$ is associated with an arity $\mathrm{ar}(a)$. Formulas of CBLL are defined by:

$$A, B ::= \quad a(p_1, \cdots, p_{\mathrm{ar}(a)}) \mid A \otimes B \mid A \multimap B \mid !_{x<p} A$$

where p in the formula $!_{x<p}A$ satisifes $x \notin \mathrm{FV}(p)$.

[10, Definition 2.6] Each occurrence of a free resource variable in a formula is classified into *positive* or *negative*. Below we inductively define a *positive occurrence* of a resource variable. An occurrence of x in:

- $a(p_1, \cdots, p_{\mathrm{ar}(a)})$ is always positive.
- $A \otimes B$ is positive iff it is in A and positive, or so in B.

$$\frac{A \sqsubseteq_{\mathscr{C}} B}{A \vdash_{\mathscr{C}} B}\ (\mathrm{Ax}) \qquad \frac{\Gamma \vdash_{\mathscr{C}} A \quad \mathscr{D} \vDash \mathscr{C}}{\Gamma \vdash_{\mathscr{D}} A}\ (\mathrm{Str}) \qquad \frac{\Gamma \vdash_{\mathscr{C}} B}{\Gamma, !_{x<0} A \vdash_{\mathscr{C}} B}\ (!\mathrm{W})$$

$$\frac{A\{0/x\}, \Gamma \vdash_{\mathscr{C}} B}{!_{x<1} A, \Gamma \vdash_{\mathscr{C}} B}\ (!\mathrm{D}) \qquad \frac{\Gamma, !_{x<p} A, !_{y<q} A\{p+y/x\} \vdash_{\mathscr{C}} B}{\Gamma, !_{x<p+q} A \vdash_{\mathscr{C}} B}\ (!\mathrm{C})$$

$$\frac{A_1, \cdots, A_n \vdash_{\mathscr{C} \cup \{x<p\}} B \quad x \notin \mathrm{FV}(\mathscr{C})}{!_{x<p} A_1, \cdots, !_{x<p} A_n \vdash_{\mathscr{C}} !_{x<p} B}\ (!\mathrm{P})$$

$$\frac{!_{y<p} !_{z<q\{y/w\}} A\left\{\left(z+\sum_{w<y} q(w)\right)/x\right\}, \Gamma \vdash_{\mathscr{C}} B}{!_{x<\sum_{w<p} q(w)} A, \Gamma \vdash_{\mathscr{C}} B}\ (!\mathrm{N})$$

Fig. 2. Inference Rules for CBLL ($\otimes$ and $\multimap$ are omitted)

- $A \multimap B$ is positive iff it is in A and negative, or it is in B and positive.
- $!_{x'<p} A$ is positive iff it is in A and positive. We remark that an occurrence of a free resource variable in p is counted as negative in $!_{x'<p} A$.

[10, Definition 2.8] We extend the order $\sqsubseteq_{\mathscr{C}}$ on resource polynomials to the one on CBLL formulas.

$$\begin{aligned}
&a(p_1, \cdots, p_{\mathrm{ar}(a)}) \sqsubseteq_{\mathscr{C}} a(q_1, \cdots, q_{\mathrm{ar}(a)}) \text{ iff } \forall i. p_i \sqsubseteq_{\mathscr{C}} q_i \\
&A \otimes B \sqsubseteq_{\mathscr{C}} C \otimes D \text{ iff } (A \sqsubseteq_{\mathscr{C}} C) \wedge (B \sqsubseteq_{\mathscr{C}} D) \\
&A \multimap B \sqsubseteq_{\mathscr{C}} C \multimap D \text{ iff } (C \sqsubseteq_{\mathscr{C}} A) \wedge (B \sqsubseteq_{\mathscr{C}} D) \\
&!_{x<p} A \sqsubseteq_{\mathscr{C}} !_{x<q} B \text{ iff } (q \sqsubseteq_{\mathscr{C}} p) \wedge (x \notin \mathrm{FV}(\mathscr{C})) \wedge (A \sqsubseteq_{\mathscr{C} \cup \{x<q\}} B)
\end{aligned} \tag{3.1}$$

[10, Section 2.3] A CBLL *judgment* is an expression $\Gamma \vdash_{\mathscr{C}} A$, where $\mathscr{C}$ is a constraint set, Γ is a multiset of formulas and A is a formula. A judgment $\Gamma \vdash_{\mathscr{C}} A$ means that A is a consequence of Γ under the constraints $\mathscr{C}$.

Inference rules (Fig. 2) are almost the same as those of QBAL; we omit the rules for $\otimes, \multimap$ and Cut. Note that weakening is restricted to !-formulas. Every BLL proof of $\Gamma \vdash A$ can be translated to a CBLL proof of $\Gamma \vdash_{\emptyset} A$.

3.3 Translation into GBAL$^+$

As mentioned at the beginning of Section 3, we will give a translation from CBLL to GBAL$^+$. When translating a CBLL proof $\Gamma \vdash_{\mathscr{C}} A$, we also need to supply a set F of free resource variables satisfying $F \supseteq \mathrm{FV}(\Gamma) \cup \mathrm{FV}(A) \cup \mathrm{FV}(\mathscr{C})$. Then the translation of the proof of $\Gamma \vdash_{\mathscr{C}} A$ yields a proof of $V_{\mathscr{C}}(F) \mid [\Gamma]^{(F;\mathscr{C})} \vdash [A]^{(F;\mathscr{C})}$ in GBAL$^+$.

For Constraints We define an *environment* over a finite set F of resource variables to be a function from F to $\mathbb{N}$; by $V(F)$ we mean the set of environments over F. Given an environment $\rho \in V(F)$ and a resource variable $x \notin F$ and $n \in \mathbb{N}$, by $\rho\{x \mapsto n\}$ we mean the environment over $F \cup \{x\}$ that extends ρ

with a mapping $x \mapsto n$. Given a resource polynomial p such that $FV(p) \subseteq F$, by $[\![p]\!] : V(F) \to \mathbb{N}$ we mean the function that evaluates the resource polynomial p under a given environment. For resource polynomials $p_1, \cdots, p_n$ such that $\mathrm{FV}(p_i) \subseteq F$, we give a function $\langle p_1, \cdots, p_n \rangle : (V(F)) \to \mathbb{N}^n$ by $\langle p_1, \cdots, p_n \rangle \rho = ([\![p_1]\!]\rho, \cdots, [\![p_n]\!]\rho)$.

Let $\rho \vDash p \leq q$ denote $[\![p]\!]\rho \leq [\![q]\!]\rho$ for a constraint $p \leq q$ with a set F of free resource variables (such that $\mathrm{FV}(p) \cup \mathrm{FV}(q) \subseteq F$) and for an environment $\rho \in V(F)$. For a subset $S \subset V(F)$ and for a constraint set $\mathscr{C}$, $S \vDash \mathscr{C}$ is also defined similarly: for every $\rho \in S$ and for every $p \leq q \in \mathscr{C}$, $\rho \vDash p \leq q$. Given a constraint set $\mathscr{C}$ and a set F of resource variables such that $\mathrm{FV}(\mathscr{C}) \subseteq F$, let a set $V_{\mathscr{C}}(F)$ and a function $\iota_{F,\mathscr{C}} : V_{\mathscr{C}}(F) \to V(F)$ be given by:

$$V_{\mathscr{C}}(F) \stackrel{\mathrm{def}}{=} \{\rho \in V(F) \mid \rho \vDash \mathscr{C}\}, \qquad \iota_{F,\mathscr{C}}(\rho) \stackrel{\mathrm{def}}{=} \rho.$$

For a resource polynomial p, a free resource variable x such that $x \notin \mathrm{FV}(p)$, a constraint set $\mathscr{C}$ and a set F of resource variables such that $\mathrm{FV}(p) \cup \mathrm{FV}(\mathscr{C}) \subseteq F$, we introduce a map $[x < p]_{(F,\mathscr{C})} : V_{\mathscr{C}}(F) \to V_{\mathscr{C} \cup \{x<p\}}(F \cup \{x\})^*$ by

$$[x < p]_{(F,\mathscr{C})}\rho \stackrel{\mathrm{def}}{=} \rho\{x \mapsto 0\}, \rho\{x \mapsto 1\}, \cdots, \rho\{x \mapsto ([\![p]\!]\rho - 1)\}$$

For Formulas Given a CBLL formula A, a constraint set $\mathscr{C}$ and a set of resource variables F such that $F \supseteq \mathrm{FV}(A) \cup \mathrm{FV}(\mathscr{C})$, the translation $[A]^{(F;\mathscr{C})}$ of a well-formed formula $V_{\mathscr{C}}(F) \vdash A$ is defined inductively as follows:

$$[a(p_1, ..., p_n)]^{(F;\mathscr{C})} \stackrel{\mathrm{def}}{=} [a] \star (\langle p_1, ..., p_n \rangle \circ \iota_{F,\mathscr{C}})$$
$$[A \otimes B]^{(F;\mathscr{C})} \stackrel{\mathrm{def}}{=} [A]^{(F;\mathscr{C})} \otimes [B]^{(F;\mathscr{C})}$$
$$[A \multimap B]^{(F;\mathscr{C})} \stackrel{\mathrm{def}}{=} [A]^{(F;\mathscr{C})} \multimap [B]^{(F;\mathscr{C})}$$
$$[!_{x<p}A]^{(F;\mathscr{C})} \stackrel{\mathrm{def}}{=} !_{[x<p]_{(F,\mathscr{C})}} [A]^{(F\cup\{x\};\mathscr{C}\cup\{x<p\})}$$

For Proofs To give a translation of proofs, we define another notation. For a resource polynomial p, q, a set F of resource variables and a constraint set $\mathscr{C}$ such that $\mathrm{FV}(p) \cup \mathrm{FV}(\mathscr{C}) \subseteq F$, a set $[p, q)^{(F,\mathscr{C})}$ of environments is defined by

$$[p, q)^{(F,\mathscr{C})} = \{\rho \in V(F \cup \{t\}) \mid \rho \vDash \mathscr{C}, [\![p]\!](\rho) \leq \rho(t) < [\![p + q]\!]\rho\}$$

here t is a "fresh" resource variable such that $t \notin F$.

Given a proof $\pi \triangleright \Gamma \vdash_{\mathscr{C}} A$, a translation $[\pi]^{(F;\mathscr{C})} \triangleright V_{\mathscr{C}}(F) \mid [\Gamma]^{(F;\mathscr{C})} \vdash [A]^{(F;\mathscr{C})}$ is defined inductively on the structure of the proof:

- For Axiom rule, we can prove $V_{\mathscr{C}}(F) \mid [A]^{(F;\mathscr{C})} \vdash [B]^{(F;\mathscr{C})}$ for formulas A, B such that $A \sqsubseteq_{\mathscr{C}} B$.
- For rules (Cut), ($\otimes$L), ($\otimes$R), ($\multimap$L), ($\multimap$R) and (!W), the translation is simple replacement of each formula A with $[A]^{(F;\mathscr{C})}$.

- For (Str) rule, we have a map $r \in \mathbf{Set}(V_{\mathscr{D}}(F), V_{\mathscr{C}}(F))$. Then the translation is given as reindexed proof $[\pi']^{(F;\mathscr{C})}|_r$ of the translation $[\pi']^{(F;\mathscr{C})}$ of the premise.
- For (!D) rule, the premise is translated to $V_{\mathscr{C}}(F) \mid A', [\Gamma]^{(F;\mathscr{C})} \vdash [B]^{(F;\mathscr{C})}$, where $A' = [A]^{(F\cup\{x\};\mathscr{C}\cup\{x<1\})}|_r$ and r is a map such that $Jr = [x < 1]_{(F,\mathscr{C})}$.
- For (!C) rule, we define a morphism $s_{p,q}^{(F;\mathscr{C})}$ in $\mathbf{Idx}_a$ and functions $r_{p,q}^{(F;\mathscr{C})}$, $i_1^{(p,q;F;\mathscr{C})}$, $i_2^{(p,q;F;\mathscr{C})}$ (s, r, i_1 and i_2 for short) by

$$
\begin{aligned}
s_{p,q}^{(F;\mathscr{C})} : \quad & V_{\mathscr{C}}(F) \to [p,q)^{(F,\mathscr{C})} \\
& \rho \mapsto \rho\{t \mapsto [\![p]\!]\rho\}, \cdots, \rho\{t \mapsto ([\![p+q]\!]\rho - 1)\} \\
r_{p,q}^{(F;\mathscr{C})} : \quad & [p,q)^{(F,\mathscr{C})} \xrightarrow{\sim} V_{\mathscr{C}\cup\{y<q\}}(F \cup \{y\}) \\
& \rho\{t \mapsto ([\![p]\!]\rho + k)\} \mapsto \rho\{y \mapsto k\} \\
i_1^{(p,q;F;\mathscr{C})} : \quad & V_{\mathscr{C}\cup\{x<p\}}(F \cup \{x\}) \to V_{\mathscr{C}\cup\{x<p+q\}}(F \cup \{x\}) \\
& \rho\{x \mapsto k\} \mapsto \rho\{x \mapsto k\} \\
i_2^{(p,q;F;\mathscr{C})} : \quad & [p,q)^{(F,\mathscr{C})} \to V_{\mathscr{C}\cup\{x<p+q\}}(F \cup \{x\}) \\
& \rho\{t \mapsto ([\![p]\!]\rho + k)\} \mapsto \rho\{x \mapsto [\![p]\!]\rho + k\}
\end{aligned}
$$

They satisfy $!_{[x<p]}[A]^{(F\cup\{x\};\mathscr{C}\cup\{x<p\})} = !_{[x<p]}([A]^{(F\cup\{x\};\mathscr{C}\cup\{x<p+q\})}|_{i_1})$ and $!_{[y<q]}[A\{^{p+y}/_{x}\}]^{(F\cup\{y\};\mathscr{C}\cup\{y<q\})} = !_{Jr\bullet s}([A]^{(F\cup\{x\};\mathscr{C}\cup\{x<p+q\})}|_{i_2 \circ r^{-1}})$. Then the conclusion of (!C) is obtained:

$$V_{\mathscr{C}}(F) \mid [\Gamma]^{(F;\mathscr{C})}, !_{(Ji_i\bullet[x<p])+(Ji_2\bullet s)}[A]^{(F\cup\{x\};\mathscr{C}\cup\{x<p+q\})} \vdash [B]^{(F;\mathscr{C})}.$$

- For (!P) rule, let $F' = F \cup \{x\}$ and $\mathscr{C}' = \mathscr{C} \cup \{x < p\}$. We can prove the translated conclusion from the translated premise by the following proof:

$$
\dfrac{\dfrac{V_{\mathscr{C}'}(F') \mid [A_1]^{(F';\mathscr{C}')}, \cdots, [A_n]^{(F';\mathscr{C}')} \vdash [B]^{(F';\mathscr{C}')}}{\dfrac{n \text{ times (!D)'s} \vdots}{V_{\mathscr{C}'}(F') \mid !_{\mathrm{id}}[A_1]^{(F';\mathscr{C}')}, \cdots, !_{\mathrm{id}}[A_n]^{(F';\mathscr{C}')} \vdash [B]^{(F';\mathscr{C}')}}}}{V_{\mathscr{C}}(F) \mid !_{[x<p]}[A_1]^{(F';\mathscr{C}')} \cdots !_{[x<p]}[A_n]^{(F';\mathscr{C}')} \vdash !_{[x<p]}[B]^{(F';\mathscr{C}')}}
$$

- For (!N) rule, we define index sets $\Delta_0, \Delta_1, \Delta_2$ and constraints $\mathscr{C}_0, \mathscr{C}_1, \mathscr{C}_2$ by

$$
\begin{aligned}
\mathscr{C}_0 &= \mathscr{C} \cup \{y < p\} & \Delta_0 &= V_{\mathscr{C}_0}(F \cup \{y\}) \\
\mathscr{C}_1 &= \mathscr{C} \cup \{y < p, z < q\{^y/_w\}\} & \Delta_1 &= V_{\mathscr{C}_1}(F \cup \{y, z\}) \\
\mathscr{C}_2 &= \mathscr{C} \cup \{x < \sum_{w<p} q(w)\} & \Delta_2 &= V_{\mathscr{C}_2}(F \cup \{x\})
\end{aligned}
$$

There is an isomorphism $r \in \mathbf{Set}(\Delta_1, \Delta_2)$, and it holds an equation $[z < q\{^y/_w\}]_{(F\cup\{y\},\mathscr{C}_0)} \bullet [y < p]_{(F,\mathscr{C})} = Jr^{-1} \bullet [x < \sum_{w<p} q(w)]_{(F,\mathscr{C})}$. Therefore, (!N) rule can be translated to the following provable judgment:

$$V_{\mathscr{C}}(F) | !_{[x<\sum_{w<p} q]}[A]^{(F\cup\{x\};\mathscr{C}_2)} \vdash !_{[y<p]}!_{[z<q\{^y/_w\}]}[A\{^{z+\sum_{w<y} q}/_{x}\}]^{(F\cup\{y,z\};\mathscr{C}_1)}$$

Since every BLL proof $\Gamma \vdash A$ can be translated to a CBLL proof $\Gamma \vdash_{\emptyset} A$, it can further be translated to a GBAL^{+} proof $V_{\emptyset}(F) \mid [\Gamma]^{(F;\emptyset)} \vdash [A]^{(F;\emptyset)}$.

4 Categorical Semantics for GBLL

We give a categorical semantics of GBLL. First, notice that each index set Δ determines a multiplicative linear logic under Δ. We model this situation by a *set-indexed symmetric monoidal closed categories*, given by a functor $C : \mathbf{Set}^{op} \to \mathbf{SMCC}_{\mathrm{strict}}$. That is, for each $\Delta \in \mathbf{Set}$, a symmetric monoidal closed category $C\Delta$ is given, and any function $f : \Delta \to \Delta'$ induces a strict symmetric monoidal closed functor $Cf : C\Delta' \to C\Delta$, performing renaming of indexes.

Upon this indexed symmetric monoidal closed categories, we introduce a categorical structure that models the $!_f$ modality. We call it *indexed linear exponential comonad*. This is a generalization of the *semiring-graded linear exponential comonad* studied in [13,22]. Our generalization replaces the semiring with **Idx**, which may be regarded as a many-object pseudo-semiring (Proposition 2.1).

We write $[\mathbb{C}, \mathbb{D}]_l$ for the category of symmetric lax monoidal functors from $\mathbb{C}$ to $\mathbb{D}$ and monoidal natural transformations between them. We equip it with the pointwise symmetric monoidal structure $(\dot{I}, \dot{\otimes})$ given by $\dot{I}X = I$ and $(F \dot{\otimes} G)X = FX \otimes GX$ for $X \in \mathbb{C}$.

Definition 4.1. *An* indexed linear exponential comonad *(ILEC for short) over a set-indexed SMCC C consists of:*

- *A collection of symmetric colax monoidal functors*

$$(D, w^{\Delta,\Delta'}, c^{\Delta,\Delta'}) : \mathbf{Idx}(\Delta, \Delta') \to [C\Delta', C\Delta]_l \quad (\Delta, \Delta' \in \mathbf{Set}).$$

  The symmetric lax monoidal structure of Df is denoted by $m_f : I \to DfI$ and $m_{f,A,B} : DfA \otimes DfB \to Df(A \otimes B)$.
- *Monoidal natural transformations $\epsilon^\Delta : D(\mathrm{i}_\Delta) \to \mathrm{Id}_{D\Delta}$ and $\delta_{g,f} : D(g \bullet f) \to Df \circ Dg$ satisfying axioms in Figure 3.*
- *$Cr' \circ Df \circ Cr = D(Jr \bullet f \bullet Jr')$ holds for any morphism f in* **Idx** *and r, r' in* **Set** *of appropriate type.*

The last axiom has two purposes: the equality $Cr'(DfA) = D(f \bullet Jr')A$ is to allow reindexing functions to act from outside, and the other equality $Df(CrA) = D(Cr \bullet f)A$ is to make D invariant under internal reindexing of formulas. These equalities are tied up with the formula equivalence in (2.1) and the definition of reindexing at $!_f A$ in Definition 2.1, respectively. We postpone a concrete example of ILEC to Section 4.2.

4.1 Semantics of GBLL

We interpret a well-formed formula $\Delta \vdash A$ as an object $[\![\Delta \vdash A]\!] \in C\Delta$. This is done by induction on the structure of the formula. We assume that each atomic formula $a \in \mathcal{A}(\Delta)$ comes with its interpretation as an object $[a] \in C\Delta$.

$$[\![\Delta \vdash a \star r]\!] \stackrel{\text{def}}{=} Cr[a] \qquad [\![\Delta \vdash !_f A]\!] \stackrel{\text{def}}{=} Df[\![\Delta' \vdash A]\!]$$

$$[\![\Delta \vdash A \otimes B]\!] \stackrel{\text{def}}{=} [\![\Delta \vdash A]\!] \otimes [\![\Delta \vdash B]\!] \quad [\![\Delta \vdash A \multimap B]\!] \stackrel{\text{def}}{=} [\![\Delta \vdash A]\!] \multimap [\![\Delta \vdash B]\!]$$

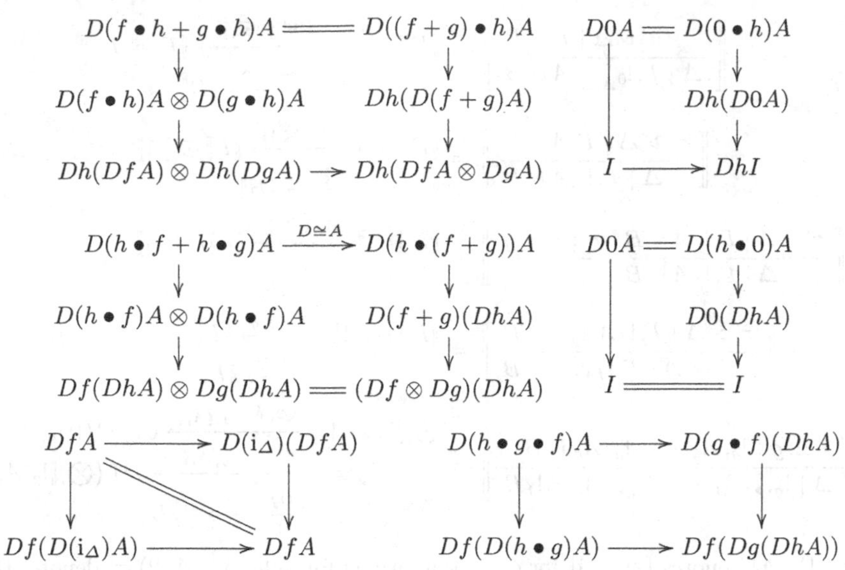

Fig. 3. Axioms of Indexed Linear Exponential Comonad

Proposition 4.1. *For any* $r \in \mathbf{Set}(\Delta, \Delta')$ *and well-formed formula* $\Delta' \vdash A$*, we have* $[\![\Delta \vdash A|_r]\!] = Cr[\![\Delta' \vdash A]\!]$.

Proposition 4.2. $[\![\Delta \vdash !_{Jr \bullet f} A]\!] = [\![\Delta' \vdash !_f (A|_r)]\!]$.

Each proof $\pi \triangleright \Delta \mid \Gamma \vdash A$ of GBLL is interpreted as a morphism $[\![\Delta \mid \Gamma \vdash A]\!] : [\![\Delta \vdash \Gamma]\!] \to [\![\Delta \vdash A]\!]$ in $C\Delta$. Here, for a sequence $\Gamma = C_1, \cdots, C_m$ of formulas, $[\![\Delta \vdash \Gamma]\!]$ denotes $[\![\Delta \vdash C_1]\!] \otimes \cdots \otimes [\![\Delta \vdash C_m]\!]$. We write out the interpretation only for the cases of modalities, because the other rules, Axiom, Exchange, Cut, $\otimes$(L, R) and $\multimap$(L, R) are interpreted similarly to the semantics of multiplicative intuitionistic linear logic. Fig. 4 shows the interpretation of rules related to $!_f$.

Theorem 4.1. *For a proof* $\pi \triangleright \Delta \mid \Gamma \vdash A$*, if* π *has a reducible cut and reduces into* π' *by a reduction step, then* $[\![\pi]\!] = [\![\pi']\!]$ *in* $C\Delta$.

4.2 Construction of an Indexed Linear Exponential Comonad

We present a construction of an indexed SMCCs $C : \mathbf{Set}^{op} \to \mathbf{SMCC}_{\text{strict}}$ and an ILEC $D : \mathbf{Idx}(\Delta, \Delta') \to [C\Delta', C\Delta]_l$ over C from a SMCC $\langle \mathbb{C}, \otimes, I, \multimap \rangle$, and a symmetric lax monoidal comonad $\langle V, m^V, m^V_{X,Y}, \epsilon, \delta \rangle$ on $\mathbb{C}$.

Construction of Indexed SMCCs First, for each index set Δ, we define the category $\Delta \pitchfork \mathbb{C}$ to be the product of Δ-many copies of $\mathbb{C}$. We represent objects and morphisms of this category by maps $\mathsf{X} : \Delta \to \mathrm{Obj}(\mathbb{C})$ and maps

$$\left[\!\left[\frac{\pi' \rhd \Delta \mid \Gamma \vdash B}{\Delta \mid \Gamma, !_{0_{\Delta,\Delta'}} A \vdash B}\right]\!\right] = [\![\Gamma, !_{0_{\Delta,\Delta'}} A]\!] \xrightarrow{\mathrm{id} \otimes w^{\Delta,\Delta'}_{[\![A]\!]}} [\![\Gamma]\!] \otimes I \xrightarrow{[\![\pi']\!] \circ \cong} [\![B]\!]$$

$$\left[\!\left[\frac{\pi' \rhd \Delta \mid \Gamma, A \vdash B}{\Delta \mid \Gamma, !_{\mathrm{id}} A \vdash B}\right]\!\right] = [\![\Gamma, !_{\mathrm{i}_\Delta} A]\!] \xrightarrow{\mathrm{id} \otimes \epsilon^{\Delta}_{[\![A]\!]}} [\![\Gamma]\!] \otimes [\![A]\!] \xrightarrow{[\![\pi']\!]} [\![B]\!]$$

$$\left[\!\left[\frac{\pi' \rhd \Delta \mid \Gamma, !_g A \vdash B \quad \sigma : f \Rightarrow g}{\Delta \mid \Gamma, !_f A \vdash B}\right]\!\right] = [\![\Gamma, !_f A]\!] \xrightarrow{\mathrm{id} \otimes (D\sigma)_{[\![A]\!]}} [\![\Gamma, !_g A]\!] \xrightarrow{[\![\pi']\!]} [\![B]\!]$$

$$\left[\!\left[\frac{\pi' \rhd \Delta \mid \Gamma, !_f A, !_g A \vdash B}{\Delta \mid \Gamma, !_{f+g} A \vdash B}\right]\!\right] = [\![\Gamma, !_{f+g} A]\!] \xrightarrow{\mathrm{id} \otimes c_{f,g}} [\![\Gamma]\!] \otimes ([\![!_f A]\!] \otimes [\![!_g A]\!]) \xrightarrow{[\![\pi']\!] \circ \cong} [\![B]\!]$$

$$\left[\!\left[\frac{\pi' \rhd \Delta, g \mid !_{g_1} A_1, \cdots, !_{g_k} A_k \vdash B}{\Delta \mid !_{g_1 \bullet f} A_1, \cdots, !_{g_k \bullet f} A_k \vdash !_f B}\right]\!\right] = \bigotimes_i [\![!_{g_i \bullet f} A_i]\!] \xrightarrow{\bigotimes_i \delta_{g_i, f, [\![A_i]\!]}} \bigotimes_i Df([\![!_{g_i} A_i]\!]) \xrightarrow{m_{f, \cdots [\![!_{g_i} A_i]\!] \cdots}} Df\left(\bigotimes_i [\![!_{g_i} A_i]\!]\right) \xrightarrow{Df([\![\pi']\!])} [\![!_f B]\!]$$

Here, 1) $[\![A]\!]$ denotes $[\![\Delta \vdash A]\!]$ for each well-formed formula $\Delta \vdash A$. 2) π' denotes the proof of the premise of each rule.

Fig. 4. Interpretations of Modal Rules.

$\mathsf{f} : \Delta \to \mathrm{Mor}(\mathbb{C})$, respectively. Since SMCCs are closed under products, $\Delta \pitchfork \mathbb{C}$ is a SMCC by the component-wise tensor product and internal hom:

$$\mathsf{I}(d) \stackrel{\mathrm{def}}{=} I, \quad \mathsf{X} \dot{\otimes} \mathsf{Y}(d) \stackrel{\mathrm{def}}{=} \mathsf{X}(d) \otimes \mathsf{Y}(d), \quad \mathsf{X} \dot{\multimap} \mathsf{Y}(d) \stackrel{\mathrm{def}}{=} \mathsf{X}(d) \multimap \mathsf{Y}(d)$$

We then define the indexed SMCCs C by $C\Delta \stackrel{\mathrm{def}}{=} \Delta \pitchfork \mathbb{C}$.

Folding Product We next introduce the *folding product* functor T; we later compose it with the symmetric lax monoidal comonad V so that we can derive various ILECs over C. Note that T itself *is* also an ILEC; set $V = \mathrm{Id}$. The type of T is $\Delta^* \times (\Delta \pitchfork \mathbb{C}) \longrightarrow \mathbb{C}$, and is defined by

$$\mathsf{T}(i_1 i_2 \cdots i_n, \mathsf{A}) \stackrel{\mathrm{def}}{=} \mathsf{A}(i_1) \otimes \mathsf{A}(i_2) \otimes \cdots \otimes \mathsf{A}(i_n), \quad \mathsf{T}((), \mathsf{A}) \stackrel{\mathrm{def}}{=} I$$

On morphisms, T maps a list permutation in the first argument to the symmetry morphism in $\mathbb{C}$. T is symmetric strong monoidal in each argument. Moreover, each strong monoidal structure interacts well with each other, concluding that it becomes a multi-symmetric strong monoidal functor in the sense of [21].

Proposition 4.3. *For $f \in \mathbf{Idx}(\Delta, \Delta')$ and $l = i_1 \cdots i_k \in \Delta^*$, let $f(l)$ denote $f(i_1) \cdots f(i_k)$. Then it holds $\mathsf{T}(f(l), \mathsf{A}) \simeq \mathsf{T}(l, \mathsf{T}(f(\_), \mathsf{A}))$ and this isomorphism is natural for A.*

Remark 4.1. Usually the !-modal formula $!A$ in linear logic is interpreted by the object consisting of many copies of the same data (referred as *uniformity* of $!A$ [8]). We leave the development of uniform folding product as a future work.

Construction of ILEC We now compose the folding product functor with the symmetric lax monoidal comonad V, to derive another ILEC. Let Δ, Δ' be index sets. We define a symmetric strong (hence colax) monoidal functor $D : \mathbf{Idx}(\Delta, \Delta') \longrightarrow [C\Delta', C\Delta]_l$ by

$$Df\mathsf{A}(i) \stackrel{\mathrm{def}}{=} \mathsf{T}(f(i), V \circ \mathsf{A}) \quad Df\mathsf{p}(i) \stackrel{\mathrm{def}}{=} \mathsf{T}(f(i), V\mathsf{p}) \quad D\alpha\mathsf{A} \stackrel{\mathrm{def}}{=} \mathsf{T}(\alpha, V \circ \mathsf{A}). \tag{4.1}$$

Here, $\mathsf{A} \in \Delta' \pitchfork \mathbb{C}$, and p and α are morphisms in $\Delta' \pitchfork \mathbb{C}$ and $\mathbf{Idx}(\Delta, \Delta')$, respectively. We also define a helper morphism $\gamma^l_{\mathsf{A}} : \mathsf{T}(l, V \circ \mathsf{A}) \to V\mathsf{T}(l, \mathsf{A})$ for $(l_1 \cdots l_k) \in \Delta^*$ and $\mathsf{A} \in \Delta \pitchfork \mathbb{C}$. It is the multiple composite of $m_{A,B}$:

$$V\mathsf{A}(l_1) \otimes \cdots \otimes V\mathsf{A}(l_k) \to V(\mathsf{A}(l_1) \otimes \cdots \otimes \mathsf{A}(l_k)).$$

It is routine to verify that this morphism is monoidal natural on l and A.

Two monoidal natural transformations $\epsilon : D\mathrm{i}_\Delta \to \mathrm{Id}_{\Delta \pitchfork \mathbb{C}}$ and $\delta_{g,f} : D(g \bullet f) \to Df \circ Dg$ are defined by:

$$\epsilon_{\mathsf{A},i} : \mathsf{T}(i, V \circ \mathsf{A}) = V\mathsf{A}(i) \tag{4.2}$$

$$\begin{aligned}\delta_{g,f;\mathsf{A};i} : &\mathsf{T}((g \bullet f)(i), V \circ \mathsf{A}) \xrightarrow{\sim} \mathsf{T}(f, \mathsf{T}(g(\_), V \circ \mathsf{A})) \\ &\xrightarrow{\mathsf{T}(f, \mathsf{T}(g(\_), \delta_{\mathsf{A}}))} \mathsf{T}(f, \mathsf{T}(g(\_), V \circ V \circ \mathsf{A})) \xrightarrow{\mathsf{T}(f, \gamma^{g(\_)}_{\mathsf{A}})} Df(Dg\mathsf{A})(i).\end{aligned} \tag{4.3}$$

Theorem 4.2. *The symmetric colax monoidal functor D (4.1) and monoidal natural transformations ϵ, δ (4.2,4.3) determine an ILEC over C.*

4.3 GBLL Semantics by Realizability Category

Hofmann et al., and also Dal Lago et al. employ a *realizability semantics* to show that the complexity of BLL proof reductions belongs to P-time [19,10]. In this section we compare their semantics and the simple semantics of GBLL constructed in the previous section.

We instantiate $\mathbb{C}$ in the previous section with the realizability category over a BCI algebra $(A, \cdot)$, which is a combinatory algebra based on B, C, I-combinators; see e.g. [2,20]. We then form the realizability category $\mathbf{Ass}(A)$ by the following data: an object is a function f into P^+A, where P^+ is the nonempty powerset construction, and a morphism from f to g is a function $h : \mathrm{dom}\, f \to \mathrm{dom}\, g$ with the following property: there exists an element $e \in A$ such that for any $x \in \mathrm{dom}\, f$ and $a \in f(x)$, we have $e \cdot a \in g(h(x))$. The category $\mathbf{Ass}(A)$ is symmetric monoidal closed; see e.g. [20, Proposition 4]. The tensor product of f and g is given by $(f \otimes g)(x, y) = \{u \boxtimes v \mid u \in f(x), v \in g(y)\}$, where $u \boxtimes v$ is the BCI-algebra element corresponding to $\lambda x.xuv$ [20, Section 2].

Next, let Δ be a set and consider the power category $\Delta \pitchfork \mathbf{Ass}(A)$. Under the axiom of choice, $\Delta \pitchfork \mathbf{Ass}(A)$ is equivalently described as follows: an object is a family of functions $\{f_i\}_{i\in\Delta}$ into P^+A, and a morphism from $\{f_i\}_{i\in\Delta}$ to $\{g_i\}_{i\in\Delta}$ is a family of functions $\{h_i : \mathrm{dom}\, f_i \to \mathrm{dom}\, g_i\}_{i\in\Delta}$ with the following property: there exists a function $e : \Delta \to A$ such that for any $i \in \Delta$, $x \in \mathrm{dom}\, f_i$ and $a \in f_i(x)$, we have $e(i) \cdot a \in g_i(h_i(x))$.

This power category is quite close to the realizability category introduced in [19, Section 4] and [10, Section 4]. A membership statement $a \in f_i(x)$ for an object $\{f_i\}_{i\in\Delta} \in \Delta \pitchfork \mathbf{Ass}(A)$ corresponds to a realizability statement $i, a \Vdash x$ in the realizability category (see [19]). The major difference between these categories is twofold: 1) In the realizability category, a computability constraint is imposed on $e : \Delta \to A$ to achieve the characterization of P-time complexity. 2) Objects in the realizability category are limited to $\Delta \pitchfork \mathbf{Ass}(A)$-objects such that all f_i share the common domain. This is to synchronize with the set-theoretic semantics *ignoring* resource polynomials [19, Section 3] [10, Section 3].

We compute the bounded !-modality using the folding product ILEC T with respect to the indexed SMCC $(\_) \pitchfork \mathbf{Ass}(A)$. Let F be a finite set of variables, $x \notin F$ be a resource variable, p be a resource polynomial and $\mathscr{C}$ be a constraint set under F. For any object X in $V_{\mathscr{C}\cup\{v\le p\}}(F\cup\{v\}) \pitchfork \mathbf{Ass}(A)$, the folding product $\mathsf{T}([v \le p]_{(F,\mathscr{C})}, \mathsf{X})$ is an object in $V_{\mathscr{C}}(F) \pitchfork \mathbf{Ass}(A)$ satisfying

$$\begin{aligned}&\mathsf{T}([v < p]_{(F,\mathscr{C})}, \mathsf{X})(i)\\ &= \lambda(x_0, \cdots, x_{[\![p]\!]i-1}) \, . \, \{a_0 \otimes \cdots \otimes a_{[\![p]\!]i-1} \mid a_j \in \mathsf{X}(i\{v \mapsto j\})(x_j)\} \qquad (4.4)\end{aligned}$$

This is different from the modality over the realizability category introduced in [19, Definition 16] and [10, Definition 4.6]:

$$(!_{v<p}\mathsf{X})(i) = \lambda x \, . \, \{a_0 \otimes \cdots \otimes a_{[\![p]\!]i-1} \mid a_j \in \mathsf{X}(i\{v \mapsto j\})(x)\};$$

it only takes a single argument. This is again because their realizability semantics is designed to synchronize with the set-theoretic semantics ignoring resource polynomials — especially it interprets $[\![!_{x\le p}A]\!] = [\![A]\!]$. On the other hand, the bounded quantification computed in (4.4) does *not* ignore resource polynomials and indexing, as the domain of (4.4) is the index-dependent product $\prod_j \mathrm{dom}(\mathsf{X}(i\{v \mapsto j\}))$. From this, we conjecture that the semantics of BLL using the ILEC T over $(\_) \pitchfork \mathbf{Ass}(A)$ realizes an *index-dependent* set-theoretic semantics of BLL — we leave this semantics as a future work.

5 Conclusion and Related Work

We introduced GBLL, a generalization of Girard et al.'s BLL. We analyzed the complexity of cut-elimination in GBLL, and gave a translation from CBLL, an extension of BLL with constraints to GBAL^+. We then introduced ILEC as a categorical structure for interpreting the !-modality of GBLL. The ILEC is a **Idx**-graded linear exponential comonad interacting well with a specified indexed SMCCs. We gave an elementary construction of ILEC using the folding product, and a technique to derive its variants by inserting symmetric monoidal comonads. We gave the semantics of BLL using the folding product on the category of assemblies of a BCI-algebra, and related with the realizability category studied in [19,10].

Girard's BLL has a great influence on the subsequent development of indexed modalities and implicit complexity theory [16]. Hofmann and Scott introduced the realizability technique to BLL and semantically proved that BLL characterizes P-time complexity [19]. Their work was further enriched and studied by Dal Lago and Hofmann [10]. Gaboardi combined the !-modality involving variable binding with PCF and showed that the combined system is relatively complete [24].

Bucciarelli and Ehrhard's *indexed linear logic with exponential* [9] is one of the closest systems to GBLL. However, the type of the !-modality is different: their system derives $\Delta \vdash !_f A$ from $\Delta' \vdash A$ and an *almost injective function* $f : \Delta' \to \Delta$; it is a function where each $f^{-1}(i)$ is finite. To relate their system and GBLL, let us use the finite powerset construction P_{fin} and convert f into its inverse $f^{-1} : \Delta \to P_{\mathrm{fin}}(\Delta')$. This exhibits the similarity with GBLL: GBLL relaxes P_{fin} to $(\_)^*$, and takes the inverse as the parameter for the !-modality. The novelty of this work to [9] is that a categorical axiomatization for the $!_f$ modality is identified as an extension of the graded linear exponential comonads [7,22]. Another novelty is to show that GBLL is enough to encode BLL.

As described in Section 1, the simple form of !-modality $!_r A$ is also widely used in various type systems and programming languages. Examples include: INTML [30], coeffect calculus [28,7] and its combination with effect systems [13], Granule language [26], bounded linear type system [14,26], type systems for the analysis of higher-order model-checking [18,17], a generic BLL-like logic $\mathrm{B}_{\mathcal{S}}\mathrm{LL}$ over semirings [6], Fuzz type system for function sensitivity and differential privacy [29,12,3], and many more. A combination of $!_r A$ with dependent type theory called QTT is also introduced in [25] and [4]. Among these systems, each of [12,26,1] supports 1) full universal and existential, 2) full universal and 3) partial universal quantification over grades, respectively.

The categorical structure corresponding to the simple form of !-modality appears in [7,13,22] and is identified as *semiring-graded linear exponential comonad.* Breuvart constructed various examples of semiring-graded linear exponential comonads on relational models of linear logic [6] using his *slicing* technique. In this work we replaced semirings to **Idx**, which may be seen as a multi-object pseudo-semiring. In the study of graded monad, Orchard et al. generalize the grading structure from ordered monoids to 2-categories [27]. The main difference from this work is that their generalized graded monad is defined over a single category, while an ILEC is defined over an *indexed* SMCCs.

References

1. Abel, A., Bernardy, J.P.: A unified view of modalities in type systems. Proc. ACM Program. Lang. **4**(ICFP) (Aug 2020). https://doi.org/10.1145/3408972

2. Abramsky, S., Lenisa, M.: Linear realizability and full completeness for typed lambda-calculi. Ann. Pure Appl. Log. **134**(2-3), 122–168 (2005). https://doi.org/10.1016/j.apal.2004.08.003
3. de Amorim, A.A., Gaboardi, M., Hsu, J., Katsumata, S., Cherigui, I.: A semantic account of metric preservation. In: Castagna, G., Gordon, A.D. (eds.) Proceedings of the 44th ACM SIGPLAN Symposium on Principles of Programming Languages, POPL 2017, Paris, France, January 18-20, 2017. pp. 545–556. ACM (2017). https://doi.org/10.1145/3009837, http://dl.acm.org/citation.cfm?id=3009890
4. Atkey, R.: Syntax and semantics of quantitative type theory. In: Dawar, A., Grädel, E. (eds.) Proceedings of the 33rd Annual ACM/IEEE Symposium on Logic in Computer Science, LICS 2018, Oxford, UK, July 09-12, 2018. pp. 56–65. ACM (2018). https://doi.org/10.1145/3209108.3209189
5. Breuvart, F.: Dissecting Denotational Semantics: From the Well-established $\mathcal{H}^*$ to the More Recent Quantitative Coeffects. Ph.D. thesis, Université Paris Diderot (2015), https://lipn.univ-paris13.fr/~breuvart/These_breuvart.pdf
6. Breuvart, F., Pagani, M.: Modelling coeffects in the relational semantics of linear logic. In: Kreutzer [23], pp. 567–581, http://www.dagstuhl.de/dagpub/978-3-939897-90-3
7. Brunel, A., Gaboardi, M., Mazza, D., Zdancewic, S.: A core quantitative coeffect calculus. In: Shao, Z. (ed.) Programming Languages and Systems. pp. 351–370. Springer Berlin Heidelberg, Berlin, Heidelberg (2014). https://doi.org/10.1007/978-3-642-54833-8_19
8. Bucciarelli, A., Ehrhard, T.: On phase semantics and denotational semantics in multiplicative-additive linear logic. Ann. Pure Appl. Log. **102**(3), 247–282 (2000). https://doi.org/10.1016/S0168-0072(99)00040-8
9. Bucciarelli, A., Ehrhard, T.: On phase semantics and denotational semantics: the exponentials. Annals of Pure and Applied Logic **109**(3), 205 – 241 (2001). https://doi.org/10.1016/S0168-0072(00)00056-7, http://www.sciencedirect.com/science/article/pii/S0168007200000567
10. Dal Lago, U., Hofmann, M.: Bounded linear logic, revisited. In: Curien, P.L. (ed.) Typed Lambda Calculi and Applications. pp. 80–94. Springer Berlin Heidelberg, Berlin, Heidelberg (2009). https://doi.org/10.1007/978-3-642-02273-9_8
11. Fujii, S., Katsumata, S., Melliès, P.: Towards a formal theory of graded monads. In: Jacobs, B., Löding, C. (eds.) Foundations of Software Science and Computation Structures - 19th International Conference, FOSSACS 2016, Held as Part of the European Joint Conferences on Theory and Practice of Software, ETAPS 2016, Eindhoven, The Netherlands, April 2-8, 2016, Proceedings. Lecture Notes in Computer Science, vol. 9634, pp. 513–530. Springer (2016). https://doi.org/10.1007/978-3-662-49630-5_30
12. Gaboardi, M., Haeberlen, A., Hsu, J., Narayan, A., Pierce, B.C.: Linear dependent types for differential privacy. In: Giacobazzi, R., Cousot, R. (eds.) The 40th Annual ACM SIGPLAN-SIGACT Symposium on Principles of Programming Languages, POPL '13, Rome, Italy - January 23 - 25, 2013. pp. 357–370. ACM (2013). https://doi.org/10.1145/2429069.2429113, http://dl.acm.org/citation.cfm?id=2429069
13. Gaboardi, M., Katsumata, S., Orchard, D., Breuvart, F., Uustalu, T.: Combining effects and coeffects via grading. SIGPLAN Not. **51**(9), 476–489 (Sep 2016). https://doi.org/10.1145/3022670.2951939
14. Ghica, D.R., Smith, A.I.: Bounded linear types in a resource semiring. In: Shao, Z. (ed.) Programming Languages and Systems. pp. 331–350. Springer Berlin Heidelberg, Berlin, Heidelberg (2014). https://doi.org/10.1007/978-3-642-54833-8_18

15. Girard, J.: Linear logic. Theor. Comput. Sci. **50**, 1–102 (1987). https://doi.org/10.1016/0304-3975(87)90045-4
16. Girard, J., Scedrov, A., Scott, P.J.: Bounded linear logic: A modular approach to polynomial-time computability. Theor. Comput. Sci. **97**(1), 1–66 (1992). https://doi.org/10.1016/0304-3975(92)90386-T
17. Grellois, C., Melliès, P.: An infinitary model of linear logic. In: Pitts, A.M. (ed.) Foundations of Software Science and Computation Structures - 18th International Conference, FoSSaCS 2015, Held as Part of the European Joint Conferences on Theory and Practice of Software, ETAPS 2015, London, UK, April 11-18, 2015. Proceedings. Lecture Notes in Computer Science, vol. 9034, pp. 41–55. Springer (2015). https://doi.org/10.1007/978-3-662-46678-0_3
18. Grellois, C., Melliès, P.: Relational semantics of linear logic and higher-order model checking. In: Kreutzer [23], pp. 260–276, http://www.dagstuhl.de/dagpub/978-3-939897-90-3
19. Hofmann, M., Scott, P.J.: Realizability models for bll-like languages. Theor. Comput. Sci. **318**(1-2), 121–137 (2004). https://doi.org/10.1016/j.tcs.2003.10.019
20. Hoshino, N.: Linear realizability. In: Duparc, J., Henzinger, T.A. (eds.) Computer Science Logic, 21st International Workshop, CSL 2007, 16th Annual Conference of the EACSL, Lausanne, Switzerland, September 11-15, 2007, Proceedings. Lecture Notes in Computer Science, vol. 4646, pp. 420–434. Springer (2007). https://doi.org/10.1007/978-3-540-74915-8_32
21. Hyland, M., Power, J.: Pseudo-commutative monads and pseudo-closed 2-categories. Journal of Pure and Applied Algebra **175**(1), 141 – 185 (2002). https://doi.org/10.1016/S0022-4049(02)00133-0, http://www.sciencedirect.com/science/article/pii/S0022404902001330, special Volume celebrating the 70th birthday of Professor Max Kelly
22. Katsumata, S.: A double category theoretic analysis of graded linear exponential comonads. In: Baier, C., Dal Lago, U. (eds.) Foundations of Software Science and Computation Structures. pp. 110–127. Springer International Publishing, Cham (2018). https://doi.org/10.1007/978-3-319-89366-2_6
23. Kreutzer, S. (ed.): 24th EACSL Annual Conference on Computer Science Logic, CSL 2015, September 7-10, 2015, Berlin, Germany, LIPIcs, vol. 41. Schloss Dagstuhl - Leibniz-Zentrum für Informatik (2015), http://www.dagstuhl.de/dagpub/978-3-939897-90-3
24. Lago, U.D., Gaboardi, M.: Linear dependent types and relative completeness. In: Proceedings of the 26th Annual IEEE Symposium on Logic in Computer Science, LICS 2011, June 21-24, 2011, Toronto, Ontario, Canada. pp. 133–142. IEEE Computer Society (2011). https://doi.org/10.1109/LICS.2011.22, https://ieeexplore.ieee.org/xpl/conhome/5968099/proceeding
25. McBride, C.: I got plenty o' nuttin'. In: Lindley, S., McBride, C., Trinder, P.W., Sannella, D. (eds.) A List of Successes That Can Change the World - Essays Dedicated to Philip Wadler on the Occasion of His 60th Birthday. Lecture Notes in Computer Science, vol. 9600, pp. 207–233. Springer (2016). https://doi.org/10.1007/978-3-319-30936-1_12
26. Orchard, D., Liepelt, V.B., Eades III, H.: Quantitative program reasoning with graded modal types. Proc. ACM Program. Lang. **3**(ICFP) (Jul 2019). https://doi.org/10.1145/3341714
27. Orchard, D., Wadler, P., Eades, H.: Unifying graded and parameterised monads. Electronic Proceedings in Theoretical Computer Science **317**, 18–38 (May 2020). https://doi.org/10.4204/eptcs.317.2

28. Petricek, T., Orchard, D., Mycroft, A.: Coeffects: Unified static analysis of context-dependence. In: Fomin, F.V., Freivalds, R., Kwiatkowska, M., Peleg, D. (eds.) Automata, Languages, and Programming. pp. 385–397. Springer Berlin Heidelberg, Berlin, Heidelberg (2013). https://doi.org/10.1007/978-3-642-39212-2_35
29. Reed, J., Pierce, B.C.: Distance makes the types grow stronger: a calculus for differential privacy. In: Hudak, P., Weirich, S. (eds.) Proceeding of the 15th ACM SIGPLAN international conference on Functional programming, ICFP 2010, Baltimore, Maryland, USA, September 27-29, 2010. pp. 157–168. ACM (2010). https://doi.org/10.1145/1863543.1863568
30. Schöpp, U.: Computation-by-interaction with effects. In: Yang, H. (ed.) Programming Languages and Systems. pp. 305–321. Springer Berlin Heidelberg, Berlin, Heidelberg (2011). https://doi.org/10.1007/978-3-642-25318-8_23

Focused Proof-search in the Logic of Bunched Implications*

Alexander Gheorghiu(✉) and Sonia Marin(✉)

University College London, London, United Kingdom
{alexander.gheorghiu.19, s.marin}@ucl.ac.uk

Abstract. The logic of Bunched Implications (BI) freely combines additive and multiplicative connectives, including implications; however, despite its well-studied proof theory, proof-search in BI has always been a difficult problem. The focusing principle is a restriction of the proof-search space that can capture various goal-directed proof-search procedures. In this paper we show that focused proof-search is complete for BI by first reformulating the traditional bunched sequent calculus using the simpler data-structure of nested sequents, following with a polarised and focused variant that we show is sound and complete via a cut-elimination argument. This establishes an operational semantics for focused proof-search in the logic of Bunched Implications.

Keywords: Logic · Proof-search · Focusing · Bunched Implications.

1 Introduction

The *Logic of Bunched Implications* (BI) [31] is well-known for its applications in systems modelling [32], especially a particular theory (of a variant of BI) called *Separation Logic* [37,23] which has found industrial use in program verification. In this work, we study an aspect of proof search in BI, relying on its well-developed and well-studied proof theory [33]. We show that a goal-directed proof-search procedure known as *focused proof-search* is complete; that is, if there is a proof then there is a focused one. Focused proofs are both interesting in the abstract, giving insight into the proof theory of the logic, and have (for other logics) been a useful modelling technology in applied settings. For example, focused proof-search forms an operational semantics of the DPLL SAT-solvers [14], logic programming [29,1,13,7], automated theorem provers [28], and has been successful in providing a meta-theoretic framework in intuitionistic, substructural, and modal logics [27,30,25].

Syntactically BI combines additive and multiplicative connectives, but unlike related logics such as Linear Logic (LL) [22], BI takes all the connectives as primitive. Indeed, it arose from a proof-theoretic investigation on the relationship between conjunction and implication. As a result, sequents in BI have a

* This work has been partially supported by the UK's EPSRC through research grant EP/S013008/1.

S. Kiefer and C. Tasson (Eds.): FOSSACS 2021, LNCS 12650, pp. 247–267, 2021.
https://doi.org/10.1007/978-3-030-71995-1_13

more complicated structure: each implication comes with an associated context-former. Therefore, in BI contexts are not lists, nor multisets, but instead are *bunches*: binary trees whose leaves are formulas and internal nodes context-formers. Additive composition $(\Gamma; \Delta)$ admits the structural rules of weakening and contraction, whereas multiplicative composition (Γ, Δ) denies them. The principal technical challenges when studying proof-search in BI arise from the interaction between the additive and multiplicative fragments. We overcome these challenges by restricting the application of structural rules in the sequent calculus LBI as well as working with a representation of bunches as nested multisets.

Throughout we use the term *sequent calculus* in a strict sense; that is, meaning a label-free internal sequent calculus, formed in the case of BI by a context (a bunch) and a consequent (a formula). The term *proof-search* is consistently understood to be read as backward reduction within such a system. Although there is an extensive body of research on systems and procedures for semantics-based calculi in BI [19,20,16,17,18], there has been comparatively little formal study on proof-search in the strict sense. One exception is the completeness result for (unit-simple) uniform proofs [2] which is partially subsumed by the results herein.

The *focusing principle* was introduced for Linear Logic [1] and is characterised by alternating *focused* and *unfocused* phases of goal-directed proof-search. The unfocused phase comprises rules which are safe to apply (i.e. rules where provability is invariant); conversely, the focused phase contains the reduction of a formula and its sub-formulas where potentially invalid sequents may arise, and backtracking may be required. During focused proof-search the unfocused phases are performed eagerly, followed by controlled goal-directed focused phases, until safe reductions are available again. We say that the focusing principle holds when every provable sequent has a focused proof. This alternation can be enforced by a mechanism based on a partition of the set of formulas into two classes, *positive* and *negative*, which correspond to safe behaviour on the left and right respectively; that is, for negative formulas provability is invariant with respect to the application of a right rule, and for positive formulas, of a left rule, but in the other cases the application may result in invalid sequents.

The original proof of the focusing principle in Linear Logic was via long and tedious permutations of rules [1]. In this paper, we use for BI a different methodology, originally presented in [24], which has since been implemented in a variety of logics [25,5,6] and proof systems [13]. The method is as follows: given a sequent calculus, first one polarises the syntax according to the positive/negative behaviours; second, one gives a focused variation of the sequent calculus where the control flow of proof-search is managed by polarisation; third, one shows that this system admits cut (the only non-analytic rule); and, finally, one shows that in the presence of cut the original sequent calculus may be simulated in the focused one. When the polarised system is complete, the focusing principle holds.

In LBI certain rules (the structural rules) have no natural placement in either the focused or the unfocused phases of proof-search. Thus, a design choice must be made: to eliminate/constrain these rules, or to permit them without restriction. The first gives a stricter control proof-search regime, but the latter typically achieves a more well-behaved proof theoretic meta-theory. In this paper, we choose the former as our motivation is to study computational behaviour of proof-search in BI, the latter being recovered by familiar admissibility results. The only case where confinement is not possible is the *exchange* rule. In standard sequent calculi the exchange rule is made implicit by working with a more convenient data-structure such as multisets as opposed to lists; however, the specific structure of bunches in BI means that a more complex alternative is required. The solution presented is to use nested multisets of two types (additive and multiplicative) corresponding to the two different context-formers/conjunctions.

In Section 2 we present the logic of Bunched Implications; in particular, Section 2.1 and Section 2.2 contain the background on BI (the syntax and sequent calculus respectfully); meanwhile, Section 2.3 gives representation of bunches as nested multisets. Section 3 contains the focused system: first, in Section 3.1 we introduce the polarised syntax; second, in Section 3.2 we introduce the focused sequents calculus and some metatheory, most importantly the cut-admissibility result; finally, in Section 3.3 we give the completeness theorem, from which the validity of the focusing principle follows as a corollary. We conclude in Section 4 with some further discussion and future directions.

2 Re-presentations of BI

2.1 Traditional Syntax

The logic BI has a well-studied metatheory admitting familiar categorical, algebraic, and truth-functional semantics which have the expected dualities [34,17,33,11,32]. In practice, it is the free combination (or, more precisely, the fibration [15,33]) of intuitionistic logic (IL) and the multiplicative fragment of intuitionistic linear logic (MILL), which imposes the presence of two distinct context-formers in its sequent presentation. That is to say, the two conjunctions $\land$ and $*$ are represented at the meta-level by context-formers ; and , in place of the usual commas for IL and MILL respectively.

Definition 1 (Formula). *Let* P *be a denumerable set of propositional letters. The* formulas *of BI, denoted by small Greek letters* $(\varphi, \psi, \chi, \ldots)$*, are defined by the following grammar, where* $A \in \mathsf{P}$*,*

$$\varphi ::= \top \mid \bot \mid \top^* \mid A \mid (\varphi \land \varphi) \mid (\varphi \lor \varphi) \mid (\varphi \to \varphi) \mid (\varphi * \varphi) \mid (\varphi \mathrel{-\!\!*} \varphi)$$

If $\circ \in \{\land, \lor, \to, \top\}$ *then it is an additive connective and if* $\circ \in \{*, \mathrel{-\!\!*}, \top^*\}$ *then it is a multiplicative connective. The set of all formulas is denoted* $\mathbb{F}$.

Definition 2 (Bunch). *A* bunch *is constructed from the following grammar, where* $\varphi \in \mathbb{F}$*,*

$$\Delta ::= \varphi \mid \varnothing_+ \mid \varnothing_\times \mid (\Delta; \Delta) \mid (\Delta, \Delta)$$

The symbols $\varnothing_+$ and $\varnothing_\times$ are the additive and multiplicative units respectively, and the symbols ; and , are the additive and multiplicative context-formers respectively. A bunch is basic *if it is a formula, $\varnothing_+$, or $\varnothing_\times$ and* complex *otherwise. The set of all bunches is denoted $\mathbb{B}$, the set of complex bunches with additive root context-former by $\mathbb{B}^+$, and the set of complex bunches with multiplicative root context-former by $\mathbb{B}^\times$.*

For two bunches $\Delta, \Delta' \in \mathbb{B}$ if Δ' is a sub-tree of Δ, it is called a *sub-bunch.* We may use the standard notation $\Delta(\Delta')$ (despite its slight inpracticality) to denote that Δ' is a sub-bunch of Δ, in which case $\Delta(\Delta'')$ is the result of replacing the occurrence of Δ' by Δ''. If δ is a sub-bunch of Δ, then the context-former $\circ$ is said to be its principal context-former in $\Delta(\Delta' \circ \delta)$ (and $\Delta(\delta \circ \Delta')$).

Example 3. Let φ, ψ and χ be formulas, and let $\Delta = (\varphi, (\chi; \varnothing_+)); (\psi; (\psi; \varnothing_\times))$. The bunch may be written for example as $\Delta(\varphi, (\chi; \varnothing_+))$ which means that we can have $\Delta(\varphi; \varphi) = (\varphi; \varphi); (\psi; (\psi; \varnothing_\times))$.

Definition 4 (Bunched Sequent). *A bunched sequent is a pair of a bunch Δ, called the context, and a formula φ, denoted $\Delta \Rightarrow \varphi$.*

Bunches are intended to be considered up-to *coherent equivalence* ($\equiv$). It is the least relation satisfying:

- Commutative monoid equations for ; with unit $\varnothing_+$,
- Commutative monoid equations for , with unit $\varnothing_\times$,
- Congruence: if $\Delta' \equiv \Delta''$ then $\Delta(\Delta') \equiv \Delta(\Delta'')$.

It will be useful to have a measure on sub-bunches which can identify their distance from the root node.

Definition 5 (Rank). *If Δ' is a sub-bunch of Δ, then $\rho(\Delta')$ is the number of alternations of additive and multiplicative context-formers between the principal context-former of Δ', and the root context-former of Δ.*

Let Δ be a complex bunch, we use $\Delta' \in \Delta$ to denote that Δ' is a (proper) top-most sub-bunch; that is, Δ is a sub-bunch satisfying $\Delta \neq \Delta'$ but $\rho(\Delta') = 0$.

Example 6. Let Δ be as in Example 3, then $\rho(\varnothing_+) = 2$ whereas $\rho(\varnothing_\times) = 0$; hence, ψ, $\varnothing_\times$ and $(\varphi, (\chi, \varnothing_\times)) \in \Delta$. Consider the parse-tree of Δ:

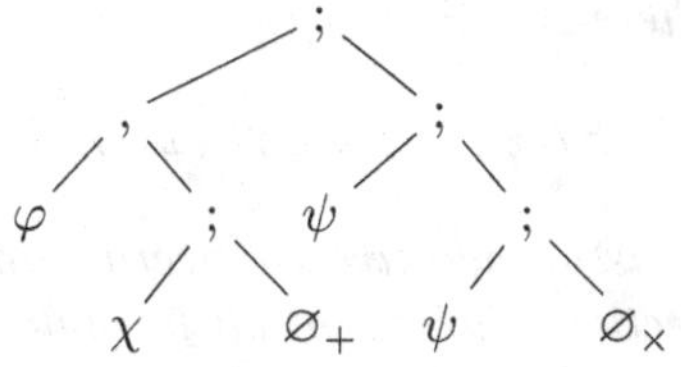

Reading upward from $\varnothing_+$ one encounters first ; which changes into , and then back to ; so the rank is 2; whereas counting up from $\varnothing_\times$ one only encounters ; so the rank is 0.

$$\frac{}{A \Rightarrow A}\,\mathsf{Ax} \qquad \frac{}{\Delta(\bot) \Rightarrow \varphi}\,\bot_{\mathsf{L}} \qquad \frac{}{\varnothing_{\times} \Rightarrow \top^*}\,\top^*_{\mathsf{R}} \qquad \frac{}{\varnothing_{+} \Rightarrow \top}\,\top_{\mathsf{R}}$$

$$\frac{\Delta' \Rightarrow \varphi \quad \Delta(\Delta'', \psi) \Rightarrow \chi}{\Delta(\Delta', \Delta'', \varphi \mathrel{-\!\!*} \psi) \Rightarrow \chi}\,{-\!\!*}_{\mathsf{L}} \qquad \frac{\Delta, \varphi \Rightarrow \psi}{\Delta \Rightarrow \varphi \mathrel{-\!\!*} \psi}\,{-\!\!*}_{\mathsf{R}}$$

$$\frac{\Delta(\varphi, \psi) \Rightarrow \chi}{\Delta(\varphi * \psi) \Rightarrow \chi}\,*_{\mathsf{L}} \qquad \frac{\Delta \Rightarrow \varphi \quad \Delta' \Rightarrow \psi}{\Delta, \Delta' \Rightarrow \varphi * \psi}\,*_{\mathsf{R}} \qquad \frac{\Delta(\varnothing_{\times}) \Rightarrow \chi}{\Delta(\top^*) \Rightarrow \chi}\,\top^*_{\mathsf{L}}$$

$$\frac{\Delta(\varphi; \psi) \Rightarrow \chi}{\Delta(\varphi \wedge \psi) \Rightarrow \chi}\,\wedge_{\mathsf{L}} \qquad \frac{\Delta \Rightarrow \varphi \quad \Delta' \Rightarrow \psi}{\Delta; \Delta' \Rightarrow \varphi \wedge \psi}\,\wedge_{\mathsf{R}} \qquad \frac{\Delta(\varnothing_{+}) \Rightarrow \chi}{\Delta(\top) \Rightarrow \chi}\,\top_{\mathsf{L}}$$

$$\frac{\Delta(\varphi) \Rightarrow \chi \quad \Delta(\psi) \Rightarrow \chi}{\Delta(\varphi \vee \psi) \Rightarrow \chi}\,\vee_{\mathsf{L}} \qquad \frac{\Delta \Rightarrow \varphi}{\Delta \Rightarrow \varphi \vee \psi}\,\vee_{\mathsf{R1}} \qquad \frac{\Delta \Rightarrow \psi}{\Delta \Rightarrow \varphi \vee \psi}\,\vee_{\mathsf{R2}}$$

$$\frac{\Delta' \Rightarrow \varphi \quad \Delta(\Delta''; \psi) \Rightarrow \chi}{\Delta(\Delta'; \Delta''; \varphi \to \psi) \Rightarrow \chi}\,\to_{\mathsf{L}} \qquad \frac{\Delta; \varphi \Rightarrow \psi}{\Delta \Rightarrow \varphi \to \psi}\,\to_{\mathsf{R}} \qquad \frac{\Delta(\Delta'; \Delta') \Rightarrow \chi}{\Delta(\Delta') \Rightarrow \chi}\,\mathsf{C}$$

$$\frac{\Delta(\Delta') \Rightarrow \chi}{\Delta(\Delta'; \Delta'') \Rightarrow \chi}\,\mathsf{W} \qquad \frac{\Delta \Rightarrow \chi}{\Delta' \Rightarrow \chi}\,\mathsf{E}_{(\Delta \equiv \Delta')} \qquad \frac{\Delta' \Rightarrow \varphi \quad \Delta(\varphi) \Rightarrow \chi}{\Delta(\Delta') \Rightarrow \chi}\,\mathsf{cut}$$

Fig. 1. Sequent Calculus LBI

2.2 Sequent Calculus

The proof theory of BI is well-developed including familiar Hilbert, natural deduction, sequent calculi, tableaux systems, and display calculi [33,17,3]. In the foregoing we restrict attention to the sequent calculus as it more amenable to studying proof-search as computation, having local correctness while enjoying the completeness of analytic proofs.

Definition 7 (System LBI). *The bunched sequent calculus* LBI *is composed of the rules in Figure 1.*

The classification of $\wedge$ as additive may seem dubious upon reading the $\wedge_{\mathsf{R}}$ rule, but the designation arises from the use of the structural rules; that is, the $\wedge_{\mathsf{R}}$ and $\to_{\mathsf{R}}$ rules may be replaced by *additive* variants without loss of generality. The presentation in Figure 1 is as in [33] and simply highlights the nature of the additive and multiplicative context-formers. Nonetheless, the choice of rule does affect proof-search behaviours, and the consequences are discussed in more detailed in Section 3.1.

Lemma 8 (Cut-elimination). *If φ has a* LBI*-proof, then it has a* cut*-free* LBI*-proof, i.e., a proof with no occurence of the* cut *rule.*

Throughout, unless specified otherwise, we take proof to mean cut-free proof. Moreover, if L is a sequent calculus we use $\vdash_{\mathsf{L}} \Delta \Rightarrow \varphi$ to denote that there is an L-proof of $\Delta \Rightarrow \varphi$. Further, if R is a rule, then we may denote L + R to denote the sequent calculus combining the rules of L with R.

The following result, that a generalised version of the axiom is derivable in LBI, will allow for such sequents to be used in proof-construction later on.

Lemma 9. *For any formula* φ, $\vdash_{\mathsf{LBI}} \varphi \Rightarrow \varphi$.

Proof. Follows from induction on size of φ. □

The remainder of this section is the meta-theory required to control the structural rules, which pose the main issue to the study of proof-search in BI.

Lemma 10. *The following rules are derivable in* LBI, *and replacing* W *with them does not affect the completeness of the system.*

$$\frac{}{\Delta; A \Rightarrow A}\ \mathsf{Ax}' \qquad \frac{}{\Delta; \varnothing_\times \Rightarrow \top^*}\ \top^*{}'_{\mathsf{R}} \qquad \frac{}{\Delta; \varnothing_+ \Rightarrow \top}\ \top'_{\mathsf{R}}$$

$$\frac{\Delta \Rightarrow \varphi \quad \Delta' \Rightarrow \psi}{(\Delta, \Delta'); \Delta'' \Rightarrow \varphi * \psi}\ *'_{\mathsf{R}} \qquad \frac{\Delta' \Rightarrow \varphi \quad \Delta(\Delta'', \psi) \Rightarrow \chi}{\Delta(\Delta', \Delta'', (\Delta'''; \varphi \mathrel{-\!\!*} \psi)) \Rightarrow \chi}\ \mathrel{-\!\!*}'_{\mathsf{L}}$$

Proof. We can construct in LBI derivations with the same premisses and conclusion as these rules by use of the structural rules. Let LBI′ be LBI without W but with these new rules (retaining also $*_{\mathsf{R}}$, $\mathrel{-\!\!*}_{\mathsf{L}}$, $\top^*_{\mathsf{R}}$, $\top_{\mathsf{R}}$, and Ax), then W is admissible in LBI′ using standard permutation argument. □

One may regard the above modification to LBI as forming a new calculus, but since all the new rules are derivable it is really a restriction of the calculus, in the sense that all proofs in the new system have equivalent proofs in LBI differing only by explicitly including instances of weakening.

2.3 Nested Calculus

Originally, sequents in the calculi for classical and intuitionistic logics (LK and LJ, respectively) were introduced as lists, and a formal *exchange* rule was required to permute elements when needed for a logical rule to be applied [21]. However, in practice, the exchange rule is often suppressed, and contexts are simply presented as multisets of formulas. This reduces the number of steps/choices being made during proof-search without increasing the complexity of the underlying data structure. Bunches have considerably more structure than lists, but a quotient with respect to coherent equivalence can be made resulting in two-sorted nested multisets; this was first suggested in [12], though never formally realised.

Definition 11 (Two-sorted Nest). *Nests* (Γ) *are formulas or multisets, ascribed either additive* (Σ), *or multiplicative* (Π) *kind, containing nests of the opposite kind:*

$$\Gamma := \Sigma \mid \Pi \qquad \Sigma := \varphi \mid \{\Pi_1, ..., \Pi_n\}_+ \qquad \Pi := \varphi \mid \{\Sigma_1, ..., \Sigma_n\}_\times$$

The constructors are multiset constructors which may be empty in which case the nests are denoted $\varnothing_+$ *and* $\varnothing_\times$ *respectively. No multiset is a singleton; and the set of all nests is denoted* $\mathbb{B}/\!\equiv$.

Given nests Λ and Γ, we write $\Lambda \in \Gamma$ to denote either that $\Lambda = \Gamma$, if Γ is a formula, or that Λ is an element of the multiset Γ otherwise. Furthermore, we write $\Lambda \subseteq \Gamma$ to denote $\forall \gamma \in \mathbb{B}/\equiv$ if $\gamma \in \Lambda$ then $\gamma \in \Gamma$.

We will depart from the standard, yet impractical subbunch notation, and adopt a context notation for nests instead. We write $\Gamma\{\cdot\}_+$ (resp. $\Gamma\{\cdot\}_\times$) for a nest with a hole within one of its additive (resp. multiplicative) multisets.The notation $\Gamma\{\Lambda\}_+$ (resp. $\Gamma\{\Lambda\}_\times$), denotes that Λ is a sub-nest of Γ of additive (resp. multiplicative) kind; we may use $\Gamma\{\Lambda\}$ when the kind is not specified. In either case $\Gamma\{\Lambda'\}$ denotes the substitution of Λ for Λ'. A promotion in the syntax tree may be required after a substitution either to handle a singleton or an improper alternation of constructor types.

Example 12. The following inclusions are valid,

$$\{\varphi, \chi\}_\times \in \Big\{\{\varphi, \chi\}_\times, \psi\Big\}_+ \subseteq \Big\{\{\varphi, \chi\}_\times, \psi, \psi, \varnothing_\times\Big\}_+ = \Gamma\{\{\varphi, \chi\}_\times\}_+$$

It follow that $\Gamma\{\{\varphi, \varphi\}_+\}_+ = \{\varphi, \varphi, \psi, \psi, \varnothing_\times\}_+$. Note the absence of the $\{\cdot\}_+$ constructor after substitution, this is due to a promotion in the syntax tree to avoid having two nested additive constructors. Similarly, since $\varnothing_\times$ denotes the empty multiset of multiplicative kind, substituting χ with it gives $\{\varphi, \psi, \psi, \varnothing_\times\}_+$; that is, first the improper $\{\varphi, \varnothing_\times\}_\times$ becomes $\{\varphi\}_\times$; then, the resulting singleton $\{\varphi\}_\times$ is promoted to φ.

Typically we will only be interested in fragments of sub-nests so we have the following abuse of notation, where $\circ \in \{+, \times\}$:

$$\Gamma\{\{\Pi_1, ..., \Pi_i\}_\circ, \Pi_{i+1}, .., \Pi_n\}_\circ := \Gamma\{\Pi_1, ..., \Pi_n\}_\circ$$

The notion of rank has a natural analogue in this setting.

Definition 13 (Depth, Rank). *Let $\circ \in \{+, \times\}$ be a nest, we define the depth on $\mathbb{B}$ as follows:*

$$\delta(\varphi) := 0 \qquad \delta(\{\Gamma_1, ..., \Gamma_n\}_\circ) := \max\{\delta(\Gamma_1), ..., \delta(\Gamma_n)\} + 1$$

The equivalence of the two presentations, bunches and nests, follows from a moral (in the sense that bunches are intended to be considered modulo congruence) inverse between a *nestifying* function η and a *bunching* function β. The transformation β is simply going from a tree with arbitrary branching to a binary one, and η is the reverse.

Definition 14 (Canonical Translation). *The canonical translation $\eta : \mathbb{B} \to \mathbb{B}/\equiv$ is defined recursively as follows,*

$$\eta(\Delta) := \begin{cases} \Delta & \text{if } \Delta \in \mathbb{F} \cup \{\varnothing_+, \varnothing_\times\} \\ \{\eta(\Delta') \in \mathbb{B}/\equiv \mid \rho(\Delta') = 1 \text{ and } \Delta' \in \mathbb{B}^\times\}_+ & \text{if } \Delta \in \mathbb{B}^+ \\ \{\eta(\Delta') \in \mathbb{B}/\equiv \mid \rho(\Delta') = 1 \text{ and } \Delta' \in \mathbb{B}^+\}_\times & \text{if } \Delta \in \mathbb{B}^\times \end{cases}$$

The canonical translation $\beta : \mathbb{B}/\equiv \to \mathbb{B}$ *is defined recursively as follows,*

$$\beta(\Gamma) := \begin{cases} \Gamma & \text{if } \Gamma \in \mathbb{F} \cup \{\varnothing_+, \varnothing_\times\} \\ \beta(\Pi_1); (\beta(\Pi_2); ...) & \text{if } \Gamma = \{\Pi_1, \Pi_2, ...\}_+ \\ \beta(\Sigma_1), (\beta(\Sigma_2), ...) & \text{if } \Gamma = \{\Sigma_1, \Sigma_2, ...\}_\times \end{cases}$$

Example 15. Applying η to the bunch in Example 3 gives the nest in Example 12:

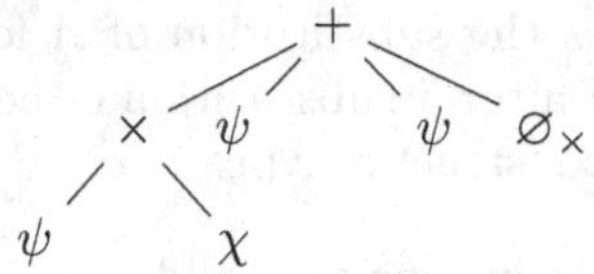

Lemma 16. *The translations are inverses up-to congruence; that is,*

1. *if* $\Delta \in \mathbb{B}$ *then* $(\beta \circ \eta)(\Delta) \equiv \Delta$;
2. *if* $\Gamma \in \mathbb{B}/\equiv$ *then* $(\eta \circ \beta)(\Gamma) \equiv \Gamma$;
3. *let* $\Delta, \Delta' \in \mathbb{B}$, *then* $\Delta \equiv \Delta'$ *if and only if* $\eta(\Delta) = \eta(\Delta')$.

Proof. The first two statements follow by induction on the depth (either for bunches or nests), where one must take care to consider the case of a context consisting entirely of units. The third statement employs the first in the forward direction, and proceeds by induction on depth in the reverse direction. □

Definition 17 (System ηLBI). *The nested sequent calculus* ηLBI *is composed of the rules in Figure 2, where the metavariables denote possibly empty nests.*

Observe the use of metavariable Γ' instead of Π (resp. Σ) as sub-contexts in Figure 2. This allows classes of inferences such as

$$\frac{\{\Sigma_0, ..., \Sigma_i\}_\times \Rightarrow \varphi \quad \{\Sigma_{i+1}, ..., \Sigma_n\}_\times \Rightarrow \varphi}{\{\Sigma_0, ..., \Sigma_n\}_\times \Rightarrow \varphi * \psi} *_\mathsf{R}$$

to be captured by a single figure. In practice it implements the abuse of notation given above:

$$\{\{\Sigma_0, ..., \Sigma_i\}_\times, \{\Sigma_{i+1}, ..., \Sigma_n\}_\times\}_\times \Rightarrow \varphi * \psi$$

This system is a new and very convenient presentation of LBI, not *per se* a development of the proof theory for the logic.

Lemma 18 (Soundness and Completeness of ηLBI). *Systems* LBI *and* ηLBI *are equivalent:*

Soundness: If $\vdash_{\eta\mathsf{LBI}} \Gamma \Rightarrow \varphi$ *then* $\vdash_{\mathsf{LBI}} \beta(\Gamma) \Rightarrow \varphi$;
Completeness: If $\vdash_{\mathsf{LBI}} \Delta \Rightarrow \varphi$ *then* $\vdash_{\eta\mathsf{LBI}} \eta(\Delta) \Rightarrow \varphi$.

Proof. Each claim follows by induction on the context, appealing to Lemma 16 to organise the data structure for the induction hypothesis, without loss of generality.

$$\dfrac{}{\{\Gamma, A\}_+ \Rightarrow A}\,\mathsf{Ax} \qquad \dfrac{}{\Gamma\{\bot\} \Rightarrow \chi}\,\bot_{\mathsf{L}} \qquad \dfrac{}{\varnothing_\times \Rightarrow \top^*}\,\top^*_{\mathsf{R}} \qquad \dfrac{}{\Gamma \Rightarrow \top}\,\top_{\mathsf{R}}$$

$$\dfrac{\Gamma' \Rightarrow \varphi \quad \Gamma\{\Gamma'', \psi\}_\times \Rightarrow \chi}{\Gamma\{\Gamma', \Gamma'', \{\Gamma''', \varphi \mathrel{-\!\!*} \psi\}_+\}_\times \Rightarrow \chi}\,\mathrel{-\!\!*}_{\mathsf{L}} \qquad \dfrac{\{\Gamma, \varphi\}_\times \Rightarrow \psi}{\Gamma \Rightarrow \varphi \mathrel{-\!\!*} \psi}\,\mathrel{-\!\!*}_{\mathsf{R}}$$

$$\dfrac{\Gamma\{\{\varphi, \psi\}_\times\} \Rightarrow \chi}{\Gamma\{\varphi * \psi\} \Rightarrow \chi}\,*_{\mathsf{L}} \qquad \dfrac{\Gamma \Rightarrow \varphi \quad \Gamma' \Rightarrow \psi}{\{\{\Gamma, \Gamma'\}_\times, \Gamma''\}_+ \Rightarrow \varphi * \psi}\,*_{\mathsf{R}} \qquad \dfrac{\Gamma\{\varnothing_\times\} \Rightarrow \chi}{\Gamma\{\top^*\} \Rightarrow \chi}\,\top^*_{\mathsf{L}}$$

$$\dfrac{\Gamma\{\{\varphi, \psi\}_+\} \Rightarrow \chi}{\Gamma\{\varphi \wedge \psi\} \Rightarrow \chi}\,\wedge_{\mathsf{L}} \qquad \dfrac{\Gamma \Rightarrow \varphi \quad \Gamma \Rightarrow \psi}{\Gamma \Rightarrow \varphi \wedge \psi}\,\wedge_{\mathsf{R}} \qquad \dfrac{\Gamma\{\varnothing_+\} \Rightarrow \chi}{\Gamma\{\top\} \Rightarrow \chi}\,\top_{\mathsf{L}}$$

$$\dfrac{\Gamma\{\varphi\} \Rightarrow \chi \quad \Gamma\{\psi\} \Rightarrow \chi}{\Gamma\{\varphi \vee \psi\} \Rightarrow \chi}\,\vee_{\mathsf{L}} \qquad \dfrac{\Gamma \Rightarrow \varphi}{\Gamma \Rightarrow \varphi \vee \psi}\,\vee_{\mathsf{R1}} \qquad \dfrac{\Gamma \Rightarrow \psi}{\Gamma \Rightarrow \varphi \vee \psi}\,\vee_{\mathsf{R2}}$$

$$\dfrac{\Gamma' \Rightarrow \varphi \quad \Gamma\{\Gamma', \psi\}_+ \Rightarrow \chi}{\Gamma\{\Gamma', \varphi \to \psi\}_+ \Rightarrow \chi}\,\to_{\mathsf{L}} \qquad \dfrac{\{\Gamma, \varphi\}_+ \Rightarrow \psi}{\Gamma \Rightarrow \varphi \to \psi}\,\to_{\mathsf{R}} \qquad \dfrac{\Gamma\{\Gamma', \Gamma'\}_+ \Rightarrow \chi}{\Gamma\{\Gamma'\}_+ \Rightarrow \chi}\,\mathsf{C}$$

Fig. 2. Sequent Calculus $\eta\mathsf{LBI}$

Example 19. The following is a proof in $\eta\mathsf{LBI}$.

$$\dfrac{\dfrac{\dfrac{\dfrac{\dfrac{\dfrac{}{A \Rightarrow A}\,\mathsf{Ax} \quad \dfrac{}{\{B, C\}_+ \Rightarrow B}\,\mathsf{Ax}}{\{A, \{B, C\}_+\}_\times \Rightarrow A * B}\,*_{\mathsf{R}}}{\{A, (B \wedge C)\}_\times \Rightarrow A * B}\,\wedge_{\mathsf{L}} \quad \dfrac{\dfrac{}{A \Rightarrow A}\,\mathsf{Ax} \quad \dfrac{\dfrac{}{\{B, C\}_+ \Rightarrow C}\,\mathsf{Ax}}{B \wedge C \Rightarrow C}\,\wedge_{\mathsf{L}}}{\{A, (B \wedge C)\}_\times \Rightarrow A * C}\,*_{\mathsf{R}}}{\{A, (B \wedge C)\}_\times \Rightarrow (A * B) \wedge (A * C)}\,\wedge_{\mathsf{R}}}{A * (B \wedge C) \Rightarrow (A * B) \wedge (A * C)}\,*_{\mathsf{L}}}{\varnothing_\times \Rightarrow (A * (B \wedge C)) \mathrel{-\!\!*} ((A * B) \wedge (A * C))}\,\mathrel{-\!\!*}_{\mathsf{R}}$$

We expect no obvious difficulty in studying focused proof-search with bunches instead of nested multisets; the design choice is simply to reduce the complexity of the argument by pushing all uses of exchange (E) to Lemma 18, rather than tackle it at the same time as focusing itself. In particular, working without the nested system would mean working with a weaker notion of focusing since the exchange rule must then be permissible during both focused and unfocused phases of reduction.

3 A Focused System

At no point in this section will we refer to bunches, thus the variable Δ, so far reserved for elements of $\mathbb{B}$, is re-appropriated as an alternative to Γ.

3.1 Polarisation

Polarity in the focusing principle is determined by the invariance of provability under application of a rule, that is, by the proof rules themselves. One way the distinction between positive and negative connectives is apparent is when their rule behave either *synchronously* or *asynchronously*. For example, the $*_{\mathsf{R}}$ and $-\!\!*_{\mathsf{L}}$ highlight the synchronous behaviour of the multiplicative connectives since the structure of the context affects the applicability of the rule. Displaying such a synchronous behaviour on the left makes $-\!\!*$ a negative connective, while having it on the right makes $*$ a positive connective.

Another way to characterise the polarity of a connective is the study of the inveribility properties of the corresponding rules. For example, consider the inverses of the $\vee_{\mathsf{L}}$ rule,

$$\frac{\Gamma\{\varphi \vee \psi\} \Rightarrow \chi}{\Gamma\{\varphi\} \Rightarrow \chi}\ \vee^{\mathsf{inv}}_{\mathsf{L1}} \qquad \frac{\Gamma\{\varphi \vee \psi\} \Rightarrow \chi}{\Gamma\{\psi\} \Rightarrow \chi}\ \vee^{\mathsf{inv}}_{\mathsf{L2}}$$

They are derivable in LBI with cut (below – the left branch being closed using Lemma 9) and therefore admissible in LBI without cut (by Lemma 8).

$$\frac{\dfrac{\varphi \Rightarrow \varphi}{\varphi \Rightarrow \varphi \vee \psi}\ \vee_{\mathsf{R}} \quad \Gamma\{\varphi \vee \psi\} \Rightarrow \chi}{\Gamma\{\varphi\} \Rightarrow \chi}\ \mathsf{cut} \qquad \frac{\dfrac{\psi \Rightarrow \psi}{\psi \Rightarrow \varphi \vee \psi}\ \vee_{\mathsf{R}} \quad \Gamma\{\varphi \vee \psi\} \Rightarrow \chi}{\Gamma\{\psi\} \Rightarrow \chi}\ \mathsf{cut}$$

This means that provability is invariant in general upon application of $\vee_{\mathsf{L}}$ since it can always be reverted if needed, as follows

$$\frac{\dfrac{\Gamma\{\varphi \vee \psi\} \Rightarrow \chi}{\Gamma\{\varphi\} \Rightarrow \chi}\ \vee^{\mathsf{inv}}_{\mathsf{L1}} \quad \dfrac{\Gamma\{\varphi \vee \psi\} \Rightarrow \chi}{\Gamma\{\psi\} \Rightarrow \chi}\ \vee^{\mathsf{inv}}_{\mathsf{L2}}}{\Gamma\{\varphi \vee \psi\} \Rightarrow \chi}\ \vee_{\mathsf{L}}$$

Note however that dual connectives do not necessarily have dual behaviours in terms of provability invariance, on the left and on the right. For example, consider all the possible rules for $\wedge$, of which some qualify as positive and others as positive.

$$\frac{\Gamma\{\varphi\} \Rightarrow \chi}{\Gamma\{\varphi \wedge \psi\} \Rightarrow \chi}\ \wedge^{-}_{\mathsf{L1}} \quad \frac{\Gamma\{\psi\} \Rightarrow \chi}{\Gamma\{\varphi \wedge \psi\} \Rightarrow \chi}\ \wedge^{-}_{\mathsf{L2}} \qquad \frac{\Gamma \Rightarrow \varphi \quad \Gamma \Rightarrow \psi}{\Gamma \Rightarrow \varphi \wedge \psi}\ \wedge^{-}_{\mathsf{R}}$$

$$\frac{\Gamma\{\{\varphi, \psi\}_{+}\} \Rightarrow \chi}{\Gamma\{\varphi \wedge \psi\} \Rightarrow \chi}\ \wedge^{+}_{\mathsf{L}} \qquad \frac{\Gamma \Rightarrow \varphi \quad \Gamma\{\{\varphi \wedge \psi\}_{+}\} \Rightarrow \chi}{\{\Gamma, \Delta\}_{+} \Rightarrow \varphi \wedge \psi}\ \wedge^{+}_{\mathsf{R}}$$

All of these rules are sound, and replacing the conjunction rules in LBI with any pair of a left and right rule will result in a sound and complete system. Indeed, the rules are inter-derivable when the structural rules are present, but otherwise they can be paired to form two sets of rules which have essentially different proof-search behaviours. That is, the rules in the top-row make $\wedge$ negative while the bottom row make $\wedge$ positive. Each conjunction also comes with an associated unit, that is, $\top^{-}$ for negative conjunctio and $\top^{+}$ for positive conjunction. We choose to add all of them to our system in order to have access to those different proof search behaviours at will.

Finally, the polarity of the propositional letters can be assigned arbitrarily as long as only once for each.

Definition 20 (Polarised Syntax). *Let $\mathsf{P}^+ \sqcup \mathsf{P}^-$ be a partition of P, and let $A^+ \in \mathsf{P}^+$ and $A^- \in \mathsf{P}^-$, then the polarised formulas are defined by the following grammar,*

$$P, Q ::= L \mid P \vee Q \mid P * Q \mid P \wedge^+ Q \mid \top^+ \mid \top^* \mid \bot \qquad L ::= \downarrow N \mid A^+$$
$$N, M ::= R \mid P \to N \mid P \mathrel{-\!\!*} N \mid N \wedge^- M \mid \top^- \qquad R ::= \uparrow P \mid A^-$$

The set of positive formulas P is denoted $\mathbb{F}^+$; the set of negative formulas N is denoted $\mathbb{F}^-$; and the set of all polarised formulas is denoted $\mathbb{F}^\pm$. The sub-classifications L and R are left-neutral and right-neutral formulas respectfully.

The shift operators have no logical meaning; they simply mediate the exchange of polarity, and thus the *shifting* into a new phase of proof-search. Consequently, to reduces cases in subsequent proofs, we will consider formulas of the form $\uparrow\downarrow N$ and $\downarrow\uparrow P$, but not $\downarrow\uparrow\downarrow N$, $\downarrow\uparrow\downarrow\uparrow P$, etc.

Definition 21 (Depolarisation). *Let $\circ \in \{\vee, *, \to, \mathrel{-\!\!*}\}$, and let $A^+ \in \mathsf{P}^+$ and $A^- \in \mathsf{P}^-$, then the depolarisation function $\lfloor\cdot\rfloor : \mathbb{F}^\pm \to \mathbb{F}$ is defined as follows:*

$$\lfloor \top^+ \rfloor := \lfloor \top^- \rfloor := \top \qquad \lfloor \bot \rfloor := \bot \qquad \lfloor \top^* \rfloor := \top^*$$
$$\lfloor A^+ \rfloor := \lfloor A^- \rfloor := A \qquad \lfloor \uparrow\varphi \rfloor := \lfloor \downarrow\varphi \rfloor := \lfloor \varphi \rfloor$$
$$\lfloor \varphi \circ \psi \rfloor := \lfloor \varphi \rfloor \circ \lfloor \psi \rfloor \qquad \lfloor \varphi \wedge^+ \psi \rfloor := \lfloor \varphi \wedge^- \psi \rfloor := \lfloor \varphi \rfloor \wedge \lfloor \psi \rfloor$$

Since proof-search is controlled by polarity, the construction of sequents in the focused system must be handled carefully to avoid ambiguity.

Definition 22 (Polarised Sequents). Positive *and* neutral *nests, denoted by Γ and $\vec{\Gamma}$ resp., are defined according to the following grammars*

$$\Gamma := \Sigma \mid \Pi \qquad \Sigma := P \mid \{\Pi_1, ..., \Pi_n\}_+ \qquad \Pi := P \mid \{\Sigma_1, ..., \Sigma_n\}_\times$$
$$\vec{\Gamma} := \vec{\Sigma} \mid \vec{\Pi} \qquad \vec{\Sigma} := L \mid \{\vec{\Pi}_1, ..., \vec{\Pi}_n\}_+ \qquad \vec{\Pi} := L \mid \{\vec{\Sigma}_1, ..., \vec{\Sigma}_n\}_\times$$

A pair of a polarised nest and a polarised formula is a polarised sequent *if it falls into one of the following cases*

$$\Gamma \Rightarrow N \quad \mid \quad \vec{\Gamma} \Rightarrow \langle P \rangle \quad \mid \quad \vec{\Gamma}\{\langle N \rangle\} \Rightarrow R$$

The decoration $\langle\varphi\rangle$ indicates that the formula is in focus; that is, it is a positive formula on the right, or a negative formula on the left. Of the three possible cases for well-formed polarised sequents, the first may be called *unfocused*, with the particular case of being *neutral* when of the form $\vec{\Gamma} \Rightarrow R$; and the latter two may be called *focused*.

Definition 23 (Depolarised Nest). *The depolarisation map extends to polarised nests $\lfloor\cdot\rfloor : \mathbb{B}/\equiv^\pm \to \mathbb{B}/\equiv$ as follows:*

$$\lfloor \{\Pi_1, ..., \Pi_n\}_+ \rfloor = \{\lfloor \Pi_1 \rfloor, ..., \lfloor \Pi_n \rfloor\}_+ \qquad \lfloor \{\Sigma_1, ..., \Sigma_n\}_\times \rfloor = \{\lfloor \Sigma_1 \rfloor, ..., \lfloor \Sigma_n \rfloor\}_\times$$

Focused

$$\frac{}{\{\Gamma, A_+\}_+ \Rightarrow \langle A_+ \rangle}\,\mathsf{Ax}^+ \qquad \frac{}{\{\Gamma, \langle A_- \rangle\}_+ \Rightarrow A_-}\,\mathsf{Ax}^- \qquad \frac{}{\overrightarrow{\Gamma} \Rightarrow \langle \top^+ \rangle}\,\top^+_{\mathsf{R}}$$

$$\frac{}{\{\overrightarrow{\Gamma}, \downarrow\uparrow P\}_+ \Rightarrow \langle P \rangle}\,\mathsf{P} \qquad \frac{}{\{\overrightarrow{\Gamma}, \langle N \rangle\}_+ \Rightarrow \uparrow\downarrow N}\,\mathsf{N} \qquad \frac{}{\{\overrightarrow{\Gamma}, \varnothing_\times\}_+ \Rightarrow \langle \top^* \rangle}\,\top^*_{\mathsf{R}}$$

$$\frac{\overrightarrow{\Gamma}\{\langle N_i \rangle\}_+ \Rightarrow R}{\overrightarrow{\Gamma}\{\langle N_1 \wedge^- N_2 \rangle\}_+ \Rightarrow R}\,\wedge^-_{\mathsf{L}i} \qquad \frac{\overrightarrow{\Gamma} \Rightarrow \langle P_i \rangle}{\overrightarrow{\Gamma} \Rightarrow \langle P_1 \vee P_2 \rangle}\,\vee_{\mathsf{R}i} \qquad \frac{\overrightarrow{\Gamma}\{\varnothing_+\} \Rightarrow R}{\overrightarrow{\Gamma}\{\langle \top^- \rangle\} \Rightarrow R}\,\top^-_{\mathsf{L}}$$

$$\frac{\overrightarrow{\Gamma} \Rightarrow \langle P \rangle \quad \overrightarrow{\Gamma}' \Rightarrow \langle Q \rangle}{\{\overrightarrow{\Gamma}, \overrightarrow{\Gamma}'\}_+ \Rightarrow \langle P \wedge^+ Q \rangle}\,\wedge^+_{\mathsf{R}} \qquad \frac{\overrightarrow{\Delta} \Rightarrow \langle P \rangle \quad \overrightarrow{\Gamma}\{\overrightarrow{\Delta}, \langle N \rangle\}_+ \Rightarrow R}{\overrightarrow{\Gamma}\{\overrightarrow{\Delta}, \langle P \to N \rangle\}_+ \Rightarrow R}\,\to_{\mathsf{L}}$$

$$\frac{\overrightarrow{\Gamma} \Rightarrow \langle P \rangle \quad \overrightarrow{\Gamma}' \Rightarrow \langle Q \rangle}{\{\{\overrightarrow{\Gamma}, \overrightarrow{\Gamma}'\}_\times, \overrightarrow{\Gamma}''\}_+ \Rightarrow \langle P * Q \rangle}\,*_{\mathsf{R}} \qquad \frac{\overrightarrow{\Delta} \Rightarrow \langle P \rangle \quad \overrightarrow{\Gamma}\{\overrightarrow{\Delta}', \langle N \rangle\}_\times \Rightarrow R}{\overrightarrow{\Gamma}\{\overrightarrow{\Delta}, \overrightarrow{\Delta}', \{\overrightarrow{\Delta}'', \langle P \mathrel{-\!\!*} N \rangle\}_+\}_\times \Rightarrow R}\,\mathrel{-\!\!*}_{\mathsf{L}}$$

Neutral

$$\frac{\overrightarrow{\Gamma} \Rightarrow \langle P \rangle}{\overrightarrow{\Gamma} \Rightarrow \uparrow P}\,\uparrow_{\mathsf{R}} \qquad \frac{\overrightarrow{\Gamma}\{P\} \Rightarrow R}{\overrightarrow{\Gamma}\{\langle \uparrow P \rangle\} \Rightarrow R}\,\uparrow_{\mathsf{L}} \qquad \frac{\overrightarrow{\Gamma} \Rightarrow N}{\overrightarrow{\Gamma} \Rightarrow \langle \downarrow N \rangle}\,\downarrow_{\mathsf{R}} \qquad \frac{\overrightarrow{\Gamma}\{\langle N \rangle\} \Rightarrow R}{\overrightarrow{\Gamma}\{\downarrow N\} \Rightarrow R}\,\downarrow_{\mathsf{L}}$$

$$\frac{\overrightarrow{\Gamma}\{\{\overrightarrow{\Delta}, \overrightarrow{\Delta}\}_+\} \Rightarrow R}{\overrightarrow{\Gamma}\{\overrightarrow{\Delta}\} \Rightarrow R}\,\mathsf{C}$$

Unfocused

$$\frac{}{\Gamma \Rightarrow \top^-}\,\top^-_{\mathsf{R}} \qquad \frac{}{\Gamma\{\bot\} \Rightarrow N}\,\bot_{\mathsf{L}}$$

$$\frac{\Gamma \Rightarrow N \quad \Gamma \Rightarrow M}{\Gamma \Rightarrow N \wedge^- M}\,\wedge^-_{\mathsf{R}} \qquad \frac{\Gamma\{P\} \Rightarrow N \quad \Gamma\{Q\} \Rightarrow N}{\Gamma\{P \vee Q\} \Rightarrow N}\,\vee_{\mathsf{L}}$$

$$\frac{\Gamma\{\{P, Q\}_+\} \Rightarrow N}{\Gamma\{P \wedge^+ Q\} \Rightarrow N}\,\wedge^+_{\mathsf{L}} \qquad \frac{\{\Gamma, P\}_+ \Rightarrow N}{\Gamma \Rightarrow P \to N}\,\to_{\mathsf{R}} \qquad \frac{\Gamma\{\varnothing_+\} \Rightarrow N}{\Gamma\{\top^+\} \Rightarrow N}\,\top^+_{\mathsf{L}}$$

$$\frac{\Gamma\{\{P, Q\}_\times\} \Rightarrow N}{\Gamma\{P * Q\} \Rightarrow N}\,*_{\mathsf{L}} \qquad \frac{\{\Gamma, P\}_\times \Rightarrow N}{\Gamma \Rightarrow P \mathrel{-\!\!*} N}\,\mathrel{-\!\!*}_{\mathsf{R}} \qquad \frac{\Gamma\{\varnothing_\times\} \Rightarrow N}{\Gamma\{\top^*\} \Rightarrow N}\,\top^*_{\mathsf{L}}$$

Fig. 3. System fBI

3.2 Focused Calculus

We may now give the focused system. That is, the operational semantics for focused proof-search in LBI. All the rules, with the exception of P and N, are polarised versions of the rules from ηLBI.

Definition 24 (System fBI). *The focused system* fBI *is composed of the rules on Figure 3.*

Note the absence of a cut-rule, this is because the above system is intended to encapsulate precisely *focused* proof-search. Below we show that a cut-rule is indeed admissible, but proofs in fBI + cut are not necessarily focused themselves. Here the distinction between the methodologies for establishing the focusing principle becomes present since one may show completeness without leaving fBI by a permutation argument instead of a cut-elimination one.

The P and N rules will allow us to move a formula from one side to another during the proof of the completeness of fBI + cut (Lemma 34).The depolarised version are not directly present in LBI, but are derivable in LBI (Lemma 9). However, the way they are focused renders them not provable in fBI because it forces one to begin with a potentially *bad* choice; for example, $A \vee B \Rightarrow A \vee B$ has no proof beginning with $\vee_\mathsf{R}$. In practice, they are a feature rather than a bug since they allow one to terminate proof-search early, without unnecessary further expansion of the axiom. In related works, such as [6,5], the analogous rules are eliminated by initially working with a weaker notion of focused proof-search, and it is reasonable to suppose that the same may be true for BI. We leave this to future investigation.

Note also that, although it is perhaps proof-theoretically displeasing to incorporate weakening into the operational rules as in $-\!*'_\mathsf{L}$ and $*'_\mathsf{R}$, it has good computational behaviour during focused proof-search since the reduction of $\varphi -\!* \psi$ can only arise out of an explicit choice made earlier in the computation.

Soundness follows immediately from the depolarisation map; that is, the interpretation of polarised sequents as nested sequents, and hence proofs in fBI actually are focused proofs in ηLBI.

Theorem 25 (Soundness of fBI). *Let Γ be a polarised nest and N a negative formula. If $\vdash_{\mathsf{fBI}} \Gamma \Rightarrow N$ then $\vdash_{\eta\mathsf{LBI}} \lfloor \Gamma \rceil \Rightarrow \lfloor N \rfloor$*

Proof. Every rule in fBI except the shift rules, as well as the P and N axioms, become a rule in ηBI when the antecedent(s) and consequent are depolarised. Instance of the shift rule can be ignored since the depolarised versions of the consequent and antecedents are the same. Finally, the depolarised versions of P and N follow from Lemma 9 with the use of some weakening. □

Example 26. Consider the following proof in fBI, we suppose here that propositional letters A and C are negative, but B is positive.

$$
\begin{prooftree}
\AxiomC{}
\RightLabel{Ax^-}
\UnaryInfC{$\langle A\rangle \Rightarrow A$}
\RightLabel{$\downarrow_\mathsf{L}$}
\UnaryInfC{$\downarrow A \Rightarrow A$}
\RightLabel{$\downarrow_\mathsf{R}$}
\UnaryInfC{$\downarrow A \Rightarrow \langle \downarrow A\rangle$}
\AxiomC{}
\RightLabel{Ax^+}
\UnaryInfC{$B \Rightarrow \langle B\rangle$}
\RightLabel{$*_\mathsf{R}$}
\BinaryInfC{$\{\downarrow A, B\}_\times \Rightarrow \langle \downarrow A * B\rangle$}
\RightLabel{$\uparrow_\mathsf{R}$}
\UnaryInfC{$\{\downarrow A, B\}_\times \Rightarrow \uparrow(\downarrow A * B)$}
\RightLabel{$\uparrow_\mathsf{L}$}
\UnaryInfC{$\{\downarrow A, \langle \uparrow B\rangle\}_\times \Rightarrow \uparrow(\downarrow A * B)$}
\RightLabel{$\wedge^-_{\mathsf{L1}}$}
\UnaryInfC{$\{\downarrow A, \langle \uparrow B \wedge^- C\rangle\}_\times \Rightarrow \uparrow(\downarrow A * B)$}
\RightLabel{$\uparrow_\mathsf{L}$ (1)}
\UnaryInfC{$\{\downarrow A, \downarrow(\uparrow B \wedge^- C)\}_\times \Rightarrow \uparrow(\downarrow A * B)$}
\AxiomC{}
\RightLabel{Ax^-}
\UnaryInfC{$\langle A\rangle \Rightarrow A$}
\RightLabel{$\downarrow_\mathsf{L}$}
\UnaryInfC{$\downarrow A \Rightarrow A$}
\RightLabel{$\downarrow_\mathsf{R}$}
\UnaryInfC{$\downarrow A \Rightarrow \langle \downarrow A\rangle$}
\AxiomC{}
\RightLabel{Ax^-}
\UnaryInfC{$\langle C\rangle \Rightarrow C$}
\RightLabel{$\wedge^-_{\mathsf{L2}}$}
\UnaryInfC{$\langle \uparrow B \wedge^- C\rangle \Rightarrow C$}
\RightLabel{$\downarrow_\mathsf{L}$}
\UnaryInfC{$\downarrow(\uparrow B \wedge^- C) \Rightarrow C$}
\RightLabel{$\downarrow_\mathsf{R}$}
\UnaryInfC{$\downarrow(\uparrow B \wedge^- C)\}_\times \Rightarrow \langle \downarrow C\rangle$}
\RightLabel{$*_\mathsf{R}$}
\BinaryInfC{$\{\downarrow A, \downarrow(\uparrow B \wedge^- C)\}_\times \Rightarrow \langle \downarrow A * \downarrow C\rangle$}
\RightLabel{$\uparrow_\mathsf{R}$ (2)}
\UnaryInfC{$\{\downarrow A, \downarrow(\uparrow B \wedge^- C)\}_\times \Rightarrow \uparrow(\downarrow A * \downarrow C)$}
\RightLabel{$\wedge^-_\mathsf{R}$}
\BinaryInfC{$\{\downarrow A, \downarrow(\uparrow B \wedge^- C)\}_\times \Rightarrow \uparrow(\downarrow A * B) \wedge^- \uparrow(\downarrow A * \downarrow C)$}
\RightLabel{$*_\mathsf{L}$}
\UnaryInfC{$\downarrow A * \downarrow(\uparrow B \wedge^- C) \Rightarrow \uparrow(\downarrow A * B) \wedge^- \uparrow(\downarrow A * \downarrow C)$}
\RightLabel{$-\!*_\mathsf{R}$}
\UnaryInfC{$\varnothing_\times \Rightarrow (\downarrow A * \downarrow(\uparrow B \wedge^- C)) -\!* (\uparrow(\downarrow A * B) \wedge^- \uparrow(\downarrow A * \downarrow C))$}
\end{prooftree}
$$

It is a focused version of the proof given in Example 19. Observe that the only non-deterministic choices are which formula to focus on, such as in steps (1) and

(2), where different choices have been made for the sake of demonstration. The point of focusing is that *only* at such points do choices that affect termination occur. The assignment of polarity to the propositional letters is what forced the shape of the proof; for example, if B had been negative the above would not have been well-formed. This phenomenon is standardly observed in focused systems (e.g. [7]).

We now introduce the tool which will allow us to show that if there is a proof of a sequent (*a priori* unstructured), then there is necessarily a focused one.

Definition 27. *All instances of the following rule where the sequents are well-formed are instances of* cut, *where* $\overrightarrow{\varphi}$ *denotes that* φ *is possibly prenexed with an additional shift*

$$\frac{\Delta \Rightarrow \varphi \quad \Gamma\{\overrightarrow{\varphi}\} \Rightarrow \chi}{\Gamma\{\Delta\} \Rightarrow \chi}\ \mathsf{cut}$$

Admissibility follows from the usual argument, but within the focused system; that is, through the upward permutation of cuts until they are eliminated in the axioms or are reduced in some other measure.

Definition 28 (Good and Bad Cuts). *Let* $\mathcal{D}$ *be a* $\mathsf{fBI}+\mathsf{cut}$ *proof, a cut is a quadruple* $\langle \mathcal{L}, \mathcal{R}, \mathcal{C}, \varphi \rangle$ *where* $\mathcal{L}$ *and* $\mathcal{R}$ *are the premises to a* cut *rule, concluding* $\mathcal{C}$ *in* $\mathcal{D}$, *and* φ *is the* cut*-formula. They are classified as follows:*

Good *- If* φ *is principal in both* $\mathcal{L}$ *and* $\mathcal{R}$.
Bad *- If* φ *is not principal in one of* $\mathcal{L}$ *and* $\mathcal{R}$.
Type 1: If φ *is not principal in* $\mathcal{L}$.
Type 2: If φ *is not principal in* $\mathcal{R}$.

Definition 29 (Cut Ordering). *The* cut*-rank of a cut* $\langle \mathcal{L}, \mathcal{R}, \mathcal{C}, \varphi \rangle$ *in a proof is the triple* ⟨cut*-complexity,* cut*-duplicity,* cut*-level*⟩, *where the* cut*-complexity is the size of* φ, *the* cut*-duplicity is the number of contraction instances above the cut, the* cut*-level is the sum of the heights of the sub-proofs concluding* $\mathcal{L}$ *and* $\mathcal{R}$.

Let $\mathcal{D}$ *and* $\mathcal{D}'$ *be two* $\mathsf{fBI}+\mathsf{cut}$ *proofs, let* σ *and* σ' *denote their multiset of cuts respectively. Proofs are ordered by* $\mathcal{D} \prec \mathcal{D}' \iff \sigma < \sigma'$, *where* $<$ *is the multiset ordering derived from the lexicographic ordering on* cut*-rank.*

It follows from a result in [10] that the ordering on proofs is a well-order, since the ordering on cuts is a well-order.

Lemma 30 (Good Cuts Elimination). *Let* $\mathcal{D}$ *be a* $\mathsf{fBI}+\mathsf{cut}$ *proof of* S; *there is a* $\mathsf{fBI}+\mathsf{cut}$ *proof* $\mathcal{D}'$ *of* S *containing no good cuts such that* $\mathcal{D}' \preceq \mathcal{D}$.

Proof. Let $\mathcal{D}$ be as in hypothesis, if it contains no good cuts then $\mathcal{D} = \mathcal{D}'$ gives the desired proof. Otherwise, there is at least one good cut $\langle \mathcal{L}, \mathcal{R}, \mathcal{C}, \varphi \rangle$. Let ∂ be the sub-proof in $\mathcal{D}$ concluding $\mathcal{C}$, then there is a transformation $\partial \mapsto \partial'$ where ∂' is a $\mathsf{fBI}+\mathsf{cut}$ proof of S with $\partial' \prec \partial$ such that the multiset of good cuts in ∂' is smaller (with respect to $\prec$) than the multiset of good cuts in ∂. Since $\prec$

is a well-order indefinitely replacing ∂ with ∂' in $\mathcal{D}$ for various cuts yields the desired $\mathcal{D}'$.

The key step is that a cut of a certain cut-complexity is replaced by cuts of lower cut-complexity, possibly increasing the cut-duplicity or cut-level of other cuts in the proof, but not modifying their complexity.

$$\dfrac{\dfrac{}{\{\vec{\Gamma}', A^+\}_+ \Rightarrow \langle A^+\rangle}\,\mathsf{Ax}^+ \qquad \vec{\Gamma}\{A^+\} \Rightarrow \langle A^+\rangle}{\vec{\Gamma}\{\{\vec{\Gamma}', A^+\}_+\} \Rightarrow \langle A^+\rangle}\,\mathsf{cut} \quad \mapsto \quad \dfrac{\vec{\Gamma}\{A^+\} \Rightarrow \langle A^+\rangle}{\vec{\Gamma}\{\{\vec{\Gamma}', A^+\}_+\} \Rightarrow \langle A^+\rangle}\,\mathsf{W}$$

$$\dfrac{\dfrac{\{\vec{\Delta}''', P\}_\times \Rightarrow N}{\vec{\Delta}''' \Rightarrow P \mathrel{-\!\!*} N}\,\mathrel{-\!\!*}_{\mathsf{R}} \qquad \dfrac{\vec{\Delta} \Rightarrow \langle P\rangle \quad \Gamma\{\vec{\Delta}', \langle N\rangle\}_\times \Rightarrow R}{\vec{\Gamma}\{\vec{\Delta}, \vec{\Delta}', \{\vec{\Delta}'', \langle P \mathrel{-\!\!*} N\rangle\}_+\}_\times \Rightarrow R}\,\mathrel{-\!\!*}_{\mathsf{L}}}{\vec{\Gamma}\{\vec{\Delta}, \vec{\Delta}', \{\vec{\Delta}'', \vec{\Delta}'''\}_+\}_\times \Rightarrow R}\,\mathsf{cut}$$

$$\mapsto \quad \dfrac{\dfrac{\vec{\Delta} \Rightarrow \langle P\rangle \qquad \dfrac{\{\vec{\Delta}'', P\}_\times \Rightarrow N \quad \vec{\Gamma}\{\vec{\Delta}', \langle N\rangle\}_\times \Rightarrow R}{\vec{\Gamma}\{\vec{\Delta}, \vec{\Delta}'', P\}_\times \Rightarrow R}\,\mathsf{cut}}{\vec{\Gamma}\{\vec{\Delta}, \vec{\Delta}', \vec{\Delta}''\}_\times \Rightarrow R}\,\mathsf{cut}}{\vec{\Gamma}\{\vec{\Delta}, \vec{\Delta}', \{\vec{\Delta}'', \vec{\Delta}'''\}_+\}_\times \Rightarrow R}\,\mathsf{W}$$

We denote by a double-line the fact that we do not actually use a weakening, but only the fact that it is admissible in fBI by construction (Lemma 10). □

Lemma 31 (Bad Cuts Elimination). *Let $\mathcal{D}$ be a* fBI + cut *proof of S that contains only one cut which is bad, then there is a* fBI + cut *proof $\mathcal{D}'$ of S such that $\mathcal{D}' \prec \mathcal{D}$.*

Proof. Without loss of generality suppose the cut is the last inference in the proof, then it may be replaced by other cuts whose cut-level or cut-duplicity is smaller, but with same cut-complexity.

First we consider bad cuts when $\mathcal{L}$ and $\mathcal{R}$ are both axioms. There are no Type 1 bad cuts on axioms as the formula is always principal, meanwhile the Type 2 bad cuts can trivially be permuted upwards or ignored; for example,

$$\dfrac{\dfrac{}{\{\vec{\Delta}''', A_+\}_+ \Rightarrow \langle A_+\rangle}\,\mathsf{Ax}^+ \qquad \dfrac{\vec{\Delta} \Rightarrow \langle P\rangle \quad \vec{\Gamma}\{\vec{\Delta}', \langle N\rangle\}_\times \Rightarrow R}{\vec{\Gamma}\{\vec{\Delta}, \vec{\Delta}', \{\vec{\Delta}'', A_+, \langle P \mathrel{-\!\!*} N\rangle\}_+\}_\times \Rightarrow R}\,\mathrel{-\!\!*}_{\mathsf{L}}}{\vec{\Gamma}\{\vec{\Delta}, \vec{\Delta}', \{\vec{\Delta}'', \vec{\Delta}''', A_+, \langle P \mathrel{-\!\!*} N\rangle\}_+\}_\times \Rightarrow R}\,\mathsf{cut}$$

$$\mapsto \quad \dfrac{\dfrac{\vec{\Delta} \Rightarrow \langle P\rangle \quad \vec{\Gamma}\{\vec{\Delta}', \langle N\rangle\}_\times \Rightarrow R}{\vec{\Gamma}\{\vec{\Delta}, \vec{\Delta}', \{\vec{\Delta}'', A_+, \langle P \mathrel{-\!\!*} N\rangle\}_+\}_\times \Rightarrow R}\,\mathrel{-\!\!*}_{\mathsf{L}}}{\vec{\Gamma}\{\vec{\Delta}, \vec{\Delta}', \{\vec{\Delta}'', \vec{\Delta}''', A_+, \langle P \mathrel{-\!\!*} N\rangle\}_+\}_\times \Rightarrow R}\,\mathsf{W}$$

Here again we are using an appropriate version of Lemma 10.

For the remaining cases the cuts are commutative in the sense that they may be permuted upward thereby reducing the cut-level. An example is given below.

$$\dfrac{\dfrac{\vec{\Delta}\{\langle N_1\rangle\} \Rightarrow M}{\vec{\Delta}\{\langle N_1 \wedge^- N_2\rangle\} \Rightarrow M}\,\wedge^-_{\mathsf{L1}} \qquad \vec{\Gamma}\{M\} \Rightarrow R}{\vec{\Gamma}\{\vec{\Delta}\{\langle N_1 \wedge^- N_2\rangle\}\} \Rightarrow R}\,\mathsf{cut} \quad \mapsto \quad \dfrac{\dfrac{\Delta\{\langle N_1\rangle\} \Rightarrow M \quad \Gamma\{M\} \Rightarrow R}{\Gamma\{\Delta\{\langle N_1\rangle\}\} \Rightarrow R}\,\mathsf{cut}}{\Gamma\{\Delta\{\langle N_1 \wedge^- N_2\rangle\}\} \Rightarrow R}\,\wedge^-_{\mathsf{L1}}$$

The exceptional case is the interaction with contraction where the cut is replaced by cuts of possibly equal cut-level, but cut-duplicity decreases.

$$\dfrac{\vec{\Delta}' \Rightarrow \langle L\rangle \quad \dfrac{\vec{\Gamma}\{\{\vec{\Delta}\{L\}, \vec{\Delta}\{L\}\}_+\} \Rightarrow R}{\vec{\Gamma}\{\vec{\Delta}\{L\}\} \Rightarrow R}\,\mathsf{C}}{\vec{\Gamma}\{\vec{\Delta}\{\vec{\Delta}'\}\} \Rightarrow R}\,\mathsf{cut}$$

$$\mapsto \quad \dfrac{\dfrac{\vec{\Delta}' \Rightarrow \langle L\rangle \quad \dfrac{\vec{\Delta}' \Rightarrow \langle L\rangle \quad \vec{\Gamma}\{\{\vec{\Delta}\{L\}, \vec{\Delta}\{L\}\}_+\} \Rightarrow R}{\vec{\Gamma}\{\{\vec{\Delta}\{\vec{\Delta}'\}, \vec{\Delta}\{L\}\}_+\} \Rightarrow R}\,\mathsf{cut}}{\vec{\Gamma}\{\{\vec{\Delta}\{\vec{\Delta}'\}, \vec{\Delta}\{\vec{\Delta}'\}\}_+\} \Rightarrow R}\,\mathsf{cut}}{\vec{\Gamma}\{\vec{\Delta}\{\vec{\Delta}'\}\} \Rightarrow R}\,\mathsf{C}$$

□

Theorem 32 (Cut-elimination in fBI). *Let Γ be a positive nest and N a negative formula. Then, $\vdash_{\mathsf{fBI}} \Gamma \Rightarrow N$ if and only if $\vdash_{\mathsf{fBI+cut}} \Gamma \Rightarrow N$.*

Proof. ($\Rightarrow$) Trivial as any fBI-proof is a fBI + cut-proof. ($\Leftarrow$) Let $\mathcal{D}$ be a fBI + cut-proof of $\Gamma \Rightarrow N$, if it has no cuts then it is a fBI-proof so we are done. Otherwise, there is at least one cut, and we proceed by well-founded induction on the ordering of proofs and sub-proofs of $\mathcal{D}$ with respect to $\prec$.

Base Case. Assume $\mathcal{D}$ is minimal with respect to $\prec$ with at least one cut; without loss of generality, by Lemma 30, assume the cut is bad. It follows from Lemma 31 that there is a proof strictly smaller in $\prec$-ordering, but this proof must be cut-free as $\mathcal{D}$ is minimal.

Inductive Step. Let $\mathcal{D}$ be as in the hypothesis, then by Lemma 30 there is a proof ∂ of $\Gamma \Rightarrow N$ containing no good cuts such that $\mathcal{D}' \preceq \mathcal{D}$. Either $\mathcal{D}'$ is cut-free and we are done, or it contains bad cuts. Consider the topmost cut, and denote the sub-proof by ∂, it follows from Lemma 31 that there is a proof ∂' of the same sequent such that $\partial' \prec \partial$. Hence, by inductive hypothesis, there is a cut-free proof the sequent and replacing ∂ by this proof in $\mathcal{D}$ gives a proof of $\Gamma \Rightarrow \varphi$ strictly smaller in $\prec$-ordering, thus by inductive hypothesis there is a cut-free proof as required. □

3.3 Completeness of fBI

The completeness theorem of the focused system, the operational semantics, is with respect to an interpretation (i.e. a polarisation). Indeed, any polarisation may be considered; for example, both $(\downarrow A^- * B^+) \wedge^+ \downarrow A^-$ and $\downarrow(A^+ * \downarrow B^-) \wedge^+ A^+$ are correct polarised versions of the formulas $(A * B) \wedge A$. Taking arbitrary φ the process is as follows: first, fix a polarised syntax (i.e. a partition of the propositional letters into positive and negative sets), then assign a polarity to φ with the following steps:

- If φ is a propositional atom, it must be polarised by default;
- If $\varphi = \top$, then *choose* polarisation $\top^+$ or $\top^-$;
- If $\varphi = \psi_1 \wedge \psi_2$, first polarise ψ_1 and ψ_2, then *choose* an additive conjunction and combine accordingly, using shifts to ensure the formula is well-formed;

– If $\varphi = \psi_1 \circ \psi_2$ where $\circ \in \{*, -\!\!*, \to, \vee\}$, then polarise ψ_1 and ψ_2 and combine with $\circ$ accordingly, using shifts where necessary.

Example 33. Suppose A is negative and B is positive, then $(A * B) \wedge A$ may be polarised by choosing the additive conjunction to be positive resulting in $(\downarrow A * B) \wedge^+ \downarrow A$ (when $\downarrow(A * \downarrow B) \wedge^+ A)$ would not be well-formed). Choosing to shift one can ascribe a negative polarisation $\uparrow((\downarrow A * B) \wedge^+ \downarrow A)$.

The above generates the set of all such polarised formulas when all possible choices are explored. The free assignment of polarity to formulas means several distinct focusing procedures are captured by the completeness theorem.

Lemma 34 (Completeness of $\mathsf{fBI}+\mathsf{cut}$). *For any unfocused sequent $\Gamma \Rightarrow N$, if $\vdash_{\eta\mathsf{LBI}} \lfloor \Gamma \Rightarrow N \rfloor$ then $\vdash_{\mathsf{fBI+cut}} \Gamma \Rightarrow N$.*

Proof. We show that every rule in $\eta\mathsf{LBI}$ is derivable in $\mathsf{fBI}+\mathsf{cut}$, consequently every proof in $\eta\mathsf{LBI}$ may be simulated; hence, every provable sequent has a focused proof. For unfocused rules $\to_{\mathsf{R}}, -\!\!*_{\mathsf{R}}, \wedge^-_{\mathsf{R}}, \wedge^+_{\mathsf{L}}, \vee_{\mathsf{L}}, *_{\mathsf{L}}, \bot_{\mathsf{L}}, \top^-_{\mathsf{R}}, \top^+_{\mathsf{L}}, \top^*_{\mathsf{L}}$, this is immediate; as well as for Ax and C. Below we give an example on how to simulate a focused rule.

Where it does not matter (e.g. in the case of inactive nests), we do not distinguish the polarised and unpolarised versions; each of the simulations can be closed thanks to the presence of the P and N rules in fBI.

$\dfrac{\Gamma \Rightarrow \varphi \quad \Delta \Rightarrow \psi}{\{\{\Gamma, \Delta\}_\times, \Delta'\}_+ \Rightarrow \varphi * \psi}\,*_{\mathsf{R}}$ in $\eta\mathsf{LBI}$ is simulated in $\mathsf{fBI}+\mathsf{cut}$ by

$$\dfrac{\dfrac{\dfrac{\Gamma \Rightarrow \uparrow\varphi^+}{\Gamma \Rightarrow \langle\downarrow\uparrow\varphi^+\rangle}\,\downarrow_{\mathsf{R}} \quad \dfrac{\Delta \Rightarrow \uparrow\psi^+}{\Delta \Rightarrow \langle\downarrow\uparrow\psi^+\rangle}\,\downarrow_{\mathsf{R}}}{\{\Gamma, \Delta\}_\times \Rightarrow \langle\downarrow\uparrow\varphi^+ * \downarrow\uparrow\psi^+\rangle}\,*_{\mathsf{R}} \quad \dfrac{\dfrac{\dfrac{\dfrac{}{\downarrow\uparrow\varphi^+ \Rightarrow \langle\varphi^+\rangle}\,\mathsf{P} \quad \dfrac{}{\downarrow\uparrow\psi^+ \Rightarrow \langle\psi^+\rangle}\,\mathsf{P}}{\{\{\downarrow\uparrow\varphi^+, \downarrow\uparrow\psi^+\}_\times, \Delta'\}_+ \Rightarrow \langle\varphi^+ * \psi^+\rangle}\,*_{\mathsf{R}}}{\{\{\downarrow\uparrow\varphi^+, \downarrow\uparrow\psi^+\}_\times, \Delta'\}_+ \Rightarrow \uparrow(\varphi^+ * \psi^+)}\,\uparrow_{\mathsf{R}}}{\{\downarrow\uparrow\varphi^+ * \downarrow\uparrow\psi^+, \Delta'\}_+ \Rightarrow \uparrow(\varphi^+ * \psi^+)}\,*_{\mathsf{L}}}{\{\{\Gamma, \Delta\}_\times, \Delta'\}_+ \Rightarrow \uparrow(\varphi^+ * \psi^+)}\,\mathsf{cut}$$

□

Theorem 35 (Completeness of fBI). *For any unfocused $\Gamma \Rightarrow N$, if $\vdash_{\eta\mathsf{LBI}} \lfloor \Gamma \Rightarrow N \rfloor$ then $\vdash_{\mathsf{fBI}} \Gamma \Rightarrow N$.*

Proof. It follows from Lemma 34 that there is a proof of $\Gamma \Rightarrow N$ in $\mathsf{fBI}+\mathsf{cut}$, and then it follows from Lemma 32 that there is a proof of $\Gamma \Rightarrow N$ in fBI. □

Given an arbitrary sequent the above theorem guarantees the existence of a focused proof, thus the focusing principle holds for $\eta\mathsf{LBI}$ and therefore for LBI.

4 Conclusion

By proving the completeness of a focused sequent calculus for the logic of Bunched Implications, we have demonstrated that it satisfies the focusing principle; that is, any polarisation of a BI-provable sequent can be proved following a focused search procedure. This required a careful analysis of how to restrict the usage of structural rules. In particular, we had to fully develop the congruence-invariant representation of bunches as nested multisets (originally proposed in [12]) to treat the exchange rule within bunched structures.

Proof-theoretically the completeness of the focused systems suggests a syntactic orderliness of LBI, though the P and N rules leave something to be desired. Computationally, these axioms are unproblematic as during search it makes sense to terminate a branch as soon as possible; however, unless they may be eliminated it means that the focusing principle holds in BI only up to a point. In related works (c.f. [6]) the analogous problem is overcome by first considering a *weak* focused system; that is, one where the structural rules are not controlled and unfocused rules may be performed inside focused phases if desired. Completeness of (strong) focusing is achieved by appealing to a *synthetic* system. It seems reasonable to suppose the same can be done for BI, resulting in a more proof-theoretically satisfactory focused calculus, exploring this possibility is a natural extension of the work on fBI.

The methodology employed for proving the focusing principle can be interpreted as soundness and completeness of an operational semantics for goal-directed search. The robustness of this technique is demonstrated by its efficacy in modal [6,5] and substructural logics [26], including now bunched ones. Although BI may be the most employed bunched logic, there are a number of others, such as the family of relevant logics [36], and the family of bunched logics [11], for which the focusing principle should be studied. However, without the presence of a cut-free sequent calculus goal-directed search becomes unclear, and currently such calculi do not exist for the two main variants of BI: Boolean BI [33] and Classical BI [4]. On the other hand, large families of bunched and substructural logics have been given hypersequent calculi [8,9]. Effective proof-search procedures have been established for the hypersequent calculi in the substructural case [35], but not the bunched one, and focused proof-search for neither. There is a technical challenge in focusing these systems as one must not only decide which formula to reduce, but also which sequent.

In the future it will be especially interesting to see how focused search, when combined with the expressiveness of BI, increases its modelling capabilities. Indeed, the dynamics of proof-search can be used to represent models of computation within (propositional) logics; for example, the undecidability of Linear Logic involves simulating two-counter machines [26]. One particularly interesting direction is to see how focused proof-search in BI may prove valuable within the context of Separation Logic. Focused systems in particular have been used to emulate proofs for other logics [27]; and to give structural operational semantics for systems used in industry, such as algorithms for solving constraint satisfaction problems [14]. A more immediate possibility though is the formulation of a theorem prover; we leave providing specific implementation or benchmarks to future research.

References

1. Andreoli, J.: Logic programming with focusing proofs in linear logic. Journal of Logic and Computation **2**, 297–347 (1992)
2. Armelin, P.: Bunched Logic Programming. Ph.D. thesis, Queen Mary College, University of London (2002)
3. Brotherston, J.: A unified display proof theory for bunched logic. Electronic Notes in Theoretical Computer Science **265**, 197 – 211 (2010), proceedings of the 26th Conference on the Mathematical Foundations of Programming Semantics (MFPS 2010)
4. Brotherston, J., Calcagno, C.: Classical BI: Its semantics and proof theory. Logical Methods in Computer Science **6** (05 2010). https://doi.org/10.2168/LMCS-6(3:3)2010
5. Chaudhuri, K., Marin, S., Straßburger, L.: Focused and synthetic nested sequents. In: Foundations of Software Science and Computation. pp. 390–407 (04 2016)
6. Chaudhuri, K., Marin, S., Straßburger, L.: Modular focused proof systems for intuitionistic modal logics. In: FSCD 2016 - 1st International Conference on Formal Structures for Computation and Deduction (2016)
7. Chaudhuri, K., Pfenning, F., Price, G.: A logical characterization of forward and backward chaining in the inverse method. In: Furbach, U., Shankar, N. (eds.) Automated Reasoning. pp. 97–111. Springer Berlin Heidelberg, Berlin, Heidelberg (2006)
8. Ciabattoni, A., Galatos, N., Terui, K.: Algebraic proof theory for substructural logics: Cut-elimination and completions. Annals of Pure and Applied Logic **163**(3), 266–290 (Mar 2012). https://doi.org/10.1016/j.apal.2011.09.003, http://dx.doi.org/10.1016/j.apal.2011.09.003
9. Ciabattoni, A., Ramanayake, R.: Bunched hypersequent calculi for distributive substructural logics. In: Eiter, T., Sands, D. (eds.) LPAR-21. 21st International Conference on Logic for Programming, Artificial Intelligence and Reasoning. EPiC Series in Computing, vol. 46, pp. 417–434. EasyChair (2017). https://doi.org/10.29007/ngp3, https://easychair.org/publications/paper/sr2D
10. Dershowitz, N., Manna, Z.: Proving termination with multiset orderings. In: Maurer, H.A. (ed.) Automata, Languages and Programming. pp. 188–202. Springer Berlin Heidelberg, Berlin, Heidelberg (1979)
11. Docherty, S.: Bunched Logics: A Uniform Approach. Ph.D. thesis, University College London (2019)
12. Donnelly, K., Gibson, T., Krishnaswami, N., Magill, S., Park, S.: The inverse method for the logic of bunched implications. In: Baader, F., Voronkov, A. (eds.) Logic for Programming, Artificial Intelligence, and Reasoning. pp. 466–480. Springer Berlin Heidelberg, Berlin, Heidelberg (2005)
13. Dyckhoff, R., Lengrand, S.: LJQ: A strongly focused calculus for intuitionistic logic. In: Beckmann, A., Berger, U., Löwe, B., Tucker, J.V. (eds.) Logical Approaches to Computational Barriers. pp. 173–185. Springer Berlin Heidelberg, Berlin, Heidelberg (2006)
14. Farooque, M., Graham-Lengrand, S., Mahboubi, A.: A bisimulation between DPLL(T) and a proof-search strategy for the focused sequent calculus. In: Proceedings of the Eighth ACM SIGPLAN international workshop on Logical frameworks & meta-languages: theory & practice. p. 3–14. LFMTP '13, Association for Computing Machinery, New York, NY, USA (2013). https://doi.org/10.1145/2503887.2503892

15. Gabbay, D.: Fibring Logics. Oxford Logic Guides, Clarendon Press (1998), https://books.google.co.uk/books?id=mpA1uUV-uYsC
16. Galmiche, D., Méry, D.: Semantic labelled tableaux for propositional BI. J. Log. Comput. **13**, 707–753 (2003)
17. Galmiche, D., Méry, D., Pym, D.: The semantics of BI and resource tableaux. Mathematical Structures in Computer Science **15**(6), 1033–1088 (2005)
18. Galmiche, D., Marti, M., Méry, D.: Relating labelled and label-free bunched calculi in BI logic. In: International Conference on Automated Reasoning with Analytic Tableaux and Related Methods. pp. 130–146. Springer (2019)
19. Galmiche, D., Méry, D.: Proof-search and countermodel generation in propositional BI Logic - extended abstract -. In: N. Kobayashi, B.P. (ed.) 4th International Symposium on Theoretical Aspects of Computer Software - TACS 2001. Lecture Notes in Computer Science, vol. 2215, pp. 263–282. Springer, Sendai, Japan (2001)
20. Galmiche, D., Méry, D.: Connection-based proof search in propositional BI logic. In: Voronkov, A. (ed.) 18th International Conference on Automated Deduction - CADE-18. Lecture Notes in Computer Science, vol. 2392, pp. 111–128. Springer Verlag, Copenhagen/Denmark (2002)
21. Gentzen, G.: The Collected Papers of Gerhard Gentzen. Amsterdam: North-Holland Pub. Co. (1969)
22. Girard, J.Y.: Linear logic. Theoretical Computer Science **50**(1), 1 – 101 (1987)
23. Ishtiaq, S., O'Hearn, P.W.: BI as an assertion language for mutable data structures. SIGPLAN Not. **46**(4), 84–96 (May 2011). https://doi.org/10.1145/1988042.1988050, https://doi.org/10.1145/1988042.1988050
24. Laurent, O.: A proof of the focalization property of linear logic (04 2004), https://perso.ens-lyon.fr/olivier.laurent/llfoc.pdf
25. Liang, C., Miller, D.: Focusing and polarization in linear, intuitionistic, and classical logics. Journal of Theoretical Computer Science. **410**(46), 4747–4768 (Nov 2009)
26. Lincoln, P., Mitchell, J., Scedrov, A., Shankar, N.: Decision problems for propositional linear logic. Annals of Pure and Applied Logic **56**(1), 239 – 311 (1992)
27. Marin, S., Miller, D., Volpe, M.: A focused framework for emulating modal proof systems. In: 11th conference on" Advances in Modal Logic". pp. 469–488. College Publications (2016)
28. McLaughlin, S., Pfenning, F.: Imogen: Focusing the polarized focused inverse method for intuitionistic propositional logic. In: Cervesato, I., Veith, H., Voronkov, A. (eds.) 15th International Conference on Logic, Programming, Artificial Intelligence and Reasoning (LPAR). vol. 5330, pp. 174–181 (Nov 2008)
29. Miller, D., Nadathur, G., Pfenning, F., Scedrov, A.: Uniform proofs as a foundation for logic programming. Annals of Pure and Applied Logic **51**(1), 125 – 157 (1991)
30. Miller, D., Pimentel, E.: A formal framework for specifying sequent calculus proof systems. Theoretical Compututer Science **474**, 98–116 (Feb 2013)
31. O'Hearn, P., Pym, D.: The logic of bunched implications. The Bulletin of Symbolic Logic **5**(2), 215–244 (1999)
32. Pym, D.: Resource semantics: Logic as a modelling technology. ACM SIGLOG News **6**(2), 5–41 (Apr 2019). https://doi.org/10.1145/3326938.3326940, https://doi.org/10.1145/3326938.3326940
33. Pym, D.J.: The Semantics and Proof Theory of the Logic of Bunched Implications, Applied Logic Series, vol. 26. Springer Netherlands, Dordrecht (2002)

34. Pym, D.J., O'Hearn, P.W., Yang, H.: Possible Worlds and Resources: the Semantics of BI. Theoretical Computer Science **315**(1), 257 – 305 (2004). https://doi.org/https://doi.org/10.1016/j.tcs.2003.11.020, http://www.sciencedirect.com/science/article/pii/S0304397503006248
35. Ramanayake, R.: Extended Kripke lemma and decidability for hypersequent substructural logics. In: Proceedings of the 35th Annual ACM/IEEE Symposium on Logic in Computer Science. p. 795–806. LICS '20, Association for Computing Machinery, New York, NY, USA (2020). https://doi.org/10.1145/3373718.3394802
36. Read, S.: Relevant Logic: A Philosophical Examination of Inference. B. Blackwell (1988)
37. Reynolds, J.C.: Separation logic: a logic for shared mutable data structures. In: Proceedings 17th Annual IEEE Symposium on Logic in Computer Science. pp. 55–74 (2002)

Interpolation and Amalgamation for Arrays with MaxDiff

Silvio Ghilardi[1], Alessandro Gianola[2(✉)], and Deepak Kapur[3⋆]

[1] Dipartimento di Matematica, Università degli Studi di Milano, Milano, Italy
[2] Faculty of Computer Science, Free University of Bozen-Bolzano, Bolzano, Italy
gianola@inf.unibz.it
[3] Department of Computer Science, University of New Mexico, Albuquerque, USA

Abstract. In this paper, the theory of McCarthy's extensional arrays enriched with a maxdiff operation (this operation returns the biggest index where two given arrays differ) is proposed. It is known from the literature that a diff operation is required for the theory of arrays in order to enjoy the Craig interpolation property at the quantifier-free level. However, the diff operation introduced in the literature is merely instrumental to this purpose and has only a purely formal meaning (it is obtained from the Skolemization of the extensionality axiom). Our maxdiff operation significantly increases the level of expressivity; however, obtaining interpolation results for the resulting theory becomes a surprisingly hard task. We obtain such results via a thorough semantic analysis of the models of the theory and of their amalgamation properties. The results are modular with respect to the index theory and it is shown how to convert them into concrete interpolation algorithms via a hierarchical approach.

Keywords: Interpolation · Arrays · Amalgamation · SMT

1 Introduction

Since McMillan's seminal papers [31,32], interpolation has been successfully applied in software model checking, also in combination with orthogonal techniques like PDR [38] or k-induction [29]. The reason why interpolation techniques are so attractive is because they allow to discover in a completely *automatic* way new atoms (improperly often called 'predicates') that might contribute to the construction of invariants. In fact, software model-checking problems are typically infinite state, so invariant synthesis may require introducing formulae whose search is not finitely bounded. One way to discover them is to analyze spurious error traces; for instance, if the system under examination (described by a transition formula $Tr(\underline{x}, \underline{x}')$) cannot reach in n-step an error configuration in $U(\underline{x})$ starting from an initial configuration in $In(\underline{x})$, this means that the formula

$$In(\underline{x}_0) \wedge Tr(\underline{x}_0, \underline{x}_1) \wedge \cdots \wedge Tr(\underline{x}_{n-1}, \underline{x}_n) \wedge U(\underline{x}_n)$$

⋆ The third author has been partially supported by the National Science Foundation CCF award 1908804.

S. Kiefer and C. Tasson (Eds.): FOSSACS 2021, LNCS 12650, pp. 268–288, 2021.
https://doi.org/10.1007/978-3-030-71995-1_14

is inconsistent (modulo a suitable theory T). From the inconsistency proof, by computing an interpolant, say at the i-th iteration, one can produce a formula $\phi(\underline{x})$ such that, modulo T, we have

$$In(\underline{x}_0) \wedge \bigwedge_{j=0}^{i} Tr(\underline{x}_{j-1}, \underline{x}_j) \models \phi(\underline{x}_i) \quad \text{and} \quad \phi(\underline{x}_i) \wedge \bigwedge_{j=i+1}^{n} Tr(\underline{x}_{j-1}, \underline{x}_j) \wedge U(\underline{x}_n) \models \bot. \tag{1}$$

This formula (and the atoms it contains) can contribute to the refinement of the current candidate loop invariant guaranteeing safey. This fact can be exploited in very different ways during invariant search, depending on the various techniques employed. It should be noticed however that interpolants are not unique and that different interpolation algorithms may return interpolants of different quality: all interpolants restrict search, but not all of them might be conclusive.

This new application of interpolation is different from the role of interpolants for analyzing proof theories of various logics starting with the pioneering works of [15,24,34]. It should be said however that Craig interpolation theorem in first order logic does not give by itself any information on the shape the interpolant can have when a specific theory is involved. Nevertheless, this is crucial for the applications: when we extract an interpolant from a trace like (1), we are typically handling a theory which might be undecidable, but whose quantifier-free fragment is decidable for satisfiability (usually within a somewhat 'reasonable' computational complexity). Thus, it is desirable (although not always possible) that the interpolant is quantifier-free, a fact which is not guaranteed in the general case. This is why a lot of effort has been made in analyzing *quantifier-free* interpolation, also exploiting its connection to semantic properties like amalgamation and strong amalgamation (see [9] for comprehensive results in the area).

The specific theories we want to analyze in this paper are variants of *McCarthy's theory of arrays* [30] *with extensionality* (see Section 3 below for a detailed description). The main operations considered in this theory are the *write* operation (i.e. the array update) and the *read* operation (i.e., the access to the content of an array cell). As such, this theory is suitable to formalize programs over arrays, like standard copying, comparing, searching, sorting, etc. functions; verification problems of this kind are collected in the SV-COMP benchmarks category "ReachSafety-Arrays"[4], where safety verification tasks involving arrays of *finite but unknown length* are considered.

By itself, the theory of arrays with extensionality does not have quantifier free interpolation [28][5]; however, in [8] it was shown that quantifier-free interpolation is restored if one enriches the language with a binary function skolemizing the extensionality axiom (the result was confirmed - via different interpolation algorithms - in [23,37]). Such a Skolem function, applied to two array variables

---

[4] https://sv-comp.sosy-lab.org/2020/benchmarks.php

[5] This is the counterexample (due to R. Jhala): the formula $x = wr(y, i, e)$ is inconsistent with the formula $rd(x, j) \neq rd(y, j) \wedge rd(x, k) \neq rd(y, k) \wedge j \neq k$, but all possible interpolants require quantifiers to be written (with diff symbols, instead, it is possible to write down an interpolant without quantifiers, as shown in [8]).

a, b, returns an index $\mathtt{diff}(a, b)$ where a, b differ (it returns an arbitrary value if a is equal to b). This semantics for the `diff` operation is very undetermined and does not have a significant interpretation in concrete programs. That is why we propose to modify it in order to give it a defined and natural meaning: we ask for $\mathtt{diff}(a, b)$ to return the *biggest* index where a, b differ (in case $a = b$ we ask for $\mathtt{diff}(a, b)$ to be the minimum index 0). Since it is natural to view arrays as functions defined on initial intervals of the nonnegative integers, this choice has a clear semantic motivation. The expressive power of the theory of arrays so enriched becomes bigger: for instance, if we also add to the language a constant symbol ϵ for the undefined array constantly equal to some 'undefined' value $\perp$ (where $\perp$ is meant to be different from the values $a[i]$ actually in use), then we can define $|a|$ as $\mathtt{diff}(a, \epsilon)$. In this way we can model the fact that a is undefined outside the interval $[\,0, |a|\,]$ - this is useful to formalize the above mentioned SV-COMP benchmarks.

The effectiveness of quantifier-free interpolation in the theory of arrays with maxdiff is exemplified in the simple example of Figure 1: the invariant certifying the assert in line 7 of the `Strcpy` algorithm can be obtained taking a suitable quantifier-free interpolant out of the spurious trace (1) already for $n = 2$. In more realistic examples, as witnessed by current research [2,3,4,5,16,22,25,13], it is quite clear that useful invariants require universal quantifiers to be expressed and if undecidable fragments are invaded, incomplete solvers must be used. However, even in such circumstances, quantifier-free interpolation does not lose its interest: for instance, the tool BOOSTER [5][6] synthesizes universally quantified invariants out of quantifer-free interpolants (quantifier-free interpolation problems are generated by negating and skolemizing universally quantified formulae arising during invariants search, see [4] for details).

```
1 int a[N];
2 int b[N];
3 int I = 0;
4 while I < N do
5     b[I] = a[I];
6     I++;
7 assert(a = b);
```

- $In(a, b, I) \equiv I = 0 \wedge |a| = N-1 \wedge |b| = N-1 \wedge N > 0$
- $Tr(a, b, I, a', b', I') \equiv I < N \wedge I' = I + 1 \wedge a' = a \wedge b' = wr(b, I, rd(a, I))$
- $U(a, b) \equiv a \neq b \wedge I = N$

Fig. 1. `Strcpy` function: code and associated transition system (with program counter missed in the latter for simplicity).
Loop invariant: $a = b \vee (N > \mathtt{diff}(a, b) \wedge \mathtt{diff}(a, b) \geq I)$.

Proving that the theory of arrays with the above 'maxdiff' operation enjoys quantifier-free interpolation revealed to be a surprisingly difficult task. In

[6] BOOSTER is no longer maintained, however it is still referred to in current experimental evaluations [16,13].

the end, the interpolation algorithm we obtain resembles the interpolation algorithms generated via the hierarchic locality techniques introduced in [35,36] and employed also in [37]; however, its correctness, completeness and termination proofs require a large détour going through non-trivial model-theoretic arguments (these arguments do not substantially simplify adopting the complex framework of 'amalgamation closures' and 'W-separability' of [37], and that is the reason why we preferred to supply direct proofs).

This paper concentrates on theoretical and methodological results, rather than on experimental aspects. It is almost completely dedicated to the correctness and completeness poof of our interpolation algorithm: in Subsection 3.1 we summarize our proof plan and supply basic intuitions. The paper is structured as follows: in Section 2 we recall some background, in Section 3 we introduce our theory of arrays with maxdiff; Sections 4 and 5 supply the semantic proof of the amalgamation theorem; Sections 6 and 7 are dedicated to the algorithmic aspects, whereas Section 8 analyzes complexity for the restricted case where indexes are constrained by the theory of total orders. In the final Section 9, we mention some still open problems. The main results in the paper are Theorems 2,4,5: for space reasons, *all proofs of these theorems will be only sketched*, full details are nevertheless supplied in the online available extended version [21]. This extended version contains additional material on complexity analysis and implementation. It contains also a proof about nonexistence of uniform interpolants (see [26,27,20,10,11,12] for the definition and more information on uniform interpolants).

2 Formal Preliminaries

We assume the usual syntactic (e.g., signature, variable, term, atom, literal, formula, and sentence) and semantic (e.g., structure, sub-structure, truth, satisfiability, and validity) notions of (possibly many-sorted) first-order logic. The equality symbol "=" is included in all signatures considered below. Notations like $E(\underline{x})$ mean that the expression (term, literal, formula, etc.) E contains free variables only from the tuple $\underline{x}$. A 'tuple of variables' is a list of variables without repetitions and a 'tuple of terms' is a list of terms (possibly with repetitions). Finally, whenever we use a notation like $E(\underline{x}, \underline{y})$ we implicitly assume not only that both the $\underline{x}$ and the $\underline{y}$ are pairwise distinct, but also that $\underline{x}$ and $\underline{y}$ are disjoint. A *constraint* is a conjunction of literals. A formula is *universal* (*existential*) iff it is obtained from a quantifier-free formula by prefixing it with a string of universal (existential, resp.) quantifiers.

Theories and satisfiability modulo theory. A *theory* T is a pair (Σ, Ax_T), where Σ is a signature and Ax_T is a set of Σ-sentences, called the *axioms* of T (we shall sometimes write directly T for Ax_T). The *models* of T are those Σ-structures in which all the sentences in Ax_T are true. A Σ-formula ϕ is T-*satisfiable* (or T-consistent) if there exists a model $\mathcal{M}$ of T such that ϕ is true in $\mathcal{M}$ under a suitable assignment $\mathtt{a}$ to the free variables of ϕ (in symbols, $(\mathcal{M}, \mathtt{a}) \models \phi$); it is *T-valid* (in symbols, $T \vdash \varphi$) if its negation is T-unsatisfiable or, equivalently, φ is provable from the axioms of T in a complete calculus for first-order logic. A theory $T = (\Sigma, Ax_T)$ is *universal* iff all sentences in Ax_T are universal. A formula φ_1 *T-entails* a formula φ_2 if $\varphi_1 \to \varphi_2$ is *T-valid* (in symbols, $\varphi_1 \vdash_T \varphi_2$ or simply $\varphi_1 \vdash \varphi_2$ when T is clear from the context). If Γ is a set of formulæ and ϕ a formula, $\Gamma \vdash_T \phi$ means that there are $\gamma_1, \ldots, \gamma_n \in \Gamma$ such that $\gamma_1 \wedge \cdots \wedge \gamma_n \vdash_T \phi$. The *satisfiability modulo the theory* T (SMT(T)) *problem* amounts to establishing the T-satisfiability of quantifier-free Σ-formulæ (equivalently, the T-satisfiability of Σ-constraints). A theory T admits *quantifier-elimination* iff for every formula $\phi(\underline{x})$ there is a quantifier-free formula $\phi'(\underline{x})$ such that $T \vdash \phi \leftrightarrow \phi'$.

Some theories have special names, which are becoming standard in SMT-literature; for instance, $\mathcal{EUF}(\Sigma)$ is the pure equality theory in the signature Σ (this is commonly abbreviated as $\mathcal{EUF}$ if there is no need to specify the signature Σ). More standard theory names will be recalled during the paper.

Embeddings and sub-structures The support of a structure $\mathcal{M}$ is denoted with $|\mathcal{M}|$. For a (sort, function, relation) symbol σ, we denote as $\sigma^{\mathcal{M}}$ the interpretation of σ in $\mathcal{M}$. An embedding is a homomorphism that preserves and reflects relations and operations (see, e.g., [14]). Formally, a *Σ-embedding* (or, simply, an embedding) between two Σ-structures $\mathcal{M}$ and $\mathcal{N}$ is any mapping $\mu : |\mathcal{M}| \longrightarrow |\mathcal{N}|$ satisfying the following three conditions: (a) it is a (sort-preserving) injective function; (b) it is an algebraic homomorphism, that is for every n-ary function symbol f and for every $a_1, \ldots, a_n \in |\mathcal{M}|$, we have $f^{\mathcal{N}}(\mu(a_1), \ldots, \mu(a_n)) = \mu(f^{\mathcal{M}}(a_1, \ldots, a_n))$; (c) it preserves and reflects predicates, i.e. for every n-ary predicate symbol P, we have $(a_1, \ldots, a_n) \in P^{\mathcal{M}}$ iff $(\mu(a_1), \ldots, \mu(a_n)) \in P^{\mathcal{N}}$. If $|\mathcal{M}| \subseteq |\mathcal{N}|$ and the embedding $\mu : \mathcal{M} \longrightarrow \mathcal{N}$ is just the identity inclusion $|\mathcal{M}| \subseteq |\mathcal{N}|$, we say that $\mathcal{M}$ is a *substructure* of $\mathcal{N}$ or that $\mathcal{N}$ is a *superstructure* of $\mathcal{M}$. As it is known, the truth of a universal (resp. existential) sentence is preserved through substructures (resp. superstructures).

Combinations of theories. A theory T is *stably infinite* iff every T-satisfiable quantifier-free formula (from the signature of T) is satisfiable in an infinite model of T. By compactness, it is possible to show that T is stably infinite iff every model of T embeds into an infinite one (see, e.g., [17]). A theory T is *convex* iff for every conjunction of literals δ, if $\delta \vdash_T \bigvee_{i=1}^{n} x_i = y_i$ then $\delta \vdash_T x_i = y_i$ holds for some $i \in \{1, ..., n\}$. Let T_i be a stably-infinite theory over the signature Σ_i such that the $SMT(T_i)$ problem is decidable for $i = 1, 2$ and such that Σ_1 and Σ_2 are disjoint (i.e. the only shared symbol is equality). Under these assumptions, the *Nelson-Oppen combination result* [33] says that the SMT problem for the combination $T_1 \cup T_2$ of the theories T_1 and T_2 is decidable.

Interpolation properties. Craig's interpolation theorem [14] roughly states that if a formula ϕ implies a formula ψ then there is a third formula θ, called an interpolant, such that ϕ implies θ, θ implies ψ, and every non-logical symbol in θ occurs both in ϕ and ψ. Our interest is to specialize this result to the computation of quantifier-free interpolants modulo (combinations of) theories.

Definition 1. *[Plain quantifier-free interpolation] A theory T* admits (plain) quantifier-free interpolation *(or, equivalently,* has quantifier-free interpolants*) iff for every pair of quantifier-free formulae ϕ, ψ such that $\psi \wedge \phi$ is T-unsatisfiable, there exists a quantifier-free formula θ, called an* interpolant, *such that: (i) ψ T-entails θ, (ii) $\theta \wedge \phi$ is T-unsatisfiable, and (iii) only the variables occurring in both ψ and ϕ occur in θ.*

In verification, the following extension of Definition 1 is considered more useful.

Definition 2. *[General quantifier-free interpolation] Let T be a theory in a signature Σ; we say that T has the* general quantifier-free interpolation property *iff for every signature Σ' (disjoint from Σ) and for every pair of ground $\Sigma \cup \Sigma'$-formulæ ϕ, ψ such that $\phi \wedge \psi$ is T-unsatisfiable*[7]*, there is a ground formula θ such that: (i) ϕ T-entails θ; (ii) $\theta \wedge \psi$ is T-unsatisfiable; (iv) all relations, constants and function symbols from Σ' occurring in θ also occur in ϕ and ψ.*

By replacing free variables with free constants, it should be clear that general quantifier-free interpolation (Definition 2) implies plain quantifier-free interpolation (Definition 1); however, the converse implication does not hold.

Amalgamation and strong amalgamation. Interpolation can be characterized semantically via amalgamation.

Definition 3. *A universal theory T has the* amalgamation property *iff given models $\mathcal{M}_1$ and $\mathcal{M}_2$ of T and a common submodel $\mathcal{A}$ of them, there exists a further model $\mathcal{M}$ of T (called T-amalgam) endowed with embeddings $\mu_1 : \mathcal{M}_1 \longrightarrow \mathcal{M}$ and $\mu_2 : \mathcal{M}_2 \longrightarrow \mathcal{M}$ whose restrictions to $|\mathcal{A}|$ coincide.*

A universal theory T has the strong amalgamation property *if the above embeddings μ_1, μ_2 and the above model $\mathcal{M}$ can be chosen so to satisfy the following additional condition: if, for some $m_1 \in |\mathcal{M}_1|, m_2 \in |\mathcal{M}_2|$, $\mu_1(m_1) = \mu_2(m_2)$ holds, then there exists an element a in $|\mathcal{A}|$ such that $m_1 = a = m_2$.*

The first statement of the following theorem is an old result due to [6]; the second statement is proved in [9] (where it is also suitably reformulated for theories which are not universal):

Theorem 1. *Let T be a universal theory. Then*
(i) *T has the amalgamation property iff it admits quantifier-free interpolants;*
(ii) *T has the strong amalgamation property iff it has the general quantifier-free interpolation property.*

We underline that, in presence of stable infiniteness, strong amalgamation is a modular property (in the sense that it transfers to signature-disjoint unions of theories), whereas amalgamation is not (see again [9] for details).

[7] By this (and similar notions) we mean that $\phi \wedge \psi$ is unsatisfiable in all Σ'-structures whose Σ-reduct is a model of T.

3 Arrays with MaxDiff

The *McCarthy theory of arrays* [30] has three sorts ARRAY, ELEM, INDEX (called "array", "element", and "index" sort, respectively) and two function symbols rd ("read") and wr ("write") of appropriate arities; its axioms are:

$$\forall y, i, e.\ rd(wr(y, i, e), i) = e$$
$$\forall y, i, j, e.\ i \neq j \rightarrow rd(wr(y, i, e), j) = rd(y, j).$$

The McCarthy theory of *arrays with extensionality* has the further axiom

$$\forall x, y. x \neq y \rightarrow (\exists i.\ rd(x, i) \neq rd(y, i)), \tag{2}$$

called the 'extensionality' axiom. The theory of arrays with extensionality is not universal and quantifier-free interpolation fails for it [28]. In [8] a variant of the McCarthy theory of arrays with extensionality, obtained by Skolemizing the axioms of extensionality, is introduced. This variant of the theory turns out to be universal and to enjoy quantifier-free interpolation. However, the Skolem function introduced in [8] is generic, here we want to make it more informative, so as to return the biggest index where two different arrays differ. To locate our contribution in the general context, we need the notion of an index theory.

Definition 4. *An* index theory T_I *is a mono-sorted theory (let* INDEX *be its sort) satisfying the following conditions:*
- T_I *is universal, stably infinite and has the general quantifier-free interpolation property (i.e. it is strongly amalgamable, see Theorem 1);*
- $SMT(T_I)$ *is decidable;*
- T_I *extends the theory* TO *of linear orderings with a distinguished element* 0.

We recall that TO is the theory whose only proper symbols (beside equality) are a binary predicate $\leq$ and a constant 0 subject to the axioms saying that $\leq$ is reflexive, transitive, antisymmetric and total (the latter means that $i \leq j \vee j \leq i$ holds for all i, j). Thus, the signature of an index theory T_I contains at least the binary relation symbol $\leq$ and the constant 0. In the paper, by a T_I-term, T_I-atom, T_I-formula, etc. we mean a term, atom, formula in the signature of T_I. Below, we use the abbreviation $i < j$ for $i \leq j \wedge i \neq j$. The constant 0 is meant to separate 'formally positive' indexes - those satisfying $0 \leq i$ - from the remaining 'formally negative' ones.

Examples of index theories are TO itself, integer difference logic $\mathcal{IDL}$, integer linear arithmetic $\mathcal{LIA}$, and real linear arithmetics $\mathcal{LRA}$. In order to match the requirements of Definition 4, one must however make a careful choice of the language, see [9] for details: the most important detail is that integer (resp. real) division by all positive integers should be added to the language of $\mathcal{LIA}$ (resp. $\mathcal{LRA}$). For most applications, $\mathcal{IDL}$ (namely the theory of integer numbers with 0, ordering, successor and predecessor) [8] suffices as in this theory one can model counters for scanning arrays.

[8] The name 'integer difference logic' comes from the fact that atoms in this theory are equivalent to formulæ of the kind $S^n(i) \bowtie j$ (where $\bowtie \in \{\leq, \geq, =\}$), thus they represent difference bound constraints of the kind $j - i \bowtie n$ for $n \geq 0$.

Given an index theory T_I, we now introduce our *array theory with maxdiff* $\mathcal{ARD}(T_I)$ (parameterized by T_I) as follows. We still have three sorts ARRAY, ELEM, INDEX; the language includes the symbols of T_I, the read and write operations rd, wr, a binary function diff of type ARRAY $\times$ ARRAY $\rightarrow$ INDEX, as well as constants ϵ and $\perp$ of sorts ARRAY and ELEM, respectively. The constant $\perp$ models an undetermined (e.g. undefined, not-in-use, not coming from appropriate initialization, etc.) value and ε models the totally undefined array; the term $\texttt{diff}(x, y)$ returns the maximum index where x and y differ and returns 0 if x and y are equal. [9] Formally, the axioms of $\mathcal{ARD}(T_I)$ include, besides the axioms of T_I, the following ones:

$$\forall y, i, e.\ i \geq 0 \rightarrow rd(wr(y, i, e), i) = e \tag{3}$$

$$\forall y, i, j, e.\ i \neq j \rightarrow rd(wr(y, i, e), j) = rd(y, j) \tag{4}$$

$$\forall x, y.\ x \neq y \rightarrow rd(x, \texttt{diff}(x, y)) \neq rd(y, \texttt{diff}(x, y)) \tag{5}$$

$$\forall x, y, i.\ i > \texttt{diff}(x, y) \rightarrow rd(x, i) = rd(y, i) \tag{6}$$

$$\forall x.\ \texttt{diff}(x, x) = 0 \tag{7}$$

$$\forall x.i\ \ i < 0 \rightarrow rd(x, i) = \perp \tag{8}$$

$$\forall i.\ rd(\varepsilon, i) = \perp \tag{9}$$

In the read-over-write axiom (3), we put the proviso $i \geq 0$ because we want all our arrays to be undefined on negative indexes (negative updates makes no sense and have no effect: by axiom (8), reading a negative index always produces $\perp$).

We call $\mathcal{AR}_{\text{ext}}(T_I)$ (the 'theory of arrays with extensionality parameterized by T_I') the theory obtained from $\mathcal{ARD}(T_I)$ by removing the symbol diff and by replacing the axioms (5)-(7) by the extensionality axiom (2). Since the extensionality axioms follows from axiom (5), $\mathcal{ARD}(T_I)$ is an extension of $\mathcal{AR}_{\text{ext}}(T_I)$.

As an effect of the above axioms, we have that an array x is undefined outside the interval $[0, |x|]$, where $|x|$ is defined as $|x| := \texttt{diff}(x, \varepsilon)$. Typically, this interval is finite and in fact our proof of Theorem 3 below shows that any satisfiable constraint is satisfiable in a model where all such intervals (relatively to the variables involved in the constraint) are finite.

The next lemma is immediate from the axiomatization of $\mathcal{ARD}(T_I)$:

Lemma 1. *An atom of the form $a = b$ is equivalent (modulo $\mathcal{ARD}(T_I)$) to*

$$\texttt{diff}(a, b) = 0 \wedge rd(a, 0) = rd(b, 0)\ . \tag{10}$$

An atom of the form $a = wr(b, i, e)$ is equivalent (modulo $\mathcal{ARD}$) to

$$(i \geq 0 \rightarrow rd(a, i) = e)\ \wedge\ \forall h\ (h \neq i \rightarrow rd(a, h) = rd(b, h))\ . \tag{11}$$

An atom of the form $\texttt{diff}(a, b) = i$ is equivalent (modulo $\mathcal{ARD}(T_I)$) to

$$i \geq 0 \wedge \forall h\ (h > i \rightarrow rd(a, h) = rd(b, h))\ \wedge\ (i > 0 \rightarrow rd(a, i) \neq rd(b, i))\ . \tag{12}$$

[9] Notice that it might well be the case that $\texttt{diff}(x, y) = 0$ for different x, y, but in that case 0 is the only index where x, y differ.

For our interpolation algorithm in Section 7, we need to introduce iterated $\mathtt{diff}$ operations, similarly to [37]. As we know $\mathtt{diff}(a, b)$ returns the biggest index where a and b differ (it returns 0 if $a = b$). Now we want an operator that returns the last-but-one index where a, b differ (0 if a, b differ in at most one index), an operator that returns the last-but-two index where a, b differ (0 is they differ in at most two indexes), etc. Our language is already enough expressive for that, so we can introduce such operators explicitly as follows. Given array variables a, b, we define by mutual recursion the sequence of array terms $b_1, b_2, \dots$ and of index terms $\mathtt{diff}_1(a, b), \mathtt{diff}_2(a, b), \dots$:

$$b_1 := b; \qquad \mathtt{diff}_1(a,b) := \mathtt{diff}(a, b_1);$$
$$b_{k+1} := wr(b_k, \mathtt{diff}_k(a,b), rd(a, \mathtt{diff}_k(a,b))); \quad \mathtt{diff}_{k+1}(a,b) := \mathtt{diff}(a, b_{k+1})$$

Intuitively, b_{k+1} is the same as b except for all k-last indexes on which a and b differ, in correspondence of which b_{k+1} has the same value as a. A useful fact is that conjunctions of formulae of the kind $\bigwedge_{j<l} \mathtt{diff}_j(a, b) = k_j$ can be eliminated in favor of universal clauses in a language whose only symbol for array variables is rd. In detail:

Lemma 2. *A formula like*

$$\mathtt{diff}_1(a,b) = k_1 \wedge \cdots\cdots \wedge \mathtt{diff}_l(a,b) = k_l \tag{13}$$

is equivalent modulo $\mathcal{ARD}$ to the conjunction of the following five formulae:

$$k_1 \geq k_2 \wedge \cdots \wedge k_{l-1} \geq k_l \wedge k_l \geq 0 \tag{14}$$
$$\bigwedge_{j<l} (k_j > k_{j+1} \to rd(a, k_j) \neq rd(b, k_j)) \tag{15}$$
$$\bigwedge_{j<l} (k_j = k_{j+1} \to k_j = 0) \tag{16}$$
$$\bigwedge_{j\leq l} (rd(a, k_j) = rd(b, k_j) \to k_j = 0) \tag{17}$$
$$\forall h \; (h > k_l \to rd(a,h) = rd(b,h) \vee h = k_1 \vee \cdots \vee h = k_{l-1}) \tag{18}$$

3.1 Our roadmap

The main result of the paper is that, for every index theory T_I, the array theory with maxdiff $\mathcal{ARD}(T_I)$ indexed by T_I *enjoys quantifier-free interpolation* and that *interpolants can be computed hierarchically* by relying on a black-box quantifier-free interpolation algorithm for the weaker theory $T_I \cup \mathcal{EUF}$ (the latter theory has quantifier free interpolation because T_I is strongly amalgamable and because of Theorem 1). In this subsection, we supply intuitions and we give a qualitative high-level view to our proofs: more technical details and full proofs can be found in [21].

The algorithm.

By general easy transformations (recalled in Section 7 below), it is sufficient to be able to extract a quantifier-free interpolant out of a pair of quantifier-free formulae A, B such that (i) $A \wedge B$ is $\mathcal{ARD}(T_I)$-inconsistent; (ii) both A and B are conjunctions of flat literals, i.e. of literals which are equalities between variables, disequalities between variables or literals of the form $R(\underline{x}), \neg R(\underline{x}), f(\underline{x}) = y$ (where $\underline{x}, y$ are variables, R is a predicate symbol and f a function symbol).

Let us call *common* the variables occurring in both A and B. The fact that a quantifier-free interpolant exists intuitively means that there are two reasoners (an A-reasoner operating on formulae involving only the variables occurring in A and a B-reasoner operating on formulae involving only the variables occurring in B) that are able to discover the inconsistency of $A \wedge B$ by exchanging information on the common language, i.e. by communicating each other only the entailed quantifier-free formulae involving the common variables.

A problem that can be addressed when designing an interpolation algorithm, is that there are infinitely many common terms that can be built up out of finitely many common variables and it may happen that some uncommon terms can be recognized to be equal to some common terms during the deductions performed by the A-reasoner and the B-reasoner.

As an example, suppose that A contains the literals $c_1 = wr(c_2, i, e), c_1 \neq c_2, a = wr(c_3, i, e)$, where only c_1, c_2, c_3 are common (i.e. only these variables occur in B). Then using diff operations, we can deduce $i = \mathtt{diff}(c_1, c_2), e = rd(c_1, i)$ so that in the end we can conclude that a is also 'common', being definable in term of common variables. Thus, the A-reasoner must communicate (via a defining common term or in some other indirect way) to the B-reasoner any fact it discovers about a, although a was not listed among the common variables since the very beginning. In more sophisticated examples, iterated diff operations are needed to discover 'hidden' common facts.

To cope with the above problem, our algorithm *gives names* $i_k = \mathtt{diff}_k(c_1, c_2)$ to all the iterated diffs of common array variables c_1, c_2 (the newly introduced names i_k are considered common and can be replaced back with their defining terms when the interpolants are computed at the end of the algorithm).

The second component of our algorithm is *instantiation.* Both the A- and the B-reasoner use the content of Lemmas 1 and 2 in order to handle atoms of the kind $a = b$, $a_1 = wr(a_2, i, e)$, $i = \mathtt{diff}_k(a_1, a_2)$. Whenever they come across such atoms, the equivalent formulæ supplied by these lemmas are taken into consideration; in fact, whenever the lemmas produce universally quantified clauses of the kind $\forall h\, C$, they replace in C the universally quantified index variable h by *all possible instantiations* with their own index terms (these are the terms built up from index variables occurring in A for the A-reasoner and occurring in B for the B-reasoner respectively). Such instantiations can be read as *clauses in the language of* $T_I \cup \mathcal{EUF}$ if we replace every array variable a by a fresh unary function symbol f_a and read terms like $rd(a, i)$ as $f_a(i)$.

Of course both the production of names for iterated diff-terms and the instantiation with owned index terms need to be repeated (possibly, infinitely many times); we prove however (this is the content of our main Theorem 4 below) that *if* $A \wedge B$ *is* $\mathcal{ARD}(T_I)$*-inconsistent, then sooner or later the union of the sets of the clauses deduced by the* A*-reasoner and the* B*-reasoner in the restricted* *signature of* $T_I \cup \mathcal{EUF}$ *is* $T_I \cup \mathcal{EUF}$*-inconsistent*, i.e., the instantiation process terminates. This means that an interpolant can be extracted, using a black-box quantifier-free interpolation algorithm for the weaker theory $T_I \cup \mathcal{EUF}$. In the simple case where T_I is just the theory TO of total orders, we shall prove in Section 8 that a *quadratic* number of instantiations always suffices. In the general case, however, the situation is similar to the statement of Herbrand theorem: finitely many instantiations suffice to get an inconsistency proof in the weaker logical formalism, but a bound cannot be given.

The proof.

Theorem 4 is proved in a contrapositive way: we show that *if a* $T_I \cup \mathcal{EUF}$*-inconsistency never arises, then* $A \wedge B$ *is* $\mathcal{ARD}(T_I)$*-consistent.* This is proved in two steps: if $T_I \cup \mathcal{EUF}$-inconsistency does not arise, we produce two $\mathcal{ARD}(T_I)$-models $\mathcal{A}$ and $\mathcal{B}$, where $\mathcal{A}$ satisfies A and $\mathcal{B}$ satisfies B. Moreover, $\mathcal{A}$ and $\mathcal{B}$ are built up in such a way that they share the same $\mathcal{ARD}(T_I)$-substructure. In the second step, we prove the amalgamation theorem for $\mathcal{ARD}(T_I)$, so that the amalgamated model will produce the desired model of $A \wedge B$. In fact, the two steps are inverted in our exposition: we first prove the amalgamation theorem in Section 5 (Theorem 2) and then our main theorem in Section 7 (Theorem 4).

4 Embeddings

We preliminarily discuss the class of models of $\mathcal{ARD}(T_I)$ and we make important clarifications about embeddings between such models. A model $\mathcal{M}$ of $\mathcal{AR}_{\text{ext}}(T_I)$ or of $\mathcal{ARD}(T_I)$ is *functional* when the following conditions are satisfied:

(i) $\mathtt{ARRAY}^{\mathcal{M}}$ is a subset of the set of all positive-support functions from $\mathtt{INDEX}^{\mathcal{M}}$ to $\mathtt{ELEM}^{\mathcal{M}}$ (a function a is *positive-support* iff $a(i) = \bot$ for every $i < 0$);
(ii) rd is function application;
(iii) wr is the point-wise update operation (i.e., for $i \geq 0$, the function $wr(a, i, e)$ returns the same values as the function a, except at the index i where it returns the element e).

Because of the extensionality axiom, it can be shown that every model is *isomorphic to a functional one.* For an array $a \in \mathtt{INDEX}^{\mathcal{M}}$ in a functional model $\mathcal{M}$ and for $i \in \mathtt{INDEX}^{\mathcal{M}}$, since a is a function, we interchangeably use the notations $a(i)$ and $rd(a, i)$. A functional model $\mathcal{M}$ is said to be *full* iff $\mathtt{ARRAY}^{\mathcal{M}}$ consists of *all* the positive-support functions from $\mathtt{INDEX}^{\mathcal{M}}$ to $\mathtt{ELEM}^{\mathcal{M}}$.

Let a, b be elements of $\mathtt{ARRAY}^{\mathcal{M}}$ in a model $\mathcal{M}$. We say that *a and b are cardinality dependent* (in symbols, $\mathcal{M} \models \|a - b\| < \omega$) iff $\{i \in \mathtt{INDEX}^{\mathcal{M}} \mid \mathcal{M} \models rd(a, i) \neq rd(b, i)\}$ is finite. Cardinality dependency in $\mathcal{M}$ is obviously an equivalence relation, that we sometimes denote as $\sim_{\mathcal{M}}$.

Passing to $\mathcal{ARD}(T_I)$, a further remark is in order: in a functional model $\mathcal{M}$ of $\mathcal{ARD}(T_I)$, the index $\mathtt{diff}(a, b)$ (if it exists) is uniquely determined: it must be the maximum index where a, b differ (it is 0 if $a = b$). We say that $\mathtt{diff}(a, b)$ is *defined* iff there is a maximum index where a, b differ (or if $a = b$).

An embedding $\mu : \mathcal{M} \longrightarrow \mathcal{N}$ between $\mathcal{AR}_{\rm ext}(T_I)$-models is said to be $\texttt{diff}$-faithful iff whenever $\texttt{diff}(a, b)$ is defined so is $\texttt{diff}(\mu(a), \mu(b))$ and it is equal to $\mu(\texttt{diff}(a, b))$. Since there might not be a maximum index where a, b differ, in principle it is not always possible to expand a functional model of $\mathcal{AR}_{\rm ext}(T_I)$ to a functional model of $\mathcal{ARD}(T_I)$, keeping the set of indexes unchanged. Indeed, in order to do that in a $\texttt{diff}$-faithful way, one needs to explicitly add to $\texttt{INDEX}^{\mathcal{M}}$ new indexes including at least indexes representing the missing maximum indexes where two given array differ. This idea is used in the following lemma (proved in the online available extended version [21]):

Lemma 3. *For every index theory T_I, every model of $\mathcal{AR}_{\rm ext}(T_I)$ has a* $\texttt{diff}$*-faithful embedding into a model of $\mathcal{ARD}(T_I)$.*

5 Amalgamation

We now sketch the proof of the amalgamation property for $\mathcal{ARD}(T_I)$. We recall that strong amalgamation holds for models of T_I (see Definition 4).

Theorem 2. *$\mathcal{ARD}(T_I)$ enjoys the amalgamation property.*

Proof. Take two embeddings $\mu_1 : \mathcal{N} \longrightarrow \mathcal{M}_1$ and $\mu_2 : \mathcal{N} \longrightarrow \mathcal{M}_2$. As we know, we can suppose—w.l.o.g.—that $\mathcal{N}, \mathcal{M}_1, \mathcal{M}_2$ are functional models; in addition, via suitable renamings, we can freely suppose that μ_1, μ_2 restricts to inclusions for the sorts $\texttt{INDEX}$ and $\texttt{ELEM}$, and that $(\texttt{ELEM}^{\mathcal{M}_1} \setminus \texttt{ELEM}^{\mathcal{N}}) \cap (\texttt{ELEM}^{\mathcal{M}_2} \setminus \texttt{ELEM}^{\mathcal{N}}) = \emptyset$, $(\texttt{INDEX}^{\mathcal{M}_1} \setminus \texttt{INDEX}^{\mathcal{N}}) \cap (\texttt{INDEX}^{\mathcal{M}_2} \setminus \texttt{INDEX}^{\mathcal{N}}) = \emptyset$. To build the amalgamated model of $\mathcal{ARD}(T_I)$, we first build a full model $\mathcal{M}$ of $\mathcal{AR}_{\rm ext}(T_I)$ with $\texttt{diff}$-faithful embeddings $\nu_1 : \mathcal{M}_1 \longrightarrow \mathcal{M}$ and $\nu_2 : \mathcal{M}_2 \longrightarrow \mathcal{M}$ such that $\nu_1 \circ \mu_1 = \nu_2 \circ \mu_2$. If we succeed, the claim follows by Lemma 3: indeed, thanks to that lemma, we can embed in a $\texttt{diff}$-faithful way $\mathcal{M}$ (which is a model of $\mathcal{AR}_{\rm ext}(T_I)$) to a model $\mathcal{M}'$ of $\mathcal{ARD}(T_I)$, which is the required $\mathcal{ARD}(T_I)$-amalgam.

We take the T_I-reduct of $\mathcal{M}$ to be a model supplied by the strong amalgamation property of T_I (again, we can freely assume that the T_I-reducts of $\mathcal{M}_1, \mathcal{M}_2$ identically include in it); we let $\texttt{ELEM}^{\mathcal{M}}$ to be $\texttt{ELEM}^{\mathcal{M}_1} \cup \texttt{ELEM}^{\mathcal{M}_2}$. We need to define $\nu_i : \mathcal{M}_i \longrightarrow \mathcal{M}$ $(i = 1, 2)$ in such a way that ν_i is $\texttt{diff}$-faithful and $\nu_1 \circ \mu_1 = \nu_2 \circ \mu_2$. We take the $\texttt{INDEX}$ and the $\texttt{ELEM}$-components of ν_1, ν_2 to be just identical inclusions. The only relevant point is the action of ν_i on $\texttt{ARRAY}^{\mathcal{M}_i}$: since we have strong amalgamation for indexes, in order to define it, it is sufficient to extend any $a \in \texttt{ARRAY}^{\mathcal{M}_i}$ to all the indexes $k \in (\texttt{INDEX}^{\mathcal{M}} \setminus \texttt{INDEX}^{\mathcal{M}_i})$. For indexes $k \in (\texttt{INDEX}^{\mathcal{M}} \setminus (\texttt{INDEX}^{\mathcal{M}_1} \cup \texttt{INDEX}^{\mathcal{M}_2}))$ we can just put $\nu_i(a)(k) = \perp$. If $k \in (\texttt{INDEX}^{\mathcal{M}} \setminus \texttt{INDEX}^{\mathcal{M}_i})$ and $k \in (\texttt{INDEX}^{\mathcal{M}_1} \cup \texttt{INDEX}^{\mathcal{M}_2})$, then $k \in (\texttt{INDEX}^{\mathcal{M}_{3-i}} \setminus \texttt{INDEX}^{\mathcal{N}})$; the definition for such k is as follows:

(*) we let $\nu_i(a)(k)$ be equal to $\mu_{3-i}(c)(k)$, where c is any array $c \in \texttt{ARRAY}^{\mathcal{N}}$ for which there is $a' \in \texttt{ARRAY}^{\mathcal{M}_i}$ such that $a \sim_{\mathcal{M}_i} a'$ and such that the relation $k > \texttt{diff}^{\mathcal{M}_i}(a', \mu_i(c))$ holds in $\texttt{INDEX}^{\mathcal{M}}$;[10] if such c does not exist, then we put $\nu_i(a)(k) = \perp$.

[10] This should be properly written as $k > \nu_i(\texttt{diff}^{\mathcal{M}_i}(a', \mu_i(c)))$, however recall that the $\texttt{INDEX}$-component of ν_i is identity, so the simplified notation is nevertheless correct.

Definition (*) is forced by some constraints that $\nu_i(a)(k)$ must satisfy. Of course, definition (*) itself needs to be justified: besides showing that it enjoys the required properties, we must also prove that it is well-given (i.e. that it does not depend on the selected c and a'). It is easy to see that, if the definition is correct, then we have $\nu_1 \circ \mu_1 = \nu_2 \circ \mu_2$; also, it is clear that ν_i preserves read and write operations (hence, it is a homomorphism) and is injective. For (i) justifying the definition of ν_i and (ii) showing that it is also diff-faithful, we need to show the following two claims (the proof is not easy, see the extended version [21] for details) for arrays $a_1, a_2 \in \mathtt{ARRAY}_1^{\mathcal{M}}$, for an index $k \in (\mathtt{INDEX}^{\mathcal{M}_2} \setminus \mathtt{INDEX}^{\mathcal{N}})$ and for arrays $c_1, c_2 \in \mathtt{ARRAY}^{\mathcal{N}}$ (checking the same facts in $\mathcal{M}_2$ is symmetrical):

(i) if $a_1 \sim_{\mathcal{M}_1} a_2$ and $k > \mathtt{diff}^{\mathcal{M}_1}(a_1, \mu_1(c_1))$, $k > \mathtt{diff}^{\mathcal{M}_1}(a_2, \mu_1(c_2))$, then $\mu_2(c_1)(k) = \mu_2(c_2)(k)$.

(ii) if $k > \mathtt{diff}^{\mathcal{M}_1}(a_1, a_2)$, then $\nu_1(a_1)(k) = \nu_1(a_2)(k)$. ⊣

6 Satisfiability

The key step of the interpolation algorithm that will be proposed in Section 7 depends upon the problem of checking satisfiability (modulo $\mathcal{ARD}(T_I)$) of quantifier-free formulæ; this will be solved in the present section by adapting instantiation techniques, like those from [7].

We define the *complexity* $c(t)$ of a term t as the number of function symbols occurring in t (thus variables and constants have complexity 0). A *flat* literal L is a formula of the kind $x_1 = t$ or $x_1 \neq x_2$ or $R(x_1, \ldots, x_n)$ or $\neg R(x_1, \ldots, x_n)$, where the x_i are variables, R is a relation symbol, and t is a term of complexity less or equal to 1. If $\mathcal{I}$ is a set of T_I-terms, an $\mathcal{I}$-*instance* of a universal formula of the kind $\forall i\, \phi$ is a formula of the kind $\phi(t/i)$ for some $t \in \mathcal{I}$.

A pair of sets of quantifier-free formulae $\Phi = (\Phi_1, \Phi_2)$ is a *separated pair* iff

(1) Φ_1 contains equalities of the form $\mathtt{diff}_k(a, b) = i$ and $a = wr(b, i, e)$; moreover if it contains the equality $\mathtt{diff}_k(a, b) = i$, it must also contain an equality of the form $\mathtt{diff}_l(a, b) = j$ for every $l < k$;

(2) Φ_2 contains Boolean combinations of T_I-atoms and of atoms of the forms:

$$rd(a, i) = rd(b, j), \;\; rd(a, i) = e, \;\; e_1 = e_2, \tag{19}$$

where a, b, i, j, e, e_1, e_2 are variables or constants of the appropriate sorts.

The separated pair is said to be finite iff Φ_1 and Φ_2 are both finite.

In practice, in a separated pair $\Phi = (\Phi_1, \Phi_2)$, reading $rd(a, i)$ as a functional application, it turns out that *the formulæ from Φ_2 can be translated into quantifier-free formulæ of the combined theory $T_I \cup \mathcal{EUF}$* (the array variables occurring in Φ_2 are converted into free unary function symbols). $T_I \cup \mathcal{EUF}$ enjoys the decidability of the quantifier-free fragment and has quantifier-free interpolation because T_I is an index theory (see Nelson-Oppen results [33] and Theorem 1): we adopt a hierarchical approach (similar to [35,36]) and *we rely on satisfiability and interpolation algorithms for such a theory as black boxes.*

Let $\mathcal{I}$ be a set of T_I-terms and let $\Phi = (\Phi_1, \Phi_2)$ be a separated pair; we let $\Phi(\mathcal{I}) = (\Phi_1(\mathcal{I}), \Phi_2(\mathcal{I}))$ be the smallest separated pair satisfying the following conditions:

- $\Phi_1(\mathcal{I})$ is equal to Φ_1 and $\Phi_2(\mathcal{I})$ contains Φ_2;
- $\Phi_2(\mathcal{I})$ contains all $\mathcal{I}$-instances of the two formulæ

$$\forall i\, rd(\varepsilon, i) = \bot,\ \forall i\, (i < 0 \rightarrow rd(a, i) = \bot),$$

  where a is any array variable occurring in Φ_1 or Φ_2;
- if Φ_1 contains the atom $a = wr(b, i, e)$ then $\Phi_2(\mathcal{I})$ contains *all the $\mathcal{I}$-instances of the formulae* (11);
- if Φ_1 contains the conjunction $\bigwedge_{i=1}^{l} \mathtt{diff}_i(a, b) = k_i$, then $\Phi_2(\mathcal{I})$ contains the formulae (14), (15), (16), (17) as well as *all $\mathcal{I}$-instances of the formula* (18).

For $M \in \mathbb{N} \cup \{\infty\}$, the *$M$-instantiation* of $\Phi = (\Phi_1, \Phi_2)$ is the separated pair $\Phi(\mathcal{I}_\Phi^M) = (\Phi_1(\mathcal{I}_\Phi^M), \Phi_2(\mathcal{I}_\Phi^M))$, where $\mathcal{I}_\Phi^M$ is the set of T_I-terms of complexity at most M built up from the index variables occurring in Φ_1, Φ_2. The *full instantiation* of $\Phi = (\Phi_1, \Phi_2)$ is the separated pair $\Phi(\mathcal{I}_\Phi^\infty) = (\Phi_1(\mathcal{I}_\Phi^\infty), \Phi_2(\mathcal{I}_\Phi^\infty))$ (which is usually not finite). A separated pair $\Phi = (\Phi_1, \Phi_2)$ is *M-instantiated* iff $\Phi = \Phi(\mathcal{I}_\Phi^M)$; it is $\mathcal{ARD}(T_I)$-satisfiable iff so it is the formula $\bigwedge \Phi_1 \wedge \bigwedge \Phi_2$[11]

Example 1. Let Φ_1 contain the four atoms

$$\{\ \mathtt{diff}(a, c_1) = i_1,\ \mathtt{diff}(b, c_2) = i_1,\ a = wr(a_1, i_3, e_3),\ a_1 = wr(b, i_1, e_1)\ \}$$

and let Φ_2 be empty. Then (Φ_1, Φ_2) is a separated pair; 0-instantiating it adds to Φ_2 the following formulae (we delete those which are redundant)

$$\begin{array}{ll}
i_1 \geq 0 & \\
rd(a, i_1) = rd(c_1, i_1) \rightarrow i_1 = 0 & rd(b, i_1) = rd(c_2, i_1) \rightarrow i_1 = 0 \\
i_3 > i_1 \rightarrow rd(a, i_3) = rd(c_1, i_3) & i_3 > i_1 \rightarrow rd(b, i_3) = rd(c_2, i_3) \\
i_3 \geq 0 \rightarrow rd(a, i_3) = e_3 & i_1 \geq 0 \rightarrow rd(a_1, i_1) = e_1 \\
i_1 \neq i_3 \rightarrow rd(a, i_1) = rd(a_1, i_1) & i_1 \neq i_3 \rightarrow rd(a_1, i_3) = rd(b, i_3)
\end{array}$$

The following results are proved in the extended version [21]:

Lemma 4. *Let ϕ be a quantifier-free formula; then it is possible to compute finitely many finite separation pairs $\Phi^1 = (\Phi_1^1, \Phi_2^1), \ldots, \Phi^n = (\Phi_1^n, \Phi_2^n)$ such that ϕ is $\mathcal{ARD}(T_I)$-satisfiable iff so is one of the Φ^i.*

Lemma 5. *The following conditions are equivalent for a finite separation pair $\Phi = (\Phi_1, \Phi_2)$:*
(i) *Φ is $\mathcal{ARD}(T_I)$-satisfiable;*
(ii) *$\bigwedge \Phi_2(\mathcal{I}_\Phi^0)$ is $T_I \cup \mathcal{EUF}$-satisfiable.*

Theorem 3. *The SMT($\mathcal{ARD}(T_I)$) problem is decidable for every index theory T_I (i.e. for every theory satisfying Definition 4).*

[11] This might be an infinitary formula if Φ is not finite. In such a case, satisfiability obviously means that there is a model $\mathcal{M}$ where we can assign values to all variables occurring in the formulæ from $\Phi_1 \cup \Phi_2$ in such a way that such formulæ become simultaneously true.

Concerning the complexity of the above procedure, notice that the satisfiability of the quantifier-free fragment of common index theories (like $\mathcal{IDL}$, $\mathcal{LIA}$, $\mathcal{LRA}$) is decidable in NP; as a consequence, from the above proof we get (for such index theories) also an NP bound for our $SMT(\mathcal{ARD}(T_I)))$-problems because 0-instantiation is clearly finite and polynomial. The fact that 0-instantiation suffices is a common feature of the above satisfiability procedure and of the satisfiability procedures from [7]. Unfortunately, when coming to interpolation algorithms in the next section, there is no evidence that 0-instantiation suffices.

7 An interpolation algorithm

Since amalgamation is equivalent to quantifier-free interpolation for universal theories like $\mathcal{ARD}(T_I)$ (see Theorem 1), Theorem 2 ensures that $\mathcal{ARD}(T_I)$ has the quantifier-free interpolation property. However, the proof of Theorem 2 is not constructive, so in order to compute an interpolant for an $\mathcal{ARD}(T_I)$-unsatisfiable conjunction like $\psi(\underline{x}, \underline{y}) \wedge \phi(\underline{y}, \underline{z})$, one should enumerate all quantifier-free formulæ $\theta(\underline{y})$ which are logical consequences of ϕ and are inconsistent with ψ (modulo $\mathcal{ARD}(T_I)$). Since the quantifier-free fragment of $\mathcal{ARD}(T_I)$ is decidable by Theorem 3, this is an effective procedure and, since interpolants of jointly unsatisfiable pairs of formulæ exist, it also terminates. However, such kind of an algorithm is not practical.

In this section, we improve the situation by supplying a better algorithm based on instantiation (à-la-Herbrand). In the next section, using the results of the present section, for the special case where T_I is just the theory of linear orders, we identify a complexity bound for this algorithm.

Our problem is the following: given two quantifier-free formulae A and B such that $A \wedge B$ is not satisfiable (modulo $\mathcal{ARD}(T_I)$), to compute a quantifier-free formula C such that $\mathcal{ARD}(T_I) \models A \to C$, $\mathcal{ARD}(T_I) \models C \wedge B \to \bot$ and such that C contains only the variables (of sort `INDEX`, `ARRAY`, `ELEM`) which occur both in A and in B.

We call the variables occurring in both A and B *common variables*, whereas the variables occurring in A (resp. in B) are called *A-variables* (resp. *B-variables*). The same terminology applies to terms, atoms and formulae: e.g., a term t is an A-term (B-term, common term) iff it is built up from A-variables (B-variables, common variables, resp.).

The following operations can be freely performed (see [9] or [8] for details):

(i) pick an A-term t and a fresh variable a (of appropriate sort) and conjoin A to $a = t$ (a will be considered an A-variable from now on);

(ii) pick a B-term t and a fresh variable b (of appropriate sort) and conjoin B to $b = t$ (b will be considered a B-variable from now on);

(iii) pick a common term t and a fresh variable c (of appropriate sort) and conjoin both A and B to $c = t$ (c will be considered a common variable from now on);

(iv) conjoin A with some quantifier-free A-formula which is implied (modulo $\mathcal{ARD}(T_I)$) by A;

(v) conjoin B with some quantifier-free B-formula which is implied (modulo $\mathcal{ARD}(T_I)$) by B.

Operations (i)-(v) either add logical consequences or explicit definitions that can be eliminated (if desired) after the final computation of the interpolant. In addition, notice that if A is the form $A' \vee A''$ (resp. B is of the form $B' \vee B''$) then from interpolants of $A' \wedge B$ and $A'' \wedge B$ (resp. of $A \wedge B'$ and $A \wedge B''$), we can recover an interpolant of $A \wedge B$ by taking disjunction (resp. conjunction).

Because of the above remarks, using the procedure in the proof of Lemma 4, both A and B are assumed to be given in the form of finite separated pairs. Thus A is of the form $\bigwedge A_1 \wedge \bigwedge A_2$, B is of the form $\bigwedge B_1 \wedge \bigwedge B_2$, for separated pairs (A_1, A_2) and (B_1, B_2). Also, by (iv)-(v) above, A and B are assumed to be both 0-instantiated. We call A (resp. B) the separated pair (A_1, A_2) (resp. (B_1, B_2)). We also use the letters A_1, A_2, B_1, B_2 both for sets of formulae and for the corresponding conjunctions; similarly, A represent both the pair (A_1, A_2) and the conjunction $\bigwedge A_1 \wedge \bigwedge A_2$ (and similarly for B).

The formulæ from A_2 and B_2 are formulæ from the signature of $T_I \cup \mathcal{EUF}$ (after rewriting terms of the kind $rd(a, i)$ to $f_a(i)$, where the f_a are free function symbols). Of course, if $A_2 \wedge B_2$ is $T_I \cup \mathcal{EUF}$-inconsistent, *we can get our quantifier-free interpolant by using our black box algorithm for interpolation in the weaker theory* $T_I \cup \mathcal{EUF}$: recall that $T_I \cup \mathcal{EUF}$ has quantifier-free interpolation because T_I is an index theory and for Theorem 1. The remarkable fact is that $A_2 \wedge B_2$ always becomes $T_I \cup \mathcal{EUF}$-inconsistent if *sufficiently many* `diff`*s among common array variables are introduced* and *sufficiently many instantiations are performed.*

Formally, we shall *apply the loop below until $A_2 \wedge B_2$ becomes inconsistent*: the loop is justified by (i)-(v) above and Theorem 4 guarantees that $A_2 \wedge B_2$ eventually becomes inconsistent modulo $T_I \cup \mathcal{EUF}$, if $A \wedge B$ was originally inconsistent modulo $\mathcal{ARD}(T_I)$. When $A_2 \wedge B_2$ becomes inconsistent modulo $T_I \cup \mathcal{EUF}$, we can get our interpolant using the interpolation algorithm for $T_I \cup \mathcal{EUF}$. [Of course, in the interpolant returned by $T_I \cup \mathcal{EUF}$, the extra variables introduced by the explicit definitions from (iii) above need to be eliminated.] We need a counter M recording how many times the Loop below has been executed (initially $M = 0$).

Loop *(to be repeated until $A_2 \wedge B_2$ becomes inconsistent modulo $T_I \cup \mathcal{EUF}$). Pick two distinct common* `ARRAY`*-variables c_1, c_2 and $n \geq 1$ and s.t. no conjunct of the kind $\mathtt{diff}_n(c_1, c_2) = k$ occurs in both A_1 and B_1 for some $n \geq 1$ (but s.t. for every $l < n$ there is a conjunct of the form $\mathtt{diff}_l(a, b) = k$ occurring in both A_1 and B_1). Pick also a fresh* `INDEX` *constant k_n; conjoin $\mathtt{diff}_n(c_1, c_2) = k_n$ to both A_1 and B_1; then M-instantiate both A and B. Increase M to $M + 1$.*

Notice that the fresh index constants k_n introduced during the loop are considered common constants (they come from explicit definitions like (iii) above) and so they are considered in the M-instantiation of both A and B.

Example 2. Let A be the formula $\bigwedge \Phi_1$ from Example 1 and let B be

$$i_1 < i_2 \;\wedge\; i_2 < i_3 \;\wedge\; rd(c_1, i_2) \neq rd(c_2, i_2)$$

B is 0-instantiated; 0-instantiating A produces the formulæ shown in Example 1. The loop needs to be executed twice; it adds the literals $\mathtt{diff}_0(c_1, c_2) =$

k_0, $\texttt{diff}_1(c_1, c_2) = k_1$; 0-instantiation produces formulae A_2, B_2 whose conjunction is $T_I \cup \mathcal{EUF}$-inconsistent (inconsistency can be tested via an SMT-solver like Z3 *or* MathSat, *see the ongoing implementation [1]). The related $T_I \cup \mathcal{EUF}$-interpolant (once k_0 and k_1 are replaced by $\texttt{diff}_0(c_1, c_2)$ and $\texttt{diff}_1(c_1, c_2)$, respectively) gives our $\mathcal{ARD}(T_I)$-interpolant.* ⊣

Theorem 4. *If $A \wedge B$ is $\mathcal{ARD}(T_I)$-inconsistent, then the above loop terminates.*

Proof. Suppose that the loop does not terminate and let $A' = (A'_1, A'_2)$ and $B' = (B'_1, B'_2)$ be the separated pairs obtained after infinitely many executions of the loop (they are the union of the pairs obtained in each step). Notice that both A' and B' are fully instantiated.[12] We claim that (A', B') is $\mathcal{ARD}(T_I)$-consistent (contradicting the assumption that (A, B) was already $\mathcal{ARD}(T_I)$-inconsistent).

Since no contradiction was found, by compactness of first-order logic, $A'_2 \cup B'_2$ has a $T_I \cup \mathcal{EUF}$-model $\mathcal{M}$ (below we treat index and element variables occurring in A, B as free constants and the array variables occurring in A, B as free unary function symbols). $\mathcal{M}$ is a two-sorted structure (the sorts are INDEX and ELEM) endowed for every array variable a occurring in A, B of a function $a^{\mathcal{M}} : \texttt{INDEX}^{\mathcal{M}} \longrightarrow \texttt{ELEM}^{\mathcal{M}}$. In addition, $\texttt{INDEX}^{\mathcal{M}}$ is a model of T_I. We build three $\mathcal{ARD}(T_I)$-structures $\mathcal{A}, \mathcal{B}, \mathcal{C}$ and two embeddings $\mu_1 : \mathcal{C} \longrightarrow \mathcal{A}$, $\mu_2 : \mathcal{C} \longrightarrow \mathcal{B}$ such that $\mathcal{A} \models A'$, $\mathcal{B} \models B'$ and such that for every common variable x we have $\mu_1(x^{\mathcal{C}}) = x^{\mathcal{A}}$ and $\mu_2(x^{\mathcal{C}}) = x^{\mathcal{B}}$. The consistency of $A' \cup B'$ then follows from the amalgamation Theorem 2. The two structures $\mathcal{A}, \mathcal{B}$ are obtained by taking the full functional model induced by the restriction of $\mathcal{M}$ to the interpretation of A-terms and B-terms (respectively) of sort INDEX, ELEM and then by applying Lemma 3; the construction of $\mathcal{C}$ requires some subtleties, to be detailed in the extended version [21], where the full proof of the theorem is provided. ⊣

8 When indexes are just a total order

Comparing the results from Sections 7 and 6, a striking difference emerges: whereas variable and constant instantiations are sufficient for satisfiability checking, our interpolation algorithm requires full instantiation over all common terms. Such a full instantiation might be quite impractical, especially in index theories like $\mathcal{LIA}$ and $\mathcal{LRA}$ (it is less annoying in theories like $\mathcal{IDL}$: here all terms are of the kind $S^n(x)$ or $P^n(x)$, where x is a variable or 0 and S, P are the successor and the predecessor functions). The problem disappears in simpler theories like the theory of linear orders TO, where all terms are variables (or the constant 0). Still, even in the case of TO, the proof of Theorem 4 does not give a bound for termination of the interpolation algorithm: we know that sooner or later an inconsistency will occur, but we do not know how many times we need to execute the main loop. We now improve the proof of Theorem 4 by supplying the missing bound. In this section, the index theory is fixed to be TO and we abbreviate $\mathcal{ARD}(TO)$ as $\mathcal{ARD}$. The full proof of the theorem below is in [21].

[12] On the other hand, the joined pair $(A'_1 \cup B'_1, A'_2 \cup B'_2)$ is not even 0-instantiated.

Theorem 5. *If $A \wedge B$ is inconsistent modulo $\mathcal{ARD}$, then the above loop terminates in at most $(\frac{m^2-m}{2}) \cdot (n+1)$ steps, where n is the number of the index variables occurring in A, B and m is the number of the common array variables.*

Proof. We sketch a proof of the theorem: the idea is that if after $N := (\frac{m^2-m}{2}) \cdot (n+1)$ steps no inconsistency occurs, then we can run the algorithm for infinitely many further steps without finding an inconsistency either. Let $A^N = (A_1^N, A_2^N)$ and $B^N = (B_1^N, B_2^N)$ be obtained after N-executions of the loop and let $\mathcal{M}$ be a $\mathcal{TO} \cup \mathcal{EUF}$-model of $A_2^N \wedge B_2^N$. Fix a pair of distinct common array variables c_1, c_2 to be handled in Step $N+1$; since all pairs of common array variables have been examined in a fair way, A_1^N and B_1^N contain the atom $\mathtt{diff}_{n+1}(c_1, c_2) = k_{n+1}$ (in fact $N := (\frac{m^2-m}{2}) \cdot (n+1)$ and $(\frac{m^2-m}{2})$ is the number of distinct unordered pairs of common array variables, so the pair (c_1, c_2) has been examined more than n times). In $\mathcal{M}$, some index variable k_l for $l \leq k_{n+1}$, if not assigned to 0, is assigned to an element x which is different from the elements assigned to the n variables occurring in A, B. This allows us to enlarge $\mathcal{M}$ to a superstructure which is a model of $A_2^{N+1} \wedge B_2^{N+1}$ by 'duplicating' x. Continuing in this way, we produce a chain of $\mathcal{TO} \cup \mathcal{EUF}$-models witnessing that we can run infinitely many steps of the algorithm without finding an inconsistency. ⊣

9 Conclusions and further work

We studied an extension of McCarthy theory of arrays with a maxdiff symbol. This symbol produces a much more expressive theory than the theory of plain diff symbol already considered in the literature [8,37].

We have also considered another strong enrichment, namely the combination with arithmetic theories like $\mathcal{IDL}, \mathcal{LIA}, \mathcal{LRA}, \ldots$ (all such theories are encompassed by the general notion of an 'index theory'). Such a combination is non trivial because it is a non disjoint combination (the ordering relation is in the shared signature) and does not fulfill the T_0-compatibility requirements of [17,19,18] needed in order to modularly import satisfiability and interpolation algorithms from the component theories.

The above enrichments come with a substantial cost: although decidability of satisfiability of quantifier-free formulae is not difficult to obtain, quantifier-free interpolation becomes challenging. In this paper, we proved that quantifier-free interpolants indeed do exist: the interpolation algorithm is indeed rather simple, but its justification comes via a complicated détour involving semantic investigations on amalgamation properties.

The interpolation algorithm is based on hierarchic reduction to general quantifier-free interpolation in the index theory. The reduction requires the introduction of iterated diff terms and a finite number of instantiations of the universal clauses associated to write and diff-atoms. For the simple case where the index theory is just the theory of total orders, we were able to polynomially bound the depth of the iterated diff terms to be introduced as well as the number of instantiations needed. The main open problem we leave for future is the determination of analogous bounds for richer index theories.

References

1. AXDInterpolator, https://github.com/typesAreSpaces/AXDInterpolator, accessed: 2020-10-12
2. Alberti, F., Bruttomesso, R., Ghilardi, S., Ranise, S., Sharygina, N.: Lazy abstraction with interpolants for arrays. In: Proc. of LPAR-18. LNCS, vol. 7180, pp. 46–61. Springer (2012). https://doi.org/10.1007/978-3-642-28717-6_7
3. Alberti, F., Bruttomesso, R., Ghilardi, S., Ranise, S., Sharygina, N.: SAFARI: SMT-based abstraction for arrays with interpolants. In: Proc. of CAV. LNCS, vol. 7358, pp. 679–685. Springer (2012). https://doi.org/10.1007/978-3-642-31424-7_49
4. Alberti, F., Bruttomesso, R., Ghilardi, S., Ranise, S., Sharygina, N.: An extension of lazy abstraction with interpolation for programs with arrays. Formal Methods Syst. Des. **45**(1), 63–109 (2014)
5. Alberti, F., Ghilardi, S., Sharygina, N.: Booster: An acceleration-based verification framework for array programs. In: Proc. of ATVA. LNCS, vol. 8837, pp. 18–23. Springer (2014). https://doi.org/10.1007/978-3-319-11936-6_2
6. Bacsich, P.D.: Amalgamation properties and interpolation theorems for equational theories. Algebra Universalis **5**, 45–55 (1975)
7. Bradley, A.R., Manna, Z., Sipma, H.B.: What's decidable about arrays? In: Proc. of VMCAI. LNCS, vol. 3855, pp. 427–442. Springer (2006). https://doi.org/10.1007/11609773_28
8. Bruttomesso, R., Ghilardi, S., Ranise, S.: Quantifier-free interpolation of a theory of arrays. Log. Methods Comput. Sci. **8**(2) (2012)
9. Bruttomesso, R., Ghilardi, S., Ranise, S.: Quantifier-free interpolation in combinations of equality interpolating theories. ACM Trans. Comput. Log. **15**(1), 5:1–5:34 (2014)
10. Calvanese, D., Ghilardi, S., Gianola, A., Montali, M., Rivkin, A.: Model completeness, covers and superposition. In: Proc. of CADE. LNCS (LNAI), vol. 11716, pp. 142–160. Springer (2019). https://doi.org/10.1007/978-3-030-29436-6_9
11. Calvanese, D., Ghilardi, S., Gianola, A., Montali, M., Rivkin, A.: Combined covers and Beth definability. In: Proc. of IJCAR. LNCS (LNAI), vol. 12166, pp. 181–200. Springer (2020). https://doi.org/10.1007/978-3-030-51074-9_11
12. Calvanese, D., Ghilardi, S., Gianola, A., Montali, M., Rivkin, A.: Model completeness, uniform interpolants and superposition calculus (with applications to verificaton of data-aware processes). J. Autom. Reasoning (To appear)
13. Chakraborty, S., Gupta, A., Unadkat, D.: Verifying array manipulating programs with full-program induction. In: Proc. of TACAS. LNCS, vol. 12078, pp. 22–39. Springer (2020). https://doi.org/10.1007/978-3-030-45190-5_2
14. Chang, C.C., Keisler, H.J.: Model Theory. North-Holland Publishing Co., Amsterdam-London, third edn. (1990)
15. Craig, W.: Three uses of the Herbrand-Gentzen theorem in relating model theory and proof theory. J. Symbolic Logic **22**, 269–285 (1957)
16. Fedyukovich, G., Prabhu, S., Madhukar, K., Gupta, A.: Quantified invariants via syntax-guided synthesis. In: Proc. of CAV. LNCS, vol. 11561, pp. 259–277. Springer (2019). https://doi.org/10.1007/978-3-030-25540-4_14
17. Ghilardi, S.: Model theoretic methods in combined constraint satisfiability. J. Autom. Reasoning **33**(3-4), 221–249 (2004)
18. Ghilardi, S., Gianola, A.: Interpolation, amalgamation and combination (the non-disjoint signatures case). In: Proc. of FroCoS. LNCS (LNAI), vol. 10483, pp. 316–332. Springer (2017). https://doi.org/10.1007/978-3-319-66167-4_18

19. Ghilardi, S., Gianola, A.: Modularity results for interpolation, amalgamation and superamalgamation. Ann. Pure Appl. Logic **169**(8), 731–754 (2018)
20. Ghilardi, S., Gianola, A., Kapur, D.: Computing uniform interpolants for EUF via (conditional) DAG-based compact representations. In: Proc. of CILC. CEUR Workshop Proceedings, vol. 2710, pp. 67–81. CEUR-WS.org (2020)
21. Ghilardi, S., Gianola, A., Kapur, D.: Interpolation and amalgamation for Arrays with MaxDiff (extended version). Technical Report arXiv:2010.07082, arXiv.org (2020), https://arxiv.org/abs/2010.07082
22. Gurfinkel, A., Shoham, S., Vizel, Y.: Quantifiers on demand. In: Proc. of ATVA. LNCS, vol. 11138, pp. 248–266. Springer (2018). https://doi.org/10.1007/978-3-030-01090-4_15
23. Hoenicke, J., Schindler, T.: Efficient interpolation for the theory of arrays. In: Proc. of IJCAR. LNCS (LNAI), vol. 10900, pp. 549–565. Springer (2018). https://doi.org/10.1007/978-3-319-94205-6_36
24. Huang, G.: Constructing Craig interpolation formulas. In: Computing and Combinatorics *COCOON*. LNCS, vol. 959, pp. 181–190. Springer (1995). https://doi.org/10.1007/BFb0030832
25. Ish-Shalom, O., Itzhaky, S., Rinetzky, N., Shoham, S.: Putting the squeeze on array programs: Loop verification via inductive rank reduction. In: Proc. of VMCAI. LNCS, vol. 11990, pp. 112–135. Springer (2020). https://doi.org/10.1007/978-3-030-39322-9_6
26. Kapur, D.: Nonlinear polynomials, interpolants and invariant generation for system analysis. In: Proc. of the 2nd International Workshop on Satisfiability Checking and Symbolic Computation co-located with ISSAC (2017)
27. Kapur, D.: Conditional congruence closure over uninterpreted and interpreted symbols. J. Systems Science & Complexity **32**(1), 317–355 (2019)
28. Kapur, D., Majumdar, R., Zarba, C.G.: Interpolation for Data Structures. In: Proc. of SIGSOFT-FSE. pp. 105–116. ACM (2006)
29. Krishnan, H.G.V., Vizel, Y., Ganesh, V., Gurfinkel, A.: Interpolating strong induction. In: Proc. of CAV. LNCS, vol. 11562, pp. 367–385. Springer (2019). https://doi.org/10.1007/978-3-030-25543-5_21
30. McCarthy, J.: Towards a Mathematical Science of Computation. In: IFIP Congress. pp. 21–28 (1962)
31. McMillan, K.L.: Interpolation and SAT-based model checking. In: Proc. of CAV. LNCS, vol. 2725, pp. 1–13. Springer (2003). https://doi.org/10.1007/978-3-540-45069-6_1
32. McMillan, K.L.: Lazy abstraction with interpolants. In: Proc. of CAV. LNCS, vol. 4144, pp. 123–136. Springer (2006). https://doi.org/10.1007/11817963_14
33. Nelson, G., Oppen, D.C.: Simplification by Cooperating Decision Procedures. ACM Transactions on Programming Languages and Systems **1**(2), 245–57 (1979)
34. Pudlák, P.: Lower bounds for resolution and cutting plane proofs and monotone computations. J. Symb. Log. **62**(3), 981–998 (1997)
35. Sofronie-Stokkermans, V.: Interpolation in local theory extensions. Log. Methods Comput. Sci. **4**(4) (2008)
36. Sofronie-Stokkermans, V.: On interpolation and symbol elimination in theory extensions. Log. Methods Comput. Sci. **14**(3) (2018)
37. Totla, N., Wies, T.: Complete instantiation-based interpolation. J. Autom. Reasoning **57**(1), 37–65 (2016)
38. Vizel, Y., Gurfinkel, A.: Interpolating property directed reachability. In: Proc. of CAV. LNCS, vol. 8559, pp. 260–276. Springer (2014). https://doi.org/10.1007/978-3-319-08867-9_17

Adjoint Reactive GUI Programming

Christian Uldal Graulund[1] [✉], Dmitrij Szamozvancev[2], and Neel Krishnaswami[2]

[1] IT University of Copenhagen, 2300 Copenhagen, DK cgra@itu.dk
[2] University of Cambridge, Cambridge CB3 0FD, UK
nk480@cl.cam.ac.uk,ds709@cl.cam.ac.uk

Abstract. Most interaction with a computer is via graphical user interfaces. These are traditionally implemented imperatively, using shared mutable state and callbacks. This is efficient, but is also difficult to reason about and error prone. Functional Reactive Programming (FRP) provides an elegant alternative which allows GUIs to be designed in a declarative fashion. However, most FRP languages are synchronous and continually check for new data. This means that an FRP-style GUI will "wake up" on each program cycle. This is problematic for applications like text editors and browsers, where often nothing happens for extended periods of time, and we want the implementation to sleep until new data arrives. In this paper, we present an *asynchronous* FRP language for designing GUIs called $\lambda_{\mathsf{Widget}}$. Our language provides a novel semantics for widgets, the building block of GUIs, which offers both a natural Curry–Howard logical interpretation and an efficient implementation strategy.

Keywords: Linear Types · FRP · Asynchrony · GUIs

Introduction

Many programs, like compilers, can be thought of as functions – they take a single input (a source file) and then produce an output (such as a type error message). Other programs, like embedded controllers, video games, and integrated development environments (IDEs), engage in a dialogue with their environment: they receive an input, produce an output, and then wait for a new input that depends on the prior input, and produce a new output which is in turn potentially based on the whole history of prior inputs.

The usual techniques for programming interactive applications are often confusing, since different parts of the program are not written to interact via structured control flow (e.g., by passing and return values from functions). Instead, they communicate indirectly, via state-manipulating callbacks which are implicitly invoked by an event loop. This makes program reasoning very challenging, since each of aliased mutable state, higher-order functions, and concurrency is tricky on its own, and interactive programs rely upon their *combination*.

This challenge has led to a great deal of work on better abstractions for programming reactive systems. Two of the main lines of work on this problem are *synchronous dataflow* and *functional reactive programming*. The synchronous

S. Kiefer and C. Tasson (Eds.): FOSSACS 2021, LNCS 12650, pp. 289–309, 2021.
https://doi.org/10.1007/978-3-030-71995-1_15

dataflow languages, like Esterel [5], Lustre [9], and Lucid Synchrone [28], feature a programming model inspired by Kahn networks. Programs are networks of stream-processing nodes which communicate with each other, each node consuming and producing a fixed number of primitive values at each clock tick. The first-order nature of these languages makes them strongly analysable, which lets them offer powerful guarantees on space and time usage. This means they see substantial use in embedded and safety-critical contexts.

Functional reactive programming, introduced by Elliott and Hudak [13], also uses time-indexed values, dubbed signals, rather than mutable state as its basic primitive. However, FRP differs from synchronous dataflow by sacrificing static analysability in favour of a much richer programming model. Signals are true first-class values, and can be used freely, including in higher-order functions and signal-valued signals. This permits writing programs with a dynamically-varying dataflow network, which simplifies writing programs (such as GUIs) in which the available signals can change as the program executes. Over the past decade, a long line of work has refined FRP via the Curry–Howard correspondence [21,18,17,19,20,10,1]. This approach views functional reactive programs as the programming counterpart for proofs of formulas in linear temporal logic [27], and has enabled the design of calculi which can rule out spacetime leaks [20] or can enforce temporal safety and liveness properties [10].

However, both synchronous dataflow and FRP (in both original and modal flavours) have a *synchronous* (or "pull") model of time – time passes in ticks, and the program wakes up on every tick to do a little bit more computation. This is suitable for applications in which something new happens at every time step (e.g., video games), but many GUI programs like text editors and spreadsheets spend most of their time doing nothing. That is, even at each event, most of the program will continue doing nothing, and we only want to wake up a component when an event directly relevant to it occurs. This is important both from a performance point of view, as well as for saving energy (and extending battery life). Because of this need, most GUI programs continue to be written in the traditional callbacks-on-mutable-state style.

In this paper, we give a reactive programming language whose type system both has a very straightforward logical reading, and which can give natural types to stateful widgets and the event-based programming model they encourage. We also derive a denotational semantics of the language, by first working out a semantics of widgets in terms of the operations that can be performed upon them and the behaviour they should exhibit. Then, we find the categorical setting in which the widget semantics should live, and by studying the structure this setup has, we are able to interpret all of the other types of the programming language.

Contributions The contributions of this paper are:

– We give a descriptive semantics for widgets in GUI programming, and show that this semantics correctly models a variety of expected behaviours. For example, our semantics shows that a widget which is periodically re-set to the colour red is different from a widget that was only persistently set to the colour red at the first timestep. Our semantic model can show that as long as neither one is updated, they look the same, but that they differ if they are ever set to blue – the first will return to red at reset time, and the second will remain blue.

- From this semantics, we find a categorical model within which the widget semantics naturally fits. This model is a Kripke–Joyal presheaf semantics, which is morally a "proof-relevant" Kripke model of temporal logic.
- We give a concrete calculus for event-based reactive programming, which can be implemented in terms of the standard primitives for modern GUI programming, scene graphs (or DOM) which are updated via callbacks invoked upon events. We then show that our model can soundly interpret the types of our calculus in an entirely standard way, showing that the types of our reactive programming language can be interpreted as time-varying sets.
- Furthermore, this calculus has an entirely standard logical reading in terms of the Curry–Howard correspondence. It is a "linear temporal linear logic", with the linear part of the language corresponding to the Benton–Wadler [3] LNL calculus for linear logic, and the temporal part of the language corresponding to S4.3 linear temporal logic. We also give a proof term for the $S_t4.3$ axiom enforcing the linearity of time, and show that it corresponds to the `select` primitive of concurrent programming.

The Language

We now present $\lambda_{\mathsf{Widget}}$ through the API of the Widget type. This API mirrors how one would work with a GUI at the browser level. An important feature of a well-designed GUI is that it should not do anything when not in use. In particular, it should not check for new inputs in each program cycle (*pull*-based reactive programming), but rather sleep until new data arrives (*push*-based reactive programming). Many FRP languages are *synchronous* languages and have some internal notion of a timestep. These languages are mostly pull-based, whereas more traditional imperative reactive languages are push-based. The former have clear semantics and are easy to reason about, the latter have efficient implementations. In $\lambda_{\mathsf{Widget}}$ we would like to combine these aspects and get a language that is easy to reason about with an efficient implementation.

In general, we think of a widget as a *state through time*, i.e., at each timestep, the widget is in some state which is presented to the user. The widget is modified by *commands*, which can update the state. To program with widgets, the programmer applies commands at various times.

The proper type system for a language of widgets should thus be a system with both state and time. If we consider what a *logic* for widgets should be, there are two obvious choices. A logic for state is linear logic [14], and a logic for time is linear temporal logic [27]. The combination of these two is the correct setting for a language of widgets, and, going through Curry–Howard, the corresponding type theory is a linear, linear temporal type theory.

Widget API To work with widgets, we define a API which mirrors how one would work with a browser level GUI:

$$\begin{aligned}
&\mathsf{newWidget} : \mathsf{I} \multimap \exists\,(i : \mathsf{Id}), \mathsf{Widget}\ i\\
&\mathsf{dropWidget} : \forall\,(i : \mathsf{Id}), \mathsf{Widget}\ i \multimap \mathsf{I}\\
&\mathsf{setColor} \quad : \forall\,(i : \mathsf{Id}), \mathsf{F\ Color} \otimes \mathsf{Widget}\ i \multimap \mathsf{Widget}\ i\\
&\mathsf{onClick} \quad : \forall\,(i : \mathsf{Id}), \mathsf{Widget}\ i \multimap \mathsf{Widget}\ i \otimes \Diamond\,\mathsf{I}\\
&\mathsf{onKeypress} : \forall\,(i : \mathsf{Id}), \mathsf{Widget}\ i \multimap \mathsf{Widget}\ i \otimes \Diamond\,(\mathsf{F\ Char})\\
&\mathsf{out} \quad : \Diamond\,A \multimap \exists\,(n : \mathsf{Time}), A\,@\,n\\
&\mathsf{into} \quad : \exists\,(n : \mathsf{Time}), A\,@\,n \multimap \Diamond\,A\\
&\mathsf{split} \quad : \forall\,(i : \mathsf{Id})\,(t : \mathsf{Time}), \mathsf{Widget}\ i \multimap \mathsf{Prefix}\ i\ t \otimes (\mathsf{Widget}\ i)\,@\,t\\
&\mathsf{join} \quad : \forall\,(i : \mathsf{Id})\,(t : \mathsf{Time}), \mathsf{Prefix}\ i\ t \otimes (\mathsf{Widget}\ i)\,@\,t \multimap \mathsf{Widget}\ i
\end{aligned}$$

The first two commands creates and deletes widgets, respectively. The $\multimap$ should be understood as *state passing.* We read the type of newWidget as "consuming no state, produce a new identifier index and a widget with that identifier index". The identifier indices are used to ensure the correct behavior when using the split and join commands explained below. The existential quantification describes the *non-deterministic* creation of an identifier index. The use of non-determinism is crucial in our language and will be explaining in further detail in section 1. Since $\lambda_{\mathsf{Widget}}$ has a linear type system, we need an explicit construction to delete state. For widgets, this is dropWidget. The type is read as "for any identifier index, consume a widget with that identifier index and produce nothing".

The first command that modifies the state of a widget is setColor. Here we see the adjoint nature of the calculus with F Color. A color is itself *not* a linear thing, and as such, to use it in the linear setting, we apply F, which moves from the non-linear (Cartesian) fragment and into the linear fragment. The second new thing is the linear product $\otimes$. This differs from the regular non-linear product in that we do not have projection maps. Again, because of the linearity of our language, we cannot just discard state. We can now read the type of setColor as "Given a color and a identified widget, consume both and produce a new widget". The produced widget is the same as the consumed widget, but with the color attribute updated.

The next two commands, onClick and onKeypress, are roughly similar. Both register a handle on the widget, for a mouse click and a key press, respectively. Here we see the first use of the $\Diamond$ modality, which represents an *event.* The type $\Diamond A$ represents that *at some point in the future* we will receive something of type A. Importantly, because of the asynchronous nature of $\lambda_{\mathsf{Widget}}$, we do not know *when* it happens. We can then read the type of onClick as "Consuming an identified widget, produce an updated widget together with a mouse click event". The same holds for onKeypress except a key press event is produced.

The two commands out and into allows us to work with events in a more precise way. Given an event, we can use out to "unfold" it into an existential. The @ connective describes a type that is only available at a certain timestep, i.e., $A\,@\,n$ means "at the timestep n, a term of type A will be available". The into commands is the reverse of out and turns an existential and an @ into an event.

Note the besides the above ways of constructing events, we can also turn any value into an event using the evt construction which is part of the core calculus. Given some element $a : A$, we get evt $a : \Diamond A$ which represents the event that returns immediately.

So far, we have only applied commands to a widget in the current timestep, but to program appropriately with widgets, we should be able to react to events and apply commands "in the future". This is exactly what the split and join commands allows us to do. The type of split is read as "Given any time step and any identified widget, split the widget into all the states *before* that time and the widget *at* that time". We denote the collection of states before a given time a *prefix* and give it the type Prefix. Given the state of the widget at a given timestep, we can now apply commands *at that timestep*. Note that both the prefix and the widget is indexed by the same identifier index. This is to ensure that when we use join, we combined the correct prefix and future.

Widget Programming To see the API in action, we now proceed with several examples of widget programming. For each example, we will add a comment on each line with the type of variables, and then explain the example in text afterwards.

One of the simplest things we can do with a widget is to perform some action when the widget is clicked. In the following example, we register a handler for mouse clicks, and then we use the click event to change the color of the widget to red at the time of the click. To do this, we use the out map to get the time of the event, then we split the widget and apply setColor at that point in the future.

```
1  turnRedOnClick : ∀ (i : Id), Widget i ⊸ Widget i
2  turnRedOnClick i w0 =
3    let (w1, c0)          = onClick i w0 in      -- w1 : Widget i, c0 : ◇I
4    let unpack (x, c1)    = out c0 in            -- x : Time, c1 : I @ x
5    let c2 @ x            = c1 in                -- c2 : I at x
6    let ⟨⟩ @ x            = c2 in
7    let (p, w2)           = split i x w1 in      -- p : Prefix i x, w2 : Widget i @ x
8    let w3 @ x            = w2 in                -- w3 : Widget i at time x
9    let w4                =                      -- w4 : Widget i @ x
10       (setColor (F Red) w3) @ x in
11   join i x (p, w4)
```

To see why this type checks, we go through the example line by line. In line 3, we register a handle for a mouse click on the widget. In line 4, we turn the click event into an existential. In line 5, we get c_2 which is a binding that is only available at the timestep x. Since we only need the *time* of the click, we discharge the click itself in line 6. In line 7 and 8, we split the widget using the timestep x and bind w_3 to the state of the widget at that timestep. In line 9-10, we change the color of the widget to red at x and in line 11 we recompose the widget.

In general, we will allow pattern matching in eliminations and since widget identity indices can always be inferred, we will omit them. In this style, the above example become:

```
turnRedOnClick : ∀ (i : Id), Widget i ⊸ Widget i
turnRedOnClick w0 =
  let (w1, c0)              = onClick w1 in   -- w1 : Widget i, c0 : ◇I
  let unpack (x, ⟨⟩ @ x) = out c0 in        -- x : Time
  let (p, w2 @ x)           = split x w1 in   -- p : Prefix i x, w2 : Widget i at time x
  join x (p, (setColor (F Red) w2) @ x)
```

We will use the same sugared style throughout the rest of the examples.

The above example turns a widget red exactly at the time of the mouse click, but will not do anything with successive clicks. To also handle further mouse clicks, we must register an event handler *recursively*. This is a simple modification of the previous code:

```
keepTurningRed : ∀ (i : Id), Widget i ⊸ Widget i
keepTurningRed w0 =
  let (w1, c0)              = onClick w1 in   -- w1 : Widget i, c0 : ◇I
  let unpack (x, ⟨⟩ @ x) = out c0 in        -- x : Time
  let (p, w2 @ x)           = split x w1 in   -- p : Prefix i x, w2 : Widget i at time x
  join (p, (setColor (F Red) (keepTurningRed w2) @ x))
```

By calling itself recursively, this function will make sure a widget will always turn red on a mouse click.

To understand the difference between two above examples, consider the code *turnBlueOnClick*(*keepTurningRed* w), where w is some widget. On the first click, the widget will turn blue, on the second click it will turn red and on any subsequent click, it will keep turning red, i.e., stay red unless further modified.

When working with widgets, we will often register multiple handlers on a single widget. For example, a widget should have one behavior for a click and another behavior for a key press. To choose between two events, we use the select construction. This construction is central to our language and how to think about a push-based reactive language.

Given two events, $t_1 : \Diamond A, t_2 : \Diamond B$, there are three possible behaviors: Either t_1 returns first, and we wait for t_2 or t_2 returns first and we wait for t_1 or they return at the same time. In general, we want to select between n events, but if we need to handle all possible cases, this will give 2^n cases, so to keep the syntax linear in size, we will omit the last case. In the case events *actually* return at the same time, we do a non-deterministic choice between them. The syntax for select is

$$\mathsf{select}\ (t_1\ \mathsf{as}\ x \mapsto t_1' \mid t_2\ \mathsf{as}\ y \mapsto t_2')$$

where $x : A, y : B, t_1' : A \multimap \Diamond B \multimap \Diamond C$ and $t_2' : B \multimap \Diamond A \multimap \Diamond C$. The second important thing to understand when working with select is that given we are working with events, we do not actually know at which timestep the events will trigger, and hence, we do not know what the (linear) context contains. Thus, when using select, we will *only* know either $a : A, t_2 : \Diamond B$ or $t_1 : \Diamond A, b : B$. We can think of the select rule a *case-expression* that must respect time.

In the following example, we register two handlers, one for clicks and one for key presses, and change the color of the widget based on which returns first. We will only annotate the new parts.

```
1  widgetSelect : ∀ (i : Id), Widget i ⊸ Widget i
2  widgetSelect w0 =
3    let (w1, c)        = onClick w0 in        -- c : ◇I.
4    let (w2, k)        = onKeypress w1 in     -- k : ◇(F char).
5    let col            =                      -- col : ◇(F Color)
6      select
7        ( c as x → let ⟨⟩ = x in              -- x : I, k : ◇(F Color).
8                   let unpack (t, ⟨⟩ @ t)
9                      = out (mapE (fun F (_) → ⟨⟩) k) in
10                  evt (F Red)
11       | k as y → let F k′ = y in            -- y : F char, c : ◇I
12                  let unpack (t, ⟨⟩ @ t) = c in
13                  evt (F Blue))
14   let unpack (x, col′ @ x) = out col in     -- col′ : F Color at time x.
15   let (p, w3 @ x)          = split x w2 in
16   join (p, (setColor col′ w3) @ x)
```

In line 3 and 4, we register the two handlers. In line 5-13, we use the select construction. In the first case, the click happens first and we return the color red. In the second case, the key press happens first and we return the color blue. In both cases, because of the linear nature of the language, we need to discharge the unit and char, respectively, and the event that does not return first. In line 14, we turn the color event into an existential. In line 15, we use the timestep of the color event to split the widget, and in line 16, we change the color of the widget at that time and recompose it.

To see how $\lambda_{\mathsf{Widget}}$ differs from more traditional synchronous FRP languages, we will examine how to encode a kind of streams. Since our language is *asynchronous*, the stream type must be encoded as

$$\mathsf{Str}\ A := \nu\alpha.\Diamond(A \otimes \alpha)$$

This asynchronous stream will *at some point in the future* give a head and a tail. We do not know when the first element of the stream will arrive, and after each element of the stream is produced, we will wait an indeterminate amount of time for the next element. The reason why the stream type in $\lambda_{\mathsf{Widget}}$ must be like this is essentially that we want a *push-based* language, i.e., we do not want to wake up and check for new data in each program cycle. Instead, the program should sleep until new data arrives.

To show the difference between the asynchronous stream and the more traditional synchronous stream, we will look at some examples. With a traditional stream, a standard operation is zipping two streams: that is, given $\mathsf{Str}\ A$ and $\mathsf{Str}\ B$, we can produce $\mathsf{Str}\ A \times B$, which should be the element-wise pairing of the two streams. It should be clear that this is not possible for our asynchronous streams. Given two streams, we can wait until the first stream produces an element, but the second stream may only produce an element after a long period of time. Hence, we would need to buffer the first element, which is not supported in general. Remember, when using select, we can not use any already defined linear variables, since we do not know if they will be available in the future.

Rather than zipping stream, we can instead do a kind of *interleaving* as shown below. We use fold and unfold to denote the folding and unfolding of the fixpoint.

```
interleave : Str A ⊸ Str B ⊸ Str (A ⊕ B)
interleave xs ys = fold (
  select
    ( unfold xs as xs' →
        let (x, xs'') = xs' in    -- xs' : A ⊗ Str A, x : A, xs'' : Str A
        evt (inl x, interleave xs'' ys)
    | unfold ys as ys' →
        let (y, ys'') = ys' in    -- ys' : B ⊗ Str B, y : B, ys'' : Str A
        evt (inr y, interleave xs ys'')))
```

Here, we use select to choose between which stream returns first, and then we let that element be the first element of the new stream.

On the other hand, some of the traditional FRP functions on streams can be translated. For instance, we can map of function over a stream, given that *it is available at each step in time*:

```
map : F (G (A ⊸ B)) ⊸ Str A ⊸ Str B
map f0 xs =
  let F f1          = f0 in    -- f1 : G(A ⊸ B)
  let (y, (x, xs') @ y) =      -- y : Time, x : A, xs' : ◇Str A at time y
    out (unfold xs) in
  fold (evt ((runG f1) x, map f0 xs'))
```

The type $\mathsf{F}(\mathsf{G}(A \multimap B))$ is read as a linear function with no free variables that can be used in a non-linear fashion, i.e., duplicated. This restriction to such "globally available functions" is reminiscent of the "box" modality in Bahr et al. [1] and Krishnaswami [20], and the F and G construction can be understood as decomposing the box modality into two separate steps. This relationship will be made precise in the logical interpretation of $\lambda_{\mathsf{Widget}}$ in section 1

As a final example, we will show how to dynamically update the GUI, i.e., how to add new widgets on the fly. Before we can give the example, we need to extend our widget API, to allow composition of widgets. To that end, we add the vAttach command to our API.

$$\mathsf{vAttach} : \forall (i, j : \mathsf{Id}), \mathsf{Widget}\ i \multimap \mathsf{Widget}\ j \multimap \mathsf{Widget}\ i$$

This command should be understood as an abstract version the div tag in HTML. In the following example, we think of the widget as a simple button that when clicked, will create a new button. When *any* of the buttons gets clicked, a new button gets attached.

```
1 buttonStack : ∀ i, Widget i ⊸ Widget i
2 buttonStack w0 =
3   let (w1, c)       = onClick w0 in
4   let (x, ⟨⟩ @ x)   = out e in
5   let (p, w2 @ x)   = split x w1 in
6   let w3            = (let (y, w) = newWidget ⟨⟩ in
7                        vAttach w2 (buttonStack w)) @ x in
8   join (p, w3)
```

The important step here is in line 6 and 7. Here the new button is attached at the time of the mouse click, and *buttonStack* is called recursively on the newly created button.

Formal Calculus

This sections gives the rules, meta-theory and logical interpretation of $\lambda_{\mathsf{Widget}}$. Briefly, the language is a mixed linear-non-linear adjoint calculus in the style of Benton–Wadler [4,3]. The non-linear fragment, also called Cartesian in the following, is a minimal simply typed lambda calculus whereas the linear fragment contains several non-standard judgments used for widget programming.

Contexts and Typing Judgments We have three typing judgments: one for indices, one for Cartesian (non-linear) terms, and one for linear terms. These are distinguished by a subscript on the turnstile, i for indices, c for Cartesian terms and l for linear terms. These depend on different contexts. The index judgment depends only on a index context, whereas the Cartesian and linear judgments depends on both an index and a linear and/or a Cartesian context. The rules for context formation is given in Figure 1. These are mostly standard except for the dependence on a previously defined context and the fact that the linear context contains variables of the form $a :_\tau A$, i.e., temporal variables. The judgment $a :_\tau A$ is read as "a has the type A at the timestep τ". In the linear setting we will write $a : A$ instead of $a :_0 A$, i.e., a judgment in the current timestep.

Indices: $$\frac{}{\vdash_i \cdot} \qquad \frac{\vdash_i \Theta \qquad s \notin \mathsf{dom}(\Theta) \qquad \sigma \in \{\mathsf{Id}, \mathsf{Time}\}}{\vdash_i \Theta, s : \sigma}$$

Cartesian: $$\frac{}{\cdot \vdash_c} \qquad \frac{\Theta \vdash_c \Gamma \qquad x \notin \mathsf{dom}(\Gamma) \qquad \Theta \vdash_c X}{\Theta \vdash_c \Gamma, x : X}$$

Linear: $$\frac{}{\cdot \vdash_l} \qquad \frac{\Theta \vdash_l \Delta \qquad x \notin \mathsf{dom}(\Delta) \qquad \Theta \vdash_l A \qquad \Theta \vdash_i \tau : \mathsf{Time}}{\Theta \vdash_l \Delta, a :_\tau A}$$

Fig. 1. Context Formation

The index judgment describes how to introduce indices. The typing rules are given in Figure 2. The judgment $\Theta \vdash_i \tau : \sigma$ contains a single context, Θ, for index variables. There are only two sorts of indices, identifiers and timesteps.

Index Judgments:

$$\frac{\tau \in \mathsf{Time}}{\Theta \vdash_i \tau : \mathsf{Time}}\ \text{TIME} \qquad \frac{\iota \in \mathsf{Id}}{\Theta \vdash_i \iota : \mathsf{Id}}\ \text{ID} \qquad \frac{i : \sigma \in \Theta}{\Theta \vdash_i i : \sigma}\ \text{VAR}$$

Fig. 2. Index Typing rules

The Cartesian judgment describes the Cartesian, or non-linear, fragment. This is a minimal simply typed lambda calculus with the addition of the G type, used for moving between the linear and Cartesian fragment, and explained further below. The judgment $\Theta; \Gamma \vdash_c t : A$ has two contexts; Θ for indices and Γ for Cartesian variables.

The linear fragment is most of the language, and a selection of typing rules is given in Figure 3. The judgment is done w.r.t three contexts, Θ for index variables, Γ for Cartesian variables and Δ for linear variables. Many of the rules are standard for a linear calculus, except for the presence of the additional contexts. We will not describe the standard rules any further.

The first non-standard rule is for $\Diamond$. The introduction and elimination rules follow from the fact that $\Diamond$ is a non-strong monad. More interesting is the select rule. Here we see the formal rule corresponding to the informal explanation in section 1. The important thing here is that we can not use any previously defined linear variable when typing t_1' and t_2', since we do not actually know *when* the typing happens. Note, we can see the select rule as a binary version of the $\Diamond$ let-binding. This could be extended to an n-ary version, but we do not do this in our core calculus. The rules for $A \mathbin{@} \tau$ shows how to move between the judgment $t : A \mathbin{@} \tau$ and $t :_\tau A$. That is, moving from knowing in the current timestep that t will have the type A at time τ and knowing at time τ that t has type A. The (F -I), (F -E), (G -I) and (G -E) rules show the adjoint structure of the language. The (G -I) rule takes a closed linear term of type A and gives it the Cartesian type G A. Note, because it has no free linear variables, it is safe to duplicate. The (G -E) rule lets us get an A without needing any linear resources. Conversely, the (F -I) rule embeds a intuitionistic term into the linear fragment and the (F -E) rule binds an intuitionistic variable to let us freely use the value. The (Delay) rule shows what happens when we actually *know* the timestep. The important part is $\Delta' = \Delta \downarrow^\tau$ which means two things. One, all the variables in Δ are on the form $a :_\tau A$, i.e., judgments at time τ and two, we shift Δ into the future such that all the variables of Δ' is of the form $a : A$. The way to understand this is, if all the variables in Δ are typed at time τ and the conclusion is at time τ, it is enough to "move to" time τ and then type w.r.t that timestep. Finally, we have (I_τ-E) and ($\otimes_\tau$-E). These allow us to work with linear unit and products at time τ. These are added explicitly since they can not be derived by the other rules, and are needed for typing certain kinds of programs.

Unfolding Events to Exists The type system as given above contains both $\Diamond A$ and $A \mathbin{@} k$, as two distinct ways to handle time. The former means that

$$\frac{\Theta \vdash_i \tau : \mathsf{Time} \qquad \Theta; \Gamma; \Delta \vdash_l t :_\tau A}{\Theta; \Gamma; \Delta \vdash_l t\,@\,\tau : A\,@\,\tau}\ (@\text{-I})$$

$$\frac{\Theta \vdash_i t : \mathsf{Time} \qquad \Theta; \Gamma; \Delta_1 \vdash_l t_1 : A\,@\,\tau \qquad \Theta; \Gamma; \Delta_2, a :_\tau A \vdash_l t_2 : B}{\Theta; \Gamma; \Delta_1, \Delta_2 \vdash_l \mathsf{let}\ a\,@\,\tau = t_1\ \mathsf{in}\ t_2 : B}\ (@\text{-E})$$

$$\frac{\Theta; \Gamma \vdash_c e : \mathsf{G}\,A}{\Theta; \Gamma; \cdot \vdash_l \mathsf{runG}\ e : A}\ (\mathsf{G}\text{-E}) \qquad \frac{\Theta; \Gamma \vdash_c e : X}{\Theta; \Gamma; \cdot \vdash_l \mathsf{F}\ e : \mathsf{F}\ x}\ (\mathsf{F}\text{-I})$$

$$\frac{\Theta; \Gamma; \Delta_1 \vdash_l t_1 : \mathsf{F}\ X \qquad \Theta; \Gamma, x : X; \Delta_2 \vdash_l t_2 : B}{\Theta; \Gamma; \Delta_1, \Delta_2 \vdash_l \mathsf{let}\ \mathsf{F}\ x = t_1\ \mathsf{in}\ t_2 : B}\ (\mathsf{F}\text{-E})$$

$$\frac{\Theta, i : \sigma; \Gamma; \Delta \vdash_l t : A}{\Theta; \Gamma; \Delta \vdash_l \Lambda(i : \sigma).t : \forall(i : \sigma).A}\ (\forall\text{-I}) \qquad \frac{\Theta \vdash_i s : \sigma \qquad \Theta; \Gamma; \Delta \vdash_l t : \forall(i : \sigma).A}{\Theta; \Gamma; \Delta \vdash_l t_s : \{s/i\}A}\ (\forall\text{-E})$$

$$\frac{\Theta \vdash_i s : \sigma \qquad \Theta; \Gamma; \Delta \vdash_l t : \{s/i\}A}{\Theta; \Gamma; \Delta \vdash_l \{s, t\} : \exists(i : \sigma).A}\ (\exists\text{-I})$$

$$\frac{\Theta; \Gamma; \Delta_1 \vdash_l t_1 : \exists(i : \sigma).A \qquad \Theta, s : \sigma; \Gamma; \Delta_2, a : \{s/i\}A \vdash_l t_2 : B}{\Theta; \Gamma; \Delta_1, \Delta_2 \vdash_l \mathsf{let\ unpack}\ \{s, a\} = t_1\ \mathsf{in}\ t_2 : B}\ (\exists\text{-E})$$

$$\frac{\begin{array}{cc}\Theta; \Gamma; \Delta_1 \vdash_l t_1 : \Diamond A & \Theta; \Gamma; \Delta_2 \vdash_l t_2 : \Diamond B \\ \Theta; \Gamma; a : A, t_2 : \Diamond B \vdash_l t_1' : \Diamond C & \Theta; \Gamma; b : B, t_1 : \Diamond A \vdash_l t_2' : \Diamond C\end{array}}{\Theta; \Gamma; \Delta_1, \Delta_2 \vdash_l \mathsf{select}\ (t_1\ \mathsf{as}\ a \mapsto t_1' \mid t_2\ \mathsf{as}\ b \mapsto t_2') : \Diamond C}\ (\textsc{select})$$

$$\frac{\Theta \vdash_i \tau : \mathsf{Time} \qquad \Delta' = \Delta \downarrow^\tau \qquad \Theta; \Gamma; \Delta' \vdash_l t : A}{\Theta; \Gamma; \Delta \vdash_l t :_\tau A}\ (\textsc{delay})$$

$$\frac{\Theta \vdash_i \tau : \mathsf{Time} \qquad \Theta; \Gamma; \Delta_1 \vdash_l t_1 :_\tau \mathsf{I} \qquad \Theta; \Gamma; \Delta_2 \vdash_l t_2 : B}{\Theta; \Gamma; \Delta_1, \Delta_2 \vdash_l \mathsf{let}\ \langle\rangle\,@\,\tau = t_1\ \mathsf{in}\ t_2 : B}\ (\mathsf{I}_\tau\text{-E})$$

$$\frac{\begin{array}{c}\Theta \vdash_i \tau : \mathsf{Time} \\ \Theta; \Gamma; \Delta_1 \vdash_l t_1 :_\tau A \otimes B \qquad \Theta; \Gamma; \Delta_2, a :_\tau A, b :_\tau B \vdash_l t_2 : C\end{array}}{\Theta; \Gamma; \Delta_1, \Delta_2 \vdash_l \mathsf{let}\ (a, b)\,@\,\tau = t_1\ \mathsf{in}\ t_2 : C}\ (\otimes_\tau\text{-E})$$

Fig. 3. Selected Linear Typing rules

something of type A will arrive at *some* point in the future, whereas the latter means an A arrives at a *specific* point in the future. The strength of $\Diamond$ is that is gives easy and concise typing rules, whereas the strength of $A @ k$ is that it allows for a more precise usage of time. To connect these two, we add the linear isomorphism $\Diamond A \cong \exists k. A @ k$ to our language, which is witnessed by out and into, as part of the widget API. This isomorphism is true semantically, but can not be derived in the type system. In particular, this isomorphism allows the select rule to be given with $\Diamond$, while still allowing the use timesteps when working with the resulting event. If we were to give the equivalent definition using timesteps, one would need to have some sort of *constraint system* for deciding which events happens first. Avoiding such constraints also allows for a simpler implementation, as everything is our type system can be inferred.

Meta-theory of Substitution The meta-theory of $\lambda_{\mathsf{Widget}}$ is given in the form of a series of substitution lemmas. Since we have three different contexts, we will end up with six different substitutions into terms. The Cartesian to Cartesian, Cartesian to linear and linear to linear are the usual notion of mutual recursive substitution. More interesting is the substitution of indices into Cartesian and linear terms and types. We prove the following lemma, showing that typing is preserved under index substitution:

Lemma 1 (Preservation of Typing under Index Substitution).

$$\frac{\zeta : \Theta' \to \Theta \qquad \Theta; \Gamma \vdash_c e : X}{\Theta'; \zeta(\Gamma) \vdash_c \zeta(e) : \zeta(X)} \qquad \frac{\zeta : \Theta' \to \Theta \qquad \Theta; \Gamma; \Delta \vdash_l t :_\tau A}{\Theta'; \zeta(\Gamma); \zeta(\Delta) \vdash_l \zeta(t) :_\tau \zeta(A)}$$

Both are these (and all other cases for substitution) are proved by a lengthy but standard induction over the typing tree. See the technical appendix for full proofs of all six substitution lemmas.

Logical Interpretation Our language has a straightforward logical interpretation. The logic corresponding to the Cartesian fragment is a propositional intuitionistic logic, following the usual Curry–Howard interpretation. The logic corresponding to the substructural part of the language is a linear, linear temporal logic. The single-use condition on variables means that the syntax and typing rules correspond to the rules of intuitionistic linear logic (i.e., the first occurrence of linear in "linear, linear temporal"). However, we do not have a comonadic exponential modality $!A$ as a primitive. Instead, we follow the Benton–Wadler approach [4,3] and decompose the exponential into the composition of a pair of adjoint functors mediating between the Cartesian and linear logic.

In addition to the Benton–Wadler rules, we have a temporal modality $\Diamond A$, which corresponds to the eventually modality of linear temporal logic (i.e., the second occurrence of "linear" in "linear, linear temporal logic"). This connective is usually written $F\,A$ in temporal logic, but that collides with the F modality of the Benton–Wadler calculus. Therefore we write it as $\Diamond A$ to reflect its nature as a possibility modality (or monad). In our calculus, the axioms of S4.3 are derivable:

$$(T) : A \multimap \Diamond A$$
$$(4) : \Diamond\Diamond A \multimap \Diamond A$$
$$(.3) : \Diamond(A \otimes B) \multimap \Diamond((\Diamond A \otimes B) \oplus \Diamond(A \otimes \Diamond B) \oplus \Diamond(A \otimes B))$$

Since the ambient logic is linear, intuitionistic implication $X \to Y$ is replaced with the linear implication $A \multimap B$, and intuitionistic conjunction $X \wedge Y$ is replaced with the linear tensor product $A \otimes B$. It is easy to see that the first two axiom corresponds to the monadic structure of $\Diamond$, and the .3 axiom corresponds to the select rule (with our syntax for select corresponding to immediately waiting for and then pattern-matching on the sum type). In the literature, the .3 axiom is often written in terms of the box modality $\Box A$ [8], but we present it here in a (classically) equivalent formulation mentioning the eventually modality $\Diamond A$. We do not need to an explicit box modality $\Box A$, since the decomposition of the exponential $\mathsf{F}(\mathsf{G}A)$ from the linear-non-linear calculus serves that role.

In our system, *we do not offer* the next-step operator $\rhd A$. Since we model asynchronous programs, we do not let programmers write programs which wake up in a specified amount of time. We only offer an iterated version of this connective, $A \,@\, n$, which can be interpreted as $\rhd^n A$, and our term syntax has no numeric constants which can be used to demand a specific delay.

Finally, the universal and existential quantifiers (in both the intuitionistic and linear fragments) are the usual quantifier rules for first-order logic.

Semantics

In this section we give a denotational model for $\lambda_{\mathsf{Widget}}$. It is a linear-non-linear (LNL) hyperdoctrine [24,16] with the non-linear part being Set and the linear part being the category of internal relations over a suitable "reactive" category. The hyperdoctrine structure is used to interpret the quantification over indices. This model is nearly entirely standard: the most interesting thing is the reactive base category and the interpretation of widgets. It is well known that any symmetric monoidal closed category (SMCC) models multiplicative intuitionistic linear logic (MILL), and it is similarly well known that the category of relations over Set can be give the structure of a SMCC by using the Cartesian product as both the monoidal product and monoidal exponential. This construction lift directly to any category of internal relations over a category that is suitably "Set-like", i.e., a topos. Our base category is a simple presheaf category, and hence, we use this construction to model the linear fragment of $\lambda_{\mathsf{Widget}}$.

The Base Reactive Category The base reactive category is where the notion of time will arise and is it this notion that will be lifted all the way up to the LNL hyperdoctrine. The simplest model of "time" is $\mathsf{Set}^{\mathbb{N}}$, which can be understood as "sets through time" [23]. This can indeed by used as a model for a reactive setting, but for our purposes it is too simple, and further, depending on which ordering is considered for $\mathbb{N}$, may have undesirable properties for the reactive setting. Instead, we use the only slightly more complicated $\mathsf{Set}^{\mathbb{N}+1}$, henceforth denoted $\mathcal{R}$, where the ordering on $\mathbb{N}+1$ is the discrete ordering on $\mathbb{N}$ and 1 is related to everything else. Adding this "point at infinity" allows global reasoning about objects, an intuition that is further supported by the definition of the sub-object classifier below. Further, this model is known to be able to differentiate between least and greatest fixpoints [15], and even though we do not use this for $\lambda_{\mathsf{Widget}}$, we consider it a useful property for further work (see section 1). Objects in $\mathcal{R}$ can be visualized as

$$A = \begin{array}{ccccc} & & A_\infty & & \\ & \swarrow^{\pi_1} & \downarrow^{\pi_2} & \searrow & \\ A_0 & & A_1 & & \cdots \end{array}$$

We can think of A_∞ as the global view of the object and A_n as the local view of the object at each timestep. Morphisms are natural transformations between such diagrams and the naturality condition means that having a map from A_∞ to B_∞ must also come with coherent maps at each timestep.

In $\mathcal{R}$ we define two endofunctors, which can be seen as describing the passage of time:

Definition 1. *We define the* later *and* previous *endofunctors on $\mathcal{R}$, denoted $\triangleright$ and $\triangleleft$, respectively:*

$$(\triangleright A)_n := \begin{cases} 1 & n = 0 \\ A_{n'} & n = n' + 1 \\ A_\infty & n = \infty \end{cases} \qquad (\triangleleft A)_n := \begin{cases} A_{n+1} & n \neq \infty \\ A_\infty & n = \infty \end{cases}$$

Note that when we apply the later functor, the global view does not change, but the local views are shifted forward in time.

Theorem 1. *The later and previous endofunctors form an adjunction.*

Definition 2. *The sub-object classifier, denoted Ω, in $\mathcal{R}$ is the object*

$$\Omega_\infty = \mathcal{P}(\mathbb{N}) + 1 \qquad\qquad \Omega_n = \{0, 1\}$$

For each $n \in \mathbb{N}$, Ω_n denotes whether a given proposition is true at the nth timestep. Ω_∞ gives the "global truth" of a given proposition. The left injection is some subset of $\mathbb{N}$ that denotes at which points in time something is true. The right injection denotes that something is true "at the limit", and in particular, also at all timesteps. Note, a proposition can be true at all timesteps but not at the limit. This extra point at infinity is precisely what allows us the differentiate between least and greatest fixpoints.

The Category of Internal Relations To interpret the linear fragment of the language, we will use the category of internal relations on $\mathcal{R}$. Given two objects A and B in $\mathcal{R}$, an *internal relation* is a sub-object of the product $A \times B$. This can equivalently by understood as a map $A \times B \to \Omega$. The category of internal relations in the category where the objects are the objects of $\mathcal{R}$ and the morphisms $A \to B$ are internal relations $A \times B \to \Omega$ in $\mathcal{R}$. We denote the category of internal relations as $\mathsf{Rel}_{\mathcal{R}}$.

Theorem 2. *Using $A \otimes B = A \times B$ and $A \multimap B = A \times B$ as monoidal product and exponential, respectively, $\mathsf{Rel}_{\mathcal{R}}$ is a symmetric monoidal closed category.*

Theorem 3. *There is an adjunction $\lhd \vdash \rhd$ in $\mathsf{Rel}_{\mathcal{R}}$ where $\lhd$ and $\rhd$ are the lifting of the previous and later functors from $\mathcal{R}$ to $\mathsf{Rel}_{\mathcal{R}}$.*

Definition 3. *We define the* iterated later modality *or the "at" connective as a successive application of the later modality.*

$$\rhd^0 A = A$$
$$\rhd^{(k+1)} A = \rhd(\rhd^k A)$$

and we will alternatively write $A \mathbin{@} k$ to mean $\rhd^k A$.

Definition 4. *We define the* event *functor on $\mathsf{Rel}_{\mathcal{R}}$ as an iterated later.*

$$\Diamond A : \mathsf{Rel}_{\mathcal{R}} \to \mathsf{Rel}_{\mathcal{R}}$$
$$(\Diamond A)_\infty = A_\infty$$
$$(\Diamond A)_n = \Sigma(k : \mathbb{N}).(\rhd^k A)_n$$

The event functor additionally carries a monadic structure (see [29] and the technical appendix).

Theorem 4. *We have the isomorphism $\Diamond A \cong \Sigma(n : \mathbb{N}).A \mathbin{@} n$ for any A*

Theorem 5. *We have the following adjunctions between* Set, $\mathcal{R}$ *and* $\mathsf{Rel}_{\mathcal{R}}$*:*

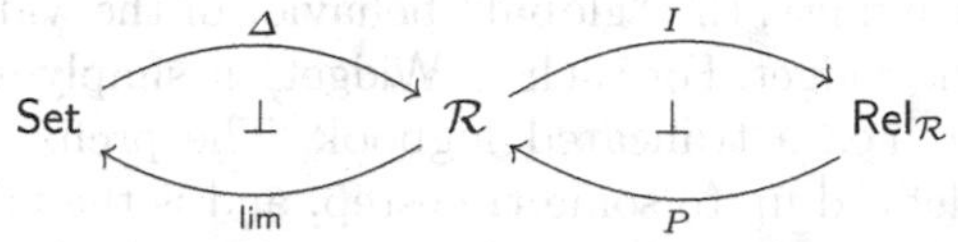

where Δ is the constant functor, lim *is the limit functor, I is the inclusion functor and P is the image functor. This induces an adjunction between* Set *and* $\mathsf{Rel}_{\mathcal{R}}$.

The Widget Object One of the most important objects in $\mathsf{Rel}_{\mathcal{R}}$ is the *widget* object. This object is used to interpret widgets and prefixes. The widget object will be defined with respect to an ambient notion of identifiers, which we will denote Id. These will be part of the hyperdoctrine structure define below, and for now, we will just assume such an object to exists. We will also use a notion of timesteps internal to the widget object. Note that this timestep is different from the abstract timestep used for defining $\mathsf{Rel}_{\mathcal{R}}$, but are related as defined below. We denote the abstract timesteps with Time.

Before we can define the widget object, we need to define an appropriate object of commands. In our minimal Widget API, the only *semantic* commands will be setColor, onClick and onKeypress. The rest of the API is defined as morphisms on the widget object itself. To work with the semantics commands, we additionally need a *compatibility* relation. This relation describes what commands can be applied at the same time. In our setting this relation is minimal, but can in principle be used to encode whatever restrictions is needed for a given API.

Definition 5. *We define the command object as*

$$\mathsf{Cmd} = \{(\mathsf{setColor}, \mathit{color}), \mathsf{onClick}, \mathsf{onKeypress}\}$$

where color is an element of a "color" object. The compatibility relations are:

$$(op, arg) \bowtie (op', arg') \text{ iff } (op = op' \Rightarrow arg = arg')$$

The only non-compatible combination of commands is two application of the setColor command, the idea being that you can not set the color twice in the same timestep.

We can now define the widget and prefix objects

Definition 6. *The widget object, denoted* Widget, *is indexed by* $i \in \mathsf{Id}$ *and is defined as*

$$\begin{aligned} \mathsf{Widget}_\infty\ i &= \{(w, i) \mid w \in \mathcal{P}(\mathsf{Time} \times \mathsf{Cmd}), (t, c) \in w \wedge (t, c') \in w \rightarrow c \bowtie c'\} \\ \mathsf{Widget}_n\ i &= \{(w, i) \subset \mathsf{Widget}_\infty\ i \mid \forall (t, c) \in w, t \leqslant n\} \end{aligned}$$

The prefix object, denoted Prefix, *is indexed by* $i \in \mathsf{Id}$ *and* $t \in \mathsf{Time}$ *and is:*

$$\mathsf{Prefix}_\infty\ i\ t = \{(P, i) \subset \mathsf{Widget}_\infty\ i \mid \forall (t', c) \in P, t' \leqslant t\}$$

$$\mathsf{Prefix}_n\ i\ t = \begin{cases} \{(P, i) \subset \mathsf{Prefix}_\infty\ i\ t \mid \forall (t', c) \in P, t' \leqslant n\} & n < t \\ \mathsf{I} & \textit{otherwise} \end{cases}$$

The widget object is a collection of times and commands keeping track of what has happened to it at various times – imagine a *logbook* with entries for each time step. At the point at infinity, the "global" behavior of the widget is defined, i.e., the full logbook of the widget. For each n, Widget_n is simply what has happened to the widget so far, i.e., a truncated logbook. The prefix object is a widget object that is only defined up to some timestep, and is the unit after that. This yields a semantic difference between the widget where the color is set only once, and the widget where the color is set at every timestep. This reflects a real difference in actual widget behavior: if *turnRedOnClick* w later set to be blue, it will remain blue, but *keepTurningRed* w will turn back to being red.

To manipulate widgets we define two "restriction" maps.

Definition 7. *We define the following on widgets and prefixes*

$$\mathsf{shift}\ t : \mathsf{Widget}\ i \rightarrow_{\mathsf{Rel}_\mathcal{R}} \mathsf{Widget}\ i$$
$$(\mathsf{shift}\ t\ W)_n = \{(t' - t, c) \mid (t', c) \in W \wedge t \leqslant t'\}$$

$$\mathsf{prefix}\ t\ i : \mathsf{Widget}\ i \rightarrow_{\mathsf{Rel}_\mathcal{R}} \mathsf{Prefix}\ i\ t$$

$$(\mathsf{prefix}\ t\ i\ W)_n = \begin{cases} \{(t', c) \in W \mid t' < t\} & n < t \\ \mathsf{I} & n \geqslant t \end{cases}$$

The intuition behind these is that prefix $t\ i$ "cuts off" the widget after t, giving a prefix, whereas shift t shifts forward all entries in the widget by t.

Using the above, we can now define the split and join morphisms. These are again given w.r.t ambient Id and Time objects, which will be part of the full hyperdoctrine structure:

Definition 8. *We define the following morphisms on the widget object*

$$\mathsf{split}\ i\ t : \mathsf{Widget}\ i \to_{\mathsf{Rel}_{\mathcal{R}}} \mathsf{Prefix}\ i\ t \otimes \mathsf{Widget}\ i \mathbin{@} t$$
$$(\mathsf{split}\ i\ t\ w)_n = (\mathsf{prefix}\ t\ i\ w, \mathsf{shift}\ t\ w)_n$$

$$\mathsf{join}\ i\ t : \mathsf{Prefix}\ i\ t \otimes \mathsf{Widget}\ i \mathbin{@} t \to_{\mathsf{Rel}_{\mathcal{R}}} \mathsf{Widget}\ i$$
$$(\mathsf{join}\ i\ t\ (p, w))_n = \begin{cases} p_n & n < t \\ w_{n-t} & n \geqslant t \end{cases}$$

Linear-non-linear Hyperdoctrine So far we have not explained in details how to model the quantifiers in our system. To do this, we use the notion of a *hyperdoctrine* [22]. For first-order logic, this is a functor from a category of contexts and substitutions to the category of Cartesian closed categories, with the idea that we have one CCC for each valuation of the free first-order variables.

As our category of contexts, we use a Cartesian category to interpret our index objects, Time and Id. The former is interpreted as $\mathbb{N} + 1$ and the latter as $\mathbb{N}$. In our case, both Set and $\mathsf{Rel}_{\mathcal{R}}$ are themselves hyperdoctrines w.r.t to this category of contexts, the former a first-order hyperdoctrine and the latter a multiplicative intuitionistic linear logic (MILL) hyperdoctrine. Together these form a linear-non-linear hyperdoctrine through the adjunction given in Theorem 5.

Definition 9. *A linear-non-linear hyperdoctrine is a MILL hyperdoctrine L together with a first-order hyperdoctrine C and a fiber-wise monoidal adjunction $F : L \leftrightarrows C : G$.*

Theorem 6. *The categories* Set *and* $\mathsf{Rel}_{\mathcal{R}}$ *form a linear-non-linear hyperdoctrine w.r.t the interpretation of the indices objects, with the adjunction given as in Theorem 5.*

We refer the reader to the accompanying technical appendix for the full details.

Denotational Semantics We the above, we have enough structure to give an interpretation of $\lambda_{\mathsf{Widget}}$. Again, most of this interpretation is standard in the use of the hyperdoctrine structure, and we interpret $\Diamond$ in the obvious way using the linear hyperdoctrine structure on $\mathsf{Rel}_{\mathcal{R}}$. As an example, we sketch the interpretation of the widget object and the setColor command below.

Definition 10. *We interpret the* Widget i *and* Prefix i *types using the widget and prefix objects:*

$$[\![\Theta \vdash \mathsf{Widget}\ i]\!] = \mathsf{Widget}\ [\![\Theta \vdash_s i : \mathsf{Id}]\!]$$
$$[\![\Theta \vdash \mathsf{Prefix}\ i\ t]\!] = \mathsf{Prefix}\ [\![\Theta \vdash_s i : \mathsf{Id}]\!]\ [\![\Theta \vdash_s t : \mathsf{Time}]\!]$$

and we interpret the setColor *commands as:*

$$[\![\mathsf{setColor} : \forall (i : \mathsf{Id}), \mathsf{Widget}\ i \otimes \mathsf{F}\ \mathsf{Color} \multimap \mathsf{Widget}\ i]\!] =$$
$$\{w \cup_W \{(0, (\mathsf{setColor}, col))\} \mid w \in [\![\mathsf{Widget}\ i]\!], col \in [\![\mathsf{Color}]\!]\}$$

where $\cup_W$ is a "widget union", which is a union of sets such that identifiers indices and compatibility of commands are respected

This interpretation shows that a widget is indeed a logbook of events. Using the setColor command simply adds an entry to the logbook of the widget. Note we only set the color in the current timestep. To set the color in the future, we combine the above with appropriate uses of splits and joins. The interpretation of split and join are done using their semantic counterparts, and the interpretation of onClick and onKeypress are done, using our non-deterministic semantics, by associating a widget with *all possible occurrences* of the corresponds event.

Soundness of Substitution Finally, we prove that semantic substitution is sound w.r.t syntactic substitution. As with the proofs of type preservation for syntactic substitution, there are several cases for the different kinds of substitution, but the main results is again concerned with substitution of indices:

Theorem 7. *Given* $\zeta : \Theta' \to \Theta, \Theta; \Gamma \vdash_c e : X$ *and* $\Theta; \Gamma; \Delta \vdash_l t : A$ *then*

$$[\![\zeta]\!] \, [\![\Theta; \Gamma \vdash_c e : X]\!] = [\![\Theta'; \zeta(\Gamma) \vdash_c \zeta(e) : \zeta(X)]\!]$$
$$[\![\zeta]\!] \, [\![\Theta; \Gamma; \Delta \vdash_l t : A]\!] = [\![\Theta'; \zeta(\Gamma); \zeta(\Delta) \vdash_l \zeta(t) : \zeta(A)]\!]$$

Proofs for all six substitutions lemmas can be found in the technical appendix.

Related and Future Work

Much work has aimed at a logical perspective on FRP via the Curry–Howard correspondence [21,18,17,19,20,10,1]. As mentioned earlier, most of this work has focused on calculi that have a Nakano-style later modality [25], but this has the consequence that it makes it easy to write programs which wake up on every clock tick. In this paper, we remove the explicit next-step modality from the calculus, which opens the door to a more efficient implementation style based on the so-called "push" (or event/notification-based) implementation style. Elliott [12] also looked at implementing a push-based model, but viewed it as an optimization rather than a first-class feature in its own right. In future work, we plan on implementing a language based upon this calculus, with the idea that we can compile to Javascript, and represent widgets with DOM nodes, and represent the $\Diamond A$ and $A \mathbin{@} n$ temporal connectives using doubly-negated callback types (in Haskell notation, `Event A = (A -> IO ()) -> IO ()`). This should let us write GUI programs in functional style, while generating imperative, callback-based code in the same style that a handwritten GUI program would use.

Our model, in terms of $\mathsf{Set}^{\mathbb{N}+1}$, enriches LTL's semantics from time-indexed truth-values to time-indexed sets. The addition of the global view or point at infinity enables our model to distinguishes between least and greatest fixed points [15] (i.e., inductive and coinductive types), unlike in models of guarded recursion where guarded types are bilimit-compact [6]. This lets us encode temporal liveness and safety properties using inductive and coinductive types [10,2].

A recent development for comonadic modalities is the introduction of the so-called 'Fitch-style' calculi [7,11] as an alternative to the Pfenning–Davies pattern-style elimination [26]. These calculi have been used successfully for FRP [1], and one interesting question is whether they extend to adjoint calculi as well – i.e., can the $\mathsf{F}\ (X)$ modality support a direct-style eliminator?

References

1. Bahr, P., Graulund, C.U., Møgelberg, R.E.: Simply RaTT: a Fitch-style modal calculus for reactive programming without space leaks. Proceedings of the ACM on Programming Languages **3**(ICFP), 1–27 (2019)
2. Bahr, P., Graulund, C.U., Møgelberg, R.: Diamonds are not forever: Liveness in reactive programming with guarded recursion (2020)
3. Benton, N., Wadler, P.: Linear logic, monads and the lambda calculus. Proceedings 11th Annual IEEE Symposium on Logic in Computer Science pp. 420–431 (1996)
4. Benton, P.N.: A mixed linear and non-linear logic: Proofs, terms and models. In: Pacholski, L., Tiuryn, J. (eds.) Computer Science Logic. pp. 121–135. Springer Berlin Heidelberg, Berlin, Heidelberg (1995)
5. Berry, G., Cosserat, L.: The ESTEREL synchronous programming language and its mathematical semantics. In: Brookes, S.D., Roscoe, A.W., Winskel, G. (eds.) Seminar on Concurrency. pp. 389–448. Springer Berlin Heidelberg, Berlin, Heidelberg (1985)
6. Birkedal, L., Møgelberg, R.E., Schwinghammer, J., Støvring, K.: First steps in synthetic guarded domain theory: Step-indexing in the topos of trees. In: In Proc. of LICS (2011)
7. Bizjak, A., Grathwohl, H.B., Clouston, R., Møgelberg, R.E., Birkedal, L.: Guarded dependent type theory with coinductive types. In: International Conference on Foundations of Software Science and Computation Structures. pp. 20–35. Springer (2016)
8. Blackburn, P., de Rijke, M., Venema, Y.: Modal Logic. Cambridge Tracts in Theoretical Computer Science, Cambridge University Press (2002)
9. Caspi, P., Pilaud, D., Halbwachs, N., Plaice, J.A.: LUSTRE: A Declarative Language for Real-time Programming. In: Proceedings of the 14th ACM SIGACT-SIGPLAN Symposium on Principles of Programming Languages. pp. 178–188. POPL '87, ACM, New York, NY, USA (1987). https://doi.org/10.1145/41625.41641
10. Cave, A., Ferreira, F., Panangaden, P., Pientka, B.: Fair Reactive Programming. In: Proceedings of the 41st ACM SIGPLAN-SIGACT Symposium on Principles of Programming Languages. pp. 361–372. POPL '14, ACM, San Diego, California, USA (2014). https://doi.org/10.1145/2535838.2535881
11. Clouston, R.: Fitch-style modal lambda calculi. In: Baier, C., Dal Lago, U. (eds.) Foundations of Software Science and Computation Structures. pp. 258–275. Springer International Publishing, Cham (2018)
12. Elliott, C.: Push-pull functional reactive programming. In: Haskell Symposium (2009), http://conal.net/papers/push-pull-frp
13. Elliott, C., Hudak, P.: Functional reactive animation. In: Proceedings of the Second ACM SIGPLAN International Conference on Functional Programming. pp. 263–273. ICFP '97, ACM, New York, NY, USA (1997). https://doi.org/10.1145/258948.258973
14. Girard, J.Y.: Linear logic. Theoretical Computer Science **50**(1), 1 – 101 (1987). https://doi.org/10.1016/0304-3975(87)90045-4
15. Graulund, C.: Lambda Calculus for Reactive Programming. Master's thesis, IT University of Copenhagen (2018)
16. Haim, M., Malherbe, O.: Linear Hyperdoctrines and Comodules. arXiv e-prints arXiv:1612.06602 (Dec 2016)

17. Jeffrey, A.: LTL types FRP: linear-time temporal logic propositions as types, proofs as functional reactive programs. In: Proceedings of the sixth workshop on Programming Languages meets Program Verification, PLPV 2012, Philadelphia, PA, USA, January 24, 2012. pp. 49–60. Philadelphia, PA, USA (2012). https://doi.org/10.1145/2103776.2103783
18. Jeltsch, W.: Towards a common categorical semantics for linear-time temporal logic and functional reactive programming. Electronic Notes in Theoretical Computer Science **286**, 229–242 (2012)
19. Jeltsch, W.: Temporal Logic with "Until", Functional Reactive Programming with Processes, and Concrete Process Categories. In: Proceedings of the 7th Workshop on Programming Languages Meets Program Verification. pp. 69–78. PLPV '13, ACM, New York, NY, USA (2013). https://doi.org/10.1145/2428116.2428128
20. Krishnaswami, N.R.: Higher-order Functional Reactive Programming Without Spacetime Leaks. In: Proceedings of the 18th ACM SIGPLAN International Conference on Functional Programming. pp. 221–232. ICFP '13, ACM, Boston, Massachusetts, USA (2013). https://doi.org/10.1145/2500365.2500588
21. Krishnaswami, N.R., Benton, N.: Ultrametric semantics of reactive programs. In: 2011 IEEE 26th Annual Symposium on Logic in Computer Science. pp. 257–266. IEEE Computer Society, Washington, DC, USA (June 2011). https://doi.org/10.1109/LICS.2011.38
22. Lawvere, F.W.: Adjointness in foundations. Dialectica **23**(3-4), 281–296 (1969). https://doi.org/10.1111/j.1746-8361.1969.tb01194.x
23. MacLane, S., Moerdijk, I.: Sheaves in Geometry and Logic: A First Introduction to Topos Theory. Universitext, Springer New York (1994). https://doi.org/10.1007/978-1-4612-0927-0
24. Maietti, M.E., de Paiva, V., Ritter, E.: Categorical models for intuitionistic and linear type theory. In: Tiuryn, J. (ed.) Foundations of Software Science and Computation Structures. pp. 223–237. Springer Berlin Heidelberg, Berlin, Heidelberg (2000)
25. Nakano, H.: A modality for recursion. In: Proceedings Fifteenth Annual IEEE Symposium on Logic in Computer Science (Cat. No.99CB36332). pp. 255–266. IEEE Computer Society, Washington, DC, USA (June 2000). https://doi.org/10.1109/LICS.2000.855774
26. Pfenning, F., Davies, R.: A judgmental reconstruction of modal logic. Mathematical Structures in Computer Science **11**(4), 511–540 (2001). https://doi.org/10.1017/S0960129501003322
27. Pnueli, A.: The temporal logic of programs. In: 18th Annual Symposium on Foundations of Computer Science (sfcs 1977). pp. 46–57. IEEE (1977)
28. Pouzet, M.: Lucid Synchrone, version 3. Tech. rep., Laboratoire d'Informatique de Paris (2006)
29. Szamozvancev, D.: Semantics of temporal type systems. Master's thesis, University of Cambridge (2018)

On the Expressiveness of Büchi Arithmetic

Christoph Haase[1]* (✉) and Jakub Różycki[2]

[1] Department of Computer Science, University of Oxford, Oxford, UK
christoph.haase@cs.ox.ac.uk

[2] Institute of Mathematics, University of Warsaw, Warsaw, Poland

Abstract. We show that the existential fragment of Büchi arithmetic is strictly less expressive than full Büchi arithmetic of any base, and moreover establish that its Σ_2-fragment is already expressively complete. Furthermore, we show that regular languages of polynomial growth are definable in the existential fragment of Büchi arithmetic.

Keywords: logical theories · logical definability · quantifier elimination · automatic structures · regular languages

1 Introduction

This paper studies the expressive power of Büchi arithmetic, an extension of Presburger arithmetic, the first-order theory of the structure $\langle \mathbb{N}, 0, 1, + \rangle$. Büchi arithmetic additionally allows for expressing restricted divisibility properties while retaining decidability. Given an integer $p \geq 2$, *Büchi arithmetic of base* p is the first-order theory of the structure $\langle \mathbb{N}, 0, 1, +, V_p \rangle$, where V_p is a binary predicate such that $V_p(a, b)$ holds if and only if a is the largest power of p dividing b without remainder, i.e., $a = p^k$, $a \mid b$ and $p \cdot a \nmid b$.

Presburger arithmetic admits quantifier-elimination in the extended structure $\langle \mathbb{N}, 0, 1, +, \{c|\cdot\}_{c>1} \rangle$ additionally consisting of unary divisibility predicates $c|\cdot$ for every $c > 1$ [10]. It follows that the existential fragment of Presburger arithmetic is expressively complete, since any predicate $c|\cdot$ can be expressed using an additional existentially quantified variable. We study the analogous question for Büchi arithmetic and show, as the main result of this paper, that its existential fragment is, in any base, strictly less expressive than full Büchi arithmetic. Notably, this result implies that there does not exist a quantifier-elimination result *à la* Presburger for Büchi arithmetic, i.e., any extension of Büchi arithmetic with additional predicates definable in existential Büchi arithmetic does not admit quantifier elimination.

A central result about Büchi arithmetic is that it is an automatic structure: a set $M \subseteq \mathbb{N}^n$ is definable in Büchi arithmetic of base p if and only if M is recognizable by a finite-state automaton under a base p encoding of the natural

* Parts of this research were carried out while the first author was affiliated with the Department of Computer Science, University College London, UK.

S. Kiefer and C. Tasson (Eds.): FOSSACS 2021, LNCS 12650, pp. 310–323, 2021.
https://doi.org/10.1007/978-3-030-71995-1_16

numbers. Equivalently, M is *p-regular*. This result was first stated by Büchi [4], albeit in an incorrect form, and later correctly stated and proved by Bruyère [2], see also [3]. Villemaire showed that the Σ_3-fragment of Büchi arithmetic is expressively complete [13, Cor. 2.4]. He established this result by showing how to construct a Σ_3-formula defining the language of a given finite-state automaton. We observe that Villemaire's construction can actually be improved to a Σ_2-formula and thus obtain a full characterization of the expressive power of Büchi arithmetic in terms of the number of quantifier alternations.

Our approach to separating the expressiveness of existential Büchi arithmetic from full Büchi arithmetic in base p is based on a counting argument. Given a set $M \subseteq \mathbb{N}$, define the counting function $d_M(n) := \#(M \cap \{p^{n-1}, \ldots, p^n - 1\})$ which counts the numbers of bit-length n in base p in M. If M is definable in existential Büchi arithmetic of base p, we show that d_M is either $O(n^c)$ for some $c \geq 0$, or at least $c \cdot p^n$ for some constant $c > 0$ and infinitely many $n \in \mathbb{N}$. Since, for instance, for $M_p \subseteq \mathbb{N}$ defined as the set of numbers with p-ary expansion in the regular language $\{10, 01\}^*$, we have $d_{M_p}(n) = \Theta(2^{n/2})$, and hence M_p is not definable in existential Büchi arithmetic of base p. However, M_p being p-regular implies that M_p is definable by a Σ_2-formula of Büchi arithmetic of base p.

We also show that existential Büchi arithmetic defines all regular languages of polynomial density, encoded as sets of integers. Given a language $L \subseteq \Sigma^*$, let the counting function $d_L \colon \mathbb{N} \to \mathbb{N}$ be such that $d_L(n) := \#(L \cap \Sigma^n)$. Szilard et al. [11] say that L has *polynomial density* whenever $d_L(n)$ is $O(n^c)$ for some non-negative integer c. If moreover L is regular then Szilard et al. show that L is represented as a finite union of regular expressions of the form $v_0 w_1^* v_1 \cdots w_k^* v_k$ such that $0 \leq k \leq c+1$, $v_0, w_1, v_1, \ldots, v_k, w_k \in \Sigma^*$ [11, Thm. 3]. We show that existential Büchi arithmetic defines any language represented by a regular expression $v_0 w_1^* v_1 \cdots w_k^* v_k$, which implies that existential Büchi arithmetic defines all regular languages of polynomial density.

2 Preliminaries

Given $\boldsymbol{v} = (v_1, \ldots, v_d) \in \mathbb{Z}^d$, we denote by $\|\boldsymbol{v}\|_\infty$ the maximum norm of $\boldsymbol{v}$, i.e., $\|\boldsymbol{v}\|_\infty = \max\{|v_1|, \ldots, |v_d|\}$. For a matrix $\mathbf{A} \in \mathbb{Z}^{m \times d}$ with entries $a_{i,j}$, $1 \leq i \leq m$, $1 \leq j \leq d$, we denote by $\|\mathbf{A}\|_{1,\infty}$ the one-infinity norm of $\mathbf{A}$, i.e., $\|\mathbf{A}\|_{1,\infty} = \max\{|a_{i,1}| + \cdots + |a_{i,d}| : 1 \leq i \leq m\}$.

Let Σ be an alphabet and $w \in \Sigma^*$, we denote by $|w|$ the length of w. Given a set $U \subseteq \mathbb{N}$, we denote by $w^U := \{w^u : u \in U\}$. Thus, for example, $w^* = w^{\mathbb{N}}$.

For an integer $p \geq 2$, let $\Sigma_p := \{0, \ldots, p-1\}$. We view words over Σ_p as numbers encoded in p-ary most-significant bit first encoding. Tuples of numbers of dimension n can be encoded as words over the alphabet Σ_p^n. For $w = \boldsymbol{v}_m \cdots \boldsymbol{v}_0 \in (\Sigma_p^n)^{m+1}$, we denote by $[\![w]\!]_p \in \mathbb{N}^n$ the n-tuple

$$[\![w]\!]_p := \sum_{i=0}^{m} \boldsymbol{v}_i \cdot p^i \,.$$

We furthermore define $[\![\varepsilon]\!]_p := 0$. Note that $[\![\cdot]\!]_p$ is not injective since, e.g., 01 and 001 both encode the number one. Given $L \subseteq (\Sigma_p^n)^*$, we define

$$[\![L]\!]_p := \{[\![w]\!]_p : w \in L\} \subseteq \mathbb{N}^n .$$

Automata. A *deterministic automaton* is a tuple $A = (Q, \Sigma, \delta, q_0, F)$, where

- Q is a set of *states*,
- Σ is a finite alphabet,
- $\delta \colon Q \times \Sigma \to Q \cup \{\bot\}$, where $\bot \notin Q$, is the *transition function*,
- $q_0 \in Q$ is the *initial state*, and
- $F \subseteq Q$ is the set of *final states.*

For states $q, r \in Q$ and $u \in \Sigma$, we write $q \xrightarrow{u} r$ if $\delta(q, u) = r$, and extend $\to$ inductively to words by stipulating, for $w \in \Sigma^*$ and $u \in \Sigma$, that $q \xrightarrow{w \cdot u} r$ if there is $s \in Q$ such that $q \xrightarrow{w} s \xrightarrow{u} r$. The *language of A* is defined as $L(A) = \{w \in \Sigma^* : q_0 \xrightarrow{w} q_f, q_f \in F\}$.

Note that *a priori* we allow automata to have infinitely many states and to have partially defined transition functions (due to the presence of $\bot$ in the co-domain of δ). If Q is finite then we call A a *deterministic finite automaton (DFA)*, and if in addition $\Sigma = \Sigma_p^n$ for some $p \geq 2$ and $n \geq 1$ then A is called a *p-automaton.* Throughout this paper, we assume, without loss of generality, that all states of a DFA are live, i.e., every state is reachable from the initial state and can reach an accepting state.

Arithmetic theories. As stated in the introduction, Presburger arithmetic is the first-order theory of the structure $\langle \mathbb{N}, 0, 1, + \rangle$, and Büchi arithmetic of base p the first-order theory of the extended structure $\langle \mathbb{N}, 0, 1, +, V_p \rangle$. We write atomic formulas of Presburger arithmetic as $\boldsymbol{a} \cdot \boldsymbol{x} = c$, where $\boldsymbol{a} = (a_1, \ldots, a_d)^\intercal$ with $a_i \in \mathbb{Z}$, $c \in \mathbb{Z}$, and $\boldsymbol{x} = (x_1, \ldots, x_d)$ is a vector of unknowns. In Büchi arithmetic we additionally have atomic formulas $V_p(x, y)$ for the unknowns x and y. For technical convenience, we assert that $V_p(x, 0)$ never holds.[3] We write $\Phi(x)$ or $\Phi(\boldsymbol{x})$ to indicate that x or a vector of unknowns $\boldsymbol{x}$ occurs free in Φ. If there are further free variables in Φ, we assume them to be implicitly existentially quantified.

We may without loss of generality assume that no negation symbol occurs in a formula of Büchi arithmetic. First, we have $\neg(\boldsymbol{a} \cdot \boldsymbol{x} = c) \equiv \boldsymbol{a} \cdot \boldsymbol{x} \leq c - 1 \vee \boldsymbol{a} \cdot \boldsymbol{x} \geq c + 1$, and the order relation $\leq$ can easily be expressed by introducing an additionally existentially quantified variable. Moreover, we have

$$\neg V_p(x, y) \equiv y = 0 \vee \exists z \colon V_p(z, y) \wedge \neg(x = z) .$$

Finally, $P_p(x) := V_p(x, x)$ denotes the macro asserting that x is a power of p.

Given a formula $\Phi(\boldsymbol{x})$ of Büchi arithmetic of base p, we define

$$[\![\Phi(\boldsymbol{x})]\!]_p := \{\boldsymbol{m} \in \mathbb{N}^d : \Phi[\boldsymbol{m}/\boldsymbol{x}] \text{ is valid}\} ,$$

[3] Other conventions are possible, e.g., asserting that $V_p(x, 0)$ holds if and only if $x = 1$ as in [3], but this does not change the sets of numbers definable in Büchi arithmetic.

where, for $\boldsymbol{m} = (m_1, \dots, m_d)$ and $\boldsymbol{x} = (x_1, \dots, x_d)$, $\Phi[\boldsymbol{m}/\boldsymbol{x}]$ is the formula obtained from replacing every x_i by m_i in Φ. The set of sets of numbers definable in Presburger arithmetic is denoted by

$$\mathbf{PA} := \{[\![\Phi(x)]\!] : \Phi(x) \text{ is a formula of Presburger arithmetic}\}.$$

Analogously, we define the sets of numbers definable in fragments of Büchi arithmetic of base p with a fixed number of quantifier-alternations as

$$\Sigma_i\text{-}\mathbf{BA}_p := \{[\![\Phi(x)]\!]_p : \Phi(x) \text{ is a } \Sigma_i\text{-formula of Büchi arithmetic of base } p\}.$$

Finally, $\mathbf{BA}_p := \bigcup_{i \geq 1} \Sigma_i\text{-}\mathbf{BA}_p$ denotes the sets of numbers definable in Büchi arithmetic of base p.

For separating existential Büchi arithmetic from full Büchi arithmetic, we employ some tools from enumerative combinatorics. As defined in [15], a formula of *parametric Presburger arithmetic* with parameter t is a formula of Presburger arithmetic Φ_t in which atomic formulas are of the form $\boldsymbol{a} \cdot \boldsymbol{x} = c(t)$, where $c(t)$ is a univariate polynomial with indeterminate t and coefficients in $\mathbb{Z}$. For $n \in \mathbb{N}$, we denote by Φ_n the formula of Presburger arithmetic obtained from replacing $c(t)$ in every atomic formula of Φ_t by the value of $c(n)$. We associate to a formula $\Phi_t(\boldsymbol{x})$ the counting function $\#\Phi_t(\boldsymbol{x}) \colon \mathbb{N} \to \mathbb{N} \cup \{\infty\}$ such that

$$\#\Phi_t(\boldsymbol{x})(n) := \#[\![\Phi_n(\boldsymbol{x})]\!].$$

Throughout this paper, we constraint ourselves to formulas $\Phi_t(\boldsymbol{x})$ of parametric Presburger arithmetic in which $c(t)$ is the identity function and $\#\Phi_t(\boldsymbol{x})(n)$ is finite for all $n \in \mathbb{N}$.

Definition 1. *A function $f \colon \mathbb{N} \to \mathbb{Q}$ is an* eventual quasi-polynomial *if there exist a threshold $t \in \mathbb{N}$ and polynomials $p_0, \dots, p_{m-1} \in \mathbb{Q}[x]$ such that for all $n > t$, $f(n) = p_i(n)$ whenever $n \equiv i \bmod m$.*

Given an eventual quasi-polynomial f with threshold t and $n > t$, we denote by f_n the polynomial p_i such that $n \equiv i \bmod m$. We say that the polynomials $p_0, \dots, p_{m-1}$ *constitute* the eventual quasi-polynomial f. A result by Woods [15, Thm. 3.5(b)] shows that the counting functions associated to parametric Presburger formulas as defined above are eventual quasi-polynomial.

Proposition 1 (Woods). *Let $\Phi_t(\boldsymbol{x})$ be a formula of parametric Presburger arithmetic. Then $\#\Phi_t(\boldsymbol{x})$ is an eventual quasi-polynomial.*

Semi-linear sets. A result by Ginsburg and Spanier establishes that the sets of numbers definable in Presburger arithmetic are semi-linear sets [7]. A *linear set* in dimension d is given by a base vector $\boldsymbol{b} \in \mathbb{N}^d$ and a finite set of period vectors $P = \{\boldsymbol{p}_1, \dots, \boldsymbol{p}_n\} \subseteq \mathbb{N}^d$ and defines the set

$$L(\boldsymbol{b}, P) := \{\boldsymbol{b} + \lambda_1 \cdot \boldsymbol{p}_1 + \cdots + \lambda_n \cdot \boldsymbol{p}_n : \lambda_i \in \mathbb{N}, 1 \leq i \leq n\}.$$

A *semi-linear set* is a finite union of linear sets. For a finite $B \subseteq \mathbb{N}^d$, we write $L(B, P)$ for $\bigcup_{\boldsymbol{b} \in B} L(\boldsymbol{b}, P)$. Semi-linear sets of the form $L(B, P)$ are called hybrid linear sets in [5], and it is known that the set of non-negative integer solutions of a system of linear Diophantine inequalities $S\colon \mathbf{A}\cdot\boldsymbol{x}\geq\boldsymbol{c}$ is a hybrid linear set [5].

Semi-linear sets in dimension one are also known as *ultimately periodic sets.* In this paper, we represent an ultimately periodic set as a four-tuple $U = (t,\ell,B,R)$, where $t\geq 0$ is a *threshold*, $\ell>0$ is a *period*, $B\subseteq\{0,\ldots,t-1\}$ and $R\subseteq\{0,\ldots,\ell-1\}$, and U defines the set

$$[\![U]\!] := B\cup\{t+r+\ell\cdot i : r\in R, i\geq 0\}\,.$$

3 The inexpressiveness of existential Büchi arithmetic

We now establish the main result of this paper and show that the existential fragment of Büchi arithmetic is strictly less expressive than general Büchi arithmetic.

Theorem 1. *For any base $p\geq 2$, Σ_1-$\mathbf{BA}_p\neq\mathbf{BA}_p$. In particular, there exists a fixed regular language $L\subseteq\{0,1\}^*$ such that $[\![L]\!]_p\in\mathbf{BA}_p\setminus\Sigma_1$-$\mathbf{BA}_p$ for every base $p\geq 2$.*

Given a set $M\subseteq\mathbb{N}$, recall that for a fixed base $p\geq 2$, $d_M(n)$ counts the numbers of bit-length n in base p in M. As already discussed in the introduction, we prove Theorem 1 by characterizing the growth of d_M for sets M definable in Büchi arithmetic.

For any formula $\Phi(x)$ of existential Büchi arithmetic in prenex normal form, we can with no loss of generality assume that its matrix is in disjunctive normal form, i.e., a disjunction of *systems of linear Diophantine equations with valuation constraints*, each of the form

$$\mathbf{A}\cdot\boldsymbol{x}=\boldsymbol{c}\wedge\bigwedge_{i\in I}V_p(x_i,y_i),$$

where the x_i and y_i are unknowns from the vector of unknowns $\boldsymbol{x}$. For $M = [\![\Phi(x)]\!]_p$, in order to determine the growth of d_M, it suffices to determine the maximum growth occurring in any of its systems of linear Diophantine equations with valuation constraints in the matrix of $\Phi(x)$, which in turn can be obtained by analyzing the growth of the number of words accepted by a p-automaton defining the set of solutions of such a system.

Let $S\colon \mathbf{A}\cdot\boldsymbol{x}=\boldsymbol{c}$ be a system of linear Diophantine equations such that, throughout this section, $\mathbf{A}$ is an $m\times d$ integer matrix, and fix a base $p\geq 2$. Following Wolper and Boigelot [14], we define an automaton $A := (Q,\Sigma_p^d,\delta,\boldsymbol{q}_0,F)$ whose language encodes all solutions of S over the alphabet Σ_p:

- $Q:=\mathbb{Z}^m$,
- $\delta(\boldsymbol{q},\boldsymbol{u}) := p\cdot\boldsymbol{q}+\mathbf{A}\cdot\boldsymbol{u}$ for all $\boldsymbol{q}\in Q$ and $\boldsymbol{u}\in\Sigma_p^d$,
- $\boldsymbol{q}_0 := \mathbf{0}$, and
- $F := \{\boldsymbol{c}\}$.

As discussed in [14], see also [8], only states $\boldsymbol{q}$ such that $\|\boldsymbol{q}\|_\infty \leq \|\mathbf{A}\|_{1,\infty}$ and $\|\boldsymbol{q}\|_\infty \leq \|\boldsymbol{c}\|_\infty$ can reach the accepting state. Hence, all words $w \in (\Sigma_p^d)^*$ such that $\mathbf{A} \cdot [\![w]\!] = \boldsymbol{c}$ only visit a finite number of states of A, and to obtain the p-automaton $A(S)$ defining the sets of solutions of S we subsequently restrict Q to only such states. The following lemma recalls an algebraic characterization of the reachability relation of $A(S)$ established in the proof of Proposition 14 in [8].

Lemma 1. *Let $\boldsymbol{q}, \boldsymbol{r} \in \mathbb{Z}^m$ be states of $A(S)$, $w \in (\Sigma_p^d)^n$ and $\boldsymbol{x} = [\![w]\!]_p$. Then $\boldsymbol{q} \xrightarrow{w} \boldsymbol{r}$ if and only if there is $y \in \mathbb{N}$ such that*

$$\boldsymbol{q} = \boldsymbol{r} \cdot y + \mathbf{A} \cdot \boldsymbol{x}, \ \|\boldsymbol{x}\|_\infty < y, \ y = p^n.$$

Let x be a distinguished variable of $\boldsymbol{x}$. For a word $w \in (\Sigma_p^d)^*$ encoding solutions of S, denote by $\pi_x(w)$ the word $v \in \Sigma_p^*$ obtained from projecting w onto the component of w corresponding to x. Let q be a state of a p-automaton A, define the counting function $C_{q,x} \colon \mathbb{N} \to \mathbb{N}$ as

$$C_{q,x}(n) := \# \left\{ \pi_x(w) : q \xrightarrow{w} q, w \in (\Sigma_p^d)^n \right\}.$$

We now show that for p-automata arising from systems of linear Diophantine equations, $C_{q,x}$ can be obtained from an eventual quasi-polynomial.

Lemma 2. *For the p-automaton $A(S)$ associated to $S\colon \mathbf{A} \cdot \boldsymbol{x} = \boldsymbol{c}$ with states Q and all $q \in Q$, there is an eventual quasi-polynomial f such that $C_{q,x}(n) = f(p^n)$ for all $n \in \mathbb{N}$. Moreover, for all sufficiently large $n \in \mathbb{N}$, f_{p^n} is a linear polynomial.*

Proof. Let $q = \boldsymbol{q} \in \mathbb{Z}^d$. By Lemma 1, $\boldsymbol{q} \xrightarrow{w} \boldsymbol{q}$ for $w \in (\Sigma_p^d)^n$ if and only if there is a $y \in \mathbb{N}$ such that

$$\boldsymbol{q} = \boldsymbol{q} \cdot y + \mathbf{A} \cdot \boldsymbol{x}, \ \|\boldsymbol{x}\|_\infty < y, \ y = p^n,$$

where $\boldsymbol{x} = [\![w]\!]_p$. The set of solutions of $S'\colon \mathbf{A} \cdot \boldsymbol{x} + \boldsymbol{q} \cdot y = \boldsymbol{q}, \|\boldsymbol{x}\|_\infty < y$ is a hybrid linear set $L(D, R) \subseteq \mathbb{N}^{d+1}$. Let $L(B, P) \subseteq \mathbb{N}^2$ be obtained from $L(D, R)$ by projecting onto the components corresponding to x and y, and assume that x corresponds to the first and y to the second component of $L(B, P)$. Let $M_t := \mathbb{N} \times \{t\}$ and

$$f(t) := \#(L(B, P) \cap M_t).$$

Observe that $C_{q,x}(n) = f(p^n)$ and that $f(n)$ is finite for all $n \in \mathbb{N}$ due to the constraint $x < y$. Let $P = \{\boldsymbol{p}_1, \ldots, \boldsymbol{p}_k\}$, the following formula of parametric Presburger arithmetic defines $L(B, P) \cap M_t$:

$$\Phi_t(x, y) := \exists z_1 \cdots \exists z_k \colon \bigvee_{\boldsymbol{b} \in B} \begin{pmatrix} x \\ y \end{pmatrix} = \boldsymbol{b} + \sum_{i=1}^{k} \boldsymbol{p}_i \cdot z_i \wedge y = t$$

Thus, $f = \#\Phi_t(x, y)$ and, by application of Proposition 1, f is an eventual quasi-polynomial.

Since $C_{q,x}(n) \leq p^n - 1$ for all $n \in \mathbb{N}$, we in particular have that all polynomials f_{p^n} constituting f are linear as they would otherwise outgrow $C_{q,x}$. □

The next step is to lift Lemma 2 to systems of linear Diophantine equations with valuation constraints. To this end, we define a DFA whose language encodes the set of all solutions of predicates of the form $V_p(x, y)$. Formally, for $S \colon V_p(x, y)$ we define $A(S) := (Q, \Sigma_p^d, \delta, q_0, F)$ such that

- $Q := \{0, 1\}$,
- $\delta(0, \boldsymbol{u}) := 0$ for all $\boldsymbol{u} \in \Sigma_p^d$ such that $\pi_x(\boldsymbol{u}) = 0$,
- $\delta(0, \boldsymbol{u}) := 1$ for all $\boldsymbol{u} \in \Sigma_p^d$ such that $\pi_x(\boldsymbol{u}) = 1$ and $\pi_y(\boldsymbol{u}) > 0$,
- $\delta(1, \boldsymbol{u}) := 1$ for all $\boldsymbol{u} \in \Sigma_p^d$ such that $\pi_x(\boldsymbol{u}) = \pi_y(\boldsymbol{u}) = 0$,
- $q_0 := 0$, and
- $F := \{1\}$.

For $S \colon \mathbf{A} \cdot \boldsymbol{x} = \boldsymbol{c} \wedge \bigwedge_{1 \le i \le \ell} V_p(x_i, y_i)$, we denote by $A(S)$ the DFA that can be obtained from the standard product construction on all DFA for the atomic formulas of S. Hence, the set of states of $A(S)$ is a finite subset of $\mathbb{Z}^m \times \{0,1\}^\ell$. We now show that the number of words along a cycle of $A(S)$ can also be obtained from an eventual quasi-polynomial.

Lemma 3. *Let S be a system of linear Diophantine equations with valuation constraints with the associated DFA $A(S)$ with states Q, and let $q \in Q$. There is an eventual quasi-polynomial f such that $C_{q,x}(n) = f(p^n)$. Moreover, f_{p^n} is a linear polynomial for all $n \in \mathbb{N}$.*

Proof. Let $S \colon \mathbf{A} \cdot \boldsymbol{x} = \boldsymbol{c} \wedge \bigwedge_{1 \le i \le \ell} V_p(x_i, y_i)$, we have $Q \subseteq \mathbb{Z}^m \times \{0,1\}^\ell$ and thus $q = (\boldsymbol{q}, b_1, \ldots, b_\ell) \in Q$. Any self-loop $q \xrightarrow{w}_S q$ with $q = (\boldsymbol{q}, b_1, \ldots, b_\ell)$ is a self-loop for the DFA induced by the system of linear Diophantine equations $\mathbf{A} \cdot \boldsymbol{x} = \boldsymbol{c}$ with the additional requirement that $\pi_{x_i}([\![w]\!]_p) = 0$ for all $1 \le i \le \ell$ and furthermore $\pi_{y_i}([\![w]\!]_p) = 0$ whenever $b_i = 1$. Thus $(\boldsymbol{q}, \boldsymbol{0}) \xrightarrow{w}_{S'} (\boldsymbol{q}, \boldsymbol{0})$ where

$$S' \colon \mathbf{A} \cdot \boldsymbol{x} = \boldsymbol{c} \wedge \bigwedge_{1 \le i \le \ell} x_i = 0 \wedge \bigwedge_{1 \le i \le \ell, b_i = 1} y_i = 0\,.$$

Conversely, $(\boldsymbol{q}, \boldsymbol{0}) \xrightarrow{w}_{S'} (\boldsymbol{q}, \boldsymbol{0})$ immediately gives $q \xrightarrow{w}_S q$. The statement is now an immediate consequence of the application of Lemma 2 to S'. □

We will from now on implicitly apply Lemma 3. As a first application, we show that Lemma 3 allows us to classify the DFA associated to a system of linear Diophantine equations with valuation constraints.

Lemma 4. *The DFA $A(S)$ associated to a system of linear Diophantine equations with valuation constraints S with states Q has either of the following properties:*

(i) there is $q \in Q$ such that $C_{q,x}$ is an eventual quasi-polynomial f and f_{p^n} is a non-constant polynomial for infinitely many $n \in \mathbb{N}$; or
(ii) there is a constant $d \ge 0$ such that $C_{q,x}(n) \le d$ for all $q \in Q$ and $n \in \mathbb{N}$.

Proof. Suppose $A(S)$ has Property (i). For a contradiction, suppose $d \geq 0$ exists. Let f be the eventual quasi-polynomial from Property (i). Every non-constant polynomial f_{p^n} constituting f is of the form $a \cdot x + b$ with $a > 0$. As there are infinitely many such n, there is some linear polynomial $g(x) = a \cdot x + b$ such that $g = f_{p^n}$ for infinitely many $n \in \mathbb{N}$. Hence $g(p^n) > d$ for some sufficiently large $n \in \mathbb{N}$.

For the converse, suppose that $A(S)$ does not have Property (i). Then there are $\ell, m > 0$ such that all f_{p^n} are constant polynomials bounded by some value $m \in \mathbb{N}$ for all $n \geq \ell$, $q \in Q$ and $f = C_{q,x}$. Hence we can choose $d = \max(\{C_{q,x}(n) : q \in Q, 0 < n \leq \ell\} \cup \{m\})$. □

We are now in a position to prove a dichotomy of the growth of the number of words accepted by a DFA corresponding to a system of linear Diophantine equations with valuation constraints.

Lemma 5. *Let S be a fixed system of linear Diophantine equations with valuation constraints with the associated DFA $A(S)$. Let $L = \pi_x(L(A(S)))$, then either*

(i) $d_L(n) \geq c \cdot p^n$ for some fixed constant $c > 0$ and infinitely many $n \in \mathbb{N}$; or
(ii) $d_L(n) = O(n^c)$ for some fixed constant $c \geq 0$.

Proof. Let $A(S)$ have the set of states Q, initial state q_0 and final state q_f. The DFA $A(S)$ has one of the two properties stated in Lemma 4.

If $A(S)$ has the Property (i) of Lemma 4 then consider $q \in Q$ such that $C_{q,x}$ is an eventual quasi-polynomial f such that f_{p^n} is non-constant for infinitely many $n \in \mathbb{N}$, and let $i_1 < i_2 < \ldots \in \mathbb{N}$ be such that all $f_{p^{i_j}}$ are the same non-constant polynomial $a \cdot x + b$. Consider v and w such that $q_0 \xrightarrow{v} q \xrightarrow{w} q_f$. Then for all sufficiently large j we have

$$d_L(i_j + |v| + |w|) \geq a \cdot p^{i_j} + b \geq c \cdot p^{(i_j+|v|+|w|)}$$

for some fixed constant $c > 0$.

Otherwise, $A(S)$ has the Property (ii) of Lemma 4, and there is some fixed $d \geq 0$ such that $C_{q,x}(n) \leq d$ for all $n \in \mathbb{N}$ and $q \in Q$. Every $w \in L$ such that $|w| = n$ can uniquely be decomposed as $w = v_0 w_1 v_1 w_2 \cdots w_k v_k$ for some $k \leq |Q|$ such that

$$q_0 \xrightarrow{v_0} q_{a_1} \xrightarrow{w_1} q_{a_1} \xrightarrow{v_1} q_{a_2} \xrightarrow{w_2} q_{a_2} \xrightarrow{v_2} q_{a_3} \cdots \xrightarrow{w_k} q_{a_k} \xrightarrow{v_k} q_{a_{k+1}}, \tag{1}$$

where $q_{a_{k+1}} = q_f$, $q_{a_i} \neq q_{a_j}$ for all $i \neq j$ and each $q_{a_i} \xrightarrow{v_i} q_{a_{i+1}}$ corresponds to a loop-free path in $A(S)$. Since $C_{q,x} \leq d$, there are at most $d^k \leq d^{(\#Q)}$ words $u \in L$ of length n that have the same sequence of states in the decomposition of Eq. (1) at the same position where they occur in w. Moreover, there are at most $\binom{n}{2k} \leq \binom{n}{2 \cdot \#Q} \leq n^{(2 \cdot \#Q)}$ possibilities at which the states q_{a_i} can appear in any $u \in L$ of length n for any particular sequence of states in the decomposition of Eq. (1). Finally, there are at most $(\#Q)^{(\#Q)}$ such sequences. We thus derive

$$d_L(n) \leq (\#Q)^{\#Q} \cdot n^{(2 \cdot \#Q)} \cdot d^{(\#Q)} = O(n^c)$$

for some constant $c \geq 0$. □

Corollary 1. *Let $\Phi(x)$ be a fixed formula of existential Büchi arithmetic of base $p \geq 2$. Let $M = [\![\Phi(x)]\!]_p$, then either:*

(i) $d_M(n) \geq c \cdot p^n$ for some fixed constant $c > 0$ and infinitely many $n \in \mathbb{N}$; or
(ii) $d_M(n) = O(n^c)$ for some fixed constant $c \geq 0$.

Proof. Without loss of generality we may assume that $\Phi(x)$ is in disjunctive normal form such that $\Phi(x) = \bigvee_{i \in I} \Phi_i(x)$ and each $\Phi_i(x)$ is a system of linear Diophantine equations with valuation constraints S_i. For $M_i = [\![\Phi_i(x)]\!]_p$, we obtain d_{M_i} by application of Lemma 5. If there is a constant $c \geq 0$ such that $d_{M_i} = O(n^c)$ for all $i \in I$ then $d_M = O(n^c)$. Otherwise, if there is some $i \in I$ such that $d_{M_i}(n) \geq c \cdot p^n$ for some constant $c > 0$ and infinitely many $n \in \mathbb{N}$ then $d_M(n) \geq c \cdot p^n$ for infinitely many $n \in \mathbb{N}$. □

As an immediate consequence of Corollary 1, we obtain:

Corollary 2. *Let $p \geq 2$ and $M \subseteq \mathbb{N}$ such that $f = o(d_M)$ for any $f = O(n^c)$, $c \geq 0$, and $d_M = o(p^n)$. Then $M \notin \Sigma_1$-$\mathbf{BA}_p$.*

For any $p \geq 2$, consider $L = \{01, 10\}^* \subseteq \Sigma_p^*$ and $M = [\![L]\!]_p$. We have $d_M(n) = \Theta(2^{n/2})$, and thus Corollary 2 yields $M \notin \Sigma_1$-$\mathbf{BA}_p$. However, since M is p-regular, we have $M \in \mathbf{BA}_p$. This concludes the proof of Theorem 1.

4 Expressive completeness of the Σ_2-fragment of Büchi arithmetic

For a regular language $L \subseteq (\Sigma_p^d)^*$ given by a DFA, Villemaire shows in the proof of Theorem 2.2 in [13] how to construct a Σ_3-formula of Büchi arithmetic $\Phi_L(\boldsymbol{x})$ such that $[\![\Phi_L(\boldsymbol{x})]\!]_p = [\![L]\!]_p$. This construction is modularized and relies on an existential formula $\Phi_{p,j}(x, y)$ expressing that *"x is a power of p and the coefficient of this power of p in the representation of y in base p is j"*:

$$\Phi_{p,j}(x, y) \equiv P_p(x) \wedge \exists t\, \exists u\, \exists z \colon (y = z + j \cdot x + t) \wedge (z < x) \wedge \\ \wedge ((V_p(u, t) \wedge x < u) \vee t = 0)\,.$$

The only reason why $\Phi_L(\boldsymbol{x})$ in [13] is a Σ_3-formula is that $\Phi_{p,j}(x, y)$ appears in an implication both as antecedent and as consequent inside an existential formula. Thus, if one could additionally define $\Phi_{p,j}(x, y)$ by a Π_1-formula then $\Phi_L(\boldsymbol{x})$ immediately becomes a Σ_2-formula. That is, however, not difficult to achieve by defining:

$$\widetilde{\Phi}_{p,j}(x, y) := P_p(x) \wedge \forall s\, \forall t\, \forall u\, \forall z \colon \\ \Big(\neg(s = z + j \cdot x + t) \vee (z \geq x) \vee (\neg V_p(u, t) \vee x \geq u) \wedge \neg(t = 0))\Big) \to \neg(s = y)\,.$$

Note that the order relation can also be expressed by a universal formula: $x \leq y$ if and only if $\forall z \colon (y + z = x) \to (z = 0)$. Thus, $\widetilde{\Phi}_{p,j}(x, y)$ is indeed a Π_1 formula.

Combining $\widetilde{\Phi}_{p,j}(x, y)$ with the results in [13], we obtain that the Σ_2-fragment of Büchi arithmetic is expressively complete.

Theorem 2. *For any base $p \geq 2$, Σ_2-$\mathbf{BA}_p = \mathbf{BA}_p$.*

5 Existential Büchi arithmetic defines regular languages of polynomial growth

For a language $L \subseteq \Sigma^*$, Szilard et al. [11] say that L has *polynomial growth* if $d_L(n) = O(n^c)$ for some constant $c \geq 0$ and all $n \in \mathbb{N}$. One of the main results of [11] is that a regular language L has polynomial growth if and only if L can be represented as a finite union of regular expressions of the form

$$v_0 w_1^* v_1 \cdots v_{k-1} w_k^* v_k \,. \tag{2}$$

Denote by

$$\mathbf{PREG}_p := \{[\![L]\!]_p : L \subseteq \Sigma_p^*,\ L \text{ is a regular language of polynomial growth}\}$$

the numerical encoding of all regular languages of polynomial growth in base p. We show in this section that existential Büchi arithmetic defines any regular language of the form in Eq. (2). This immediately gives the following theorem.

Theorem 3. *For any base $p \geq 2$, $\mathbf{PREG}_p \subseteq \Sigma_1\text{-}\mathbf{BA}_p$.*

We first require a couple of abbreviations. Define

$$W_p(x, y) := P_p(y) \wedge x < y \leq p \cdot x,$$

which expresses that y is the smallest power of p strictly greater than x.

Let $\ell > 0$, Lohrey and Zetzsche introduce in [9] the predicate $S_\ell(x, y)$ which holds whenever

$$x = p^r \text{ and } y = p^{r+\ell\cdot i} \text{ for some } i, r \geq 0\,.$$

They show that $S_\ell(x, y)$ is definable in existential Büchi arithmetic. Since $y = p^{\ell\cdot i}\cdot x$ if and only if $y \equiv x \bmod (p^\ell - 1)$, one can obtain S_ℓ as

$$S_\ell(x, y) := P_p(x) \wedge P_p(y) \wedge \exists z\colon (y - x = (p^\ell - 1)\cdot z) \wedge y \geq x\,.$$

We slightly generalize S_ℓ. Let $U \subseteq \mathbb{N}$, define the predicate $S_U(x, y)$ to hold whenever

$$x = p^r \text{ and } y = p^{r+u} \text{ for some } r \geq 0 \text{ and } u \in U\,.$$

Lemma 6. *For any ultimately periodic set $U \subseteq \mathbb{N}$, the predicate $S_U(x, y)$ is definable in existential Büchi arithmetic*

Proof. Suppose that U is given as (t, ℓ, B, R), we define

$$S_U(x, y) := P_p(x) \wedge P_p(y) \wedge \bigvee_{b\in B} y = p^b \cdot x \vee \bigvee_{r\in R} S_\ell(p^{t+r}\cdot x, y)\,.$$

□

Towards proving Theorem 3, we now show that we can define $[\![w^*]\!]_p$ for any $w \in \Sigma_p$.

Lemma 7. *For any $w \in \Sigma_p^*$, $[\![w^*]\!]_p$ is definable by a formula of existential Büchi arithmetic $\Phi_{w^*}(x)$.*

Proof. Let $m = p^\ell$ be the smallest power of p greater than $[\![w]\!]_p$. Then for any $k > 0$,

$$[\![w^k]\!]_p = [\![w]\!]_p \cdot \sum_{i=0}^{k-1} m^i = [\![w]\!]_p \cdot \frac{m^k - 1}{m - 1}.$$

It follows that $[\![w^*]\!]_p$ is defined by

$$\Phi_{w^*}(x) := x = 0 \vee \exists y \colon S_\ell(m, y) \wedge (m-1) \cdot x = [\![w]\!]_p \cdot (y - 1).$$

□

Building upon Lemma 7, we now show that, for any $w \in \Sigma_p$, we can define $[\![w^+]\!]_p$ shifted to the left by a number of zeros specified by an ultimately periodic set.

Lemma 8. *Let $w \in \Sigma_p^*$ and U be an ultimately periodic set. Then $[\![w^+0^U]\!]_p$ is definable by a formula of existential Büchi arithmetic $\Phi_{U,w^+}(x)$.*

Proof. The case $w \in 0^*$ is trivial. Thus, let $w = w' \cdot w_0$ such that $w' \in \Sigma_p^* \cdot (\Sigma_p \setminus \{0\})$ and $w_0 \in 0^*$. Observe that for $i < j$, $[\![w^j]\!]_p - [\![w^i]\!]_p = [\![w^{j-i}0^i]\!]_p$. We define

$$\Phi_{U,w^+}(x) := \exists y \, \exists z \colon y < z \wedge \Phi_{w^*}(y) \wedge \Phi_{w^*}(z) \wedge \bigvee_{0 \le i < |w|} x = p^i \cdot (z - y) \wedge$$
$$\wedge \, \exists s \, \exists t \colon S_U(1, s) \wedge V_p(t, x) \wedge t = p^{|w_0|+1} \cdot s.$$

The first line defines the set $[\![w^+0^*]\!]_p$, whereas the second line ensures that the tailing number of zeros is in the set $U + |w_0|$. □

We have now all the ingredients to prove the following key proposition.

Proposition 2. *Let $L = v_0 w_1^* v_1 \cdots v_{k-1} w_k^* v_k$. Then $[\![L]\!]_p$ is definable in existential Büchi arithmetic.*

Proof. The proposition follows from showing the statement for languages of the form

$$L' = v_0 w_1^+ v_1 \cdots v_{k-1} w_k^+ v_k.$$

We show the statement by induction on k. The induction base case $k = 0$ is trivial. For the induction step, assume that for $M = v_1 w_2^+ v_2 \cdots v_{k-1} w_k^+ v_k$, $[\![M]\!]_p$ is defined by a formula $\Phi_k(x)$ of existential Büchi arithmetic, and let $v_0, w_1 \in \Sigma_p^*$.

We first show how to define $N = w_1^+ v_1 w_2^+ v_2 \cdots v_{k-1} w_k^+ v_k$. To this end, factor $M = M_0 \cdot M'$, where $M_0 \subseteq 0^*$ and $M \subseteq (\Sigma_p \setminus \{0\}) \cdot \Sigma_p^*$. Observe that $[\![M']\!]_p = [\![\Phi_k(x)]\!]_p$, and that both $U = \{|w| : w \in M\}$ and $V = \{|w| : w \in M_0\}$ are ultimately periodic sets, cf. [6,12]. We moreover assume that $w_1 \notin 0^*$, otherwise we are done. Factor $w_1 = w' \cdot w_0$ such that $w' \in \Sigma_p^* \cdot (\Sigma_p \setminus \{0\})$ and $w_0 \in 0^*$.

Recall that $W_p(x, y)$ holds if and only if y is the smallest power of p strictly greater than x, and define

$$\Psi_{k+1}(x) := \exists y \, \exists z \colon \Phi_k(y) \land \Phi_{U,w^+}(z) \land x = y + z \land \\ \land \exists s \, \exists t \colon W_p(y, s) \land S_V(s, t) \land V_p(p^{|w_0|+1} \cdot t, z) \, .$$

The first line composes x as the sum of some $y \in [\![M]\!]_p$ and $z \in [\![w^+ 0^U]\!]_p$. The second line ensures that the number of zeros between the leading bit of y and the last non-zero digit of z in their p-ary expansion is in $V + |w_0|$. Thus, $[\![N]\!]_p = [\![\Psi_{k+1}(x)]\!]$.

We now show how to define L' along similar lines. To this end, factor $N = N_0 \cdot N'$ such that $N_0 \subseteq 0^*$ and $N' \subseteq (\Sigma_p \setminus \{0\}) \cdot \Sigma_p^*$, and let $T = \{|w| : w \in N_0\}$, which is an ultimately periodic set. We now obtain the desired formula of existential Büchi arithmetic as

$$\Phi_{k+1}(x) := \exists y \, \exists z \colon x = y + p \cdot z \cdot [\![v_0]\!]_p \land \Psi_{k+1}(y) \land \exists s \colon W_p(y, s) \land S_T(s, z) \, .$$

□

Since we can define any regular language of the form (2) in existential Büchi arithmetic via Proposition 2, we can define a finite union of such languages and thus define all regular languages of polynomial growth in existential Büchi arithmetic. This completes the proof of Theorem 3.

Note that $\mathbf{PREG}_p \not\subseteq \mathbf{PA}$ for any base $p \geq 2$: since $M = [\![\Phi(x)]\!]$ is ultimately periodic for any formula $\Phi(x)$ of Presburger arithmetic, whenever $[\![\Phi(x)]\!]$ is infinite it follows that $d_M(n) = \Omega(p^n)$, i.e., not of polynomial growth.

6 Conclusion

The main result of this paper is that existential Büchi arithmetic is strictly less expressive than full Büchi arithmetic of any base. This is in contrast to Presburger arithmetic, for which it is known that its existential fragment is expressively complete.

When considered as the first-order theory of the structure $\langle \mathbb{N}, 0, 1, + \rangle$, Presburger arithmetic does not have a quantifier elimination procedure. The extended structure $\langle \mathbb{N}, 0, 1, +, \{c|\cdot\}_{c>1} \rangle$, however, admits quantifier elimination. Those additional divisibility predicates are definable in existential Presburger arithmetic. Our main result shows that even if we extended the structure underlying Büchi arithmetic with predicates definable in existential Büchi arithmetic, the resulting first-order theory would not admit quantifier-elimination. On the positive side, Benedikt et al. [1, Thm. 3.1] give an extension of Büchi arithmetic which has quantifier elimination.

We conclude this paper with an interesting yet likely challenging open problem: Is it decidable whether a set definable in Büchi arithmetic is definable in existential Büchi arithmetic?

Acknowledgments. We would like to thank Dmitry Chistikov and Alex Fung for inspiring discussions on the topics of this paper, and the FoSSaCS'21 reviewers for their comments and suggestions.

This work is part of a project that has received funding from the European Research Council (ERC) under the European Union's Horizon 2020 research and innovation programme (Grant agreement No. 852769, ARiAT).

References

1. Benedikt, M., Libkin, L., Schwentick, T., Segoufin, L.: Definable relations and first-order query languages over strings. J. ACM **50**(5), 694–751 (2003). https://doi.org/10.1145/876638.876642
2. Bruyère, V.: Entiers et automates finis. Mémoire de fin d'études (1985)
3. Bruyère, V., Hansel, G., Michaux, C., Villemaire, R.: Logic and *p*-recognizable sets of integers. Bull. Belg. Math. Soc. Simon Stevin **1**(2), 191–238 (1994). https://doi.org/doi:10.36045/bbms/1103408547
4. Büchi, J.: Weak second-order arithmetic and finite automata. Math. Logic Quart. **6**(1-6), 66–92 (1960). https://doi.org/10.1002/malq.19600060105
5. Chistikov, D., Haase, C.: The taming of the semi-linear set. In: Automata, Languages, and Programming, ICALP. LIPIcs, vol. 55, pp. 128:1–128:13. Schloss Dagstuhl - Leibniz-Zentrum fuer Informatik (2016). https://doi.org/10.4230/LIPIcs.ICALP.2016.128
6. Chrobak, M.: Finite automata and unary languages. Theor. Comput. Sci. **47**(3), 149–158 (1986). https://doi.org/10.1016/0304-3975(86)90142-8
7. Ginsburg, S., Spanier, E.: Bounded ALGOL-like languages. T. Am. Math. Soc. pp. 333–368 (1964). https://doi.org/10.2307/1994067
8. Guépin, F., Haase, C., Worrell, J.: On the existential theories of Büchi arithmetic and linear p-adic fields. In: Logic in Computer Science, LICS. pp. 1–10. IEEE (2019). https://doi.org/10.1109/LICS.2019.8785681
9. Lohrey, M., Zetzsche, G.: Knapsack and the power word problem in solvable Baumslag-Solitar groups. In: Mathematical Foundations of Computer Science, MFCS. LIPIcs, vol. 170, pp. 67:1–67:15. Schloss Dagstuhl - Leibniz-Zentrum für Informatik (2020). https://doi.org/10.4230/LIPIcs.MFCS.2020.67
10. Presburger, M.: Über die Vollständigkeit eines gewissen Systems der Arithmetik ganzer Zahlen, in welchem die Addition als einzige Operation hervortritt. In: Comptes Rendus du I congres de Mathematiciens des Pays Slaves, pp. 92–101 (1929)
11. Szilard, A., Yu, S., Zhang, K., Shallit, J.: Characterizing regular languages with polynomial densities. In: Mathematical Foundations of Computer Science, MFCS. Lect. Notes Comp. Sci., vol. 629, pp. 494–503. Springer (1992). https://doi.org/10.1007/3-540-55808-X_48
12. To, A.: Unary finite automata vs. arithmetic progressions. Inf. Process. Lett. **109**(17), 1010–1014 (2009). https://doi.org/10.1016/j.ipl.2009.06.005
13. Villemaire, R.: The theory of $\langle \mathbb{N}, +, V_k, V_l \rangle$ is undecidable. Theor. Comput. Sci. **106**(2), 337–349 (1992). https://doi.org/10.1016/0304-3975(92)90256-F
14. Wolper, P., Boigelot, B.: On the construction of automata from linear arithmetic constraints. In: Tools and Algorithms for the Construction and Analysis of Systems, TACAS. Lect. Notes Comp. Sci., vol. 1785, pp. 1–19. Springer (2000). https://doi.org/10.1007/3-540-46419-0_1

15. Woods, K.: The unreasonable ubiquitousness of quasi-polynomials. Elect. J. Combin. **21**(1), P1.44 (2014). https://doi.org/10.37236/3750

Parametricity for Primitive Nested Types

Patricia Johann ✉, Enrico Ghiorzi, and Daniel Jeffries
Appalachian State University, Boone, NC, USA
{johannp,ghiorzie,jeffriesd}@appstate.edu

Abstract. This paper considers parametricity and its resulting free theorems for nested data types. Rather than representing nested types via their Church encodings in a higher-kinded or dependently typed extension of System F, we adopt a functional programming perspective and design a Hindley-Milner-style calculus with primitives for constructing nested types directly as fixpoints. Our calculus can express all nested types appearing in the literature, including truly nested types. At the term level, it supports primitive pattern matching, map functions, and fold combinators for nested types. Our main contribution is the construction of a parametric model for our calculus. This is both delicate and challenging: to ensure the existence of semantic fixpoints interpreting nested types, and thus to establish a suitable Identity Extension Lemma for our calculus, our type system must explicitly track functoriality of types, and cocontinuity conditions on the functors interpreting them must be appropriately threaded throughout the model construction. We prove that our model satisfies an appropriate Abstraction Theorem and verifies all standard consequences of parametricity for primitive nested types.

1 Introduction

Algebraic data types (ADTs), both built-in and user-defined, have long been at the core of functional languages such as Haskell, ML, Agda, Epigram, and Idris. ADTs, such as that of natural numbers, can be unindexed. But they can also be indexed over other types. For example, the ADT of lists (here coded in Agda)

```
data List (A : Set) : Set where
    nil  : List A
    cons : A → List A → List A
```

is indexed over its element type `A`. The instance of `List` at index `A` depends only on itself, and so is independent of `List B` for any other index `B`. That is, `List`, like all other ADTs, defines a *family of inductive types*, one for each index type.

Over time, there has been a notable trend toward data types whose non-regular indexing can capture invariants and other sophisticated properties that can be used for program verification and other applications. A simple example of such a type is given by Bird and Meertens' [4] prototypical nested type

```
data PTree (A : Set) : Set where
    pleaf : A → PTree A
    pnode : PTree (A × A) → PTree A
```

of perfect trees, which can be thought of as constraining lists to have lengths that are powers of 2. The above code makes clear that perfect trees at index type `A` are defined in terms of perfect trees at index type `A × A`. This is typical of nested types, one type instance of which can depend on others, so that the entire family

S. Kiefer and C. Tasson (Eds.): FOSSACS 2021, LNCS 12650, pp. 324–343, 2021.
https://doi.org/10.1007/978-3-030-71995-1_17

of types must actually be defined at once. A nested type thus defines not a family of inductive types, but rather an *inductive family of types.* Nested types include simple nested types, like perfect trees, none of whose recursive occurrences occur below another type constructor; "deep" nested types [18], such as the nested type

```
data PForest (A : Set) : Set where
    fempty : PForest A
    fnode : A → PTree (PForest A) → PForest A
```

of perfect forests, whose recursive occurrences appear below type constructors for other nested types; and truly nested types, such as the nested type

```
data Bush (A : Set) : Set where
    bnil : Bush A
    bcons : A → Bush (Bush A) → Bush A
```

of bushes, whose recursive occurrences appear below their own type constructors.

Relational parametricity encodes a powerful notion of type-uniformity, or representation independence, for data types in polymorphic languages. It formalizes the intuition that a polymorphic program must act uniformly on all of its possible type instantiations by requiring that every such program preserves all relations between pairs of types at which it is instantiated. Parametricity was originally put forth by Reynolds [24] for System F [11], the calculus at the core of all polymorphic functional languages. It was later popularized as Wadler's "theorems for free" [27], so called because it can deduce properties of programs in such languages solely from their types, i.e., with no knowledge whatsoever of the text of the programs involved. Most of Wadler's free theorems are consequences of naturality for polymorphic list-processing functions. However, parametricity can also derive results that go beyond just naturality, such as correctness for ADTs of the program optimization known as *short cut fusion* [10,14].

But what about nested types? Does parametricity still hold if such types are added to polymorphic calculi? More practically, can we justifiably reason type-independently about (functions over) nested types in functional languages?

Type-independent reasoning about ADTs in functional languages is usually justified by first representing ADTs by their Church encodings, and then reasoning type-independently about these encodings. This is typically justified by constructing a parametric model — i.e, a model in which polymorphic functions preserve relations *á la* Reynolds — for a suitable fragment of System F, demonstrating that an initial algebra exists for the positive type constructor corresponding to the functor underlying an ADT of interest, and showing that each such initial algebra is suitably isomorphic to its corresponding Church encoding. In fact, this isomorphism of initial algebras and their Church encodings is one of the "litmus tests" for the goodness of a parametric model.

This approach works well for ADTs, which are always fixpoints of *first-order* functors, and whose Church encodings, which involve quantification over only type variables, are always expressible in System F. For example, $\mathtt{List\,A}$ is the fixpoint of the first-order functor $F\,X = 1 + A \times X$ and has Church encoding $\forall \alpha.\, \alpha \to (\mathtt{A} \to \alpha \to \alpha) \to \alpha$. But despite Cardelli's [7] claim that "virtually any basic type of interest can be encoded within F_2" — i.e., within System

F — non-ADT nested types cannot. Not even our prototypical nested type of perfect trees has a Church encoding expressible in System F! Indeed, `PTree A` cannot be represented as the fixpoint of any *first-order* functor. However, it can be seen as the instance at index `A` of the fixpoint of the *higher-order* functor $H\,F\,A = (A \rightarrow F\,A) \rightarrow (F\,(A \times A) \rightarrow F\,A) \rightarrow F\,A$. It thus has Church encoding $\forall f.\,(\forall \alpha.\,\alpha \rightarrow f\alpha) \rightarrow (\forall \alpha.\,f(\alpha \times \alpha) \rightarrow f\alpha) \rightarrow \forall \alpha.\,f\alpha$, which requires quantification at the higher kind $* \rightarrow *$ for f. A similar situation obtains for any (non-ADT) nested type. Unfortunately, higher-kinded quantification is not available in System F, so if we want to reason type-independently about nested types in a language based on it we have only two options: *i)* move to an extension of System F, such as the higher-kinded calculus F_ω or a dependent type theory, and reason via their Church encodings in a known parametric model for that extension, or *ii)* add nested types to System F as primitives — i.e., as primitive type-level fixpoints — and construct a parametric model for the result.

Since the type systems of F_ω and dependent type theories are designed to extend System F with far more than non-ADT data types, it seems like serious overkill to pass to their parametric models to reason about nested types in System F. Indeed, such calculi support fundamentally new features that add complexity to their models that is entirely unnecessary for reasoning about nested types. This paper therefore pursues the second option above. We first design a Hindley-Milner-style calculus supporting primitive nested types, together with primitive types of natural transformations representing morphisms between them. Our calculus can express all nested types appearing in the literature, including truly nested types. At the term-level, it supports primitive pattern matching, map functions, and fold combinators for nested types.[1] Our main contribution is the construction of a parametric model for our calculus. This is both delicate and challenging. To ensure the existence of semantic fixpoints interpreting nested types, and thus to establish a suitable Identity Extension Lemma, our type system must explicitly track functoriality of types, and cocontinuity conditions on the functors interpreting them must be appropriately threaded throughout the model construction. Our model validates all standard consequences of parametricity in the presence of primitive nested types, including the isomorphism of primitive ADTs and their Church encodings, and correctness of short cut fusion for nested types. The relationship between naturality and parametricity has long been of interest, and our inclusion of a primitive type of natural transformations allows us to clearly delineate those consequences of parametricity that follow from naturality, from those, such as short cut fusion for nested types, that require the full power of parametricity.

[1] We leave incorporating general term-level recursion to future work because, as Pitts [23] reminds us, "it is hard to construct models of both impredicative polymorphism and fixpoint recursion". In fact, as the development in this paper shows, constructing a parametric model even for our predicative calculus with primitive nested types — and even without term-level fixpoints — is already rather involved. On the other hand, our calculus is strongly normalizing, so it perhaps edges us toward the kind of provably total practical programming language proposed in [27].

Structure of this Paper We introduce our calculus in Section 2. Its type system is based on the level-2-truncation of the higher-kinded grammar from [17], augmented with a primitive type of natural transformations. (Since [17] contains no term calculus, the issue of parametricity could not even be raised there.) In Section 3 we give set and relational interpretations of our types. Set interpretations are possible precisely because our calculus is predicative — as ensured by our primitive natural transformation types — and [17] guarantees that local finite presentability of Set makes it suitable for interpreting nested types. As is standard in categorical models, types are interpreted as functors from environments interpreting their type variable contexts to sets or relations, as appropriate. To ensure that these functors satisfy the cocontinuity properties needed for the semantic fixpoints interpreting nested types to exist, set environments must map k-ary type constructor variables to appropriately cocontinuous k-ary functors on sets, relation environments must map k-ary type constructor variables to appropriately cocontinuous k-ary relation transformers, and these cocontinuity conditions must be threaded through our type interpretations in such a way that an Identity Extension Lemma (Theorem 1) can be proved. Properly propagating the cocontinuity conditions requires considerable care, and Section 4, where it is done, is (apart from tracking functoriality in the calculus so that it is actually possible) where the bulk of the work in constructing our model lies.

In Section 5, we give set and relational interpretations for the terms of our calculus. As usual in categorical models, terms are interpreted as natural transformations from interpretations of their term contexts to interpretations of their types, and these must cohere in what is essentially a fibred way. In Section 6.1 we prove a scheme deriving free theorems that are consequences of naturality of polymorphic functions over nested types. This scheme is very general, and is parameterized over both the data type and the type of the polymorphic function at hand. It has, for example, analogues for nested types of Wadler's map-rearrangement free theorems as instances. In Section 6.2 we prove that our model satisfies an Abstraction Theorem (Theorem 4), which we use to derive other parametricity results that go beyond naturality. We conclude in Section 7.

Related Work There is a long line of work on categorical models of parametricity for System F; see, e.g., [3,6,8,9,12,13,20,26]. To our knowledge, all such models treat ADTs via their Church encodings, verifying in the just-constructed parametric model that each ADT is isomorphic to its encoding. This paper draws on this rich tradition of categorical models of parametricity for System F, but modifies them to treat nested types (and thus ADTs) as primitive data types. The only other extensions we know of System F with primitive data types are those in [19,21,22,23,27]. Wadler [27] treats full System F, and sketches parametricity for its extension with lists. Martin and Gibbons [21] outline a semantics for a grammar of primitive nested types similar to that in [17], but treat only polynomial nested types. Unfortunately, the model suggested in [21] is not entirely correct (see [17]), and parametricity is nowhere mentioned. Matthes [19] treats System F with non-polynomial ADTs and nested types, but focuses on expressivity of generalized Mendler iteration for them. He gives no semantics.

In [23], Pitts adds list ADTs to full System F with a term-level fixpoint primitive. Other ADTs are included in [22], but nested types are not expressible in either syntax. Pitts constructs parametric models for his calculi based on operational, rather than categorical, semantics. A benefit of using operational semantics to build parametric models is that it avoids needing to work in a suitable metatheory to accommodate System F's impredicativity. It is well-known that there are no set-based parametric models of System F [25], so parametric models for it and its extensions are often constructed in a syntactic metatheory such as the impredicative Calculus of Inductive Constructions (iCIC). By adding primitive nested types to a Hindley-Milner-style calculus and working in a categorical setting we side-step such metatheoretic distractions. It is important to note that different consequences of parametricity are available in syntactic and semantic metatheories. Consequences of parametricity are possible for both closed and open System F terms in a syntactic metatheory — although not all that can be formulated can be always proved; see, e.g., the end of Section 7 of [4]. By contrast, in a categorical metatheory consequences of parametricity are expressible only for *closed* terms. For this reason, validating the standard consequences of parametricity for closed terms is — going all the way back to Reynolds [24] — all that is required for a model of parametricity to be considered good.

Atkey [2] treats parametricity for arbitrary higher kinds, constructing a parametric model for System F_ω within iCIC, rather than in a semantic category. His construction is in some ways similar to ours, but he represents (now higher-kinded) data types using Church encodings rather than as primitives. Moreover, the *fmap* functions associated to Atkey's functors must be *given*, presumably by the programmer, together with their underlying type constructors. This absolves him of imposing cocontinuity conditions on his model to ensure that fixpoints of his functors exist, but, unfortunately, he does not indicate which type constructors support *fmap* functions. We suspect explicitly spelling out which types can be interpreted as strictly positive functors would result in a full higher-kinded extension of a calculus akin to that presented here.

2 The Calculus

2.1 Types

For each $k \geq 0$, we assume countable sets $\mathbb{T}^k$ of *type constructor variables of arity* k (i.e., of kind $* \to ... \to * \to *$, with k arrows and $k+1$ $*$s in this sequence) and $\mathbb{F}^k$ of *functorial variables of arity* k, all mutually disjoint. The sets of all type constructor variables and functorial variables are $\mathbb{T} = \bigcup_{k\geq 0} \mathbb{T}^k$ and $\mathbb{F} = \bigcup_{k\geq 0} \mathbb{F}^k$, respectively, and a *type variable* is any element of $\mathbb{T} \cup \mathbb{F}$. We use lower case Greek letters for type variables, writing ϕ^k to indicate that $\phi \in \mathbb{T}^k \cup \mathbb{F}^k$, and omitting the arity indicator k when convenient. Letters from the beginning of the alphabet denote type variables of arity 0, i.e., elements of $\mathbb{T}^0 \cup \mathbb{F}^0$. We write $\overline{\phi}$ for either a set $\{\phi_1, ..., \phi_n\}$ of type constructor variables or a set of functorial variables when the cardinality n of the set is unimportant or clear from context. If V is a set of type variables we write $V, \overline{\phi}$ for $V \cup \overline{\phi}$ when $V \cap \overline{\phi} = \emptyset$. We omit the vector notation for a singleton set, thus writing ϕ, instead of $\overline{\phi}$, for $\{\phi\}$.

If Γ is a finite subset of $\mathbb{T}$, Φ is a finite subset of $\mathbb{F}$, $\overline{\alpha}$ is a finite subset of $\mathbb{F}^0$ disjoint from Φ, and $\phi^k \in \mathbb{F}^k \setminus \Phi$, then the set $\mathcal{F}$ of well-formed types is given in Definition 1. The notation there entails that type application $\phi F_1 ... F_k$ is allowed only when ϕ is a type variable of arity k, or ϕ is a subexpression of the form $\mu\psi^k.\lambda\alpha_1...\alpha_k.F'$. Moreover, if ϕ has arity k then ϕ must be applied to exactly k arguments. Accordingly, an overbar indicates a sequence of subexpressions whose length matches the arity of the type applied to it. Requiring that types are always in such *η-long normal form* avoids having to consider β-conversion of types. In a subexpression $\mathsf{Nat}^{\overline{\alpha}} F\, G$, the Nat operator binds all occurrences of the variables in $\overline{\alpha}$ in F and G; intuitively, $\mathsf{Nat}^{\overline{\alpha}} F\, G$ represents the type of a natural transformation in $\overline{\alpha}$ from the functor F to the functor G. In a subexpression $\mu\phi^k.\lambda\overline{\alpha}.F$, the μ operator binds all occurrences of the variable ϕ, and the λ operator binds all occurrences of the variables in $\overline{\alpha}$, in the body F.

A *type constructor*, or *non-functorial, context* is a finite set Γ of type constructor variables, and a *functorial context* is a finite set Φ of functorial variables. In Definition 1, a judgment of the form $\Gamma;\Phi \vdash F$ indicates that the type F is intended to be functorial in the variables in Φ but not necessarily in those in Γ.

Definition 1. *The formation rules for the set $\mathcal{F}$ of* (well-formed) types *are*

$$\frac{}{\Gamma;\Phi \vdash \mathbb{0}} \qquad \frac{}{\Gamma;\Phi \vdash \mathbb{1}} \qquad \frac{\Gamma;\Phi \vdash F \qquad \Gamma;\Phi \vdash G}{\Gamma;\Phi \vdash F + G} \qquad \frac{\Gamma;\Phi \vdash F \qquad \Gamma;\Phi \vdash G}{\Gamma;\Phi \vdash F \times G}$$

$$\frac{\Gamma;\overline{\alpha^0} \vdash F \qquad \Gamma;\overline{\alpha^0} \vdash G}{\Gamma;\emptyset \vdash \mathsf{Nat}^{\overline{\alpha^0}} F\, G} \qquad \frac{\phi^k \in \Gamma \cup \Phi \qquad \overline{\Gamma;\Phi \vdash F}}{\Gamma;\Phi \vdash \phi^k \overline{F}}$$

$$\frac{\Gamma;\overline{\alpha^0},\phi^k \vdash F \qquad \overline{\Gamma;\Phi \vdash G}}{\Gamma;\Phi \vdash (\mu\phi^k.\lambda\overline{\alpha^0}.\,F)\,\overline{G}}$$

We write $\vdash F$ for $\emptyset;\emptyset \vdash F$. Definition 1 ensures that the expected weakening rules for well-formed types hold (but weakening does not change the contexts in which types can be formed). If $\Gamma;\emptyset \vdash F$ and $\Gamma;\emptyset \vdash G$, then our rules allow formation of $\Gamma;\emptyset \vdash \mathsf{Nat}^{\emptyset} F\, G$, which represents the arrow type $\Gamma \vdash F \to G$ in our calculus. The type $\Gamma;\emptyset \vdash \mathsf{Nat}^{\overline{\alpha}}\, \mathbb{1}\, F$ represents the $\forall$-type $\Gamma;\emptyset \vdash \forall\overline{\alpha}.F$. Some System F types, such as $\forall\alpha.\,(\alpha \to \alpha) \to \alpha$, are not representable in our calculus.

Since the body F of a type $(\mu\phi.\lambda\overline{\alpha}.F)\overline{G}$ can only be functorial in ϕ and the variables in $\overline{\alpha}$, the representation of *List* α as the ADT $\mu\beta.\,\mathbb{1} + \alpha \times \beta$ cannot be functorial in α. By contrast, if *List* α is represented as the nested type $(\mu\phi.\lambda\beta.\,\mathbb{1} + \beta \times \phi\beta)\,\alpha$ then we can choose α to be a functorial variable or not when forming the type. This observation holds for other ADTs as well; for example, if *Tree* $\alpha\,\gamma = \mu\beta.\alpha + \beta \times \gamma \times \beta$, then $\alpha,\gamma;\emptyset \vdash$ *Tree* $\alpha\,\gamma$ is well-formed, but $\emptyset;\alpha,\gamma \vdash$ *Tree* $\alpha\,\gamma$ is not. It also applies to some non-ADT types, such as *GRose* $\phi\,\alpha = \mu\beta.\mathbb{1} + \alpha \times \phi\beta$, in which ϕ and α must both be non-functorial variables. It is in fact possible to allow "extra" 0-ary functorial variables in the body of μ-types (functorial variables of higher arity are the real problem). This would allow the first-order representations of ADTs to be functorial, but doing so requires some changes to the formation rule for μ-types, as well as the delicate threading of some additional conditions throughout our model construction. But since we can always use an ADT's (semantically equivalent) second-order representation when functoriality is needed, disallowing such "extra" variables does not negatively impact the expressivity of our calculus. We therefore pursue the simpler syntax here.

Definition 1 allows well-formed types to be functorial in no variables. Functorial variables can also be demoted to non-functorial status: if $F[\phi :== \psi]$ is the textual replacement of ϕ in F, then $\Gamma, \psi^k; \Phi \vdash F[\phi^k :== \psi^k]$ is derivable whenever $\Gamma; \Phi, \phi^k \vdash F$ is. In addition to textual replacement, we also have substitution for types. If $\Gamma; \Phi \vdash F$ is a type, if Γ and Φ contain only type variables of arity 0, and if $k = 0$ for every occurrence of ϕ^k bound by μ in F, then we say that F is *first-order*; otherwise we say that F is *second-order*. Substitution for first-order types is the usual capture-avoiding textual substitution. We write $F[\alpha := \sigma]$ for the result of substituting σ for α in F, and $F[\alpha_1 := F_1, ..., \alpha_k := F_k]$, or $F[\overline{\alpha := F}]$ when convenient, for $F[\alpha_1 := F_1][\alpha_2 := F_2, ..., \alpha_k := F_k]$. The operation $(\cdot)[\phi :=_{\overline{\alpha}} F]$ of *second-order type substitution along* $\overline{\alpha}$ is defined by induction on types exactly as expected. The only interesting clause is that for type application, which defines $(\psi\overline{G})[\phi :=_{\overline{\alpha}} F]$ to be $F[\overline{\alpha := G[\phi :=_{\overline{\alpha}} F]}]$ if $\psi = \phi$ and $\overline{G[\phi :=_{\overline{\alpha}} F]}$ otherwise. Of course, $(\cdot)[\phi^0 :=_{\emptyset} F]$ coincides with first-order substitution. We omit $\overline{\alpha}$ when convenient, but note that it is not correct to substitute along non-functorial variables. It is not hard to see that if $\Gamma; \Phi, \phi^k \vdash H$ and $\Gamma; \Phi, \overline{\alpha} \vdash F$ with $|\overline{\alpha}| = k$, then $\Gamma; \Phi \vdash H[\phi :=_{\overline{\alpha}} F]$. Similarly, if $\Gamma, \phi^k; \Phi \vdash H$, and if $\Gamma; \overline{\psi}, \overline{\alpha} \vdash F$ with $|\overline{\alpha}| = k$ and $\Phi \cap \overline{\psi} = \emptyset$, then $\Gamma, \overline{\psi}'; \Phi \vdash H[\phi :=_{\overline{\alpha}} F[\overline{\psi :== \psi'}]]$.

2.2 Terms

Assume an infinite set $\mathcal{V}$ of term variables disjoint from $\mathbb{T}$ and $\mathbb{F}$. If Γ is a type constructor context and Φ is a functorial context, then a *term context for* Γ *and* Φ is a finite set of bindings of the form $x : F$, where $x \in \mathcal{V}$ and $\Gamma; \Phi \vdash F$. We adopt the above conventions for disjoint unions and vectors in term contexts. If Δ is a term context for Γ and Φ then the formation rules for the set of *well-formed terms over* Δ are given in Figure 1. An expression $L_{\overline{\alpha}}x.t$ binds all occurrences of the type variables in $\overline{\alpha}$ in the types of x and t, as well as all occurrences of x in t. In the rule for $t_{\overline{K}}s$ there is one functorial expression in $\overline{K}$ for every variable in $\overline{\alpha}$. In the rule for $\mathsf{map}_H^{\overline{F},\overline{G}}$ there is one functorial expression in $\overline{F}$ and one functorial expression in $\overline{G}$ for each variable in $\overline{\phi}$. Moreover, for each ϕ^k in $\overline{\phi}$ the number of variables in $\overline{\beta}$ in the judgments for functorial expresssions in $\overline{F}$ and $\overline{G}$ is k. In the rules for in_H and fold_H^F, the variables in $\overline{\beta}$ are fresh with respect to H, and there is one β for every α. Substitution for terms is the obvious extension of the usual capture-avoiding textual substitution, and weakening is respected.

The "extra" functorial variables in $\overline{\gamma}$ in the rules for $\mathsf{map}_H^{\overline{F},\overline{G}}$ (i.e., those variables not affected by the substitution of ϕ) allow us to map polymorphic functions over nested types. Suppose, for example, that we want to map the polymorphic function *flatten* : $\mathsf{Nat}^\beta(\mathit{PTree}\,\beta)\,(\mathit{List}\,\beta)$ over lists. The map term for this is typeable as follows:

$$\frac{\Gamma; \alpha, \gamma \vdash \mathit{List}\,\alpha \qquad \Gamma; \gamma \vdash \mathit{PTree}\,\gamma \qquad \Gamma; \gamma \vdash \mathit{List}\,\gamma}{\Gamma; \emptyset \mid \emptyset \vdash \mathsf{map}_{\mathit{List}\,\alpha}^{\mathit{PTree}\,\gamma, \mathit{List}\,\gamma} : \mathsf{Nat}^\emptyset(\mathsf{Nat}^\gamma(\mathit{PTree}\,\gamma)\,(\mathit{List}\,\gamma))\,(\mathsf{Nat}^\gamma\,(\mathit{List}\,(\mathit{PTree}\,\gamma))\,(\mathit{List}\,(\mathit{List}\,\gamma)))}$$

$$\frac{\Gamma;\Phi \vdash F}{\Gamma;\Phi \mid \Delta, x : F \vdash x : F} \qquad \frac{\Gamma;\Phi \mid \Delta \vdash t : \mathbb{0} \qquad \Gamma;\Phi \vdash F}{\Gamma;\Phi \mid \Delta \vdash \bot_F t : F} \qquad \frac{}{\Gamma;\Phi \mid \Delta \vdash \top : \mathbb{1}}$$

$$\frac{\Gamma;\Phi \mid \Delta \vdash s : F}{\Gamma;\Phi \mid \Delta \vdash \mathsf{inL}\; s : F + G} \qquad \frac{\Gamma;\Phi \mid \Delta \vdash t : G}{\Gamma;\Phi \mid \Delta \vdash \mathsf{inR}\; t : F + G}$$

$$\frac{\Gamma;\Phi \vdash F, G \qquad \Gamma;\Phi \mid \Delta \vdash t : F + G \qquad \Gamma;\Phi \mid \Delta, x : F \vdash l : K \qquad \Gamma;\Phi \mid \Delta, y : G \vdash r : K}{\Gamma;\Phi \mid \Delta \vdash \mathsf{case}\, t\, \mathsf{of}\, \{x \mapsto l;\, y \mapsto r\} : K}$$

$$\frac{\Gamma;\Phi \mid \Delta \vdash s : F \qquad \Gamma;\Phi \mid \Delta \vdash t : G}{\Gamma;\Phi \mid \Delta \vdash (s,t) : F \times G} \qquad \frac{\Gamma;\Phi \mid \Delta \vdash t : F \times G}{\Gamma;\Phi \mid \Delta \vdash \pi_1 t : F} \qquad \frac{\Gamma;\Phi \mid \Delta \vdash t : F \times G}{\Gamma;\Phi \mid \Delta \vdash \pi_2 t : G}$$

$$\frac{\Gamma;\overline{\alpha} \vdash F \qquad \Gamma;\overline{\alpha} \vdash G \qquad \Gamma;\overline{\alpha} \mid \Delta, x : F \vdash t : G}{\Gamma;\emptyset \mid \Delta \vdash L_{\overline{\alpha}} x.t : \mathsf{Nat}^{\overline{\alpha}}\, F\, G}$$

$$\frac{\overline{\Gamma;\Phi \vdash K} \qquad \Gamma;\emptyset \mid \Delta \vdash t : \mathsf{Nat}^{\overline{\alpha}}\, F\, G \qquad \Gamma;\Phi \mid \Delta \vdash s : F[\overline{\alpha := K}]}{\Gamma;\Phi \mid \Delta \vdash t_{\overline{K}} s : G[\overline{\alpha := K}]}$$

$$\frac{\Gamma;\overline{\phi},\overline{\gamma} \vdash H \qquad \overline{\Gamma;\overline{\beta},\overline{\gamma} \vdash F} \qquad \overline{\Gamma;\overline{\beta},\overline{\gamma} \vdash G}}{\Gamma;\emptyset \mid \emptyset \vdash \mathsf{map}_H^{\overline{F},\overline{G}} : \mathsf{Nat}^{\emptyset}\, (\overline{\mathsf{Nat}^{\overline{\beta},\overline{\gamma}}\, F\, G})\, (\mathsf{Nat}^{\overline{\gamma}}\, H[\overline{\phi :=_{\overline{\beta}} F}]\; H[\overline{\phi :=_{\overline{\beta}} G}])}$$

$$\frac{\Gamma;\phi,\overline{\alpha} \vdash H}{\Gamma;\emptyset \mid \emptyset \vdash \mathsf{in}_H : \mathsf{Nat}^{\overline{\beta}} H[\phi :=_{\overline{\beta}} (\mu\phi.\lambda\overline{\alpha}.H)\overline{\beta}][\overline{\alpha := \beta}]\, (\mu\phi.\lambda\overline{\alpha}.H)\overline{\beta}}$$

$$\frac{\Gamma;\phi,\overline{\alpha} \vdash H \qquad \Gamma;\overline{\beta} \vdash F}{\Gamma;\emptyset \mid \emptyset \vdash \mathsf{fold}_H^F : \mathsf{Nat}^{\emptyset}\, (\mathsf{Nat}^{\overline{\beta}}\, H[\phi :=_{\overline{\beta}} F][\overline{\alpha := \beta}]\, F)\, (\mathsf{Nat}^{\overline{\beta}}\, (\mu\phi.\lambda\overline{\alpha}.H)\overline{\beta}\, F)}$$

Fig. 1. Well-formed terms

However, this derivation would not possible without the "extra" variable γ.

Our calculus is expressive enough to define, e.g., a function *reversePTree* : $\mathsf{Nat}^{\alpha}\, (\mathit{PTree}\,\alpha)(\mathit{PTree}\,\alpha)$ that reverses the order of the leaves in a perfect tree. It maps the perfect tree $((1,2),(3,4))$ to $((4,3),(2,1))$. Unfortunately, we cannot define recursive functions — such as a concatenation function for perfect trees or a zip function for bushes — that take as inputs a nested type and an argument of another type, both of which are parameterized over the same variable. The fundamental issue is that recursion is expressible only via fold, which produces natural transformations in some variables $\overline{\alpha}$ from μ-types to other functors F. The restrictions on Nat-types entail that F cannot itself be a Nat-type containing $\overline{\alpha}$, so, e.g., $\mathsf{Nat}^{\alpha}\, (\mathit{PTree}\,\alpha)(\mathsf{Nat}^{\emptyset}\, (\mathit{PTree}\,\alpha)(\mathit{PTree}\,(\alpha \times \alpha)))$ is not well-typed. Uncurrying gives $\mathsf{Nat}^{\alpha}\, (\mathit{PTree}\,\alpha \times \mathit{PTree}\,\alpha)(\mathit{PTree}\,(\alpha \times \alpha))$, which is well-typed, but fold cannot produce a term of this type because $\mathit{PTree}\,\alpha \times \mathit{PTree}\,\alpha$ is not a μ-type. Our calculus can, however, express types of recursive functions that take multiple nested types as arguments, provided they are parameterized over disjoint sets of type variables and the return type of the function is parameterized over only the variables occurring in the type of its final argument. Even for ADTs there is a difference between which folds over them we can type when they are viewed as ADTs (i.e., as fixpoints of first-order functors) versus as proper nested types (i.e., as fixpoints of higher-order functors). This is because, in the return type of fold, the arguments of the μ-type must be variables bound by Nat. For ADTs, the μ-type takes no arguments, making it possible to write recursive functions, such as a concatenation function for lists of type $\alpha; \emptyset \vdash \mathsf{Nat}^{\emptyset}\,(\mu\beta.\mathbb{1}+\alpha\times\beta)\,(\mathsf{Nat}^{\emptyset}(\mu\beta.\mathbb{1}+\alpha\times\beta)\,(\mu\beta.\mathbb{1}+\alpha\times\beta))$. This is not possible for nested types — even when they are semantically equivalent to ADTs.

Interestingly, even some recursive functions of a single proper nested type — e.g., a reverse function for bushes that is a true involution — cannot be expressed as folds because the algebra arguments needed to define them are again recursive functions with types of the same problematic form as the type of, e.g., a zip function for perfect trees. Expressivity of folds for nested types has long been a vexing issue, and this is naturally inherited by our calculus. Adding more expressive recursion combinators — e.g., generalized folds or Mendler iterators — could help, but since this is orthogonal to the issue of parametricity in the presence of primitive nested types we do not consider it further here.

3 Interpreting Types

We denote the category of sets and functions by Set. The category Rel has as objects triples (A, B, R), where R is a relation between sets A and B. It has as morphisms from (A, B, R) to (A', B', R') pairs $(f : A \to A', g : B \to B')$ of morphisms in Set such that $(fa, g\,b) \in R'$ if $(a, b) \in R$. We may write $R : \mathsf{Rel}(A, B)$ for (A, B, R). If $R : \mathsf{Rel}(A, B)$ we write $\pi_1 R$ and $\pi_2 R$ for the *domain* A of R and the *codomain* B of R, respectively, and assume π_1 and π_2 are surjective. We write $\mathsf{Eq}_A = (A, A, \{(x, x) \mid x \in A\})$ for the *equality relation* on the set A.

The key idea underlying Reynolds' parametricity is to give each type $F(\alpha)$ with one free variable α a *set interpretation* F_0 taking sets to sets and a *relational interpretation* F_1 taking relations $R : \mathsf{Rel}(A, B)$ to relations $F_1(R) : \mathsf{Rel}(F_0(A), F_0(B))$, and to interpret each term $t(\alpha, x) : F(\alpha)$ with one free term variable $x : G(\alpha)$ as a map t_0 associating to each set A a function $t_0(A) : G_0(A) \to F_0(A)$. These interpretations are given inductively on the structures of F and t in such a way that they imply two fundamental theorems. The first is an *Identity Extension Lemma*, which states that $F_1(\mathsf{Eq}_A) = \mathsf{Eq}_{F_0(A)}$, and is the essential property that makes a model relationally parametric rather than just induced by a logical relation. The second is an *Abstraction Theorem*, which states that, for any $R : \mathsf{Rel}(A, B)$, $(t_0(A), t_0(B))$ is a morphism in Rel from $(G_0(A), G_0(B), G_1(R))$ to $(F_0(A), F_0(B), F_1(R))$. The Identity Extension Lemma is similar to the Abstraction Theorem except that it holds for *all* elements of a type's interpretation, not just those that interpret terms. Similar theorems are required for types and terms with any number of free variables.

The key to proving our Identity Extension Lemma is a familiar "cutting down" of the interpretations of universally quantified types to include only the "parametric" elements; the relevant types here are Nat types. This requires that the set interpretations of types (Section 3.1) are defined simultaneously with their relational interpretations (Section 3.2). While set interpretations are relatively straightforward, relational interpretations are less so because of the cocontinuity conditions needed to know they are well-defined. We develop these conditions in Sections 3.1 and 3.2. This separates our set and relational interpretations in space, but has no other impact on the mutually inductive definitions.

$$
\begin{aligned}
[\![\Gamma;\Phi \vdash \mathbb{0}]\!]^{\mathsf{Set}}\rho &= 0\\
[\![\Gamma;\Phi \vdash \mathbb{1}]\!]^{\mathsf{Set}}\rho &= 1\\
[\![\Gamma;\emptyset \vdash \mathsf{Nat}^{\overline{\alpha}}\, F\, G]\!]^{\mathsf{Set}}\rho &= \{\eta : \lambda\overline{A}.\,[\![\Gamma;\overline{\alpha} \vdash F]\!]^{\mathsf{Set}}\rho[\overline{\alpha := A}] \Rightarrow \lambda\overline{A}.\,[\![\Gamma;\overline{\alpha} \vdash G]\!]^{\mathsf{Set}}\rho[\overline{\alpha := A}]\\
&\qquad \mid \forall \overline{A},\overline{B} : \mathsf{Set}.\forall \overline{R} : \mathsf{Rel}(\overline{A},\overline{B}).\\
&\qquad (\eta_{\overline{A}},\eta_{\overline{B}}) : [\![\Gamma;\overline{\alpha} \vdash F]\!]^{\mathsf{Rel}}\mathsf{Eq}_\rho[\overline{\alpha := R}] \to [\![\Gamma;\overline{\alpha} \vdash G]\!]^{\mathsf{Rel}}\mathsf{Eq}_\rho[\overline{\alpha := R}]\}\\
[\![\Gamma;\Phi \vdash \phi\overline{F}]\!]^{\mathsf{Set}}\rho &= (\rho\phi)\,\overline{[\![\Gamma;\Phi \vdash F]\!]^{\mathsf{Set}}\rho}\\
[\![\Gamma;\Phi \vdash F+G]\!]^{\mathsf{Set}}\rho &= [\![\Gamma;\Phi \vdash F]\!]^{\mathsf{Set}}\rho + [\![\Gamma;\Phi \vdash G]\!]^{\mathsf{Set}}\rho\\
[\![\Gamma;\Phi \vdash F\times G]\!]^{\mathsf{Set}}\rho &= [\![\Gamma;\Phi \vdash F]\!]^{\mathsf{Set}}\rho \times [\![\Gamma;\Phi \vdash G]\!]^{\mathsf{Set}}\rho\\
[\![\Gamma;\Phi \vdash (\mu\phi.\lambda\overline{\alpha}.H)\overline{G}]\!]^{\mathsf{Set}}\rho &= (\mu T^{\mathsf{Set}}_{H,\rho})\overline{[\![\Gamma;\Phi \vdash G]\!]^{\mathsf{Set}}\rho}\\
\text{where } T^{\mathsf{Set}}_{H,\rho}F &= \lambda\overline{A}.[\![\Gamma;\phi,\overline{\alpha} \vdash H]\!]^{\mathsf{Set}}\rho[\phi := F][\overline{\alpha := A}]\\
\text{and } T^{\mathsf{Set}}_{H,\rho}\eta &= \lambda\overline{A}.[\![\Gamma;\phi,\overline{\alpha} \vdash H]\!]^{\mathsf{Set}}id_\rho[\phi := \eta][\overline{\alpha := id_A}]
\end{aligned}
$$

Fig. 2. Set interpretation

3.1 Interpreting Types as Sets

We interpret types in our calculus as ω-cocontinuous functors on locally finitely presentable categories [1]. Since functor categories of locally finitely presentable categories are again locally finitely presentable, this ensures that the fixpoints interpreting μ-types in Set and Rel exist, and thus that both the set and relational interpretations of all of the types in Definition 1 are well-defined [17]. To bootstrap this process, we interpret type variables as ω-cocontinuous functors. If $\mathcal{C}$ and $\mathcal{D}$ are locally finitely presentable categories, we write $[\mathcal{C},\mathcal{D}]$ for the category of ω-cocontinuous functors from $\mathcal{C}$ to $\mathcal{D}$.

A *set environment* maps each type variable in $\mathbb{T}^k \cup \mathbb{F}^k$ to an element of $[\mathsf{Set}^k, \mathsf{Set}]$. A morphism $f : \rho \to \rho'$ for set environments ρ and ρ' with $\rho|_{\mathbb{T}} = \rho'|_{\mathbb{T}}$ maps each type constructor variable $\psi^k \in \mathbb{T}$ to the identity natural transformation on $\rho\psi^k = \rho'\psi^k$ and each functorial variable $\phi^k \in \mathbb{F}$ to a natural transformation from the k-ary functor $\rho\phi^k$ on Set to the k-ary functor $\rho'\phi^k$ on Set. Composition of morphisms on set environments is componentwise, with the identity morphism mapping each one to itself. This gives a category of set environments and morphisms between them, denoted SetEnv. We identify a functor in $[\mathsf{Set}^0, \mathsf{Set}]$ with its value on $*$, and consider a set environment to map a type variable of arity 0 to a set. If $\overline{\alpha} = \{\alpha_1, ..., \alpha_k\}$ and $\overline{A} = \{A_1, ..., A_k\}$, then we write $\rho[\overline{\alpha := A}]$ for the set environment ρ' such that $\rho'\alpha_i = A_i$ for $i = 1, ..., k$ and $\rho'\alpha = \rho\alpha$ if $\alpha \notin \{\alpha_1, ..., \alpha_k\}$. If $\rho \in \mathsf{SetEnv}$ we write Eq_ρ for the relation environment (see Section 3) such that $\mathsf{Eq}_\rho v = \mathsf{Eq}_{\rho v}$ for every type variable v. The *set interpretation* $[\![\cdot]\!]^{\mathsf{Set}} : \mathcal{F} \to [\mathsf{SetEnv}, \mathsf{Set}]$ is defined in Figure 2. The relational interpretations in the second clause of Figure 2 are given in full in Figure 3.

If $\rho \in \mathsf{SetEnv}$ and $\vdash F$ we write $[\![\vdash F]\!]^{\mathsf{Set}}$ for $[\![\vdash F]\!]^{\mathsf{Set}}\rho$ since the environment is immaterial. The third clause of Figure 2 does indeed define a set: local finite presentability of Set and ω-cocontinuity of $[\![\Gamma;\overline{\alpha} \vdash F]\!]^{\mathsf{Set}}\rho$ ensure that the set of natural transformations $\{\eta : [\![\Gamma;\overline{\alpha} \vdash F]\!]^{\mathsf{Set}}\rho \Rightarrow [\![\Gamma;\overline{\alpha} \vdash G]\!]^{\mathsf{Set}}\rho\}$ (which contains $[\![\Gamma;\emptyset \vdash \mathsf{Nat}^{\overline{\alpha}}\, F\, G]\!]^{\mathsf{Set}}\rho$) is a subset of $\big\{([\![\Gamma;\overline{\alpha} \vdash G]\!]^{\mathsf{Set}}\rho[\overline{\alpha := S}])^{([\![\Gamma;\overline{\alpha} \vdash F]\!]^{\mathsf{Set}}\rho[\overline{\alpha := S}])} \mid \overline{S} = (S_1, ..., S_{|\overline{\alpha}|})$, and S_i is a finite set for $i = 1, ..., |\overline{\alpha}|\big\}$. There are countably many tuples $\overline{S}$, each giving a morphism from $[\![\Gamma;\overline{\alpha}\vdash F]\!]^{\mathsf{Set}}\rho[\overline{\alpha := S}]$ to $[\![\Gamma;\overline{\alpha}\vdash G]\!]^{\mathsf{Set}}\rho[\overline{\alpha := S}]$, and only Set-many such morphisms since Set is locally small. In addition, $[\![\Gamma;\emptyset\vdash \mathsf{Nat}^{\overline{\alpha}} F\,G]\!]^{\mathsf{Set}}$ is ω-cocontinuous since it is constant on ω-directed sets. Interpretations of Nat types ensure that $[\![\Gamma\vdash F\to G]\!]^{\mathsf{Set}}$ and $[\![\Gamma\vdash \forall\overline{\alpha}.F]\!]^{\mathsf{Set}}$ are as expected in parametric models.

To make sense of the last clause in Figure 2, we need to know that, for each $\rho\in\mathsf{SetEnv}$, $T^{\mathsf{Set}}_{H,\rho}$ is an ω-cocontinuous endofunctor on $[\mathsf{Set}^k,\mathsf{Set}]$, and thus admits a fixpoint. Since $T^{\mathsf{Set}}_{H,\rho}$ is defined in terms of $[\![\Gamma;\phi,\overline{\alpha}\vdash H]\!]^{\mathsf{Set}}$, interpretations of types must be such functors, which entails that the actions of set interpretations of types on objects and on morphisms in SetEnv are intertwined. We know from [17] that, for every $\Gamma;\overline{\alpha}\vdash G$, $[\![\Gamma;\overline{\alpha}\vdash G]\!]^{\mathsf{Set}}$ is actually in $[\mathsf{Set}^k,\mathsf{Set}]$ where $k=|\overline{\alpha}|$, so that, for each $[\![\Gamma;\phi^k,\overline{\alpha}\vdash H]\!]^{\mathsf{Set}}$, the corresponding operator T^{Set}_H can be extended to a *functor* from SetEnv to $[[\mathsf{Set}^k,\mathsf{Set}],[\mathsf{Set}^k,\mathsf{Set}]]$. The action of T^{Set}_H on an object $\rho\in\mathsf{SetEnv}$ is given by the higher-order functor $T^{\mathsf{Set}}_{H,\rho}$, whose actions on objects (functors in $[\mathsf{Set}^k,\mathsf{Set}]$) and morphisms between them are given in Figure 2. Its action on a morphism $f:\rho\to\rho'$ is the higher-order natural transformation $T^{\mathsf{Set}}_{H,f}:T^{\mathsf{Set}}_{H,\rho}\to T^{\mathsf{Set}}_{H,\rho'}$ whose action on $F:[\mathsf{Set}^k,\mathsf{Set}]$ is the natural transformation $T^{\mathsf{Set}}_{H,f}\,F:T^{\mathsf{Set}}_{H,\rho}\,F\to T^{\mathsf{Set}}_{H,\rho'}\,F$ whose component at $\overline{A}$ is $(T^{\mathsf{Set}}_{H,f}\,F)_{\overline{A}}=[\![\Gamma;\phi,\overline{\alpha}\vdash H]\!]^{\mathsf{Set}}f[\phi:=id_F][\overline{\alpha:=id_A}]$. The next definition uses T^{Set}_H to define the functorial action of set interpretation.

Definition 2. *The action of* $[\![\Gamma;\Phi\vdash F]\!]^{\mathsf{Set}}$ *on* $f:\rho\to\rho'$ *in* SetEnv *is given by:*

- $[\![\Gamma;\Phi\vdash \mathbb{0}]\!]^{\mathsf{Set}}f=id_0$
- $[\![\Gamma;\Phi\vdash \mathbb{1}]\!]^{\mathsf{Set}}f=id_1$
- $[\![\Gamma;\emptyset\vdash \mathsf{Nat}^{\overline{\alpha}}\,F\,G]\!]^{\mathsf{Set}}f=id_{[\![\Gamma;\emptyset\vdash\mathsf{Nat}^{\overline{\alpha}}\,F\,G]\!]^{\mathsf{Set}}\rho}$
- $[\![\Gamma;\Phi\vdash\phi\overline{F}]\!]^{\mathsf{Set}}f:[\![\Gamma;\Phi\vdash\phi\overline{F}]\!]^{\mathsf{Set}}\rho\to[\![\Gamma;\Phi\vdash\phi\overline{F}]\!]^{\mathsf{Set}}\rho'=(\rho\phi)\overline{[\![\Gamma;\Phi\vdash F]\!]^{\mathsf{Set}}\rho}\to(\rho'\phi)\overline{[\![\Gamma;\Phi\vdash F]\!]^{\mathsf{Set}}\rho'}$ *is defined by* $[\![\Gamma;\Phi\vdash\phi\overline{F}]\!]^{\mathsf{Set}}f=(f\phi)_{\overline{[\![\Gamma;\Phi\vdash F]\!]^{\mathsf{Set}}\rho'}}\circ(\rho\phi)\overline{[\![\Gamma;\Phi\vdash F]\!]^{\mathsf{Set}}f}=(\rho'\phi)\overline{[\![\Gamma;\Phi\vdash F]\!]^{\mathsf{Set}}f}\circ(f\phi)_{\overline{[\![\Gamma;\Phi\vdash F]\!]^{\mathsf{Set}}\rho}}$. *This holds since* $\rho\phi$ *and* $\rho'\phi$ *are functors and* $f\phi:\rho\phi\to\rho'\phi$ *is a natural transformation.*
- $[\![\Gamma;\Phi\vdash F+G]\!]^{\mathsf{Set}}f$ *is defined by* $[\![\Gamma;\Phi\vdash F+G]\!]^{\mathsf{Set}}f(\mathsf{inL}\;x)=\mathsf{inL}\,([\![\Gamma;\Phi\vdash F]\!]^{\mathsf{Set}}fx)$ *and* $[\![\Gamma;\Phi\vdash F+G]\!]^{\mathsf{Set}}f(\mathsf{inR}\;y)=\mathsf{inR}\,([\![\Gamma;\Phi\vdash G]\!]^{\mathsf{Set}}fy)$
- $[\![\Gamma;\Phi\vdash F\times G]\!]^{\mathsf{Set}}f=[\![\Gamma;\Phi\vdash F]\!]^{\mathsf{Set}}f\times[\![\Gamma;\Phi\vdash G]\!]^{\mathsf{Set}}f$
- $[\![\Gamma;\Phi\vdash(\mu\phi.\lambda\overline{\alpha}.H)\overline{G}]\!]^{\mathsf{Set}}f\;:\;[\![\Gamma;\Phi\vdash(\mu\phi.\lambda\overline{\alpha}.H)\overline{G}]\!]^{\mathsf{Set}}\rho\to[\![\Gamma;\Phi\vdash(\mu\phi.\lambda\overline{\alpha}.H)\overline{G}]\!]^{\mathsf{Set}}\rho'=(\mu T^{\mathsf{Set}}_{H,\rho})\overline{[\![\Gamma;\Phi\vdash G]\!]^{\mathsf{Set}}\rho}\to(\mu T^{\mathsf{Set}}_{H,\rho'})\overline{[\![\Gamma;\Phi\vdash G]\!]^{\mathsf{Set}}\rho'}$ *is defined by* $(\mu T^{\mathsf{Set}}_{H,f})\overline{[\![\Gamma;\Phi\vdash G]\!]^{\mathsf{Set}}\rho'}\circ(\mu T^{\mathsf{Set}}_{H,\rho})\overline{[\![\Gamma;\Phi\vdash G]\!]^{\mathsf{Set}}f}=(\mu T^{\mathsf{Set}}_{H,\rho'})\overline{[\![\Gamma;\Phi\vdash G]\!]^{\mathsf{Set}}f}\circ(\mu T^{\mathsf{Set}}_{H,f})\overline{[\![\Gamma;\Phi\vdash G]\!]^{\mathsf{Set}}\rho}$. *This holds since* $\mu T^{\mathsf{Set}}_{H,\rho}$ *and* $\mu T^{\mathsf{Set}}_{H,\rho'}$ *are functors and* $\mu T^{\mathsf{Set}}_{H,f}:\mu T^{\mathsf{Set}}_{H,\rho}\to\mu T^{\mathsf{Set}}_{H,\rho'}$ *is a natural transformation.*

3.2 Interpreting Types as Relations

A *k-ary relation transformer* F is a triple (F^1,F^2,F^*), where $F^1,F^2:[\mathsf{Set}^k,\mathsf{Set}]$ and $F^*:[\mathsf{Rel}^k,\mathsf{Rel}]$ are functors, if $\overline{R_i:\mathsf{Rel}(A_i,B_i)}$ for $i=1,...,k$ then $F^*\overline{R}:\mathsf{Rel}(F^1\overline{A},F^2\overline{B})$, and if $\overline{(\alpha_i,\beta_i)\in\mathsf{Hom}_{\mathsf{Rel}}(R_i,S_i)}$ for $i=1,...,k$, then $F^*\overline{(\alpha,\beta)}=$

$(F^1\overline{\alpha}, F^2\overline{\beta})$. We define $F\overline{R}$ to be $F^*\overline{R}$ and $F\overline{(\alpha,\beta)}$ to be $F^*\overline{(\alpha,\beta)}$. The last clause above expands to: if $\overline{(a,b) \in R}$ implies $\overline{(\alpha\, a, \beta\, b) \in S}$ then $(c,d) \in F^*\overline{R}$ implies $(F^1\overline{\alpha}\, c, F^2\overline{\beta}\, d) \in F^*\overline{S}$. We identify a 0-ary relation transformer (A, B, R) with $R : \mathsf{Rel}(A, B)$, and write $\pi_1 F$ for F^1 and $\pi_2 F$ for F^2. Below we extend these conventions to relation environments in the obvious ways.

The category RT_k of k-ary relation transformers is given by the following data: an object of RT_k is a k-ary relation transformer; a morphism $\delta : (G^1, G^2, G^*) \to (H^1, H^2, H^*)$ in RT_k is a pair of natural transformations (δ^1, δ^2) where $\delta^1 : G^1 \to H^1$, $\delta^2 : G^2 \to H^2$ such that, for all $\overline{R : \mathsf{Rel}(A, B)}$, if $(x, y) \in G^*\overline{R}$ then $(\delta^1_{\overline{A}} x, \delta^2_{\overline{B}} y) \in H^*\overline{R}$; and identity morphisms and composition are inherited from the category of functors on Set. An endofunctor H on RT_k is a triple $H = (H^1, H^2, H^*)$, where H^1 and H^2 are functors from $[\mathsf{Set}^k, \mathsf{Set}]$ to $[\mathsf{Set}^k, \mathsf{Set}]$; H^* is a functor from RT_k to $[\mathsf{Rel}^k, \mathsf{Rel}]$; for all $\overline{R : \mathsf{Rel}(A, B)}$, $\pi_1((H^*(\delta^1, \delta^2))_{\overline{R}}) = (H^1\delta^1)_{\overline{A}}$ and $\pi_2((H^*(\delta^1, \delta^2))_{\overline{R}}) = (H^2\delta^2)_{\overline{B}}$; the action of H on objects is given by $H\,(F^1, F^2, F^*) = (H^1F^1, H^2F^2, H^*(F^1, F^2, F^*))$; and the action of H on morphisms is given by $H\,(\delta^1, \delta^2) = (H^1\delta^1, H^2\delta^2)$ for $(\delta^1, \delta^2) : (F^1, F^2, F^*) \to (G^1, G^2, G^*)$. Since applying an endofunctor H to k-ary relation transformers and morphisms between them must give k-ary relation transformers and morphisms between them, this definition implicitly requires the following three conditions to hold: *i)* $H^*(F^1, F^2, F^*)\overline{R} : \mathsf{Rel}(H^1F^1\overline{A}, H^2F^2\overline{B})$ if $R_1 : \mathsf{Rel}(A_1, B_1), ..., R_k : \mathsf{Rel}(A_k, B_k)$; *ii)* $H^*(F^1, F^2, F^*)\,\overline{(\alpha,\beta)} = (H^1F^1\overline{\alpha}, H^2F^2\overline{\beta})$ if $(\alpha_1, \beta_1) \in \mathsf{Hom}_{\mathsf{Rel}}(R_1, S_1), ..., (\alpha_k, \beta_k) \in \mathsf{Hom}_{\mathsf{Rel}}(R_k, S_k)$; and *iii)* if $(\delta^1, \delta^2) : (F^1, F^2, F^*) \to (G^1, G^2, G^*)$ and $R_1 : \mathsf{Rel}(A_1, B_1), ..., R_k : \mathsf{Rel}(A_k, B_k)$, then $((H^1\delta^1)_{\overline{A}} x, (H^2\delta^2)_{\overline{B}} y) \in H^*(G^1, G^2, G^*)\overline{R}$ if $(x, y) \in H^*(F^1, F^2, F^*)\overline{R}$. Note, however, that this last condition is automatically satisfied because it is implied by the third condition on functors on relation transformers.

If H and K are endofunctors on RT_k, then a *natural transformation* $\sigma : H \to K$ is a pair $\sigma = (\sigma^1, \sigma^2)$, where $\sigma^1 : H^1 \to K^1$ and $\sigma^2 : H^2 \to K^2$ are natural transformations between endofunctors on $[\mathsf{Set}^k, \mathsf{Set}]$ and the component of σ at $F \in RT_k$ is given by $\sigma_F = (\sigma^1_{F^1}, \sigma^2_{F^2})$. This definition entails that $\sigma^i_{F^i}$ is natural in $F^i : [\mathsf{Set}^k, \mathsf{Set}]$, and, for every F, both $(\sigma^1_{F^1})_{\overline{A}}$ and $(\sigma^2_{F^2})_{\overline{A}}$ are natural in $\overline{A}$. Moreover, since the results of applying σ to k-ary relation transformers must be morphisms of k-ary relation transformers, it implicitly requires that $(\sigma_F)_{\overline{R}} = ((\sigma^1_{F^1})_{\overline{A}}, (\sigma^2_{F^2})_{\overline{B}})$ is a morphism in Rel for any k-tuple of relations $\overline{R : \mathsf{Rel}(A, B)}$, i.e., that if $(x, y) \in H^*F\overline{R}$, then $((\sigma^1_{F^1})_{\overline{A}} x, (\sigma^2_{F^2})_{\overline{B}} y) \in K^*F\overline{R}$.

Critically, we can compute ω-directed colimits in RT_k. Indeed, if $\mathcal{D}$ is an ω-directed set then $\varinjlim_{d \in \mathcal{D}} (F^1_d, F^2_d, F^*_d) = (\varinjlim_{d \in \mathcal{D}} F^1_d, \varinjlim_{d \in \mathcal{D}} F^2_d, \varinjlim_{d \in \mathcal{D}} F^*_d)$. We define an endofunctor $T = (T^1, T^2, T^*)$ on RT_k to be *ω-cocontinuous* if T^1 and T^2 are ω-cocontinuous endofunctors on $[\mathsf{Set}^k, \mathsf{Set}]$ and T^* is an ω-cocontinuous functor from RT_k to $[\mathsf{Rel}^k, \mathsf{Rel}]$, i.e., is in $[RT_k, [\mathsf{Rel}^k, \mathsf{Rel}]]$. Now, for any k, any $A : \mathsf{Set}$, and any $R : \mathsf{Rel}(A, B)$, let K^{Set}_A be the constantly A-valued functor from Set^k to Set and K^{Rel}_R be the constantly R-valued functor from Rel^k to Rel. Also let 0 denote the initial object of either Set or Rel, as appropriate. Observing that, for every k, K^{Set}_0 is initial in $[\mathsf{Set}^k, \mathsf{Set}]$, and K^{Rel}_0 is initial in $[\mathsf{Rel}^k, \mathsf{Rel}]$, we have that, for each k, $K_0 = (K^{\mathsf{Set}}_0, K^{\mathsf{Set}}_0, K^{\mathsf{Rel}}_0)$ is initial in RT_k. Thus, if

$T = (T^1, T^2, T^*) : RT_k \to RT_k$ is an endofunctor on RT_k we can define the relation transformer μT to be $\varinjlim_{n\in\mathbb{N}} T^n K_0 = (\mu T^1, \mu T^2, \varinjlim_{n\in\mathbb{N}} (T^n K_0)^*)$. If $T : [RT_k, RT_k]$ then μT is a fixpoint for T, i.e., $\mu T \cong T(\mu T)$. The isomorphism is given by $(in_1, in_2) : T(\mu T) \to \mu T$ and $(in_1^{-1}, in_2^{-1}) : \mu T \to T(\mu T)$ in RT_k. The latter is always a morphism in RT_k, but the former need not be if T is not ω-cocontinuous. Since μT's third component is the colimit in $[\mathsf{Rel}^k, \mathsf{Rel}]$ of third components of relation transformers, rather than a fixpoint of an endofunctor on $[\mathsf{Rel}^k, \mathsf{Rel}]$, there is an asymmetry between μT's first two and third components.

A *relation environment* maps each type variable in $\mathbb{T}^k \cup \mathbb{F}^k$ to a k-ary relation transformer. A morphism $f : \rho \to \rho'$ between relation environments ρ and ρ' with $\rho|_{\mathbb{T}} = \rho'|_{\mathbb{T}}$ maps each $\psi^k \in \mathbb{T}$ to the identity morphism on $\rho\psi^k = \rho'\psi^k$ and each $\phi^k \in \mathbb{F}$ to a morphism from the k-ary relation transformer $\rho\phi$ to the k-ary relation transformer $\rho'\phi$. Composition of morphisms on relation environments is componentwise, with the identity morphism mapping each to itself; this gives a category RelEnv of relation environments and their morphisms. We identify a 0-ary relation transformer with its codomain, and consider a relation environment to map a type variable of arity 0 to a relation. We write $\rho[\overline{\alpha := R}]$ for the relation environment ρ' such that $\rho'\alpha_i = R_i$ for $i = 1, ..., k$ and $\rho'\alpha = \rho\alpha$ if $\alpha \notin \{\alpha_1, ..., \alpha_k\}$. If $\rho \in \mathsf{RelEnv}$ we write $\pi_1\rho$ and $\pi_2\rho$ for the set environments mapping each type variable ϕ to the functors $(\rho\phi)^1$ and $(\rho\phi)^2$, respectively.

For each k, an ω-cocontinuous functor $H : [\mathsf{RelEnv}, RT_k]$ is a triple $H = (H^1, H^2, H^*)$, where $H^1, H^2 : [\mathsf{SetEnv}, [\mathsf{Set}^k, \mathsf{Set}]]$; $H^* : [\mathsf{RelEnv}, [\mathsf{Rel}^k, \mathsf{Rel}]]$; for all $\overline{R : \mathsf{Rel}(A, B)}$ and morphisms f in RelEnv, $\pi_1(H^* f\, \overline{R}) = H^1(\pi_1 f)\, \overline{A}$ and $\pi_2(H^* f\, \overline{R}) = H^2(\pi_2 f)\, \overline{B}$; the action of H on ρ in RelEnv is given by $H\rho = (H^1(\pi_1\rho), H^2(\pi_2\rho), H^*\rho)$; and the action of H on morphisms $f : \rho \to \rho'$ in RelEnv is given by $Hf = (H^1(\pi_1 f), H^2(\pi_2 f))$. The last two points above give: *i)* if $\overline{R_i : \mathsf{Rel}(A_i, B_i)}$, $i = 1, ..., k$, then $H^*\rho\,\overline{R} : \mathsf{Rel}(H^1(\pi_1\rho)\,\overline{A}, H^2(\pi_2\rho)\,\overline{B})$; *ii)* if $\overline{(\alpha_i, \beta_i) \in \mathrm{Hom}_{\mathsf{Rel}}(R_i, S_i)}$, $i = 1, ..., k$, then $H^*\rho\,\overline{(\alpha, \beta)} = (H^1(\pi_1\rho)\,\overline{\alpha}, H^2(\pi_2\rho)\,\overline{\beta})$; and *iii)* if $f : \rho \to \rho'$ and $\overline{R_i : \mathsf{Rel}(A_i, B_i)}$, $i = 1, ..., k$, then if $(x, y) \in H^*\rho\,\overline{R}$ then $(H^1(\pi_1 f)\,\overline{A}\,x, H^2(\pi_2 f)\,\overline{B}\,y) \in H^*\rho'\,\overline{R}$.

Computation of ω-directed colimits in RT_k extends componentwise to colimits in RelEnv. Similarly, ω-cocontinuity for endofunctors on RT_k extends to functors from RelEnv to RT_k. Our relational interpretation $[\![\cdot]\!]^{\mathsf{Rel}} : \mathcal{F} \to [\mathsf{RelEnv}, \mathsf{Rel}]$ is given in Figure 3. It ensures that $[\![\Gamma \vdash F \to G]\!]^{\mathsf{Rel}}$ and $[\![\Gamma \vdash \forall\overline{\alpha}.F]\!]^{\mathsf{Rel}}$ are as expected. As for set interpretations, $[\![\Gamma; \emptyset \vdash \mathsf{Nat}^{\overline{\alpha}} F\, G]\!]^{\mathsf{Rel}}$ is ω-cocontinuous because it is constant on ω-directed sets. If $\rho \in \mathsf{RelEnv}$ we write $[\![\vdash F]\!]^{\mathsf{Rel}}$ for $[\![\vdash F]\!]^{\mathsf{Rel}}\rho$. For the last clause in Figure 3 to be well-defined we need $T_{H,\rho}$ to be an ω-cocontinuous endofunctor on RT, so that it admits a fixpoint. Since $T_{H,\rho}$ is defined in terms of $[\![\Gamma; \phi^k, \overline{\alpha} \vdash H]\!]^{\mathsf{Rel}}$, this means that relational interpretations of types must be ω-cocontinuous functors from RelEnv to RT_0, which in turn entails that the actions of relational interpretations of types on objects and on morphisms in RelEnv are intertwined. We know from [17] that, for every $\Gamma; \overline{\alpha} \vdash F$, $[\![\Gamma; \overline{\alpha} \vdash F]\!]^{\mathsf{Rel}}$ is actually in $[\mathsf{Rel}^k, \mathsf{Rel}]$ where $k = |\overline{\alpha}|$. We first define the actions of each of these functors on morphisms between relation environments, and then argue that they are well-defined and have the required properties. To do this, we

$$
\begin{aligned}
[\![\Gamma;\Phi \vdash \mathbb{0}]\!]^{\mathsf{Rel}}\rho &= 0\\
[\![\Gamma;\Phi \vdash \mathbb{1}]\!]^{\mathsf{Rel}}\rho &= 1\\
[\![\Gamma;\emptyset \vdash \mathsf{Nat}^{\overline{\alpha}}\,F\,G]\!]^{\mathsf{Rel}}\rho &= \{\eta : \lambda\overline{R}.\,[\![\Gamma;\overline{\alpha} \vdash F]\!]^{\mathsf{Rel}}\rho[\overline{\alpha := R}] \Rightarrow \lambda\overline{R}.\,[\![\Gamma;\overline{\alpha} \vdash G]\!]^{\mathsf{Rel}}\rho[\overline{\alpha := R}]\}\\
&= \{(t,t') \in [\![\Gamma;\emptyset \vdash \mathsf{Nat}^{\overline{\alpha}}\,F\,G]\!]^{\mathsf{Set}}(\pi_1\rho) \times [\![\Gamma;\emptyset \vdash \mathsf{Nat}^{\overline{\alpha}}\,F\,G]\!]^{\mathsf{Set}}(\pi_2\rho) \mid\\
&\qquad \forall R_1 : \mathsf{Rel}(A_1,B_1) \ldots R_k : \mathsf{Rel}(A_k,B_k).\\
&\qquad\quad (t_{\overline{A}}, t'_{\overline{B}}) \in ([\![\Gamma;\overline{\alpha} \vdash G]\!]^{\mathsf{Rel}}\rho[\overline{\alpha := R}])^{[\![\Gamma;\overline{\alpha}\vdash F]\!]^{\mathsf{Rel}}\rho[\overline{\alpha := R}]}\}\\
[\![\Gamma;\Phi \vdash \phi\overline{F}]\!]^{\mathsf{Rel}}\rho &= (\rho\phi)\overline{[\![\Gamma;\Phi \vdash F]\!]^{\mathsf{Rel}}\rho}\\
[\![\Gamma;\Phi \vdash F + G]\!]^{\mathsf{Rel}}\rho &= [\![\Gamma;\Phi \vdash F]\!]^{\mathsf{Rel}}\rho + [\![\Gamma;\Phi \vdash G]\!]^{\mathsf{Rel}}\rho\\
[\![\Gamma;\Phi \vdash F \times G]\!]^{\mathsf{Rel}}\rho &= [\![\Gamma;\Phi \vdash F]\!]^{\mathsf{Rel}}\rho \times [\![\Gamma;\Phi \vdash G]\!]^{\mathsf{Rel}}\rho\\
[\![\Gamma;\Phi \vdash (\mu\phi.\lambda\overline{\alpha}.H)\overline{G}]\!]^{\mathsf{Rel}}\rho &= (\mu T_{H,\rho})\overline{[\![\Gamma;\Phi \vdash G]\!]^{\mathsf{Rel}}\rho}\\
\text{where } T_{H,\rho} &= (T^{\mathsf{Set}}_{H,\pi_1\rho}, T^{\mathsf{Set}}_{H,\pi_2\rho}, T^{\mathsf{Rel}}_{H,\rho})\\
\text{and } T^{\mathsf{Rel}}_{H,\rho}\,F &= \lambda\overline{R}.[\![\Gamma;\phi,\overline{\alpha} \vdash H]\!]^{\mathsf{Rel}}\rho[\phi := F][\overline{\alpha := R}]\\
\text{and } T^{\mathsf{Rel}}_{H,\rho}\,\delta &= \lambda\overline{R}.[\![\Gamma;\phi,\overline{\alpha} \vdash H]\!]^{\mathsf{Rel}}id_\rho[\phi := \delta][\overline{\alpha := id_R}]
\end{aligned}
$$

Fig. 3. Relational interpretation

extend T_H to a *functor* from RelEnv to $[[\mathsf{Rel}^k, \mathsf{Rel}], [\mathsf{Rel}^k, \mathsf{Rel}]]$. Its action on an object $\rho \in \mathsf{RelEnv}$ is given by the higher-order functor $T_{H,\rho}$ whose actions on objects and morphisms are given in Figure 3. Its action on a morphism $f : \rho \to \rho'$ is the higher-order natural transformation $T_{H,f} : T_{H,\rho} \to T_{H,\rho'}$ whose action on any $F : [\mathsf{Rel}^k, \mathsf{Rel}]$ is the natural transformation $T_{H,f}\,F : T_{H,\rho}\,F \to T_{H,\rho'}\,F$ whose component at $\overline{R}$ is $(T_{H,f}\,F)_{\overline{R}} = [\![\Gamma;\phi,\overline{\alpha} \vdash H]\!]^{\mathsf{Rel}} f[\phi := id_F][\overline{\alpha := id_R}]$.

Using T_H, we can define the functorial action of relational interpretation. The action $[\![\Gamma;\Phi \vdash F]\!]^{\mathsf{Rel}} f$ of $[\![\Gamma;\Phi \vdash F]\!]^{\mathsf{Rel}}$ on $f : \rho \to \rho'$ in RelEnv is given as in Definition 2, except that all interpretations are relational interpretations and all occurrences of $T^{\mathsf{Set}}_{H,f}$ are replaced by $T_{H,f}$. For this definition and Figure 3 to be well-defined we need that, for every H, $T_{H,\rho}\,F$ is a relation transformer, and $T_{H,f}\,F : T_{H,\rho}\,F \to T_{H,\rho'}\,F$ is a morphism of relation transformers, whenever F is a relation transformer and $f : \rho \to \rho'$ is in RelEnv. This is immediate from

$$[\![\Gamma;\Phi \vdash F]\!] = ([\![\Gamma;\Phi \vdash F]\!]^{\mathsf{Set}}, [\![\Gamma;\Phi \vdash F]\!]^{\mathsf{Set}}, [\![\Gamma;\Phi \vdash F]\!]^{\mathsf{Rel}}) \in [\mathsf{RelEnv}, RT_0] \quad (1)$$

The proof is a straightforward induction on the structure of F, using an appropriate result from [17] to deduce ω-cocontinuity of $[\![\Gamma;\Phi \vdash F]\!]$ in each case.

We can prove by simultaneous induction that set and relational interpretations of types respect demotion of functorial variables to non-functorial ones and, for $\mathsf{D} \in \{\mathsf{Set}, \mathsf{Rel}\}$, $[\![\Gamma;\Phi \vdash G[\overline{\alpha := K}]]\!]^{\mathsf{D}}\rho = [\![\Gamma;\Phi,\overline{\alpha} \vdash G]\!]^{\mathsf{D}}\rho[\overline{\alpha := [\![\Gamma;\Phi \vdash K]\!]^{\mathsf{D}}\rho}]$, and $[\![\Gamma;\Phi \vdash G[\overline{\alpha := K}]]\!]^{\mathsf{D}} f = [\![\Gamma;\Phi,\overline{\alpha} \vdash G]\!]^{\mathsf{D}} f[\overline{\alpha := [\![\Gamma;\Phi \vdash K]\!]^{\mathsf{D}} f}]$, and $[\![\Gamma;\Phi \vdash F[\phi := H]]\!]^{\mathsf{D}}\rho = [\![\Gamma;\Phi,\phi \vdash F]\!]^{\mathsf{D}}\rho[\phi := \lambda\overline{A}.\,[\![\Gamma;\Phi,\overline{\alpha} \vdash H]\!]^{\mathsf{D}}\rho[\overline{\alpha := A}]]$, and, finally, $[\![\Gamma;\Phi \vdash F[\phi := H]]\!]^{\mathsf{D}} f = [\![\Gamma;\Phi,\phi \vdash F]\!]^{\mathsf{D}} f[\phi := \lambda\overline{A}.\,[\![\Gamma;\Phi,\overline{\alpha} \vdash H]\!]^{\mathsf{D}} f[\overline{\alpha := id_A}]]$.

4 The Identity Extension Lemma

In most treatments of parametricity, equality relations are taken as *given*, either directly as diagonal relations or perhaps via reflexive graphs. By contrast, we give a categorical definition of graph relations for natural transformations and *construct* equality relations as particular such relations. Our definitions specialize to the usual ones for morphisms between sets and equality relations on sets.

The standard definition $(x, y) \in \langle f \rangle$ iff $fx = y$ of the graph $\langle f \rangle$ of a morphism $f : A \to B$ in Set naturally generalizes to associate to each natural transformation between k-ary functors on Set a k-ary relation transformer. Indeed, if $F, G : \mathsf{Set}^k \to \mathsf{Set}$ and $\alpha : F \to G$ is a natural transformation, then the functor $\langle \alpha \rangle^* : \mathsf{Rel}^k \to \mathsf{Rel}$ is defined as follows. Given $R_1 : \mathsf{Rel}(A_1, B_1), ..., R_k : \mathsf{Rel}(A_k, B_k)$, let $\iota_{R_i} : R_i \hookrightarrow A_i \times B_i$, for $i = 1, ..., k$, be the inclusion of R_i as a subset of $A_i \times B_i$, let $h_{\overline{A \times B}}$ be the unique morphism making the left diagram below commute, and let $h_{\overline{R}} : F\overline{R} \to F\overline{A} \times G\overline{B}$ be $h_{\overline{A \times B}} \circ F\overline{\iota_R}$. Further, let $\alpha^\wedge \overline{R}$ be the subobject through which $h_{\overline{R}}$ is factorized by the mono-epi factorization system in Set, as in the right diagram below. Then $\alpha^\wedge \overline{R} : \mathsf{Rel}(F\overline{A}, G\overline{B})$ by construction, so the action of $\langle \alpha \rangle^*$ on objects can be given by $\langle \alpha \rangle^* \overline{(A, B, R)} = (F\overline{A}, G\overline{B}, \iota_{\alpha^\wedge \overline{R}} \alpha^\wedge \overline{R})$. Its action on morphisms is given by $\langle \alpha \rangle^* \overline{(\beta, \beta')} = (F\overline{\beta}, G\overline{\beta'})$.

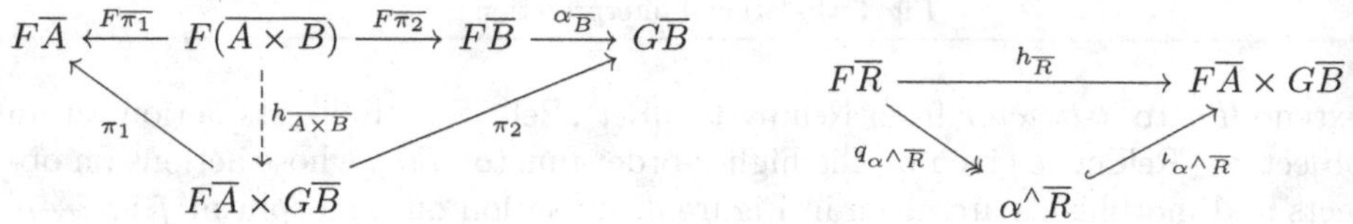

Lemma 1. *If $F, G : [\mathsf{Set}^k, \mathsf{Set}]$, and if $\alpha : F \to G$ is a natural transformation, then the* graph relation transformer for α *defined by $\langle \alpha \rangle = (F, G, \langle \alpha \rangle^*)$ is in RT_k.*

The action of a graph relation transformer on a graph relation can be computed explicitly: if $\alpha : F \to G$ is a morphism in $[\mathsf{Set}^k, \mathsf{Set}]$ and $f_1 : A_1 \to B_1, ..., f_k : A_k \to B_k$, then $\langle \alpha \rangle^* \langle \overline{f} \rangle = \langle G\overline{f} \circ \alpha_{\overline{A}} \rangle = \langle \alpha_{\overline{B}} \circ F\overline{f} \rangle$.

To prove the IEL we also need to know that equality relation transformers preserve equality relations. The *equality relation transformer* on $F : [\mathsf{Set}^k, \mathsf{Set}]$ is $\mathsf{Eq}_F = \langle id_F \rangle = (F, F, \langle id_F \rangle^*)$. The above definition then gives that, for all $\overline{A : \mathsf{Set}}$, $\mathsf{Eq}_F^* \overline{\mathsf{Eq}_A} = \langle id_F \rangle^* \langle id_{\overline{A}} \rangle = \langle F id_{\overline{A}} \circ (id_F)_{\overline{A}} \rangle = \langle id_{F\overline{A}} \circ id_{F\overline{A}} \rangle = \langle id_{F\overline{A}} \rangle = \mathsf{Eq}_{F\overline{A}}$. In addition, if $\rho, \rho' \in \mathsf{SetEnv}$ and $f : \rho \to \rho'$, then the *graph relation environment* $\langle f \rangle$ is defined pointwise by $\langle f \rangle \phi = \langle f \phi \rangle$ for every ϕ. This entails that $\pi_1 \langle f \rangle = \rho$ and $\pi_2 \langle f \rangle = \rho'$. The *equality relation environment* Eq_ρ is defined to be $\langle id_\rho \rangle$. Our IEL is thus:

Theorem 1 (IEL). *If $\rho \in \mathsf{SetEnv}$, then $[\![\Gamma; \Phi \vdash F]\!]^{\mathsf{Rel}} \mathsf{Eq}_\rho = \mathsf{Eq}_{[\![\Gamma; \Phi \vdash F]\!]^{\mathsf{Set}} \rho}$.*

The IEL's highly non-trivial proof is by induction on the structure of F. Only the Nat, application, and fixpoint cases are non-routine. The latter two explicitly calculate actions of graph relation transformers as above. The fixpoint case also uses that, for every $n \in \mathbb{N}$, the following intermediate results can be proved by simultaneous induction with Theorem 1: for any H, ρ, $\overline{A}$, and subformula J of H, both $T^n_{H, \mathsf{Eq}_\rho} K_0 \, \overline{\mathsf{Eq}_A} = (\mathsf{Eq}_{(T^{\mathsf{Set}}_{H,\rho})^n K_0})^* \overline{\mathsf{Eq}_A}$ and $[\![\Gamma; \Phi, \phi, \overline{\alpha} \vdash J]\!]^{\mathsf{Rel}} \mathsf{Eq}_\rho[\phi := T^n_{H, \mathsf{Eq}_\rho} K_0]\overline{[\alpha := \mathsf{Eq}_A]} = [\![\Gamma; \Phi, \phi, \overline{\alpha} \vdash J]\!]^{\mathsf{Rel}} \mathsf{Eq}_\rho[\phi := \mathsf{Eq}_{(T^{\mathsf{Set}}_{H,\rho})^n K_0}]\overline{[\alpha := \mathsf{Eq}_A]}$ hold.

$$
\begin{array}{ll}
\llbracket \Gamma;\Phi \,|\, \Delta, x : F \vdash x : F \rrbracket^{\mathsf{D}} \rho & = \pi_{|\Delta|+1} \\
\llbracket \Gamma;\emptyset \,|\, \Delta \vdash L_{\overline{\alpha}} x.t : \mathsf{Nat}^{\overline{\alpha}}\, F\, G \rrbracket^{\mathsf{D}} \rho & = \mathsf{curry}(\llbracket \Gamma;\overline{\alpha} \,|\, \Delta, x : F \vdash t : G \rrbracket^{\mathsf{D}} \rho[\overline{\alpha := \_}]) \\
\llbracket \Gamma;\Phi \,|\, \Delta \vdash t_{\overline{K}} s : G[\overline{\alpha := K}] \rrbracket^{\mathsf{D}} \rho & = \mathsf{eval} \circ \langle \lambda d.\, (\llbracket \Gamma;\emptyset \,|\, \Delta \vdash t : \mathsf{Nat}^{\overline{\alpha}}\, F\, G \rrbracket^{\mathsf{D}} \rho\, d)_{\overline{\llbracket \Gamma;\Phi \vdash K \rrbracket^{\mathsf{D}} \rho}}, \\
& \qquad \llbracket \Gamma;\Phi \,|\, \Delta \vdash s : F[\overline{\alpha := K}] \rrbracket^{\mathsf{D}} \rho \rangle \\
\llbracket \Gamma;\Phi \,|\, \Delta \vdash \bot_F t : F \rrbracket^{\mathsf{D}} \rho & = !^{\mathbb{0}}_{\llbracket \Gamma;\Phi \vdash F \rrbracket^{\mathsf{D}} \rho} \circ \llbracket \Gamma;\Phi \mid \Delta \vdash t : \mathbb{0} \rrbracket^{\mathsf{D}} \rho, \text{ where} \\
& \quad !^{\mathbb{0}}_{\llbracket \Gamma;\Phi \vdash F \rrbracket^{\mathsf{D}} \rho} \text{ is the unique morphism from } 0 \\
& \quad \text{to } \llbracket \Gamma;\Phi \vdash F \rrbracket^{\mathsf{D}} \rho \\
\llbracket \Gamma;\Phi \,|\, \Delta \vdash \top : \mathbb{1} \rrbracket^{\mathsf{D}} \rho & = !_{\mathbb{1}}^{\llbracket \Gamma;\Phi \vdash \Delta \rrbracket^{\mathsf{D}} \rho}, \text{ where } !_{\mathbb{1}}^{\llbracket \Gamma;\Phi \vdash \Delta \rrbracket^{\mathsf{D}} \rho} \\
& \quad \text{is the unique morphism from } \llbracket \Gamma;\Phi \vdash \Delta \rrbracket^{\mathsf{D}} \rho \text{ to } 1 \\
\llbracket \Gamma;\Phi \,|\, \Delta \vdash (s,t) : F \times G \rrbracket^{\mathsf{D}} \rho & = \llbracket \Gamma;\Phi \,|\, \Delta \vdash s : F \rrbracket^{\mathsf{D}} \rho \times \llbracket \Gamma;\Phi \,|\, \Delta \vdash t : G \rrbracket^{\mathsf{D}} \rho \\
\llbracket \Gamma;\Phi \,|\, \Delta \vdash \pi_1 t : F \rrbracket^{\mathsf{D}} \rho & = \pi_1 \circ \llbracket \Gamma;\Phi \,|\, \Delta \vdash t : F \times G \rrbracket^{\mathsf{D}} \rho \\
\llbracket \Gamma;\Phi \,|\, \Delta \vdash \pi_2 t : G \rrbracket^{\mathsf{D}} \rho & = \pi_2 \circ \llbracket \Gamma;\Phi \,|\, \Delta \vdash t : F \times G \rrbracket^{\mathsf{D}} \rho \\
\llbracket \Gamma;\Phi \mid \Delta \vdash \mathsf{case}\, t\, \mathsf{of}\, \{x \mapsto l;\, y \mapsto r\} : K \rrbracket^{\mathsf{D}} \rho & = \mathsf{eval} \circ \langle \mathsf{curry}\, [\llbracket \Gamma;\Phi \,|\, \Delta, x : F \vdash l : K \rrbracket^{\mathsf{D}} \rho, \\
& \qquad \llbracket \Gamma;\Phi \,|\, \Delta, y : G \vdash r : K \rrbracket^{\mathsf{D}} \rho], \\
& \qquad \llbracket \Gamma;\Phi \,|\, \Delta \vdash t : F + G \rrbracket^{\mathsf{D}} \rho \rangle \\
\llbracket \Gamma;\Phi \,|\, \Delta \vdash \mathsf{inL}\; s : F + G \rrbracket^{\mathsf{D}} \rho & = \mathsf{inL} \circ \llbracket \Gamma;\Phi \,|\, \Delta \vdash s : F \rrbracket^{\mathsf{D}} \rho \\
\llbracket \Gamma;\Phi \,|\, \Delta \vdash \mathsf{inR}\; t : F + G \rrbracket^{\mathsf{D}} \rho & = \mathsf{inR} \circ \llbracket \Gamma;\Phi \,|\, \Delta \vdash t : G \rrbracket^{\mathsf{D}} \rho \\
\llbracket \Gamma;\emptyset \,|\, \emptyset \vdash \mathsf{map}_H^{\overline{F},\overline{G}} : \mathsf{Nat}^{\emptyset}(\mathsf{Nat}^{\overline{\beta},\overline{\gamma}} F\, G) & = \lambda d\, \overline{\eta}\, \overline{C}.\, \llbracket \Gamma;\overline{\phi},\overline{\gamma} \vdash H \rrbracket^{\mathsf{D}} id_{\rho[\overline{\gamma := C}]} [\overline{\phi := \lambda \overline{B}.\eta_{\overline{B}\,\overline{C}}}] \\
\quad (\mathsf{Nat}^{\overline{\gamma}}\, H[\overline{\phi :=_{\overline{\beta}} F}]\, H[\overline{\phi :=_{\overline{\beta}} G}]) \rrbracket^{\mathsf{D}} \rho & \\
\llbracket \Gamma;\emptyset \,|\, \emptyset \vdash \mathsf{in}_H : Nat^{\overline{\beta}}\, H[\phi := (\mu\phi.\lambda\overline{\alpha}.H)\overline{\beta}][\overline{\alpha := \beta}] & = \lambda d.\, in_{T^X_{H,\rho}} \quad \text{where } X \text{ is } \mathsf{Set} \text{ when} \\
\quad (\mu\phi.\lambda\overline{\alpha}.H)\overline{\beta} \rrbracket^{\mathsf{D}} \rho & \qquad \mathsf{D} = \mathsf{Set} \text{ and not present when } \mathsf{D} = \mathsf{Rel} \\
\llbracket \Gamma;\emptyset \,|\, \emptyset \vdash \mathsf{fold}_H^F : \mathsf{Nat}^{\emptyset}\, (\mathsf{Nat}^{\overline{\beta}}\, H[\phi :=_{\overline{\beta}} F][\overline{\alpha := \beta}]\, F) & = \lambda d.\, fold_{T^X_{H,\rho}} \\
\quad (\mathsf{Nat}^{\overline{\beta}}\, (\mu\phi.\lambda\overline{\alpha}.H)\overline{\beta}\, F) \rrbracket^{\mathsf{D}} \rho & \qquad \text{where } X \text{ is as above}
\end{array}
$$

Fig. 4. Term semantics

The case of the proof when F and J are both μ-types makes clear that if functorial variables of arity greater than 0 were allowed to appear in the bodies of μ-types, then the IEL would fail.

With the IEL in hand we can prove a Graph Lemma for our setting:

Lemma 2. *If $\rho, \rho' \in \mathsf{SetEnv}$ and $f : \rho \to \rho'$ then*

$$\langle \llbracket \Gamma;\Phi \vdash F \rrbracket^{\mathsf{Set}} f \rangle = \llbracket \Gamma;\Phi \vdash F \rrbracket^{\mathsf{Rel}} \langle f \rangle$$

5 Interpreting Terms

If $\Delta = x_1 : F_1, ..., x_n : F_n$ is a term context for Γ and Φ, define $\llbracket \Gamma;\Phi \vdash \Delta \rrbracket^{\mathsf{D}} = \llbracket \Gamma;\Phi \vdash F_1 \rrbracket^{\mathsf{D}} \times ... \times \llbracket \Gamma;\Phi \vdash F_n \rrbracket^{\mathsf{D}}$, where D is Set or Rel as appropriate. Then every well-formed term has a set (resp., relational) interpretation as a natural transformation from the set (resp., relational) interpretation of its term context to that of its type. These interpretations, given in Figure 4, respect weakening, so that $\llbracket \Gamma;\Phi \,|\, \Delta, x : F \vdash t : G \rrbracket^{\mathsf{D}} \rho = (\llbracket \Gamma;\Phi \,|\, \Delta \vdash t : G \rrbracket^{\mathsf{D}} \rho) \circ \pi_\Delta$, where $\rho \in \mathsf{SetEnv}$ or $\rho \in \mathsf{RelEnv}$, and π_Δ is the projection $\llbracket \Gamma;\Phi \vdash \Delta, x : F \rrbracket^{\mathsf{D}} \to \llbracket \Gamma;\Phi \vdash \Delta \rrbracket^{\mathsf{D}}$.

The return type for the semantic fold is $\llbracket \Gamma;\overline{\beta} \vdash F \rrbracket^{\mathsf{D}} \rho[\overline{\beta := B}]$. This interpretation gives $\llbracket \Gamma;\emptyset \,|\, \Delta \vdash \lambda x.t : F \to G \rrbracket^{\mathsf{D}} \rho = \mathsf{curry}(\llbracket \Gamma;\emptyset \,|\, \Delta, x : F \vdash t : G \rrbracket^{\mathsf{D}} \rho)$ and $\llbracket \Gamma;\emptyset \,|\, \Delta \vdash st : G \rrbracket^{\mathsf{D}} \rho = \mathsf{eval} \circ \langle \llbracket \Gamma;\emptyset \,|\, \Delta \vdash s : F \to G \rrbracket^{\mathsf{D}} \rho, \llbracket \Gamma;\emptyset \,|\, \Delta \vdash t : F \rrbracket^{\mathsf{D}} \rho \rangle$, so it specializes to the standard interpretations for System F terms. If t is closed, i.e., if $\emptyset;\emptyset \mid \emptyset \vdash t : F$, then we write $\llbracket \vdash t : F \rrbracket^{\mathsf{D}}$ instead of $\llbracket \emptyset;\emptyset \mid \emptyset \vdash t : F \rrbracket^{\mathsf{D}}$. In addition, term interpretation respects substitution for both functorial and non-functorial type variables, as well as term substitution. Direct calculation reveals that interpretations of terms also satisfy $\llbracket \Gamma;\Phi \mid \Delta \vdash (L_{\overline{\alpha}} x.t)_{\overline{K}} s \rrbracket^{\mathsf{D}} =$

$[\![\Gamma;\Phi \mid \Delta \vdash t[\overline{\alpha := K}][x := s]]\!]^{\mathsf{D}}$. Term extensionality for both types and terms — i.e., $[\![\Gamma;\Phi \vdash (L_\alpha x.t)_\alpha \top : F]\!]^{\mathsf{D}} = [\![\Gamma;\Phi \vdash t : F]\!]^{\mathsf{D}}$ and $[\![\Gamma;\Phi \vdash (L_\alpha x.t)_\alpha x : F]\!]^{\mathsf{D}} = [\![\Gamma;\Phi \vdash t : F]\!]^{\mathsf{D}}$ — follow (when both sides of these equations are defined).

6 Free Theorems for Nested Types

6.1 Consequences of Naturality

Define, for $\Gamma;\overline{\alpha} \vdash F$, the term id_F to be $\Gamma;\emptyset \mid \emptyset \vdash L_{\overline{\alpha}}x.x : \mathsf{Nat}^{\overline{\alpha}} F\,F$ and, for terms $\Gamma;\emptyset \mid \Delta \vdash t : \mathsf{Nat}^{\overline{\alpha}} F\,G$ and $\Gamma;\emptyset \mid \Delta \vdash s : \mathsf{Nat}^{\overline{\alpha}} G\,H$, the *composition* $s \circ t$ of t and s to be $\Gamma;\emptyset \mid \Delta \vdash L_{\overline{\alpha}}x.s_{\overline{\alpha}}(t_{\overline{\alpha}}x) : \mathsf{Nat}^{\overline{\alpha}} F\,H$. Then $[\![\Gamma;\emptyset \mid \emptyset \vdash id_F : \mathsf{Nat}^{\overline{\alpha}} F\,F]\!]^{\mathsf{Set}}\rho* = id_{\lambda\overline{A}.[\![\Gamma;\overline{\alpha}\vdash F]\!]^{\mathsf{Set}}\rho[\overline{\alpha:=A}]}$ for any set environment ρ and $[\![\Gamma;\emptyset \mid \Delta \vdash s \circ t : \mathsf{Nat}^{\overline{\alpha}} F\,H]\!]^{\mathsf{Set}} = [\![\Gamma;\emptyset \mid \Delta \vdash s : \mathsf{Nat}^{\overline{\alpha}} G\,H]\!]^{\mathsf{Set}} \circ [\![\Gamma;\emptyset \mid \Delta \vdash t : \mathsf{Nat}^{\overline{\alpha}} F\,G]\!]^{\mathsf{Set}}$. Also, terms of Nat type behave as natural transformations with respect to their source and target types:

Theorem 2. *If* $\Gamma;\emptyset \mid \Delta \vdash s : \mathsf{Nat}^{\overline{\alpha},\overline{\gamma}} F\,G$ *and* $\overline{\Gamma;\emptyset \mid \Delta \vdash t : \mathsf{Nat}^{\overline{\gamma}} K\,H}$*, then*

$$[\![\Gamma;\emptyset \mid \Delta \vdash ((\mathsf{map}_G^{\overline{K},\overline{H}})_\emptyset\,\overline{t}) \circ (L_{\overline{\gamma}}z.s_{\overline{K},\overline{\gamma}}z) : \mathsf{Nat}^{\overline{\gamma}} F[\overline{\alpha := K}]\,G[\overline{\alpha := H}]]\!]^{\mathsf{Set}}$$
$$= [\![\Gamma;\emptyset \mid \Delta \vdash (L_{\overline{\gamma}}z.s_{\overline{H},\overline{\gamma}}z) \circ ((\mathsf{map}_F^{\overline{K},\overline{H}})_\emptyset\,\overline{t}) : \mathsf{Nat}^{\overline{\gamma}} F[\overline{\alpha := K}]\,G[\overline{\alpha := H}]]\!]^{\mathsf{Set}}$$

Theorem 2 gives rise to an entire family of free theorems that are consequences of naturality, and thus do not require the full power of parametricity. In particular, we can prove that the interpretation of every map_H is a functor, and that map is itself a higher-order functor. For example, the former property can be stated as: if $\Gamma;\overline{\alpha},\overline{\gamma} \vdash H$, $\overline{\Gamma;\emptyset \mid \Delta \vdash g : \mathsf{Nat}^{\overline{\gamma}} F\,G}$, and $\overline{\Gamma;\emptyset \mid \Delta \vdash f : \mathsf{Nat}^{\overline{\gamma}} G\,K}$, then

$$[\![\Gamma;\emptyset \mid \Delta \vdash (\mathsf{map}_H^{\overline{F},\overline{K}})_\emptyset\,\overline{(f \circ g)} : \mathsf{Nat}^{\overline{\gamma}} H[\overline{\alpha := F}]\,H[\overline{\alpha := K}]]\!]^{\mathsf{Set}}$$
$$= [\![\Gamma;\emptyset \mid \Delta \vdash (\mathsf{map}_H^{\overline{G},\overline{K}})_\emptyset\,\overline{f} \circ (\mathsf{map}_H^{\overline{F},\overline{G}})_\emptyset\,\overline{g} : \mathsf{Nat}^{\overline{\gamma}} H[\overline{\alpha := F}]\,H[\overline{\alpha := K}]]\!]^{\mathsf{Set}}$$

We can also prove the expected properties of map, in, and fold, and their interpretations, e.g., uniqueness and the universal property of the interpretation of fold, and the interpretation of in is an isomorphism.

6.2 The Abstraction Theorem

To get consequences of parametricity that are not merely consequences of naturality, we prove an Abstraction Theorem (Theorem 4). As usual for such theorems, we prove a more general result (Theorem 3) for open terms, and recover our Abstraction Theorem as its special case for closed terms of closed type.

Theorem 3. *Every well-formed term* $\Gamma;\Phi \mid \Delta \vdash t : F$ *induces a natural transformation from* $[\![\Gamma;\Phi \vdash \Delta]\!]$ *to* $[\![\Gamma;\Phi \vdash F]\!]$*, i.e., a triple of natural transformations* $([\![\Gamma;\Phi \mid \Delta \vdash t : F]\!]^{\mathsf{Set}}, [\![\Gamma;\Phi \mid \Delta \vdash t : F]\!]^{\mathsf{Set}}, [\![\Gamma;\Phi \mid \Delta \vdash t : F]\!]^{\mathsf{Rel}})$*, where, for* $\mathsf{D} \in \{\mathsf{Set}, \mathsf{Rel}\}$*, and for* $\rho \in \mathsf{SetEnv}$ *or* $\rho \in \mathsf{RelEnv}$ *as appropriate,* $[\![\Gamma;\Phi \mid \Delta \vdash t : F]\!]^{\mathsf{D}} : [\![\Gamma;\Phi \vdash \Delta]\!]^{\mathsf{D}} \to [\![\Gamma;\Phi \vdash F]\!]^{\mathsf{D}}$ *has component* $[\![\Gamma;\Phi \mid \Delta \vdash t : F]\!]^{\mathsf{D}}\rho : [\![\Gamma;\Phi \vdash \Delta]\!]^{\mathsf{D}}\rho \to [\![\Gamma;\Phi \vdash F]\!]^{\mathsf{D}}\rho$ *at* ρ*. Moreover, for all* $\rho \in \mathsf{RelEnv}$*, we have* $[\![\Gamma;\Phi \mid \Delta \vdash t : F]\!]^{\mathsf{Rel}}\rho = ([\![\Gamma;\Phi \mid \Delta \vdash t : F]\!]^{\mathsf{Set}}(\pi_1\rho), [\![\Gamma;\Phi \mid \Delta \vdash t : F]\!]^{\mathsf{Set}}(\pi_2\rho))$.

The proof is by induction on t. It requires showing that set and relational interpretations of term judgments are natural transformations, and that all set interpretations of terms of Nat-types satisfy the appropriate equality preservation conditions from Figure 2. For the interesting cases of abstraction, application, map, in, and fold terms, propagating the naturality conditions is somewhat involved; the latter two especially require some delicate diagram chasing. That it is possible provides strong evidence that our development is sensible, natural, and at an appropriate level of abstraction.

Using Theorem 3 we can prove that our calculus admits no terms with the type $\mathsf{Nat}^{\alpha}\mathbb{1}\,\alpha$ of the polymorphic bottom, and every closed term g of type $\mathsf{Nat}^{\alpha}\alpha\,\alpha$ denotes the polymorphic identity function. Moreover, an immediate consequence of Theorem 3 is that if $\rho \in \mathsf{RelEnv}$, and $(a,b) \in [\![\Gamma;\Phi \vdash \Delta]\!]^{\mathsf{Rel}}\rho$, then $([\![\Gamma;\Phi\,|\,\Delta \vdash t : F]\!]^{\mathsf{Set}}(\pi_1\rho)\,a\,,[\![\Gamma;\Phi\,|\,\Delta \vdash t : F]\!]^{\mathsf{Set}}(\pi_2\rho)\,b) \in [\![\Gamma;\Phi \vdash F]\!]^{\mathsf{Rel}}\rho$. Its instantiation to closed terms of closed type gives

Theorem 4 (Abstraction Theorem). $([\![\vdash t : F]\!]^{\mathsf{Set}}, [\![\vdash t : F]\!]^{\mathsf{Set}}) \in [\![\vdash F]\!]^{\mathsf{Rel}}$

Using Theorem 4 we can recover free theorems, such as that for the type of the standard *filter* function for lists, that go beyond mere naturality, and extend them to those nested types for which analogous functions can be defined. In particular, we can extend short cut fusion for lists [10] to nested types, thereby formally proving correctness of the categorically inspired theorem from [16]. As shown there, replacing $\mathbb{1}$ with any type $\emptyset;\alpha \vdash C$ generalizes Theorem 5 to a free theorem whose conclusion is $\mathit{fold}_H\, B\, \circ\, G\,\mu H\; \mathit{in}_H = G\,[\![\emptyset;\alpha \vdash K]\!]^{\mathsf{Set}}\,B$.

Theorem 5. *If* $\emptyset;\phi,\alpha \vdash F$, $\emptyset;\alpha \vdash K$, $H : [\mathsf{Set},\mathsf{Set}] \to [\mathsf{Set},\mathsf{Set}]$ *is defined by* $H f x = [\![\emptyset;\phi,\alpha \vdash F]\!]^{\mathsf{Set}}[\phi := f][\alpha := x]$, *and* $G = [\![\phi;\emptyset\,|\,\emptyset \vdash g : \mathsf{Nat}^{\emptyset}\,(\mathsf{Nat}^{\alpha}\,F\,(\phi\alpha))\,(\mathsf{Nat}^{\alpha}\,\mathbb{1}\,(\phi\alpha))]\!]^{\mathsf{Set}}$ *for some g, then for every* $B \in H[\![\emptyset;\alpha \vdash K]\!]^{\mathsf{Set}} \to [\![\emptyset;\alpha \vdash K]\!]^{\mathsf{Set}}$ *we have* $\mathit{fold}_H\, B\,(G\,\mu H\; \mathit{in}_H) = G\,[\![\emptyset;\alpha \vdash K]\!]^{\mathsf{Set}}\,B$.

7 Conclusion and Directions for Future Work

We have constructed a parametric model for a calculus supporting primitive nested types, and used its Abstraction Theorem to derive free theorems for these types. This was not possible before [17] because these types were not previously known to have well-defined interpretations in locally finitely presentable categories (here, Set and Rel), and, to our knowledge, no term calculus for them existed either. We naturally hope (some appropriate variant of) the construction elaborated here will generalize to more advanced data types. For example, GADTs can be represented using left Kan extensions, and it was shown in [17] that adding a Lan construct to a calculus such as ours preserves the λ-cocontinuity needed for the data types it defines to have well-defined interpretations in locally λ-presentable categories. (Interestingly, $\lambda > \aleph_1$ is required to interpret even common GADTs.) This suggests carrying out our model construction in locally λ-presentable cartesian closed categories (lpcccs) $\mathcal{C}$ whose categories of (abstract) relations, obtained by pullback as in [13], are also lpcccs and are appropriately fibred over $\mathcal{C}$. Adding term-level fixpoints further requires our semantic categories not just to be locally λ-presentable, but to support some kind of domain structure as well.

References

1. Adámek, J., Rosický, J.: Locally Presentable and Accessible Categories. Cambridge University Press (1994)
2. Atkey, R.: Relational Parametricity for Higher Kinds. In: Computer Science Logic, pp. 46–61. Schloss Dagstuhl–Leibniz-Zentrum fuer Informatik (2012)
3. Bainbridge, E. S., Freyd, P. J., Scedrov, A., Scott, P. J.: Functorial Polymorphism. Theoretical Computer Science 70, 35–64 (1990)
4. Bird, R., Meertens, L.: Nested datatypes. In: Mathematics of Program Construction, pp. 52–67. Springer (1998)
5. Bird, R., Paterson, R.: Generalised folds for nested datatypes. Formal Aspects of Computing 11, 200–222 (1999)
6. Birkedal, L., Møgelberg, R. E.: Categorical models for Abadi and Plotkin's logic for parametricity. Mathematical Structures in Computer Science 15, 709–772 (2005)
7. Cardelli, L: Type Systems. In: CRC Handbook of Computer Science and Engineering, pp. 2208–2236. CRC Press (1984)
8. Dunphy, B., Reddy, U.: Parametric Limits. In: Logic in Computer Science, pp. 242–252. IEEE (2004)
9. Ghani, N., Johann, P., Nordvall Forsberg, F., Orsanigo, F., Revell, T.: Bifibrational Functorial Semantics for Parametric Polymorphism. Electronic Notes in Theoretical Computer Science 319, 165–181. (2015)
10. Gill, A., Launchbury, J., Peyton Jones, S. L.: A short cut to deforestation. In: Functional Programming Languages and Computer Architecture, Proceedings, pp. 223–232. Association for Computing Machinery (1993)
11. Girard, J.-Y.: Interprétation fonctionnelle et élimination des coupures de l'arithmétique d'ordre supérieur. PhD thesis, University of Paris (1972)
12. Hasegawa, R.: Categorical data types in parametric polymorphism. Mathematical Structures in Computer Science 4, 71–109 (1994)
13. Jacobs, B.: Categorical Logic and Type Theory. Elsevier (1999)
14. Johann, P.: A Generalization of Short-Cut Fusion and Its Correctness Proof. Higher-Order and Symbolic Computation 15, 273–300 (2002)
15. Johann, P., Ghani, N.: Foundations for Structured Programming with GADTs. In: Principles of Programming Languages, pp. 297–308. Association for Computing Machinery (2008)
16. Johann, P., Ghani, N.: Haskell Programming with Nested Types: A Principled Approach Higher-Order and Symbolic Computation 22(2), 155–189 (2010)
17. Johann, P., Polonsky, A.: Higher-kinded data types: Syntax and Semantics. In: Logic in Computer Science, pp. 1–13. IEEE (2019)
18. Johann, P., Polonsky, A.: Deep Induction: Induction Rules for (Truly) Nested Types. In: Foundations of Software Science and Computation Structures, pp. 339–358. Springer (2020)
19. Matthes, R.: Map Fusion for Nested Datatypes in Intensional Type Theory. Science of Computer Programming 76(3), 204–224 (2011)
20. Ma, Q., Reynolds, J. C.: Types, abstraction, and parametric polymorphism, part 2. In: Mathematical Foundations of Program Semantics, pp. 1–40. Springer-Verlag (1992)
21. Martin, C., Gibbons, J.: On the semantics of nested datatypes. Information Processing Letters 80(5), 233–238 (2001)
22. Pitts, A.: Parametric polymorphism, recursive types, and operational equivalence. (1998)

23. Pitts, A.: Parametric polymorphism and operational equivalence. Mathematical Structures in Computer Science 10, 321–359 (2000)
24. Reynolds, J. C.: Types, abstraction, and parametric polymorphism. Information Processing 83(1), 513–523 (1983)
25. Reynolds, J. C.: Polymorphism is not set-theoretic. Semantics of Data Types, 145–156 (1984)
26. Robinson, E., Rosolini, G.: Reflexive graphs and parametric polymorphism. In: Logic in Computer Science, pp. 364–371. IEEE (1994)
27. Wadler, P.: Theorems for free!. In: Functional Programming Languages and Computer Architecture, Proceedings, pp. 347–359. Association for Computing Machinery (1989)

The Spirit of Node Replication

Delia Kesner[1,2], Loïc Peyrot[1], and Daniel Ventura[3]*

[1] Université de Paris, CNRS, IRIF, Paris, France
{kesner,lpeyrot}@irif.fr
[2] Institut Universitaire de France, France
[3] Univ. Federal de Goiás, Goiânia, Brazil
ventura@ufg.br

Abstract. We define and study a term calculus implementing higher-order node replication. It is used to specify two different (weak) evaluation strategies: call-by-name and fully lazy call-by-need, that are shown to be observationally equivalent by using type theoretical technical tools.

1 Introduction

Computation in the λ-calculus is based on higher-order substitution, a complex operation being able to erase and copy terms during evaluation. Several formalisms have been proposed to model higher-order substitution, going from explicit substitutions (ES) [1] (see a survey in [41]) and labeled systems [15] to pointer graphs [60] or optimal sharing graphs [49]. The model of copying behind each of these formalisms is not the same.

Indeed, suppose one wants to substitute all the free occurrences of some variable x in a term t by some term u. We can imagine at least four ways to do that. (1) A drastic solution is a one-shot substitution, called *non-linear* (or *full*) *substitution*, based on simultaneously replacing *all* the free occurrences of x in t by the whole term u. This notion is generally defined by induction on the structure of the term t. (2) A refined method substitutes *one* free occurrence of x at a time, the so-called *linear* (or *partial*) *substitution*. This notion is generally defined by induction on the number of free occurrences of x in the term t. An orthogonal approach can be taken by replicating *one* term-constructor of u *at a time*, instead of replicating u as a whole, called here *node replication*. This notion can be defined by induction on the structure of the term u, and also admits two versions: (3) non-linear, *i.e.* by simultaneously replacing all the occurrences of x in t, or (4) linear. The linear version of the node replication approach can be formally defined by combining (2) and (3).

It is not surprising that different notions of substitution give rise to different evaluation strategies. Indeed, linear substitution is the common model in well-known abstract machines for call-by-name and call-by-value (see *e.g.* [3]), while (linear) node replication is used to implement fully lazy sharing [60]. However, node replication, originally introduced to implement optimal graph reduction in

* Supported by CNPq grant Universal 430667/2016-7.

S. Kiefer and C. Tasson (Eds.): FOSSACS 2021, LNCS 12650, pp. 344–364, 2021.
https://doi.org/10.1007/978-3-030-71995-1_18

a graphical formalism, has only been studied from a Curry-Howard perspective by means of a term language known as the atomic λ-calculus [33].

The Atomic Lambda-Calculus. The Curry-Howard isomorphism uncovers a deep connection between logical systems and term calculi. It is then not surprising that different methods to implement substitution correspond to different ways to normalize logical proofs. Indeed, full substitution (1) can be explained in terms of natural deduction, while partial substitution (2) corresponds to cut elimination in Proof-Nets [2]. Replication of nodes (3)-(4) is based on a Curry-Howard interpretation of deep inference [32,33]. Indeed, the logical aspects of intuistionistic deep inference are captured by the atomic λ-calculus [33], where copying of terms proceeds *atomically*, *i.e.* node by node, similar to the optimal graph reduction of Lamping [49].

The atomic λ-calculus is based on *explicit control of resources* such as erasure and duplication. Its operational semantics explicitly handles the structural constructors of weakening and contraction, as in the calculus of resources λ`lxr` [43,44]. As a result, comprehension of the meta-properties of the term-calculus, in a higher-level, and its application to concrete implementations of reduction strategies in programming languages, turn out to be quite difficult. In this paper, we take one step back, by studying the paradigm of *node replication* based on *implicit*, rather than *explicit*, weakening and contraction. This gives a new concise formulation of node replication which is simple enough to model different programming languages based on reduction strategies.

Call-by-Name, Call-by-Value, Call-by-Need. *Call-by-name* is used to implement programming languages in which arguments of functions are first copied, then evaluated. This is frequently expensive, and may be improved by *call-by-value*, in which arguments are evaluated first, then consumed. The difference can be illustrated by the term $t = \Delta(\mathtt{II})$, where $\Delta = \lambda x.xx$ and $\mathtt{I} = \lambda z.z$: call-by-name first duplicates the argument `II`, so that its evaluation is also duplicated, while call-by-value first reduces `II` to (the value) `I`, so that duplications of the argument do not cause any duplicated evaluation. It is not always the best solution, though, because evaluating erasable arguments is useless.

Call-by-need, instead, takes the best of call-by-name and call-by-value: as in call-by-name, erasable arguments are not evaluated at all, and as in call-by-value, reduction of arguments occurs at most once. Furthermore, call-by-need implements a *demand-driven* evaluation, in which erasable arguments are never needed (so they are not evaluated), and non-erasable arguments are evaluated only if needed. Technically, some sharing mechanism is necessary, for example by extending the λ-calculus with explicit substitutions/let constructs [7]. Then β-reduction is decomposed in at least two steps: one creating an explicit (pending) substitution, and the other ones (linearly) substituting *values*. Thus for example, $(\lambda x.xx)(\mathtt{II})$ reduces to $(xx)[x\backslash\mathtt{II}]$, and the substitution argument is thus evaluated in order to find a value before performing the linear substitution.

Even when adopting this wise evaluation scheme, there are still some unnecessary copies of redexes: while only *values* (*i.e.* abstractions) are duplicated, they may contain redexes as subterms, *e.g.* $\lambda z.z(\mathtt{II})$ whose subterm $\mathtt{II}$ is a redex. Duplication of such values might cause redex duplications in *weak* (*i.e.* when evaluation is forbidden inside abstractions) call-by-need. This happens in particular in the *confluent* variant of weak reduction in [52].

Full laziness. Alas, it is not possible to keep all values shared forever, typically when they potentially contribute to the creation of a future β-reduction step. The key idea to gain in efficiency is then to keep the subterm $\mathtt{II}$ as a *shared* redex. Therefore, the (full) value $\lambda z.z(\mathtt{II})$ to be copied is split into two separate parts. The first one, called *skeleton*, contains the minimal information preserving the bound structure of the value, *i.e.* the linked structure between the binder and each of its (bound) variables. In our example, this is the term $\lambda z.zy$, where y is a fresh variable. The second one is a multiset of *maximal free expressions* (MFE), representing all the shareable expressions (here only the term $\mathtt{II}$). Only the skeleton is then copied, while the problematic redex $\mathtt{II}$ remains shared:

$$(\lambda x.xx)(\lambda z.z(\mathtt{II})) \to (xx)[x\backslash\lambda z.z(\mathtt{II})] \to ((\lambda z.zy)x)[x\backslash\lambda z.zy][y\backslash\mathtt{II}]$$

When the subterm $\mathtt{II}$ is needed ahead, it is first reduced inside the ES, as it is usual in (standard) call-by-need, thus avoiding to compute the redex twice. This optimization is called *fully lazy sharing* and is due to Wadsworth [60].

In the confluent weak setting evoked earlier [52], the fully lazy optimization is even optimal in the sense of Lévy [51]. This means that the strategy reaches the weak normal form in the same number of β-steps as the shortest possible weak reduction sequence in the usual λ-calculus without sharing. Thus, fully lazy sharing turns out to be a *decidable* optimal strategy, in contrast to other weak evaluation strategies in the λ-calculus without sharing, which are also optimal but not decidable [11].

Contributions. The first contribution of this paper is a term calculus implementing (full) node replication and internally encoding skeleton extraction (Sec. 2). We study some of its main operational properties: termination of the substitution calculus, confluence, and its relation with the λ-calculus.

Our second contribution is the use of the node replication paradigm to give an alternative specification of two evaluation strategies usually described by means of full or linear substitution: call-by-name (Sec. 4.1) and weak fully lazy reduction (Sec. 4.2), based on the key notion of skeleton. The former can be related to (weak) head reduction, while the latter is a fully lazy version of (weak) call-by-need. In contrast to other implementations of fully lazy reduction relying on (external) meta-level definitions, our implementation is based on formal operations internally defined over the term syntax of the calculus.

Furthermore, while it is known that call-by-name and call-by-need specified by means of full/linear substitution are observationally equivalent [7], it was not clear at first whether the same property would hold in our case. Our third contribution is a proof of this result (Sec. 6) using semantical tools coming from proof theory –notably intersection types. This proof technique [42] considerably simplifies other approaches [7,54] based on syntactical tools. Moreover, the use of intersection types has another important consequence: standard call-by-name and call-by-need turn out to be observationally equivalent to call-by-name and call-by-need with node replication, as well as to the more semantical notion of neededness (see [45]).

Intersection types provide quantitative information about fully lazy evaluation so that a fourth contribution of this work is a measure based on type derivations which turns out to be an upper bound to the length of reduction sequences to normal forms in a fully lazy implementation.

More generally, our work bridges the gap between the Curry-Howard theoretical understanding of node replication and concrete implementations of fully lazy sharing. Related works are presented in the concluding Sec. 7.

2 A Calculus for Node Replication

We now present the syntax and operational semantics of the $\lambda\mathtt{R}$-calculus ($\mathtt{R}$ for Replication), as well as a notion of *level* playing a key role in the next sections.

Syntax. Given a countably infinite set $\mathcal{X}$ of variables $x, y, z, \ldots$, we consider the following grammars.

(Terms)	$t, u ::= x \mid \lambda x.t \mid tu \mid t[x\backslash u] \mid t[x\backslash\!\!\backslash \lambda y.u]$
(Pure Terms)	$p, q ::= x \mid \lambda x.p \mid pq$
(Term Contexts)	$\mathtt{C} ::= \Box \mid \lambda x.\mathtt{C} \mid \mathtt{C}t \mid t\mathtt{C} \mid \mathtt{C}[x\backslash t] \mid \mathtt{C}[x\backslash\!\!\backslash \lambda y.u] \mid t[x\backslash \mathtt{C}] \mid t[x\backslash\!\!\backslash \lambda y.\mathtt{C}]$
(List Contexts)	$\mathtt{L} ::= \Box \mid \mathtt{L}[x\backslash u] \mid \mathtt{L}[x\backslash\!\!\backslash \lambda y.u]$

The set of terms (resp. **pure** terms) is denoted by $\Lambda_{\mathtt{R}}$ (resp. Λ). We write $|t|$ for the **size** of t, *i.e.* for its number of constructors. We write $\mathtt{I}$ for the identity function $\lambda x.x$. The construction $[x\backslash u]$ is an **explicit substitution (ES)**, and $[x\backslash\!\!\backslash \lambda y.u]$ an **explicit distributor**: the first one is used to copy arbitrary terms, while the second one is used specifically to duplicate abstractions. We write $[x \triangleleft u]$ to denote an **explicit cut** in general, which is either $[x\backslash u]$ or $[x\backslash\!\!\backslash u]$ when u is $\lambda y.u'$, typically to factorize some definitions and proofs where they behave similarly in both cases. When using the general notation $t[x \triangleleft u]$, we define $x(\triangleleft) = 1$ if the term is an ES, and $x(\triangleleft) = 0$ otherwise.

We use two notions of **contexts**. Term contexts $\mathtt{C}$ extend those of the λ-calculus to explicit cuts. List contexts $\mathtt{L}$ denote an arbitrary list of explicit cuts. They will be used to implement reduction *at a distance* in the operational semantics defined ahead.

Free/bound variables of terms are defined as usual, notably $\mathtt{fv}(t[x\triangleleft u]) := \mathtt{fv}(t)\backslash\{x\} \cup \mathtt{fv}(u)$. These notions are extended to contexts as expected, in particular $\mathtt{fv}(\Box) := \emptyset$. The **domain** of a **list context** is given by $\mathtt{dlc}(\Box) := \emptyset$ and $\mathtt{dlc}(\mathtt{L}[x \triangleleft u]) := \mathtt{dlc}(\mathtt{L}) \cup \{x\}$. α-conversion [13] is extended to $\lambda\mathtt{R}$-terms as expected and used to avoid capture of free variables. We write $t\{x\backslash u\}$ for the meta-level (capture-free) substitution simultaneously replacing all the free occurrences of the variable x in t by the term u.

The **application of a context** $\mathtt{C}$ **to a term** t, written $\mathtt{C}\langle t\rangle$, replaces the hole $\Box$ of $\mathtt{C}$ by t. For instance, $\Box\langle t\rangle = t$ and $(\lambda x.\Box)\langle t\rangle = \lambda x.t$. This operation is not defined modulo α-conversion, so that capture of variables eventually happens. Thus, we also consider another kind of application of contexts to terms, denoted with double brackets, which is only defined if there is no capture of variables. For instance, $(\lambda y.\Box)\langle\langle x\rangle\rangle = \lambda y.x$ while $(\lambda x.\Box)\langle\langle x\rangle\rangle$ is undefined.

Operational semantics. ES may block some expected *meaningful* (*i.e.* non-structural) reductions. For instance, β-reduction is blocked in $(\lambda x.t)[y\backslash v]u$ because an ES lies between the function and its argument. This kind of stuck redexes do not happen in graphical representations (*e.g.* [28]), but it is typical in the sequential structure of *term* syntaxes.

There are at least two ways to handle this issue. The first one is based on *structural/permutation* rules, as in [33], where the substitution is first pushed outside the application node, as $(\lambda x.t)[y\backslash v]u \to ((\lambda x.t)u)[y\backslash v]$, so that β-reduction is finally unblocked. The second, less elementary, possibility is given by an operational semantics *at a distance* [6,4], where the β-rule can be fired by a rule like $\mathtt{L}\langle\lambda x.t\rangle u \to \mathtt{L}\langle t[x\backslash u]\rangle$, $\mathtt{L}$ being an arbitrary list context. The distance paradigm is therefore used to gather meaningful and permutation rules in only one reduction step. In $\lambda\mathtt{R}$, we combine these two technical tools. First, we consider the following permutation rules, all of them are constrained by the condition $x \notin \mathtt{fv}(t)$.

$$\begin{array}{llll} \lambda x.u[y\lhd t] \mapsto_\pi (\lambda x.u)[y\lhd t] & & v[x\lhd u]t & \mapsto_\pi (vt)[x\lhd u] \\ tv[x\lhd u] \quad \mapsto_\pi (tv)[x\lhd u] & & t[y\lhd v[x\lhd u]] & \mapsto_\pi t[y\lhd v][x\lhd u] \end{array}$$

The reduction relation $\to_\pi$ is defined as the closure of the rules $\mapsto_\pi$ under *all* contexts. It does not hold any computational content, only a structural one that unblocks redexes by moving explicit cuts out.

In order to highlight the computational content of node replication we combine distance and permutations within the **λR-calculus**, given by the closure of the following rules by all the contexts.

$$\begin{array}{llll} \mathtt{L}\langle\lambda x.t\rangle u & \mapsto_{\mathtt{dB}} & \mathtt{L}\langle t[x\backslash u]\rangle & \\ t[x\backslash\mathtt{L}\langle uv\rangle] & \mapsto_{\mathtt{app}} & \mathtt{L}\langle t\{x\backslash yz\}[y\backslash u][z\backslash v]\rangle & \text{where } y \text{ and } z \text{ are fresh} \\ t[x\backslash\mathtt{L}\langle\lambda y.u\rangle] & \mapsto_{\mathtt{dist}} & \mathtt{L}\langle t[x\backslash\!\backslash\lambda y.z[z\backslash u]]\rangle & \text{where } z \text{ is fresh} \\ t[x\backslash\!\backslash\lambda y.u] & \mapsto_{\mathtt{abs}} & \mathtt{L}\langle t\{x\backslash\lambda y.p\}\rangle & \text{where } u \to_\pi^* \mathtt{L}\langle p\rangle \text{ and } y \notin \mathtt{fv}(\mathtt{L}) \\ t[x\backslash\mathtt{L}\langle y\rangle] & \mapsto_{\mathtt{var}} & \mathtt{L}\langle t\{x\backslash y\}\rangle & \end{array}$$

Notice in the five rules above that the (meta-level) substitution is *full* (it is performed simultaneously on all free occurrences of the variable x), and the list context $\mathtt{L}$ is always pushed outside the term t. We will highlight in *green* such list contexts in the forthcoming examples to improve readability. Apart from rule $\mathtt{dB}$ used to fire β-reductions, there are four substitution rules used to copy abstractions, applications and variables, pushing outside all the cuts surrounding the node to be copied. Rule $\mathtt{app}$ copies one application node, while rule $\mathtt{var}$ copies one variable node. The case of abstractions is more involved as explained below.

The specificity in copying an abstraction $\lambda y.u$ is due to the (binding) relation between λy and all the free occurrences of y in its body u. Abstractions are thus copied in two stages. The first one is implemented by the rule dist, creating a distributor in which a potentially replaceable abstraction is placed, while moving its body inside a new ES. There are then two ways to replicate nodes of the body. Either they can be copied inside the distributor (where the binding relation between λy and the bound occurrences of y is kept intact), or they can be pushed outside the distributor, by means of the (non-deterministic) rule abs. In the second case, however, free occurrences of y cannot be pushed outside the abstraction (with binder y) to be duplicated, at the risk of breaking consistency: only shared components without y links can be then pushed outside. These components are gathered together into a list context L, which is pushed outside by using permutation rules, before performing the substitution of the pure body containing all the bound occurrences of y. Specifying this operation using only distance is hard, thus permutation rules are also used in our rule abs.

The s-substitution relation $\rightarrow_{\mathtt{s}}$ (resp. distant Beta relation $\rightarrow_{\mathtt{dB}}$) is defined as the closure of $\mapsto_{\mathtt{app}} \cup \mapsto_{\mathtt{dist}} \cup \mapsto_{\mathtt{abs}} \cup \mapsto_{\mathtt{var}}$ (resp. $\mapsto_{\mathtt{dB}}$) under *all* contexts, and the reduction relation $\rightarrow_{\mathtt{R}}$ is the union of $\rightarrow_{\mathtt{s}}$ and $\rightarrow_{\mathtt{dB}}$.

Example 1. Let $t_0 = (\lambda x_1.x_1)(\lambda y.\mathtt{I}y)$. In what follows, we underline the term where the reduction is performed:

$$t_0 \rightarrow_{\mathtt{dB}} \underline{x_1[x_1\backslash\lambda y.\mathtt{I}y]} \rightarrow_{\mathtt{dist}} x_1[x_1\backslash\!\!\backslash\lambda y.\underline{z[z\backslash\mathtt{I}y]}] \rightarrow_{\mathtt{app}} x_1[x_1\backslash\!\!\backslash\lambda y.\underline{(z_1z_2)[z_1\backslash\mathtt{I}][z_2\backslash y]}]$$
$$\rightarrow_{\mathtt{dist}} \underline{x_1[x_1\backslash\!\!\backslash\lambda y.(z_1z_2)[z_1\backslash\!\!\backslash\lambda x_3.z_3[z_3\backslash x_3]][z_2\backslash y]]}$$
$$\rightarrow_{\mathtt{var}} \underline{x_1[x_1\backslash\!\!\backslash\lambda y.(z_1y)\,[z_1\backslash\!\!\backslash\lambda x_3.z_3[z_3\backslash x_3]]\,]} \rightarrow_{\mathtt{abs}} (\lambda y.z_1y)[z_1\backslash\!\!\backslash\lambda x_3.z_3[z_3\backslash x_3]]$$

Let $\mathcal{R}$ be any reduction relation. We write $\rightarrow^*_{\mathcal{R}}$ for the reflexive-transitive closure of $\rightarrow_{\mathcal{R}}$. A term t is said to be **$\mathcal{R}$-confluent** *iff* $t \rightarrow^*_{\mathcal{R}} u$ and $t \rightarrow^*_{\mathcal{R}} s$ implies there is t' such that $u \rightarrow^*_{\mathcal{R}} t'$ and $s \rightarrow^*_{\mathcal{R}} t'$. The relation $\mathcal{R}$ is **confluent** *iff* every term is $\mathcal{R}$-confluent. A term t is said to be in **$\mathcal{R}$-normal form** (written also $\mathcal{R}$-nf) *iff* there is no t' such that $t \rightarrow_{\mathcal{R}} t'$. A term t is said to be **$\mathcal{R}$-terminating** or **$\mathcal{R}$-normalizing** *iff* there is no infinite $\mathcal{R}$-sequence starting at t. The reduction $\mathcal{R}$ is said to be **terminating** *iff* every term is $\mathcal{R}$-terminating.

Levels. The notion of level plays a key role in this work. Intuitively, the level of a variable in a term indicates the maximal depth of its free occurrences w.r.t. ES (and not w.r.t. explicit distributors). However, in order to keep soundness w.r.t. the permutation rules, levels are computed along *linked chains* of ES. For instance, the level of w in both $x[x\backslash y[y\backslash w]]$ and $x[x\backslash y][y\backslash w]$ is 2. Formally, the **level** of a variable z in a term t is defined by (structural) induction, while assuming by α-conversion that z is not a bound variable in t:

$$\mathtt{lv}_z(x) := 0 \quad \mathtt{lv}_z(t_1t_2) := \max(\mathtt{lv}_z(t_1), \mathtt{lv}_z(t_2)) \quad \mathtt{lv}_z(\lambda y.t) := \mathtt{lv}_z(t)$$
$$\mathtt{lv}_z(t[x \triangleleft u]) := \begin{cases} \mathtt{lv}_z(t) & \text{if } z \notin \mathtt{fv}(u) \\ \max(\mathtt{lv}_z(t), \mathtt{lv}_x(t) + \mathtt{lv}_z(u) + x(\triangleleft)) & \text{otherwise} \end{cases}$$

Notice that $\mathtt{lv}_w(t) = 0$ whenever $w \notin \mathtt{fv}(t)$ or t is pure. We illustrate the concept of level by an example. Consider $t = x[x\backslash z[y\backslash w]][w\backslash w']$, then $\mathtt{lv}_z(t) = 1$, $\mathtt{lv}_{w'}(t) = 3$ and $\mathtt{lv}_y(t) = 0$ because $y \notin \mathtt{fv}(t)$. This notion is also extended to contexts as expected, *i.e.* $\mathtt{lv}_\Box(\mathtt{C}) = \mathtt{lv}_z(\mathtt{C}\langle\langle z\rangle\rangle)$, where z is a fresh variable.

Lemma 2. *Let $t \in \Lambda_{\mathtt{R}}$. If $t_0 \rightarrow_{\pi,\mathtt{s}} t_1$, then $\mathtt{lv}_w(t_0) \geq \mathtt{lv}_w(t_1)$ for any $w \in \mathcal{X}$.*

It is worth noticing that there are two cases when the level of a variable in a term may decrease: using a permutation rule to push an explicit cut out of another cut when the first one is a void cut, or using rule $\mapsto_{\mathtt{var}}$.

Hence, levels alone are not enough to prove termination of $\rightarrow_{\mathtt{s}}$. We then define a decreasing measure for $\rightarrow_{\mathtt{s}}$ in which not only variables are indexed by a level, but also constructors. For instance, in $t[x\backslash\lambda y.yz]$, we can consider that the level of *all* the constructors of $\lambda y.yz$ have level $\mathtt{lv}_x(t)$. This will ensure that the level of an abstraction will decrease when applying rule $\mathtt{dist}$, as well as the level of an application when applying rule $\mathtt{app}$. This is what we do next.

3 Operational Properties

We now prove three key properties of the $\lambda\mathtt{R}$-calculus: termination of the reduction system $\rightarrow_{\mathtt{s}}$, relation between $\lambda\mathtt{R}$ and the λ-calculus, and confluence of the reduction system $\rightarrow_{\lambda\mathtt{R}}$.

Termination of $\rightarrow_{\mathtt{s}}$. Some (rather informal) arguments are provided in [33] to justify termination of the substitution subrelation of their whole calculus. We expand these ideas into an alternative full formal proof adapted to our case, which is based on a measure being strictly decreasing w.r.t. $\rightarrow_{\mathtt{s}}$.

We consider a set $\mathcal{O}$ of objects of the form $\mathtt{a}(k,n)$ or $\mathtt{b}(k)$ ($k,n \in \mathbb{N}$), which is equipped with the following ordering $>^{\mathcal{O}}$:

$$\begin{array}{llll} \mathtt{a}(k,n) >^{\mathcal{O}} \mathtt{a}(k',n) & \text{if } k > k', \text{ or } (k = k' \text{ and } n > n') & \mathtt{b}(k) >^{\mathcal{O}} \mathtt{a}(k',n) & \text{if } k \geq k' \\ \mathtt{a}(k,n) >^{\mathcal{O}} \mathtt{b}(k') & \text{if } k > k' & \mathtt{b}(k) >^{\mathcal{O}} \mathtt{b}(k') & \text{if } k > k' \end{array}$$

Lemma 3. *The order $>^{\mathcal{O}}$ on the set $\mathcal{O}$ is well-founded.*

We write $>^{\mathcal{O}}_{\mathtt{MUL}}$ for the multiset extension of the order $>^{\mathcal{O}}$ on $\mathcal{O}$, which turns out to be well-founded [8] by Lem. 3. We are now ready to (inductively) define our **cuts level** measure $\mathsf{C}(\_)$ on terms, where the following operation on multisets is used $p \cdot M := [\mathtt{a}(p+k,n) \mid \mathtt{a}(k,n) \in M] \sqcup [\mathtt{b}(p+k) \mid \mathtt{b}(k) \in M]$, where $\sqcup$ denotes multiset union.

$$\begin{array}{l} \mathsf{C}(x) := [\,] \qquad \mathsf{C}(\lambda x.t) := \mathsf{C}(t) \qquad \mathsf{C}(tu) := \mathsf{C}(t) \sqcup \mathsf{C}(u) \\ \mathsf{C}(t[x\backslash u]) := \mathsf{C}(t) \sqcup (\mathtt{lv}_x(t)+1) \cdot \mathsf{C}(u) \sqcup [\mathtt{a}(\mathtt{lv}_x(t)+1, |u|)] \\ \mathsf{C}(t[x\backslash\!\backslash u]) := \mathsf{C}(t) \sqcup \mathtt{lv}_x(t) \cdot \mathsf{C}(u) \sqcup [\mathtt{b}(\mathtt{lv}_x(t))] \end{array}$$

Intuitively, the integer k in $\mathtt{a}(k,n)$ and $\mathtt{b}(k)$ counts the level of variables bound by explicit cuts, while n counts the size of terms to be substituted by an ES. Remark that for every pure term p we have $\mathsf{C}(p) = [\,]$. Moreover:

Lemma 4. *Let* $t_0 \in \Lambda_{\mathtt{R}}$. *Then* $t_0 \to_\pi t_1$ *(resp.* $t_0 \to_{\mathtt{s}} t_1$*) implies* $\mathsf{C}(t_0) \geq^{\mathcal{O}}_{\mathtt{MUL}} \mathsf{C}(t_1)$ *(resp.* $\mathsf{C}(t_0) >^{\mathcal{O}}_{\mathtt{MUL}} \mathsf{C}(t_1)$*).*

As an example, consider the following reduction sequence:

$$\begin{array}{llll} t_0 = (yy)[y\backslash(\lambda z.x)w] & \to_{\mathtt{app}} (y_1y_2)(y_1y_2)[y_1\backslash\lambda z.x][y_2\backslash w] = t_1 \to_{\mathtt{var}} \\ t_2 = (y_1w)(y_1w)[y_1\backslash\lambda z.x] \to_{\mathtt{dist}} (y_1w)(y_1w)[y_1\backslash\!\backslash\lambda z.r[r\backslash x]] & = t_3 \end{array}$$

We have $\mathsf{C}(t_0) = [\mathsf{a}(1,4)]$, $\mathsf{C}(t_1) = [\mathsf{a}(1,1), \mathsf{a}(1,2)]$, $\mathsf{C}(t_2) = [\mathsf{a}(1,2)]$, $\mathsf{C}(t_3) = [\mathsf{a}(1,1), \mathsf{b}(0)]$. So $\mathsf{C}(t_i) >_{\mathtt{MUL}} \mathsf{C}(t_{i+1})$ for $i = 0, 1, 2, 3$.

Corollary 5. *The reduction relation* $\to_{\mathtt{s}}$ *is terminating.*

Simulations. We show the relation between $\lambda\mathtt{R}$ and the λ-calculus, as well as the atomic λ-calculus. For that, we introduce a projection from $\lambda\mathtt{R}$-terms to λ-terms implementing the unfolding of all the explicit cuts: $x^{\downarrow} := x$, $(\lambda x.t)^{\downarrow} := \lambda x.t^{\downarrow}$, $(tu)^{\downarrow} := t^{\downarrow}u^{\downarrow}$, $(t[x \lhd u])^{\downarrow} := t^{\downarrow}\{x\backslash u^{\downarrow}\}$. Thus *e.g.* $x[x\backslash z][y\backslash w][w\backslash w']^{\downarrow} = z$.

Lemma 6. *Let* $t_0 \in \Lambda_{\mathtt{R}}$. *If* $t_0 \to_{\mathtt{R}} t_1$, *then* $t_0^{\downarrow} \to^*_\beta t_1^{\downarrow}$. *In particular, if either* $t_0 \to_\pi t_1$ *or* $t_0 \to_{\mathtt{s}} t_1$, *then* $t_0^{\downarrow} = t_1^{\downarrow}$.

The relation $\to_{\mathtt{s}}$ enjoys **full composition** on *pure* terms, namely, for any $p \in \Lambda$, $t[x\backslash p] \to^+_{\mathtt{s}} t\{x\backslash p\}$. This property does not hold in general. Indeed, if $t = xx$, then $(xx)[x\backslash z[z\backslash w]]$ does not $\mathtt{s}$-reduce to $(z[z\backslash w])(z[z\backslash w])$, but to $(zz)[z\backslash w]$. However, full composition restricted to pure terms is sufficient to prove simulation of the λ-calculus.

Lemma 7 (Simulation of the λ-calculus). *Let* $p_0 \in \Lambda$. *If* $p_0 \to_\beta p_1$, *then* $p_0 \to_{\mathtt{dB}}\to^+_{\mathtt{s}} p_1$.

The previous results have an important consequence relating the original atomic λ-calculus and the $\lambda\mathtt{R}$-calculus. Indeed, it can be shown that reduction in the atomic λ-calculus is captured by $\lambda\mathtt{R}$, and vice-versa. More precisely, the $\lambda\mathtt{R}$-calculus can be simulated into the atomic λ-calculus by Lem. 6 and [33], while the converse holds by [33] and Lem. 7.

A more structural correspondence between $\lambda\mathtt{R}$ and the atomic λ-calculus could also be established. Indeed, $\lambda\mathtt{R}$ can be first refined into a (non-linear) calculus *without* distance, let say $\lambda\mathtt{R}'$, so that permutation rules are integrated in the intermediate calculus as independent rules. Then a structural relation can be established between $\lambda\mathtt{R}$ and $\lambda\mathtt{R}'$ on one side, and $\lambda\mathtt{R}'$ and the atomic λ-calculus on the other side (as for example done in [43] for the λ-calculus).

Confluence. By Cor. 5 the reduction relation $\to_{\mathtt{s}}$ is terminating. It is then not difficult to conclude confluence of $\to_{\mathtt{s}}$ by using the unfolding function $\_^{\downarrow}$. Therefore, by termination of $\to_{\mathtt{s}}$ any $t \in \Lambda_{\mathtt{R}}$ has an $\mathtt{s}$-nf, and by confluence this $\mathtt{s}$-nf is unique (and computed by the unfolding function). Using the interpretation method [35] together with Lem. 6, Cor. 5, and Lem. 7, one obtains:

Theorem 8. *The reduction relation* $\to_{\mathtt{R}}$ *is confluent.*

4 Encoding Evaluation Strategies

In the theory of programming languages [56], the notion of *calculus* is usually based on a non-deterministic rewriting relation, providing an equational system of calculation, while the deterministic notion of *strategy* is associated to a concrete machinery being able to implement a specific evaluation procedure. Typical evaluation strategies are call-by-name, call-by-value, call-by-need, etc.

Although the atomic λ-calculus was introduced as a technical tool to implement full laziness, only its (non-deterministic) equational theories was studied. In this paper we bridge the gap between the theoretical presentation of the atomic λ-calculus and concrete specifications of evaluation strategies. Indeed, we use the λR-calculus to investigate two concrete cases: a call-by-name strategy implementing weak head reduction, based on full substitution, and the call-by-need fully lazy strategy, which uses linear substitution.

In both cases, explicit cuts can in principle be placed anywhere in the distributors, thus demanding to dive deep in such terms to deal with them. We then restrict the set of terms to a subset $\mathtt{U}$, which simplifies the formal reasoning of explicit cuts inside distributors. Indeed, distributors will all be of the shape $[x\backslash\!\!\backslash \lambda y.\mathtt{L}\langle p\rangle]$, where p is a pure term (and $\mathtt{L}$ is a *commutative list* defined below). We argue that this restriction is natural in a weak implementation of the λ-calculus: it is true on pure terms and is preserved through evaluation. We consider the following grammars.

$$\begin{array}{lll}
\textbf{(Linear Cut Values)} & \mathtt{T} & ::= \lambda x.\mathtt{LL}\langle p\rangle \text{ where } y \in \mathtt{dlc}(\mathtt{LL}) \implies |p|_y = 1 \\
\textbf{(Commutative Lists)} & \mathtt{LL} & ::= \Box \mid \mathtt{LL}[x\backslash p] \mid \mathtt{LL}[x\backslash\!\!\backslash \mathtt{T}] \text{ where } |\mathtt{LL}|_x = 0 \\
\textbf{(Values)} & v & ::= \lambda x.p \\
\textbf{(Restricted Terms)} & \mathtt{U} & ::= x \mid v \mid \mathtt{UU} \mid \mathtt{U}[x\backslash \mathtt{U}] \mid \mathtt{U}[x\backslash\!\!\backslash \mathtt{T}]
\end{array}$$

A term t generated by any of the grammars G defined above is written $t \in G$. Thus *e.g.* $\lambda x.(yz)[y\backslash \mathtt{I}][z\backslash \mathtt{I}] \in \mathtt{T}$ but $\lambda x.(yy)[y\backslash \mathtt{I}] \notin \mathtt{T}$, $\Box[x\backslash yz][x'\backslash \mathtt{I}] \in \mathtt{LL}$ but $\Box[x\backslash yz][y\backslash \mathtt{I}] \notin \mathtt{LL}$, and $(yz)[y\backslash\!\!\backslash \mathtt{I}] \in \mathtt{U}$ but $(yz)[y\backslash\!\!\backslash \lambda x.(yy)[y\backslash \mathtt{I}]] \notin \mathtt{U}$.

The set $\mathtt{T}$ is stable by the relation $\rightarrow_{\mathtt{s}}$, but $\mathtt{U}$ is clearly not stable under the whole $\rightarrow_{\mathtt{R}}$ relation, where dB-reductions may occur under abstractions. However, $\mathtt{U}$ is stable under both weak strategies to be defined: call-by-name and call-by-need. We factorize the proofs by proving stability for a more general relation $\rightarrow_{\mathtt{R}'}$, defined as the relation $\rightarrow_{\mathtt{R}}$ with dB-reductions forbidden under abstractions and inside distributors.

Lemma 9 (Stability of the Grammar by $\rightarrow_{\mathtt{s}}/\rightarrow_{\mathtt{R}'}$).

1. *If $t \in \mathtt{T}$ and $t \rightarrow_{\mathtt{s}} t'$, then $t' \in \mathtt{T}$.*
2. *If $t \in \mathtt{U}$ and $t \rightarrow_{\mathtt{R}'} t'$, then $t' \in \mathtt{U}$.*

4.1 Call-by-name

The **call-by-name** (CBN) strategy $\rightarrow_{\mathtt{name}}$ (Fig. 1) is defined on the set of terms $\mathtt{U}$ as the union of the following relations $\rightarrow_{\mathtt{ndb}}$ and $\rightarrow_{\mathtt{ns}}$. The strategy is *weak* as there is no reduction under abstractions. It is also worth noticing (as a particular case of Lem. 9) that $t \in \mathtt{U}$ and $t \rightarrow_{\mathtt{name}} t'$ implies $t' \in \mathtt{U}$.

$$\frac{t \mapsto_{\mathtt{dB}} t'}{t \to_{\mathtt{ndb}} t'}\ (\mathtt{dB}) \qquad \frac{t \to_{\mathtt{ndb}} t'}{tu \to_{\mathtt{ndb}} t'u}\ (\mathtt{app\_dB}) \qquad \frac{t \to_{\mathtt{ndb}} t'}{t[x \lhd u] \to_{\mathtt{ndb}} t'[x \lhd u]}\ (\mathtt{sub\_dB})$$

$$\frac{t \mapsto_{\mathtt{s}} t'}{t \to_{\mathtt{ns}} t'}\ (\mathtt{s}) \qquad \frac{t \to_{\mathtt{ns}} t'}{tu \to_{\mathtt{ns}} t'u}\ (\mathtt{app\_s}) \qquad \frac{t \to_{\mathtt{ns}} t'}{u[x \backslash\!\backslash \lambda y.t] \to_{\mathtt{ns}} u[x \backslash\!\backslash \lambda y.t']}\ (\mathtt{sub\_s})$$

Fig. 1. Call-by-Name Strategy

Example 10. Let $t_0 = (\lambda x_1.\mathtt{I}(x_1\mathtt{I}))(\lambda y.\mathtt{I}y)$. Then,

$$t_0 \to_{\mathtt{dB}}^{(\mathtt{dB})} \underline{(\mathtt{I}(x_1\mathtt{I}))[x_1 \backslash \lambda y.\mathtt{I}y]} \to_{\mathtt{dist}}^{(\mathtt{s})} (\mathtt{I}(x_1\mathtt{I}))[x_1 \backslash\!\backslash \lambda y.\underline{z[z\backslash \mathtt{I}y]}] \to_{\mathtt{app}}^{(\mathtt{sub\_s})}$$
$$(\mathtt{I}(x_1\mathtt{I}))[x_1 \backslash\!\backslash \lambda y.\underline{(z_1z_2)[z_1\backslash\mathtt{I}][z_2\backslash y]}] \to_{\mathtt{var}}^{(\mathtt{sub\_s})} \underline{(\mathtt{I}(x_1\mathtt{I}))[x_1 \backslash\!\backslash \lambda y.(z_1y)\ [z_1\backslash\mathtt{I}]\ ]} \to_{\mathtt{abs}}^{(\mathtt{s})}$$
$$\underline{(\mathtt{I}((\lambda y.z_1y)\mathtt{I}))}[z_1\backslash\mathtt{I}] \to_{\mathtt{dB}}^{(\mathtt{sub\_dB})} x_2[x_2\backslash(\lambda y.z_1y)\mathtt{I}][z_1\backslash\mathtt{I}]$$

Although the strategy $\to_{\mathtt{name}}$ is not deterministic, it enjoys the remarkable *diamond* property, guaranteeing in particular that all reduction sequences starting from t and ending in a normal form have the same length.

It is worth noticing that simulation lemmas also hold between call-by-name in the λ-calculus, known as weak head reduction and denoted by $\to_{\mathtt{whr}}$, and the $\lambda\mathtt{R}$-calculus. Indeed, $\to_{\mathtt{whr}}$ is defined as the β-reduction rule closed by contexts $\mathtt{E} ::= \Box \mid \mathtt{E}t$. Then, as a consequence of Lem. 7, we have that $p_0 \to_{\mathtt{whr}} p_1$ implies $p_0 \to_{\mathtt{R}}^* p_1$, and as a consequence of Lem. 6, we have that $t_0 \to_{\mathtt{name}} t_1$ implies $t_0^{\downarrow} \to_{\beta}^* t_1^{\downarrow}$. More importantly, call-by-name in the λ-calculus and call-by-name in the $\lambda\mathtt{R}$-calculus are also related. Indeed,

Lemma 11 (Relating Call-by-Name Strategies).

- *Let* $p_0 \in \Lambda$. *If* $p_0 \to_{\mathtt{whr}} p_1$ *then* $p_0 \to_{\mathtt{name}}^+ p_1$.
- *Let* $t_0 \in \mathcal{U}$. *If* $t_0 \to_{\mathtt{name}} t_1$ *then* $t_0^{\downarrow} \to_{\mathtt{whr}}^* t_1^{\downarrow}$.

4.2 Call-by-need

We now specify a deterministic strategy `flneed` implementing demand-driven computations and only linearly replicating nodes of *values* (*i.e.* pure abstractions). Given a value $\lambda x.p$, only the piece of structure containing the paths between the binder λx and all the free occurrences of x in p, named *skeleton*, will be copied. All the other components of the abstraction will remain shared, thus avoiding some future duplications of redexes, as explained in the introduction. By copying only the smallest possible substructure of the abstraction, the strategy `flneed` implements an optimization of call-by-need called `f`*ully* `l`*azy sharing* [60]. First, we formally define the key notions we are going to use.

A **free expression** [39,9] of a *pure* term p is a strict subterm q of p such that every free occurrence of a variable in q is also a free occurrence of the variable in p. A **free expression** of p is **maximal** if it is not a subterm of another free expression of p. From now on, we will consider the multiset of all maximal free expressions (**MFE**) of a term. Thus *e.g.* the MFEs of $\lambda y.p$, where $p = (\mathtt{I}y)\mathtt{I}(\lambda z.zyw)$, is given by the multiset $[\mathtt{I}, \mathtt{I}, w]$.

An **n-ary context** ($n \geq 0$) is a term with n holes $\square$. A skeleton is an n-ary pure context where the maximal free expressions w.r.t. a variable set θ are replaced with holes. Formally, the **θ-skeleton** $\{\{p\}\}^\theta$ of a pure term p, where $\theta = \{x_1 \dots x_n\}$, is the n-ary pure context $\{\{p\}\}^\theta$ such that $\{\{p\}\}^\theta \langle q_1, \dots, q_n \rangle = p$, for $[q_1, \dots, q_n]$ the maximal free expressions of $\lambda x_1. \dots \lambda x_n.p$ [4]. Thus, for the same p as before, $\lambda y.\{\{p\}\}^y = \lambda y.(\square y)\square(\lambda z.zy\square)$.

The Splitting Operation. Splitting a term into a skeleton and a multiset of MFEs is at the core of full laziness. This can naturally be implemented in the node replication model, as observed in [33]. Here, we define a (small-step) strategy $\rightarrow_{\mathtt{st}}$ on the set of terms $\mathtt{T}$ to achieve it (Fig. 2), which is indeed a subset of the reduction relation $\lambda\mathtt{R}$[5]. The relation $\rightarrow_{\mathtt{st}}$ makes use of four basic rules which are parameterized by the variable y upon which the skeleton is built, written $\mapsto^y$. There are also two contextual (inductive) rules.

$$\frac{}{t[x\backslash y] \mapsto^y_{\mathtt{var}} t\{x\backslash y\}} \qquad \frac{y \in \mathtt{fv}(p_1p_2)}{t[x\backslash p_1p_2] \mapsto^y_{\mathtt{app}} t\{x\backslash x_1x_2\}[x_1\backslash p_1][x_2\backslash p_2]}$$

$$\frac{y \in \mathtt{fv}(\lambda z.p)}{t[x\backslash \lambda z.p] \mapsto^y_{\mathtt{dist}} t[x\backslash\!\backslash \lambda z.w[w\backslash p]]} \qquad \frac{y \in \mathtt{fv}(\lambda z.\mathtt{LL}\langle p\rangle) \quad z \notin \mathtt{fv}(\mathtt{LL})}{t[x\backslash\!\backslash \lambda z.\mathtt{LL}\langle p\rangle] \mapsto^y_{\mathtt{abs}} \mathtt{LL}\langle t\{x\backslash \lambda z.p\}\rangle}$$

$$\frac{t \mapsto^y t' \quad y \in \mathtt{fv}(t) \quad y \notin \mathtt{fv}(\mathtt{LL})}{\lambda y.\mathtt{LL}\langle t\rangle \rightarrow_{\mathtt{st}} \lambda y.\mathtt{LL}\langle t'\rangle}\,\mathtt{ctx}_1 \qquad \frac{t \rightarrow_{\mathtt{st}} t' \quad y \in \mathtt{fv}(t) \quad y \notin \mathtt{fv}(\mathtt{LL})}{\lambda y.\mathtt{LL}\langle u[x\backslash\!\backslash t]\rangle \rightarrow_{\mathtt{st}} \lambda y.\mathtt{LL}\langle u[x\backslash\!\backslash t']\rangle}\,\mathtt{ctx}_2$$

Fig. 2. Relation $\rightarrow_{\mathtt{st}}$: Splitting Skeleton and MFEs in Small-Step Semantics

Example 12. Let $y, z \notin \mathtt{fv}(t)$, so that t is the MFE of $\lambda y.x[x\backslash \lambda z.(yt)z]$. Then,

$$\lambda y.\underline{x[x\backslash \lambda z.(yt)z]} \rightarrow^y_{\mathtt{dist}} \lambda y.x[x\backslash\!\backslash \lambda z.\underline{w[w\backslash(yt)z]}] \rightarrow^z_{\mathtt{app}}$$
$$\lambda y.x[x\backslash\!\backslash \lambda z.\underline{(w_1w_2)[w_1\backslash yt][w_2\backslash z]}] \rightarrow^z_{\mathtt{var}} \lambda y.\underline{x[x\backslash\!\backslash \lambda z.(w_1z)\ [w_1\backslash yt]\ ]} \rightarrow^y_{\mathtt{abs}}$$
$$\lambda y.\underline{(\lambda z.w_1z)[w_1\backslash yt]} \rightarrow^y_{\mathtt{app}} \lambda y.\underline{(\lambda z.(x_1x_2)z)[x_1\backslash y]}[x_2\backslash t] \rightarrow^y_{\mathtt{var}} \lambda y.(\lambda z.(yx_2)z)[x_2\backslash t]$$

Notice that the focused variable changes from y to z, then back to y. This is because $\rightarrow_{\mathtt{st}}$ constructs the innermost skeletons first.

Lemma 13. *The reduction relation $\rightarrow_{\mathtt{st}}$ is confluent and terminating.*

Thus, from now on, we denote by $\Downarrow_{\mathtt{st}}$ the function relating a term of $\mathtt{T}$ to its unique $\mathtt{st}$-nf.

[4] The order of variables in the set θ is indeed irrelevant.

[5] Since $\rightarrow_{\mathtt{st}}$ acts only on terms in $\mathtt{T}$, it is handled by linear substitution.

Lemma 14 (Correctness of $\to_{\mathtt{st}}$). *Let $p \in \Lambda$ and $q_1, \ldots, q_n$ be the MFEs of $\lambda y.p$. Then $\lambda y.z[z\backslash p] \Downarrow_{\mathtt{st}} \lambda y.\{\{p\}\}^{\{y\}}\langle x_1, \ldots, x_n\rangle[x_i\backslash q_i]_{i\leq n}$ where the variables $x_1, \ldots, x_n$ are fresh and pairwise distinct.*

Since the small-step semantics is contained in the λR-calculus, we use it to build our call-by-need strategy of λR.

The strategy. The **call-by-need strategy** $\to_{\mathtt{flneed}}$ (Fig. 3) is defined on the set of terms U, by using closure under the *need contexts*, given by the grammar $\mathtt{N} ::= \Box \mid \mathtt{N}t \mid \mathtt{N}[x \lhd t] \mid \mathtt{N}\langle\langle x\rangle\rangle[x\backslash\mathtt{N}]$, where $\mathtt{N}\langle\langle\_\rangle\rangle$ denotes capture-free application of contexts (Sec. 2). As for call-by-name (Sec. 4.1), the call-by-need strategy is *weak*, because no *meaningful* reduction steps are performed under abstractions.

$$\begin{array}{lll} \mathtt{L}\langle\lambda x.p\rangle u & \mapsto_{\mathtt{dB}} \mathtt{L}\langle p[x\backslash u]\rangle & \\ \mathtt{N}\langle\langle x\rangle\rangle[x\backslash\mathtt{L}\langle\lambda y.p\rangle] & \mapsto_{\mathtt{spl}} \mathtt{L}\langle\mathtt{LL}\langle\mathtt{N}\langle\langle x\rangle\rangle[x\backslash\!\backslash\lambda y.p']\rangle\rangle & \text{if } \lambda y.z[z\backslash p] \Downarrow_{\mathtt{st}} \lambda y.\mathtt{LL}\langle p'\rangle \\ \mathtt{N}\langle\langle x\rangle\rangle[x\backslash\!\backslash v] & \mapsto_{\mathtt{sub}} \mathtt{N}\langle\langle v\rangle\rangle[x\backslash\!\backslash v] & \end{array}$$

Fig. 3. Call-by-Need Strategy

Rule dB is the same one used to define name. Although rules spl and sub could have been presented in a unique rule of the form $\mathtt{N}\langle\langle x\rangle\rangle[x\backslash\mathtt{L}\langle\lambda y.p\rangle] \mapsto \mathtt{L}\langle\mathtt{LL}\langle\mathtt{N}\langle\langle\lambda y.p'\rangle\rangle[x\backslash\!\backslash\lambda y.p']\rangle\rangle$, we prefer to keep them separate since they represent different stages in the strategy. Indeed, rule spl only uses node replication operations to compute the skeleton of the abstraction, while rule sub implements one-shot *linear* substitution.

Notice that as a particular case of Lem. 9, $t \in \mathtt{U}$ and $t \to_{\mathtt{flneed}} t'$ implies $t' \in \mathtt{U}$. Another interesting property is that $t \to_{\mathtt{sub}} t'$ implies $\mathtt{lv}_z(t) \geq \mathtt{lv}_z(t')$. Moreover, $\to_{\mathtt{flneed}}$ is deterministic.

Example 15. Let $t_0 = (\lambda x.(\mathtt{I}(\mathtt{I}x)))\lambda y.y\mathtt{I}$. Needed variable occurrences are highlighted in *orange*.

$$\begin{array}{l} t_0 \to_{\mathtt{dB}} \underline{(\mathtt{I}(\mathtt{I}x))}[x\backslash\lambda y.y\mathtt{I}] \to_{\mathtt{dB}} x_1[x_1\backslash\underline{\mathtt{I}x}][x\backslash\lambda y.y\mathtt{I}] \\ \to_{\mathtt{dB}} x_1[x_1\backslash x_2[x_2\backslash x]][x\backslash\lambda y.y\mathtt{I}] \to_{\mathtt{spl}} x_1[x_1\backslash x_2[x_2\backslash x]][x\backslash\!\backslash\lambda y.yz_1][z_1\backslash\mathtt{I}] \\ \to_{\mathtt{sub}} x_1[x_1\backslash x_2[x_2\backslash\lambda y.yz_1]][x\backslash\!\backslash\lambda y.yz_1][z_1\backslash\mathtt{I}] \\ \to_{\mathtt{spl}} x_1[x_1\backslash x_2[x_2\backslash\!\backslash\lambda y.yz_2][z_2\backslash z_1]][x\backslash\!\backslash\lambda y.yz_1][z_1\backslash\mathtt{I}] \\ \to_{\mathtt{sub}} x_1[x_1\backslash(\lambda y.yz_2)[x_2\backslash\!\backslash\lambda y.yz_2][z_2\backslash z_1]][x\backslash\!\backslash\lambda y.yz_1][z_1\backslash\mathtt{I}] \\ \to_{\mathtt{spl}} x_1[x_1\backslash\!\backslash\lambda y.yz_3][z_3\backslash z_2][x_2\backslash\!\backslash\lambda y.yz_2][z_2\backslash z_1][x\backslash\!\backslash\lambda y.yz_1][z_1\backslash\mathtt{I}] \\ \to_{\mathtt{sub}} (\lambda y.yz_3)[x_1\backslash\!\backslash\lambda y.yz_3][z_3\backslash z_2][x_2\backslash\!\backslash\lambda y.yz_2][z_2\backslash z_1][x\backslash\!\backslash\lambda y.yz_1][z_1\backslash\mathtt{I}] \end{array}$$

5 A Type System for the λR-calculus

This section introduces a quantitative type system $\mathcal{V}$ for the λR-calculus. Non-idempotent intersection [26] has one main advantage over the idempotent model

[14]: it gives *quantitative* information about the length of reduction sequences to normal forms [21]. Indeed, not only typability and normalization can be proved to be equivalent, but a measure based on type derivations provides an *upper bound* to normalizing reduction sequences. This was extensively investigated in different logical/computational frameworks [5,18,20,25,42,47]. However, no quantitative result based on types exists in the literature for the node replication model, including the attempts done for deep inference [30]. The typing rules of our system are in themselves not surprising (see [46]), but they provide a handy quantitative characterization of fully lazy normalization (Sec. 6).

Types are built on the following grammar of types and multi-types, where α ranges over a set of base types and a is a special type constant used to type terms reducing to normal abstractions.

$$\textbf{(Types)}\ \sigma := \mathsf{a} \mid \alpha \mid \mathcal{M} \to \sigma \qquad \textbf{(Multi-Types)}\ \mathcal{M} := [\sigma_i]_{i \in I}$$

We write $|\mathcal{M}|$ to denote the **size of a multi-type** $\mathcal{M}$. **Typing contexts,** written Γ, Δ, Σ are functions from variables to multiset types, assigning the empty multiset to all but a finite set of variables. The domain of Γ is given by $\mathtt{dom}(\Gamma) := \{x \mid \Gamma(x) \neq [\,]\}$. The **union of contexts**, written $\Gamma + \Delta$, is defined by $(\Gamma + \Delta)(x) := \Gamma(x) \sqcup \Delta(x)$, where $\sqcup$ denotes multiset union. An example is $(x : [\sigma], y : [\tau]) + (x : [\sigma], z : [\tau]) = (x : [\sigma, \sigma], y : [\tau], z : [\tau])$. This notion is extended to several contexts as expected, so that $+_{i \in I} \Gamma_i$ denotes a finite union of contexts, and the empty context when $I = \emptyset$. We write $\Gamma; \Delta$ for $\Gamma + \Delta$ when $\mathtt{dom}(\Gamma) \cap \mathtt{dom}(\Delta) = \emptyset$. **Type judgments** have the form $\Gamma \vdash t : \sigma$, where Γ is a typing context, t is a term and σ is a type.

$$\frac{}{x : [\sigma] \vdash x : \sigma}\,(\mathtt{ax}) \qquad \frac{\Gamma; x : \mathcal{M} \vdash t : \sigma}{\Gamma \vdash \lambda x.t : \mathcal{M} \to \sigma}\,(\mathtt{abs}) \qquad \frac{\Gamma \vdash t : \mathcal{M} \to \sigma \quad \Delta \vdash u : \mathcal{M}}{\Gamma + \Delta \vdash t\,u : \sigma}\,(\mathtt{app})$$

$$\frac{}{\vdash \lambda x.t : \mathsf{a}}\,(\mathtt{ans}) \qquad \frac{(\Gamma_i \vdash t : \sigma_i)_{i \in I}}{+_{i \in I} \Gamma_i \vdash t : [\sigma_i]_{i \in I}}\,(\mathtt{many}) \qquad \frac{\Gamma; x : \mathcal{M} \vdash t : \sigma \qquad \Delta \vdash u : \mathcal{M}}{\Gamma + \Delta \vdash t[x \triangleleft u] : \sigma}\,(\mathtt{cut})$$

Fig. 4. Typing System $\mathcal{V}$

A **(typing) derivation** is a tree obtained by applying the (inductive) typing rules of system $\mathcal{V}$ (Fig. 4), introduced in [46]. The notation $\Phi \triangleright \Gamma \vdash t : \sigma$ means there is a derivation named Φ of the judgment $\Gamma \vdash t : \sigma$ in system $\mathcal{V}$. A term t is typable in system $\mathcal{V}$, or $\mathcal{V}$-typable, iff there is a context Γ and a type σ such that $\Phi \triangleright \Gamma \vdash t : \sigma$. The **size of a type derivation** $\mathbf{sz}(\Phi)$ is defined as the number of its $\mathtt{abs}$, $\mathtt{app}$ and $\mathtt{ans}$ rules. The typing system is **relevant** in the sense that $\Phi \triangleright \Gamma \vdash t : \sigma$ implies $\mathtt{dom}(\Gamma) \subseteq \mathtt{fv}(t)$.

Type derivations can be measured by 3-tuples. We use a + operation on 3-tuples as pointwise addition: $(a, b, c) + (e, f, g) = (a + e, b + f, c + g)$. These 3-tuples are computed by a **weighted derivation level** function defined on typing derivations as $\mathsf{D}(\Phi) := \mathsf{M}(\Phi, 1)$, where $\mathsf{M}(-, -)$ is inductively defined below. In the cases (abs), (app) and (cut), we let Φ_t (resp. Φ_u) be the subderivation of the type of t (resp. Φ_u) and in (many) we let Φ_t^i be the i-th derivation of the type of t for each $i \in I$.

- For (ax), $\mathsf{M}(\Phi_x, m) = (0, 0, 1)$,
- For (abs), $\mathsf{M}(\Phi_{\lambda x.t}, m) = \mathsf{M}(\Phi_t, m) + (1, m, 0)$.
- For (ans), $\mathsf{M}(\Phi_{\lambda x.t}, m) = (1, m, 0)$.
- For (app), $\mathsf{M}(\Phi_{tu}, m) = \mathsf{M}(\Phi_t, m) + \mathsf{M}(\Phi_u, m) + (1, m, 0)$.
- For (cut), $\mathsf{M}(\Phi_{t[x \triangleleft u]}, m) = \mathsf{M}(\Phi_t, m) + \mathsf{M}(\Phi_u, m + \mathtt{lv}_x(t) + x(\triangleleft))$.
- For (many), $\mathsf{M}(\Phi_t, m) = \sum_{i \in I} \mathsf{M}(\Phi_t^i, m)$.

Notice that the first and the third components of any 3-tuple $\mathsf{M}(\Phi, m)$ do not depend on m. Intuitively, the first (resp. third) component of the 3-tuple counts the number of application/abstraction (resp. (ax)) rules in the typing derivation. The second one takes into account the number of application/abstraction rules as well, but *weighted* by the level of the constructor. The 3-tuples are ordered lexicographically.

Example 16. Let $\sigma = [\tau] \to \tau$. Consider the following type derivation Φ:

$$\dfrac{\dfrac{}{x : [\tau] \vdash x : \tau}\,(\mathtt{ax}) \quad \dfrac{\dfrac{\dfrac{}{y : [\sigma] \vdash y : \sigma}\,(\mathtt{ax}) \quad \dfrac{\dfrac{}{z : [\tau] \vdash z : \tau}\,(\mathtt{ax})}{z : [\tau] \vdash z : [\tau]}\,(\mathtt{many})}{y : [\sigma], z : [\tau] \vdash yz : \tau}\,(\mathtt{app})}{y : [\sigma], z : [\tau] \vdash yz : [\tau]}\,(\mathtt{many})}{y : [\sigma], z : [\tau] \vdash x[x \backslash yz] : \tau}\,(\mathtt{cut})$$

This gives $\mathsf{D}(\Phi) = (1, 2, 3)$. Moreover, for $x[x \backslash yz] \to_{\mathtt{app}} (x_1 x_2)[x_1 \backslash y][x_2 \backslash z]$ we have $\Phi' \rhd y : [\sigma], z : [\tau] \vdash (x_1 x_2)[x_1 \backslash y][x_2 \backslash z] : \tau$ and $\mathsf{D}(\Phi') = (1, 1, 4)$.

6 Observational Equivalence

The type system $\mathcal{V}$ characterizes normalization of both `name` and `flneed` strategies as follows: every typable term normalizes and every normalisable term is typable. In this sense, system $\mathcal{V}$ can be seen as a (quantitative) *model* [17] of our call-by-name and call-by-need strategies. We prove these results by studying the appropriate lemmas, notably weighted subject reduction and weighted subject expansion. We then deduce observational equivalence between the `name` and the `flneed` strategies from the fact that their associated normalization properties are both fully characterized by the same typing system.

Soundness. Soundness of system $\mathcal{V}$ w.r.t. both $\to_{\mathtt{name}}$ and $\to_{\mathtt{flneed}}$ is investigated in this section. More precisely, we show that typable terms are normalizing for both strategies. In contrast to reducibility techniques needed to show this kind of result for simple types [34], soundness is achieved here by relatively simple combinatorial arguments based again on decreasing measures. We start by studying the interaction between system $\mathcal{V}$ and linear as well as full substitution.

Lemma 17 (Partial Substitution). *Let $\Phi \rhd \Gamma; x : \mathcal{M} \vdash \mathtt{C}\langle\langle x \rangle\rangle : \sigma$ and $\sqsubseteq$ denote multiset inclusion. Then, there exists $\mathcal{N} \sqsubseteq \mathcal{M}$ such that for every $\Phi_u \rhd \Delta \vdash u : \mathcal{N}$ we have $\Psi \rhd \Gamma + \Delta; x : \mathcal{M} \setminus \mathcal{N} \vdash \mathtt{C}\langle\langle u \rangle\rangle : \sigma$ and, for every $m \in \mathbb{N}$, $\mathsf{M}(\Psi, m) = \mathsf{M}(\Phi, m) + \mathsf{M}(\Phi_u, m + \mathtt{lv}_{\Box}(\mathtt{C})) - (0, 0, |\mathcal{N}|)$.*

Corollary 18 (Substitution). *If $\Phi_t \rhd \Gamma; x : \mathcal{M} \vdash t : \sigma$ and $\Phi_u \rhd \Delta \vdash u : \mathcal{M}$, then $\Phi \rhd \Gamma + \Delta \vdash t\{x \backslash u\} : \sigma$, and for all $m \in \mathbb{N}$ we have $\mathsf{M}(\Phi, m) \leq \mathsf{M}(\Phi_t, m) + \mathsf{M}(\Phi_u, m + \mathtt{lv}_x(t))$. Moreover, $|\mathcal{M}| > 0$ iff the inequality is strict.*

The key idea to show soundness is that the measure $\mathsf{D}(\_)$ decreases w.r.t. the reduction relations $\rightarrow_{\mathtt{name}}$ and $\rightarrow_{\mathtt{flneed}}$:

Lemma 19 (Weighted Subject Reduction). *Let $\Phi_{t_0} \rhd \Gamma \vdash t_0 : \sigma$.*

1. *If $t_0 \rightarrow_{\pi} t_1$, then there exists $\Phi_{t_1} \rhd \Gamma \vdash t_1 : \sigma$ such that $\mathsf{D}(\Phi_{t_0}) = \mathsf{D}(\Phi_{t_1})$.*
2. *If $t_0 \rightarrow_{\mathtt{s}} t_1$, then there exists $\Phi_{t_1} \rhd \Gamma \vdash t_1 : \sigma$ such that $\mathsf{D}(\Phi_{t_0}) \geq \mathsf{D}(\Phi_{t_1})$.*
3. *If $t_0 \rightarrow_{\mathtt{ndb}} t_1$, then there exists $\Phi_{t_1} \rhd \Gamma \vdash t_1 : \sigma$ such that $\mathsf{D}(\Phi_{t_0}) > \mathsf{D}(\Phi_{t_1})$.*
4. *If $t_0 \rightarrow_{\mathtt{flneed}} t_1$, then there exists $\Phi_{t_1} \rhd \Gamma \vdash t_1 : \sigma$ such that $\mathsf{D}(\Phi_{t_0}) > \mathsf{D}(\Phi_{t_1})$.*

Proof. By induction on $\mathtt{r} \in \{\pi, \mathtt{s}, \mathtt{ndb}, \mathtt{flneed}\}$, using Lem. 17 and Cor. 18.

Theorem 20 (Typability implies name-Normalization). *Let $\Phi_t \rhd \Gamma \vdash t : \sigma$. Then t is* name*-normalizing.*

Proof. Suppose t is not name-normalizing. Since $\rightarrow_{\mathtt{s}}$ is terminating by Cor. 5, then every infinite $\rightarrow_{\mathtt{name}}$-reduction sequence starting at t must necessarily have an infinite number of dB-steps. Moreover, all terms in such an infinite sequence are typed by Lem 19. Therefore, Lem. 19:3 (resp. Lem. 19:2) guarantees that all dB (resp. s) reduction steps involved in such $\rightarrow_{\mathtt{name}}$-reduction sequence strictly decrease (resp. do not increase) the measure $\mathsf{D}(\_)$. This leads to a contradiction because the order $>$ on 3-tuples $\mathsf{D}(\_)$ is well-founded. Then t is necessarily name-normalizing.

Theorem 21 (Typability implies flneed-Normalization). *Let $\Phi_t \rhd \Gamma \vdash t : \sigma$. Then t is* flneed*-normalizing. Moreover, $\mathsf{D}(\Phi_t)$ is an upper bound to the length of the* flneed*-reduction evaluation to* flneed*-nf.*

Proof. The property trivially holds by Lem. 19:4 since the lexicographic order on 3-tuples is well-founded.

Completeness. We address here completeness of system $\mathcal{V}$ with respect to $\rightarrow_{\mathtt{name}}$ and $\rightarrow_{\mathtt{flneed}}$. More precisely, we show that normalizing terms in each strategy are typable. The basic property in showing that consists in guaranteeing that normal forms are typable.

The following lemma makes use of a notion of **needed variable**: $\mathtt{nv}(x) := \{x\}$, $\mathtt{nv}(tu) := \mathtt{nv}(t)$, $\mathtt{nv}(t[x \backslash\!\backslash u]) := \mathtt{nv}(t)$, $\mathtt{nv}(\lambda x.t) := \emptyset$, $\mathtt{nv}(t[y \backslash u]) := (\mathtt{nv}(t) \setminus \{y\}) \cup \mathtt{nv}(u)$ if $y \in \mathtt{nv}(t)$ and $\mathtt{nv}(t[y \backslash u]) := \mathtt{nv}(t)$ otherwise.

Lemma 22 (flneed-nfs are Typable). *Let t be in* flneed*-nf. Then there exists a derivation $\Phi \rhd \Gamma \vdash t : \tau$ such that for any $x \notin \mathtt{nv}(t)$, $\Gamma(x) = [\,]$.*

Because $\mathtt{name}$-nfs are also $\mathtt{flneed}$-nfs, we infer the following corollary for free.

Corollary 23 (name-nfs are Typable). *Let t be in* $\mathtt{name}$*-nf. Then there is a derivation $\Phi \rhd \Gamma \vdash t : \tau$.*

Now we need lemmas stating the behavior of partial and full (anti-)substitution w.r.t. typing.

Lemma 24 (Partial Anti-Substitution). *Let $\mathtt{C}\langle\langle x\rangle\rangle$, u be terms s.t. $x \notin \mathtt{fv}(u)$ and $\Phi \rhd \Gamma \vdash \mathtt{C}\langle\langle u\rangle\rangle : \sigma$. Then $\exists\Gamma'$, $\exists\Delta$, $\exists\mathcal{M}$, $\exists\Phi'$, $\exists\Phi_u$ s.t. $\Gamma = \Gamma' + \Delta$, $\Phi' \rhd \Gamma' + x : \mathcal{M} \vdash \mathtt{C}\langle\langle x\rangle\rangle : \sigma$ and $\Phi_u \rhd \Delta \vdash u : \mathcal{M}$.*

Corollary 25 (Anti-Substitution). *Let u be a term s.t. $x \notin \mathtt{fv}(u)$ and $\Phi \rhd \Gamma \vdash t\{x\backslash u\} : \sigma$. Then $\exists\Gamma'$, $\exists\Delta$, $\exists\mathcal{M}$, $\exists\Phi'$, $\exists\Phi_u$ s.t. $\Gamma = \Gamma' + \Delta$, $\Phi' \rhd \Gamma'; x : \mathcal{M} \vdash t : \sigma$ and $\Phi_u \rhd \Delta \vdash u : \mathcal{M}$.*

To achieve completeness, we show that typing is preserved by anti-reduction. We decompose the property as follows:

Lemma 26 (Subject Expansion). *Let $\Phi_{t_1} \rhd \Gamma \vdash t_1 : \sigma$. If $t_0 \to_{\mathtt{r}} t_1$, where $\mathtt{r} \in \{\pi, \mathtt{s}, \mathit{ndb}, \mathtt{flneed}\}$, then there exists $\Phi_{t_0} \rhd \Gamma \vdash t_0 : \sigma$.*

Proof. The proof is by induction on $\to_{\mathtt{r}}$ and uses Lem. 24 and Cor. 25.

Theorem 27 (name-Normalization implies Typability). *Let t be a term. If t is* $\mathtt{name}$*-normalizing, then t is $\mathcal{V}$-typable.*

Proof. Let t be $\mathtt{name}$-normalizing. Then $t \to^n_{\mathtt{name}} u$ and u is a $\mathtt{name}$-nf. We reason by induction on n. If $n = 0$, then $t = u$ is typable by Cor. 23. Otherwise, we have $t \to_{\mathtt{name}} t' \to^{n-1}_{\mathtt{name}} u$. By the *i.h.* t' is typable and thus by Lem. 26 (because $\to_{\mathtt{ns}}$ is included in $\to_{\mathtt{s}}$), t turns out to be also typable.

Theorem 28 (flneed-Normalization implies Typability). *Let t be a term. If t is* $\mathtt{flneed}$*-normalizing, then t is $\mathcal{V}$-typable.*

Proof. Similar to the previous proof but using Lem. 22 instead of Cor. 23.

Summing up, Thms. 20, 27, 21 and 28 give:

Theorem 29. *Let t be a $\lambda\mathtt{R}$-term. t is* $\mathtt{name}$*-normalizing iff t is* $\mathtt{flneed}$*-normalizing iff t is $\mathcal{V}$-typable.*

All the technical tools are now available to conclude observational equivalence between our two evaluation strategies based on node replication. Let $\mathcal{R}$ be any reduction notion on $\Lambda_{\mathtt{R}}$. Then, two terms $t, u \in \Lambda_{\mathtt{R}}$ are said to be **$\mathcal{R}$-observationally equivalent**, written $t \equiv u$, if for any context $\mathtt{C}$, $\mathtt{C}\langle t\rangle$ is $\mathcal{R}$-normalizing *iff* $\mathtt{C}\langle u\rangle$ is $\mathcal{R}$-normalizing.

Theorem 30. *For all terms $t, u \in \Lambda_{\mathtt{R}}$, t and u are* $\mathtt{name}$*-observationally equivalent iff t and u are* $\mathtt{flneed}$*-observationally equivalent.*

Proof. By Thm. 29, $t \equiv_{\mathtt{name}} u$ means that $\mathtt{C}\langle t\rangle$ is $\mathcal{V}$-typable *iff* $\mathtt{C}\langle u\rangle$ is $\mathcal{V}$-typable, for all $\mathtt{C}$. By the same theorem, this is also equivalent to say that $\mathtt{C}\langle t\rangle$ is $\mathtt{flneed}$-normalizing *iff* $\mathtt{C}\langle u\rangle$ is $\mathtt{flneed}$-normalizing for any $\mathtt{C}$, *i.e.* $t \equiv_{\mathtt{flneed}} u$.

7 Related Works and Conclusion

Several calculi with ES bridge the gap between formal higher-order calculi and concrete implementations of programming languages (see a survey in [40]). The first of such calculi, *e.g.* [1,16], were all based on *structural* substitution, in the sense that the ES operator is syntactically propagated step-by-step through the term structure until a variable is reached, when the substitution finally takes place. The correspondence between ES and Linear Logic Proof-Nets [24] led to the more recent notion of calculi *at a distance* [6,4,2], enlightening a natural and new application of the Curry-Howard interpretation. These calculi implement linear/partial substitution *at a distance*, where the search of variable occurrences is abstracted out with context-based rewriting rules, and thus no ES propagation rules are necessary. A third model was introduced by the seminal work of Gundersen, Heijltjes, and Parigot [33,34], introducing the atomic λ-calculus to implement node replication.

Inspired by the last approach we introduced the $\lambda\texttt{R}$-calculus, capturing the essence of node replication. In contrast to [33], we work with an implicit (structural) mechanism of weakening and contraction, a design choice which aims at focusing and highlighting the node replication model, which is the core of our calculus, so that we obtain a rather simple and natural formalism used in particular to specify evaluation strategies. Indeed, besides the proof of the main operational meta-level properties of our calculus (confluence, termination of the substitution calculus, simulations), we use linear and non-linear versions of $\lambda\texttt{R}$ to specify evaluation strategies based on node replication, namely call-by-name and call-by-need evaluation strategies.

The first description of call-by-need was given by Wadsworth [60], where reduction is performed on *graphs* instead of terms. Weak call-by-need on *terms* was then introduced by Ariola and Felleisen [7], and by Maraist, Odersky and Wadler [54,53]. Reformulations were introduced by Accattoli, Barenbaum and Mazza [3] and by Chang and Felleisen [22]. Our call-by-need strategy is inspired by the calculus in [3], which uses the distance paradigm [6] to gather together meaningful and permutation rules, by clearly separating *multiplicative* from *exponential* rules, in the sense of Linear Logic [27].

Full laziness has been formalized in different ways. Pointer graphs [60,59] are DAGs allowing for an elegant representation of sharing. Labeled calculi [15] implement pointer graphs by adding annotations to λ-terms, which makes the syntax more difficult to handle. Lambda-lifting [38,39] implements full laziness by resorting to translations from λ-terms to supercombinators. In contrast to all the previous formalisms, our calculus is defined on standard λ-terms with explicit cuts, without the use of any complementary syntactical tool. So is Ariola and Felleisen's call-by-need [7], however, their notion of full laziness relies on external (ad-hoc) meta-level operations used to extract the skeleton. Our specification of call-by-need enjoys fully lazy sharing, where the skeleton extraction operation is internally encoded in the term calculus operational semantics. Last but not least, our calculus has strong links with proof-theory, notably deep inference.

Balabonski [10,9] relates many formalisms of full laziness and shows that they are equivalent when considering the number of β-steps to a normal form. It would then be interesting to understand if his unified approach, (abstractly) stated by means of the theory of residuals [50,51], applies to our own strategy.

We have also studied the calculus from a semantical point of view, by means of intersection types. Indeed, the type system can be seen as a model of our implementations of call-by-name and call-by-need, in the sense that typability and normalization turn out to be equivalent.

Intersection types go back to [23] and have been used to provide characterizations of qualitative [14] as well as quantitative [21] models of the λ-calculus, where typability and normalization coincide. Quantitative models specified by means of non-idempotent types [26,48] were first applied to the λ-calculus (see a survey in [19]) and to several other formalisms ever since, such as call-by-value [25,20], call-by-need [42,5], call-by-push-value [31,18] and classical logic [47]. In the present work, we achieve for the first time a quantitative characterization of fully lazy normalization, which provides upper bounds for the length of reduction sequences to normal forms.

The characterizations provided by intersection type systems sometimes lead to observational equivalence results (*e.g.* [42]). In this work we succeed to prove observational equivalence related to a fully lazy implementation of weak call-by-need, a result which would be extremely involved to prove by means of syntactical tools of rewriting, as done for weak call-by-need in [7]. Moreover, our result implies that our node replication implementation of full laziness is observationally equivalent to standard call-by-name and to weak call-by-need (see [42]), as well as to the more semantical notion of neededness (see [45]).

A Curry-Howard interpretation of the logical *switch* rule of deep inference is given in [58,57] as an end-of-scope operator, thus introducing the *spinal atomic λ-calculus*. The calculus implements a refined optimization of call-by-need, where only the *spine* of the abstraction (tighter than the skeleton) is duplicated. It would be interesting to adapt the λR-calculus to spine duplication by means of an appropriate end-of-scope operator, such as the one in [37]. Further optimizations might also be considered.

Finally, this paper only considers weak evaluation strategies, *i.e.* with reductions forbidden under abstractions, but it would be interesting to extend our notions to full (strong) evaluations too [29,12]. Extending full laziness to classical logic would be another interesting research direction, possibly taking preliminary ideas from [36]. We would also like to investigate (quantitative) *tight* types for our fully lazy strategy, as done for weak call-by-need in [5], which does not seem evident in our node replication framework.

References

1. Abadi, M., Cardelli, L., Curien, P., Lévy, J.: Explicit substitutions. In: POPL. pp. 31–46. ACM Press (1990)
2. Accattoli, B.: Proof nets and the linear substitution calculus. In: ICTAC. Lecture Notes in Computer Science, vol. 11187, pp. 37–61. Springer (2018)
3. Accattoli, B., Barenbaum, P., Mazza, D.: Distilling abstract machines. In: ICFP. pp. 363–376. ACM (2014)
4. Accattoli, B., Bonelli, E., Kesner, D., Lombardi, C.: A nonstandard standardization theorem. In: POPL. pp. 659–670. ACM (2014)
5. Accattoli, B., Guerrieri, G., Leberle, M.: Types by need. In: ESOP. Lecture Notes in Computer Science, vol. 11423, pp. 410–439. Springer (2019)
6. Accattoli, B., Kesner, D.: The structural *lambda*-calculus. In: CSL. Lecture Notes in Computer Science, vol. 6247, pp. 381–395. Springer (2010)
7. Ariola, Z.M., Felleisen, M.: The call-by-need lambda calculus. J. Funct. Program. **7**(3), 265–301 (1997)
8. Baader, F., Nipkow, T.: Term rewriting and all that. Cambridge University Press (1998)
9. Balabonski, T.: La plein paresse, une certain optimalité : partage de sous-termes et stratégies de réduction en réécriture d'ordre supérieur. Ph.D. thesis, Paris 7 (2012), http://www.theses.fr/2012PA077198
10. Balabonski, T.: A unified approach to fully lazy sharing. In: POPL. pp. 469–480. ACM (2012)
11. Balabonski, T.: Weak optimality, and the meaning of sharing. In: ICFP. pp. 263–274. ACM (2013)
12. Balabonski, T., Barenbaum, P., Bonelli, E., Kesner, D.: Foundations of strong call by need. Proc. ACM Program. Lang. **1**(ICFP), 20:1–20:29 (2017)
13. Barendregt, H.P.: The lambda calculus - its syntax and semantics, Studies in logic and the foundations of mathematics, vol. 103. North-Holland (1985)
14. Barendregt, H.P., Dekkers, W., Statman, R.: Lambda Calculus with Types. Perspectives in logic, Cambridge University Press (2013)
15. Blanc, T., Lévy, J., Maranget, L.: Sharing in the weak lambda-calculus. In: Processes, Terms and Cycles. Lecture Notes in Computer Science, vol. 3838, pp. 70–87. Springer (2005)
16. Bloo, R., Rose, K.: Preservation of strong normalization in named lambda calculi with explicit substitution and garbage collection. In: CSN. pp. 62–72. Netherlands Computer Science Research Foundation (1995)
17. Bucciarelli, A., Ehrhard, T.: On phase semantics and denotational semantics: the exponentials. Ann. Pure Appl. Log. **109**(3), 205–241 (2001)
18. Bucciarelli, A., Kesner, D., Ríos, A., Viso, A.: The bang calculus revisited. In: FLOPS. Lecture Notes in Computer Science, vol. 12073, pp. 13–32. Springer (2020)
19. Bucciarelli, A., Kesner, D., Ventura, D.: Non-idempotent intersection types for the lambda-calculus. Log. J. IGPL **25**(4), 431–464 (2017)
20. Carraro, A., Guerrieri, G.: A semantical and operational account of call-by-value solvability. In: FoSSaCS. Lecture Notes in Computer Science, vol. 8412, pp. 103–118. Springer (2014)
21. de Carvalho, D.: Sémantiques de la logique linéaire et temps de calcul. Ph.D. thesis, Université Aix-Marseille II (2007), https://www.theses.fr/2007AIX22066
22. Chang, S., Felleisen, M.: The call-by-need lambda calculus, revisited. In: ESOP. Lecture Notes in Computer Science, vol. 7211, pp. 128–147. Springer (2012)

23. Coppo, M., Dezani-Ciancaglini, M.: A new type assignment for λ-terms. Arch. Math. Log. **19**(1), 139–156 (1978)
24. Di Cosmo, R., Kesner, D., Polonowski, E.: Proof nets and explicit substitutions. Math. Struct. Comput. Sci. **13**(3), 409–450 (2003)
25. Ehrhard, T.: Collapsing non-idempotent intersection types. In: CSL. LIPIcs, vol. 16, pp. 259–273. Schloss Dagstuhl - Leibniz-Zentrum für Informatik (2012)
26. Gardner, P.: Discovering needed reductions using type theory. In: TACS. Lecture Notes in Computer Science, vol. 789, pp. 555–574. Springer (1994)
27. Girard, J.: Linear logic. Theor. Comput. Sci. **50**, 1–102 (1987)
28. Girard, J.Y.: Proof-nets: The parallel syntax for proof-theory, Lecture Notes in Pure Applied Mathematics, vol. 180, p. 97–124. Marcel Dekker (1996)
29. Grégoire, B., Leroy, X.: A compiled implementation of strong reduction. In: ICFP. pp. 235–246. ACM (2002)
30. Guerrieri, G., Heijltjes, W.B., Paulus, J.W.N.: A deep quantitative type system. In: CSL. LIPIcs, vol. 183, pp. 24:1–24:24. Schloss Dagstuhl - Leibniz-Zentrum für Informatik (2021)
31. Guerrieri, G., Manzonetto, G.: The bang calculus and the two girard's translations. In: Linearity-TLLA@FLoC. EPTCS, vol. 292, pp. 15–30 (2018)
32. Guglielmi, A., Gundersen, T., Parigot, M.: A proof calculus which reduces syntactic bureaucracy. In: RTA. LIPIcs, vol. 6, pp. 135–150. Schloss Dagstuhl - Leibniz-Zentrum für Informatik (2010)
33. Gundersen, T., Heijltjes, W., Parigot, M.: Atomic lambda calculus: A typed lambda-calculus with explicit sharing. In: LICS. pp. 311–320. IEEE Computer Society (2013)
34. Gundersen, T., Heijltjes, W., Parigot, M.: A proof of strong normalisation of the typed atomic lambda-calculus. In: LPAR. Lecture Notes in Computer Science, vol. 8312, pp. 340–354. Springer (2013)
35. Hardin, T.: Confluence results for the pure strong categorical logic CCL: lambda-calculi as subsystems of CCL. Theor. Comput. Sci. **65**(3), 291–342 (1989)
36. He, F.: The Atomic Lambda-Mu Calculus. Ph.D. thesis, University of Bath (Jan 2018)
37. Hendriks, D., van Oostrom, V.: $\curlywedge$. In: CADE. Lecture Notes in Computer Science, vol. 2741, pp. 136–150. Springer (2003)
38. Hughes, J.: The design and implementation of programming languages. Tech. Rep. PRG-40, Oucl (Jul 1983)
39. Jones, S.L.P.: The Implementation of Functional Programming Languages. Prentice-Hall (1987)
40. Kesner, D.: The theory of calculi with explicit substitutions revisited. In: CSL. Lecture Notes in Computer Science, vol. 4646, pp. 238–252. Springer (2007)
41. Kesner, D.: A theory of explicit substitutions with safe and full composition. Log. Methods Comput. Sci. **5**(3) (2009)
42. Kesner, D.: Reasoning about call-by-need by means of types. In: FoSSaCS. Lecture Notes in Computer Science, vol. 9634, pp. 424–441. Springer (2016)
43. Kesner, D., Lengrand, S.: Resource operators for lambda-calculus. Inf. Comput. **205**(4), 419–473 (2007)
44. Kesner, D., Renaud, F.: A prismoid framework for languages with resources. Theor. Comput. Sci. **412**(37), 4867–4892 (2011)
45. Kesner, D., Ríos, A., Viso, A.: Call-by-need, neededness and all that. In: FoSSaCS. Lecture Notes in Computer Science, vol. 10803, pp. 241–257. Springer (2018)

46. Kesner, D., Ventura, D.: Quantitative types for the linear substitution calculus. In: IFIP TCS. Lecture Notes in Computer Science, vol. 8705, pp. 296–310. Springer (2014)
47. Kesner, D., Vial, P.: Non-idempotent types for classical calculi in natural deduction style. Log. Methods Comput. Sci. **16**(1) (2020)
48. Kfoury, A.J.: A linearization of the lambda-calculus and consequences. J. Log. Comput. **10**(3), 411–436 (2000)
49. Lamping, J.: An algorithm for optimal lambda calculus reduction. In: POPL. pp. 16–30. ACM Press (1990)
50. Lévy, J.J.: Réductions correctes et optimales dans le lambda-calcul. Ph.D. thesis, Université Paris VII (1978)
51. Lévy, J.J.: Optimal reductions in the lambda-calculus. In: To H.B. Curry: Essays on Combinatory Logic, Lambda Calculus and Formalisms, pp. 159–191. Academic Press Inc (1980)
52. Lévy, J., Maranget, L.: Explicit substitutions and programming languages. In: FSTTCS. Lecture Notes in Computer Science, vol. 1738, pp. 181–200. Springer (1999)
53. Maraist, J., Odersky, M., Turner, D.N., Wadler, P.: Call-by-name, call-by-value, call-by-need and the linear lambda calculus. Theor. Comput. Sci. **228**(1-2), 175–210 (1999)
54. Maraist, J., Odersky, M., Wadler, P.: The call-by-need lambda calculus. J. Funct. Program. **8**(3), 275–317 (1998)
55. Paulus, J.W.N., Heijltjes, W.: Deep-inference intersection types. In: Int. Workshop: Twenty Years of Deep Inference. pp. 1–3 (2018), http://t-news.cn/Floc2018/FLoC2018-pages/proceedings_paper_652.pdf
56. Plotkin, G.D.: Call-by-name, call-by-value and the lambda-calculus. Theor. Comput. Sci. **1**(2), 125–159 (1975)
57. Sherratt, D., Heijltjes, W., Gundersen, T., Parigot, M.: Spinal atomic lambda-calculus. In: FoSSaCS. Lecture Notes in Computer Science, vol. 12077, pp. 582–601. Springer (2020)
58. Sherratt, D.R.: A lambda-calculus that achieves full laziness with spine duplication. Ph.D. thesis, University of Bath (Mar 2019)
59. Shivers, O., Wand, M.: Bottom-up beta-reduction: Uplinks and lambda-dags. Fundam. Informaticae **103**(1-4), 247–287 (2010)
60. Wadsworth, C.P.: Semantics and Pragmatics of the Lambda Calculus. Ph.D. thesis, Oxford University (1971)

Nondeterministic and co-Nondeterministic Implies Deterministic, for Data Languages

Bartek Klin[†], Sławomir Lasota[‡], and Szymon Toruńczyk[§](✉)

University of Warsaw, Warsaw, Poland
{klin,sl,szymtor}@mimuw.edu.pl

Abstract. We prove that if a data language and its complement are both recognized by nondeterministic register automata (without guessing), then they are also recognized by deterministic ones.

Keywords: Data languages, register automata, determinizability, deterministic separability, sets with atoms, orbit-finite sets, nominal sets

1 Introduction

Register automata are finite-state automata equipped with a finite number of registers that can store values from an infinite data domain. When processing an input string, an automaton compares the current input data value to its registers and, based on this comparison and on the current control state, it chooses its next control state and possibly stores the input value in one of its registers. In the original model, introduced over 25 year ago by Francez and Kaminski [15], data values can only be compared for equality and not for any other property. Subsequent extensions of the model allow for comparing data values with respect to some fixed relations such as a total order, or introduce alternation, variations on the allowed form of nondeterminism, etc.

It appears that register automata lack most of the good properties known from the classical theory of finite automata. For example, while languages of nondeterministic register automata are closed under unions and intersections, they are not closed under complement, and they do not determinize. Moreover, the expressivity of register automata is very sensitive to natural variants and extensions. Any of the following relaxations of the model leads to a strict increase of expressive power (see [15,23,1] for details):

- increasing the number of registers (when this number is bounded),
- extension from one-way to two-way automata,
- extension from deterministic to unambiguous, nondeterministic or alternating ones,

[†] Supported by the European Research Council (ERC) under the EU Horizon 2020 programme (ERC consolidator grant LIPA, agreement no. 683080).
[‡] Supported by the NCN grant 2019/35/B/ST6/02322.
[§] Supported by the NCN grant 2017/26/D/ST6/00201.

S. Kiefer and C. Tasson (Eds.): FOSSACS 2021, LNCS 12650, pp. 365–384, 2021.
https://doi.org/10.1007/978-3-030-71995-1_19

– adding the capability to nondeterministically guess data values.

In fact, almost every combination of these extensions leads to a different class of recognized languages. Furthermore, no satisfactory characterizations of languages of register automata in terms of regular expressions [17,20] or logic [23,12] are known. There are a few positive results: a simulation of two-way nondeterministic automata by one-way alternating automata with guessing [1], a Myhill-Nerode characterization of languages of deterministic automata [16,4,5], and the well-behaved class of languages definable by orbit-finite monoids [2], which admits equivalent characterisations in terms of logic [11] and a syntactic subclass of deterministic automata [7]. Nevertheless, register automata satisfy almost no semantic equivalences that hold for classical finite automata.

Contribution. Our primary contribution is a collapse result: if a language and its complement are both recognized by nondeterministic register automata (NRA), then they are both recognized by deterministic ones (DRA). In symbols, we prove the following equality of language classes:

$$\mathsf{NRA} \cap \text{co-}\mathsf{NRA} \quad = \quad \mathsf{DRA}.$$

This result is shown under the assumption that the data values can be compared only for equality, and it turns out to be quite fragile. For instance, it fails if the automata can compare data values using a total order relation. It also fails if NRA are additionally equipped with the capability of guessing fresh data values, even when data values can only be compared for equality.

Our secondary contribution is a collapse result for NRA with 1 register only (1-NRA), but over an arbitrary data domain that *admits well quasi-order* (WQO), meaning roughly that finite induced substructures of the data domain, ordered by embeddings, form a WQO. This includes both equality and ordered data domains. In short, we prove the following inclusion of language classes:

$$1\text{-}\mathsf{NRA} \cap \text{co-}1\text{-}\mathsf{NRA} \quad \subseteq \quad \mathsf{DRA}.$$

The inclusion is strict, as some DRA languages are not recognizable by 1-NRA.

Our proofs are mostly self-contained, but use basic notions and results about sets with atoms [1], also known as nominal sets [24]. In particular, automorphisms of the data domain play a central role in our arguments, and we extensively use notions such as finite support and orbit-finiteness of sets. In both results, we prove that for every data language $L \in \mathsf{NRA} \cap \text{co-}\mathsf{NRA}$ the set of derivative languages $w^{-1}L$ is orbit-finite, i.e., finite up to automorphism of data values. The collapse then follows from an orbit-finite version of the Myhill-Nerode theorem.

In our primary contribution, orbit-finiteness of the set of derivative languages is a consequence of a key technical result (Lem. 1), an abstract observation about orbit-finite families of sets, which we believe may be of independent interest. As another example application of this lemma, we give a new proof of decidability of universality for unambiguous register automata (URA).

Relation to other work. Our primary result partially confirms a conjecture of Thomas Colcombet [10], according to which every two disjoint languages of NRA with guessing are separable by a language recognized by an URA. Working in the special case when the NRA are complementing and have no guessing, we show more: both languages are then recognized not only by an URA but by a DRA.

NRA do not have good algorithmic properties: while the emptiness problem is PSPACE-complete [14], the universality problem (does a given automaton accept all data words?) is undecidable [15] (it is decidable only for 1-NRA [14]). Universality becomes decidable for URA, as shown recently in [22] (2-EXPSPACE upper bound, improved to 2-EXPTIME upper bound in [8]), and language containment and equality for URA reduce polynomially to universality (see [8, Lemma 8]). As mentioned above, our results allow us to re-prove this decidability result.

Register automata have been intensively investigated, with respect both to their foundational properties [15,25,17,23] and to their applications to XML databases and logics [14] (see [26] for a survey). There are several other ways to extend finite-state machines with a capability to recognize languages over infinite alphabets. These include, apart from register automata: their abstract version – nominal automata or automata over atoms [4,5,1]; symbolic automata [13]; pebble automata [21]; and data automata [3,6].

Acknowledgments. We thank Lorenzo Clemente for posing the collapse question studied in this paper, and Joanna Ochremiak and Radek Piórkowski for valuable discussions.

2 Data languages and register automata

The model of register automata, as considered in this paper, is parametrized by an underlying relational structure ATOMS over a finite vocabulary Σ. This structure constitutes a *data domain*; its elements are called *atoms*. A register automaton processes sequences of atoms, possibly coupled with labels from a fixed finite set. It may store atoms read from the input in its registers, and compare them with previously stored atoms using relations in Σ (equality included).

Here are some example data domains:

- *Equality atoms*: natural numbers with equality $(\mathbb{N}, =)$. Since equality is the only available relation, any other countably infinite set could be used instead.
- *Dense order atoms*: rational numbers with the standard order $(\mathbb{Q}, \leqslant)$. Again, any countably infinite dense order without endpoints could be used instead.
- *Nested equality atoms* (universal equivalence relation): $(\mathbb{N}^2, =_1, =)$ where $=_1$ is the equality on the first coordinate: $(n_1, n_2) =_1 (m_1, m_2)$ if $n_1 = m_1$.

In the following we consider input alphabets of the form $S \times$ ATOMS, where S is a finite set of labels. A *data word* is a finite sequence $w \in (S \times \text{ATOMS})^*$, and a *data language* is a set of data words.

A *nondeterministic register automaton* (NRA) $\mathcal{A}$ consists of:

- an input alphabet of the form $S \times$ ATOMS, for some finite set S,

- a positive integer $r \in \mathbb{N}$ (the number of registers),
- a finite set of control states (locations) Q,
- subsets $I, F \subseteq Q$ of initial resp. accepting states,
- a finite set Δ of transition rules of the form

$$(p, s, \varphi, \mathrm{ST}, q) \in \Delta, \tag{1}$$

where $p, q \in Q$, $s \in S$, $\varphi(x_1, \ldots, x_r, x)$ is a quantifier-free Σ-formula with free variables in $\{x_1, \ldots, x_r, x\}$, and $\mathrm{ST} \in \{1, \ldots, r, \mathrm{NONE}\}$.

Intuitively, φ defines a condition which needs to be satisfied by the register contents $(x_1, \ldots, x_r)$ and by the current atom (x) for a transition to happen, and ST specifies the register in which the input atom is stored after the transition, $\mathrm{ST} = \mathrm{NONE}$ meaning that it is not to be stored in any register.

An NRA $\mathcal{A}$ is *deterministic* (DRA) if it has exactly one initial state and if for every two transition rules

$$(p, s, \varphi_1, \mathrm{ST}_1, q_1),\ (p, s, \varphi_2, \mathrm{ST}_2, q_2) \in \Delta,$$

such that $\varphi_1 \wedge \varphi_2$ is satisfiable in ATOMS, we have $\mathrm{ST}_1 = \mathrm{ST}_2$ and $q_1 = q_2$. We write r-NRA, resp. r-DRA, when the number of registers r is fixed.

A configuration $q(\mathbf{a}) \in Q \times (\mathrm{ATOMS} \cup \{\bot\})^r$ of $\mathcal{A}$ consists of a control state $q \in Q$ and a content of registers $\mathbf{a} \in (\mathrm{ATOMS} \cup \{\bot\})^r$, where $\bot$ means that the content of a register is undefined (i.e., the register is empty). A rule (1) induces a transition $p(\mathbf{a}) \xrightarrow{(s,a)} q(\mathbf{b})$ from a configuration $p(\mathbf{a})$ to a configuration $q(\mathbf{b})$ if:

- $\mathrm{ATOMS}, (\mathbf{a}, a) \models \varphi$ (by definition, this fails if φ refers to any variable that has the undefined value $\bot$ in $\mathbf{a}$), and
- $\mathbf{b}$ is obtained from $\mathbf{a}$ by placing a on coordinate ST if $\mathrm{ST} \neq \mathrm{NONE}$, and $\mathbf{b} = \mathbf{a}$ otherwise.

A *run* of $\mathcal{A}$ on a data word $w = (s_1, a_1) \cdots (s_n, a_n)$ is a sequence

$$q_0(\mathbf{a}_0) \xrightarrow{(s_1,a_1)} q_1(\mathbf{a}_1) \xrightarrow{(s_2,a_2)} \ldots \xrightarrow{(s_n,a_n)} q_n(\mathbf{a}_n),$$

where q_0 is an initial state and $\mathbf{a}_0$ is a tuple where the content of all registers is undefined. We then say that the configuration $q_n(\mathbf{a}_n)$ is *reachable along* w. The finite set of all configurations reachable along w is finite, and it is denoted $\mathcal{A}(w)$.

A run is *accepting* if it ends in a configuration with an accepting state. A data word w is *accepted* by $\mathcal{A}$ if there is an accepting run of $\mathcal{A}$ on w. A NRA is *unambiguous* (URA) if every word has at most one accepting run.

The *language* of $\mathcal{A}$, denoted $L(\mathcal{A})$, is the set of all data words accepted by $\mathcal{A}$.

3 Examples

In all our examples, the finite component S of data alphabets will be a singleton set. We will therefore omit S when describing automata, so (1) will simplify to

$$(p, \varphi, \mathrm{ST}, q) \in \Delta.$$

Graphically, a transition rule like this will be presented as

if ST $= n$, and

if ST $=$ NONE.

Furthermore, $\longrightarrow$(p) means that p is initial and ((q)) means that q is accepting.

Example 1. For the equality atoms, consider the language $L \subseteq$ ATOMS* of those words where the first letter appears at some later position:

$$L = \{a_1 \dots a_n \mid n > 1, a_1 = a_i \text{ for some } i > 1\}.$$

This language is recognized by a DRA with one register and three control states:

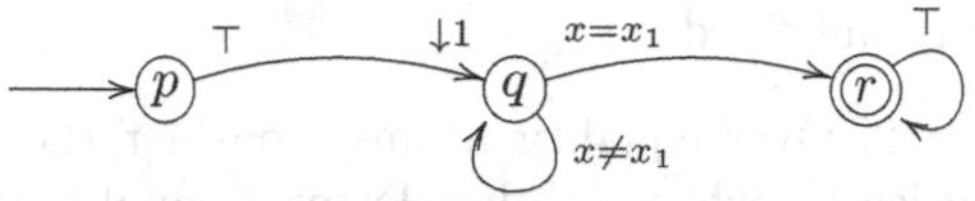

This automaton stores the first letter in its only register and then remains in the (non-accepting) state q until the letter is encountered again; then it moves to the accepting state r and stays there.

Example 2. Still for the equality atoms, consider the *reverse* of the language from Example 1, i.e., the language of those words where the last letter appears at some earlier position. This language is not recognized by any DRA, but it is recognized by a NRA with one register and three control states:

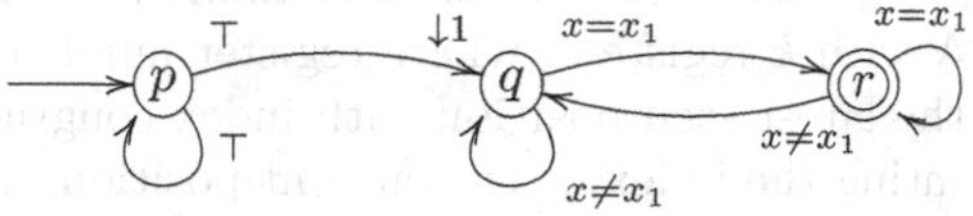

This automaton nondeterministically decides to store a letter in its register and then checks that the last letter is equal to the stored one.

Example 3. Still for the equality atoms, consider the *complement* of the language from Example 2, i.e., the language L of those words where the last letter does *not* appear at any earlier position. (In particular, we consider the empty word and all length-one words to be in this language.)

The language L is not recognized by any NRA. However, it becomes recognizable if automata are additionally equipped with the ability of *guessing*, that is, of updating the contents of their registers with arbitrary atoms, possibly different from the one that comes with the current input letter. Unlike NRA without guessing, those with guessing are closed under reversal [18, Def. 3 and Corollary 31], and the reversal of the language L is even recognized by a DRA.

Example 4. Automata from Ex. 1-3 work just as well over the dense order domain: the formulas in their transition rules simply do not use the order relation. However, over densely ordered atoms something more happens: the language from Ex. 3 is recognizable by a NRA without guessing.

The automaton has two registers. The idea is that, at any moment in an accepting run where these registers store atoms $a_1 < a_2$:

(a) in the part of the word read so far, no letter is in the open interval (a_1, a_2),
(b) the last letter of the word will belong to that open interval.

Condition (a) can be ensured easily: upon reading a letter a that belongs to the open interval (a_1, a_2), the automaton will (enter an accepting state for the moment and) put a in one of the two registers. The register is chosen nondeterministically so that condition (b) remains true. If the currently input letter is not in the interval (a_1, a_2), the automaton enters a rejecting state for the moment, with the registers kept unchanged.

Special treatment is needed to deal with situations where the last letter of the word will be larger than (or smaller than) all the letters encountered so far. These are taken care of by introducing special control states where one of the two registers remains undefined.

Example 5. Fix $k \geqslant 2$. Over equality atoms, consider the language L_k of all words w of length at least k whose kth last letter is equal to the last letter. Then L_k is recognised by a NRA with one register and $k+1$ states, depicted below:

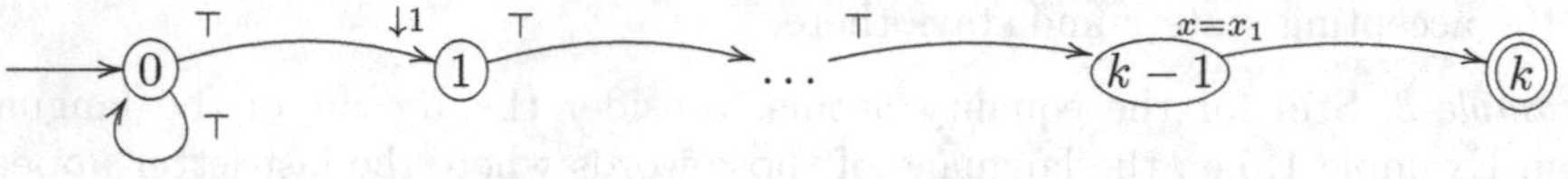

The complement of L_k is also recognised by an NRA, similar to the one above, but with $x \neq x_1$ in place of $x = x_1$ in the last transition, and with an additional component for accepting words of length smaller than k. The language L_k is also recognised by a DRA with k registers, where register number i stores the letter which appeared on the latest seen position with index congruent to i, mod k. It has k states, for counting the index of the current position, mod k.

4 Main results

Our primary contribution is:

Theorem 1. *Over equality atoms, if a data language and its complement are both recognizable by nondeterministic register automata, then they are both recognizable by deterministic register automata.*

Note that this result fails if automata with guessing are considered (see Ex. 3). Indeed, the language from Ex. 2 is recognized by a 1-NRA, and its complement in Ex. 3 is recognized by a 1-NRA with guessing, but they are not deterministically recognizable.

Moreover, the result fails (even without guessing) for densely ordered atoms. The counterexample is the same: the language from Ex. 2 is recognized by a 1-NRA, and its complement is recognized by a 2-NRA over densely ordered atoms as explained in Ex. 4, but they are not deterministically recognizable. Here the use of two registers in NRA is necessary, due to our secondary contribution: for a wide range of data domains, if a data language and its complement are both recognized by 1-NRA, then they are recognized by DRA.

We prove this for any data domain ATOMS which admits WQO in the following sense. A *well quasi-order* (WQO) is a quasi-order $(Z, \leqslant)$ such that for every infinite sequence $z_1, z_2, \ldots \in Z$ there are $1 \leqslant i < j$ with $z_i \leqslant z_j$. For a finite set X, an X-labeled substructure of ATOMS is a set $\mathcal{B} \subseteq$ ATOMS together with a labelling $\ell_{\mathcal{B}} : \mathcal{B} \to X$. For two X-labeled substructures $\mathcal{B}$ and $\mathcal{C}$ of ATOMS, we say that $\mathcal{B}$ embeds into $\mathcal{C}$ (written $\mathcal{B} \preceq \mathcal{C}$) if some automorphism π of ATOMS, restricted to $\mathcal{B}$, yields a label-preserving injection from $\mathcal{B}$ to $\mathcal{C}$, so that $\ell_{\mathcal{B}} = \ell_{\mathcal{C}} \circ \pi \restriction_{\mathcal{B}}$. Let $\text{AGE}_X(\text{ATOMS})$ be the set of all finite labeled substructures of ATOMS, partially ordered by $\preceq$. We say that ATOMS *admits* WQO if for every finite set X, the quasi-order $(\text{AGE}_X(\text{ATOMS}), \preceq)$ is a WQO. All data domains listed in Section 2 admit WQO [19]. They are also *oligomorphic* (see Sec. 5 below).

Theorem 2. *Over any oligomorphic atoms that admit* WQO*, if a data language and its complement are both recognizable by nondeterministic register automataa with one register, then they are recognizable by deterministic register automata.*

The rest of the paper consists of the proofs of Thms. 1 and 2, in Sec. 6 and 8, respectively, preceded by Sec. 5 that recalls basic definitions of the setting of sets with atoms which are used in the proofs. Our main technical lemma is proved in Sec. 6. Besides proving Thm. 1, in Sec. 7 we explain how it implies decidability of universality for unambiguous register automata.

5 Orbit-finite automata

Our proofs rely on some basic notions and results of the theory of sets with atoms [1], also known as nominal sets [24]. In this section we recall what is necessary to follow our arguments; this is part of a uniform abstract approach to register automata developed in [4,5,1].

Let Aut(ATOMS) denote the group of all automorphisms of a relational structure ATOMS. (For the equality atoms $(\mathbb{N}, =)$ this means the group of all bijections; for the densely ordered atoms $(\mathbb{Q}, \leqslant)$, the group of monotone bijections.) We consider sets equipped with an action of this group, typically, ATOMS^n for some $n \geqslant 0$ or ATOMS^* with the componentwise action.

Group actions. A (left) action of a group G on a set X is a mapping $\_ \cdot \_ :$ $G \times X \to X$ such that $1 \cdot x = x$ and $\sigma\pi \cdot x = \sigma \cdot (\pi \cdot x)$ for all $\sigma, \pi \in G$ and $x \in X$. We then say that G *acts* on X, or that X is a *G-set.* For $x \in X$, we call the set $\{\pi \cdot x \mid \pi \in G\}$ the orbit *of* x; or an orbit *in* X. The orbits in X partition X into disjoint sets. We call X *orbit-finite* if it has finitely many orbits.

Group actions canonically extend along familiar set-theoretic constructions: if X and Y are G-sets then the cartesian product $X \times Y$, the disjoint union $X \uplus Y$, the set of sequences X^*, the powerset $\mathcal{P}(X)$ etc. are all G-sets, in the expected way. For example, G acts componentwise on $X \times Y$ via $\pi \cdot (x, y) = (\pi \cdot x, \pi \cdot y)$.

Oligomorphicity. A structure ATOMS is *oligomorphic* if for every $n \in \mathbb{N}$, the componentwise action of Aut(ATOMS) on ATOMS^n induces finitely many orbits. All structures considered in this paper are oligomorphic; an example of a non-oligomorphic structure is the total order of integers.

Supports. Let $\mathrm{Aut}(\textsc{Atoms})$ act on a set X and let $x \in X$. A *support* of x is any set $S \subseteq \textsc{Atoms}$ such that the following implication holds for all $\pi \in \mathrm{Aut}(\textsc{Atoms})$:

if $\pi(s) = s$ for all $s \in S$ then $\pi \cdot x = x$.

An element $x \in X$ is *finitely supported* if it has some finite support.

For many structures $\textsc{Atoms}$, finite supports of a fixed element are always closed under intersections. Then every finitely supported x has *the least support*, denoted $\sup(x)$. This happens in particular for the equality atoms (as proved in [24, Prop. 2.3] or in [5, Cor. 9.4]) and for the dense order atoms (as proved in [5, Prop. 9.5]). It is easy to prove that taking least supports commutes with group actions: $\pi \cdot \sup(x) = \sup(\pi \cdot x)$ for every $x \in X$ and $\pi \in \mathrm{Aut}(\textsc{Atoms})$.

Equivariance. An element (or a subset, relation, function...) of an $\mathrm{Aut}(\textsc{Atoms})$-set is called *equivariant* if it is supported by the empty set; equivalently, it is fixed by every automorphism of $\textsc{Atoms}$. For example:

- a subset Z of an $\mathrm{Aut}(\textsc{Atoms})$-set X is equivariant if and only if it is a union of orbits in X (indeed, it is then equivariant as an element of $\mathcal{P}(X)$);
- a relation $R \subseteq X \times Y$ is equivariant if and only if $xRy \leftrightarrow (\pi \cdot x)R(\pi \cdot y)$ for all $x \in X$, $y \in Y$ and $\pi \in \mathrm{Aut}(\textsc{Atoms})$. An equivariant function is a function whose graph is an equivariant relation.

Standard set-theoretic relations such as set membership, or set containment, are equivariant. Indeed, $x \in Z \leftrightarrow (\pi \cdot x) \in (\pi \cdot Z)$, etc.

If $\sim$ is an equivariant equivalence relation on X then $\mathrm{Aut}(\textsc{Atoms})$ acts on the set $X/_\sim$, by $\pi \cdot C = \{\pi \cdot x \mid x \in C\}$ for each $\sim$-equivalence class $C \subseteq X$.

Register automata. Fix a structure $\textsc{Atoms}$ and let $\mathcal{R}$ be an NRA with input alphabet $S \times \textsc{Atoms}$, control states Q, and with r registers. The group $\mathrm{Aut}(\textsc{Atoms})$ acts on all the components of $\mathcal{R}$:

- on the input alphabet $A := S \times \textsc{Atoms}$, via $\pi \cdot (s, a) = (s, \pi(a))$;
- on the set $C := Q \times (\textsc{Atoms} \uplus \{\perp\})^r$ of all configurations of $\mathcal{R}$, via

$$\pi \cdot q(a_1, \ldots, a_r) = q(\pi(a_1), \ldots, \pi(a_r)) \qquad \text{(where } \pi(\perp) = \perp\text{)};$$

- the set of initial configurations and the set of accepting configurations are both equivariant subsets of C;
- the set of transitions of $\mathcal{R}$ is an equivariant relation: if $p(\mathbf{a}) \xrightarrow{(s,a)} q(\mathbf{a}')$ is a transition of $\mathcal{R}$, then so is $\pi \cdot p(\mathbf{a}) \xrightarrow{(s,\pi(a))} \pi \cdot q(\mathbf{a}')$.

Furthermore, each of these components is orbit-finite, and each of its elements has a finite support. Using the terminology of [5], this means that register automata are a special case of *orbit-finite automata.*

By equivariance of all the components above, the language $L(\mathcal{R})$ of a register automaton is an equivariant subset of $A^* = (S \times \textsc{Atoms})^*$, considered with the componentwise action of $\mathrm{Aut}(\textsc{Atoms})$ on A^*, i.e.

$$\pi \cdot ((s_1, a_1), \ldots, (s_n, a_n)) = ((s_1, \pi \cdot a_1), \ldots, (s_n, \pi \cdot a_n)).$$

Myhill-Nerode theorem. In order to prove that a language is deterministically recognizable, we use the following Myhill-Nerode characterization.

For an alphabet $A = S \times \text{ATOMS}$ and data language $L \subseteq A^*$, consider its Myhill-Nerode equivalence $\sim_L \subseteq A^* \times A^*$, defined by

$$u \sim_L v \qquad \text{if and only if} \qquad uw \in L \leftrightarrow vw \in L \quad \text{for all } w \in A^*.$$

Theorem 3. *[5, Thm. 3.8 and Thm. 6.4] Let* ATOMS *be oligomorphic and $L \subseteq (S \times \text{ATOMS})^*$ be an equivariant language. Then L is deterministically recognizable if and only if $(S \times \text{ATOMS})^*/_{\sim_L}$ is orbit-finite.*

Among other things, this theorem immediately implies that the language from Ex. 2 is not deterministically recognizable, neither for the equality atoms nor for the total order atoms. Indeed, two words are Myhill-Nerode equivalent with respect to that language if and only if they contain the same set of letters. Therefore, the language cannot be deterministically recognizable, since automorphisms of ATOMS preserve the number of distinct letters in a word.

6 Proof of Theorem 1

In the proof, we will make use of an abstract notion of a split of a family of sets.

For any family $\mathcal{F}$ of subsets of a set X, a *split* of $\mathcal{F}$ is a pair (U, V) of sets which partition X: $X = U \uplus V$, such that both U and V are *finite* unions of elements of $\mathcal{F}$. Obviously, for any splits to exist, $X = \bigcup \mathcal{F}$ must hold.

In the following lemma, ATOMS is the equality atoms.

Lemma 1. *For any* Aut(ATOMS)*-set X with finitely supported elements, and any equivariant, orbit-finite family $\mathcal{F}$ of finitely supported subsets of X, the set $\mathcal{G}$ of splits of $\mathcal{F}$ is orbit-finite. Moreover, a bound on the number of orbits of $\mathcal{G}$ and the maximal size of the support of an element in $\mathcal{G}$ are computable from the analogous bounds for $\mathcal{F}$.*

As should be clear after reading Sec. 5, the set of splits of $\mathcal{F}$ is considered with the natural action of Aut(ATOMS): $\pi \cdot (U, V) = (\pi \cdot U, \pi \cdot V)$, where $\pi \cdot W = \{\pi \cdot x \mid x \in W\}$ for $W \subseteq X$.

We will prove Lem. 1 in Sec. 6.2. For now, let us show how the lemma implies Thm. 1.

Let $\mathcal{A}$ and $\mathcal{B}$ be two NRA over an alphabet $A = S \times \text{ATOMS}$ such that $L(\mathcal{A})$ and $L(\mathcal{B})$ partition A^*. We will show that the Myhill-Nerode equivalence of $L = L(\mathcal{A})$ has orbit-finitely many classes. Together with Thm. 3, this will prove that L is deterministically recognizable.

Let C be the set of configurations of $\mathcal{A} \uplus \mathcal{B}$ (the disjoint union of $\mathcal{A}$ and $\mathcal{B}$.) Hence, C consists of tuples of the form $q(\mathbf{a})$ where q is either a state of $\mathcal{A}$ or a state of $\mathcal{B}$ (but not both), and $\mathbf{a}$ is a tuple of elements of $\text{ATOMS} \uplus \{\bot\}$ of appropriate length. For $c \in C$ denote

$$L_c := \{w \in A^* \mid \mathcal{A} \uplus \mathcal{B} \text{ accepts } w \text{ from configuration } c\},$$

and let $\mathcal{F} = \{L_c \mid c \in C\}$. Since C is equivariant and orbit-finite, so is $\mathcal{F}$. Moreover, if $c = q(\mathbf{a})$ then L_c is finitely supported by the atoms in $\mathbf{a}$. Clearly, every word $(s_1, a_1) \cdots (s_n, a_n) \in A^*$ is supported by $\{a_1, \ldots, a_n\}$. This means that $\mathcal{F}$ and $X = A^*$ satisfy the assumptions of Lem. 1, therefore $\mathcal{F}$ has only orbit-finitely many splits.

Every word $v \in A^*$ induces a partition of A^* into two disjoint sets:

$$U_v = \{w \in A^* \mid vw \in L\} \qquad \text{and} \qquad V_v = \{w \in A^* \mid vw \notin L\}.$$

Moreover, the sets U_v and V_v are finite unions of sets from $\mathcal{F}$, namely

$$U_v = \bigcup_{c \in \mathcal{A}(v)} L_c \qquad \text{and} \qquad V_v = \bigcup_{c \in \mathcal{B}(v)} L_c.$$

These unions are finite because automata $\mathcal{A}$ and $\mathcal{B}$ allow no guessing and so $\mathcal{A}(v)$ and $\mathcal{B}(v)$, the sets of configurations reachable in $\mathcal{A}$ resp. $\mathcal{B}$ by reading the word v, are finite. Therefore, (U_v, V_v) is a split of $\mathcal{F}$, for any word v.

By definition, $u \sim_L v$ if and only if $U_u = U_v$. Consider any two words $v, w \in A^*$ such that the splits (U_v, V_v) and (U_w, V_w) are in the same orbit, i.e., $U_w = \pi \cdot U_v$ (and therefore also $V_w = \pi \cdot V_v$) for some automorphism π. Since L is an equivariant language, we have $\pi \cdot U_v = U_{\pi \cdot v}$ and so $w \sim_L \pi \cdot v$. Theorem 1 now follows from Thm. 3.

6.1 Examples

Before proving Lem. 1, we give some examples of families of splits, which may be helpful in developing some intuitions.

The first example shows that the number of orbits of splits may grow as fast as double-exponentially, relative to the least supports of elements of $\mathcal{F}$.

Example 6. For the equality atoms, fix $k \geqslant 1$ and let X be the set of all k-tuples of pairwise distinct atoms. For each $S \subseteq \text{Atoms}$ with $|S| = k$, let $S^{(k)} = S^k \cap X$ and let $M_S = X \setminus S^{(k)}$. Note that $S^{(k)}$ is finite, with $k!$ elements.

The family $\mathcal{F} \subseteq \mathcal{P}(X)$ of all singletons in X and all sets M_S as above is equivariant and has two orbits. Each set in $\mathcal{F}$ has a support of size k.

For any $K \subseteq S^{(k)}$, consider the partition of X into K and $X \setminus K$. Then $(K, X \setminus K)$ is a split of $\mathcal{F}$, as $K = \bigcup_{v \in K} \{v\}$ and $X \setminus K = M_S \cup \bigcup_{v \in S^{(k)} \setminus K} \{v\}$.

Moreover, every split (U, V) of $\mathcal{F}$ is of the form $(K, X \setminus K)$ or $(X \setminus K, K)$ for some S and K as above. Indeed, suppose $U = \bigcup \mathcal{U}$ and $V = \bigcup \mathcal{V}$ for some finite $\mathcal{U}, \mathcal{V} \subseteq \mathcal{F}$. As $U \cup V = X$ is infinite, $\mathcal{U} \cup \mathcal{V}$ must contain M_S for some set S of k atoms. Suppose without loss of generality that $M_S \in \mathcal{U}$. By disjointness of U and V, the set $\mathcal{V} \subseteq \mathcal{F}$ may only contain singletons $\{v\}$, for $v \in S^{(k)}$. Then $(U, V) = (X \setminus K, K)$, where $K = \bigcup \mathcal{V}$.

For $K, K' \subseteq S^{(k)}$, the splits defined by K and K' are in the same orbit only if there is an automorphism π that fixes S as a set, such that $\pi \cdot K = K'$. Since there are only $k!$ bijections on S, the set of splits of $\mathcal{F}$ has at least $\frac{2^{k!}}{k!}$ orbits. $\square$

The next example shows the difference between splits and the finite subfamilies of $\mathcal{F}$ that define those splits: the set of those families may be orbit-infinite.

Example 7. Let X be the set of all finite sets of equality atoms. For any distinct atoms a, b, define $E_{a,b}, D_{a,b} \subseteq X$ by:

$$E_{a,b} = \{F \in X \mid a \in F \leftrightarrow b \in F\} \qquad D_{a,b} = X \setminus E_{a,b}$$

And let $\mathcal{F}$ contain all sets $E_{a,b}$ and $D_{a,b}$. This $\mathcal{F}$ has two orbits.

Obviously, $(U, V) = (X, \emptyset)$ is a split of $\mathcal{F}$; it is enough to take $\mathcal{U} = \{D_{a,b}, E_{a,b}\}$ and $\mathcal{V} = \emptyset$ for any fixed a, b. However, there are many more minimal families $\mathcal{U}$ and $\mathcal{V}$ that achieve the same effect. Indeed, for any number n, and for any pairwise distinct atoms $a_1, \ldots, a_n$, consider:

$$\mathcal{U} = \{D_{a_1,a_2}, D_{a_2,a_3}, \ldots, D_{a_{n-1},a_n}, E_{a_1,a_n}\} \qquad \mathcal{V} = \emptyset$$

It is easy to check that $\bigcup \mathcal{U} = X$. All such families are minimal (in fact, removing any element from $\mathcal{U}$ would prevent it from being the part of any split of $\mathcal{F}$), and for each n these families form a separate orbit. □

The following example shows that the statement of Lem. 1 fails if the atoms are $(\mathbb{Q}, \leqslant)$. It is obtained from Ex. 4 via the translation given in the proof of Thm. 1, and a simplification replacing each word by its last letter.

Example 8. The atoms are $(\mathbb{Q}, \leqslant)$. Let $X = \mathbb{Q}$ and let $\mathcal{F} \subseteq \mathcal{P}(X)$ consist of:

- singletons $\{q\} \subseteq X$, for $q \in \mathbb{Q}$;
- open intervals $(p, q) \subseteq X$, for $p < q$ in $\mathbb{Q} \cup \{-\infty, +\infty\}$.

Then $\mathcal{F}$ has five orbits (here $\pm\infty$ are fixed under the action of Aut(Atoms)). For any finite set $K \subseteq X$, consider the partition of X into K and $X \setminus K$. Then $K = \bigcup_{q \in K} \{q\}$ whereas $X \setminus K$ is the union of all intervals (p, q), where $p < q$ are consecutive elements in $K \cup \{-\infty, +\infty\}$. Hence, $(K, X \setminus K)$ is a split of $\mathcal{F}$. In particular, the set of all splits of $\mathcal{F}$ has infinitely many orbits, because the set of finite subsets of X has infinitely many orbits. □

6.2 Proof of Lemma 1

We prove by induction a stronger statement, where the atoms are assumed to be an expansion of $(\mathbb{N}, =)$ by finitely many constants. In other words, in this section we will assume that Atoms is a structure over a vocabulary that consists of (equality and) a finite number of constant symbols; the universe of Atoms is $\mathbb{N}$, with the constants interpreted as some pairwise distinct numbers. The group Aut(Atoms) then consists of all bijections of Atoms which fix every constant.

If Atoms is such a structure and T is a finite set of atoms all different from the constants, then by Atoms_T we denote the structure, over an extended vocabulary, that arises from Atoms by interpreting all the atoms in T as additional constants. Obviously, Aut(Atoms_T) is a subgroup of Aut(Atoms), so every action of Aut(Atoms) on a set X restricts to an action of Aut(Atoms_T).

This restriction preserves and reflects the existence of finite supports: an element $x \in X$ is supported by some S in the action of Aut(ATOMS) if and only if it is supported by $S \setminus T$ in the restricted action of Aut(ATOMS_T). In particular, if ATOMS is an expansion of $(\mathbb{N}, =)$ by finitely many constants, then every finitely supported element x has a least support $\sup(x)$. Note that $\sup(x)$ never contains any constants, since those can always be safely removed from any support.

For a subset $\mathcal{U}$ of an orbit-finite equivariant set $\mathcal{F}$, its *dimension* $\dim(\mathcal{U})$ is the maximum size of the least support of an element of $\mathcal{U}$. This makes sense even if $\mathcal{U}$ is infinite, because $\mathcal{F}$ is orbit-finite and sets from the same orbit have least supports of the same size. In particular, $\dim(\mathcal{F})$ is well defined.

The following lemma says that adding constants to atoms preserves orbit-finiteness. It is a standard result in the theory of sets with atoms, see e.g. [1, Lem. 3.19] or [24, Lem. 5.22], indeed it is a fundamental property of oligomorphic structures, but we re-prove it here to extract explicit bounds:

Lemma 2. *Fix a finite set $T \subseteq$ ATOMS. For any orbit-finite* Aut(ATOMS)*-set $\mathcal{F}$ with l orbits, the corresponding action of* Aut(ATOMS_T) *on $\mathcal{F}$ is also orbit-finite, with at most $l \cdot (|T|+1)^{\dim(\mathcal{F})}$ orbits.*

Proof. Assume first that $\mathcal{F}$ has only one orbit in the Aut(ATOMS)-action, i.e., that $l = 1$. Let $d = \dim(\mathcal{F})$. Let Y denote the set of d-tuples of pairwise distinct atoms different from the constants in ATOMS. This is a single-orbit set under the componentwise action of Aut(ATOMS). Pick any $x_0 \in \mathcal{F}$. Let $y_0 = (a_1, \ldots, a_d) \in Y$ be an enumeration of $\sup(x_0)$. There is a unique equivariant surjection $f \colon Y \to X$ such that $f(\pi \cdot y_0) = \pi \cdot x_0$ for all $\pi \in$ Aut(ATOMS). (The function f is total since Y has one orbit; it is well defined because y_0 enumerates a support of x_0, and it is surjective since X has one orbit.) Two tuples in Y are in the same orbit in the action of Aut(ATOMS_T) if and only if they contain the same arrangement of atoms from T at the same positions. There are at most $(|T|+1)^d$ such arrangements, (in fact fewer than this if $d > 1$, because tuples in Y are pairwise distinct), so Y has at most $(|T|+1)^d$ such orbits. X is an image of the equivariant function $f \colon Y \to X$, so the same bound applies to X. For a set $\mathcal{F}$ with l orbits, each of dimension at most d, the bound simply multiplies by l. □

From now on consider ATOMS as described above, and let X and $\mathcal{F}$ be as in the statement of Lem. 1. The following key lemma says that every split of $\mathcal{F}$ has a support of a bounded size.

Lemma 3. *Let $U \uplus V$ be a split of $\mathcal{F}$ and let $\mathcal{U}, \mathcal{V}$ be finite subfamilies of $\mathcal{F}$ such that $\bigcup \mathcal{U} = U$ and $\bigcup \mathcal{V} = V$. Then U and V each have a support of size at most N, for some bound N computable only from* $\dim(\mathcal{U}), \dim(\mathcal{V}), \dim(\mathcal{F})$ *and the number of orbits in $\mathcal{F}$.*

The crux of this lemma is that the number N does not depend on the split $U \uplus V$. It only depends on the number of orbits in $\mathcal{F}$, its dimension $\dim(\mathcal{F})$, and on $\dim(\mathcal{U})$ and $\dim(\mathcal{V})$ (which, anyway, are bounded from above by $\dim(\mathcal{F})$).

Proof (of Lem. 3). We proceed by induction on $k = \dim(\mathcal{U}) + \dim(\mathcal{V})$. Fix $k \geqslant 0$ and assume that the statement of the lemma holds for all smaller values of k. Without loss of generality, we may assume that $\emptyset$ does not belong to $\mathcal{U}$ nor $\mathcal{V}$ (as it can be safely removed from each of them).

For a finitely supported set $F \subseteq X$ define

$$F^\sharp := \{\pi \cdot y \mid \pi \in \mathrm{Aut}(\textsc{Atoms}), y \in F, \sup(y) \cap \sup(F) = \emptyset\}.$$

Intuitively, $F^\sharp$ arises by taking all elements of F that are "fresh for F", i.e., ones whose supports share no atoms with the support of F, and then by applying arbitrary atom automorphisms to those elements. Note that that $F^\sharp$ is equivariant and $F^\sharp = (\pi \cdot F)^\sharp$ for any automorphism π.

Claim 1 $X = \bigcup_{F \in \mathcal{U} \cup \mathcal{V}} F^\sharp$.

Proof. Take any $x \in X$. Let $S = \bigcup_{F \in \mathcal{U} \cup \mathcal{V}} \sup(F)$. Since $\mathcal{U}$ and $\mathcal{V}$ are finite, S is a finite set. Pick an automorphism π such that its inverse π^{-1} maps $\sup(x)$ to a set disjoint with S. Consider the element $y = \pi^{-1} \cdot x \in X$. Since $U \cup V = X$, there must be some $F \in \mathcal{U} \cup \mathcal{V}$ such that $y \in F$. Then $x \in F^\sharp$. □

Let us first prove the lemma for the **special case** where $X = F^\sharp$ for some $F \in \mathcal{U} \cup \mathcal{V}$. Suppose that $X = F^\sharp$ for some $F \in \mathcal{U}$ (the case $F \in \mathcal{V}$ is symmetric).

Claim 2 Every $y \in X$ with $\sup(y) \cap \sup(F) = \emptyset$ belongs to F.

Proof. Take any y as above. As $X = F^\sharp$, there is some π and $x \in F$ such that $y = \pi \cdot x$ and $\sup(x) \cap \sup(F) = \emptyset$. Pick an automorphism θ such that:

- θ agrees with π on $\sup(x)$, mapping it bijectively to $\sup(y)$,
- θ fixes $\sup(F)$ pointwise.

Such a θ exists since $\sup(x)$ and $\sup(y)$ are both disjoint from F. Then $\theta \cdot x = \pi \cdot x = y$ by the first property above, and $\theta \cdot x \in \theta \cdot F = F$ by the second property. Altogether, $y \in F$. □

Claim 3 For every $G \in \mathcal{V}$, $\sup(F) \cap \sup(G) \neq \emptyset$.

Proof. We show that if $\sup(G)$ is disjoint from $\sup(F)$ then G must be empty, contradicting our previous assumption.

Suppose $x \in G$. Pick an automorphism π which fixes $\sup(G)$ pointwise and maps $\sup(x)$ to a set disjoint with $\sup(F)$. Such a π exists because $\sup(G)$ and $\sup(F)$ are disjoint. Letting $y := \pi \cdot x$, we have $y \in F$ by Claim 2, and moreover $y = \pi \cdot x \in \pi \cdot G = G$. Then $y \in F \cap G \subseteq U \cap V = \emptyset$, a contradiction. This proves $G = \emptyset$, which in turn contradicts the assumption that $\emptyset \notin \mathcal{V}$. □

Denote $T = \sup(F)$. If $T = \emptyset$ then by Claim 3, $\mathcal{V}$ has dimension 0 and therefore V is supported by the empty set. So we may assume that $T \neq \emptyset$. For the same reason we may assume that the family $\mathcal{V}$ is not empty.

Let ATOMS_T be obtained from ATOMS by including the elements of T as new constants. Hence, ATOMS_T extends ATOMS by at most r constants, where $r := \dim(\mathcal{F})$.

Let l be the number of orbits in $\mathcal{F}$. By Lem. 2, the family $\mathcal{F}$, treated as a family of sets over the atoms ATOMS_T, is still orbit-finite, with the number of orbits l' depending only on l and r. Clearly, $U \uplus V$ remains a split of $\mathcal{F}$. Note that if $F \in \mathcal{F}$ is supported by some set S over ATOMS, then F is supported by S, indeed even by $S \setminus T$, over ATOMS_T. In particular, the dimension of $\mathcal{F}$ does not increase by moving from ATOMS to ATOMS_T. More interestingly, by Claim 3, the least supports of all the elements in $\mathcal{V}$ actually *decrease* when considering ATOMS_T as atoms. Since $\mathcal{V}$ is not empty, the dimension of $\mathcal{V}$ strictly decreases and it follows that $\dim(\mathcal{U}) + \dim(\mathcal{V}) < k$ over ATOMS_T. Applying the inductive assumption yields a set T' of size N', depending on $k-1$ and l', such that T' supports V over ATOMS_T. By construction, V is supported by $T \cup T'$ over ATOMS. Note that

$$|T \cup T'| \leqslant N'' := N' + r.$$

This concludes the proof in the special case when $X = F^\sharp$ for some $F \in \mathcal{U} \cup \mathcal{V}$. In the **general case**, for each $F \in \mathcal{U} \cup \mathcal{V}$ define:

$$\mathcal{F}_F := \{G \cap F^\sharp \mid G \in \mathcal{F}\}$$

$$\mathcal{U}_F := \{G \cap F^\sharp \mid G \in \mathcal{U}\} \qquad \mathcal{V}_F := \{G \cap F^\sharp \mid G \in \mathcal{V}\}$$
$$U_F := U \cap F^\sharp = \bigcup \mathcal{U}_F \qquad V_F := V \cap F^\sharp = \bigcup \mathcal{V}_F.$$

Then $\bigcup \mathcal{F}_F = F^\sharp$ and (U_F, V_F) is a split of $\mathcal{F}_F$ which falls into the special case considered above. Hence, U_F has some support S_F of size at most N''.

Then U is supported by $S := \bigcup_{F \in \mathcal{U} \cup \mathcal{V}} S_F$. Note that S_F only depends on the orbit of F, as $F^\sharp = (\pi \cdot F)^\sharp$ for any automorphism π. As there are l such orbits contained in $\mathcal{F}$, it follows that S has size at most $N := N''l$. This concludes the inductive step, and the proof of Lem. 3. □

Using Lem. 3, we now proceed to prove Lem. 1.

Proof (of Lemma 1). Consider an equivariant set X and an equivariant, orbit finite family $\mathcal{F}$ of finitely supported subsets of X. Let $((U_i, V_i))_{i \in I}$ be a family of splits of $\mathcal{F}$. By Lem. 3, each one of these splits is supported by some set of a bounded size. Applying suitable automorphisms to each of these splits, we can obtain a family of splits $((U'_i, V'_i))_{i \in I}$ such that, for all $i \in I$:

- U'_i and U_i are in the same orbit, and
- each U'_i is supported by the same set S.

It is now enough to show that there are only finitely many subsets $U \subseteq X$ supported by a fixed set S, which are unions of elements of $\mathcal{F}$.

By Lem. 2 it follows that $\mathcal{F}$ has finitely many orbits under the action of the group $\text{Aut}(\text{ATOMS}_S)$ of all automorphisms which fix S pointwise. (Here, as in the statement of Lem. 1, ATOMS are the pure equality atoms without any constants.) If a set $U \subseteq X$ supported by S contains some $F \in \mathcal{F}$ as a subset, then it contains $\pi \cdot F$ for every $\pi \in \mathrm{Aut}(\mathrm{ATOMS}_S)$. In other words, U contains (the union of) the entire orbit in $\mathcal{F}$ under the action of $\mathrm{Aut}(\mathrm{ATOMS}_S)$. Since we assume that U is a union of elements of $\mathcal{F}$, it is a union of (the unions of) orbits in $\mathcal{F}$, and there are only finitely many of these.

This completes the proof of Lem. 1. □

7 Application to Unambiguous Register Automata

Lemma 1 is interesting in its own right and its applications are not limited to the ones mentioned in Sec. 4. We shall now show how it can be used to decide universality (and hence also language containment and equality, cf. [8, Lem. 8]) of URA over the pure equality atoms ATOMS.

Theorem 4. *[22, Thm. 14] The language containment and equality problems are decidable for unambiguous register automata.*

As an application of Lem. 1, we give an alternative decidability proof for the universality problem of URA. First, we prove a consequence of Lem. 1.

Lemma 4. *Let X be an equivariant set over equality atoms, and let $\mathcal{F}$ be an equivariant, orbit-finite family of finitely supported subsets of X. There is a bound M, computable from $\dim(\mathcal{F})$ and the number of orbits in $\mathcal{F}$, such that every $\mathcal{P} \subseteq \mathcal{F}$ which is a partition of X has size at most M.*

Proof. Let $\mathcal{G} = \{U \mid (U, V) \text{ is a split of } \mathcal{F}\}$. By Lem. 1, $\mathcal{G}$ is orbit-finite. Moreover, its elements are finitely supported. Let $\mathcal{P} \subseteq \mathcal{F}$ be a partition of X into nonempty subsets. For each $\mathcal{U} \subseteq \mathcal{P}$, the union $\bigcup \mathcal{U}$ belongs to $\mathcal{G}$; in particular, we have $2^{|\mathcal{P}|}$ elements of $\mathcal{G}$, each containing different sets in $\mathcal{P}$. The proof is completed by the following counting argument.[1]

Let $S = \bigcup_{F \in \mathcal{P}} \sup(F)$. An *S-orbit* in $\mathcal{G}$ is an orbit in $\mathcal{G}$ with respect to the action of those atom permutations which fix S pointwise. Equivalently, it is an orbit in $\mathcal{G}$ viewed as a $\mathrm{Aut}(\mathrm{ATOMS}_S)$-set. By Lem. 2, for any finite $S \subseteq \mathrm{ATOMS}$, the number of S-orbits in $\mathcal{G}$ is bounded by $l \cdot (|S| + 1)^k$, where k and l are computable from $\dim(\mathcal{F})$ and the number of orbits of $\mathcal{F}$.

Two splits $G, G' \in \mathcal{G}$ in the same S-orbit contain the same elements of $\mathcal{P}$: if $G' = \pi \cdot G$ then by equivariance of $\mathcal{F}$ and $\mathcal{G}$, for each $F \in \mathcal{P}$ we have $F \subseteq G$ if and only if $\pi \cdot F \subseteq \pi \cdot G$, but $\pi \cdot F = F$ when π fixes S pointwise. Hence, for any two distinct $\mathcal{U}, \mathcal{U}' \subseteq \mathcal{P}$, their unions $\bigcup \mathcal{U}$ and $\bigcup \mathcal{U}'$ belong to different S-orbits in $\mathcal{G}$, so there are least $2^{|\mathcal{P}|}$ such orbits. As $|S| \leqslant \dim(\mathcal{F}) \cdot |\mathcal{P}|$, we get:

$$2^{|\mathcal{P}|} \leqslant l \cdot (|S| + 1)^k \leqslant l \cdot (\dim(\mathcal{F}) \cdot |\mathcal{P}| + 1)^k.$$

It follows that $|\mathcal{P}|$ is bounded by some M computable from k, l, and $\dim(\mathcal{F})$. □

[1] It exhibits the well-known fact that equality atoms have the *NIP property* studied in model theory.

Lemma 4 has the following corollary, which is a strong restriction on the structure of universal URA and easily yields Thm. 4.

Call a configuration c of a NRA $\mathcal{A}$ *nonempty* if the NRA accepts some word from this configuration, i.e., the following language is nonempty:

$$L_c := \{w \in A^* \mid \mathcal{A} \text{ accepts } w \text{ from } c\}$$

Since NRA emptiness is decidable, it is not difficult to modify any given NRA to one with only nonempty configurations. This transformation preserves URA, so we may safely assume that we only consider URA with this property.

Corollary 1. *Let $\mathcal{A}$ be a* URA *with nonempty configurations and which accepts every input word. Then there is a computable bound M such that $\mathcal{A}$ may reach at most M different configurations when reading any given input word.*

Proof. Let $\mathcal{A}$ be an URA over an input alphabet $A = S \times \text{Atoms}$. Let C be the set of configurations of $\mathcal{A}$ and let $\mathcal{F} := \{L_c \mid c \in C\}$. Note that $\dim(\mathcal{F})$ is not larger than the number of registers r of $\mathcal{A}$, and the number of orbits in $\mathcal{F}$ is not larger than the number of orbits of configurations in $\mathcal{A}$, which in turn is equal to the number of control states in $\mathcal{A}$ times the number of orbits in $(\text{Atoms} \uplus \{\bot\})^r$ (equal to the $r+1$-st Bell number).

For each $w \in A^*$, the set $\mathcal{A}(w) \subseteq C$ of configurations reachable when reading w is finite, since $\mathcal{A}$ has no guessing. Unambiguity of $\mathcal{A}$ implies that the family

$$\mathcal{P}_w := \{L_c \mid c \in \mathcal{A}(w)\} \subseteq \mathcal{F}$$

consists of pairwise disjoint sets. If additionally $L(\mathcal{A}) = A^*$, then $\mathcal{P}_w$ forms a partition of A^*, so $|\mathcal{P}_w| \leqslant M$ where M is the bound from Lemma 4. As $|\mathcal{A}(w)| \leqslant |\mathcal{P}_w|$, this yields the corollary. □

Decidability of universality of URA now follows using standard ideas.

Proof (of Thm. 4, sketch). We use the notation of the proof of Cor. 1. The idea is to construct the truncated powerset automaton whose states are sets of at most M states of $\mathcal{A}$.

Let C' denote the family of subsets of C of size at most M; then C' is orbit-finite. We define a deterministic automaton $\mathcal{A}'$ with an infinite, but orbit-finite state space C'. Its transitions are $X \xrightarrow{a} Y$, for $X, Y \in C'$ such that

$$Y = \left\{ y \in C \;\middle|\; x \xrightarrow{a} y \text{ in } \mathcal{A}, x \in X \right\}.$$

The initial state of $\mathcal{A}'$ is the set $C_0 \subseteq C$ of initial configurations of $\mathcal{A}$ (unless $|C_0| > M$, but then $L(\mathcal{A}) \neq A^*$ by the corollary). Accepting states are all states $X \in C'$ which contain an accepting configuration of $\mathcal{A}$. All the ingredients of $\mathcal{A}'$ are equivariant, orbit-finite sets, so $\mathcal{A}'$ is an *orbit-finite deterministic automaton*, and can be effectively constructed given $\mathcal{A}$ and M. Its language $L(\mathcal{A}')$ is defined as usual. By construction,

– $L(\mathcal{A}') \subseteq L(\mathcal{A}) \subseteq A^*$;

– if $L(\mathcal{A}) = A^*$ then $L(\mathcal{A}') = A^*$, by Cor. 1.

Hence, $\mathcal{A}'$ is universal if and only if $\mathcal{A}$ is universal. Since $\mathcal{A}'$ is orbit-finite, universality of $\mathcal{A}'$ can be effectively decided, using standard techniques for orbit-finite automata [1,5]: by first complementing and then testing emptiness. □

8 Proof of Theorem 2

Towards proving Thm. 2, assume $\mathcal{A}$ and $\mathcal{B}$ are two complementing 1-NRA over an alphabet $A = S \times \text{ATOMS}$ and that ATOMS admit WQO.

Recall that configurations of a 1-NRA are either of the form $q(a)$ where q is a control state and $a \in \text{ATOMS}$ is the register value, or of the form $q(\bot)$ when the register value is still undefined. We assume, without losing generality, that both register automata $\mathcal{A}$ and $\mathcal{B}$ immediately update their register, i.e., every transition rule outgoing from an initial state updates the register.

Let Q and Q' denote sets of control states of $\mathcal{A}$ and $\mathcal{B}$, respectively, and assume without losing generality that Q and Q' are disjoint.

For every nonempty data word $w \in A^+$, the set $\mathcal{A}(w) \cup \mathcal{B}(w)$ of configurations of $\mathcal{A}$ and $\mathcal{B}$ reachable along w is finite, since NRA have no guessing, and contains no undefined configurations of the form $q(\bot)$ due to the immediate update assumption. For every $w \in A^+$ define a finite induced substructure $\mathcal{C}_w$ of ATOMS, labeled with the finite set $P = \mathcal{P}(Q \cup Q')$, as follows. The elements of $\mathcal{C}_w$ are the atoms that appear in configurations in $\mathcal{A}(w) \cup \mathcal{B}(w)$:

$$\mathcal{C}_w = \{a \in \text{ATOMS} \mid (q, a) \in \mathcal{A}(w) \cup \mathcal{B}(w) \text{ for some state } q.\}$$

The labeling $\ell_w \colon \mathcal{C}_w \to P$ of $\mathcal{C}_w$ maps $a \in \mathcal{C}_w$ to the set of all control states which appear in $\mathcal{A}(w) \cup \mathcal{B}(w)$ together with a:

$$\ell_w(a) = \{q \in Q \mid (q, a) \in \mathcal{A}(w)\} \cup \{q \in Q' \mid (q, a) \in \mathcal{B}(w)\}.$$

Let $L = L(\mathcal{A})$. For each $v \in A^*$ define the partition of A^* into:

$$U_v = \{w \in A^* \mid vw \in L\} \quad \text{and} \quad V_v = \{w \in A^* \mid vw \notin L\}.$$

Recall that $u \sim_L v$ if and only if $U_u = U_v$.

Claim. Let $u, v \in A^+$. If $\mathcal{C}_u \preceq \mathcal{C}_v$ then $\pi \cdot u \sim_L v$ for some automorphism π.

Proof. By definition of $\preceq$, there is some $\pi \in \text{Aut}(\text{ATOMS})$ which maps $\mathcal{C}_u$ to a substructure of $\mathcal{C}_v$, so that $\pi \cdot \mathcal{C}_u \subseteq \mathcal{C}_v$ and

$$\ell_u(a) = \ell_v(\pi(a)) \qquad \text{for } a \in \mathcal{C}_u. \tag{2}$$

Let $u' = \pi \cdot u$. By equivariance of register automata, if $\mathcal{A}$ reaches a configuration (q, a) when reading u, then it reaches the configuration $(q, \pi(a))$ when reading $u' = \pi \cdot u$. Hence, $\mathcal{C}_{u'} \subseteq \mathcal{C}_v$ and $\ell_u(a) = \ell_{u'}(\pi(a))$ for $a \in \mathcal{C}_u$. Together with (2) we get $\ell_{u'}(a) = \ell_v(a)$ for all $a \in \mathcal{C}_{u'}$.

We show that this implies $U_{u'} = U_v$, which will yield the claim as $u' = \pi \cdot u$. Towards proving $U_{u'} \subseteq U_v$ take any $w \in U_{u'}$; then $u'w \in L$. Pick an accepting run of $\mathcal{A}$ on $u'w$. Let $q(a)$ be the configuration of $\mathcal{A}$ in this run reached after reading the (nonempty) prefix u'. In particular, $\mathcal{A}$ accepts w starting from the configuration $q(a)$. Moreover, $a \in \mathfrak{C}_{u'}$ and $q \in \ell_u(a)$. As $\mathfrak{C}_{u'} \subseteq \mathfrak{C}_v$ and $\ell_{u'}(a) = \ell_v(a)$, it follows that $\mathcal{A}$ may reach the configuration $q(a)$ after reading v. As w is accepted by $\mathcal{A}$ from this configuration, it follows that $\mathcal{A}$ accepts vw, so $w \in U_v$.

The inclusion $V_{u'} \subseteq V_v$ is proved by a similar argument, using $\mathcal{B}$ instead of $\mathcal{A}$, since $L(\mathcal{B}) = A^* \setminus L(\mathcal{A}) = A^* \setminus L$. As $U_{u'} = A^* \setminus V_{u'}$ and $V_v = A^* \setminus U_v$, the inclusion $V_{u'} \subseteq V_v$ implies $U_{u'} \supseteq U_v$. Altogether, $U_{u'} = U_v$, so $u' \sim_L v$, yielding the claim. □

Theorem 2 now follows easily: assume towards a contradiction that $A^*/_{\sim_L}$ is not orbit-finite. Then there is an infinite set $X \subseteq A^+$ such that $\pi(u) \not\sim_L v$ for all distinct $u, v \in X$ and $\pi \in \mathrm{Aut}(\textsc{Atoms})$. As $\textsc{Atoms}$ admits $\textsc{wqo}$, there are distinct $u, v \in X$ such that $\mathfrak{C}_u \preceq \mathfrak{C}_v$. The claim above yields a contradiction. □

9 Final remarks

We have studied a deterministic collapse for NRA: if a language and its complement are both recognized by NRA then they are also recognized by DRA. We have proved this for register automata over equality atoms; and for automata with one register only, over any atoms that admit WQO. We have also applied our key technical observation, namely orbit-finiteness of the set of splits of an orbit-finite family of sets, in order to re-prove decidability of universality of URA.

The assumed form $A = S \times \textsc{Atoms}$ of the input alphabets is not important; the results apply to arbitrary orbit-finite input alphabets A.

The proof of our main result (also of decidability of universality of URA) is effective, with elementary bounds. In particular, given two NRA with complementing languages the equivalent DRA from Thm. 1 has an exponential number of registers and a doubly-exponential number of orbits of states. The same bounds apply to a DRA constructed in our proof of Thm. 4. Moreover, assuming $\textsc{Atoms}$ satisfy standard effectiveness assumptions, like decidability of their first-order theory, one can also compute an equivalent DRA from Thm. 2.

Concerning possible generalisations of our results, we believe that Thm. 1 holds not only for equality atoms, but for arbitrary oligomorphic ω-stable atoms. These include e.g. the nested equality atoms mentioned in Sec. 2. On the other hand Thm. 1 does not extend to disjoint but non-complementing NRA languages: it is not true that for every two disjoint NRA languages there is a DRA language that *separates* them, i.e., includes one of them and is disjoint from the other. The corresponding decision problem (given two disjoint NRA, does a separating DRA exist?) is decidable when the number of registers of a separating automaton is fixed [9], and open in general.

An intriguing open question (not unlike the WQO Dichotomy Conjecture [19]) is whether it is necessary for $\textsc{Atoms}$ to admit WQO for Thm. 2 to hold.

References

1. M. Bojańczyk. Slightly infinite sets. A draft of a book available at https://www.mimuw.edu.pl/~bojan/paper/atom-book.
2. M. Bojanczyk. Data monoids. In *Proc. STACS 2011*, volume 9 of *LIPIcs*, pages 105–116. Schloss Dagstuhl - Leibniz-Zentrum für Informatik, 2011.
3. M. Bojańczyk, C. David, A. Muscholl, T. Schwentick, and L. Segoufin. Two-variable logic on data words. *ACM Trans. Comput. Log.*, 12(4):27:1–27:26, 2011.
4. M. Bojańczyk, B. Klin, and S. Lasota. Automata with group actions. In *Proc. LICS 2011*, pages 355–364, 2011.
5. M. Bojańczyk, B. Klin, and S. Lasota. Automata theory in nominal sets. *Log. Methods Comput. Sci.*, 10(3), 2014.
6. M. Bojańczyk and S. Lasota. An extension of data automata that captures XPath. *Log. Methods Comput. Sci.*, 8(1), 2012.
7. M. Bojańczyk and R. Stefański. Single-use automata and transducers for infinite alphabets. In *Proc. ICALP 2020*, volume 168 of *LIPIcs*, pages 113:1–113:14. Schloss Dagstuhl - Leibniz-Zentrum für Informatik, 2020.
8. L. Clemente and C. Barloy. Bidimensional linear recursive sequences and universality of unambiguous register automata. Submited for publication, 2020.
9. L. Clemente, S. Lasota, and R. Piórkowski. Timed games and deterministic separability. In *Proc. ICALP 2020*, volume 168 of *LIPIcs*, pages 121:1–121:16, 2020.
10. T. Colcombet. Forms of Determinism for Automata. In *STACS'12 (29th Symposium on Theoretical Aspects of Computer Science)*, volume 14, pages 1–23. LIPIcs, 2012.
11. T. Colcombet, C. Ley, and G. Puppis. Logics with rigidly guarded data tests. *Log. Methods Comput. Sci.*, 11(3), 2015.
12. T. Colcombet and A. Manuel. Generalized data automata and fixpoint logic. In *Proc. FSTTCS 2014*, volume 29 of *LIPIcs*, pages 267–278. Schloss Dagstuhl - Leibniz-Zentrum für Informatik, 2014.
13. L. D'Antoni and M. Veanes. Minimization of symbolic automata. In *Proc. POPL '14*, pages 541–554. ACM, 2014.
14. S. Demri and R. Lazic. LTL with the freeze quantifier and register automata. *ACM Trans. Comput. Log.*, 10(3):16:1–16:30, 2009.
15. N. Francez and M. Kaminski. Finite-memory automata. *Theor. Comput. Sci.*, 134(2):329–363, 1994.
16. N. Francez and M. Kaminski. An algebraic characterization of deterministic regular languages over infinite alphabets. *Theor. Comput. Sci.*, 306(1-3):155–175, 2003.
17. M. Kaminski and T. Tan. Regular expressions for languages over infinite alphabets. *Fundam. Informaticae*, 69(3):301–318, 2006.
18. M. Kaminski and D. Zeitlin. Finite-memory automata with non-deterministic reassignment. *Int. J. Found. Comput. Sci.*, 21(5):741–760, 2010.
19. S. Lasota. Decidability border for Petri nets with data: WQO dichotomy conjecture. In *Proc. PETRI NETS 2016*, volume 9698 of *Lecture Notes in Computer Science*, pages 20–36. Springer, 2016.
20. L. Libkin, T. Tan, and D. Vrgoc. Regular expressions for data words. *J. Comput. Syst. Sci.*, 81(7):1278–1297, 2015.
21. T. Milo, D. Suciu, and V. Vianu. Typechecking for XML transformers. *J. Comput. Syst. Sci.*, 66(1):66–97, 2003.
22. A. Mottet and K. Quaas. The containment problem for unambiguous register automata. In *Proc. STACS 2019*, volume 126 of *LIPIcs*, pages 53:1–53:15, 2019.

23. F. Neven, T. Schwentick, and V. Vianu. Finite state machines for strings over infinite alphabets. *ACM Trans. Comput. Log.*, 5(3):403–435, 2004.
24. A. M. Pitts. *Nominal Sets: Names and Symmetry in Computer Science*, volume 57 of *Cambridge Tracts in Theoretical Computer Science*. Cambridge University Press, 2013.
25. H. Sakamoto and D. Ikeda. Intractability of decision problems for finite-memory automata. *Theor. Comput. Sci.*, 231(2):297–308, 2000.
26. L. Segoufin. Automata and logics for words and trees over an infinite alphabet. In *Proc. CSL 2006*, volume 4207 of *Lecture Notes in Computer Science*, pages 41–57. Springer, 2006.

Certifying Inexpressibility*

Orna Kupferman[1] and Salomon Sickert[1,2] (✉)

[1] School of Computer Science and Engineering,
The Hebrew University, Jerusalem, Israel.
orna@cs.huji.ac.il, salomon.sickert@mail.huji.ac.il
[2] Technische Universität München, Munich, Germany.
s.sickert@tum.de

Abstract Different classes of automata on infinite words have different expressive power. Deciding whether a given language $L \subseteq \Sigma^\omega$ can be expressed by an automaton of a desired class can be reduced to deciding a game between Prover and Refuter: in each turn of the game, Refuter provides a letter in Σ, and Prover responds with an annotation of the current state of the run (for example, in the case of Büchi automata, whether the state is accepting or rejecting, and in the case of parity automata, what the color of the state is). Prover wins if the sequence of annotations she generates is correct: it is an accepting run iff the word generated by Refuter is in L. We show how a winning strategy for Refuter can serve as a simple and easy-to-understand certificate to inexpressibility, and how it induces additional forms of certificates. Our framework handles all classes of deterministic automata, including ones with structural restrictions like weak automata. In addition, it can be used for refuting *separation* of two languages by an automaton of the desired class, and for finding automata that *approximate* L and belong to the desired class.

Keywords: Automata on infinite words · Expressive power · Games.

1 Introduction

Finite *automata on infinite objects* were first introduced in the 60's, and were the key to the solution of several fundamental decision problems in mathematics and logic [8,33,41]. Today, automata on infinite objects are used for specification, verification, and synthesis of nonterminating systems. The automata-theoretic approach reduces questions about systems and their specifications to questions about automata [28,49], and is at the heart of many algorithms and tools. Industrial-strength property-specification languages such as the IEEE 1850

* The full version of this article is available from [27]. Orna Kupferman is supported in part by the Israel Science Foundation, grant No. 2357/19. Salomon Sickert is supported in part by the Deutsche Forschungsgemeinschaft (DFG) under project numbers 436811179 and 317422601 ("Verified Model Checkers"), and in part funded by the European Research Council (ERC) under the European Union's Horizon 2020 research and innovation programme under grant agreement No. 787367 (PaVeS).

S. Kiefer and C. Tasson (Eds.): FOSSACS 2021, LNCS 12650, pp. 385–405, 2021.
https://doi.org/10.1007/978-3-030-71995-1_20

Standard for Property Specification Language (PSL) [14] include regular expressions and/or automata, making specification and verification tools that are based on automata even more essential and popular.

A run r of an automaton on infinite words is an infinite sequence of states, and acceptance is determined with respect to the set of states that r visits infinitely often. For example, in *Büchi* automata, some of the states are designated as accepting states, denoted by α, and a run is accepting iff it visits states from the accepting set α infinitely often [8]. Dually, in *co-Büchi* automata, a run is accepting if it visits the set α only finitely often. Then, in *parity* automata, the acceptance condition maps each state to a color in some set $C = \{j, \ldots, k\}$, for $j \in \{0, 1\}$ and some *index* $k \geq 0$, and a run is accepting if the maximal color it visits infinitely often is odd.

The different classes of automata have different *expressive power*. For example, while deterministic parity automata can recognize all ω-regular languages, deterministic Büchi automata cannot [29]. We use DBW, DCW, and DPW to denote a deterministic Büchi, co-Büchi, and parity word automaton, respectively, or (this would be clear from the context) the set of languages recognizable by the automata in the corresponding class. There has been extensive research on expressiveness of automata on infinite words [48,20]. In particular, researchers have studied two natural expressiveness hierarchies induced by different classes of deterministic automata. The first hierarchy is the *Mostowski Hierarchy*, induced by the index of parity automata [35,50]. Formally, let DPW$[0, k]$ denote a DPW with $C = \{0, \ldots, k\}$, and similarly for DPW$[1, k]$ and $C = \{1, \ldots, k\}$. Clearly, DPW$[0, k] \subseteq$ DPW$[0, k + 1]$, and similarly DPW$[1, k] \subseteq$ DPW$[1, k + 1]$. The hierarchy is infinite and strict. Moreover, DPW$[0, k]$ complements DPW$[1, k+1]$, and for every $k \geq 0$, there are languages L_k and L'_k such that $L_k \in$ DPW$[0, k] \setminus$ DPW$[1, k + 1]$ and $L'_k \in$ DPW$[1, k + 1] \setminus$ DPW$[0, k]$. At the bottom of this hierarchy, we have DBW and DCW. Indeed, DBW=DPW$[0, 1]$ and DCW=DPW$[1, 2]$.

While the Mostowski Hierarchy refines DPWs, the second hierarchy, which we term the *depth hierarchy*, refines deterministic *weak* automata (DWWs). Weak automata can be viewed as a special case of Büchi or co-Büchi automata in which every strongly connected component in the graph induced by the structure of the automaton is either contained in α or is disjoint from α, where α is depending on the acceptance condition the set of accepting or rejecting states. The structure of weak automata captures the alternation between greatest and least fixed points in many temporal logics, and they were introduced in this context in [36]. DWWs have been used to represent vectors of real numbers [6], and they have many appealing theoretical and practical properties [32,21]. In terms of expressive power, DWW $=$ DCW $\cap$ DBW.

The depth hierarchy is induced by the depth of alternation between accepting and rejecting components in DWWs. For this, we view a DWW as a DPW in which the colors visited along a run can only increase. Accordingly, each run eventually gets trapped in a single color, and is accepting iff this color is odd. We use DWW$[0, k]$ and DWW$[1, k]$ to denote weak-DPW$[0, k]$ and weak-

DPW[1, k], respectively. The picture obtained for the depth hierarchy is identical to that of the Mostowski hierarchy, with DWW[j, k] replacing DPW[j, k] [50]. At the bottom of the depth hierarchy we have *co-safety* and *safety* languages [2]. Indeed, co-safety languages are DWW[0, 1] and safety are DWW[1, 2].

Beyond the theoretical interest in expressiveness hierarchies, their study is motivated by the fact many algorithms, like synthesis and probabilistic model checking, need to operate on deterministic automata [5,3]. The lower the automata are in the expressiveness hierarchy, the simpler are algorithms for reasoning about them. Simplicity goes beyond complexity, which typically depends on the parity index [16], and involves important practical considerations like minimization and canonicity (exists only for DWWs [32]), circumvention of Safra's determinization [26], and symbolic implementations [47]. Of special interest is the characterization of DBWs. For example, it is shown in [25] that given a *linear temporal logic* formula ψ, there is an *alternation-free μ-calculus* formula equivalent to $\forall\psi$ iff ψ can be recognized by a DBW. Further research studies *typeness* for deterministic automata, examining the ability to define a weaker acceptance condition on top of a given automaton [19,21].

Our goal in this paper is to provide a simple and easy-to-understand explanation to inexpressibility results. The need to accompany results of decision procedures by an explanation (often termed "certificate") is not new, and includes certification of a "correct" decision of a model checker [24,44], reachability certificates in complex multi-agent systems [1], and explainable reactive synthesis [4]. To the best of our knowledge, our work is the first to provide certification to inexpressibility results.

The underlying idea is simple: Consider a language L and a class γ of deterministic automata. We consider a turn-based two-player game in which one player (Refuter) provides letters in Σ, and the second player (Prover) responds with letters from a set A of annotations that describe states in a deterministic automaton. For example, when we consider a DBW, then $A = \{\text{ACC}, \text{REJ}\}$, and when we consider a DPW[0, k], then $A = \{0, \ldots, k\}$. Thus, during the interaction, Refuter generates a word $x \in \Sigma^\omega$ and Prover responds with a word $y \in A^\omega$. Prover wins if for all words $x \in \Sigma^\omega$, we have that $x \in L$ iff y is accepting according to γ. Clearly, if there is a deterministic γ automaton for L, then Prover can win by following its run on x. Dually, a finite-state winning strategy for Prover induces a deterministic γ automaton for L. The game-based approach is not new, and has been used for deciding the membership of given ω-regular languages in different classes of deterministic automata [26]. Further, the game-based formulation is used in descriptive set theory to classify sets into hierarchies, see for example [39, Chapters 4 and 5] for an introduction that focuses on ω-regular languages. Our contribution is a study of strategies for Refuter. Indeed, since the above described game is determined [9] and the strategies are finite-state, Refuter has a winning strategy iff no deterministic γ automaton for L exists, and this winning strategy can serve as a certificate for inexpressibility.

Example 1. Consider the language $L_{\neg\infty a} \subseteq \{a, b\}^\omega$ of all words with only finitely many a's. It is well known that L cannot be recognized by a DBW [29]. In Fig-

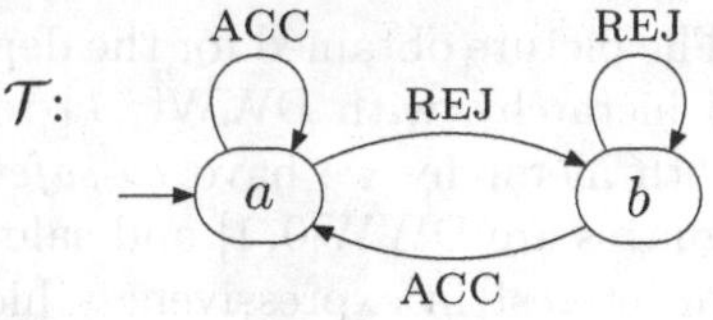

Figure 1. A refuter for DBW-recognizability of "only finitely many a's".

ure 1 we describe what we believe to be the neatest proof of this fact. The figure describes a transducer $\mathcal{R}$ with inputs in {ACC,REJ} and outputs in $\{a,b\}$ – the winning strategy of Refuter in the above described game. The way to interpret $\mathcal{R}$ is as follows. In each round of the game, Prover tells Refuter whether the run of her DBW for $L_{\neg\infty a}$ is in an accepting or a rejecting state, and Refuter uses $\mathcal{R}$ in order to respond with the next letter in the input word. For example, if Prover starts with ACC, namely declaring that the initial state of her DBW is accepting, then Refuter responds with a, and if Prover continues with REJ, namely declaring that the state reachable with a is rejecting, then Refuter responds with b. If Prover continues with REJ forever, then Prover continues with b forever. Thus, together Prover and Refuter generate two words: $y \in \{\text{ACC},\text{REJ}\}^\omega$ and $x \in \{a,b\}^\omega$. Prover wins whenever $x \in L_{\neg\infty a}$ iff y contains infinitely many ACC's. If Prover indeed has a DBW for $L_{\neg\infty a}$, then she can follow its transition function and win the game. By following the refuter $\mathcal{R}$, however, Refuter can always fool Prover and generate a word x such that $x \in L_{\neg\infty a}$ iff y contains only finitely many ACC's. ∎

We first define refuters for DBW-recognizability, and study their construction and size for languages given by deterministic or nondeterministic automata. Our refuters serve as a first inexpressibility certificate. We continue and argue that each DBW-refuter for a language L induces three words $x \in \Sigma^*$ and $x_1, x_2 \in \Sigma^*$, such that $x \cdot (x_1 + x_2)^* \cdot x_1^\omega \subseteq L$ and $x \cdot (x_1^* \cdot x_2)^\omega \cap L = \emptyset$. The triple $\langle x, x_1, x_2 \rangle$ is an additional certificate for L not being in DBW. Indeed, we show that a language L is not in DBW iff it has a certificate as above. For example, the language $L_{\neg\infty a}$ has a certificate $\langle \epsilon, b, a \rangle$. In fact, we show that Landweber's proof for $L_{\neg\infty a}$ can be used as is for all languages not in DBW, with x_1 replacing b, x_2 replacing a, and adding x as a prefix.

We then generalize our results on DBW-refutation and certification in two orthogonal directions. The first is an extension to richer classes of deterministic automata, in particular all classes in the two hierarchies discussed above, as well as all deterministic Emerson-Lei automata (DELWs) [17]. For the depth hierarchy, we add to the winning condition of the game a *structural restriction*. For example, in a weak automaton, Prover loses if the sequence $y \in A^\omega$ of annotations she generates includes infinitely many alternations between ACC and REJ. We show how structural restrictions can be easily expressed in our framework.

The second direction is an extension of the recognizability question to the questions of *separation* and *approximation*: We say that a language $L \subseteq \Sigma^\omega$ is a *separator* for two languages $L_1, L_2 \subseteq \Sigma^\omega$ if $L_1 \subseteq L$ and $L \cap L_2 = \emptyset$. Studies of separation include a search for regular separators of general languages [11], as well as separation of regular languages by weaker classes of languages, e.g., FO-definable languages [40] or piecewise testable languages [12]. In the context of ω-regular languages, [2] presents an algorithm computing the smallest safety language containing a given language L_1, thus finding a safety separator for L_1 and L_2. As far as we know, besides this result there has been no systematic study of separation of ω-regular languages by deterministic automata.

In addition to the interest in separators, we use them in the context of recognizability in two ways. First, a third type of certificate that we suggest for DBW-refutation of a language L are "simple" languages L_1 and L_2 such that $L_1 \subseteq L$, $L \cap L_2 = \emptyset$, and $\langle L_1, L_2 \rangle$ are not DBW-separable. Second, we use separability in order to approximate languages that are not in DBW. Consider such a language $L \subseteq \Sigma^\omega$. A user may be willing to approximate L in order to obtain DBW-recognizability. Specifically, we assume that there are languages $I_\downarrow \subseteq L$ and $I_\uparrow \subseteq \Sigma^\omega \setminus L$ of words that the user is willing to under- and over-approximate L with. Thus, the user searches for a language that is a separator for $L \setminus I_\downarrow$ and $\Sigma^\omega \setminus (L \cup I_\uparrow)$. We study DBW-separability and DBW-approximation, namely separability and approximation by languages in DBW. In particular, we are interested in finding "small" approximating languages $I_\downarrow$ and $I_\uparrow$ with which L has a DBW-approximation, and we show how certificates that refute DBW-separation can direct the search to for successful $I_\downarrow$ and $I_\uparrow$. Essentially, as in *counterexample guided abstraction-refinement* (CEGAR) for model checking [10], we use certificates for non-DBW-separability in order to suggest interesting *radius languages*. While in CEGAR the refined system excludes the counterexample, in our setting the approximation of L excludes the certificate. As has been the case with recognizability, we extend our results to all classes of deterministic automata.

2 Preliminaries

2.1 Transducers and Realizability

Consider two finite alphabets Σ and A. It is convenient to think about Σ as the "main" alphabet, and about A as an alphabet of annotations. For two words $x = x_0 \cdot x_1 \cdot x_2 \cdots \in \Sigma^\omega$ and $y = y_0 \cdot y_1 \cdot y_2 \cdots \in A^\omega$, we define $x \oplus y$ as the word in $(\Sigma \times A)^\omega$ obtained by merging x and y. Thus, $x \oplus y = (x_0, y_0) \cdot (x_1, y_1) \cdot (x_2, y_2) \cdots$.

A (Σ/A)-*transducer* models a finite-state system that responds with letters in A while interacting with an environment that generates letters in Σ. Formally, a (Σ/A)-transducer is $\mathcal{T} = \langle \Sigma, A, \iota, S, s_0, \rho, \tau \rangle$, where $\iota \in \{sys, env\}$ indicates who initiates the interaction – the system or the environment, S is a set of states, $s_0 \in S$ is an initial state, $\rho : S \times \Sigma \to S$ is a transition function, and $\tau : S \to A$ is a labelling function on the states. Consider an input word $x = x_0 \cdot x_1 \cdot x_2 \cdots \in \Sigma^\omega$. The *run* of $\mathcal{T}$ on x is the sequence $s_0, s_1, s_2 \ldots$ such that for all $j \geq 0$, we have that $s_{j+1} = \rho(s_j, x_j)$. The *annotation of x by $\mathcal{T}$*, denoted $\mathcal{T}(x)$, depends on ι. If $\iota = sys$, then $\mathcal{T}(x) = \tau(s_0) \cdot \tau(s_1) \cdot \tau(s_2) \cdots \in A^\omega$. Note that the first letter in A is the output of $\mathcal{T}$ in s_0. This reflects the fact that the system initiates the interaction. If $\iota = env$, then $\mathcal{T}(x) = \tau(s_1) \cdot \tau(s_2) \cdot \tau(s_3) \cdots \in A^\omega$. Note that now, the output in s_0 is ignored, reflecting the fact that the environment initiates the interaction.

Consider a language $L \subseteq (\Sigma \times A)^\omega$. Let $comp(L)$ denote the complement of L. Thus, $comp(L) = (\Sigma \times A)^\omega \setminus L$. We say that a language $L \subseteq (\Sigma \times A)^\omega$ is *(Σ/A)-realizable by the system* if there is a (Σ/A)-transducer $\mathcal{T}$ with $\iota = sys$ such that for every word $x \in \Sigma^\omega$, we have that $x \oplus \mathcal{T}(x) \in L$. Then, L is *(A/Σ)-realizable by the environment* if there is an (A/Σ)-transducer $\mathcal{T}$ with $i = env$ such that for every word $y \in A^\omega$, we have that $\mathcal{T}(y) \oplus y \in L$. When the language L is regular, realizability reduces to deciding a game with a regular winning condition. Then, by determinacy of games and due to the existence of finite-memory winning strategies [9], we have the following.

Proposition 1. *For every ω-regular language $L \subseteq (\Sigma \times A)^\omega$, exactly one of the following holds.*

1. *L is (Σ/A)-realizable by the system.*
2. *$comp(L)$ is (A/Σ)-realizable by the environment.*

2.2 Automata

A *deterministic word automaton* over a finite alphabet Σ is $\mathcal{A} = \langle \Sigma, Q, q_0, \delta, \alpha \rangle$, where Q is a set of states, $q_0 \in Q$ is an initial state, $\delta : Q \times \Sigma \to Q$ is a transition function, and α is an acceptance condition. We extend δ to words in Σ^* in the expected way, thus for $q \in Q$, $w \in \Sigma^*$, and letter $\sigma \in \Sigma$, we have that $\delta(q, \epsilon) = q$ and $\delta(q, w\sigma) = \delta(\delta(q, w), \sigma)$. A *run* of $\mathcal{A}$ on an infinite word $\sigma_0, \sigma_1, \dots \in \Sigma^\omega$ is the sequence of states $r = q_0, q_1, \dots$, where for every position $i \geq 0$, we have that $q_{i+1} = \delta(q_i, \sigma_i)$. We use $inf(r)$ to denote the set of states that r visits infinitely often. Thus, $inf(r) = \{q : q_i = q \text{ for infinitely many } i \geq 0\}$.

The acceptance condition α refers to $inf(r)$ and determines whether the run r is accepting. For example, in the *Büchi*, acceptance condition, we have that $\alpha \subseteq Q$, and a run is accepting iff it visits states in α infinitely often; that is, $\alpha \cap inf(r) \neq \emptyset$. Dually, in *co-Büchi*, $\alpha \subseteq Q$, and a run is accepting iff it visits states in α only finitely often; that is, $\alpha \cap inf(r) = \emptyset$. The language of $\mathcal{A}$, denoted $L(\mathcal{A})$, is then the set of words w such that the run of $\mathcal{A}$ on w is accepting.

A parity condition is $\alpha : Q \to \{0, \dots, k\}$, for $k \geq 0$, termed the *index* of α. A run r satisfies α iff the maximal color $i \in \{0, \dots, k\}$ such that $\alpha^{-1}(i) \cap inf(r) \neq \emptyset$ is odd. That is, r is accepting iff the maximal color that r visits infinitely often is odd. Then, a Rabin condition is $\alpha = \{\langle G_1, B_1 \rangle, \dots, \langle G_k, B_k \rangle\}$, with $G_i, B_i \subseteq Q$, for all $0 \leq i \leq k$. A run r satisfies α iff there is $1 \leq i \leq k$ such that $inf(r) \cap G_i \neq \emptyset$ and $inf(r) \cap B_i = \emptyset$. Thus, there is a pair $\langle G_i, B_i \rangle$ such that r visits states in G_i infinitely often and visits states in B_i only finitely often.

All the acceptance conditions above can be viewed as special cases of the *Emerson-Lei acceptance condition* (EL-condition, for short) [17], which we define below. Let $\mathbb{M}$ be a finite set of marks. Given an infinite sequence $\pi = M_0 \cdot M_1 \cdots \in (2^{\mathbb{M}})^\omega$ of subsets of marks, let $inf(\pi)$ be the set of marks that appear infinitely often in sets in π. Thus, $inf(\pi) = \{m \in \mathbb{M} : \text{there exist infinitely many } i \geq 0 \text{ such that } m \in M_i\}$. An EL-condition is a Boolean assertion over atoms in $\mathbb{M}$. For simplicity, we consider assertions in positive normal form, where negation is applied only to atoms. Intuitively, marks that appear positively should repeat infinitely often and marks that appear negatively should repeat only finitely often. Formally, a deterministic EL-automaton is $\mathcal{A} = \langle \Sigma, Q, q_0, \delta, \mathbb{M}, \tau, \theta \rangle$, where $\tau : Q \to 2^{\mathbb{M}}$ maps each state to a set of marks, and θ is an EL-condition over $\mathbb{M}$. A run r of a $\mathcal{A}$ is accepting if $inf(\tau(r))$ satisfies θ.

For example, a Büchi condition $\alpha \subseteq Q$ can be viewed as an EL-condition with $\mathbb{M} = \{\text{ACC}\}$ and $\tau(q) = \{\text{ACC}\}$ for $q \in \alpha$ and $\tau(q) = \emptyset$ for $q \notin \alpha$. Then, the assertion $\theta = \text{ACC}$ is satisfied by sequences π induced by runs r with $inf(r) \cap \alpha \neq \emptyset$. Dually, the assertion $\theta = \neg\text{REJ}$ with $\mathbb{M} = \{\text{REJ}\}$ is satisfied by sequences π induced by runs r with $inf(r) \cap \alpha = \emptyset$, and thus corresponds to a co-Büchi condition. In the case of a parity condition $\alpha : Q \to \{0, \ldots, k\}$, it is not hard to see that α is equivalent to an EL-condition in which $\mathbb{M} = \{0, 1, \ldots, k\}$, for every state $q \in Q$, we have that $\tau(q) = \{\alpha(q)\}$, and θ expresses the parity condition. Lastly, a Rabin condition $\alpha = \{\langle G_1, B_1 \rangle, \ldots, \langle G_k, B_k \rangle\}$ is equivalent to an EL-condition with $\mathbb{M} = \{G_1, B_1, \ldots, G_k, B_k\}$ and $\tau(q) = \{m \in \mathbb{M} : q \in m\}$. Note that now, the mapping τ is not to singletons, and each state is marked by all sets in α in which it is a member. Then, $\theta = \bigvee_{1 \leq i \leq k} (G_i \wedge \neg B_i)$.

We use DBW, DCW, DPW, DRW, DELW to denote deterministic Büchi, co-Büchi, parity, Rabin, and EL word automata, respectively. For parity automata, we also use DPW$[0, k]$ and DPW$[1, k]$, for $k \geq 0$, to denote DPWs in which the colours are in $\{0, \ldots, k\}$ and $\{1, \ldots, k\}$, respectively. For Rabin automata, we use DRW$[k]$, for $k \geq 0$, to denote DRWs that have at most k elements in α. Finally, we use DELW$[\theta]$, to denote DELWs with EL-condition θ. We sometimes use the above acronyms in order to refer to the set of languages that are recognizable by the corresponding class of automata. For example, we say that a language L is in DBW if L is *DBW-recognizable*, thus there is a DBW $\mathcal{A}$ such that $L = L(\mathcal{A})$. Note that DBW = DPW$[0, 1]$, DCW = DPW$[1, 2]$, and DRW$[1]$ = DPW$[0, 2]$. In fact, in terms of expressiveness, DRW$[k]$ = DPW$[0, 2k]$ [43,31].

Consider a directed graph $G = \langle V, E \rangle$. A *strongly connected set* of G (SCS) is a set $C \subseteq V$ of vertices such that for every two vertices $v, v' \in C$, there is a path from v to v'. An SCS C is *maximal* if it cannot be extended to a larger SCS. Formally, for every nonempty $C' \subseteq V \setminus C$, we have that $C \cup C'$ is not an SCS. The maximal strongly connected sets are also termed *strongly connected components* (SCC). An automaton $\mathcal{A} = \langle \Sigma, Q, Q_0, \delta, \alpha \rangle$ induces a directed graph $G_{\mathcal{A}} = \langle Q, E \rangle$ in which $\langle q, q' \rangle \in E$ iff there is a letter σ such that $q' \in \delta(q, \sigma)$. When we talk about the SCSs and SCCs of $\mathcal{A}$, we refer to those of $G_{\mathcal{A}}$. Consider a run r of an automaton $\mathcal{A}$. It is not hard to see that the set $inf(r)$ is an SCS. Indeed, since every two states q and q' in $inf(r)$ are visited infinitely often, the state q' must be reachable from q.

3 Refuting DBW-Recognizability

Let $A = \{\text{ACC}, \text{REJ}\}$. We use ∞ACC to denote the subset $\{a_0 \cdot a_1 \cdot a_2 \cdots \in A^\omega :$ there are infinitely many $j \geq 0$ with $a_j = \text{ACC}\}$ and $\neg\infty\text{ACC} = comp(\infty\text{ACC}) = \{a_0 \cdot a_1 \cdot a_2 \cdots \in A^\omega :$ there are only finitely many $j \geq 0$ with $a_j = \text{ACC}\}$.

A DBW $\mathcal{A} = \langle \Sigma, Q, q_0, \delta, \alpha \rangle$ can be viewed as a (Σ/A)-transducer $\mathcal{T}_\mathcal{A} = \langle \Sigma, A, sys, Q, q_0, \delta, \tau \rangle$, where for every state $q \in Q$, we have that $\tau(q) = \text{ACC}$ if $q \in \alpha$, and $\tau(q) = \text{REJ}$ otherwise. Then, for every word $x \in \Sigma^\omega$, we have that $x \in L(\mathcal{A})$ iff $\mathcal{T}_\mathcal{A}(x) \in \infty\text{ACC}$.

For a language $L \subseteq \Sigma^\omega$, we define the language $\text{DBW}(L) \subseteq (\Sigma \times A)^\omega$ of words with correct annotations. Thus,

$$\text{DBW}(L) = \{x \oplus y : x \in L \text{ iff } y \in \infty\text{ACC}\}.$$

Note that $comp(\text{DBW}(L))$ is the language

$$\text{NoDBW}(L) = \{x \oplus y : (x \in L \text{ and } y \notin \infty\text{ACC}) \text{ or } (x \notin L \text{ and } y \in \infty\text{ACC})\}.$$

A *DBW-refuter for L* is an (A/Σ)-transducer with $\iota = env$ realizing $\text{NoDBW}(L)$.

Example 2. For every language $R \subseteq \Sigma^*$ of finite words, the language $R^\omega \subseteq \Sigma^\omega$ consists of infinite concatenations of words in R. It was recently shown that R^ω may not be in DBW [30]. The language used in [30] is $R = \$ + (0 \cdot \{0, 1, \$\}^* \cdot 1)$. In Figure 2 below we describe a DBW-refuter for R^ω.

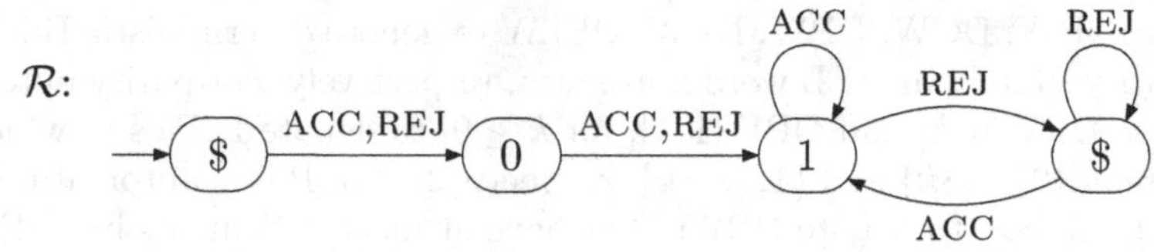

Figure 2. A DBW-refuter for $(\$ + (0 \cdot \{0, 1, \$\}^* \cdot 1))^\omega$.

Following $\mathcal{R}$, Refuter starts by generating a prefix $0 \cdot 1$ and then responds to ACC with 1 and responds with \$ to REJ. Accordingly, if Prover generates a rejecting run, Prover generates a word in $0 \cdot 1 \cdot (1 + \$)^* \cdot \$^\omega$, which is in R^ω. Also, if Prover generates an accepting run, Prover generates a word in $0 \cdot 1 \cdot (1^+ \cdot \$^*)^\omega$, which has a single 0 and infinitely many 1's, and is therefore not in R^ω. ■

By Proposition 1, we have the following.

Proposition 2. *Consider a language* $L \subseteq \Sigma^\omega$. *Let* $A = \{\text{ACC}, \text{REJ}\}$. *Exactly one of the following holds:*

- *L is in DBW, in which case the language* $\text{DBW}(L)$ *is* (Σ/A)*-realizable by the system, and a finite-memory winning strategy for the system induces a DBW for L.*
- *L is not in DBW, in which case the language* $\text{NoDBW}(L)$ *is* (A/Σ)*-realizable by the environment, and a finite-memory winning strategy for the environment induces a DBW-refuter for L.*

3.1 Complexity

In this section we analyze the size of refuters. We start with the case where the language L is given by a DPW.

Theorem 1. *Consider a DPW $\mathcal{A}$ with n states. Let $L = L(\mathcal{A})$. One of the following holds.*

1. *There is a DBW for L with n states.*
2. *There is a DBW-refuter for L with $2n$ states.*

Proof. If L is in DBW, then, as DPWs are Büchi type [19], a DBW for L can be defined on top of the structure of $\mathcal{A}$, and so it has n states. If L is not in DBW, then by Proposition 2, there is a DBW-refuter for L, namely a $(\{\text{ACC}, \text{REJ}\}/\Sigma)$-transducer that realizes NoDBW(L). We show we can define a DRW $\mathcal{U}$ with $2n$ states for NoDBW(L). The result then follows from the fact a realizable DRW is realized by a transducer of the same size as the DRW [15].

We construct $\mathcal{U}$ by taking the union of the acceptance conditions of a DRW $\mathcal{U}_1$ for $\{x \oplus y : x \in L \text{ and } y \notin \infty\text{ACC}\}$ and a DRW $\mathcal{U}_2$ for $\{x \oplus y : x \notin L \text{ and } y \in \infty\text{ACC}\}$. We obtain both DRWs by taking the product of $\mathcal{A}$, extended to the alphabet $\Sigma \times \{\text{ACC}, \text{REJ}\}$, with a 2-state automaton for ∞ACC, again extended to the alphabet $\Sigma \times \{\text{ACC}, \text{REJ}\}$.

We describe the construction in detail. Let $\mathcal{A} = \langle \Sigma, Q, q_0, \delta, \alpha \rangle$. Then, the state space of $\mathcal{U}_1$ is $Q \times \{\text{ACC}, \text{REJ}\}$ and its transition on a letter $\langle \sigma, a \rangle$ follows δ when it reads σ, with a determining whether $\mathcal{U}_1$ moves to the ACC or REJ copy. Let α_1 be the Rabin condition equivalent to α. We obtain the acceptance condition of $\mathcal{U}_1$ by replacing each pair $\langle G, B \rangle$ in α_1 by $\langle G \times \{\text{REJ}\}, B \times \{\text{REJ}\} \cup Q \times \{\text{ACC}\} \rangle$. It is not hard to see that a run of $\mathcal{U}_1$ satisfies the latter pair iff its projection on Q satisfies the pair $\langle G, B \rangle$ and its projection on $\{\text{ACC}, \text{REJ}\}$ has only finitely many ACC. The construction of $\mathcal{U}_2$ is similar, with α_2 being a Rabin condition that complements α, and then replacing each pair $\langle G, B \rangle$ in α_2 by $\langle G \times \{\text{ACC}\}, B \times \{\text{ACC}, \text{REJ}\} \rangle$. Since $\mathcal{U}_1$ and $\mathcal{U}_2$ have the same state space, and we only have to take the union of the pairs in their acceptance conditions, the $2n$ bound follows. □

Now, when L is given by an NBW, an exponential bound follows from the exponential blow up in determinization [42]. If we are also given an NBW for *comp*(L), the complexity can be tightened. Formally, we have the following.

Theorem 2. *Given NBWs with n and m states, for L and comp(L), respectively, one of the following holds.*

1. *There is a DBW for L with* $\min\{(1.65n)^n, 3^m\}$ *states.*
2. *There is a DBW-refuter for L with* $\min\{2 \cdot (1.65n)^n, 2 \cdot (1.65m)^m\}$ *states.*

Proof. If L is in DBW, then a DBW for L can be defined on top of a DPW for L, which has at most $(1.65n)^n$ states [45], or by dualizing a DCW for *comp*(L). Since the translation of an NBW with m states to a DCW, when it exists, results in a DCW with 3^m states [7], we are done. If L is not in DBW, then we proceed as in the proof of Theorem 1, defining $\mathcal{U}$ on the top of a DPW for either L or *comp*(L). □

3.2 Certifying DBW-Refutation

Consider a DBW-refuter $\mathcal{R} = \langle\{\text{ACC}, \text{REJ}\}, \Sigma, env, S, s_0, \rho, \tau\rangle$. We say that a path $s_0, \ldots, s_m$ in $\mathcal{R}$ is an REJ^+-path if it contains at least one transition and all the transitions along it are labeled by REJ; thus, for all $0 \leq j < m$, we have that $s_{j+1} = \rho(s_j, \text{REJ})$. Then, a path $s_0, \ldots, s_m$ in $\mathcal{R}$ is an ACC-path if it contains at least one transition and its first transition is labeled by ACC. Thus, $s_1 = \rho(s_0, \text{ACC})$.

Lemma 1. *Consider a DBW-refuter* $\mathcal{R} = \langle\{\text{ACC}, \text{REJ}\}, \Sigma, env, S, s_0, \rho, \tau\rangle$. *Then there exists a state* $s \in S$, *a (possibly empty) path* $p = s_0, s_1, \ldots s_m$, *a* REJ^+*-cycle* $p_1 = s_0^1, s_1^1 \ldots s_{m_1}^1$, *and an* ACC*-cycle* $p_2 = s_0^2, s_1^2 \ldots s_{m_2}^2$, *such that* $s_m = s_0^1 = s_{m_1}^1 = s_0^2 = s_{m_2}^2 = s$.

Proof. Let $s_i \in S$ be a reachable state that belongs to an ergodic component in the graph of $\mathcal{R}$ (that is, $s_i \in C$, for a set C of strongly connected states that can reach only states in C). Since $\mathcal{R}$ is responsive, in the sense it can read in each round both ACC and REJ, we can read from s_i the input sequence REJ^ω. Hence, $\mathcal{R}$ has a REJ^+-path $s_i, \ldots, s_l, \ldots, s_k$ with $s_l = s_k$, for $l < k$. It is easy to see that the claim holds with $s = s_l$. In particular, since $\mathcal{R}$ is responsive and C is strongly connected, there exists an ACC-cycle from s_l to itself. □

Theorem 3. *An* ω*-regular language* L *is not in DBW iff there exist three finite words* $x \in \Sigma^*$ *and* $x_1, x_2 \in \Sigma^+$, *such that* $x \cdot (x_1 + x_2)^* \cdot x_1^\omega \subseteq L$ *and* $x \cdot (x_1^* \cdot x_2)^\omega \cap L = \emptyset$.

Proof. Assume first that L is not in DBW. Then, by Theorem 2, there exists a DBW-refuter $\mathcal{R}$ for it. Let $p = s_0, s_1, \ldots s_m$, $p_1 = s_0^1, s_1^1, \ldots, s_{m_1}^1$, and $p_2 = s_0^2, s_1^2, \ldots, s_{m_2}^2$, be the path, REJ^+-cycle, and ACC-cycle that are guaranteed to exist by Lemma 1. Let x, x_1, and x_2 be the outputs that $\mathcal{R}$ generates along them. Formally, $x = \tau(s_1) \cdot \tau(s_2) \cdots \tau(s_m)$, $x_1 = \tau(s_1^1) \cdot \tau(s_2^1) \cdots \tau(s_{m_1}^1)$, and $x_2 = \tau(s_1^2) \cdot \tau(s_1^2) \cdots \tau(s_{m_2}^2)$. Note that as the environment initiates the interaction, the first letter in the words x, x_1, and x_2, are the outputs in the second states in p, p_1, and p_2. The final step, i.e., that x, x_1, and x_2 satisfy the two conditions of the theorem, can be found in the full version of this article [27].

For the other direction, we adjust Landweber's proof [29] for the non-DBW-recognizability of $\neg\infty a$ to L. Essentially, $\neg\infty a$ can be viewed as a special case of $x \cdot (x_1 + x_2)^* \cdot x_1^\omega$, with $x = \epsilon$, $x_1 = b$, and $x_2 = a$. Assume by way of contradiction that there is a DBW $\mathcal{A}$ with $L(\mathcal{A}) = L$. Let $\mathcal{A} = \langle \Sigma, Q, q_0, \delta, \alpha\rangle$. Consider the infinite word $w_0 = x \cdot x_1^\omega$. Since $w_0 \in x \cdot (x_1 + x_2)^* \cdot x_1^\omega$, and so $w \in L$, the run of $\mathcal{A}$ on w_0 is accepting. Thus, there is $i_1 \geq 0$ such that $\mathcal{A}$ visits α when it reads the x_1 suffix of $x \cdot x_1^{i_1}$. Consider now the infinite word $w_1 = x \cdot x_1^{i_1} \cdot x_2 \cdot x_1^\omega$. Since w_1 is also in L, the run of $\mathcal{A}$ on w_1 is accepting. Thus, there is $i_2 \geq 0$ such that $\mathcal{A}$ visits α when it reads the x_1 suffix of $x \cdot x_1^{i_1} \cdot x_2 \cdot x_1^{i_2}$. In a similar fashion we can continue to find indices $i_1, i_2, \ldots$ such for all $j \geq 1$, we have that $\mathcal{A}$ visits α when it reads the x_1 suffix of $x \cdot x_1^{i_1} \cdot x_2 \cdot x_1^{i_2} \cdot x_2 \cdots x_2 \cdot x_1^{i_j}$. Since Q is finite, we can construct a word $w \in x \cdot (x_1^* \cdot x_2)^\omega$ that is accepted, but we assumed that

$x \cdot (x_1^* \cdot x_2)^\omega \cap L = \emptyset$, and thus we have reached a contradiction. The details of this step are given in [27]. □

We refer to a triple $\langle x, x_1, x_2 \rangle$ of words that satisfy the conditions in Theorem 3 as a *certificate* to the non-DBW-recognizability of L.

Example 3. In Example 2, we described a DBW-refuter for $L = (\$+(0\cdot\{0,1,\$\}^*\cdot 1))^\omega$. A certificate to its non-DBW-recognizability is $\langle x, x_1, x_2 \rangle$, with $x = 01$, $x_1 = \$$, and $x_2 = 1$. Indeed, $01 \cdot (\$ + 1)^* \cdot \$^\omega \subseteq L$ and $01 \cdot (\$^* \cdot 1)^\omega \cap L = \emptyset$. ■

Note that obtaining certificates according to the proof of Theorem 3 may not give us the shortest certificate. For example, for L in Example 3, the proof would give us $x = 01\$$, $x_1 = \$$, and $x_2 = 1\$$, with $01\$ \cdot (\$ + 1\$)^* \cdot \$^\omega \subseteq L$ and $01\$ \cdot (\$^* \cdot 1\$)^\omega \cap L = \emptyset$. The problem of generating smallest certificates is related to the problem of finding smallest witnesses to DBW non-emptiness [22] and is harder. Formally, defining the length of a certificate $\langle x, x_1, x_2 \rangle$ as $|x|+|x_1|+|x_2|$, we have the following (see proof in [27]):

Theorem 4. *Consider a DPW $\mathcal{A}$ and a threshold $l \geq 1$. The problem of deciding whether there is a certificate of length at most l for non-DBW-recognizability of $L(\mathcal{A})$ is NP-complete, for l given in unary or binary.*

Remark 1. **[Relation with existing characterizations]** By [29], the language of a DPW $\mathcal{A} = \langle \Sigma, Q, q_0, \delta, \alpha \rangle$ is in DBW iff for every accepting SCS $C \subseteq Q$ and SCS $C' \supseteq C$, we have that C' is accepting. The proof of Landweber relies on a complicated analysis of the structural properties of $\mathcal{A}$. As we elaborate in the full version [27], Theorem 3, which relies instead on determinacy of games, suggests an alternative proof. Similarly, [50] examines the structure of a deterministic Muller automaton, and Theorem 3 can be viewed as a special case of Lemma 14 there, with a proof based on the game setting.

■

Being an (A/Σ)-transducer, every DBW-refuter $\mathcal{R}$ is responsive and may generate many different words in Σ^ω. Below we show that we can leave $\mathcal{R}$ responsive and yet let it generate only words induced by a certificate. Formally, we have the following.

Lemma 2. *Given a certificate $\langle x, x_1, x_2 \rangle$ to non-DBW-recognizability of a language $L \subseteq \Sigma^\omega$, we can define a refuter $\mathcal{R}$ for L such that for every $y \in A^\omega$, if $y \models \infty\text{ACC}$, then $\mathcal{R}(y) \in x \cdot (x_1^* \cdot x_2)^\omega$, and if $y \models \neg\infty\text{ACC}$, then $\mathcal{R}(y) \in x \cdot (x_1 + x_2)^* \cdot x_1^\omega$.*

Proof. Intuitively, $\mathcal{R}$ first ignores the inputs and outputs x. It then repeatedly outputs either x_1 or x_2, according to the following policy: in the first iteration, $\mathcal{R}$ outputs x_1. If during the output of x_1 all inputs are REJ, then $\mathcal{R}$ outputs x_1 also in the next iteration. If an input ACC has been detected, thus the prover tries to accept the constructed word, the refuter outputs x_2 in the next iteration, again keeping track of an ACC input. If no ACC has been input, $\mathcal{R}$ switches back to outputting x_1. The formal definition of $\mathcal{R}$ can be found in [27]. □

By Theorem 3, every language not in DBW has a certificate $\langle x, x_1, x_2\rangle$. As we argue below, these certificates are linear in the number of states of the refuters.

Lemma 3. *Let $\mathcal{R}$ be a DBW-refuter for $L \subseteq \Sigma^\omega$ with n states. Then, L has a certificate of the form $\langle x, x_1, x_2\rangle$ such that $|x| + |x_1| + |x_2| \leq 2 \cdot n$.*

Proof. The paths p, p_1, and p_2 that induce x, x_1 and x_2 in the proof of Theorem 3 are simple, and so they are all of length at most n. Also, while these paths may share edges, we can define them so that each edge appears in at most two paths. Indeed, if an edge appears in all three path, we can shorten p. Hence, $|x| + |x_1| + |x_2| \leq 2 \cdot n$, and we are done. □

Theorem 5. *Consider a language $L \subseteq \Sigma^\omega$ not in DBW. The length of a certificate for the non-DBW-recognizability of L is linear in a DPW for L and is exponential in an NBW for L. These bounds are tight.*

Proof. The upper bounds follow from Theorem 1 and Lemma 3, and the exponential determinization of NBWs. The lower bound in the NBW case follows from the exponential lower bound on the size of shortest non-universality witnesses for non-deterministic finite word automata (NFW) [34]. We sketch the reduction: Let $L_n \subseteq \{0,1\}^*$ be a language such that the shortest witness for non-universality of L_n is exponential in n, but L_n has a polynomial sized NFW. We then define $L'_n = (L_n \cdot \$ \cdot (0^* \cdot 1)^\omega) + ((0+1)^* \cdot \$ \cdot (0+1)^* \cdot 0^\omega)$. It is clear that L'_n has a NBW polynomial in n and is not DBW-recognizable. Note that for every word $w \in L_n$, we have $w \cdot \$ \cdot (0+1)^\omega \subseteq L'_n$. Thus, in order to satisfy Theorem 3, every certificate $\langle x, x_1, x_2\rangle$ needs to have $w \cdot \$$ as prefix of x, for some $w \notin L_n$. Hence, it is exponential in the size of the NBW. □

Remark 2. [**LTL**] When the language L is given by an LTL formula φ, then $\text{DBW}(\varphi) = \varphi \leftrightarrow \mathbf{GF}\textsc{Acc}$ and thus an off-the-shelf LTL synthesis tool can be used to extract a DBW-refuter, if one exists. As for complexity, a doubly-exponential upper bound on the size of a DPW for NoDBW(L), and then also on the size of DBW-refuters and certificates, follows from the double-exponential translation of LTL formulas to DPWs [49,42]. The length of certificates, however, and then, by Lemma 2, also the size of a minimal refuter, is related to the *diameter* of the DPW for NoDBW(L), and we leave its tight bound open. ■

4 Separability and Approximations

Consider three languages $L_1, L_2, L \subseteq \Sigma^\omega$. We say that L is a *separator* for $\langle L_1, L_2\rangle$ if $L_1 \subseteq L$ and $L_2 \cap L = \emptyset$. We say that a pair of languages $\langle L_1, L_2\rangle$ is *DBW-separable* iff there exists a language L in DBW such that L is a separator for $\langle L_1, L_2\rangle$.

Example 4. Let $\Sigma = \{a, b\}$, $L_1 = (a+b)^* \cdot b^\omega$, and $L_2 = (a+b)^* \cdot a^\omega$. By [29], L_1 and L_2 are not in DBW. They are, however, DBW-separable. A witness for this is $L = (a^* \cdot b)^\omega$. Indeed, $L_1 \subseteq L$, $L \cap L_2 = \emptyset$, and L is DBW-recognizable. ■

Consider a language $L \subseteq \Sigma^\omega$, and suppose we know that L is not in DBW. A user may be willing to approximate L in order to obtain DBW-recognizability. Specifically, we assume that there is a language $I \subseteq \Sigma^\omega$ of words that the user is *indifferent* about. Formally, the user is satisfied with a language in DBW that agrees with L on all words that are not in I. Formally, we say that a language L' *approximates* L *with radius* I if $L \setminus I \subseteq L' \subseteq L \cup I$. It is easy to see that, equivalently, L' is a separator for $\langle L \setminus I, comp(L \cup I)\rangle$. Note that the above formulation embodies the case where the user has in mind different over- and under-approximation radiuses, thus separating $\langle L \setminus I_\downarrow, comp(L \cup I_\uparrow)\rangle$ for possibly different $I_\downarrow$ and $I_\uparrow$. Indeed, by defining $I = (I_\downarrow \cap L) \cup (I_\uparrow \setminus L)$, we get $\langle L \setminus I, comp(L \cup I)\rangle = \langle L \setminus I_\downarrow, comp(L) \setminus I_\uparrow\rangle\rangle$.

It follows that by studying DBW-separability, we also study DBW-approximation, namely approximation by a language that is in DBW, possibly with different over- and under-approximation radiuses.

Remark 3. **[From recognizability to separation]** It is easy to see that DBW-separability generalizes DBW-recognizability, as L is in DBW iff $\langle L, comp(L)\rangle$ is DBW-separable. Given $L \subseteq \Sigma^\omega$, we say that a pair of languages $\langle L_1, L_2\rangle$ is a *no-DBW-witness* for L if L is a separator for $\langle L_1, L_2\rangle$ and $\langle L_1, L_2\rangle$ is not DBW-separable. Note that the latter indeed implies that L is not in DBW.

A simple no-DBW witness for L can be obtained as follows. Let $\mathcal{R}$ be a DBW refuter for L. Then, we define $L_1 = \{\mathcal{R}(y) : y \in \neg\infty\text{ACC}\}$ and $L_2 = \{\mathcal{R}(y) : y \in \infty\text{ACC}\}$. By the definition of DBW-refuters, we have $L_1 \subseteq L$ and $L_2 \cap L = \emptyset$, and so $\langle L_1, L_2\rangle$ is a no-DBW witness for L. It is simple, in the sense that when we describe L_1 and L_2 by a tree obtained by pruning the Σ^*-tree, then each node has at most two children – these that correspond to the responses of $\mathcal{R}$ to ACC and REJ. ■

4.1 Refuting Separability

For a pair of languages $\langle L_1, L_2\rangle$, we define the language $\text{SepDBW}(L) \subseteq (\Sigma \times A)^\omega$ of words with correct annotations for separation. Thus,

$$\text{SepDBW}(L_1, L_2) = \{x \oplus y : (x \in L_1 \to y \in \infty\text{ACC}) \wedge (x \in L_2 \to y \notin \infty\text{ACC})\}.$$

Note that $comp(\text{SepDBW}(L_1, L_2))$ is then the language

$$\text{NoSepDBW}(L_1, L_2) = \{x \oplus y : (x \in L_1 \wedge y \notin \infty\text{ACC}) \vee (x \in L_2 \wedge y \in \infty\text{ACC})\}.$$

A *DBW-sep-refuter for* $\langle L_1, L_2\rangle$ is an (A/Σ)-transducer with $\iota = env$ that realizes $\text{NoSepDBW}(L_1, L_2)$.

Example 5. Consider the language $L_{\neg\infty a} = (a + b)^* \cdot b^\omega$, which is not DBW. Let $I = a^* \cdot b^\omega + b^* \cdot a^\omega$, thus we are indifferent about words with only one alternation between a and b. In Figure 3 we describe a DBW-sep refuter for $\langle L_{\neg\infty a} \setminus I, comp(L_{\neg\infty a} \cup I)\rangle$. Note that the refuter generates only words in $a \cdot b \cdot a \cdot (a + b)^\omega$, whose intersection with I is empty. Consequently, the refutation is similar to the DBW-refutation of $L_{\neg\infty a}$. ■

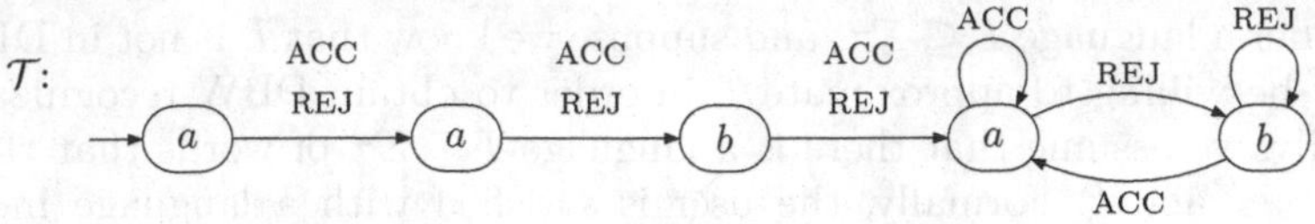

Figure 3. A DBW-sep refuter for $\langle L_{\neg\infty a} \setminus I, comp(L_{\neg\infty a} \cup I)\rangle$.

By Proposition 1, we have the following extension of Proposition 2.

Proposition 3. *Consider two languages $L_1, L_2 \subseteq \Sigma^\omega$. Let $A = \{\text{ACC}, \text{REJ}\}$. Exactly one of the following holds:*

- *$\langle L_1, L_2\rangle$ is DBW-separable, in which case the language* SepDBW(L_1, L_2) *is (Σ/A)-realizable by the system, and a finite-memory winning strategy for the system induces a DBW for a language L that separates L_1 and L_2.*
- *$\langle L_1, L_2\rangle$ is not DBW-separable, in which case the language* NoSepDBW(L) *is (A/Σ)-realizable by the environment, and a finite-memory winning strategy for the environment induces a DBW-sep-refuter for $\langle L_1, L_2\rangle$.*

As for complexity, the construction of the game for SepDBW(L_1, L_2) is similar to the one described in Theorem 1. Here, however, the input to the problem includes two DPWs. Also, the positive case, namely the construction of the separator does not follow from known results.

Theorem 6. *Consider DPWs $\mathcal{A}_1$ and $\mathcal{A}_2$ with n_1 and n_2 states, respectively. Let $L_1 = L(\mathcal{A}_1)$ and $L_2 = L(\mathcal{A}_2)$. One of the following holds.*

1. *There is a DBW $\mathcal{A}$ with $2 \cdot n_1 \cdot n_2$ states such that $L(\mathcal{A})$ DBW-separates $\langle L_1, L_2\rangle$.*
2. *There is a DBW-sep-refuter for $\langle L_1, L_2\rangle$ with $2 \cdot n_1 \cdot n_2$ states.*

Proof. We show that SepDBW(L_1, L_2) and NoSepDBW(L_1, L_2) can be recognised by DRWs with at most $2 \cdot n_1 \cdot n_2$ states. Then, by [15], we can construct a DBW or a DBW-sep-refuter with at most $2 \cdot n_1 \cdot n_2$ states. The construction is similar to the one described in the proof of Theorem 1. The only technical challenge is the fact SepDBW(L_1, L_2) is defined as the intersection, rather than union, of two languages. For this, we observe that we can define SepDBW(L_1, L_2) also as $\{x \oplus y : (y \in \infty\text{ACC} \text{ and } x \notin L_2) \text{ or } (y \notin \infty\text{ACC} \text{ and } x \notin L_1)\}$. With this formulation we then can reuse the union construction as seen in Theorem 1 to obtain DRWs with at most $2 \cdot n_1 \cdot n_2$ states. □

As has been the case with DBW-recognizability, one can generate certificates from a DBW-sep-refuter. The proof is similar to that of Theorem 3, with membership in L_1 replacing membership in L and membership in L_2 replacing being disjoint from L. Formally, we have the following.

Theorem 7. *Two ω-regular languages $L_1, L_2 \subseteq \Sigma^\omega$ are not DBW-separable iff there exist three finite words $x \in \Sigma^*$ and $x_1, x_2 \in \Sigma^+$, such that $x \cdot (x_1 + x_2)^* \cdot x_1^\omega \subseteq L_1$ and $x \cdot (x_1^* \cdot x_2)^\omega \subseteq L_2$.*

We refer to a triple $\langle x, x_1, x_2\rangle$ of words that satisfy the conditions in Theorem 7 as a *certificate* to the non-DBW-separability of $\langle L_1, L_2\rangle$. Observe that the same way we generated a no-DBW witness in Remark 3, we can extract, given a DBW-sep-refuter $\mathcal{R}$ for $\langle L_1, L_2\rangle$, languages $L_1' \subseteq L_1$ and $L_2' \subseteq L_2$ that tighten $\langle L_1, L_2\rangle$ and are still not DBW-separable.

4.2 Certificate-Guided Approximation

In this section we describe a method for finding small approximating languages $I_\downarrow$ and $I_\uparrow$ such that $\langle L \setminus I_\downarrow, comp(L) \setminus I_\uparrow\rangle$ is DBW-separable. If this method terminates we obtain an approximation for L that is DBW-recognizable. As in *counterexample guided abstraction-refinement* (CEGAR) for model checking [10], we use certificates for non-DBW-separability in order to suggest interesting approximating languages. Intuitively, while in CEGAR the refined system excludes the counterexample, here the approximation of L excludes the certificate.

Consider a certificate $\langle x, x_1, x_2\rangle$ for the non-DBW-separability of $\langle L_1, L_2\rangle$. We suggest the following five approximations:

$$\begin{array}{ll} C_0 = x \cdot (x_1 + x_2)^\omega & \rightsquigarrow \langle L_1 \setminus C_0, L_2 \setminus C_0\rangle \\ C_1 = x \cdot (x_1 + x_2)^* \cdot x_1^\omega = L_1 \cap C_0 & \rightsquigarrow \langle L_1 \setminus C_1, L_2\rangle \\ C_2 = x \cdot (x_2^* \cdot x_1)^\omega \supset C_1 & \rightsquigarrow \langle L_1, L_2 \setminus C_2\rangle \\ C_3 = x \cdot (x_1^* \cdot x_2)^\omega = L_2 \cap C_0 & \rightsquigarrow \langle L_1, L_2 \setminus C_3\rangle \\ C_4 = x \cdot (x_1 + x_2)^* \cdot x_2^\omega \subset C_3 & \rightsquigarrow \langle L_1, L_2 \setminus C_4\rangle \end{array}$$

First, it is easy to verify that $\langle x, x_1, x_2\rangle$ is indeed not a certificate for the non-DBW-separability of the obtained candidate pairs $\langle L_1', L_2'\rangle$. If $\langle L_1', L_2'\rangle$ is DBW-separable, we are done (yet may try to tighten the approximation). Otherwise, we can repeat the process with a certificate for the non-DBW-separability of $\langle L_1', L_2'\rangle$. As in CEGAR, some suggestions may be more interesting than others, in some cases the process terminates, in some it does not, and the user takes part directing the search.

Example 6. Consider again the language $L = (a + b)^* \cdot b^\omega$ and the certificate $\langle x, x_1, x_2\rangle = \langle \epsilon, b, a\rangle$. Trying to approximate L by a language in DBW, we start with the pair $\langle L, comp(L)\rangle$. Our five suggestions are then as follows.

$$\begin{array}{ll} C_0 = \Sigma^\omega & \rightsquigarrow \langle L \setminus C_0, comp(L) \setminus C_0\rangle = \langle \emptyset, \emptyset\rangle \\ C_1 = (b + a)^* \cdot b^\omega & \rightsquigarrow \langle L \setminus C_1, comp(L)\rangle = \langle \emptyset, comp(L)\rangle \\ C_2 = (a^* \cdot b)^\omega & \rightsquigarrow \langle L, comp(L) \setminus C_2\rangle = \langle L, (a + b)^* \cdot a^\omega\rangle \\ C_3 = (b^* \cdot a)^\omega & \rightsquigarrow \langle L, comp(L) \setminus C_3\rangle = \langle L, \emptyset\rangle \\ C_4 = (b + a)^* \cdot a^\omega & \rightsquigarrow \langle L, comp(L) \setminus C_4\rangle = \langle L, (a + b)^* \cdot (a \cdot a^* \cdot b \cdot b^*)^\omega\rangle \end{array}$$

Candidates C_0, C_1, and C_3 induce trivial approximations. Then, C_2 suggests to over-approximate L by setting $I_\uparrow$ to $(a^* \cdot b)^\omega$, which we view as a nice solution, approximating "eventually always b" by "infinitely often b". Then, the pair derived from C_4 is not DBW-separable. We can try to approximate it. Note, however, that repeated approximations in the spirit of C_4 are going to only extend the prefix of x in the certificates, and the process does not terminate. In the full version of this article [27], we describe the process for the certificate $\langle x, x_1, x_2\rangle = \langle a, b, a\rangle$, which again might not terminate. ■

5 Other Classes of Deterministic Automata

In this section we generalise the idea of DBW-refuters to other classes of deterministic automata. For this we take again the view that a deterministic automaton is a $\langle\Sigma, A\rangle$-transducer over a suitable annotation alphabet A. We then characterize each class of deterministic automata by two languages over A:

- The language $L_{\text{acc}} \subseteq A^\omega$, describing when a run is accepting. For example, for DBWs, we have $A = \{\text{ACC}, \text{REJ}\}$ and $L_{\text{acc}} = \infty\text{ACC}$.
- The language $L_{\text{struct}} \subseteq A^\omega$, describing structural conditions on the run. For example, recall that a DWW is a DBW in which the states of each SCS are either all accepting or all rejecting, and so each run eventually get trapped in an accepting or rejecting SCS. Accordingly, the language of runs that satisfy the structural condition is $L_{\text{struct}} = A^* \cdot (\text{ACC}^\omega + \text{REJ}^\omega)$.

We now formalize this intuition. Let A be a finite set of annotations and let $\gamma = \langle L_{\text{acc}}, L_{\text{struct}}\rangle$, for $L_{\text{acc}}, L_{\text{struct}} \subseteq A^\omega$. A deterministic automaton $\mathcal{A} = \langle\Sigma, Q, q_0, \delta, \alpha\rangle$ is a deterministic γ automaton (DγW, for short) if there is a function $\tau\colon Q \to A$ that maps each state to an annotation such that a run r of $\mathcal{A}$ satisfies α iff $\tau(r) \in L_{\text{acc}}$, and all runs r satisfy the structural condition, thus $\tau(r) \in L_{\text{struct}}$. We then say that a language L is γ-recognizable if there a DγW $\mathcal{A}$ such that $L = L(\mathcal{A})$.

Before we continue to study γ-recognizability, let us demonstrate the γ-characterization of common deterministic automata. We first start with classes γ for which L_{struct} is trivial; i.e., $L_{\text{struct}} = A^\omega$.

- DBW: $A = \{\text{ACC}, \text{REJ}\}$ and $L_{\text{acc}} = \infty\text{ACC}$.
- DCW: $A = \{\text{ACC}, \text{REJ}\}$ and $L_{\text{acc}} = \neg\infty\text{ACC}$.
- DPW$[i,k]$: $A = \{i, \ldots, k\}$ and $L_{\text{acc}} = \{y \in A^\omega : \max(\mathit{inf}(y)) \text{ is odd}\}$.
- DELW$[\theta]$: $A = 2^{\mathbb{M}}$ and $L_{\text{acc}} = \{y \in A^\omega : y \models \theta\}$.

Note that the characterizations for Büchi, co-Büchi, and parity are special cases of the characterization for DELW. In a similar way, we could define a language L_{acc} for DRW$[k]$ and other common special cases of DELWs. We continue to classes in the depth hierarchy, where γ includes also a structural restriction:

- DWW: The set A and the language L_{acc} are as for DBW or DCW. In addition, $L_{\text{struct}} = A^* \cdot (\text{ACC}^\omega + \text{REJ}^\omega)$.
- DWW$[j,k]$, for $j \in \{0,1\}$: The set A and the language L_{acc} are as for DPW$[j,k]$. In addition, $L_{\text{struct}} = \{y_0 \cdot y_1 \cdots \in A^\omega : \text{for all } i \geq 0, \text{ we have that } y_i \leq y_{i+1}\}$.

- *Bounded Languages*: A language L is bounded if it is both safety and co-safety. Thus, every word $w \in \Sigma^\omega$ has a prefix $v \in \Sigma^*$ such that either for all $u \in \Sigma^\omega$ we have $v \cdot u \in L$, or for all $u \in \Sigma^\omega$ we have $v \cdot u \notin L$ [23]. To capture this, we use $A = \{\text{ACC}, \text{REJ}, ?\}$, where "?" is used for annotating states with both accepting and rejecting continuations. Then, $L_{\text{acc}} = A^* \cdot \text{ACC}^\omega$, and $L_{\text{struct}} = ?^* \cdot (\text{ACC}^\omega + \text{REJ}^\omega)$.
- *Deterministic (m, n)-Superparity Automata* [39]: $A = \{(i, j) : 0 \leq i \leq m, 0 \leq j \leq n\}$, $L_{\text{acc}} = \{y_m \oplus y_n \in A^\omega : \max(\mathit{inf}(y_m)) + \max(y_n) \text{ is odd}\}$, and $L_{\text{struct}} = \{y_m \oplus (y_0 \cdot y_1 \cdots) \in A^\omega : y_i \leq y_{i+1}, \text{ for all } i \geq 0\}$.

Let Σ be an alphabet, let A be an annotation alphabet, and let $\gamma = \langle L_{\text{acc}}, L_{\text{struct}} \rangle$, for $L_{\text{acc}}, L_{\text{struct}} \subseteq A^\omega$. We define the language $\text{Real}(L, \gamma) \subseteq (\Sigma \times A)^\omega$ of words with correct annotations.

$$\text{Real}(L, \gamma) = \{x \oplus y : y \in L_{\text{struct}} \text{ and } (x \in L \text{ iff } y \in L_{\text{acc}})\}.$$

Note that the language $\text{DBW}(L)$ can be viewed as a special case of our general framework. In particular, in cases $L_{\text{struct}} = A^\omega$, we can remove the $y \in L_{\text{struct}}$ conjunct from $\text{Real}(L, \gamma)$. Note that $\mathit{comp}(\text{Real}(L, \gamma))$ is the language

$$\text{NoReal}(L, \gamma) = \{x \oplus y : y \notin L_{\text{struct}} \text{ or } (x \in L \text{ iff } y \notin L_{\text{acc}})\}.$$

A *γ-refuter for L* is then an (A/Σ)-transducer with $\iota = \mathit{env}$ that realizes $\text{NoReal}(L, \gamma)$. We can now state the "DγW-generalization" of Proposition 2.

Proposition 4. *Consider an ω-regular language $L \subseteq \Sigma^\omega$, and a pair $\gamma = \langle L_{\text{acc}}, L_{\text{struct}} \rangle$, for ω-regular languages $L_{\text{acc}}, L_{\text{struct}} \subseteq A^\omega$. Exactly one of the following holds:*

1. *L is in* DγW, *in which case the language* $\text{Real}(L, \gamma)$ *is (Σ/A)-realizable by the system, and a finite-memory winning strategy for the system induces a* DγW *for L.*
2. *L is not in* DγW, *in which case the language* $\text{NoReal}(L, \gamma)$ *is (A/Σ)-realizable by the environment, and a finite-memory winning strategy for the environment induces a γ-refuter for L.*

Note that every DELW can be complemented by dualization, thus by changing its acceptance condition from θ to $\neg\theta$. In particular, DBW and DCW dualize each other. As we argue below, dualization is carried over to refutation. For example, the $(\{\text{ACC}, \text{REJ}\}/\Sigma)$-transducer $\mathcal{R}$ from Figure 1 is both a DBW-refuter for $\neg\infty a$ and a DCW-refuter for ∞a. Formally, we have the following.

Theorem 8. *Consider an EL-condition θ over $\mathbb{M}$. Let $A = 2^{\mathbb{M}}$. For every (A/Σ)-transducer $\mathcal{R}$ and language L, we have that $\mathcal{R}$ is a* DELW$[\theta]$*-refuter for L iff $\mathcal{R}$ is a* DELW$[\neg\theta]$*-refuter for $\mathit{comp}(L)$. In particular, for every language L and $(\{\text{ACC}, \text{REJ}\}/\Sigma)$-transducer $\mathcal{R}$, we have that $\mathcal{R}$ is a DBW-refuter for L iff $\mathcal{R}$ is a DCW-refuter for $\mathit{comp}(L)$.*

Proof. For DELW[θ]-recognizability of L, the language of correct annotations is $\{x \oplus y : (x \in L \text{ iff } y \models \theta)\}$, which is equal to $\{x \oplus y : (x \in comp(L) \text{ iff } y \models \neg\theta)\}$, which is the language of correct annotations for DELW[$\neg\theta$]-recognizability of $comp(L)$. □

While dualization is nicely carried over to refutation, this is not the case for all expressiveness results. For example, while DWW=DBW∩DCW, and in fact DBW and DCW are weak type (that is, when the language of a DBW is in DWW, an equivalent DWW can be defined on top of its structure, and similarly for DCW [21]), we describe in [27] a DWW-refuter that is neither a DBW- nor a DCW-refuter. Intuitively, this is possible as in DWW refutation, Prover loses when the input is not in $A^* \cdot (\text{ACC}^\omega + \text{REJ}^\omega)$, whereas in DBW and DCW refutation, Refuter has to respond correctly also for these inputs.

On the other hand, as every DWW is also a DBW and a DCW, every DBW-refuter or DCW-refuter is also a DWW-refuter.

It is easy to see that our results about DγW-recognizability can be extended to separability and approximation in the same way DBW-recognizability has been extended in Section 4. We describe the details in the full version [27], as well as word-certificates for the non-DγW-recognizability and -separability of several well-known types of γ.

6 Discussion and Directions for Future Research

The automation of decision procedures makes certification essential. We suggest to use the winning strategy of the refuter in expressiveness games as a certificate to inexpressibility. We show that beyond this *state-based certificate*, the strategy induces a *word-based certificate*, generated from words traversed along a "flower structure" the strategy contains, as well as a *language-based certificate*, consisting of languages that under- and over-approximate the language in question and that are not separable by automata in the desired class.

While our work considers *expressive power*, one can use similar ideas in order to question the *size* of automata needed to recognize a given language. For example, in the case of a regular language L of finite words, the Myhill-Nerode characterization [37,38] suggests to refute the existence of deterministic finite word automata (DFW) with n states for L by providing $n+1$ prefixes that are not right-congruent. Using our approach, one can alternatively consider the winning strategy of Refuter in a game in which the set of annotations includes also the state space, and L_{struct} ensures consistency of the transition relation. Even more interesting is refutation of size in the setting of automata on infinite words. Indeed, there, minimization is NP-complete [46], and there are interesting connections between polynomial certificates and possible membership in co-NP, as well as connections between size of certificates and succinctness of the different classes of automata.

Finally, while the approximation scheme we studied is based on suggested over- and under-approximating languages, it is interesting to study approximations that are based on more flexible distance measures [13,18].

References

1. Almagor, S., Lahijanian, M.: Explainable multi agent path finding. In: Proc. 19th International Conference on Autonomous Agents and Multiagent Systems. pp. 34–42 (2020)
2. Alpern, B., Schneider, F.: Recognizing safety and liveness. Distributed computing **2**, 117–126 (1987)
3. Baier, C., de Alfaro, L., Forejt, V., Kwiatkowska, M.: Model checking probabilistic systems. In: Handbook of Model Checking., pp. 963–999. Springer (2018)
4. Baumeister, T., Finkbeiner, B., Torfah, H.: Explainable reactive synthesis. In: 18th Int. Symp. on Automated Technology for Verification and Analysis (2020). https://doi.org/10.1007/978-3-030-59152-6_23
5. Bloem, R., Chatterjee, K., Jobstmann, B.: Graph games and reactive synthesis. In: Handbook of Model Checking., pp. 921–962. Springer (2018)
6. Boigelot, B., Jodogne, S., Wolper, P.: On the use of weak automata for deciding linear arithmetic with integer and real variables. In: Proc. Int. Joint Conf. on Automated Reasoning. Lecture Notes in Computer Science, vol. 2083, pp. 611–625. Springer (2001)
7. Boker, U., Kupferman, O.: Co-ing Büchi made tight and useful. In: Proc. 24th IEEE Symp. on Logic in Computer Science. pp. 245–254 (2009)
8. Büchi, J.: On a decision method in restricted second order arithmetic. In: Proc. Int. Congress on Logic, Method, and Philosophy of Science. 1960. pp. 1–12. Stanford University Press (1962)
9. Büchi, J., Landweber, L.: Solving sequential conditions by finite-state strategies. Trans. AMS **138**, 295–311 (1969)
10. Clarke, E.M., Grumberg, O., Jha, S., Lu, Y., Veith, H.: Counterexample-guided abstraction refinement for symbolic model checking. Journal of the ACM **50**(5), 752–794 (2003)
11. Czerwinski, W., Lasota, S., Meyer, R., Muskalla, S., Kumar, K., Saivasan, P.: Regular separability of well-structured transition systems. In: Proc. 29th Int. Conf. on Concurrency Theory. LIPIcs, vol. 118, pp. 35:1–35:18. Schloss Dagstuhl - Leibniz-Zentrum für Informatik (2018)
12. Czerwinski, W., Martens, W., Masopust, T.: Efficient separability of regular languages by subsequences and suffixes. In: Proc. 40th Int. Colloq. on Automata, Languages, and Programming. Lecture Notes in Computer Science, vol. 7966, pp. 150–161. Springer (2013)
13. Dimitrova, R., Finkbeiner, B., Torfah, H.: Approximate automata for omega-regular languages. In: 17th Int. Symp. on Automated Technology for Verification and Analysis. Lecture Notes in Computer Science, vol. 11781, pp. 334–349. Springer (2019)
14. Eisner, C., Fisman, D.: A Practical Introduction to PSL. Springer (2006)
15. Emerson, E., Jutla, C.: The complexity of tree automata and logics of programs. In: Proc. 29th IEEE Symp. on Foundations of Computer Science. pp. 328–337 (1988)
16. Emerson, E., Jutla, C.: Tree automata, μ-calculus and determinacy. In: Proc. 32nd IEEE Symp. on Foundations of Computer Science. pp. 368–377 (1991)
17. Emerson, E., Lei, C.L.: Modalities for model checking: Branching time logic strikes back. Science of Computer Programming **8**, 275–306 (1987)
18. Gange, G., Ganty, P., Stuckey, P.: Fixing the state budget: Approximation of regular languages with small dfas. In: 15th Int. Symp. on Automated Technology

for Verification and Analysis. Lecture Notes in Computer Science, vol. 10482, pp. 67–83. Springer (2017)

19. Krishnan, S., Puri, A., Brayton, R.: Deterministic ω-automata vis-a-vis deterministic Büchi automata. In: Algorithms and Computations. Lecture Notes in Computer Science, vol. 834, pp. 378–386. Springer (1994)
20. Kupferman, O.: Automata theory and model checking. In: Handbook of Model Checking, pp. 107–151. Springer (2018)
21. Kupferman, O., Morgenstern, G., Murano, A.: Typeness for ω-regular automata. International Journal on the Foundations of Computer Science **17**(4), 869–884 (2006)
22. Kupferman, O., Sheinvald-Faragy, S.: Finding shortest witnesses to the nonemptiness of automata on infinite words. In: Proc. 17th Int. Conf. on Concurrency Theory. Lecture Notes in Computer Science, vol. 4137, pp. 492–508. Springer (2006)
23. Kupferman, O., Vardi, M.: On bounded specifications. In: Proc. 8th Int. Conf. on Logic for Programming Artificial Intelligence and Reasoning. Lecture Notes in Computer Science, vol. 2250, pp. 24–38. Springer (2001)
24. Kupferman, O., Vardi, M.: From complementation to certification. Theoretical Computer Science **305**, 591–606 (2005)
25. Kupferman, O., Vardi, M.: From linear time to branching time. ACM Transactions on Computational Logic **6**(2), 273–294 (2005)
26. Kupferman, O., Vardi, M.: Safraless decision procedures. In: Proc. 46th IEEE Symp. on Foundations of Computer Science. pp. 531–540 (2005)
27. Kupferman, O., Sickert, S.: Certifying inexpressibility (2021), https://arxiv.org/abs/2101.08756, (Full Version)
28. Kurshan, R.: Computer Aided Verification of Coordinating Processes. Princeton Univ. Press (1994)
29. Landweber, L.: Decision problems for ω–automata. Mathematical Systems Theory **3**, 376–384 (1969)
30. Leshkowitz, O., Kupferman, O.: On repetition languages. In: 45th Int. Symp. on Mathematical Foundations of Computer Science. Leibniz International Proceedings in Informatics (LIPIcs) (2020)
31. Löding, C.: Methods for the transformation of automata: Complexity and connection to second order logic (1999), M.Sc. Thesis, Christian-Albrechts-University of Kiel
32. Löding, C.: Efficient minimization of deterministic weak ω-automata. Information Processing Letters **79**(3), 105–109 (2001)
33. McNaughton, R.: Testing and generating infinite sequences by a finite automaton. Information and Control **9**, 521–530 (1966)
34. Meyer, A., Stockmeyer, L.: The equivalence problem for regular expressions with squaring requires exponential space. In: Proc. 13th IEEE Symp. on Switching and Automata Theory. pp. 125–129 (1972)
35. Mostowski, A.: Regular expressions for infinite trees and a standard form of automata. In: Computation Theory. Lecture Notes in Computer Science, vol. 208, pp. 157–168. Springer (1984)
36. Muller, D., Saoudi, A., Schupp, P.: Alternating automata, the weak monadic theory of the tree and its complexity. In: Proc. 13th Int. Colloq. on Automata, Languages, and Programming. Lecture Notes in Computer Science, vol. 226, pp. 275 – 283. Springer (1986)
37. Myhill, J.: Finite automata and the representation of events. Tech. Rep. WADD TR-57-624, pages 112–137, Wright Patterson AFB, Ohio (1957)

38. Nerode, A.: Linear automaton transformations. Proceedings of the American Mathematical Society **9**(4), 541–544 (1958)
39. Perrin, D., Pin, J.E.: Infinite words - automata, semigroups, logic and games, Pure and applied mathematics series, vol. 141. Elsevier Morgan Kaufmann (2004)
40. Place, T., Zeitoun, M.: Separating regular languages with first-order logic. Log. Methods Comput. Sci. **12**(1) (2016)
41. Rabin, M.: Decidability of second order theories and automata on infinite trees. Transaction of the AMS **141**, 1–35 (1969)
42. Safra, S.: On the complexity of ω-automata. In: Proc. 29th IEEE Symp. on Foundations of Computer Science. pp. 319–327 (1988)
43. Safra, S.: Exponential determinization for ω-automata with strong-fairness acceptance condition. In: Proc. 24th ACM Symp. on Theory of Computing (1992)
44. S.Almagor, Chistikov, D., Ouaknine, J., Worrell, J.: O-minimal invariants for linear loops. In: Proc. 45th Int. Colloq. on Automata, Languages, and Programming. LIPIcs, vol. 107, pp. 114:1–114:14. Schloss Dagstuhl - Leibniz-Zentrum für Informatik (2018)
45. Schewe, S.: Büchi complementation made tight. In: Proc. 26th Symp. on Theoretical Aspects of Computer Science. LIPIcs, vol. 3, pp. 661–672. Schloss Dagstuhl - Leibniz-Zentrum fuer Informatik, Germany (2009)
46. Schewe, S.: Beyond Hyper-Minimisation—Minimising DBAs and DPAs is NP-Complete. In: Proc. 30th Conf. on Foundations of Software Technology and Theoretical Computer Science. Leibniz International Proceedings in Informatics (LIPIcs), vol. 8, pp. 400–411 (2010)
47. di Stasio, A., Murano, A., Vardi, M.: Solving parity games: Explicit vs symbolic. In: 23rd International Conference on Implementation and Application of Automata. Lecture Notes in Computer Science, vol. 10977, pp. 159–172. Springer (2018)
48. Thomas, W.: Automata on infinite objects. Handbook of Theoretical Computer Science pp. 133–191 (1990)
49. Vardi, M., Wolper, P.: Reasoning about infinite computations. Information and Computation **115**(1), 1–37 (1994)
50. Wagner, K.: On ω-regular sets. Information and Control **43**, 123–177 (1979)

A General Semantic Construction of Dependent Refinement Type Systems, Categorically

Satoshi Kura[1,2]

[1] National Institute of Informatics, Tokyo, Japan
[2] The Graduate University for Advanced Studies (SOKENDAI), Kanagawa, Japan
kura@nii.ac.jp

Abstract. Dependent refinement types are types equipped with predicates that specify preconditions and postconditions of underlying functional languages. We propose a general semantic construction of dependent refinement type systems from underlying type systems and predicate logic, that is, a construction of liftings of closed comprehension categories from given (underlying) closed comprehension categories and posetal fibrations for predicate logic. We give sufficient conditions to lift structures such as dependent products, dependent sums, computational effects, and recursion from the underlying type systems to dependent refinement type systems. We demonstrate the usage of our construction by giving semantics to a dependent refinement type system and proving soundness.

1 Introduction

Dependent refinement types [6] are types equipped with predicates that restrict values in the types. They are used to specify preconditions and postconditions which may depend on input values and to verify that programs satisfy the specifications. Many dependent refinement types systems are proposed [5,6,13,14,25] and implemented in, e.g., F* [23,24] and LiquidHaskell [19,26,27].

In this paper, we address the question: "How are dependent refinement type systems, underlying type systems, and predicate logic related from the viewpoint of categorical semantics?" Although most existing dependent refinement type systems are proved to be sound using operational semantics, we believe that categorical semantics is more suitable for the general understanding of their nature, especially when we consider general computational effects and various kinds of predicate logic (e.g., for relational verification). This understanding will provide guidelines to design new dependent refinement type systems.

Our answer to the question is a general semantic construction of dependent refinement type systems from underlying type systems and predicate logic. More concretely, given a closed comprehension category (CCompC for short) for interpreting an underlying type system and a fibration for predicate logic, we combine them to obtain another CCompC that can interpret a dependent refinement type system built from the underlying type system and the predicate logic.

S. Kiefer and C. Tasson (Eds.): FOSSACS 2021, LNCS 12650, pp. 406–426, 2021.
https://doi.org/10.1007/978-3-030-71995-1_21

For example, consider giving an interpretation to the term "$x : \{\mathrm{int} \mid x \geq 0\} \vdash x + 1 : \{v : \mathrm{int} \mid v = x + 1\}$" in a dependent refinement type system. Its underlying term is "$x : \mathrm{int} \vdash x + 1 : \mathrm{int}$," and we assume that it is interpreted as the successor function of $\mathbb{Z}$ in **Set**. The problem here is how to refine this interpretation with predicates. In dependent refinement types, predicates may depend on the variables in contexts. In this example, the type "$x : \{\mathrm{int} \mid x \geq 0\} \vdash \{v : \mathrm{int} \mid v = x + 1\}$" depends on the variable x. Thus, the interpretation of such types must be a predicate on the context and the type, i.e.,

$$[\![x : \{\mathrm{int} \mid x \geq 0\} \vdash \{v : \mathrm{int} \mid v = x + 1\}]\!] = \{(x, v) \in \mathbb{Z} \times \mathbb{Z} \mid x \geq 0 \wedge v = x + 1\}.$$

As a result, the term in the dependent refinement type system is interpreted as the interpretation in the underlying type system together with the property that if the input satisfies preconditions, then the output satisfies postconditions.

$$\begin{array}{ccc} \{x \in \mathbb{Z} \mid x \geq 0\} & \dashrightarrow & \{(x, v) \in \mathbb{Z} \times \mathbb{Z} \mid x \geq 0 \wedge v = x + 1\} \\ \cap\!\!\mid & & \cap\!\!\mid \\ \mathbb{Z} & \xrightarrow{\langle \mathrm{id}_{\mathbb{Z}}, (-)+1 \rangle} & \mathbb{Z} \times \mathbb{Z} \end{array} \tag{1}$$

$$\begin{array}{ccc} \{\mathbb{E} \mid \mathbb{P}\} & \longrightarrow & \mathbb{E} \\ \downarrow & & \downarrow p \\ \mathbb{P} & \xrightarrow{q} & \mathbb{B} \end{array}$$

Fig. 1. Lifting.

We formalize this refinement process as a construction of liftings of CCompCs, which are used to interpret dependent type theories. Assume that we have a pair of a CCompC $p : \mathbb{E} \to \mathbb{B}$ for interpreting underlying type systems and a fibration $q : \mathbb{P} \to \mathbb{B}$ for predicate logic satisfying certain conditions. Then we construct a CCompC $\{\mathbb{E} \mid \mathbb{P}\} \to \mathbb{P}$ for interpreting dependent refinement type systems. This construction also yields a morphism of CCompCs from $\{\mathbb{E} \mid \mathbb{P}\} \to \mathbb{P}$ to $p : \mathbb{E} \to \mathbb{B}$ in Fig. 1. Given the simple fibration $\mathbf{s}(\mathbf{Set}) \to \mathbf{Set}$ for underlying type systems and the subobject fibration $\mathbf{Sub}(\mathbf{Set}) \to \mathbf{Set}$ for predicate logic, then we get interpretations like (1).

We extend the construction of liftings of CCompCs to liftings of fibred monads [1] on CCompCs, which is motivated by the fact that many dependent refinement type systems have computational effects, e.g., exception (like division and assertion), divergence, nondeterminism [25], and probability [5]. Assume that we have a fibred monad $\hat{T}$ on $p : \mathbb{E} \to \mathbb{B}$, a monad T on $\mathbb{B}$, and a lifting $\dot{T}$ of T along $q : \mathbb{P} \to \mathbb{B}$. Under a certain condition that roughly claims that $\hat{T}$ and T represent the same computational effects, we construct a fibred monad on $\{\mathbb{E} \mid \mathbb{P}\} \to \mathbb{P}$, which is a lifting of $\hat{T}$ in the same spirit of the given lifting $\dot{T}$. This situation is rather realistic because the fibred monad $\hat{T}$ on the CCompC $p : \mathbb{E} \to \mathbb{B}$ is often induced from the monad T on the base category $\mathbb{B}$. The lifting $\dot{T}$ of the monad T along $p : \mathbb{P} \to \mathbb{B}$ specifies how to map predicates $P \in \mathbb{P}_X$ on values $X \in \mathbb{B}$ to predicates $\dot{T}P \in \mathbb{P}_{TX}$ on computations TX, which enables us to express, for example, total/partial correctness and may/must nondeterminism [1].

We explain the usage of these categorical constructions by giving semantics to a dependent refinement type system with computational effects, which is based on [4]. Our system also supports subtyping relations induced by logical implication. We prove soundness of the dependent refinement type system.

Finally, we discuss how to handle recursion in dependent refinement type systems. In [4], Ahman gives semantics to recursion in a specific model, i.e., the fibration of continuous families of ω-cpos $\mathbf{CFam(CPO)} \to \mathbf{CPO}$. We consider more general characterization of recursion by adapting Conway operators for CCompCs, which enables us to lift the structure for recursion. We show that a rule for partial correctness in our dependent refinement type system is sound under the existence of a generalized Conway operator.

Our contributions are summarized as follows.

- We provide a general construction of liftings of CCompCs from given CCompCs and posetal fibrations satisfying certain conditions, as a semantic counterpart of construction of dependent refinement type systems from underlying type systems and predicate logic. We extend this to liftings of fibred monads on the underlying CCompCs to model computational effects.
- We consider a type system (based on EMLTT [2–4]) that includes most of basic features of dependent refinement type systems and prove its soundness in the liftings of CCompCs obtained from the above construction.
- We define Conway operators for dependent type systems. This generalizes the treatment of general recursion in [4]. We prove soundness of the typing rule for partial correctness of recursion under the existence of a lifting of Conway operators.

2 Preliminaries

We review basic definitions and fix notations for comprehension categories, which are used as categorical models for dependent type theories. We assume basic knowledge of fibrations (see e.g. [10]).

Let $p : \mathbb{E} \to \mathbb{B}$ be a fibration (opfibration). We denote the cartesian (cocartesian) lifting over $u : I \to J$ by $\overline{u}(Y) : u^*Y \to Y$ ($\underline{u}(X) : X \to u_!X$) where $u^* : \mathbb{E}_J \to \mathbb{E}_I$ ($u_! : \mathbb{E}_I \to \mathbb{E}_J$) is the reindexing (coreindexing) functor. We call $p : \mathbb{E} \to \mathbb{B}$ a *posetal fibration* if p is a fibration such that each fibre category is a poset. Note that the fibration $p : \mathbb{E} \to \mathbb{B}$ is split and faithful if p is posetal.

A *comprehension category* is a functor $\mathcal{P} : \mathbb{E} \to \mathbb{B}^{\to}$ such that the composite $\mathrm{cod} \circ \mathcal{P} : \mathbb{E} \to \mathbb{B}$ is a fibration and $\mathcal{P}$ maps cartesian morphisms to pullbacks in $\mathbb{B}$. A comprehension category $\mathcal{P}$ is *full* if $\mathcal{P}$ is fully faithful.

A *comprehension category with unit* is a fibration $p : \mathbb{E} \to \mathbb{B}$ that has a fibred terminal object $1 : \mathbb{B} \to \mathbb{E}$ and a comprehension functor $\{-\} : \mathbb{E} \to \mathbb{B}$ which is a right adjoint of the fibred terminal object functor $1 \dashv \{-\}$. Projection $\pi_X : \{X\} \to pX$ is defined by $\pi_X = p\epsilon_X^{1 \dashv \{-\}}$ for each $X \in \mathbb{E}$. Intuitively, $\mathbb{E}$ represents a collection of types $\Gamma \vdash A$ in dependent type theories; $\mathbb{B}$ represents a collection of contexts Γ; $p : \mathbb{E} \to \mathbb{B}$ is the mapping $(\Gamma \vdash A) \mapsto \Gamma$; $1 : \mathbb{B} \to \mathbb{E}$ is the unit type $\Gamma \mapsto (\Gamma \vdash 1)$; and $\{-\}$ is the mapping $(\Gamma \vdash A) \mapsto \Gamma, x : A$ where x is a fresh variable.

The comprehension category with unit $p : \mathbb{E} \to \mathbb{B}$ induces several structures. It induces a comprehension category $\mathcal{P}$ defined by $\mathcal{P}X = \pi_X$. The adjunction

$1 \dashv \{-\}$ defines the bijection $s : \mathbb{E}_I(1I, X) \cong \{f : I \to \{X\} \mid \pi_X \circ f = \mathrm{id}_I\}$ between vertical morphisms in $\mathbb{E}$ and sections in $\mathbb{B}$. For each $X, Y \in \mathbb{E}_I$, we have an isomorphism $\phi : \mathbb{E}_{\{X\}}(1\{X\}, \pi_X^* Y) \cong \mathbb{E}_I(X, Y)$. Consider the pullback square $\mathcal{P}(\overline{\pi_X}(Y))$ where $X, Y \in \mathbb{E}_I$. By the universal property of pullbacks, we have the symmetry isomorphism $\sigma_{X,Y} : \{\pi_X^* Y\} \to \{\pi_Y^* X\}$ as a unique morphism $\sigma_{X,Y}$ such that $\pi_{\pi_X^* Y} = \{\overline{\pi_Y}(X)\} \circ \sigma_{X,Y}$ and $\{\overline{\pi_X}(Y)\} = \pi_{\pi_Y^* X} \circ \sigma_{X,Y}$. Similarly, we have the diagonal morphism $\delta_X : \{X\} \to \{\pi_X^* X\}$ as a unique morphism δ_X such that $\pi_{\pi_X^* X} \circ \delta_X = \{\overline{\pi_X}(X)\} \circ \delta_X = \mathrm{id}_{\{X\}}$.

Let $p : \mathbb{E} \to \mathbb{B}$ be a comprehension category with unit and $q : \mathbb{D} \to \mathbb{B}$ be a fibration. The fibration q has *p-products* if $\pi_X^* : \mathbb{D}_{pX} \to \mathbb{D}_{\{X\}}$ has a right adjoint $\pi_X^* \dashv \prod_X$ for each $X \in \mathbb{E}$ and these adjunctions satisfy the BC (Beck-Chevalley) condition for each pullback square $\mathcal{P}f$ where $\mathcal{P}$ is a comprehension category induced by p and f is a cartesian morphism in $\mathbb{E}$. Similarly, we define *p-coproducts* by $\coprod_X \dashv \pi_X^*$ and p-equality by $\mathrm{Eq}_X \dashv \delta_X^*$ plus the BC condition for each cartesian morphism (see [10, Definition 9.3.5] for detail).

A comprehension category with unit $p : \mathbb{E} \to \mathbb{B}$ admits *products* (*coproducts*) if it has p-products (p-coproducts). The coproducts are *strong* if the canonical morphism $\kappa : \{Y\} \to \{\coprod_X Y\}$ defined by $\{\overline{\pi_X}(\coprod_X Y) \circ \eta^{\pi_X^* \dashv \coprod_X}\}$ is an isomorphism for each $X \in \mathbb{E}$ and $Y \in \mathbb{E}_{\{X\}}$. A *closed comprehension category* (CCompC) is a full comprehension category with unit that admits products and strong coproducts and has a terminal object in the base category. A *split closed comprehension category* (SCCompC) is a CCompC such that p is a split fibration, and the BC condition for products and coproducts holds strictly (i.e., canonical isomorphisms are identities). For example, the simple fibration $\mathsf{s}_{\mathbb{B}} : \mathbf{s}(\mathbb{B}) \to \mathbb{B}$ on a cartesian closed category $\mathbb{B}$ is a SCCompC (see [10, Theorem 10.5.5]). Another example of SCCompCs is the family fibration $\mathsf{fam}_{\mathbf{Set}} : \mathbf{Fam}(\mathbf{Set}) \to \mathbf{Set}$.

Fibred coproducts in a comprehension category with unit $p : \mathbb{E} \to \mathbb{B}$ are *strong* if the functor $\langle \{\iota_1\}^*, \{\iota_2\}^* \rangle : \mathbb{E}_{\{X+Y\}} \to \mathbb{E}_{\{X\}} \times \mathbb{E}_{\{Y\}}$ is fully faithful where $\iota_1 : X \to X + Y$ and $\iota_2 : Y \to X + Y$ are injections for fibred coproducts. Strong fibred coproducts are used to interpret fibred coproduct types $A + B$.

3 Lifting SCCompCs and Fibred Coproducts

In this section, we give a construction of liftings of SCCompCs with strong fibred coproducts from given SCCompCs with strong fibred coproducts for underlying types and posetal fibrations for predicate logic satisfying appropriate conditions.

3.1 Lifting SCCompCs

Let $p : \mathbb{E} \to \mathbb{B}$ be a SCCompC for underlying type systems. Let $q : \mathbb{P} \to \mathbb{B}$ be a posetal fibration with fibred finite products for predicate logic.

Definition 1. We define a category $\{\mathbb{E} \mid \mathbb{P}\}$ by the pullback of $q^{\to} : \mathbb{P}^{\to} \to \mathbb{B}^{\to}$ along $\mathcal{P} : \mathbb{E} \to \mathbb{B}^{\to}$ where the comprehension category $\mathcal{P}$ is induced by $p : \mathbb{E} \to \mathbb{B}$.

$$\begin{array}{ccc} \{\mathbb{E} \mid \mathbb{P}\} & \xrightarrow{(q^{\rightarrow})^*\mathcal{P}} & \mathbb{P}^{\rightarrow} \\ {\scriptstyle \mathcal{P}^*(q^{\rightarrow})}\downarrow & \lrcorner & \downarrow{\scriptstyle q^{\rightarrow}} \\ \mathbb{E} & \xrightarrow{\mathcal{P}} & \mathbb{B}^{\rightarrow} \end{array}$$

That is, objects are tuples (X, P, Q) where $X \in \mathbb{E}$, $P \in \mathbb{P}_{pX}$, $Q \in \mathbb{P}_{\{X\}}$, and $Q \leq \pi_X^* P$; and morphisms are tuples $(f, g, h) : (X, P, Q) \to (X', P', Q')$ where $f : X \to X'$, $g : P \to P'$, $h : Q \to Q'$, $pf = qg$, and $\{f\} = qh$.

The intuition of this definition is as follows. For each object $(X, P, Q) \in \{\mathbb{E} \mid \mathbb{P}\}$, X represents a type $\Gamma \vdash A$ in the underlying type system, P represents a predicate on the context Γ, and Q represents the conjunction of a predicate on $\Gamma, v : A$ and the predicate P (thus $Q \leq \pi_X^* P$ is imposed). Note that $\mathcal{P}^*(q^{\rightarrow}) : \{\mathbb{E} \mid \mathbb{P}\} \to \mathbb{E}$ is faithful because q is faithful.

Let $\{p \mid q\} : \{\mathbb{E} \mid \mathbb{P}\} \to \mathbb{P}$ be a functor defined by $\mathrm{cod} \circ (q^{\rightarrow})^*\mathcal{P}$, that is, $(X, P, Q) \mapsto P$. The functor $\{p \mid q\}$ inherits most of the CCompC structure of $p : \mathbb{E} \to \mathbb{B}$.

Lemma 2. *The functor $\{p \mid q\} : \{\mathbb{E} \mid \mathbb{P}\} \to \mathbb{P}$ is a split fibration. The cartesian lifting of $g : P' \to P$ is given by*

$$(\overline{qg}(X), g, \overline{\{\overline{qg}(X)\}}(Q) \circ \pi') : ((qg)^* X, P', \pi_{(qg)^* X}^* P' \wedge \{\overline{qg}(X)\}^* Q) \to (X, P, Q)$$

where π' is a projection for fibred products. □

Lemma 3. *The fibration $\{p \mid q\} : \{\mathbb{E} \mid \mathbb{P}\} \to \mathbb{P}$ is a full comprehension category with unit that admits strong coproducts.*

Proof. The main idea is that the structure in the CCompC $p : \mathbb{E} \to \mathbb{B}$ can be lifted to $\{\mathbb{E} \mid \mathbb{P}\} \to \mathbb{P}$. Here, we only show the definition of (object parts of) fibred terminal objects $1 : \mathbb{P} \to \{\mathbb{E} \mid \mathbb{P}\}$, the comprehension functor $\{-\} : \{\mathbb{E} \mid \mathbb{P}\} \to \mathbb{P}$, and coproducts $\coprod_{(X,P,Q)} : \{\mathbb{E} \mid \mathbb{P}\}_Q \to \{\mathbb{E} \mid \mathbb{P}\}_P$ for each $(X, P, Q) \in \{\mathbb{E} \mid \mathbb{P}\}$.

$$1P = (1qP, P, \pi_{1qP}^* P) \quad \{(X, P, Q)\} = Q \quad \coprod_{(X,P,Q)} (Y, Q, R) = (\coprod_X Y, P, (\kappa^{-1})^* R)$$

The rest of the proof is omitted. □

The existence of products in $\{p \mid q\}$ requires additional conditions.

Lemma 4. *If $q : \mathbb{P} \to \mathbb{B}$ has fibred exponentials and p-products (in addition to fibred finite products), then $\{p \mid q\} : \{\mathbb{E} \mid \mathbb{P}\} \to \mathbb{P}$ admits products.*

Proof. We define $\prod_{(X,P,Q)} : \{\mathbb{E} \mid \mathbb{P}\}_Q \to \{\mathbb{E} \mid \mathbb{P}\}_P$ by

$$\prod_{(X,P,Q)} (Y, Q, R) = (\prod_X Y, P, \pi_{\prod_X Y}^* P \wedge \prod_{\pi_{\prod_X Y}^* X} \sigma_{\prod_X Y, X}^* (\pi_{\pi_X^* \prod_X Y}^* Q \Rightarrow \{\epsilon_Y^{\pi_X^* \dashv \prod_X}\}^* R)).$$

$$\begin{array}{ccccccc} Q \in \mathbb{P}_{\{X\}} & \xrightarrow{\pi_{\pi_X^* \prod_X Y}^*} & & & & \xrightarrow{\prod_{\pi_{\prod_X Y}^* X}} & \\ & & \mathbb{P}_{\{\pi_X^* \prod_X Y\}} & \xrightarrow{\sigma_{\prod_X Y, X}^*} & \mathbb{P}_{\{\pi_{\prod_X Y}^* X\}} & \top & \mathbb{P}_{\{\prod_X Y\}} \\ R \in \mathbb{P}_{\{Y\}} & \xrightarrow[\{\epsilon_Y^{\pi_X^* \dashv \prod_X}\}^*]{} & & & & \xleftarrow[\pi_{\pi_{\prod_X Y}^* X}^*]{} & \end{array}$$

Then, this gives products in $\{p \mid q\}$ but we omit the lengthy proof. □

As a result, we get a lifting of SCCompCs over $p : \mathbb{E} \to \mathbb{B}$.

Theorem 5. *If $p : \mathbb{E} \to \mathbb{B}$ is a SCCompC and $q : \mathbb{P} \to \mathbb{B}$ is a fibred ccc that has p-products, then $\{p \mid q\} : \{\mathbb{E} \mid \mathbb{P}\} \to \mathbb{P}$ is a SCCompC. Moreover, $(\mathcal{P}^*(q^{\to}), q) : \{p \mid q\} \to p$ is a morphism of SCCompCs, i.e., a split fibred functor that preserves the CCompC structure strictly.*

$$\begin{array}{ccc} \{\mathbb{E} \mid \mathbb{P}\} & \xrightarrow{\mathcal{P}^*(q^{\to})} & \mathbb{E} \\ \downarrow{\scriptstyle \{p|q\}} & & \downarrow{\scriptstyle p} \\ \mathbb{P} & \xrightarrow{q} & \mathbb{B} \end{array}$$

Proof. By Lemma 3 and Lemma 4. A terminal object in $\mathbb{P}$ exists because $\mathbb{B}$ has a terminal object and $q : \mathbb{P} \to \mathbb{B}$ has fibred terminal objects. It is almost obvious that $(\mathcal{P}^*(q^{\to}), q)$ preserves the structure of CCompCs. □

Example 6. Consider the simple fibration $\mathsf{s}_{\mathbf{Set}} : \mathbf{s}(\mathbf{Set}) \to \mathbf{Set}$ and the subobject fibration $\mathsf{sub}_{\mathbf{Set}} : \mathbf{Sub}(\mathbf{Set}) \to \mathbf{Set}$ (see [10, §1.3]). Objects in $\{\mathbf{s}(\mathbf{Set}) \mid \mathbf{Sub}(\mathbf{Set})\}$ are tuples $((I, X), P, Q)$ where $(I, X) \in \mathbf{s}(\mathbf{Set})$, $P \subseteq I$, and $Q \subseteq P \times X \subseteq I \times X$, and morphisms are those in $\mathbf{s}(\mathbf{Set})$ that preserve predicates. In $\{\mathsf{s}_{\mathbf{Set}} \mid \mathsf{sub}_{\mathbf{Set}}\} : \{\mathbf{s}(\mathbf{Set}) \mid \mathbf{Sub}(\mathbf{Set})\} \to \mathbf{Sub}(\mathbf{Set})$, products are given by

$$\begin{aligned} \prod_{((I,X),P,Q)} ((I \times X, Y), Q, R) = ((I, X \Rightarrow Y), P, \{(i, f) \in I \times (X \Rightarrow Y) \mid \\ i \in P \wedge \forall x \in X, (i, x) \in Q \implies ((i, x), f(x)) \in R\}). \end{aligned} \tag{2}$$

Example 7. Let $\mathsf{erel} : \mathbf{ERel} \to \mathbf{Set}$ be the fibration of endorelations defined by change-of-base from $\mathbf{Sub}(\mathbf{Set}) \to \mathbf{Set}$ along the functor $X \mapsto X \times X$. The fibration erel is a fibred ccc and has products (i.e. right adjoints of reindexing functors that satisfy the BC condition for each pullback square). Therefore, erel has p-products for any comprehension category with unit p. If we apply Theorem 5 to erel and the simple fibration $\mathsf{s}_{\mathbf{Set}} : \mathbf{s}(\mathbf{Set}) \to \mathbf{Set}$, then products are defined similarly to Example 6.

Example 8. Consider the family fibration $\mathsf{fam}_{\mathbf{Set}} : \mathbf{Fam}(\mathbf{Set}) \to \mathbf{Set}$ [10, Def 1.2.1] and the subobject fibration $\mathsf{sub}_{\mathbf{Set}} : \mathbf{Sub}(\mathbf{Set}) \to \mathbf{Set}$. Objects in $\{\mathbf{Fam}(\mathbf{Set}) \mid \mathbf{Sub}(\mathbf{Set})\}$ are tuples $((I, X), P, Q)$ where $(I, X) \in \mathbf{Fam}(\mathbf{Set})$, $P \subseteq I$, and $Q \subseteq \coprod_{i \in P} Xi \subseteq \coprod_{i \in I} Xi$. Note that subsets $Q \subseteq \coprod_{i \in I} Xi$ have a one-to-one correspondence with families of subsets $(Qi \subseteq Xi)_{i \in I}$ when we define $Qi = \iota_i^*(Q)$ where $\iota_i : Xi \to \coprod_{i \in I} Xi$ is the i-th injection. So, we often identify Q with the family of subsets $Qi \subseteq Xi$. We get products in $\{\mathsf{fam}_{\mathbf{Set}} \mid \mathsf{sub}_{\mathbf{Set}}\} : \{\mathbf{Fam}(\mathbf{Set}) \mid \mathbf{Sub}(\mathbf{Set})\} \to \mathbf{Sub}(\mathbf{Set})$ by modifying (2) for dependent functions.

3.2 Lifting Fibred Coproducts

A sufficient condition for $\{p \mid q\} : \{\mathbb{E} \mid \mathbb{P}\} \to \mathbb{P}$ to have strong fibred coproducts is given by the following lemma, which is analogous to [9, Prop. 4.5.8].

Lemma 9. *If (1) $p : \mathbb{E} \to \mathbb{B}$ is a CCompC that has strong fibred coproducts (2) for each $X, Y \in \mathbb{E}_I$, $X', Y' \in \mathbb{E}_{I'}$, $u : I \to I'$, and pair of cartesian liftings $f : X \to X'$ and $g : Y \to Y'$ over u, the following two squares are pullbacks*

$$\begin{array}{ccccc} \{X\} & \xrightarrow{\{\iota_1\}} & \{X+Y\} & \xleftarrow{\{\iota_2\}} & \{Y\} \\ {\scriptstyle \{f\}}\downarrow & \lrcorner & \downarrow{\scriptstyle \{f+g\}} & \llcorner & \downarrow{\scriptstyle \{g\}} \\ \{X'\} & \xrightarrow{\{\iota_1\}} & \{X'+Y'\} & \xleftarrow{\{\iota_2\}} & \{Y'\} \end{array}$$

(3) $q : \mathbb{P} \to \mathbb{B}$ *is a fibred distributive category (4) for each* $X, Y \in \mathbb{E}_I$ *and* $Z \in \mathbb{E}_{\{X+Y\}}$, *q has cocartesian liftings of* $\{\iota_1\} : \{X\} \to \{X+Y\}$, $\{\iota_2\} : \{Y\} \to \{X+Y\}$, $\{\overline{\{\iota_1\}}(Z)\} : \{\{\iota_1\}^*Z\} \to \{Z\}$, *and* $\{\overline{\{\iota_2\}}(Z)\} : \{\{\iota_2\}^*Z\} \to \{Z\}$ *that satisfy the BC condition for each pullback squares and Frobenius, then* $\{p \mid q\} : \{\mathbb{E} \mid \mathbb{P}\} \to \mathbb{P}$ *has strong fibred coproducts, and the fibred functor* $(\mathcal{P}^*(q^{\to}), q) : \{p \mid q\} \to p$ *strictly preserves fibred coproducts.*

Proof. We define fibred coproducts by $(X, P, Q)+(Y, P, R) = (X+Y, P, \{\iota_1\}_!Q \vee \{\iota_2\}_!R)$. We omit the rest of the proof. □

Note that if q is fibred bicartesian closed, then q is a fibred distributive category.

Example 10. Consider $s_{\mathbf{Set}} : \mathbf{s}(\mathbf{Set}) \to \mathbf{Set}$ and $\mathrm{sub}_{\mathbf{Set}} : \mathbf{Sub}(\mathbf{Set}) \to \mathbf{Set}$ (recall Example 6). This combination satisfies four conditions in Lemma 9. Fibred coproducts in $\{\mathbf{s}(\mathbf{Set}) \mid \mathbf{Sub}(\mathbf{Set})\} \to \mathbf{Sub}(\mathbf{Set})$ are defined as follows.

$$((I, X), P, Q) + ((I, Y), P, R) = ((I, X+Y), P, \{(i, x) \mid (i, x) \in Q \vee (i, x) \in R\})$$

4 Lifting Monads on SCCompCs

Suppose we have a SCCompC $p : \mathbb{E} \to \mathbb{B}$ and a posetal fibration $q : \mathbb{P} \to \mathbb{B}$ as ingredients for $\{p \mid q\} : \{\mathbb{E} \mid \mathbb{P}\} \to \mathbb{P}$ in Theorem 5. We explain how to construct a fibred monad on $\{p \mid q\} : \{\mathbb{E} \mid \mathbb{P}\} \to \mathbb{P}$ from monads on p and q.

First, we assume that a monad T on $\mathbb{B}$ and a fibred monad $\hat{T}$ on $p : \mathbb{E} \to \mathbb{B}$ are given. These monads are intended to represent the same computational effects in underlying type systems, but T is more "primitive" than $\hat{T}$, and $\hat{T}$ is induced from T in some natural way. For example, we can use the maybe monad or the powerset monad on **Set** as T and define $\hat{T}$ by $(I, X) \mapsto (I, TX)$ on the simple fibration $\mathbf{s}(\mathbf{Set}) \to \mathbf{Set}$. In such a situation, we often have an oplax monad morphism (Definition 11) $\theta : \{\hat{T}(-)\} \to T\{-\}$. Intuitively, θ extends the action of $\hat{T}$ on types to contexts, just like strengths of strong monads. We also need a lifting $\dot{T}$ of T along $q : \mathbb{P} \to \mathbb{B}$ to specify a mapping from predicates on values in $X \in \mathbb{B}$ to predicates on computations in TX [1]. Given all these ingredients and some additional conditions, we define a fibred monad on $\{p \mid q\} : \{\mathbb{E} \mid \mathbb{P}\} \to \mathbb{P}$, which is a lifting of the fibred monad $\hat{T}$ on $p : \mathbb{E} \to \mathbb{B}$.

Definition 11 (oplax monad morphism). Let $\mathbb{C}, \mathbb{D}$ be categories, $F : \mathbb{C} \to \mathbb{D}$ be a functor, and (S, η^S, μ^S), (T, η^T, μ^T) be monads on $\mathbb{C}$ and $\mathbb{D}$, respectively. A natural transformation $\theta : FS \to TF$ is an *oplax monad morphism* if θ respects units and multiplications.

$$\begin{array}{ccc} FX & & \\ {\scriptstyle F\eta^S_X}\downarrow & \searrow{\scriptstyle \eta^T_{FX}} & \\ FSX & \xrightarrow{\theta_X} & TFX \end{array} \qquad \begin{array}{ccccc} FS^2X & \xrightarrow{\theta_{SX}} & TFSX & \xrightarrow{T\theta_X} & T^2FX \\ {\scriptstyle F\mu^S_X}\downarrow & & & & \downarrow{\scriptstyle \mu^T_{FX}} \\ FSX & & \xrightarrow{\theta_X} & & TFX \end{array}$$

Theorem 12. *Let T be a monad on $\mathbb{B}$, $\hat{T}$ be a fibred monad on $p : \mathbb{E} \to \mathbb{B}$ in the 2-category $\mathbf{Fib}_{\mathbb{B}}$ of fibrations over $\mathbb{B}$, $\theta : \{\hat{T}(-)\} \to T\{-\}$ be an oplax monad morphism, and $\dot{T}$ be a fibred lifting [1] of T along $q : \mathbb{P} \to \mathbb{B}$. If*

$$\pi^*_{\hat{T}X} P \wedge \theta^*_X \dot{T} Q \leq \theta^*_X \dot{T}(\pi^*_X P \wedge Q) \tag{3}$$

holds for each $X \in \mathbb{E}$, $P \in \mathbb{P}_{pX}$ and $Q \in \mathbb{P}_{\{X\}}$, then there exists a fibred monad S on $\{p \mid q\} : \{\mathbb{E} \mid \mathbb{P}\} \to \mathbb{P}$ in $\mathbf{Fib}_{\mathbb{P}}$ such that the fibred functor $\{p \mid q\} \to p$ in Theorem 5 is a fibred monad morphism from S to $\hat{T}$.

Proof. We define $S(X, P, Q) = (\hat{T}X, P, \pi^*_{\hat{T}X} P \wedge \theta^* \dot{T} Q)$. Then the monad structure of $\hat{T}$ lifts to S. The assumption (3) is required to prove that S is fibred.

$$\begin{array}{ccc} \mathbb{P} & & \theta^* \dot{T} Q \xrightarrow{\overline{\theta}(\dot{T}Q)} \dot{T} Q \\ \downarrow q & & \\ \mathbb{B} & & \{\hat{T}X\} \xrightarrow{\ \theta\ } T\{X\} \end{array}$$

□

Example 13. Any strong monad T on a CCC $\mathbb{B}$ gives rise to a split fibred monad $\hat{T}$ on the simple fibration $\mathsf{s}_{\mathbb{B}} : \mathbf{s}(\mathbb{B}) \to \mathbb{B}$ (actually, there is a one-to-one correspondence [10, Ex.2.6.10]). The monad $\hat{T}$ is defined by $(I, X) \mapsto (I, TX)$. An oplax monad morphism $\theta : I \times TX \to T(I \times X)$ is given by the strength.

Now consider the case where $\mathbb{B} = \mathbf{Set}$. Since the strength for the monad T on $\mathbf{Set}$ is given uniquely [17, Proposition 3.4], we can prove that (3) holds for any fibred lifting of T along the subobject fibration $\mathsf{sub}_{\mathbf{Set}} : \mathbf{Sub}(\mathbf{Set}) \to \mathbf{Set}$.

Let T be the maybe monad $(-) + \{*\}$. There are two fibred liftings of T:

$$\dot{T}_1(P \subseteq I) = (P + \{*\} \subseteq I + \{*\}) \qquad \dot{T}_2(P \subseteq I) = (P \subseteq I + \{*\})$$

for each $(P \subseteq I) \in \mathbf{Sub}(\mathbf{Set})$. The lifting $\dot{T}_1$ corresponds to partial correctness, and $\dot{T}_2$ corresponds to total correctness. The fibred monads on $\{\mathsf{s}_{\mathbf{Set}} \mid \mathsf{sub}_{\mathbf{Set}}\}$ defined in Theorem 12 from $\dot{T}_1$ and $\dot{T}_2$ are given by

$$((I, X), P, Q) \mapsto ((I, X + \{*\}), P, \{(i, x) \mid (i \in P \wedge x = *) \vee (i, x) \in Q\})$$
$$((I, X), P, Q) \mapsto ((I, X + \{*\}), P, \{(i, x) \mid (i, x) \in Q\})$$

respectively. Here, we leave the left/right injection of coproducts implicit.

Example 14. For each monad T on $\mathbf{Set}$, we have a split fibred monad on the family fibration $\mathbf{Fam}(\mathbf{Set}) \to \mathbf{Set}$ defined by $\hat{T}(I, X) = (I, T \circ X)$. We have an oplax monad morphism $\theta : \coprod_{i \in I} TXi \to T \coprod_{i \in I} Xi$ defined by the cotupling $[(T\iota_i)_{i \in I}] : \coprod_{i \in I} TXi \to T \coprod_{i \in I} Xi$ where $\iota_i : Xi \to \coprod_{i \in I} Xi$ is the i-th injection. The condition (3) holds for any fibred lifting of T along the subobject fibration $\mathbf{Sub}(\mathbf{Set}) \to \mathbf{Set}$. Moreover, we have $\iota_i^* \theta^* \dot{T} Q = \dot{T} \iota_i^* Q$ for each $Q \in \mathbf{Sub}(\mathbf{Set})_{\coprod_{i \in I} Xi}$, so the monad in Theorem 12 is given by

$$((I, X), P, (Qi \subseteq Xi)_{i \in I}) \mapsto ((I, T \circ X), P, (\dot{T} Qi \subseteq TXi)_{i \in I}).$$

5 Soundness

We consider a concrete dependent refinement type system with computational effects and define sound semantics to show that the SCCompC defined in Theorem 5 has sufficient structures for dependent refinement types. Here, we consider two type systems. One is an underlying type system that is a fragment of EMLTT [2–4]. The other is a refinement of the underlying type system that has refinement types $\{v : A \mid p\}$ and a subtyping relation $\Gamma \vdash A <: B$ induced by logical implication. The two type systems share a common syntax for terms while types are more expressive in the refinement type system. We consider liftings of fibred adjunction models to interpret the refinement type system. Here, Theorem 12 can be used to obtain a lifting of fibred adjunction models via Eilenberg-Moore construction. We prove a soundness theorem that claims if a term is well-typed in the refinement type system, then the interpretation of the term has a lifting along the morphism of CCompCs defined in Theorem 5.

5.1 Underlying Type System

We define the underlying dependent type system by a slightly modified version of a fragment of EMLTT [2–4]. We remove some of the types and terms from the original for simplicity. We parameterize our type system with a set of base type constructors (ranged over by b) and a set of value constants (ranged over by c) for convenience.

We define value types $(A, B, \dots)$, computation types $(\underline{C}, \underline{D}, \dots)$, contexts $(\Gamma, \dots)$, value terms $(V, W, \dots)$, and computation terms $(M, N, \dots)$ as follows.

$$A := 1 \mid b_A(V) \mid \Sigma x{:}A.B \mid U\underline{C} \mid A + B$$

$$\underline{C} := FA \mid \Pi x{:}A.\underline{C} \qquad\qquad \Gamma := \diamond \mid \Gamma, x : A$$

$$V := x \mid * \mid c_A \mid \langle V, W\rangle_{(x:A).B} \mid \mathbf{thunk}\ M \mid \mathbf{inl}_{A+B}\ V \mid \mathbf{inr}_{A+B}\ V$$

$$M := \mathbf{return}\ V \mid M\ \mathbf{to}\ x : A\ \mathbf{in}_{\underline{C}}\ N \mid \mathbf{force}_{\underline{C}}\ V \mid \lambda x : A.M \mid M(V)_{(x:A).\underline{C}} \mid \\ \mathbf{pm}\ V\ \mathbf{as}\ \langle x : A, y : B\rangle\ \mathbf{in}_{z.\underline{C}}\ M \mid \\ \mathbf{case}\ V\ \mathbf{of}_{z.\underline{C}}\ (\mathbf{inl}\ (x : A) \mapsto M, \mathbf{inr}\ (y : B) \mapsto N)$$

We implicitly assume that variables in Γ are mutually different. We use many type annotations in the syntax of terms for a technical reason, but we might omit them if they are clear from the context. We define substitution $A[V/x]$, $\underline{C}[V/x]$, $W[V/x]$, and $M[V/x]$ as usual.

For each type constructor b, let $\arg(b)$ be a closed value type of the argument of b. We write $b : A \to \mathrm{Type}$ if $A = \arg(b)$. For each value constant c, let $\mathrm{ty}(c)$ be a closed value type of c.

We have several kinds of judgements: well-formed contexts $\vdash \Gamma$; well-formed (value or computation) types $\Gamma \vdash A$, $\Gamma \vdash \underline{C}$; well-typed (value or computation) terms $\Gamma \vdash V : A$, $\Gamma \vdash M : \underline{C}$; and definitional equalities for contexts, types and terms $\vdash \Gamma_1 = \Gamma_2$, $\Gamma \vdash A = B$, $\Gamma \vdash \underline{C} = \underline{D}$, $\Gamma \vdash V = W : A$, $\Gamma \vdash M = N : \underline{C}$.

Typing rules are basically the same as EMLTT. Rules for base type constructors and value constants are shown in Fig. 2

$$\frac{\vdash \Gamma \qquad \diamond \vdash \mathrm{ty}(c)}{\Gamma \vdash c_{\mathrm{ty}(c)} : \mathrm{ty}(c)} \qquad \frac{b : A \to \mathrm{Type} \quad \diamond \vdash A \qquad \Gamma \vdash V : A}{\Gamma \vdash b_A(V)} \qquad \frac{b : A \to \mathrm{Type} \qquad \diamond \vdash A \quad \Gamma \vdash V = W : A}{\Gamma \vdash b_A(V) = b_A(W)}$$

Fig. 2. Some typing rules for the underlying type system.

Semantics. We use fibred adjunction models to interpret terms and types. We adapt the definition for our fragment of EMLTT as follows.

Definition 15 (Fibred adjunction models). A *fibred adjunction model* is a fibred adjunction $F \dashv U : r \to p$ where $p : \mathbb{E} \to \mathbb{B}$ is a SCCompC with strong fibred coproducts and $r : \mathbb{C} \to \mathbb{B}$ is a fibration with p-products.

The Eilenberg-Moore fibration of a CCompC $p : \mathbb{E} \to \mathbb{B}$ inherits products in p [2, Theorem 4.3.24] and thus gives an example of fibred adjunction models.

Lemma 16. *Given a SCCompC $p : \mathbb{E} \to \mathbb{B}$ with strong fibred products and a split fibred monad T on p, then the Eilenberg-Moore adjunction of T is a fibred adjunction model.* □

We assume that a fibred adjunction model $F \dashv U : r \to p$ between $p : \mathbb{E} \to \mathbb{B}$ and $r : \mathbb{C} \to \mathbb{B}$ is given and that interpretations of base type constructors $[\![b]\!] \in \mathbb{E}$ and value constants $[\![c]\!] \in \mathbb{E}_1(1, X)$ (for some $X \in \mathbb{E}_1$) are given. We define a partial interpretation $[\![-]\!]$ of the following form for raw syntax.

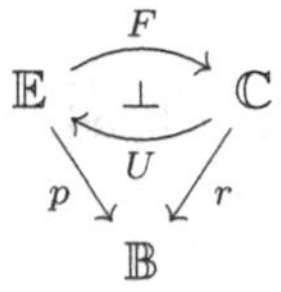

$$[\![\Gamma]\!] \in \mathbb{B} \qquad [\![\Gamma; A]\!] \in \mathbb{E}_{[\![\Gamma]\!]} \qquad [\![\Gamma; \underline{C}]\!] \in \mathbb{C}_{[\![\Gamma]\!]}$$
$$[\![\Gamma; V]\!] \in \mathbb{E}_{[\![\Gamma]\!]}(1[\![\Gamma]\!], A) \qquad \text{for some } A$$
$$[\![\Gamma; M]\!] \in \mathbb{E}_{[\![\Gamma]\!]}(1[\![\Gamma]\!], UC) \qquad \text{for some } C \in \mathbb{C}$$

Most of the definition of $[\![-]\!]$ are the same as [2]. For base type constructors b and value constants c, we define $[\![-]\!]$ as follows.

$$[\![\Gamma; b_A(V)]\!] = (s[\![\Gamma; V]\!])^* \{\overline{!_{[\![\Gamma]\!]}}([\![\diamond; A]\!])\}^* [\![b]\!] \qquad [\![\Gamma; c_A]\!] = !^*_{[\![\Gamma]\!]} [\![c]\!]$$

Here, left-hand sides are defined if right-hand sides are defined.

Proposition 17 (Soundness). *Assume that $[\![b]\!] \in \mathbb{E}_{\{[\![\diamond; A]\!]\}}$ holds for each $b :$ $A \to$ Type such that $[\![\diamond; A]\!]$ is defined, and $[\![c]\!] \in \mathbb{E}_1(1, [\![\diamond; \mathrm{ty}(c)]\!])$ holds if $[\![\diamond; \mathrm{ty}(c)]\!]$ $\in \mathbb{E}_1$ is defined. Interpretations $[\![-]\!]$ of well-formed contexts and types and well-typed terms are defined. If two contexts, types, or terms are definitionally equal, then their interpretations are equal.* □

5.2 Predicate Logic

We define syntax for logical formulas by

$$p = \top \mid p \wedge q \mid p \Rightarrow q \mid \forall x : A.p \mid V =_A W \mid a(V)$$

$$\frac{\Gamma \vdash V : A \qquad \Gamma \vdash W : A}{\Gamma \vdash V =_A W : \mathrm{Prop}} \qquad \frac{a : A \to \mathrm{Prop} \qquad \diamond \vdash A \qquad \Gamma \vdash V : A}{\Gamma \vdash a(V) : \mathrm{Prop}}$$

Fig. 3. Some rules for well-formed predicates.

where a ranges over predicate symbols. Here, we added $\top$ and $V =_A W$ for typing rule for the unique value of the unit type and variables of base types (i.e. for selfification [18]), respectively, which we describe later. However, there is a large amount of freedom to choose the syntax of logical formulas. The least requirement here is that logical formulas can be interpreted in a posetal fibration $q : \mathbb{P} \to \mathbb{B}$, and interpretations of logical formulas admit semantic weakening, substitution, and conversion in the sense of [2, Proposition 5.2.4, 5.2.6]. So, we can almost freely add or remove logical connectives and quantifiers as long as $q : \mathbb{P} \to \mathbb{B}$ admits them.

We define a standard judgement of well-formedness for logical formulas. Some of the rules for well-formedness are shown in Fig. 3

Logical formulas are interpreted in the fibration $q : \mathbb{P} \to \mathbb{B}$. We assume that interpretation $[\![a]\!] \in \mathbb{P}_{\{[\![\diamond; A]\!]\}}$ for each predicate symbol $a : A \to \mathrm{Prop}$ is given. The interpretation $[\![\Gamma \vdash p]\!] \in \mathbb{P}_{[\![\Gamma]\!]}$ is standard and defined inductively for each well-formed formulas. For example:

$$[\![\Gamma \vdash V =_A W]\!] = (s[\![\Gamma; V]\!])^*(s(\pi^*_{[\![\Gamma;A]\!]}[\![\Gamma; W]\!]))^*\mathrm{Eq}(\top\{[\![\Gamma; A]\!]\})$$
$$[\![\Gamma \vdash a(V)]\!] = s([\![\Gamma; V]\!])^*\{\overline{!_{[\![\Gamma]\!]}}([\![\diamond; A]\!])\}^*[\![a]\!]$$

where $a : A \to \mathrm{Prop}$ is a predicate symbol and s is the bijection defined in §2.

5.3 Refinement Type System

We refine the underlying type system by adding predicates to base types and the unit type. From now on, we use subscript A_u for types in the underlying type system to distinguish them from types in the refinement type system.

$$A \coloneqq \{v : b_{A_u}(V) \mid p\} \mid \{v : 1 \mid p\} \mid \Sigma x{:}A.B \mid U\underline{C} \mid A + B$$
$$\underline{C} \coloneqq FA \mid \Pi x{:}A.\underline{C} \qquad\qquad \Gamma \coloneqq \diamond \mid \Gamma, x : A$$

We use the same definition of terms as the underlying type system and the same set of base type constructors and value constants. Argument types of base type constructors $b : A_u \to \mathrm{Type}$ are also the same, but types $\mathrm{ty}(c)$ assigned to value constants c are redefined as refinement types. Given a type A (or $\underline{C}$) in the refinement type system, we define its underlying type $|A|$ (or $|\underline{C}|$) by induction where predicates are eliminated in the base cases.

$$|\{v : b_{A_u}(V) \mid p\}| = b_{A_u}(V) \qquad |\{v : 1 \mid p\}| = 1$$

Underlying contexts $|\Gamma|$ are also defined by $|\diamond| = \diamond$ and $|\Gamma, x : A| = |\Gamma|, x : |A|$.

$$\frac{b : A_u \to \mathrm{Type} \quad \vdash \Gamma \quad |\Gamma| \vdash b_{A_u}(V) \quad |\Gamma|, v : b_{A_u}(V) \vdash p : \mathrm{Prop}}{\Gamma \vdash \{v : b_{A_u}(V) \mid p\}} \qquad \frac{\vdash \Gamma \quad |\Gamma| \vdash b_{A_u}(V) = b_{A_u}(W) \quad \Gamma; v : b_{A_u}(V) \mid p \vdash q}{\Gamma \vdash \{v : b_{A_u}(V) \mid p\} <: \{v : b_{A_u}(W) \mid q\}}$$

$$\frac{\vdash \Gamma_1, x : \{v : b_{A_u}(V) \mid p\}, \Gamma_2}{\Gamma_1, x : \{v : b_{A_u}(V) \mid p\}, \Gamma_2 \vdash x : \{v : b_{A_u}(V) \mid v = x\}} \qquad \frac{\vdash \Gamma \quad \diamond \vdash \mathrm{ty}(c)}{\Gamma \vdash c_{|\mathrm{ty}(c)|} : \mathrm{ty}(c)}$$

$$\frac{\Gamma \vdash A_2 <: A_1 \quad \Gamma, x : A_1 \vdash \underline{C}_1 \quad \Gamma, x : A_2 \vdash \underline{C}_1 <: \underline{C}_2}{\Gamma \vdash \Pi x{:}A_1.\underline{C}_1 <: \Pi x{:}A_2.\underline{C}_2} \qquad \frac{\Gamma_2 \vdash V : A \quad \vdash \Gamma_1 <: \Gamma_2 \quad \Gamma_1 \vdash A <: B}{\Gamma_1 \vdash V : B} \qquad \frac{\vdash \Gamma}{\Gamma \vdash * : \{v : 1 \mid \top\}}$$

$$\frac{\vdash \Gamma \quad |\Gamma|, v : 1 \vdash p : \mathrm{Prop}}{\Gamma \vdash \{v : 1 \mid p\}} \qquad \frac{\vdash \Gamma \quad \Gamma; v : 1 \mid p \vdash q}{\Gamma \vdash \{v : 1 \mid p\} <: \{v : 1 \mid q\}}$$

Fig. 4. Some typing rules for the refinement type system.

Judgements in the refinement type system are as follows. We have judgements for well-formedness or well-typedness for contexts, types and terms in the refinement type system, which are denoted in the same way as the underlying type system. We do not consider definitional equalities for terms because they are the same as the underlying type system. Instead, we add judgements for subtyping between types and contexts. They are denoted by $\vdash \Gamma_1 <: \Gamma_2$ for context, $\Gamma \vdash A <: B$ for value types, and $\Gamma \vdash \underline{C} <: \underline{D}$ for computation types.

Most of term and type formation rules are similar to the underlying type system. We listed some of the non-trivial modifications of typing rules in Fig. 4. We add typing rules for $\{v : b_{B_u}(V) \mid p\}$ and $\{v : 1 \mid p\}$. Subtyping for these types are defined by judgements $\Gamma; v : A_u \mid p \vdash q$ for logical implication. Here, $\Gamma; v : A_u \mid p \vdash q$ means "assumptions in Γ and p implies q" where p and q are well-formed formulas in the context $|\Gamma|, v : A_u$. We do not specify derivation rules for the judgement $\Gamma; v : A_u \mid p \vdash q$ but assume soundness of the judgement (explained later). We allow "selfification" [18] for variables of base types. Subtyping for $\Sigma x{:}A.B$, $U\underline{C}$, FA, and $\Pi x{:}A.\underline{C}$ are defined covariantly except the argument type A of $\Pi x{:}A.\underline{C}$, which is contravariant. We have the rule of subsumption. Value constants are typed with a refined type assignment $\mathrm{ty}(c)$. The unique value $*$ of the unit type has type $\{v : 1 \mid \top\}$.

Lemma 18. *If we eliminate predicates in the refinement types from well-formed contexts, types and terms, then we get well-formed contexts, types and terms of the underlying type system.*

- *If* $\vdash \Gamma$, *then* $\vdash |\Gamma|$. *If* $\Gamma \vdash A$, *then* $|\Gamma| \vdash |A|$. *If* $\Gamma \vdash \underline{C}$, *then* $|\Gamma| \vdash |\underline{C}|$.
- *If* $\vdash \Gamma_1 <: \Gamma_2$, *then* $\vdash |\Gamma_1| = |\Gamma_2|$. *If* $\Gamma \vdash A <: B$, *then* $|\Gamma| \vdash |A| = |B|$. *If* $\Gamma \vdash \underline{C} <: \underline{D}$, *then* $|\Gamma| \vdash |\underline{C}| = |\underline{D}|$.

Proof. By induction on the derivation of judgements. Each typing rule in the refinement type system has a corresponding rule in the underlying system. □

Example 19. We can express conditional branching using the elimination rule of the fibred coproduct type $1 + 1$. For example, assume we have a base type constructor int : 1 → Type for integers and a value constant for comparison.

$$(\leq) : U(\Pi x{:}\mathrm{int}.\Pi y{:}\mathrm{int}.F(\{v : 1 \mid x \leq y\} + \{v : 1 \mid x > y\}))$$

We can define **if** $x \leq y$ **then** M **else** N to be a syntax sugar for

$$(x \leq' y)\ \mathbf{to}\ z\ \mathbf{in}\ (\mathbf{case}\ z\ \mathbf{of}\ (\mathbf{inl}\ v \mapsto M, \mathbf{inr}\ v \mapsto N))$$

where $(\leq') = \mathbf{force}\ (\leq)$. Note that M and N are typed in contexts that have $v : \{v : 1 \mid x \leq y\}$ or $v : \{v : 1 \mid x > y\}$ depending on the result of comparison.

5.4 Semantics

Definition 20 (lifting of fibred adjunction models). Suppose that we have two fibred adjunction models $F \dashv U : q \to p$ between $p : \mathbb{E} \to \mathbb{B}$ and $q : \mathbb{C} \to \mathbb{B}$ and $\dot{F} \dashv \dot{U} : s \to r$ between $r : \mathbb{U} \to \mathbb{P}$ and $s : \mathbb{D} \to \mathbb{P}$. The fibred adjunction model $\dot{F} \dashv \dot{U}$ is a *lifting* of $F \dashv U$ if there exists functors $u : \mathbb{U} \to \mathbb{E}$, $v : \mathbb{D} \to \mathbb{C}$, and $t : \mathbb{P} \to \mathbb{B}$ such that these functors strictly preserve all structures of $\dot{F} \dashv \dot{U}$ to those of $F \dashv U$. That is, $(u, t) : r \to p$ and $(v, t) : s \to q$ are split fibred functors, the pair of fibred functor (u, t) and (v, t) is a map of adjunctions in the 2-category **Fib**, (u, t) strictly preserves the CCompC structure and fibred coproducts, and (v, t) maps r-products to p-products in the strict sense.

We assume that a lifting of fibred adjunction models is given as follows.

$$\begin{array}{ccc} \mathbb{E} & \underset{U}{\overset{F}{\rightleftarrows}}\ (\bot) & \mathbb{C} \\ & {\scriptstyle p}\searrow \ \swarrow & \\ & \mathbb{B} & \end{array} \qquad \begin{array}{ccc} \{\mathbb{E} \mid \mathbb{P}\} & \underset{\dot{U}}{\overset{\dot{F}}{\rightleftarrows}}\ (\bot) & \mathbb{D} \\ & {\scriptstyle \{p \mid q\}}\searrow \ \swarrow & \\ & \mathbb{P} & \end{array} \qquad \begin{array}{ccc} \{\mathbb{E} \mid \mathbb{P}\} & \xrightarrow{u} & \mathbb{E} \\ \downarrow{\scriptstyle \{p \mid q\}} & & \downarrow{\scriptstyle p} \\ \mathbb{P} & \xrightarrow{q} & \mathbb{B} \end{array} \qquad \begin{array}{ccc} \mathbb{D} & \xrightarrow{v} & \mathbb{C} \\ \downarrow & & \downarrow \\ \mathbb{P} & \xrightarrow{q} & \mathbb{B} \end{array} \tag{4}$$

Here, we assume more than just a lifting of fibred adjunction models by requiring the specific SCCompC $\{p \mid q\}$ with strong fibred coproducts, and the split functor $(u, q) : \{p \mid q\} \to p$ defined in Theorem 5 and Lemma 9. The underlying fibred adjunction model $F \dashv U$ is used for the underlying type system in §5.1, and $q : \mathbb{P} \to \mathbb{B}$ is for predicate logic in §5.2. One way to obtain such liftings of fibred adjunction models is to apply the Eilenberg-Moore construction to the monad morphism in Theorem 12, but in general we do not restrict $\mathbb{C}$ and $\mathbb{D}$ to be Eilenberg-Moore categories. We further assume that q has p-equalities to interpret logical formulas of the form $V =_A W$.

We define partial interpretation of refinement types $[\![\Gamma]\!] \in \mathbb{P}$, $[\![\Gamma; A]\!] \in \{\mathbb{E} \mid \mathbb{P}\}_{[\![\Gamma]\!]}$, and $[\![\Gamma; \underline{C}]\!] \in \mathbb{D}_{[\![\Gamma]\!]}$ similarly to the underlying type system but with the following modification. Here, we make use of the definition of $\{\mathbb{E} \mid \mathbb{P}\}$.

$$\begin{aligned} [\![\Gamma; \{v : b(V) \mid p\}]\!] &= ([\![|\Gamma|; b(V)]\!], [\![\Gamma]\!], \pi^*_{[\![|\Gamma|; b(V)]\!]}[\![\Gamma]\!] \wedge [\![|\Gamma|, v : b(V) \vdash p]\!]) \\ [\![\Gamma; \{v : 1 \mid p\}]\!] &= ([\![|\Gamma|; 1]\!], [\![\Gamma]\!], \pi^*_{[\![|\Gamma|; 1]\!]}[\![\Gamma]\!] \wedge [\![|\Gamma|, v : 1 \vdash p]\!]) \end{aligned}$$

For each $(X, P, Q), (X', P', Q') \in \{\mathbb{E} \mid \mathbb{P}\}$, we define a semantic subtyping relation $(X, P, Q) <: (X', P', Q')$ by the conjunction of $X = X'$, $P = P'$, and

$Q \leq Q'$. In other words, we have $(X, P, Q) <: (X', P', Q')$ if and only if there exists a morphism $(\mathrm{id}_X, \mathrm{id}_P, h) : (X, P, Q) \to (X', P', Q')$ that is mapped to identities by $u : \{\mathbb{E} \mid \mathbb{P}\} \to \mathbb{E}$ and $\{p \mid q\} : \{\mathbb{E} \mid \mathbb{P}\} \to \mathbb{P}$.

Lemma 21. – *If $[\![\Gamma]\!]$ is defined, then $[\![|\Gamma|]\!]$ is defined and equal to $q[\![\Gamma]\!]$.*
– *If $[\![\Gamma; A]\!]$ is defined, then $[\![|\Gamma|; |A|]\!]$ is defined and equal to $u[\![\Gamma; A]\!]$.*
– *If $[\![\Gamma; \underline{C}]\!]$ is defined, then $[\![|\Gamma|; |\underline{C}|]\!]$ is defined and equal to $v[\![\Gamma; \underline{C}]\!]$.*

Proof. By simultaneous induction. The case of $\{v : A_u \mid p\}$ is obvious, and other cases follow from the definition of liftings of fibred adjunction models. □

We do not specify syntactic derivation rules for judgement for logical implication $\Gamma; v : A_u \mid p \vdash q$. Instead, we assume soundness of $\Gamma; v : A_u \mid p \vdash q$ in the following sense: $\pi^*_{[\![|\Gamma|; A_u]\!]}[\![\Gamma]\!] \wedge [\![|\Gamma|, v : A_u \vdash p]\!] \leq [\![|\Gamma|, v : A_u \vdash q]\!]$ holds in $\mathbb{P}_{[\![|\Gamma|, v:A_u]\!]}$. For example, we can define a derivation rule for logical implication $\Gamma; v : A_u \mid p \vdash q$ from derivation rules for predicate logic $\Gamma_u \mid p \vdash q$ ("p implies q in the context Γ_u"). This is done by collecting predicates in context Γ by

$$(\!|\diamond|\!) = \top \qquad (\!|\Gamma, x : A|\!) = \begin{cases} (\!|\Gamma|\!) \wedge p[x/v] & \text{if } A = \{v : A_u \mid p\} \\ (\!|\Gamma|\!) & \text{otherwise} \end{cases}$$

and defining a derivation rule for judgement for logical implication $\Gamma; v : A_u \mid p \vdash q$ by $|\Gamma|, v : A_u \mid (\!|\Gamma|\!) \wedge p \vdash q$. If the derivation rules for predicate logic $\Gamma_u \mid p \vdash q$ is sound (i.e., $\Gamma_u \mid p \vdash q$ implies $[\![\Gamma_u \vdash p]\!] \leq [\![\Gamma_u \vdash q]\!]$), then so are the derivation rule for $\Gamma; v : A_u \mid p \vdash q$. This technique is used in, e.g., [27].

Theorem 22 (Soundness). *Assume that $\Gamma; v : A_u \mid p \vdash q$ is sound in the sense described above, $[\![b]\!] \in \mathbb{E}_{\{[\![\diamond; A]\!]\}}$ holds for each $b : A \to$ Type if $[\![\diamond; A]\!]$ is defined, and $[\![c]\!] \in \{\mathbb{E} \mid \mathbb{P}\}_1(1, [\![\diamond; \mathrm{ty}(c)]\!])$ holds if $[\![\diamond; \mathrm{ty}(c)]\!] \in \{\mathbb{E} \mid \mathbb{P}\}_1$ is defined. Then we have the following.*

– *If $\vdash \Gamma$, then $[\![\Gamma]\!] \in \mathbb{P}$ is defined. If $\Gamma \vdash A$, then $[\![\Gamma; A]\!] \in \{\mathbb{E} \mid \mathbb{P}\}_{[\![\Gamma]\!]}$ is defined. If $\Gamma \vdash \underline{C}$, then $[\![\Gamma; \underline{C}]\!] \in \mathbb{D}_{[\![\Gamma]\!]}$ is defined.*
– *If $\vdash \Gamma_1 <: \Gamma_2$, then $[\![\Gamma_1]\!] \leq [\![\Gamma_2]\!]$ in a fibre category of $\mathbb{P}$.*
– *If $\Gamma \vdash A <: B$, then $[\![\Gamma; A]\!] <: [\![\Gamma; B]\!]$. If $\Gamma \vdash \underline{C} <: \underline{D}$, then $\dot{U}[\![\Gamma; \underline{C}]\!] <: \dot{U}[\![\Gamma; \underline{D}]\!]$.*
– *If $\Gamma \vdash V : A$, then there exists a lifting $[\![\Gamma; V]\!] : 1[\![\Gamma]\!] \to [\![\Gamma; A]\!]$ above $[\![|\Gamma|; V]\!]$ along $u : \{\mathbb{E} \mid \mathbb{P}\} \to \mathbb{E}$. If $\Gamma \vdash M : \underline{C}$, then there exists a lifting $[\![\Gamma; M]\!] : 1[\![\Gamma]\!] \to [\![\Gamma; \underline{C}]\!]$ above $[\![|\Gamma|; M]\!]$ along $u : \{\mathbb{E} \mid \mathbb{P}\} \to \mathbb{E}$.*

Since we have the bijection $s : \{\mathbb{E} \mid \mathbb{P}\}_P(1P, (X, P, Q)) \to \{f : P \to Q \mid \pi_{(X,P,Q)} \circ f = \mathrm{id}_P\}$ for each $(X, P, Q) \in \{\mathbb{E} \mid \mathbb{P}\}$, we obtain liftings of interpretations of terms along $q : \mathbb{P} \to \mathbb{B}$.

Corollary 23. *If $\Gamma \vdash V : A$, then $s[\![|\Gamma|; V]\!] : [\![|\Gamma|]\!] \to \{[\![|\Gamma|; A]\!]\}$ has a lifting $s[\![\Gamma; V]\!] : [\![\Gamma]\!] \to \{[\![\Gamma; A]\!]\}$ along $q : \mathbb{P} \to \mathbb{B}$ (and similarly for computation terms $\Gamma \vdash M : \underline{C}$).* □

Corollary 24. *Assume the lifting of fibred adjunction models is given by applying the Eilenberg-Moore construction to a lifting of monads in Theorem 12. If $\Gamma \vdash M : FA$, then $\theta \circ s[\![|\Gamma|; M]\!] : [\![|\Gamma|]\!] \to T\{[\![|\Gamma|; A]\!]\}$ has a lifting of type $[\![\Gamma]\!] \to \dot{T}\{[\![\Gamma; A]\!]\}$ along $q : \mathbb{P} \to \mathbb{B}$.* □

6 Toward Recursion in Refinement Type Systems

We consider how to deal with general recursion in dependent refinement type systems. In [4], Ahman used a specific model of the fibration $\mathbf{CFam}(\mathbf{CPO}) \to \mathbf{CPO}$ of continuous families of ω-cpos to extend EMLTT with recursion. However, we need to identify the structure that characterizes recursion to lift recursion from the underlying type system to dependent refinement type systems. So, we consider a generalization of Conway operators [22] and prove the soundness of the underlying and the dependent refinement type system extended with typing rules for recursion. This extension enables us to reason about partial correctness of general recursion.

Unfortunately, we still do not know an example of liftings of Conway operators, although (1) $\mathbf{CFam}(\mathbf{CPO}) \to \mathbf{CPO}$ does have a Conway operator and (2) the soundness of the refinement type system with recursion holds under the existence of a lifting of Conway operators. We leave this problem for future work.

6.1 Conway Operators

The notion of Conway operators for cartesian categories is defined in [22]. We adapt the definition for comprehension categories with unit. We allow partially defined Conway operators because we need those defined only on interpretations of computation types.

Definition 25 (Conway operator for comprehension categories with unit). Let $p : \mathbb{E} \to \mathbb{B}$ be a comprehension category with unit and $K \subseteq \mathbb{E}$ be a collection of objects. A *Conway operator* for the comprehension category with unit p defined on K is a family of mappings $(-)^{\ddagger} : \mathbb{E}_I(X, X) \to \mathbb{E}_I(1I, X)$ for each $X \in \mathbb{E}_I \cap K$ such that the following conditions are satisfied.

(Naturality) For each $X \in K$, $f \in \mathbb{E}_I(X, X)$, and $u : J \to I$, $u^* f^{\ddagger} = (u^* f)^{\ddagger}$.
(Dinaturality) For each $X, Y \in K$, $f \in \mathbb{E}_I(X, Y)$, and $g \in \mathbb{E}_I(Y, X)$, $(g \circ f)^{\ddagger} = g \circ (f \circ g)^{\ddagger}$.
(Diagonal property) For each $X \in K$ and $f \in \mathbb{E}_{\{X\}}(\pi_X^* X, \pi_X^* X)$, if $\pi_X^* X \in K$, then $(\phi(f^{\ddagger}))^{\ddagger} = (\phi(\delta_X^*(\phi^{-1}(f))))^{\ddagger}$ holds where $\phi : \mathbb{E}_{\{X\}}(1\{X\}, \pi_X^* X) \to \mathbb{E}_I(X, X)$ is the isomorphism defined in §2.

Lemma 26. *Let $\mathbb{B}$ be a cartesian category. There is a bijective correspondence between the following. (1) Conway operators $(-)^{\dagger}$ on the cartesian category $\mathbb{B}$. (2) Conway operators $(-)^{\ddagger}$ on the simple comprehension category $\mathbf{s}(\mathbb{B}) \to \mathbb{B}^{\to}$ that are defined totally on $\mathbf{s}(\mathbb{B})$.* □

Example 27. Let $K \subseteq \mathbf{CFam}(\mathbf{CPO})$ be a collection of objects defined by $K = \{(I, X) \in \mathbf{CFam}(\mathbf{CPO}) \mid \text{for each } i \in I,\ Xi \text{ has a least element}\}$. For each $(I, X) \in K$ and vertical morphism $f = (\mathrm{id}_I, (f_i)_{i \in I}) : (I, X) \to (I, X)$, we define $f^{\ddagger} = (\mathrm{id}_I, (* \mapsto \mathrm{lfp} f_i)_{i \in I}) : (I, 1) \to (I, X)$. Then $(-)^{\ddagger}$ is a Conway operator, which is implicitly used in [4].

$$\frac{\Gamma \vdash \underline{C} \qquad \Gamma, x : U\underline{C} \vdash M : \underline{C}}{\Gamma \vdash \mu x : U\underline{C}.M : \underline{C}} \qquad \frac{\Gamma \vdash \underline{C} = \underline{D} \qquad \Gamma, x : U\underline{C} \vdash M = N : \underline{C}}{\Gamma \vdash \mu x : U\underline{C}.M = \mu x : U\underline{D}.N : \underline{C}}$$

$$\frac{\Gamma \vdash \underline{C} \qquad \Gamma, x : U\underline{C} \vdash M : \underline{C}}{\begin{array}{l}\Gamma \vdash M[\mathbf{thunk}\ (\mu x : U\underline{C}.M)/x] \\ \qquad = \mu x : U\underline{C}.M : \underline{C}\end{array}} \qquad \frac{\Gamma \vdash \underline{C} \qquad \Gamma, x : U\underline{C}, y : U\underline{C} \vdash M : \underline{C}}{\begin{array}{l}\Gamma \vdash \mu x : U\underline{C}.\mu y : U\underline{C}.M \\ \qquad = \mu x : U\underline{C}.M[x/y] : \underline{C}\end{array}}$$

Fig. 5. Typing rules for general recursion.

6.2 Recursion in the Underlying Type System

Syntax. We add recursion $\mu x : U\underline{C}.M$ to the syntax of computation terms. We also add typing rules in Fig. 5.

Semantics. Assume we have a fibred adjunction model $F \dashv U : r \to p$ where $p : \mathbb{E} \to \mathbb{B}$ and $r : \mathbb{C} \to \mathbb{B}$. We need a Conway operator defined on objects in $\{[\![\Gamma; U\underline{C}]\!] \mid \Gamma \vdash \underline{C}\} \subseteq \mathbb{E}$. However, here is a circular definition because $[\![\Gamma; U\underline{C}]\!]$ may contain terms of the form $\mu x : U\underline{D}.M$, whose interpretations are defined by the Conway operator. So, we use a slightly stronger condition.

Definition 28. A *Conway operator defined on computation types* is a Conway operator defined on $K \subseteq \mathbb{E}$ such that K satisfies the following conditions. (1) $UFX \in K$ holds for each $X \in \mathbb{E}$. (2) $\prod_X Y \in K$ holds for each $X \in \mathbb{E}$ and $Y \in K \cap \mathbb{E}_{\{X\}}$. (3) For each $X \in K$ and $Y \in \mathbb{E}$, $X \cong Y$ implies $Y \in K$.

Given a Conway operator defined on computation types, we interpret $\mu x : U\underline{C}.M$ by $[\![\Gamma; \mu x : U\underline{C}.M]\!] = (\phi([\![\Gamma, x : U\underline{C}; M]\!]))^{\ddagger} : 1[\![\Gamma]\!] \to U[\![\Gamma; \underline{C}]\!]$.

Proposition 29. *Soundness (Proposition 17) holds for the underlying type system extended with general recursion.*

Proof. By induction. We can prove that the given Conway operator is defined on $\{[\![\Gamma; U\underline{C}]\!] \mid \Gamma \vdash \underline{C}\} \subseteq \mathbb{E}$ by [2, Proposition 4.1.14]. □

6.3 Recursion in Refinement Type System

Syntax. We add the typing rule for $\Gamma \vdash \mu x{:}U\underline{C}.M : \underline{C}$ in Fig. 5 to the refinement type system. Here, recall that we remove definitional equalities when we consider the refinement type system.

Semantics. We consider liftings of Conway operators to interpret recursion in the refinement type system.

Definition 30. Let $p : \mathbb{E} \to \mathbb{B}$ and $q : \mathbb{D} \to \mathbb{A}$ be comprehension categories with unit, $(u, v) : p \to q$ be a morphism of comprehension categories with unit. Assume q has a Conway operator $(-)^{\ddagger}$ defined on $K \subseteq \mathbb{D}$. A *lifting* of the Conway operator $(-)^{\ddagger}$ along (u, v) is a Conway operator $(-)^{\natural}$ for p defined on $L \subseteq \mathbb{E}$ such that $uL \subseteq K$ and $u(f^{\natural}) = (uf)^{\ddagger}$ for each $f \in \mathbb{E}_I(X, X)$ where $X \in L$.

Lemma 31. *Let (u, v) be a morphism of CCompCs defined in Theorem 5. Assume $p : \mathbb{E} \to \mathbb{B}$ has a Conway operator $(-)^{\ddagger}$ defined on $K \subseteq \mathbb{E}$. The CCompC $\{\mathbb{E} \mid \mathbb{P}\} \to \mathbb{P}$ has a lifting of the Conway operator defined on $L \subseteq \{\mathbb{E} \mid \mathbb{P}\}$ if $uL \subseteq K$ and for each $(X, P, Q) \in L$ and $f \in \{\mathbb{E} \mid \mathbb{P}\}_P((X, P, Q), (X, P, Q))$, $\{f^{\ddagger}\}$ has a lifting $\pi^*_{1pX} P \to Q$ along $q : \mathbb{P} \to \mathbb{B}$.* □

Proof. Let $(f, \mathrm{id}_P, h) : (X, P, Q) \to (X, P, Q)$ be a morphism in $\{\mathbb{E} \mid \mathbb{P}\}$ where $(X, P, Q) \in L$. We define a Conway operator by $(f, \mathrm{id}_P, h)^{\natural} = (f^{\ddagger}, \mathrm{id}_P, h') : (1pX, P, \pi^*_{1pX} P) \to (X, P, Q)$ where h' is a lifting of $\{f^{\ddagger}\}$. □

We assume that a lifting of fibred adjunction models (4) together with a lifting of Conway operators defined on computation types is given.

Theorem 32. *Soundness (Theorem 22) holds for the refinement type system extended with general recursion.* □

Consider the fibration **CFam(CPO)** $\to$ **CPO** for the underlying type system with recursion. To support recursion in our refinement type system, a natural choice of a fibration for predicate logic is the fibration of admissible subsets **Adm(CPO)** $\to$ **CPO** because the least fixed point of an ω-continuous function $f : X \to X$ is given by $\mathrm{lfp} f = \bigvee_n f^n(\bot)$. However, we cannot apply Theorem 5 because **Adm(CPO)** $\to$ **CPO** is not a fibred ccc [9, §4.3.2]. Specifically, it is not clear whether this combination admits products. We believe that our approach is quite natural but leave giving concrete examples of liftings of Conway operators for future work.

7 Related Work

Dependent refinement types. Historically, there are two kinds of refinement types. One is *datasort refinement types* [7], which are subsets of underlying types but not necessarily dependent. The other is *index refinement types* [28]. A typical example of index refinement types is a type of lists indexed by natural numbers that represent the length of lists. Nowadays, the word "refinement types" includes datasort and index refinement types, and moreover, mixtures of them.

Among a wide variety of the meaning of refinement types, we focus on types equipped with predicates that may depend on other terms [6, 20], which we call *dependent refinement types* or just *refinement types*. Dependent refinement types are widely studied [5, 13, 14, 25], and implemented in, e.g., F* [23, 24] and LiquidHaskell [19,26,27]. However, most studies focus on decidable type systems, and only a few consider categorical semantics.

We expect that some of the existing refinement type systems are combined with effect systems. For example, a dependent refinement type system for nondeterminism and partial/total correctness proposed in [25] contains types for computations indexed by quantifiers $Q_1 Q_2$ where $Q_1, Q_2 \in \{\forall, \exists\}$. Here, Q_1 represents may/must nondeterminism, and Q_2 represents total/partial correctness. It has been shown that $Q_1 Q_2$ corresponds to four cartesian liftings of the monad $P_+((-) + 1)$ [1, 12]. We conjecture that these liftings are connected by monad morphisms and hence yield a lattice-graded monad. Another example is a relational refinement type system for differential privacy [5]. Their system seems to use a graded lifting of the distribution monad where the lifting is graded by privacy parameters, as pointed out in [21]. We leave for future work combining our refinement type system with effect systems based on graded monads [8,11,15].

Categorical semantics. Our interpretation of refinement type systems is based on a morphism of CCompCs, which is a similar strategy to [16]. The difference is that our paper focuses on dependent refinement types and makes the role of predicate logic explicit by giving a semantic construction of refinement type systems from given underlying type systems and predicate logic.

Combining dependent types and computational effects is discussed in [2–4]. Although their aim is not at refinement types, their system is a basis for the design and semantics of our refinement type system with computational effects.

Semantics for types of the form $\{v : A_u \mid p\}$ are characterized categorically as right adjoints of terminal object functors in [10, Chapter 11]. Such types are called *subset types* there. They consider the situation where a given CCompC $p : \mathbb{E} \to \mathbb{B}$ is already rich enough to interpret $\{v : A_u \mid p\}$, and do not aim to interpret refinement type systems by liftings of CCompCs. Moreover, we cannot directly use the interpretations in [10] for our CCompC $\{\mathbb{E} \mid \mathbb{P}\} \to \mathbb{P}$ because we are not given a fibration for predicate logic whose base category is $\mathbb{P}$.

8 Conclusion and Future Work

We provided a general construction of liftings of CCompCs from combinations of CCompCs and posetal fibrations satisfying certain conditions. This can be seen as a semantic counterpart of constructing dependent refinement type systems from underlying type systems and predicate logic. We identified sufficient conditions for several structures in underlying type systems (e.g. products, coproducts, fibred coproducts, fibred monads, and Conway operators) to lift to dependent refinement type systems. We proved the soundness of a dependent refinement type system with computational effects with respect to interpretations in CCompCs obtained from the general construction.

We aim to extend our dependent refinement type system by combining effect systems based on graded monads [8,11,15]. We hope that this extension will give us a more expressive framework that subsumes, for example, dependent refinement type systems in [5,25]. Another direction is to define interpretations of $\{v : A_u \mid p\}$ in the style of subset types in [10, Chapter 11]. Lastly, we are interested in finding more examples of possible combinations of underlying type systems and predicate logic (especially for recursion in dependent refinement type systems but not limited to this) so that we can find a new practical application of this paper.

Acknowledgement. We thank Shin-ya Katsumata, Hiroshi Unno and the anonymous referees for helpful comments. This work was supported by JST ERATO HASUO Metamathematics for Systems Design Project (No. JPMJER1603).

References

1. Aguirre, A., Katsumata, S.: Weakest preconditions in fibrations. In: Proceedings of the Thirty-Sixth Conference on the Mathematical Foundations of Programming Semantics, MFPS 2020, Paris, France (June 2020), to appear
2. Ahman, D.: Fibred Computational Effects. PhD Thesis, University of Edinburgh (2017)
3. Ahman, D.: Handling fibred algebraic effects. Proceedings of the ACM on Programming Languages **2**, 1–29 (Jan 2018). https://doi.org/10.1145/3158095
4. Ahman, D., Ghani, N., Plotkin, G.D.: Dependent types and fibred computational effects. In: Jacobs, B., Löding, C. (eds.) Foundations of Software Science and Computation Structures, vol. 9634, pp. 36–54. Springer Berlin Heidelberg (2016). https://doi.org/10.1007/978-3-662-49630-5_3
5. Barthe, G., Gaboardi, M., Gallego Arias, E.J., Hsu, J., Roth, A., Strub, P.Y.: Higher-Order Approximate Relational Refinement Types for Mechanism Design and Differential Privacy. In: Proceedings of the 42nd Annual ACM SIGPLAN-SIGACT Symposium on Principles of Programming Languages - POPL '15. pp. 55–68. ACM Press, Mumbai, India (2015). https://doi.org/10.1145/2676726.2677000
6. Flanagan, C.: Hybrid type checking. In: Conference Record of the 33rd ACM SIGPLAN-SIGACT Symposium on Principles of Programming Languages - POPL'06. pp. 245–256. ACM Press, Charleston, South Carolina, USA (2006). https://doi.org/10.1145/1111037.1111059
7. Freeman, T., Pfenning, F.: Refinement types for ML. ACM SIGPLAN Notices **26**(6), 268–277 (Jun 1991). https://doi.org/10.1145/113446.113468
8. Fujii, S., Katsumata, S.y., Melliès, P.A.: Towards a Formal Theory of Graded Monads. In: Jacobs, B., Löding, C. (eds.) Foundations of Software Science and Computation Structures, vol. 9634, pp. 513–530. Springer Berlin Heidelberg, Berlin, Heidelberg (2016). https://doi.org/10.1007/978-3-662-49630-5_30
9. Hermida, C.: Fibrations, logical predicates and indeterminates. PhD Thesis, University of Edinburgh, UK (1993)
10. Jacobs, B.: Categorical Logic and Type Theory. No. 141 in Studies in Logic and the Foundations of Mathematics, Elsevier, paperback edn. (2001)
11. Katsumata, S.: Parametric effect monads and semantics of effect systems. In: Proceedings of the 41st ACM SIGPLAN-SIGACT Symposium on Principles of Programming Languages - POPL '14. pp. 633–645. ACM Press, San Diego, California, USA (2014). https://doi.org/10.1145/2535838.2535846
12. Katsumata, S.: private communication (2020)
13. Knowles, K., Flanagan, C.: Compositional reasoning and decidable checking for dependent contract types. In: Proceedings of the 3rd Workshop on Programming Languages Meets Program Verification - PLPV '09. p. 27. ACM Press, Savannah, GA, USA (2008). https://doi.org/10.1145/1481848.1481853
14. Lehmann, N., Tanter, É.: Gradual refinement types. ACM SIGPLAN Notices **52**(1), 775–788 (May 2017). https://doi.org/10.1145/3093333.3009856
15. McDermott, D., Mycroft, A.: Extended Call-by-Push-Value: Reasoning About Effectful Programs and Evaluation Order. In: Caires, L. (ed.) Programming Languages and Systems, vol. 11423, pp. 235–262. Springer International Publishing, Cham (2019). https://doi.org/10.1007/978-3-030-17184-1_9
16. Melliès, P.A., Zeilberger, N.: Functors are Type Refinement Systems. In: Proceedings of the 42nd Annual ACM SIGPLAN-SIGACT Symposium on Principles of Programming Languages - POPL '15. pp. 3–16. ACM Press, Mumbai, India (2015). https://doi.org/10.1145/2676726.2676970

17. Moggi, E.: Notions of computation and monads. Information and Computation **93**(1), 55–92 (Jul 1991). https://doi.org/10.1016/0890-5401(91)90052-4
18. Ou, X., Tan, G., Mandelbaum, Y., Walker, D.: Dynamic Typing with Dependent Types. In: Levy, J.J., Mayr, E.W., Mitchell, J.C. (eds.) Exploring New Frontiers of Theoretical Informatics, vol. 155, pp. 437–450. Kluwer Academic Publishers, Boston (2004). https://doi.org/10.1007/1-4020-8141-3_34
19. Rondon, P.M., Kawaguci, M., Jhala, R.: Liquid types. In: Proceedings of the 2008 ACM SIGPLAN Conference on Programming Language Design and Implementation - PLDI '08. p. 159. ACM Press, Tucson, AZ, USA (2008). https://doi.org/10.1145/1375581.1375602
20. Rushby, J., Owre, S., Shankar, N.: Subtypes for specifications: Predicate subtyping in PVS. IEEE Transactions on Software Engineering **24**(9), 709–720 (Sept/1998). https://doi.org/10.1109/32.713327
21. Sato, T., Barthe, G., Gaboardi, M., Hsu, J., Katsumata, S.y.: Approximate Span Liftings: Compositional Semantics for Relaxations of Differential Privacy. In: 2019 34th Annual ACM/IEEE Symposium on Logic in Computer Science (LICS). pp. 1–14. IEEE, Vancouver, BC, Canada (Jun 2019). https://doi.org/10.1109/LICS.2019.8785668
22. Simpson, A., Plotkin, G.: Complete axioms for categorical fixed-point operators. In: Proceedings Fifteenth Annual IEEE Symposium on Logic in Computer Science (Cat. No.99CB36332). pp. 30–41. IEEE Comput. Soc, Santa Barbara, CA, USA (2000). https://doi.org/10.1109/LICS.2000.855753
23. Swamy, N., Chen, J., Fournet, C., Strub, P.Y., Bhargavan, K., Yang, J.: Secure distributed programming with value-dependent types. Journal of Functional Programming **23**(4), 402–451 (Jul 2013). https://doi.org/10.1017/S0956796813000142
24. Swamy, N., Weinberger, J., Schlesinger, C., Chen, J., Livshits, B.: Verifying higher-order programs with the dijkstra monad. In: Proceedings of the 34th ACM SIGPLAN Conference on Programming Language Design and Implementation - PLDI '13. p. 387. ACM Press, Seattle, Washington, USA (2013). https://doi.org/10.1145/2491956.2491978
25. Unno, H., Satake, Y., Terauchi, T.: Relatively complete refinement type system for verification of higher-order non-deterministic programs. Proceedings of the ACM on Programming Languages **2**, 1–29 (Jan 2018). https://doi.org/10.1145/3158100
26. Vazou, N., Rondon, P.M., Jhala, R.: Abstract Refinement Types. In: Hutchison, D., Kanade, T., Kittler, J., Kleinberg, J.M., Mattern, F., Mitchell, J.C., Naor, M., Nierstrasz, O., Pandu Rangan, C., Steffen, B., Sudan, M., Terzopoulos, D., Tygar, D., Vardi, M.Y., Weikum, G., Felleisen, M., Gardner, P. (eds.) Programming Languages and Systems, vol. 7792, pp. 209–228. Springer Berlin Heidelberg, Berlin, Heidelberg (2013). https://doi.org/10.1007/978-3-642-37036-6_13
27. Vazou, N., Seidel, E.L., Jhala, R., Vytiniotis, D., Peyton-Jones, S.: Refinement types for Haskell. In: Proceedings of the 19th ACM SIGPLAN international conference on Functional programming - ICFP '14. pp. 269–282. ACM Press, Gothenburg, Sweden (2014). https://doi.org/10.1145/2628136.2628161
28. Xi, H., Pfenning, F.: Eliminating array bound checking through dependent types. In: Proceedings of the ACM SIGPLAN 1998 Conference on Programming Language Design and Implementation - PLDI '98. pp. 249–257. ACM Press, Montreal, Quebec, Canada (1998). https://doi.org/10.1145/277650.277732

Simple Stochastic Games with Almost-Sure Energy-Parity Objectives are in NP and coNP

Richard Mayr[1], Sven Schewe[2], Patrick Totzke[2], and Dominik Wojtczak[2(✉)]

[1] University of Edinburgh, Edinburgh, UK
[2] University of Liverpool, Liverpool, UK
{sven.schewe,totzke,d.wojtczak}@liverpool.ac.uk

Abstract. We study stochastic games with energy-parity objectives, which combine quantitative rewards with a qualitative ω-regular condition: The maximizer aims to avoid running out of energy while simultaneously satisfying a parity condition. We show that the corresponding almost-sure problem, i.e., checking whether there exists a maximizer strategy that achieves the energy-parity objective with probability 1 when starting at a given energy level k, is decidable and in $\mathsf{NP} \cap \mathsf{coNP}$. The same holds for checking if such a k exists and if a given k is minimal.

Keywords: Simple Stochastic Games, Parity Games, Energy Games

1 Introduction

Simple stochastic games (SSGs), also called *competitive Markov decision processes* [30], or $2\frac{1}{2}$-*player games* [23,22] are turn-based games of perfect information played on finite graphs. Each state is either random or belongs to one of the players (maximizer or minimizer). A game is played successively moving a pebble along the game graph, where the next state is chosen by the player who owns the current one or, in the case of random states, according to a predefined distribution. This way, an infinite run is produced. The maximizer tries to achieve an objective (in our case almost surely), while the minimizer tries to prevent this. The maximizer can be seen as a controller trying to ensure an objective in the face of both known random failure modes (encoded by the random states) and an unknown or hostile environment (encoded by the minimizer player).

Stochastic games were first introduced in Shapley's seminal work [46] in 1953 and have since then played a central role in the solution of many problems in computer science, including synthesis of reactive systems [45,42]; checking interface compatibility [27]; well-formedness of specifications [28]; verification of open systems [4]; and many others.

A huge variety of objectives for such games was already studied in the literature. We will mainly focus on three of them in this paper: parity; mean-payoff; and energy objectives. In order to define them we assume that numeric rewards are assigned to transitions, and priorities (encoded by bounded non-negative numbers) are assigned to states.

S. Kiefer and C. Tasson (Eds.): FOSSACS 2021, LNCS 12650, pp. 427–447, 2021.
https://doi.org/10.1007/978-3-030-71995-1_22

The *parity objective* simply asks that the minimal priority that appears infinitely often in a run is even. Such a condition is a canonical way to define desired behaviors of systems, such as safety, liveness, fairness, etc.; it subsumes all ω-regular objectives. The algorithmic problem of deciding the winner in non-stochastic parity games is polynomial-time equivalent to the model checking of the modal μ-calculus [49] and is at the center of the algorithmic solutions to the Church's synthesis problem [44]. But the impact of parity games goes well beyond automata theory and logic: They facilitated the solution of two long-standing open problems in stochastic planning [29] and in linear programming [32], which was done by careful adaptation of the parity game examples on which the strategy improvement algorithm [31] requires exponentially many iterations.

The parity objective can be seen as a special case of the *mean-payoff objective* that asks for the limit average reward per transition along the run to be non-negative. Mean-payoff objectives are among the first objectives studied for stochastic games and go back to a 1957 paper by Gillette [33]. They allow for reasoning about the efficiency of a system, e.g., how fast it operates once optimally controlled.

The *energy objective* [14] can be seen as a refinement of the mean-payoff objective. It asks for the accumulated reward at any point of a run not to be lower than some finite threshold. As the name suggests, it is useful when reasoning about systems with a finite initial energy level that should never become depleted. Note that the accumulated reward is not bounded a-priori, which essentially turns a finite-state game into an infinitely-state one.

In this paper we consider SSGs with *energy-parity* objectives, which requires runs to satisfy both an energy and a parity objective. It is natural to consider such an objective for systems that should not only be correct, but also energy efficient. For instance, consider a robot maintaining a nuclear power plant. We not only require the robot to correctly react to all possible chains of events (parity objective for functional correctness), but also never to run out of energy as charging it manually would be risky (energy objective).

While the complexity of games with single objectives is often in $\mathsf{NP} \cap \mathsf{coNP}$, asking for multiple objectives often makes solving games harder. Parity games are commonly viewed as the simplest of these objectives, and some traditional solutions for non-stochastic games go through simple reductions to mean-payoff or energy conditions (which are quite similar in non-stochastic games) to discounted payoff games that establishes the membership of those problems in UP and coUP [35]. However, asking for *two* parity objectives to be satisfied at the same time leads to coNP completeness [21].

We study the almost sure satisfaction of the energy-parity objective, i.e., with probability 1. Such *qualitative analysis* is important as there are many applications where we need to know whether the correct behavior arises almost-surely, e.g., in the analysis of randomized distributed algorithms (see, e.g, [43,47]) and safety-critical examples like the one from above. Moreover, the algorithms for *quantitative analysis*, i.e., computing the optimal probability of satisfaction, typically start by performing the qualitative analysis first and then solving a game with a simpler objective (see, e.g., [23,15]). Finally, there are stochastic models for which qualitative analysis is decidable but quantitative one is not (e.g., probabilistic finite automata [6]). This may also be the case for our model.

Our contributions. We consider stochastic games with energy-parity winning conditions and show that deciding whether maximizer can win almost-surely for a given initial energy level k is in $\mathsf{NP} \cap \mathsf{coNP}$. We show the same for checking if such k exists at all and checking if a given k is the smallest possible for which this holds. The proofs are considerably harder than the corresponding result for MDPs [40] (on which they are partly based), because the attainable mean-payoff value is no longer a valid criterion in the analysis (via combinations of sub-objectives). E.g., even though the stored energy might be inexorably drifting towards $+\infty$ (resp. $-\infty$), the mean-payoff value might still be zero because the minimizer (resp. maximizer) can delay payoffs for longer and longer (though not indefinitely, due to the parity condition). Moreover, the minimizer might be able to choose between different ways of losing and never commit to any particular way after any finite prefix of the play (see Example 1).

Our proof characterizes almost-sure energy-parity via a recursive combination of complex sub-objectives called *Gain* and *Bailout*, which can each eventually be solved in $\mathsf{NP} \cap \mathsf{coNP}$.

Our proof of the coNP membership is based on a result on the strategy complexity of a natural class of objectives, which is of independent interest. We show (cf. Theorem 6; based on previous work in [34]) that, if an objective $\mathcal{O}$ is such that its complement is both shift-invariant and submixing, and that every MDP admits optimal finite-memory deterministic maximizer strategies for $\mathcal{O}$, then the same is true in turn-based stochastic games.

Example 1. Fig. 1 shows an energy-parity game that the maximizer can win almost surely when starting with an energy level of ≥ 2 from the middle left node. Whenever the game is at that node with an energy level ≥ 3, then the maximizer can turn left and has at least $\frac{1}{2}$ chance that the energy level will never drop to 2 while wining the game with priority 2. This is because we can

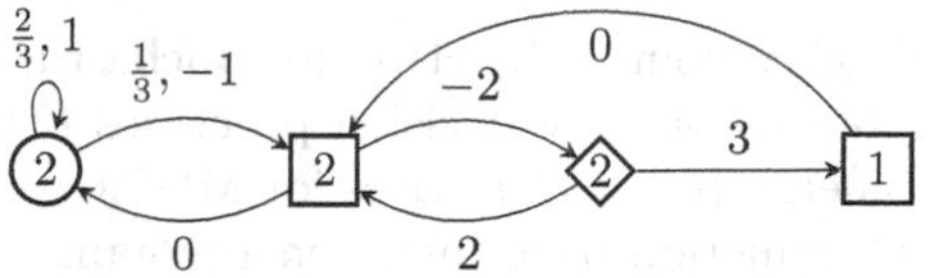

Fig. 1: A SSG with two maximizer states (□), one minimizer state (◇) and one probabilistic state (○). Each state is annotated with its priority. Each edge is annotated with a reward by which the energy level is increased after traversing it (respectively, decreased if the reward is negative). The maximizer wins if the lowest priority visited infinitely often is even and the energy level never drops below 0.

view this process as a random walk on a half line. If x_n is the probability of reaching energy level 2 when starting at energy n then these probabilities are the least point-wise positive solution to the following system of linear equations: $x_2 = 1$, $x_n = \frac{2}{3}x_{n+1} + \frac{1}{3}x_{n-1}$ for all $n \geq 3$. We then get that $x_n = \frac{1}{2^{n-2}}$ so the probability of not reaching energy level 2 is $\geq \frac{1}{2}$ for all $n \geq 3$. Always turning left guarantees that, almost surely, the parity condition holds and the limes inferior of the energy level is not $-\infty$. We call this condition *Gain*. Strategies for *Gain* can be used when the energy level is sufficiently high (at least 3 in our example) to win with a positive probability.

However, if maximizer plays for Gain and always moves left, then for every initial energy level the chance of eventually dropping the energy down to level 2 is positive, due to the negative cycle. When that happens, the only other option for the maximizer is to move right. There minimizer can 'choose how to lose', via a disjunction of two conditions that we later formalize as *Bailout*. Either minimizer goes back to the start state without changing the energy level (thus maximizer wins as the energy stays at level 2 and only the good priority 2 is seen), or minimizer turns right. In the latter case, the play visits a dominating odd priority (which is bad for maximizer) but also increases the energy by 1, which allows maximizer to switch back to playing left for the *Gain* condition until energy level 2 is reached again.

Our maximizer strategies are a complex interplay between *Bailout* and *Gain*. In the example, it is easy to see that the probability of seeing priority 1 infinitely often is zero if maximizer follows the just described strategy (the probability of requiring to go right more than n times is at most $(\frac{1}{2})^n$), so maximizer wins this energy-parity game almost surely. Note that maximizer does not win almost surely when the initial energy level is 0 or 1.

Previous work on combined objectives. Non-stochastic energy-parity games have been studied in [16]. They can be solved in $\mathsf{NP} \cap \mathsf{coNP}$ and maximizer strategies require only finite (but exponential) memory, a property that also allowed to show P-time inter-reducibility with mean-payoff parity games. More recently they were also shown to be solvable in pseudo-quasi-polynomial time [26]. Related results on non-stochastic games (e.g., mean-payoff parity) are summarized in [18].

Most existing work on combined objectives for stochastic systems, for example [17,18,9,40], is restricted to Markov decision processes (MDPs; aka $1\frac{1}{2}$-player games). Almost-sure energy-parity objectives for MDPs were first considered in [17,18], where a direct reduction to ordinary energy games was proposed. This reduction relies on the assumption that maximizer can win using finite memory if at all. Unfortunately, this assumption does not necessarily hold: it was shown in [40] that an almost sure winning strategy for energy-parity in finite MDPs may require infinite memory. Nevertheless, it was possible to recover the original result, that deciding the existence of a.s. winning strategies is in $\mathsf{NP} \cap \mathsf{coNP}$ (and pseudo-polynomial time), by showing that the existence of an a.s. winning strategy can be witnessed by the existence of two compatible, and finite-memory, winning strategies for two simpler objectives. We generalize this approach from MDPs to full stochastic games.

Stochastic mean-payoff parity games were studied in [20], where it was shown that they can be solved in $\mathsf{NP} \cap \mathsf{coNP}$. However, this does not imply a solution for stochastic energy-parity games, since, unlike in the non-stochastic case [16], there is no known reduction from energy-parity to mean-payoff parity in stochastic games. (The reduction in [16] relies on the fact that maximizer has a winning finite-memory strategy for energy-parity, which does not generally hold for stochastic games or MDPs; see above.)

A related model are the 1-counter MDPs (and stochastic games) studied in [12,11,8], since the value of the counter can be interpreted as the stored energy. These papers consider the objective of reaching counter value zero (which is dual to the energy objective of staying above zero), thus the roles of minimizer and maximizer are swapped. However, unlike in this paper, these works do not combine termination objectives with extra parity conditions.

Structure of the paper. The rest of the paper is organized as follows. We start by introducing the notation and formal definitions of games and objectives in the next section. In Section 3 we show how checking almost-sure energy-parity objectives can be characterized in terms of two newly defined auxiliary objectives: Gain and Bailout. In Sections 4 and 5, we show that almost-sure Bailout and Gain objectives, respectively, can be checked in NP and coNP. Section 6 contains our main result: NP and coNP algorithms for checking almost-sure energy-parity games with a known and unknown initial energy, as well as checking if a given initial energy is the minimal one. We conclude and point out some open problems in Section 7. Due to page restrictions, most proofs in the main body of the paper were replaced by sketches. The detailed proofs can be found in the full version of this paper [41].

2 Preliminaries

A probability distribution over a set X is a function $f : X \to [0,1]$ such that $\sum_{x \in X} f(x) = 1$. We write $\mathcal{D}(X)$ for the set of distributions over X.

Games, Strategies, Measures. A *Simple Stochastic Game (SSG)* is a directed graph $\mathcal{G} \stackrel{\text{def}}{=} (V, E, \lambda)$, where all states have an outgoing edge and the set of states is partitioned into states owned by *maximizer* ($V_\Box$), *minimizer* ($V_\Diamond$) and probabilistic states ($V_\bigcirc$). The set of *edges* is $E \subseteq V \times V$ and $\lambda : V_\bigcirc \to \mathcal{D}(E)$ assigns each probabilistic state a probability distribution over its outgoing edges. W.l.o.g., we assume that each probabilistic state has at most two successors, because one can introduce a new probabilistic state for each excess successor. We let $\lambda(ws) \stackrel{\text{def}}{=} \lambda(s)$ for all $ws \in (VE)^* V_\bigcirc$.

A *path* is a finite or infinite sequence $\rho \stackrel{\text{def}}{=} s_0 e_0 s_1 e_1 \ldots$ such that $e_i = (s_i, s_{i+1}) \in E$ holds for all indices i. A *run* is an infinite path and we write $\mathit{Runs} \stackrel{\text{def}}{=} (VE)^\omega$ for the set of all runs.

A *strategy* for maximizer is a function $\sigma : (VE)^*V_\Box \to \mathcal{D}(E)$ that assigns to each path $ws \in (VE)^*V_\Box$ a probability distribution over the outgoing edges of its target node s. That is, $\sigma(ws)(e) > 0$ implies $e = (s,t) \in E$ for some $t \in V$. A strategy is called *memoryless* if $\sigma(xs) = \sigma(ys)$ for all $x, y \in (VE)^*$ and $s \in V_\Box$, *deterministic* if $\sigma(w)$ is Dirac for all $w \in (VE)^*V_\Box$, and *finite-state* if there exists an equivalence relation $\sim$ on $(VE)^*V_\Box$ with a finite index, such that $\sigma(\rho_1) = \sigma(\rho_2)$ if $\rho_1 \sim \rho_2$. Of particular interest to us will be the class of *memoryless deterministic strategies* (*MD*) and the class of *finite-memory deterministic strategies* (*FD*). Strategies for minimizer are defined analogously and will usually be denoted by $\tau : (VE)^*V_\Diamond \to \mathcal{D}(E)$.

A maximizing (minimizing) *Markov Decision Process (MDP)* is a game in which minimizer (maximizer) has no choices, i.e., all her states have exactly one successor. We will write $\mathcal{G}[\tau]$ for the MDP resulting from fixing the strategy τ. A *Markov chain* is a game where neither player has a choice. In particular, $\mathcal{G}[\sigma, \tau]$ is a Markov chain obtained by setting, in the game $\mathcal{G}$, the strategies for maximizer and minimizer to σ and τ, respectively.

Given an initial state $s \in V$ and strategies σ and τ for maximizer and minimizer, respectively, the set of runs starting in s naturally extends to a probability space as follows. We write $Runs^{\mathcal{G}}_w$ for the *w-cylinder*, i.e., the set of all runs with prefix $w \in (VE)^*V$. We let $\mathcal{F}^{\mathcal{G}}$ be the σ-algebra generated by all these cylinders. We inductively define a probability function $\mathbb{P}^{\mathcal{G},\sigma,\tau}_s$ on all cylinders, which then uniquely extends to $\mathcal{F}^{\mathcal{G}}$ by Carathéodory's extension theorem [5], by setting $\mathbb{P}^{\mathcal{G},\sigma,\tau}_s(Runs^{\mathcal{G}}_s) \overset{\text{def}}{=} 1$ and $\mathbb{P}^{\mathcal{G},\sigma,\tau}_s(Runs^{\mathcal{G}}_w) \overset{\text{def}}{=} \prod_{i=0}^{n-1} dist_i(s_0e_0s_1e_1 \ldots s_i)(e_i)$ for $w = s_0e_0s_1e_1 \ldots e_{n-1}s_n$, where $s_0 = s$, $e_i = (s_i, s_{i+1})$ and $dist_i$ is $\sigma(\cdot)$, $\tau(\cdot)$ or $\lambda(\cdot)$, for $s_i \in V_\Box, V_\Diamond$ or $V_\bigcirc$, respectively.

Objective Functions. A (Borel) *objective* is a set $\mathsf{Obj} \in \mathcal{F}^{\mathcal{G}}$ of runs. We write $\overline{\mathsf{Obj}} \overset{\text{def}}{=} Runs \setminus \mathsf{Obj}$ for its complement. Borel objectives Obj are weakly determined [39,38], which means that

$$\sup_\sigma \inf_\tau \mathbb{P}^{\sigma,\tau}_s(\mathsf{Obj}) = \inf_\tau \sup_\sigma \mathbb{P}^{\sigma,\tau}_s(\mathsf{Obj}).$$

This quantity is called the *value* of Obj in state s, and written as $\mathsf{Val}^{\mathcal{G}}_s(\mathsf{Obj})$. We say that Obj holds *almost-surely* (abbreviated as *a.s.*) at state s iff there exists σ such that $\forall \tau, \mathbb{P}^{\mathcal{G},\sigma,\tau}_s(\mathsf{Obj}) = 1$. Let $\mathsf{AS}^{\mathcal{G}}(\mathsf{Obj})$ denote the set of states at which Obj holds almost surely. We will drop the superscript $\mathcal{G}$ and simply write *Runs*, $\mathbb{P}^{\sigma,\tau}_s$ and $\mathsf{AS}(\mathsf{Obj})$, if the game is clear from the context.

We use the syntax and semantics of operators $\mathbb{F}$ (eventually) and $\mathbb{G}$ (always) from the temporal logic LTL [25] to specify some conditions on runs.

A *reachability condition* is defined by a set of target states $T \subseteq V$. A run $\rho = s_0e_0s_1 \ldots$ satisfies the reachability condition iff there exists an $i \in \mathbb{N}$ s.t. $s_i \in T$. We write $\mathbb{F}T \subseteq Runs$ for the set of runs that satisfy this reachability condition. Given a set of states $W \subseteq V$, we lift this to a safety condition on runs and write $\mathbb{G}W \subseteq Runs$ for the set of runs $\rho = s_0e_0s_1 \ldots$ where $\forall i.\, s_i \in W$.

A *parity condition* is given by a bounded function $parity : V \to \mathbb{N}$ that assigns a priority (a non-negative integer) to each state. A run $\rho \in Runs$ satisfies the parity condition iff the minimal priority that appears infinitely often on the run is even. The *parity objective* is the subset $\mathsf{PAR} \subseteq Runs$ of runs that satisfy the parity condition.

Energy conditions are given by a function $r : E \to \mathbb{Z}$, that assigns a *reward* value to each edge. For a given initial energy value $k \in \mathbb{N}$, a run $s_0e_0s_1e_1\ldots$ satisfies the k-energy condition if, for every finite prefix of length n, the *energy level* $k + \sum_{i=0}^{n} r(e_i)$ is greater or equal to 0. Let $\mathsf{EN}(k) \subseteq Runs$ denote the k-energy objective, consisting of those runs that satisfy the k-energy condition.

The *l-storage condition* holds for a run $s_0e_0s_1e_1\ldots$ if $l + \sum_{i=m}^{n-1} r(s_i, s_{i+1}) \geq 0$ holds for every infix $s_me_ms_{m+1}\ldots s_n$. Let $\mathsf{ST}(k,l) \subseteq Runs$ denote the k-energy l-storage objective, consisting of those runs that satisfy both the k-energy and the l-storage condition. We write $\mathsf{ST}(k)$ for $\bigcup_l \mathsf{ST}(k,l)$. Clearly, $\mathsf{ST}(k) \subseteq \mathsf{EN}(k)$.

Mean-payoff and *limit-payoff conditions* are defined w.r.t. the same reward function as the energy conditions. The *mean-payoff* value of a run $\rho = s_0e_0s_1e_1\ldots$ is $MP(\rho) \stackrel{\text{def}}{=} \liminf_{n\to\infty} \frac{1}{n}\sum_{i=0}^{n-1} r(e_i)$. For $\triangle \in \{>, \geq, =, \leq, <\}$ and $c \in \mathbb{R} \cup \{-\infty, \infty\}$, the set $\mathsf{MP}(\triangle c) \subseteq Runs$ consists of all runs ρ with $MP(\rho)\triangle c$. Let $\mathsf{LimInf}(\triangle c) \subseteq Runs$ contain all runs ρ with $(\liminf_{n\to\infty} \sum_{i=0}^{n} r(e_i))\triangle c$, and likewise for $\mathsf{LimSup}(\triangle c)$.

The combined energy-parity objective $\mathsf{EN}(k) \cap \mathsf{PAR}$ is Borel and therefore weakly determined, meaning that it has a well-defined (inf sup = sup inf) value for every game [39,38]. Moreover, the almost-sure energy-parity objective (asking to win with probability 1) is even strongly determined [37]: either maximizer has a strategy to enforce the condition with probability 1 or minimizer has a strategy to prevent this.

3 Characterizing Energy-Parity via Gain and Bailout

The main theorem of this section (Theorem 5) characterizes almost sure energy-parity objectives in terms of two intermediate objectives called Gain and k-Bailout for parameters $k \geq 0$. This will form the basis of all computability results: we will show (as Theorems 14, 17 and 18) how to compute almost-sure sets for these intermediate objectives.

Definition 2. *Consider a finite SSG $\mathcal{G} = (V, E, \lambda)$, as well as reward and parity functions defining the objectives* $\mathsf{PAR}, \mathsf{LimInf}(> -\infty), \mathsf{LimSup}(= \infty)$ *as well as* $\mathsf{ST}(k,l)$ *and* $\mathsf{EN}(k)$ *for every* $k, l \in \mathbb{N}$. *We define combined objectives* Gain *and* k-Bailout $\stackrel{def}{=} \cup_l \mathsf{Bailout}(k,l)$ *where*

$$\mathsf{Gain} \stackrel{def}{=} \mathsf{LimInf}(> -\infty) \cap \mathsf{PAR}$$
$$\mathsf{Bailout}(k,l) \stackrel{def}{=} (\mathsf{ST}(k,l) \cap \mathsf{PAR}) \cup (\mathsf{EN}(k) \cap \mathsf{LimSup}(= \infty)).$$

The main idea behind these two objectives is a special witness property for energy-parity. We argue that, if maximizer has an almost-sure winning strategy for energy-parity then he also has one that combines two almost-sure winning strategies, one for Gain and one for k-Bailout.

Notice that playing an almost-sure winning strategy for Gain implies a uniformly lower-bounded strictly positive chance that the energy level never drops below zero (assuming it is sufficiently high to begin with). This fact uses the finiteness of the set of control-states and does not hold for infinite-state MDPs. In the unlikely event that the energy level does get close to zero, maximizer switches to playing an almost sure winning strategy for k-Bailout. This is a disjunction of two scenarios, and the balance might be influenced by minimizer's choices. In the first scenario ($\mathsf{ST}(k,l) \cap \mathsf{PAR}$) the energy never drops much and stays above zero (thus satisfying energy-parity). In the second scenario, ($\mathsf{EN}(k) \cap \mathsf{LimSup}(=\infty)$), the parity objective is temporarily suspended in favor of boosting (while always staying above zero) the energy to a sufficiently high level to switch back to the strategy for Gain and thus try again from the beginning. The probability of infinitely often switching between these modes is zero due to the lower-bounded chance of success in the Gain phase. Therefore, maximizer eventually wins by playing for Gain. Note that maximizer needs to remember the current energy level in order to know when to switch and consequently, this strategy uses infinite memory.

Example 3. Consider again the game in Fig. 1. The middle left state satisfies both Gain and k-Bailout objectives for all $k \geq 2$ almost-surely. The respective winning strategies are to always go left for Gain or always go right for k-Bailout when at that state. Note that it neither satisfies 0-Bailout nor 1-Bailout objectives.

We define the subset $W \subseteq V$ of states from which maximizer can almost surely win both Gain and k-Bailout (assuming sufficiently high initial energy), while at the same time ensuring that the play remains within this set of states. These are the states from which maximizer can win by freely combining individual strategies for the Gain and Bailout objectives.

Definition 4. *Given a finite SSG $\mathcal{G} = (V, E, \lambda)$, let $W \subseteq V$ be the largest subset of states satisfying the following condition*

$$W \subseteq \mathsf{AS}\left(\mathsf{Gain} \cap \mathbb{G}W\right) \cap \bigcup_k \mathsf{AS}\left(k\text{-}\mathsf{Bailout} \cap \mathbb{G}W\right)$$

This condition describes a fixed-point, and as it is easy to see that if two sets W_1 and W_2 are such fixed-points, then so is $W_1 \cup W_2$. Thus, the maximal fixed-point W is well-defined.

Our main characterization of almost-sure energy-parity objectives is the following Theorem 5. It states that maximizer can almost surely win an $\mathsf{EN}(k) \cap \mathsf{PAR}$ objective if, and only if, he can win the easier k-Bailout objective while always staying in the safe set W.

Theorem 5. *For every $k \in \mathbb{N}$,* $\mathsf{AS}\left(\mathsf{EN}(k) \cap \mathsf{PAR}\right) = \mathsf{AS}\left(k\text{-}\mathsf{Bailout} \cap \mathbb{G}W\right)$.

Our proof of this characterization theorem relies on the following claim, which allows to lift the existence of finite-memory deterministic optimal strategies from MDPs to SSGs. It applies to a fairly general class of objectives and, we believe, is of independent interest.

Recall that $\overline{\mathsf{Obj}} \stackrel{\text{def}}{=} \mathit{Runs} \setminus \mathsf{Obj}$ denotes the complement of objective Obj. For runs $a, b, c \in \mathit{Runs}$ we say that a is a *shuffle* of b and c if there exist factorizations $b = b_0 b_1 \ldots$ and $c = c_0 c_1 \ldots$ such that $a = b_0 c_0 b_1 c_1 \ldots$. An objective Obj is called *submixing* if, for every run $a \in \mathsf{Obj}$ that is a shuffle of runs b and c, either $b \in \mathsf{Obj}$ or $c \in \mathsf{Obj}$. Obj is *shift-invariant* if, for every run $s_1 e_1 s_2 e_2 \ldots$, it holds that $s_1 e_1 s_2 e_2 \ldots \in \mathsf{Obj} \iff s_2 e_2 \ldots \in \mathsf{Obj}$. Shift-invariance slightly generalizes the better-known *tail* condition (see [34] for a discussion).

Theorem 6. *Let $\mathcal{O}$ be an objective such that $\overline{\mathcal{O}}$ is both shift-invariant and submixing. If maximizer has optimal FD strategies (from any state s) for $\mathcal{O}$ for every finite MDP then maximizer has optimal FD strategies (from any state s) for $\mathcal{O}$ for every finite SSG.*

This applies in particular to the Gain objective, but not to k-Bailout objectives, as these are not shift-invariant. A proof of Theorem 6 can be found in [41]. It uses a recursive argument based on the notion of *reset strategies* from [34].

The remainder of this section is dedicated to proving Theorem 5. We will first collect the remaining technical claims about Gain, Bailout, and reachability objectives. Most notably, as Lemma 8, we show that if maximizer can almost surely win Gain in a SSG, then he can do so using a FD strategy which moreover satisfies an energy-parity objective with strictly positive (and lower-bounded) probability. This is shown in part based on Theorem 6 applied to the Gain objective. We will also need the following fact about reachability objectives in finite MDPs.

Lemma 7 ([8, Lemma 3.9]). *Let $\mathcal{M}$ be a finite MDP and Reach_T be the reachability objective with target $T \stackrel{\text{def}}{=} \{s' \mid \mathsf{Val}_{s'}(\mathsf{LimInf}(=-\infty)) = 1\}$. One can compute a rational constant $c < 1$ and an integer $h \geq 0$ such that for all states s and $i \geq h$ we have $\forall \tau . \mathbb{P}^{\tau}_{s}(\overline{\mathsf{EN}(i)} \cap \overline{\mathit{Reach}_T}) \leq \frac{c^i}{1-c}$.*

Lemma 8. *Consider a finite SSG $\mathcal{G} = (V, E, \lambda)$ where Gain holds a.s. for every state $s \in V$. Then, for every $\delta \in [0,1)$ and $s \in V$, there exists a $\hat{k} \in \mathbb{N}$ and an FD strategy $\hat{\sigma}$ s.t.*

1. *$\forall \tau . \mathbb{P}^{\hat{\sigma},\tau}_{s}(\mathsf{Gain}) = 1$, and*
2. *$\forall \tau . \mathbb{P}^{\hat{\sigma},\tau}_{s}(\mathsf{EN}(\hat{k}) \cap \mathsf{PAR}) \geq \delta$.*

Proof. Fix a $\delta \in [0,1)$ and a state $s \in V$. Both $\mathsf{LimInf}(=-\infty)$, as well as PAR objectives are *shift-invariant* and *submixing*, and therefore also the union has both these properties. It follows that $\overline{\mathsf{Gain}} = \overline{\mathsf{LimInf}(>-\infty) \cap \mathsf{PAR}} = \mathsf{LimInf}(= -\infty) \cup \overline{\mathsf{PAR}}$ is both shift-invariant and submixing, since the complement of a parity objective is also a parity objective. By Lemma 16 and Theorem 6, there exists an almost-sure winning FD strategy $\hat{\sigma}$ for maximizer for the objective Gain from s, i.e., $\forall \tau. \mathbb{P}_s^{\hat{\sigma},\tau}(\mathsf{Gain}) = 1$, thus yielding Item 1.

Let $\mathcal{M}$ be the MDP obtained from $\mathcal{G}$ by fixing the strategy $\hat{\sigma}$ for maximizer from s. Since $\mathcal{G}$ is finite and $\hat{\sigma}$ is FD, also $\mathcal{M}$ is finite. In $\mathcal{M}$ we have $\forall \tau. \mathbb{P}_s^{\tau}(\mathsf{Gain}) = 1$. In particular, in $\mathcal{M}$, the set $T \stackrel{\text{def}}{=} \{s' \mid \mathsf{Val}_{s'}(\mathsf{LimInf}(=-\infty)) = 1\}$ is not reachable, i.e., $\forall \tau. \mathbb{P}_s^{\tau}(\mathit{Reach}_T) = 0$.

By Lemma 7, in $\mathcal{M}$ there exists a horizon $h \in \mathbb{N}$ and a constant $c < 1$ such that for all $i \geq h$ we have $\forall \tau. \mathbb{P}_s^{\tau}(\overline{\mathsf{EN}(i)} \cap \overline{\mathit{Reach}_T}) \leq \frac{c^i}{1-c}$. Since T cannot be reached in $\mathcal{M}$, the condition $\overline{\mathit{Reach}_T}$ evaluates to *true* and we have $\forall \tau. \mathbb{P}_s^{\tau}(\mathsf{EN}(i)) \geq 1 - \frac{c^i}{1-c}$. Since $c < 1$ and $\delta < 1$, we can pick a sufficiently large $\hat{k} \geq h$ such that $1 - \frac{c^{\hat{k}}}{1-c} \geq \delta$ and obtain $\forall \tau. \mathbb{P}_s^{\tau}(\mathsf{EN}(\hat{k})) \geq \delta$ in $\mathcal{M}$. Moreover, the above property $\forall \tau. \mathbb{P}_s^{\tau}(\mathsf{Gain}) = 1$ in particular implies $\forall \tau. \mathbb{P}_s^{\tau}(\mathsf{PAR}) = 1$. Thus we obtain $\forall \tau. \mathbb{P}_s^{\tau}(\mathsf{EN}(\hat{k}) \cap \mathsf{PAR}) \geq \delta$ in $\mathcal{M}$.

Back in the SSG $\mathcal{G}$, we have $\forall \tau. \mathbb{P}_s^{\hat{\sigma},\tau}(\mathsf{EN}(\hat{k}) \cap \mathsf{PAR}) \geq \delta$ as required for Item 2. □

Lemma 9. $\mathsf{EN}(k) \cap \mathsf{PAR} \subseteq k$-$\mathsf{Bailout}$.

Proof. Let ρ be a run in $\mathsf{EN}(k) \cap \mathsf{PAR}$. There are two cases. In the first case we have $\rho \in \cup_l \mathsf{ST}(k,l) \cap \mathsf{PAR}$ and thus directly $\rho \in k$-$\mathsf{Bailout}$. Otherwise, $\rho \notin \cup_l \mathsf{ST}(k,l) \cap \mathsf{PAR}$. Since $\rho \in \mathsf{PAR}$, we must have $\rho \notin \cup_l \mathsf{ST}(k,l)$. Since $\rho \in \mathsf{EN}(k)$, it follows that ρ does not satisfy the l-storage condition for any $l \in \mathbb{N}$. So, for every $l \in \mathbb{N}$, there exists an infix ρ' of ρ s.t. $l + r(\rho') < 0$. Let ρ'' be the prefix of ρ before ρ'. Since $\rho \in \mathsf{EN}(k)$ we have $k + r(\rho''\rho') \geq 0$ and thus $r(\rho'') \geq -k - r(\rho') > -k + l$. To summarize, if $\rho \notin \cup_l \mathsf{ST}(k,l) \cap \mathsf{PAR}$ then, for every l, it has a prefix ρ'' with $r(\rho'') > -k + l$. Thus $\rho \in \mathsf{LimSup}(=\infty)$. Thus $\rho \in k$-$\mathsf{Bailout}$. □

We now define W' as the set of states that are almost-sure winning for energy-parity with some sufficiently high initial energy level. (W' is also called the winning set for the unknown initial credit problem.)

Definition 10. $W' \stackrel{\text{def}}{=} \bigcup_k \mathsf{AS}\,(\mathsf{EN}(k) \cap \mathsf{PAR})$.

Lemma 11.

1. $\mathsf{AS}\,(\mathsf{EN}(k) \cap \mathsf{PAR}) \subseteq \mathsf{AS}\,(\mathsf{Gain} \cap \mathbb{G}W')$
2. $\mathsf{AS}\,(\mathsf{EN}(k) \cap \mathsf{PAR}) \subseteq \mathsf{AS}\,(k\text{-}\mathsf{Bailout} \cap \mathbb{G}W')$

Proof. Let $s \in \mathsf{AS}\,(\mathsf{EN}(k) \cap \mathsf{PAR})$ and σ a strategy that witnesses this property. Except for a null-set, all runs $\rho = s e_0 s_1 e_1 \ldots e_{n-1} s_n \ldots$ from s induced by σ satisfy $\mathsf{EN}(k) \cap \mathsf{PAR}$.

Let $\rho' = s e_0 s_1 e_1 \ldots s_m$ be a finite prefix of ρ. For every $n \geq 0$ we have $k + \sum_{i=0}^{n-1} r(e_i) \geq 0$, since $\rho \in \mathsf{EN}(k)$. In particular this holds for all $n \geq m$. So, for every $n \geq m$, we have $k + \sum_{i=0}^{m-1} r(e_i) + \sum_{i=m}^{n-1} r(e_i) \geq 0$. Therefore $s_m \in \mathsf{AS}\,(\mathsf{EN}(k') \cap \mathsf{PAR})$, where $k' = k + \sum_{i=0}^{m-1} r(e_i)$, as witnessed by playing σ with history $s e_0 s_1 e_1 \ldots s_m$ from s_m. Thus $s_m \in \bigcup_k \mathsf{AS}\,(\mathsf{EN}(k) \cap \mathsf{PAR}) = W'$, i.e., almost all σ-induced runs ρ satisfy $\mathbb{G}W'$.

Towards Item 1, we have $\mathsf{EN}(k) \subseteq \mathsf{LimInf}(>-\infty)$ and thus $\mathsf{EN}(k) \cap \mathsf{PAR} \subseteq \mathsf{LimInf}(>-\infty) \cap \mathsf{PAR} = \mathsf{Gain}$. Therefore σ witnesses $s \in \mathsf{AS}\,(\mathsf{Gain} \cap \mathbb{G}W')$.

Towards Item 2, we have $\mathsf{EN}(k) \cap \mathsf{PAR} \subseteq k\text{-}\mathsf{Bailout}$ by Lemma 9. Thus σ witnesses $s \in \mathsf{AS}\,(k\text{-}\mathsf{Bailout} \cap \mathbb{G}W')$. □

Lemma 12. $W' \subseteq W$.

Proof. It suffices to show that W' satisfies the monotone condition imposed on W (cf. Definition 4), since W is defined as the largest set satisfying this condition.

Let $s \in W' = \bigcup_k \mathsf{AS}\,(\mathsf{EN}(k) \cap \mathsf{PAR})$. Then $s \in \mathsf{AS}\left(\mathsf{EN}(\hat{k}) \cap \mathsf{PAR}\right)$ for some fixed $\hat{k}$. By Lemma 11(1) we have $s \in \mathsf{AS}\,(\mathsf{Gain} \cap \mathbb{G}W')$. By Lemma 11(2) we have $s \in \mathsf{AS}\left(\hat{k}\text{-}\mathsf{Bailout} \cap \mathbb{G}W'\right) \subseteq \bigcup_k \mathsf{AS}\,(k\text{-}\mathsf{Bailout} \cap \mathbb{G}W')$. □

Proof of Theorem 5. Towards the $\subseteq$ inclusion, we have

$$\mathsf{AS}\,(\mathsf{EN}(k) \cap \mathsf{PAR}) \subseteq \mathsf{AS}\,(k\text{-}\mathsf{Bailout} \cap \mathbb{G}W') \subseteq \mathsf{AS}\,(k\text{-}\mathsf{Bailout} \cap \mathbb{G}W)$$

by Lemma 11(2) and Lemma 12.

Towards the $\supseteq$ inclusion, let $s \in \mathsf{AS}\,(k\text{-}\mathsf{Bailout} \cap \mathbb{G}W)$ and σ_1 be a strategy that witnesses this. We show that $s \in \mathsf{AS}\,(\mathsf{EN}(k) \cap \mathsf{PAR})$. We now consider the modified SSG $\mathcal{G}' = (W, E, \lambda)$ with the state set restricted to W. In particular, $s \in W$ and σ_1 witnesses $s \in \mathsf{AS}\,(k\text{-}\mathsf{Bailout})$ in $\mathcal{G}'$. We now construct a strategy σ that witnesses $s \in \mathsf{AS}\,(\mathsf{EN}(k) \cap \mathsf{PAR})$ in $\mathcal{G}'$, and thus also in $\mathcal{G}$. The strategy σ will use infinite memory to keep track of the current energy level of the run.

Apart from σ_1, we require several more strategies as building blocks for the construction of σ.

First, in $\mathcal{G}$ we had $\forall s' \in W.\, s' \in \mathsf{AS}\,(\mathsf{Gain} \cap \mathbb{G}W)$, and thus in $\mathcal{G}'$ we have $\forall s' \in W.\, s' \in \mathsf{AS}\,(\mathsf{Gain})$. For every $s' \in W$ we instantiate Lemma 8 for $\mathcal{G}'$ with $\delta = 1/2$ and obtain a number $\hat{k}_{s'}$ and a strategy $\hat{\sigma}_{s'}$ with

1. $\forall \tau.\, \mathbb{P}_{s'}^{\hat{\sigma}_{s'},\tau}(\mathsf{Gain}) = 1$, and
2. $\forall \tau.\, \mathbb{P}_{s'}^{\hat{\sigma}_{s'},\tau}(\mathsf{EN}(\hat{k}_{s'}) \cap \mathsf{PAR}) \geq 1/2$.

Let $k_1 \stackrel{\text{def}}{=} \max\{\hat{k}_{s'} \mid s' \in W\}$. The strategies $\hat{\sigma}_{s'}$ are called *gain strategies.*

Second, by the finiteness of V, there is a minimal number k_2 such that $\bigcup_k \mathsf{AS}\,(k\text{-}\mathsf{Bailout} \cap \mathbb{G}W) = \bigcup_{k \leq k_2} \mathsf{AS}\,(k\text{-}\mathsf{Bailout} \cap \mathbb{G}W)$ in $\mathcal{G}$. Therefore, in $\mathcal{G}'$ we have that

$$W \subseteq \bigcup_k \mathsf{AS}\,(k\text{-}\mathsf{Bailout}) = \bigcup_{k \leq k_2} \mathsf{AS}\,(k\text{-}\mathsf{Bailout}) = \mathsf{AS}\,(k_2\text{-}\mathsf{Bailout})\,.$$

Thus in $\mathcal{G}'$ for every $s' \in W$ there exists a strategy $\tilde{\sigma}_{s'}$ with $\forall \tau.\, \mathbb{P}_{s'}^{\tilde{\sigma}_{s'},\tau}(k_2\text{-}\mathsf{Bailout}) = 1$. The strategies $\tilde{\sigma}_{s'}$ are called *bailout strategies*. Let $k' \stackrel{\text{def}}{=} k_1 + k_2 - k + 1$. We now define the strategy σ.

Start: First σ plays like σ_1 from s. Since σ_1 witnesses $s \in \mathsf{AS}\,(k\text{-}\mathsf{Bailout})$ against every minimizer strategy τ, almost all induced runs $\rho = s e_0 s_1 e_1 \ldots$ satisfy either

(A) $(\cup_l \mathsf{ST}(k,l) \cap \mathsf{PAR})$, or
(B) $(\mathsf{EN}(k) \cap \mathsf{LimSup}(=\infty))$.

Almost all runs ρ of the latter type (B) (and potentially also some runs of type (A)) satisfy $\mathsf{EN}(k)$ and $\sum_{i=0}^{l} r(e_i) \geq k'$ eventually for some l. If we observe $\sum_{i=0}^{l} r(e_i) \geq k'$ for some prefix $se_0s_1e_1\ldots e_ls'$ of the run ρ then our strategy σ plays from s' as described in the **Gain** part below. Otherwise, if we never observe this condition, then our run ρ is of type (A) and σ continues playing like σ_1. Since property (A) implies $(\mathsf{EN}(k) \cap \mathsf{PAR})$, this is sufficient.

Gain: In this case we are in the situation where we have reached some state s' after some finite prefix ρ' of the run, where $r(\rho') \geq k'$. Our strategy σ now plays like the gain strategy $\hat{\sigma}_{s'}$, as long as $r(\rho') \geq k' - k_1$ holds for the current prefix ρ' of the run. By Item 2, this will satisfy $\forall \tau.\, \mathbb{P}_{s'}^{\hat{\sigma}_{s'},\tau}(\mathsf{EN}(\hat{k}_{s'}) \cap \mathsf{PAR}) \geq 1/2$ and thus $\forall \tau.\, \mathbb{P}_{s'}^{\hat{\sigma}_{s'},\tau}(\mathsf{EN}(k_1) \cap \mathsf{PAR}) \geq 1/2$. It follows that with probability $\geq 1/2$ we will keep playing $\hat{\sigma}_{s'}$ forever and satisfy PAR and always $r(\rho') \geq k' - k_1$ and thus $\mathsf{EN}(k)$, since $k + r(\rho') \geq k + k' - k_1 = k_2 + 1 \geq 0$.

Otherwise, if eventually $r(\rho') = k' - k_1 - 1$ then we have $k + r(\rho') = k_2$. In this case (which happens with probability $< 1/2$) we continue playing as described in the **Bailout** part below.

Bailout: In this case we are in the situation where we have reached some state $s'' \in W$ after some finite prefix ρ' of the run, where $k + r(\rho') = k_2$. Since $s'' \in W$, we can now let our strategy σ play like the bailout strategy $\tilde{\sigma}_{s''}$ and obtain $\forall \tau.\, \mathbb{P}_{s''}^{\tilde{\sigma}_{s''},\tau}(k_2\text{-}\mathsf{Bailout}) = 1$. Thus almost all induced runs $\rho'' = s''e_0s_1e_1\ldots$ from s'' satisfy either

(A) $(\cup_l \mathsf{ST}(k_2,l) \cap \mathsf{PAR})$, or
(B) $(\mathsf{EN}(k_2) \cap \mathsf{LimSup}(=\infty))$.

As long as $r(\rho') < k'$ holds for the current prefix ρ' of the run, we keep playing $\tilde{\sigma}_{s''}$. Otherwise, if eventually $r(\rho') \geq k'$ holds, then we switch back to playing the **Gain** strategy above. All the runs that never switch back to playing the **Gain** strategy must be of type (A) and thus satisfy PAR. Since we have $k_2\text{-}\mathsf{Bailout} \subseteq \mathsf{EN}(k_2)$, it follows that, for every prefix ρ'' of the run from s'', according to $\tilde{\sigma}_{s''}$ we have $k_2 + r(\rho'') \geq 0$. Thus, for every prefix ρ''' of ρ, we have $k + r(\rho''') = k + r(\rho') + r(\rho'') = k_2 + r(\rho'') \geq 0$. Therefore, the $\mathsf{EN}(k)$ objective is satisfied by all runs.

As shown above, almost all runs induced by σ that eventually stop switching between the three modes satisfy $\mathsf{EN}(k) \cap \mathsf{PAR}$. Switching from Gain/Bailout to Start is impossible, but switching from Gain to Bailout and back is possible. However, the set of runs that infinitely often switch between Gain and Bailout is a null-set, because the probability of switching from Gain to Bailout is $\leq 1/2$. Thus, σ witnesses $s \in \mathsf{AS}\,(\mathsf{EN}(k) \cap \mathsf{PAR})$. □

Remark 13. It follows from the results above that $W' = W$. The $\subseteq$ inclusion holds by Lemma 12. For the reverse inclusion we have

$$\begin{aligned} W &\subseteq \bigcup_k \mathsf{AS}\,(k\text{-}\mathsf{Bailout} \cap \mathbb{G}W) && \text{by Definition 4} \\ &= \bigcup_k \mathsf{AS}\,(\mathsf{EN}(k) \cap \mathsf{PAR}) && \text{by Theorem 5} \\ &= W' && \text{by Definition 10.} \end{aligned}$$

4 Bailout

In this section we will argue that it is possible decide, in NP and coNP, whether the bailout objective can be satisfied almost surely. More precisely, we show the existence of procedures to decide if, for a given $k \in \mathbb{N}$ and state s, there exists an $l \in \mathbb{N}$ such that s almost-surely satisfies the $\mathsf{Bailout}(k, l)$ objective

$$\mathsf{Bailout}(k,l) \stackrel{\text{def}}{=} (\mathsf{ST}(k,l) \cap \mathsf{PAR}) \cup (\mathsf{EN}(k) \cap \mathsf{LimSup}(=\infty)).$$

Recall that the idea behind the Bailout objective is that, during a game for energy-parity, maximizer is temporarily abandoning the parity (but not the energy) condition in order to increase the energy to a sufficient level (which will then allow him to try an a.s. strategy for Gain once more). However, in a stochastic game – as opposed to an MDP [40] – an opponent could possibly prevent this increase in energy level at the expense of satisfying the original energy-parity objective in the first place (cf. Example 1). The Bailout objective is designed to capture the disjunction of both outcomes, as both are favorable for the maximizer. The parameter k is the acceptable total energy drop (i.e., the initial value), and the parameter l is the acceptable energy drop on any infix of a play, which translates to the upper bound on the energy level in the second outcome.

The question can be phrased equivalently as membership of a control state s in the almost-sure set for the k-Bailout objective for a given game $\mathcal{G}$ and energy level $k \in \mathbb{N}$.

Theorem 14. *One can check in* NP, coNP *and pseudo-polynomial time if, for a given SSG* $\mathcal{G} \stackrel{\text{def}}{=} (V, E, \lambda)$, $k \in \mathbb{N}$ *and control state* $s \in V$, *maximizer can almost-surely satisfy* k-Bailout *from* s.

Moreover, there are $K, L \in \mathbb{N}$, *polynomial in* $|V|$ *and the largest absolute transition reward, so that* $\bigcup_{k\geq 0} \mathsf{AS}^{\mathcal{G}}(k\text{-}\mathsf{Bailout}) = \mathsf{AS}^{\mathcal{G}}(\mathsf{Bailout}(K, L))$. *And so, checking whether state* s *belongs to* $\bigcup_{k\geq 0} \mathsf{AS}^{\mathcal{G}}(k\text{-}\mathsf{Bailout})$ *is in* NP *and* coNP.

Proof (sketch). This is shown by a sequence of transformations of the game and ultimately reduced to a finding the winner of a non-stochastic game with an energy-parity objective, which is known to be solvable in NP, coNP and pseudo-polynomial time [19]. One important observation is that it is possible to replace, without changing the outcome, the energy $\mathsf{EN}(k)$ condition in the $\mathsf{Bailout}(k,l)$ objective by the more restrictive energy-storage $\mathsf{ST}(k,l)$ condition. See [41] for further details. □

5 Gain

In this section we will argue that it is possible to decide, in NP and coNP, whether the Gain objective (i.e., $\mathsf{LimInf}(> -\infty) \cap \mathsf{PAR}$) can be satisfied almost surely.

We start by investigating the strategy complexity of winning strategies for the Gain objective.

Lemma 15. *In every finite SSG, minimizer has optimal MD strategies for objective* Gain.

Proof. We show that maximizer has MD optimal strategies for $\overline{\mathsf{LimInf}(= -\infty) \cup \mathsf{PAR}}$. This is equivalent to the claim of the lemma because $\overline{\mathsf{LimInf}(> -\infty) \cap \mathsf{PAR}} = \mathsf{LimInf}(= -\infty) \cup \overline{\mathsf{PAR}}$ and the complement of a parity condition is itself a parity condition (with all priorities incremented by one).

We note that both $\mathsf{LimInf}(= -\infty)$, as well as parity objectives PAR are shift-invariant and submixing and therefore also that the union $\mathsf{LimInf}(= -\infty) \cup \mathsf{PAR}$ has both these properties. The claim now follows from the fact that SSGs with objectives that are both submixing and shift-invariant admit MD optimal strategies for maximizer [34, Theorem 5.2]. □

Based on the results in [40] one can show a similar claim for maximizer strategies in MDPs.

Lemma 16. *For finite MDPs, almost-sure winning maximizer strategies for* Gain *can be chosen FD.*

Using the existence of MD optimal minimizer strategies (Lemma 15) and a coNP upper bound for checking almost sure Gain in MDPs established in [40], we can derive a coNP procedure. See [41] for full details.

Theorem 17. *Checking whether a state $s \in V$ of a SSG satisfies* Gain *almost-surely is in* coNP.

The rest of this section will deal with the NP upper bound, which is the most challenging part of this paper. The crux of our proof is the observation that if maximizer has a strategy that wins almost surely against all MD minimizer strategies, then he wins almost surely. This is because one of these MD strategies is optimal due to Lemma 15. We show that, in order to witness such an almost-sure winning strategy for maximizer in SSG $\mathcal{G}$, it suffices to provide a polynomially larger SSG $\mathcal{G}_3$, together with an almost-sure winning strategy for the *storage-parity* objective (see Theorem 21 in Section 6) in $\mathcal{G}_3$. This will give us an NP algorithm, because $\mathcal{G}_3$, along with its winning strategy, can be guessed and verified in polynomial time. Formally we claim that:

Theorem 18. *Checking whether a state $s \in V$ of $\mathcal{G}$ satisfies* Gain *almost-surely is in* NP.

Proof. (sketch) For technical convenience, we will assume w.l.o.g. that every SSG henceforth is in a normal form, where every random state has only one predecessor, which is owned by the maximizer. To show the existence of $\mathcal{G}_3$, we are going to introduce two intermediate games: $\mathcal{G}_1$ and $\mathcal{G}_2$. These games are never constructed by our NP algorithm, but are just defined to break down the complex construction of $\mathcal{G}_3$ into more manageable steps.

Intuitively, $\mathcal{G}_1$ is just $\mathcal{G}$ where all rewards on edges are multiplied by a large enough factor, f, to turn strategies with a mean-payoff > 0 into ones with mean-payoff > 2. $\mathcal{G}_2$ is an extension of $\mathcal{G}_1$ where the maximizer is given a choice before every visit to a probabilistic node. He can either let the game proceed as before, or sacrifice part of his one-step reward in exchange for a more evenly balanced reward outcome, so the energy can no longer drop arbitrarily low when a probabilistic cycle is reached. As a result, in $\mathcal{G}_2$ it suffices to consider a storage-parity objective (see Theorem 21 in Section 6) instead of Gain. The number of choices maximizer is given is the number of MD minimizer strategies, which clearly can be exponential. That would not suffice for an NP algorithm. Therefore, we show that most of these choices are redundant and can be removed without impairing the almost sure wining region. As the result of that pruning, we obtain $\mathcal{G}_3$ of polynomial size. □

For the the technical details of the $\mathcal{G} \to \mathcal{G}_1 \to \mathcal{G}_2 \to \mathcal{G}_3$ constructions please see [41]. Figure 2 shows how these transformations may look like.

6 The Main Results

In this section, we prove the main results of the paper, namely that almost-sure energy parity stochastic games can be decided in NP and coNP. The proofs are straightforward and follow from the much more involved characterization of almost sure energy parity objective in terms of the Bailout and Gain objectives established in Section 3 and their computational complexity analysis in Sections 4 and 5, respectively.

Theorem 19. *Given an SSG, energy level k^*, checking if a state s is almost-sure winning for* $\mathsf{EN}(k^*) \cap \mathsf{PAR}$ *is in* $\mathsf{NP} \cap \mathsf{coNP}$.

Proof. Recall that we can compute the set W from Definition 4 by iterating

$$W_i \stackrel{\text{def}}{=} \mathsf{AS}\left(\mathsf{Gain} \cap \mathbb{G}W_{i-1}\right) \cap \bigcup_k \mathsf{AS}\left(k\text{-}\mathsf{Bailout} \cap \mathbb{G}W_{i-1}\right)$$

starting with $W_0 \stackrel{\text{def}}{=} V$, until we reach the greatest fixed point W. Note that at step i we need to solve almost sure Gain and almost sure $\bigcup_k \mathsf{AS}\,(k\text{-}\mathsf{Bailout})$, where the states of the game are restricted to W_{i-1}. There can be at most $|V|$ steps, because at least one state is removed in each iteration.

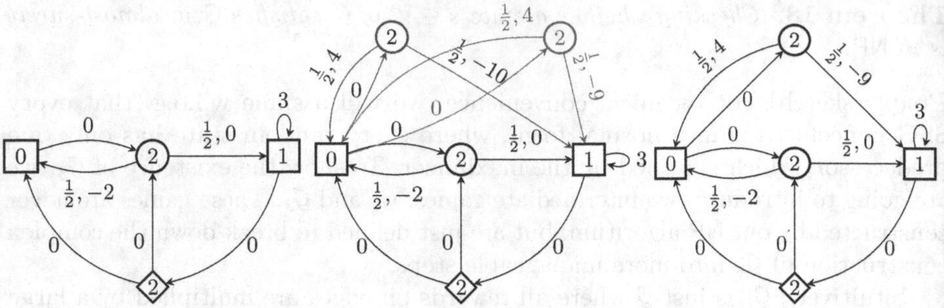

(a) The original game $\mathcal{G} = \mathcal{G}_1$ (b) The game $\mathcal{G}_2$ (c) The game $\mathcal{G}_3$

Fig. 2: An example game $\mathcal{G}$ (left) and the derived games. The strategy that always loops in the right-most state of $\mathcal{G}$ ensures a mean-payoff of 3. As this is the only MD strategy for maximizer that ensures a positive mean-payoff, a factor $f = 1$ is sufficient here and we have $\mathcal{G}_1 = \mathcal{G}$. In the derived game $\mathcal{G}_2$ in Fig. 2b there are as many trade-in options for the random state as there are MD minimizer's strategies in $\mathcal{G}_1$ (just two in this example). The blue one (top left) corresponds to minimizer going left and the red one (top right) to going up in $\mathcal{G}_1$. Maximizer almost-surely wins Gain in $\mathcal{G}$ iff he almost-surely wins a storage-parity condition (see Theorem 21) in $\mathcal{G}_3$.

It then suffices to check AS (k-Bailout $\cap\, \mathbb{G}W$) (i.e., AS (k-Bailout) for the subgame that consists only of the states of the fixed point W for $k = k^*$. Note that this step can be skipped if $k^* \geq K$, the bound from Theorem 14.

Before we discuss how to use NP and coNP procedures to construct these sets and to conduct the final test on the fixed point W, we note that the '$\cap\, \mathbb{G}W_{i-1}$' does not add anything substantial, as these are simply the same tests and procedures conducted on the subgame that only consist of the states of W_{i-1}.

To obtain an NP procedure for constructing AS (Gain)—or, as remarked above, AS (Gain $\cap\, \mathbb{G}W_{i-1}$)—we can guess and validate its membership for each state s *in* this set, using the NP result from Theorem 18, and we can guess and validate its non-membership for each state s *not in* this set in NP, using the coNP result from Theorem 17. Similarly, we can guess and validate both the membership and the non-membership in $\bigcup_k$ AS (k-Bailout $\cap\, \mathbb{G}W_{i-1}$)—and of $\bigcup_k$ AS (k-Bailout $\cap\, \mathbb{G}W_{i-1}$) by analysing the subgame with only the states in W_{i-1}—by using the NP and coNP result, respectively, from Theorem 14.

Once we can construct these sets, we can also intersect them and check if a fixed point has been reached. (One can, of course, stop when $s \notin W_i$.)

We can now conduct the final check in NP using Theorem 18.

A coNP algorithm that constructs W can be designed analogously: once W_{i-1} is known, membership and non-membership of a state s in AS (Gain $\cap\, \mathbb{G}W_{i-1}$) can be guessed and validated in coNP by Theorem 17 and by Theorem 18, respectively; and membership or non-membership of a state in $\bigcup_k$ AS (k-Bailout $\cap\, \mathbb{G}W_{i-1}$) can be guessed and validated in coNP using the coNP and NP part, respectively, of Theorem 14.

Once W is constructed, we can conduct the final check in coNP using Theorem 17. □

This result, together with the upper bound on the energy needed to win energy-parity objective, allows us to solve the "unknown initial energy problem" [7], which is to compute the minimal initial energy level required.

Corollary 20. *For any state s, checking if there is k such that* $\mathsf{AS}\,(\mathsf{EN}(k) \cap \mathsf{PAR})$ *holds is in* $\mathsf{NP} \cap \mathsf{coNP}$. *Also, for a given k^*, checking if k^* is the minimal energy level required to win almost surely is in* $\mathsf{NP} \cap \mathsf{coNP}$ *as well.*

Proof. Due to Theorem 14, if there is an energy level k for which $\mathsf{AS}\,(\mathsf{EN}(k) \cap \mathsf{PAR})$ holds, then it also holds for the bound K whose size is polynomial in the size of the game. We can then simply calculate K and then use NP and coNP algorithms from Theorem 19 for $\mathsf{AS}\,(\mathsf{EN}(K) \cap \mathsf{PAR})$.

As for the second claim, note that checking whether maximizer cannot win almost surely $\mathsf{EN}(k) \cap \mathsf{PAR}$ is also in NP and coNP as a complement of a coNP and an NP set, respectively. Therefore, for an NP/coNP upper bound it suffices to simultaneously guess certificates for almost surely $\mathsf{EN}(k^*) \cap \mathsf{PAR}$ and not almost surely $\mathsf{EN}(k^* - 1) \cap \mathsf{PAR}$ and verify them in polynomial time. □

Finally, let us mention that the slightly more restrictive *storage-parity* objectives can also be solved in $\mathsf{NP} \cap \mathsf{coNP}$. These are almost identical to energy-parity except that, in addition, there must exist some bound $l \in \mathbb{N}$ such that the energy level never drops by more than l during a run. This extra condition ensures that, if the storage-parity objective holds almost-surely, then there must exist a *finite-memory* winning strategy for maximizer.

Theorem 21. *One can check in* NP, coNP *and pseudo-polynomial time if, for a given SSG $\mathcal{H} \stackrel{def}{=} (V, E, \lambda)$, $k \in \mathbb{N}$ and control state $s \in V$, maximizer can almost-surely satisfy* $\mathsf{ST}(k) \cap \mathsf{PAR}$ *from s.*

Moreover, there is a bound $L \in \mathbb{N}$, polynomial in the number of states and the largest absolute transition reward, so that $\mathsf{ST}(k) \cap \mathsf{PAR} = \mathsf{ST}(k, L) \cap \mathsf{PAR}$.

Proof. (sketch) This result follows by a simple adaptation of the proofs showing the same computational complexity of the Bailout objective (Section 4). See [41] for further details. □

Example 22. In the game in Fig. 1, maximizer cannot ensure the storage-parity condition $\mathsf{ST}(k) \cap \mathsf{PAR}$ for any initial energy level k. This is because it would imply the existence of a finite-memory almost-surely winning strategy, which as we have already argued, cannot be true. More intuitively, to prevent an intermediate energy drop by l units, a winning maximizer strategy for storage-parity would need to stop moving left after observing the negative cycle in the leftmost state l successive times. However, when maximizer moves right, this gives minimizer the chance to visit the rightmost bad state (with dominating odd priority 1). The chance of that happening is $(1/3)^l > 0$. In particular, this probability is > 0 for any value of the intermediate energy drop l. Therefore, for any fixed l, maximizer would need to move right infinitely often to satisfy storage and lose (against an optimal minimizer strategy that moves to the rightmost state).

7 Conclusion and Outlook

We showed that several almost-sure problems for combined energy-parity objectives in simple stochastic games are in $\mathsf{NP} \cap \mathsf{coNP}$. No pseudo-polynomial algorithm is known (just like for stochastic mean-payoff parity games [20]). All these problems subsume (stochastic) parity games, by setting all rewards to 0. Thus the existence of a pseudo-polynomial algorithm would imply that (stochastic and non-stochastic) parity games are in P, which is a long-standing open problem.

It is known that maximizer already needs infinite memory to win almost-surely a combined energy-parity objective in MDPs [40]. Our results do not imply anything about the memory requirement for optimal minimizer strategies in SSGs for this objective. We conjecture that memoryless minimizer strategies suffice. If this conjecture holds (and is proven), this would greatly simplify the coNP upper bound that we established for this problem.

A natural question is whether results on mean-payoff/energy/parity games can be generalized to a setting with multi-dimensional payoffs. Non-stochastic multi-mean-payoff and multi-energy games have been studied in [48,36,1]. To the best of our knowledge, the techniques used there, e.g. upper bounds on the necessary energy levels as in [36], do not generalize to stochastic games (or MDPs).

Multiple mean-payoff objectives in MDPs have been studied in [10,24], but the corresponding multi-energy (resp. multi-energy-parity) objective has extra difficulties due to the 0-boundary condition on the energy. I.e., even on Markov chains, and without any parity condition, it subsumes problems about multi-dimensional random walks. Some partial results on Markov chains and MDPs have been obtained in [13,2,3], but the decidability of the almost-sure problem for stochastic multi-energy-parity games (and MDPs) remains open.

Acknowledgments

The work of all the authors was supported in part by EPSRC grant EP/M027287/1. Sven Schewe and Dominik Wojtczak were also supported by EPSRC grant EP/P020909/1.

References

1. Abdulla, P., Mayr, R., Sangnier, A., Sproston, J.: Solving parity games on integer vectors. In: International Conference on Concurrency Theory (CONCUR). vol. 8052 (2013)
2. Abdulla, P.A., Ciobanu, R., Mayr, R., Sangnier, A., Sproston, J.: Qualitative analysis of VASS-induced MDPs. In: International Conference on Foundations of Software Science and Computational Structures (FoSSaCS). vol. 9634 (2016)
3. Abdulla, P.A., Henda, N.B., Mayr, R.: Decisive Markov Chains. Logical Methods in Computer Science Volume 3, Issue 4 (Nov 2007)
4. Alur, R., Henzinger, T.A., Kupferman, O.: Alternating-time temporal logic. J. ACM 49(5), 672–713 (2002)
5. Billingsley, P.: Probability and Measure. Wiley (1995), third Edition
6. Blondel, V.D., Canterini, V.: Undecidable problems for probabilistic automata of fixed dimension. Theory of Computing systems 36(3) (2003)
7. Bouyer, P., Fahrenberg, U., Larsen, K.G., Markey, N., Srba, J.: Infinite runs in weighted timed automata with energy constraints. In: International Conference on Formal Modeling and Analysis of Timed Systems (FORMATS). vol. 5215, pp. 33–47 (2008)
8. Brázdil, T., Brožek, V., Etessami, K., Kučera, A.: Approximating the Termination Value of One-Counter MDPs and Stochastic Games. Information and Computation 222, 121–138 (2013)
9. Brázdil, T., Kučera, A., Novotný, P.: Optimizing the expected mean payoff in energy Markov decision processes. In: International Symposium on Automated Technology for Verification and Analysis (ATVA). vol. 9938, pp. 32–49 (2016)
10. Brázdil, T., Brožek, V., Chatterjee, K., Forejt, V., Kučera, A.: Markov decision processes with multiple long-run average objectives. Logical Methods in Computer Science 10 (2014)
11. Brázdil, T., Brožek, V., Etessami, K.: One-Counter Stochastic Games. In: IARCS Annual Conference on Foundations of Software Technology and Theoretical Computer Science (FSTTCS). vol. 8, pp. 108–119 (2010)
12. Brázdil, T., Brožek, V., Etessami, K., Kučera, A., Wojtczak, D.: One-counter Markov decision processes. In: ACM-SIAM Symposium on Discrete Algorithms (SODA). pp. 863–874 (2010)
13. Brázdil, T., Kiefer, S., Kučera, A., Novotný, P., Katoen, J.P.: Zero-reachability in probabilistic multi-counter automata. In: Proceedings of the Joint Meeting of the Twenty-Third EACSL Annual Conference on Computer Science Logic (CSL) and the Twenty-Ninth Annual ACM/IEEE Symposium on Logic in Computer Science (LICS). pp. 22:1–22:10 (2014)
14. Chakrabarti, A., De Alfaro, L., Henzinger, T.A., Stoelinga, M.: Resource interfaces. In: International Workshop on Embedded Software. pp. 117–133 (2003)
15. Chatterjee, K., De Alfaro, L., Henzinger, T.A.: The complexity of stochastic Rabin and Streett games. In: International Colloquium on Automata, Languages and Programming (ICALP). pp. 878–890 (2005)
16. Chatterjee, K., Doyen, L.: Energy parity games. In: International Colloquium on Automata, Languages and Programming (ICALP). vol. 6199, pp. 599–610 (2010)
17. Chatterjee, K., Doyen, L.: Energy and mean-payoff parity Markov decision processes. In: International Symposium on Mathematical Foundations of Computer Science (MFCS). vol. 6907, pp. 206–218 (2011)

18. Chatterjee, K., Doyen, L.: Games and Markov decision processes with mean-payoff parity and energy parity objectives. In: Mathematical and Engineering Methods in Computer Science (MEMICS). LNCS, vol. 7119, pp. 37–46. Springer (2011)
19. Chatterjee, K., Doyen, L.: Energy parity games. Theoretical Computer Science 458, 49–60 (2012)
20. Chatterjee, K., Doyen, L., Gimbert, H., Oualhadj, Y.: Perfect-information stochastic mean-payoff parity games. In: International Conference on Foundations of Software Science and Computational Structures (FoSSaCS). vol. 8412 (2014)
21. Chatterjee, K., Henzinger, T.A., Piterman, N.: Generalized parity games. In: International Conference on Foundations of Software Science and Computational Structures (FoSSaCS). pp. 153–167 (2007)
22. Chatterjee, K., Jurdziński, M., Henzinger, T.A.: Simple stochastic parity games. In: Computer Science Logic (CSL). vol. 2803, pp. 100–113. Springer (2003)
23. Chatterjee, K., Jurdziński, M., Henzinger, T.A.: Quantitative stochastic parity games. In: ACM-SIAM Symposium on Discrete Algorithms (SODA). pp. 121–130. SIAM (2004)
24. Chatterjee, K., Kretínská, Z., Kretínský, J.: Unifying two views on multiple mean-payoff objectives in Markov decision processes. Logical Methods in Computer Science 13(2) (2017)
25. Clarke, E., Grumberg, O., Peled, D.: Model Checking. MIT Press (Dec 1999)
26. Daviaud, L., Jurdziński, M., Lazić, R.: A pseudo-quasi-polynomial algorithm for mean-payoff parity games. In: Logic in Computer Science (LICS). pp. 325–334 (2018)
27. De Alfaro, L., Henzinger, T.A.: Interface automata. ACM SIGSOFT Software Engineering Notes 26(5), 109–120 (2001)
28. Dill, D.L.: Trace theory for automatic hierarchical verification of speed-independent circuits, vol. 24. MIT press Cambridge (1989)
29. Fearnley, J.: Exponential lower bounds for policy iteration. In: International Colloquium on Automata, Languages and Programming (ICALP). pp. 551–562 (2010)
30. Filar, J., Vrieze, K.: Competitive Markov Decision Processes. Springer (1997)
31. Friedmann, O.: An exponential lower bound for the parity game strategy improvement algorithm as we know it. In: Logic in Computer Science (LICS). pp. 145–156 (2009)
32. Friedmann, O., Hansen, T.D., Zwick, U.: Subexponential lower bounds for randomized pivoting rules for the simplex algorithm. In: Symposium on Theory of Computing (STOC). pp. 283–292 (2011)
33. Gillette, D.: Stochastic games with zero stop probabilities. Contributions to the Theory of Games 3, 179–187 (1957)
34. Gimbert, H., Kelmendi, E.: Two-Player Perfect-Information Shift-Invariant Submixing Stochastic Games Are Half-Positional (Jan 2014), working paper or preprint available at: https://hal.archives-ouvertes.fr/hal-00936371
35. Jurdziński, M.: Deciding the winner in parity games is in UP ∩ co-UP. Information Processing Letters 68(3), 119–124 (1998)
36. Jurdziński, M., Lazić, R., Schmitz, S.: Fixed-dimensional energy games are in pseudo-polynomial time. In: International Colloquium on Automata, Languages and Programming (ICALP). vol. 9135, pp. 260–272 (2015)
37. Kiefer, S., Mayr, R., Shirmohammadi, M., Wojtczak, D.: On strong determinacy of countable stochastic games. Logic in Computer Science (LICS) (2017)
38. Maitra, A., Sudderth, W.: Stochastic games with Borel payoffs. In: Stochastic Games and Applications, pp. 367–373. Kluwer, Dordrecht (2003)

39. Martin, D.A.: The determinacy of Blackwell games. Journal of Symbolic Logic 63(4), 1565–1581 (1998)
40. Mayr, R., Schewe, S., Totzke, P., Wojtczak, D.: MDPs with Energy-Parity Objectives. Logic in Computer Science (LICS) (2017)
41. Mayr, R., Schewe, S., Totzke, P., Wojtczak, D.: Simple Stochastic Games with Almost-Sure Energy-Parity Objectives are in NP and coNP (2021), arXiv:2101.06989
42. Pnueli, A., Rosner, R.: On the synthesis of a reactive module. In: Annual Symposium on Principles of Programming Languages (POPL). pp. 179–190 (1989)
43. Pogosyants, A., Segala, R., Lynch, N.: Verification of the randomized consensus algorithm of Aspnes and Herlihy: a case study. Distributed Computing 13(3), 155–186 (2000)
44. Rabin, M.O.: Automata on infinite objects and Church's problem, vol. 13. American Mathematical Soc. (1972)
45. Ramadge, P.J., Wonham, W.M.: Supervisory control of a class of discrete event processes. SIAM journal on control and optimization 25(1), 206–230 (1987)
46. Shapley, L.S.: Stochastic games. Proceedings of the national academy of sciences 39(10), 1095–1100 (1953)
47. Stoelinga, M.: Fun with firewire: A comparative study of formal verification methods applied to the IEEE 1394 root contention protocol. Formal aspects of computing 14(3), 328–337 (2003)
48. Velner, Y., Chatterjee, K., Doyen, L., Henzinger, T.A., Rabinovich, A., Raskin, J.F.: The complexity of multi-mean-payoff and multi-energy games. Information and Computation 241, 177 – 196 (2015)
49. Wilke, T.: Alternating tree automata, parity games, and modal mu-calculus. Bulletin of the Belgian Mathematical Society Simon Stevin 8(2), 359 (2001)

Nondeterministic Syntactic Complexity

Robert S. R. Myers, Stefan Milius[1,*], and Henning Urbat (✉)[1,**]

[1]Friedrich-Alexander-Universität Erlangen-Nürnberg, Erlangen, Germany
my.robmyers@gmail.com, {stefan.milius,henning.urbat}@fau.de

Abstract We introduce a new measure on regular languages: their *nondeterministic syntactic complexity.* It is the least degree of any extension of the 'canonical boolean representation' of the syntactic monoid. Equivalently, it is the least number of states of any *subatomic* nondeterministic acceptor. It turns out that essentially all previous structural work on nondeterministic state-minimality computes this measure. Our approach rests on an algebraic interpretation of nondeterministic finite automata as deterministic finite automata endowed with semilattice structure. Crucially, the latter form a self-dual category.

1 Introduction

Regular languages admit a plethora of equivalent representations: finite automata, finite monoids, regular expressions, formulas of monadic second-order logic, and numerous others. In many cases, the most succinct representation is given by a *nondeterministic finite automaton (nfa).* Therefore, the investigation of state-minimal nfas is of both computational and mathematical interest. However, this turns out to be surprisingly intricate; in fact, the task of minimizing an nfa, or even of deciding whether a given nfa is minimal, is known to be PSPACE-complete [23]. One intuitive reason is that minimal nfas lack structure: a language may have many non-isomorphic minimal nondeterministic acceptors, and there are no clearly identified and easily verifiable mathematical properties distinguishing them from non-minimal ones. As a consequence, all known algorithms for nfa minimization (and related problems such as inclusion or universality testing) require some form of exhaustive search [9,11,26]. This sharply contrasts the situation for minimal *deterministic finite automata (dfa)*: they can be characterized by a universal property making them unique up to isomorphism, which immediately leads to efficient minimization.

In the present paper, we work towards the goal of bringing more structure into the theory of nondeterministic state-minimality. To this end, we propose a novel algebraic perspective on nfas resting on *boolean representations* of monoids, i.e. morphisms $M \to \mathbf{JSL}(S, S)$ from a monoid M into the endomorphism monoid

* Supported by Deutsche Forschungsgemeinschaft (DFG) under projects MI 717/5-2 and MI 717/7-1, and as part of the Research and Training Group 2475 "Cybercrime and Forensic Computing" (393541319/GRK2475/1-2019)

** Supported by Deutsche Forschungsgemeinschaft (DFG) under proj. SCHR 1118/8-2

S. Kiefer and C. Tasson (Eds.): FOSSACS 2021, LNCS 12650, pp. 448–468, 2021.
https://doi.org/10.1007/978-3-030-71995-1_23

of a finite join-semilattice S. Our focus lies on quotient monoids of the free monoid Σ^* recognizing a given regular language $L \subseteq \Sigma^*$. The largest such monoid is Σ^* itself, while the smallest one is the *syntactic monoid* $\mathsf{syn}(L)$. For both of them, L induces a *canonical boolean representation*

$$\Sigma^* \to \mathbf{JSL}(\mathsf{SLD}(L), \mathsf{SLD}(L) \qquad \text{and} \qquad \mathsf{syn}(L) \to \mathbf{JSL}(\mathsf{SLD}(L), \mathsf{SLD}(L))$$

on the semilattice $\mathsf{SLD}(L)$ of all finite unions of left derivatives of L. The first representation gives rise to an algebraic characterization of minimal nfas:

Theorem. The size of a state-minimal nfa for L equals the least degree of any extension of the canonical representation of Σ^* induced by L.

Here, the *degree* of a representation refers to the number of join-irreducibles of the underlying semilattice. In the light of this result, it is natural to ask for an analogous automata-theoretic perspective on the canonical representation of $\mathsf{syn}(L)$ and its extensions. For this purpose, we introduce the class of *subatomic* nfas, a generalization of *atomic* nfas earlier introduced by Brzozowski and Tamm [6]. In order to get a handle on them, we employ an algebraic framework that interprets nfas in terms of **JSL**-*dfas*, i.e. deterministic finite automata in the category of semilattices. In this setting, the semilattice $\mathsf{SLD}(L)$ used in the canonical representations naturally arises as the *minimal* **JSL**-dfa for the language L. We shall demonstrate that much of the structure theory of (sub-)atomic nfas reduces to the observation that the category of **JSL**-dfas is *self-dual.* Our main result gives an algebraic characterization of minimal subatomic nfas:

Theorem. The size of a state-minimal subatomic nfa for L equals the least degree of any extension of the canonical representation of $\mathsf{syn}(L)$.

We call the measure suggested by the above theorem the *nondeterministic syntactic complexity* of the language L. It turns out to be extremely natural: as illustrated in Section 5, essentially all existing work on the structure of state-minimal nfas implicitly identifies classes of languages whose nondeterministic state complexity equals their nondeterministic syntactic complexity, and thus is actually concerned with computing minimal subatomic acceptors.

2 Preliminaries

We start by introducing some notation and terminology used in the paper.

Semilattices. A *(join-)semilattice* is a poset $(S, \leq_S)$ in which every finite subset $X \subseteq S$ has a least upper bound, a.k.a. join, denoted by $\bigvee X$. A *morphism* of semilattices is a map preserving all finite joins. Let **JSL** denote the category of join-semilattices and their morphisms. An element j of a semilattice S is *join-irreducible* if for all finite subsets $X \subseteq S$ with $j = \bigvee X$ one has $j \in X$. Let

$$J(S) = \{\, j \in S \ : \ j \text{ is join-irreducible} \,\}.$$

Let $2 = \{0, 1\}$ denote the two-element semilattice with $0 \leq 1$. Since $2 \cong (\mathcal{P}(1), \subseteq)$ is the free semilattice on a single generator, morphisms from 2 into a semilattice S

correspond uniquely to elements of S. Similarly, a morphism $f\colon S \to 2$ corresponds uniquely to a *prime filter* $F = f^{-1}[1] \subseteq S$, i.e. an upwards closed subset such that $\bigvee X \in F$ implies $X \cap F \neq \emptyset$ for every finite subset $X \subseteq S$. If S is finite, prime filters are precisely the sets $F = \{s \in S : s \nleq s_0\}$ for $s_0 \in S$. If S is a subsemilattice of a semilattice T, every prime filter F of S can be extended to the prime filter $T \setminus (\downarrow(S \setminus F))$ of T, where $\downarrow X = \{\, t \in T \ : \ t \leq x \text{ for some } x \in X \,\}$ denotes the down-closure of a subset $X \subseteq T$. Equivalently, every morphism $f\colon S \to 2$ can be extended to a morphism $g\colon T \to 2$. In category-theoretic terminology, this means that the semilattice 2 forms an injective object of **JSL**.

The category $\mathbf{JSL}_\mathsf{f}$ of finite semilattices is *self-dual* [25]. The equivalence functor $\mathbf{JSL}_\mathsf{f} \xrightarrow{\cong} \mathbf{JSL}_\mathsf{f}^\mathsf{op}$ sends a semilattice S to its *dual semilattice* S^op obtained by reversing the order, and a morphism $f\colon S \to T$ to the morphism $f^*\colon T^\mathsf{op} \to S^\mathsf{op}$ mapping $t \in T$ to the $\leq_S$-largest element $s \in S$ with $f(s) \leq_T t$. Note that f is *adjoint* to f^*: for $s \in S$ and $t \in T$ we have $f(s) \leq_T t$ iff $s \leq_S f^*(t)$.

Languages. A *language* is a subset L of Σ^*, the set of finite words over an alphabet Σ. We let $\overline{L} = \Sigma^* \setminus L$ denote the *complement* and $L^\mathsf{r} = \{w^\mathsf{r} : w \in L\}$ the *reverse*, where $w^\mathsf{r} = a_n \ldots a_1$ for $w = a_1 \ldots a_n$. The *left derivatives*, *right derivatives* and *two-sided derivatives* of L are, respectively, given by $u^{-1}L = \{w \in \Sigma^* : uw \in L\}$, $Lv^{-1} = \{w \in \Sigma^* : wv \in L\}$ and $u^{-1}Lv^{-1} = \{w \in \Sigma^* : uwv \in L\}$ for $u, v \in \Sigma^*$. More generally, for $U \subseteq \Sigma^*$ the language $U^{-1}L = \bigcup_{u \in U} u^{-1}L$ is called the *left quotient* of L w.r.t. U. We define the following sets of languages generated by L:

- $\mathsf{LD}(L) = \{u^{-1}L : u \in \Sigma^*\}$, the set of all left derivatives of L;
- $\mathsf{SLD}(L)$, its closure under finite union;
- $\mathsf{BLD}(L)$, its closure under all set-theoretic boolean operations;
- $\mathsf{BLRD}(L)$, its closure under all boolean operations and right derivatives.

In other words, $\mathsf{SLD}(L)$ is the $\cup$-semilattice of all left quotients of L, or equivalently, the $\cup$-subsemilattice of $\mathcal{P}(\Sigma^*)$ generated by all left derivatives. Moreover, $\mathsf{BLD}(L)$ and $\mathsf{BLRD}(L)$ form the boolean subalgebras of $\mathcal{P}(\Sigma^*)$ generated by all left derivatives and all two-sided derivatives, respectively.

3 Duality Theory of Semilattice Automata

In this section, we set up the algebraic framework in which nondeterministic automata can be studied. Since it involves considering several different types of automata, it is convenient to view them all as instances of a general categorical concept. For the rest of this paper, let Σ denote a fixed finite input alphabet.

Definition 3.1. Let $\mathscr{C}$ be a category and let $X, Y \in \mathscr{C}$ be two fixed objects. An *automaton* in $\mathscr{C}$ is a quadruple (S, δ, i, f) consisting of an object $S \in \mathscr{C}$ of *states*, a family $\delta = (\delta_a\colon S \to S)_{a \in \Sigma}$ of morphisms representing *transitions*, and two morphisms $i\colon X \to S$ and $f\colon S \to Y$ representing *initial* and *final* states (see the left-hand diagram below). A *morphism* between automata (S, δ, i, f) and (S', δ', i', f') is given by a morphism $h\colon S \to S'$ in $\mathscr{C}$ preserving transitions, initial states and final states, i.e. making the right-hand diagram below commute for all $a \in \Sigma$:

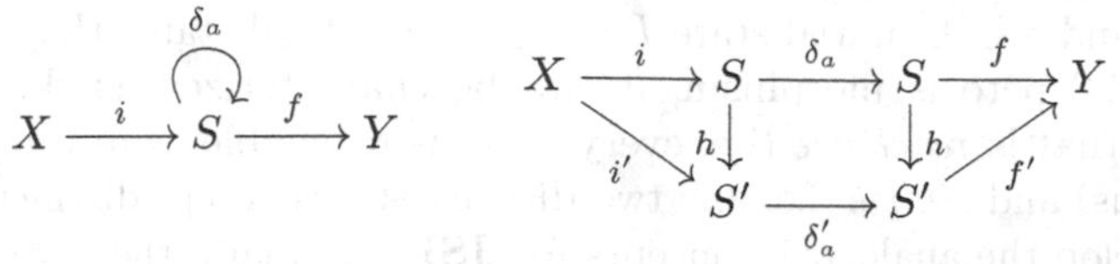

Let $\mathbf{Aut}(\mathscr{C})$ denote the category of automata in $\mathscr{C}$ and their morphisms.

Notation 3.2. We put $\delta_w := \delta_{a_n} \circ \cdots \circ \delta_{a_1}$ for $w = a_1 \dots a_n$ in Σ^*.

Example 3.3. (1) An automaton $D = (S, \delta, i, f)$ in **Set**, the category of sets and functions, with $X = 1$ and $Y = 2$, is precisely a classical *deterministic automaton.* It is called a *dfa* if S is finite. We identify the map $i\colon 1 \to S$ with an initial state $s_0 = i(*) \in S$, and the map $f\colon S \to 2$ with a set $F = f^{-1}[1] \subseteq S$ of final states. The language $L(D, s)$ *accepted* by a state $s \in S$ is the set of all words $w \in \Sigma^*$ such that $\delta_w(s) \in F$. The language $L(D)$ *accepted* by D is the language accepted by the state s_0.
(2) An automaton $N = (S, \delta, i, f)$ in **Rel**, the category of sets and relations, with $X = Y = 1$, is precisely a classical *nondeterministic automaton.* It is called an *nfa* if S is finite. We identify $i \subseteq 1 \times S$ with a set $I \subseteq S$ of initial states and $f \subseteq S \times 1$ with a set $F \subseteq S$ of final states. Thus, in our view an nfa may have multiple initial states. The language $L(N, R)$ *accepted* by a subset $R \subseteq S$ consists of all $w \in \Sigma^*$ such that $(r, s) \in \delta_w$ for some $r \in R$ and $s \in F$. The language $L(N)$ *accepted* by N is the language accepted by the set I.
(3) An automaton $A = (S, \delta, i, f)$ in **JSL** with $X = Y = 2$, shortly a **JSL**-*automaton,* is given by a semilattice S of states, a family $\delta = (\delta_a\colon S \to S)_{a\in\Sigma}$ of semilattice morphisms specifying transitions, an initial state $s_0 \in S$ (corresponding to $i\colon 2 \to S$), and a prime filter $F \subseteq S$ of final states (corresponding to $f\colon S \to 2$). It is called a **JSL**-*dfa* if S is finite. The language *accepted* by a state $s \in S$ or by the automaton A, resp., is defined as for deterministic automata.

Remark 3.4 (JSL-dfas vs. nfas). Dfas, nfas and **JSL**-dfas are expressively equivalent; they all accept precisely the regular languages. The interest of **JSL**-dfas is that they constitute an algebraic representation of nfas:

(1) Every **JSL**-dfa $A = (S, \delta, s_0, F)$ induces an equivalent nfa $J(A)$ on the set $J(S)$ of join-irreducibles of S. Given $s, t \in J(S)$ and $a \in \Sigma$, there is a transition $s \xrightarrow{a} t$ in $J(A)$ iff $t \le \delta_a(s)$; the initial states are those $s \in J(S)$ with $s \le s_0$, and the final states form the set $J(S) \cap F$.

(2) Conversely, for every nfa $N = (Q, \delta, I, F)$, the *subset construction* yields an equivalent **JSL**-dfa $\mathcal{P}(N)$ with states $\mathcal{P}(Q)$ (the $\cup$-semilattice of subsets of Q), transitions $\mathcal{P}\delta_a\colon \mathcal{P}(Q) \to \mathcal{P}(Q)$, $X \mapsto \delta_a[X]$, initial state $I \in \mathcal{P}(Q)$, and final states those subsets of Q containing some state from F. Note that $J(\mathcal{P}(Q)) \cong Q$.

It follows that the task of finding a state-minimal nfa for a given language is equivalent to finding a **JSL**-dfa with a minimum number of join-irreducibles [4]. This idea has recently been extended to a general coalgebraic framework [32, 39].

Recall that the *minimal dfa* [7] for a regular language L, denoted by $\mathsf{dfa}(L)$, has states $\mathsf{LD}(L)$ (the set of left derivatives of L), transitions $K \xrightarrow{a} a^{-1}K$ for $K \in \mathsf{LD}(L)$ and $a \in \Sigma$, initial state $L = \varepsilon^{-1}L$, and final states those $K \in \mathsf{LD}(L)$ containing ε. Up to isomorphism, it can be characterized as the unique dfa accepting L that is *reachable* (i.e. every state is reachable from the initial state via transitions) and *simple* (i.e. any two distinct states accept distinct languages). We now develop the analogous concepts for **JSL**-automata; they are instances of the categorical theory of minimality due to Arbib and Manes [3] and Goguen [15]. Let us first observe that every language has two canonical infinite **JSL**-acceptors:

Definition 3.5. Let $L \subseteq \Sigma^*$ be a language.

(1) The *initial* **JSL**-*automaton* $\mathsf{Init}(L)$ for L has states $\mathcal{P}_\mathsf{f}(\Sigma^*)$ (the $\cup$-semilattice of finite subsets of Σ^*), initial state $\{\varepsilon\}$, final states all $X \in \mathcal{P}_\mathsf{f}(\Sigma^*)$ with $X \cap L \neq \emptyset$, and transitions $X \mapsto Xa = \{xa : x \in X\}$ for $X \in \mathcal{P}_\mathsf{f}(\Sigma^*)$ and $a \in \Sigma$.

(2) The *final* **JSL**-*automaton* $\mathsf{Fin}(L)$ for L has states $\mathcal{P}(\Sigma^*)$ (the $\cup$-semilattice of all languages), initial state L, final states all languages K containing ε, and transitions $K \mapsto a^{-1}K$ for $K \in \mathcal{P}(\Sigma^*)$ and $a \in \Sigma$.

As suggested by the terminology, these automata form the initial and the final object in the category of **JSL**-automata accepting L:

Lemma 3.6 [3,15]. *For every* **JSL**-*automaton* $A = (S, \delta, s_0, F)$ *accepting the language* $L \subseteq \Sigma^*$, *there exist unique* **JSL**-*automata morphisms*

$$e_A\colon \mathsf{Init}(L) \to A \quad \text{and} \quad m_A\colon A \to \mathsf{Fin}(L).$$

The map e_A *sends* $\{w_1, \ldots, w_n\} \in \mathcal{P}_\mathsf{f}(\Sigma^*)$ *to the state* $\bigvee_{i=1}^n \delta_{w_i}(s_0)$, *and the map* m_A *sends a state* $s \in S$ *to* $L(A, s)$, *the language accepted by* s.

Definition 3.7. A **JSL**-automaton $A = (S, \delta, s_0, F)$ is called

(1) *reachable* if the unique morphism $e_A\colon \mathsf{Init}(L) \to A$ is surjective, i.e. every state is of the form $\bigvee_{i=1}^n \delta_{w_i}(s_0)$ for some $w_1, \ldots, w_n \in \Sigma^*$;

(2) *simple* if the unique morphism $m_A\colon A \to \mathsf{Fin}(L)$ in injective, i.e. any two distinct states accept distinct languages;

(3) *minimal* if it is both reachable and simple.

Remark 3.8. (1) The category **Aut(JSL)** has a factorization system given by surjective and injective morphisms. Thus, for every **JSL**-automata morphism $h\colon (S, \delta, i, f) \to (S', \delta', i', f')$ with image factorization $h = (S \overset{e}{\twoheadrightarrow} S'' \overset{m}{\rightarrowtail} S')$ in **JSL**, there exists a unique **JSL**-automaton structure $(S'', \delta'', i'', f'')$ on S'' making both e and m automata morphisms. We call e the *coimage* and m the *image* of h. *Subautomata* and *quotient automata* of **JSL**-automata are represented by injective and surjective morphisms, respectively.

(2) Every **JSL**-automaton A has a unique reachable subautomaton $\mathsf{reach}(A) \rightarrowtail A$, the *reachable part* of A. It is the smallest subautomaton of A and arises as the image of the unique morphism $e_A\colon \mathsf{Init}(L) \to A$. Thus,

$$A \text{ is reachable} \quad \text{iff} \quad A \cong \mathsf{reach}(A) \quad \text{iff} \quad A \text{ has no proper subautomaton.}$$

Let us emphasize that a state in $\mathsf{reach}(A)$ is not necessarily reachable when A is viewed as an ordinary dfa. For distinction, we thus call a state **JSL**-*reachable* if it lies in $\mathsf{reach}(A)$, and *dfa-reachable* if it is reachable in the usual sense.

(3) Dually, every **JSL**-automaton A has a unique simple quotient automaton $A \twoheadrightarrow \mathsf{simple}(A)$, the *simplification* of A. It is the smallest quotient automaton of A and arises as the coimage of the unique morphism $m_A\colon A \to \mathsf{Fin}(L)$. Thus,

$$A \text{ is simple} \quad \text{iff} \quad A \cong \mathsf{simple}(A) \quad \text{iff} \quad A \text{ has no proper quotient automaton.}$$

(4) Every language $L \subseteq \Sigma^*$ has a minimal **JSL**-automaton, unique up to isomorphism. It can be constructed as the image of the unique automata morphism $h_L\colon \mathsf{Init}(L) \to \mathsf{Fin}(L)$. Since h_L sends $\{w_1, \dots, w_n\} \in \mathcal{P}_{\mathsf{f}}(\Sigma^*)$ to the language $\bigcup_{i=1}^{n} w_i^{-1}L$, the minimal automaton of L is the subautomaton $\mathsf{SLD}(L)$ of $\mathsf{Fin}(L)$ carried by the semilattice of finite unions of left derivatives of L.

Example 3.9. The minimal **JSL**-dfa accepting $L = \{a, aa\}$ is shown below, with the dashed lines representing the partial order.

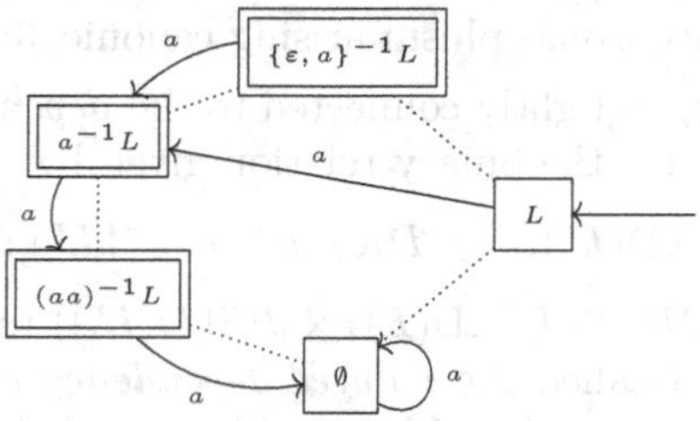

Remark 3.10. The self-duality of $\mathbf{JSL_f}$ lifts to a self-duality of the category of **JSL**-dfas. The equivalence functor $\mathbf{Aut}(\mathbf{JSL_f}) \xrightarrow{\cong} \mathbf{Aut}(\mathbf{JSL_f})^{\mathsf{op}}$ maps a **JSL**-dfa $A = (S, (\delta_a\colon S \to S)_{a\in\Sigma}, i\colon 2 \to S, f\colon S \to 2)$ to its *dual automaton*

$$A^{\mathsf{op}} = (S^{\mathsf{op}}, (\delta_a^*\colon S^{\mathsf{op}} \to S^{\mathsf{op}})_{a\in\Sigma}, f^*\colon 2 \to S^{\mathsf{op}}, i^*\colon S^{\mathsf{op}} \to 2),$$

using that $2^{\mathsf{op}} \cong 2$. Thus, the initial state of A^{op} is the $\leq_S$-largest non-final state of A, and its final states are those $s \in S$ with $s_0 \not\leq_S s$. Given $s, t \in S$ and $a \in \Sigma$, there is a transition $s \xrightarrow{a} t$ in A^{op} iff t is the $\leq_S$-largest state with $\delta_a(t) \leq_S s$.

The dualization of **JSL**-dfas can be seen as an algebraic generalization of the reversal operation on nfas. Recall that the *reverse* of an nfa N is the nfa N^{r} obtained by flipping all transitions and swapping initial and final states. If N accepts the language L, then N^{r} accepts the reverse language L^{r}.

Lemma 3.11. *For each nfa $N = (Q, \delta, I, F)$, we have the* **JSL**-*dfa isomorphism*

$$[\mathcal{P}(N)]^{\mathsf{op}} \xrightarrow{\cong} \mathcal{P}(N^{\mathsf{r}}), \qquad X \mapsto \overline{X} = Q \setminus X.$$

The following lemma summarizes some important properties of A^{op}:

Lemma 3.12. *Let $A = (S, \delta, i, f)$ be a* **JSL**-*dfa.*

(1) *For every $s \in S$, we have $L(A^{\mathsf{op}}, s) = \{ w \in \Sigma^* : \delta_{w^{\mathsf{r}}}(s_0) \not\leq_S s \}$.*

(2) *If A accepts the language L, then A^{op} accepts the reverse language L^{r}.*

(3) *We have $[\mathsf{reach}(A)]^{\mathsf{op}} \cong \mathsf{simple}(A^{\mathsf{op}})$. Thus, A is reachable iff A^{op} is simple.*

Our next goal is to give, for every regular language L, dual characterizations of $\mathsf{SLD}(L)$, $\mathsf{BLD}(L)$ and $\mathsf{BLRD}(L)$, the **JSL**-subautomata of $\mathsf{Fin}(L)$ carried by all finite unions of left derivatives, boolean combinations of left derivatives and boolean combinations of two-sided derivatives, respectively. These results form the core of our duality-based approach to (sub-)atomic nfas in the next section. The minimal **JSL**-dfa $\mathsf{SLD}(L)$ admits the following dual description:

Proposition 3.13. *For every regular language L, the minimal* **JSL**-*dfas for L and L^{r} are dual. More precisely, we have the* **JSL**-*dfa isomorphism*

$$\mathrm{dr}_L\colon [\mathsf{SLD}(L^{\mathsf{r}})]^{\mathsf{op}} \xrightarrow{\cong} \mathsf{SLD}(L), \qquad K \mapsto (\overline{K^{\mathsf{r}}})^{-1}L.$$

Remark 3.14. (1) The isomorphism dr_L induces a bijection between the *left* and *right factors* of L, i.e. the inclusion-maximal left/right solutions of $X \cdot Y \subseteq L$. Conway [10] observed that the left and right factors are respectively $\{\overline{K^{\mathsf{r}}} : K \in \mathsf{SLD}(L^{\mathsf{r}})\}$ and $\{\overline{K} : K \in \mathsf{SLD}(L)\}$ and that they biject. Backhouse [5] observed that they are dually isomorphic posets. Proposition 3.13 provides an explicit automata-theoretic lattice isomorphism arising canonically via duality.

(2) The isomorphism dr_L is tightly connected to the *dependency relation* [18,20] of a regular language L, i.e. the binary relation given by

$$\mathcal{DR}_L \subseteq \mathsf{LD}(L) \times \mathsf{LD}(L^r), \qquad \mathcal{DR}_L(u^{-1}L, v^{-1}L^r) :\iff uv^r \in L.$$

Its restriction $\mathcal{DR}_L^j := \mathcal{DR}_L \cap J(\mathsf{SLD}(L)) \times J(\mathsf{SLD}(L^r))$ to the $\cup$-irreducible left derivatives of L and L^{r} is called the *reduced dependency relation.* The following theorem shows that the semilattice of left quotients and the dependency relation are essentially the same concepts. In part (3), we use that the isomorphism dr_L restricts to a bijection between the $\cup$-irreducible derivatives of L^{r} and the meet-irreducible elements of the lattice $\mathsf{SLD}(L)$.

Theorem 3.15 (Dependency theorem).

(1) *We have the* **JSL**-*isomorphism*

$$\mathsf{SLD}(L) \xrightarrow{\cong} (\{\mathcal{DR}_L[X] : X \subseteq \mathsf{LD}(L)\}, \cup, \emptyset), \qquad K \mapsto \{v^{-1}L^{\mathsf{r}} : v \in K^{\mathsf{r}}\}.$$

Note that its codomain forms a subsemilattice of $\mathcal{P}(\mathsf{LD}(L^{\mathsf{r}}))$.

(2) *For all $u, v \in \Sigma^*$ we have $\mathcal{DR}_L(u^{-1}L, v^{-1}L^r) \iff u^{-1}L \not\subseteq \mathrm{dr}_L(v^{-1}L^r)$.*

(3) *The following diagram in* **Rel** *commutes:*

$$\begin{array}{ccc} J(\mathsf{SLD}(L^{\mathsf{r}})) & \xrightarrow[\cong]{\mathrm{dr}_L} & M(\mathsf{SLD}(L)) \\ {\scriptstyle \mathcal{DR}_L^j}\uparrow & & \uparrow{\scriptstyle \not\supseteq} \\ J(\mathsf{SLD}(L)) & =\!=\!= & J(\mathsf{SLD}(L)) \end{array}$$

Let us now turn to a dual characterization of the **JSL**-dfa $\mathsf{BLD}(L)$:

Proposition 3.16. *For every regular language L, the **JSL**-dfa $\mathsf{BLD}(L)$ is dual to the subset construction of the minimal dfa for L^{r}:*

$$[\mathsf{BLD}(L)]^{\mathsf{op}} \cong \mathcal{P}(\mathsf{dfa}(L^{\mathsf{r}})).$$

The isomorphism maps $\{w_1^{-1}L^{\mathsf{r}}, \dots, w_n^{-1}L^{\mathsf{r}}\} \in \mathcal{P}(\mathsf{dfa}(L^{\mathsf{r}}))$ to $\bigcap_{i=1}^n \overline{\mathrm{At}(w_i^{\mathsf{r}})}$, where $\mathrm{At}(x)$ is the unique atom (= join-irreducible) of $\mathsf{BLD}(L)$ containing x.

To state the dual characterization of $\mathsf{BLRD}(L)$, we recall two standard concepts from algebraic language theory [33]. The *transition monoid* of a deterministic automaton $D = (S, \delta, i, f)$ is the image $\mathsf{tm}(D) \subseteq \mathbf{Set}(S, S)$ of the morphism

$$\Sigma^* \to \mathbf{Set}(S, S), \quad w \mapsto \delta_w.$$

Thus, $\mathsf{tm}(M)$ is carried by the set of extended transition maps δ_w $(w \in \Sigma^*)$ with multiplication given by $\delta_v \bullet \delta_w = \delta_{vw}$ and unit $id_S = \delta_\varepsilon \colon S \to S$. We may view $\mathsf{tm}(D)$ as a deterministic automaton with initial state id_S, final states all δ_w such that w is accepted by D, and transitions $\delta_w \xrightarrow{a} \delta_{wa}$ for $w \in \Sigma^*$ and $a \in \Sigma$. This automaton accepts the same language as D. The *syntactic monoid* $\mathsf{syn}(L)$ of a regular language $L \subseteq \Sigma^*$ is the transition monoid of its minimal dfa:

$$\mathsf{syn}(L) = \mathsf{tm}(\mathsf{dfa}(L)).$$

Equivalently, $\mathsf{syn}(L)$ is the quotient monoid of the free monoid Σ^* modulo the *syntactic congruence* of L, i.e the monoid congruence on Σ^* given by

$$v \equiv_L w \quad \text{iff} \quad \forall x, y \in \Sigma^* : xvy \in L \iff xwy \in L.$$

The associated surjective monoid morphism $\mu_L \colon \Sigma^* \twoheadrightarrow \mathsf{syn}(L)$, mapping $w \in \Sigma^*$ to its congruence class $[w]_L \in \mathsf{syn}(L)$, is called the *syntactic morphism.*

Proposition 3.17. *For every regular language L, the **JSL**-dfa $\mathsf{BLRD}(L)$ is dual to the subset construction of $\mathsf{syn}(L^{\mathsf{r}})$, viewed as a dfa:*

$$[\mathsf{BLRD}(L)]^{\mathsf{op}} \cong \mathcal{P}(\mathsf{syn}(L^{\mathsf{r}})).$$

The isomorphism maps $\{ [w_1]_{L^{\mathsf{r}}}, \dots, [w_n]_{L^{\mathsf{r}}} \} \in \mathcal{P}(\mathsf{syn}(L^{\mathsf{r}}))$ to $\bigcap_{i=1}^n \overline{\mathrm{At}(w_i{}^{\mathsf{r}})}$, with $\mathrm{At}(x)$ denoting the unique atom of $\mathsf{BLRD}(L)$ containing x.

Our final duality result in this section concerns the *transition semiring* [35], a generalization of the transition monoid to **JSL**-automata. Note that the monoid $\mathbf{JSL}(S, S)$ of endomorphisms of a semilattice S forms an idempotent semiring with join defined pointwise: for any $f, g \colon S \to S$, the morphism $f \vee g \colon S \to S$ is given by $s \mapsto f(s) \vee g(s)$. The transition semiring of a **JSL**-automaton $A = (S, \delta, i, f)$ is the image $\mathsf{ts}(A) \subseteq \mathbf{JSL}(S, S)$ of the semiring morphism

$$\mathcal{P}_{\mathsf{f}}(\Sigma^*) \to \mathbf{JSL}(S, S), \quad \{w_1, \dots, w_n\} \mapsto \bigvee_{i=1}^{n} \delta_{w_i}.$$

Here $\mathcal{P}_f(\Sigma^*)$ is the free idempotent semiring on Σ, with composition given by concatenation of languages and join given by union. Thus, $\mathsf{ts}(A)$ is the semiring carried by all morphisms $\bigvee_{i=1}^n \delta_{w_i}$ for $w_1, \dots, w_n \in \Sigma^*$, with join given as above and multiplication $\bigvee_j \delta_{v_j} \bullet \bigvee_i \delta_{w_i} = \bigvee_{i,j} \delta_{v_j w_i}$. We view $\mathsf{ts}(A)$ as a **JSL**-automaton with initial state $id_S = \delta_\varepsilon$, final states all $\bigvee_i \delta_{w_i}$ such that some w_i is accepted by A, and transitions $\bigvee_{i=1}^n \delta_{w_i} \xrightarrow{a} \bigvee_{i=1}^n \delta_{w_i a}$ for $w_1, \dots, w_n \in \Sigma^*$ and $a \in \Sigma$. This **JSL**-automaton is reachable and accepts the same language as A. It has the following dual characterization:

Notation 3.18. Given a simple **JSL**-automaton $A = (S, \delta, i, f)$, the subautomaton of $\mathsf{Fin}(L)$ obtained by closing S (viewed as a set of languages) under right derivatives is called the *right-derivative closure* of A and denoted $\mathsf{rdc}(A)$.

Proposition 3.19. *Let A be a reachable* **JSL**-*dfa. Then the transition semiring of A, viewed as a* **JSL**-*dfa, is dual to the right-derivative closure of A^{op}:*

$$[\mathsf{ts}(A)]^{\mathsf{op}} \cong \mathsf{rdc}(A^{\mathsf{op}}).$$

Note that both $[\mathsf{ts}(A)]^{\mathsf{op}}$ and $\mathsf{rdc}(A^{\mathsf{op}})$ are simple, hence subautomata of $\mathsf{Fin}(L)$. Thus, the isomorphism just expresses that their states accept the same languages.

4 Boolean Representations and Subatomic NFAs

Based upon the duality results of the previous section, we will now introduce our algebraic approach to nondeterministic state minimality. It rests on the concept of a representation of a monoid on a finite semilattice.

Definition 4.1 (Boolean representation). Let M be a monoid.

(1) A *boolean representation* of M is given by a finite semilattice S together with a monoid morphism $\rho\colon M \to \mathbf{JSL}(S, S)$. The *degree* of ρ is

$$\deg(\rho) := |J(S)|.$$

(2) Given boolean representations $\rho_i\colon M \to \mathbf{JSL}(S_i, S_i)$, $i = 1, 2$, an *equivariant map* $f\colon \rho_1 \to \rho_2$ is a **JSL**-morphism $f\colon S_1 \to S_2$ such that

$$f(\rho_1(m)(s)) = \rho_2(m)(f(s)) \text{ for all } m \in M \text{ and } s \in S_1.$$

If f is injective, we say that the representation ρ_2 *extends* ρ_1.

Remark 4.2. (1) The above representations are called *boolean* because semilattices are precisely semimodules over the boolean semiring $2 = \{0, 1\}$ with $1 + 1 = 1$. For more on representations over general commutative semirings, see [21].

(2) The category of boolean representations of M coincides with the functor category $\mathbf{JSL}_f^M$, viewing M as a one object category.

Definition 4.3 (Canonical representation). For every regular language L, the *canonical boolean representation* of the syntactic monoid $\mathsf{syn}(L)$ is given by

$$\kappa_L \colon \mathsf{syn}(L) \to \mathbf{JSL}(\mathsf{SLD}(L), \mathsf{SLD}(L)), \quad [w]_L \mapsto \lambda K.w^{-1}K.$$

It induces the *canonical boolean presentation* of the free monoid Σ^* given by

$$\kappa_L \circ \mu_L \colon \Sigma^* \to \mathbf{JSL}(\mathsf{SLD}(L), \mathsf{SLD}(L)), \quad w \mapsto \lambda K.w^{-1}K,$$

where $\mu_L \colon \Sigma^* \twoheadrightarrow \mathsf{syn}(L)$ is the syntactic morphism.

The representation $\kappa_L \circ \mu_L$ amounts to constructing the transition semiring of the minimal **JSL**-automaton $\mathsf{SLD}(L)$, i.e. the *syntactic semiring* [35] of L.

Example 4.4. We describe the canonical boolean representation κ_{L_n} for the language $L_n := (0+1)^*1(0+1)^n$, $n \in \mathbb{N}$. Let $S := 2_\perp^{n+1}$ be the semilattice of binary words of length $n+1$, ordered pointwise, with an additional bottom element $\perp$. Then $\mathsf{SLD}(L_n)$ is isomorphic to S, as witnessed by the isomorphism

$$f \colon S \xrightarrow{\cong} \mathsf{SLD}(L_n), \quad f(\perp) = \emptyset, \quad f(w) = w^{-1}L_n.$$

Thus, κ_{L_n} is isomorphic to the representation $\rho \colon \mathsf{syn}(L_n) \to \mathbf{JSL}(S,S)$ where:
(1) $\rho([0]_{L_n}) \colon S \to S$ performs a left-shift (distinct from left-rotate);
(2) $\rho([1]_{L_n}) \colon S \to S$ performs a left-shift and sets the last bit as 1.
Finally, $\deg(\kappa_{L_n}) = \deg(\rho) = 1 + |J(2^{n+1})| = n+2$ is the number of states of the usual minimal nfa for L.

Example 4.5. We describe the canonical boolean presentation κ_L for the language $L = a_1(a_2+a_3)+a_2(a_1+a_3)+a_3(a_1+a_2)$ over $\Sigma = \{a_1,a_2,a_3\}$. Consider the $\cup$-semilattice $M_3 = \{\emptyset, \{a_1,a_2\}, \{a_1,a_3\}, \{a_2,a_3\}, \Sigma\}$. Then $\mathsf{SLD}(L)$ is isomorphic to the product semilattice $2 \times M_3 \times 2$ via the map

$$f \colon \mathsf{SLD}(L) \xrightarrow{\cong} 2 \times M_3 \times 2, \quad f(X) = (X \cap \Sigma^2, X \cap \Sigma, X \cap \{\varepsilon\}).$$

Note that the first and third component is either $\emptyset$ or one other set, i.e. it may be identified with the elements of 2. For $i = 1,2,3$ we define the following semilattice morphisms:

$$\begin{aligned} &\alpha_i \colon 2 \to M_3, && \alpha_i(1) = \Sigma \setminus \{a_i\};\\ &\beta_i \colon M_3 \to 2, && \beta_i(S) = 1 \iff a_i \in S;\\ &\gamma \colon 2 \to 2 && \gamma(1) = 0;\\ &\delta \colon M_3 \times 2 \times 2 \to 2 \times M_3 \times 2, && \delta(x,y,z) = (z,x,y). \end{aligned}$$

Then κ_L is isomorphic to $\rho \colon \mathsf{syn}(L) \to \mathbf{JSL}(2 \times M_3 \times 2,\, 2 \times M_3 \times 2)$ where

$$\rho([a_i]_L) = (\, 2 \times M_3 \times 2 \xrightarrow{\alpha_i \times \beta_i \times \gamma} M_3 \times 2 \times 2 \xrightarrow{\delta} 2 \times M_3 \times 2\,).$$

Thus, $\deg(\kappa_L) = \deg(\rho) = 1+3+1 = 5$. An analogous description of κ_L exists for any language L where each word has the same length.

The next theorem links minimal nfas and representations.

Definition 4.6. The *nondeterministic state complexity* $\mathrm{ns}(L)$ of a regular language L is the least number of states of any nfa accepting L.

Theorem 4.7. *For every regular language L, the nondeterministic state complexity* $\mathrm{ns}(L)$ *is the least degree of any boolean representation extending the canonical representation* $\kappa_L \circ \mu_L \colon \Sigma^* \to \mathbf{JSL}(\mathsf{SLD}(L), \mathsf{SLD}(L))$.

Proof (Sketch).
(1) Given a k-state nfa $N = (Q, \delta, I, F)$ accepting L, consider the subsemilattice $\mathsf{langs}(N) = \mathsf{simple}(\mathcal{P}(N))$ of $\mathcal{P}(\Sigma^*)$ on all languages accepted by subsets of Q. The embedding $\mathsf{SLD}(L) \rightarrowtail \mathsf{langs}(N)$ yields an extension of $\kappa_L \circ \mu_L$. Since the semilattice $\mathsf{langs}(N)$ is generated by the languages accepted by single states of N, this extension has degree at most k.
(2) Conversely, let $\rho\colon \Sigma^* \to \mathbf{JSL}(S, S)$ be a boolean representation of degree k extending $\kappa_L \circ \mu_L$, witnessed by an injective equivariant map $h\colon \mathsf{SLD}(L) \rightarrowtail S$. One can equip S with a **JSL**-dfa structure making h an automata morphism. Since morphisms preserve accepted languages, it follows that S accepts L. Then the nfa of join-irreducibles of S, see Remark 3.4, is a k-state nfa accepting L. □

As an application, let us return to the dependency relation $\mathcal{DR}_L$ introduced in Remark 3.14(2). Recall that a *biclique* of a relation $R \subseteq X \times Y$ (viewed as a bipartite graph) is a subset of the form $X' \times Y' \subseteq R$, where $X' \subseteq X$ and $Y' \subseteq Y$. A *biclique cover* of R is a set $\mathscr{C}$ of bicliques with $R = \bigcup \mathscr{C}$. The *bipartite dimension* $\dim(R)$ is the least cardinality of any biclique cover of R.

Theorem 4.8 (Gruber-Holzer [18]). *For every regular language L, we have*

$$\dim(\mathcal{DR}_L) \leq \mathrm{ns}(L).$$

We give a new algebraic proof of this result based on boolean representations.

Proof. (1) The task of computing biclique covers is well-known to be equivalent to the *set basis* problem. Given a family $C \subseteq \mathcal{P}(Y)$ of subsets of a finite set Y, a set basis for C is a family $B \subseteq \mathcal{P}(Y)$ such that each element of C can be expressed as a union of elements of B. A relation $R \subseteq X \times Y$ has a biclique cover of size k iff the family $C_R = \{R[x] : x \in X\} \subseteq \mathcal{P}(Y)$ of neighborhoods of nodes in X has a set basis of size k.

(2) Given an instance $C \subseteq \mathcal{P}(Y)$ of the set basis problem, consider the $\cup$-subsemilattice $\langle C \rangle \subseteq \mathcal{P}(Y)$ generated by C, i.e. the semilattice of all unions of sets in C. We claim that C has a set basis of size at most k iff there exists an extension of $\langle C \rangle$ of degree at most k, i.e. a monomorphism $\langle C \rangle \rightarrowtail S$ into some finite semilattice S with $|J(S)| \leq k$.

For the "only if" direction, suppose that $B \subseteq \mathcal{P}(Y)$ is a set basis of C of size at most k. The the embedding $\langle C \rangle \rightarrowtail \langle B \rangle$ gives an extension of $\langle C \rangle$ with the desired property: since the semilattice $\langle B\rangle$ has a set of generators with at most k elements, it has at most k join-irreducibles.

For the "if" direction, suppose that $m\colon \langle C\rangle \rightarrowtail S$ with $|J(S)| \leq k$ is given. Since the free semilattice $\mathcal{P}(Y)$ is an injective object of **JSL** [19, Corollary 2.9], there exists a morphism $f\colon S \to \mathcal{P}(Y)$ extending the embedding $\langle C\rangle \rightarrowtail \mathcal{P}(Y)$. Consider the image $S' \subseteq \mathcal{P}(Y)$ of f, leading to the commutative diagram below:

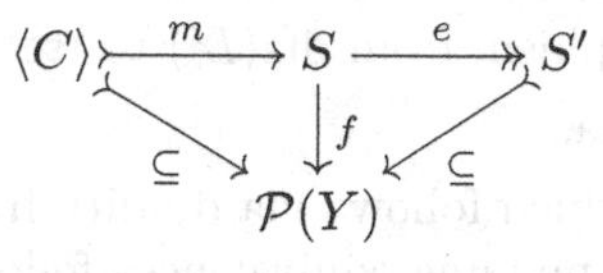

We thus have $\langle C\rangle \subseteq S' \subseteq \mathcal{P}(Y)$. Every set of generators of the semilattice S' is a basis of C. Since the morphism e is surjective, we have $|J(S')| \leq |J(S)| \leq k$, i.e. S' has a set of generators with at most k elements.

(3) Let $C_{\mathcal{DR}_L} \subseteq \mathcal{P}(\mathsf{LD}(L^{\mathsf{r}}))$ be the instance of the set basis problem corresponding to the dependency relation $\mathcal{DR}_L \subseteq \mathsf{LD}(L) \times \mathsf{LD}(L^{\mathsf{r}})$. Note that $\langle C_{\mathcal{DR}_L}\rangle$ consists of all $\mathcal{DR}_L[X]$ for $X \subseteq \mathsf{LD}(L)$. Thus, Theorem 3.15(1) shows that $\langle C_{\mathcal{DR}_L}\rangle \cong \mathsf{SLD}(L)$. In particular, every extension of the canonical boolean representation of Σ^* yields an extension of the semilattice $\langle C_{\mathcal{DR}_L}\rangle$ of the same degree. Therefore, by part (1) and (2) and Theorem 4.7, we have $\dim(\mathcal{DR}_L) \leq \mathrm{ns}(L)$, as required.

Theorem 4.7 motivates the following definition, which can be considered the key concept of our paper:

Definition 4.9. The *nondeterministic syntactic complexity* $\mathrm{n}\mu(L)$ of a regular language L is the least degree of any boolean representation of $\mathsf{syn}(L)$ extending the canonical boolean representation $\kappa_L\colon \mathsf{syn}(L) \to \mathbf{JSL}(\mathsf{SLD}(L), \mathsf{SLD}(L))$.

Just like the degrees of boolean representations of Σ^* determine the state complexity of nfas, we will provide an automata-theoretic characterization of $\mathrm{n}\mu(L)$ in terms of *subatomic* nfas in Theorem 4.14 below.

Definition 4.10. An nfa accepting the language L is called

(1) *atomic* if each state accepts a language from $\mathsf{BLD}(L)$, and

(2) *subatomic* if each state accepts a language from $\mathsf{BLRD}(L)$.

The notion of an atomic nfa goes back to Brzozowski and Tamm [6], as does the following characterization.

Notation 4.11. For any nfa N, let $\mathsf{rsc}(N)$ denote the dfa obtained via the *reachable subset construction*, i.e. the dfa-reachable part of $\mathcal{P}(N)$.

Theorem 4.12. *An nfa N is atomic iff $\mathsf{rsc}(N^{\mathsf{r}})$ is a minimal dfa.*

We present a new conceptual proof, interpreting this theorem as an instance of the self-duality of **JSL**-dfas.

Proof (Sketch). Let L be the language accepted by N. We establish the theorem by showing each of the following statements to be equivalent to the next one:
(1) N is atomic.
(2) There exists a **JSL**-automata morphism from $\mathcal{P}(N)$ to $\mathsf{BLD}(L)$.
(3) There exists a **JSL**-automata morphism from $\mathcal{P}(\mathsf{dfa}(L^{\mathsf{r}}))$ to $\mathcal{P}(N^{\mathsf{r}})$.
(4) There exists a dfa morphism from $\mathsf{dfa}(L^{\mathsf{r}})$ to $\mathcal{P}(N^{\mathsf{r}})$.
(5) There exists a dfa morphism from $\mathsf{dfa}(L^{\mathsf{r}})$ to $\mathsf{rsc}(N^{\mathsf{r}})$.
(6) $\mathsf{rsc}(N^{\mathsf{r}})$ is a minimal dfa.

The key step is (2)⇔(3), which follows via duality from Lemmas 3.11 and 3.12, and Proposition 3.16. All remaining equivalences follow from the definitions. □

The next theorem gives an analogous characterization of subatomic nfas. Again, the proof is based on duality.

Theorem 4.13. *An nfa N accepting the language L is subatomic iff the transition monoid of $\mathsf{rsc}(N^{\mathsf{r}})$ is isomorphic to the syntactic monoid $\mathsf{syn}(L^{\mathsf{r}})$.*

Proof (Sketch). Each of the following statements is equivalent to the next one:
(1) N is subatomic.
(2) There exists a **JSL**-dfa morphism from $\mathcal{P}(N)$ to $\mathsf{BLRD}(L)$.
(3) There exists a **JSL**-dfa morphism from $\mathsf{rdc}(\mathsf{simple}(P(N)))$ to $\mathsf{BLRD}(L)$.
(4) There exists a **JSL**-dfa morphism from $\mathcal{P}(\mathsf{syn}(L^{\mathsf{r}}))$ to $\mathsf{ts}(\mathsf{reach}(\mathcal{P}(N^{\mathsf{r}})))$.
(5) There exists a dfa morphism from $\mathsf{syn}(L^{\mathsf{r}})$ to $\mathsf{ts}(\mathsf{reach}(\mathcal{P}(N^{\mathsf{r}})))$.
(6) There exists a dfa morphism from $\mathsf{syn}(L^{\mathsf{r}})$ to $\mathsf{tm}(\mathsf{rsc}(N^{\mathsf{r}}))$.
(7) The monoids $\mathsf{syn}(L^{\mathsf{r}})$ and $\mathsf{tm}(\mathsf{rsc}(N^{\mathsf{r}}))$ are isomorphic.

The equivalence (3)⇔(4) follows via duality from Lemma 3.11, Proposition 3.17 and Proposition 3.19. All remaining equivalences follow from the definitions. □

We are prepared to state the main result of our paper, an automata-theoretic characterization of the nondeterministic syntactic complexity:

Theorem 4.14. *For every regular language L, the nondeterministic syntactic complexity $\mathrm{n}\mu(L)$ is the least number of states of any subatomic nfa accepting L.*

Proof (Sketch).

(1) Let N be a k-state subatomic nfa accepting the language L. As in the proof of Theorem 4.7, we consider the semilattice $\mathsf{langs}(N) = \mathsf{simple}(\mathcal{P}(N))$. Then

$$\rho\colon \mathsf{syn}(L) \to \mathbf{JSL}(\mathsf{langs}(N), \mathsf{langs}(N)), \quad [w]_L \mapsto \lambda K.w^{-1}K,$$

is a representation of $\mathsf{syn}(L)$ of degree at most k extending κ_L.

(2) Conversely, let $\rho\colon \mathsf{syn}(L) \to \mathbf{JSL}(S, S)$ be a boolean representation extending κ_L, and let $h\colon \mathsf{SLD}(Q) \rightarrowtail S$ be the embedding. As in the proof of Theorem 4.7, we can equip S with the structure of a **JSL**-dfa making h an automata morphism. Its nfa of join-irreducibles, see Remark 3.4, is a subatomic nfa accepting L with $\deg(\rho)$ states. □

We conclude this section with the observation that the state complexity of unrestricted nfas, subatomic nfas and atomic nfas generally differs:

Example 4.15 (Subatomic more succinct than atomic). Consider the language L accepted by the nfa N shown below, along with the minimal dfas for L and L^r. Each automaton has exactly one initial state, namely 0.

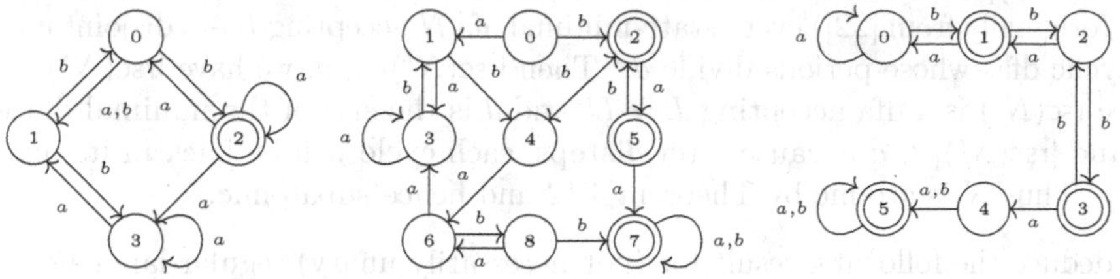

N dfa(L) dfa(L^r)

Brzozowski and Tamm [6] showed that there is no atomic nfa with four states accepting L. However, N is subatomic: one can verify that the transition monoids of dfa(L^r) and rsc(N^r) both have 22 elements. Since the former is the syntactic monoid of L^r, they are isomorphic, and so Theorem 4.13 applies.

Example 4.16 (Subatomic less succinct than general nfas). There is a regular language for which no state-minimal nfa is subatomic:

$$L := \{a^n : n \in \mathbb{N}, n \neq 5\} \subseteq \{a\}^*.$$

It is accepted by the following nfa:

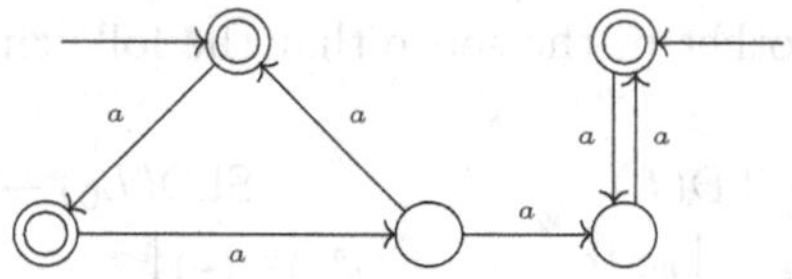

An exhaustive search shows that no subatomic nfa with five states accepts L. In fact, L is the unique (!) unary language with $\mathrm{ns}(L) \leq 5$ and $\mathrm{ns}(L) < \mathrm{n}\mu(L)$. Moreover, the above nfa and its reverse are the only state-minimal nfas for L.

5 Applications

While subatomic nfas are generally less succinct then unrestricted ones, all structural results concerning nondeterministic state complexity we have encountered in the literature are actually about nondeterministic syntactic complexity: they implicitly identify classes of languages where the two measures coincide. In the present section, we illustrate this in a few selected applications.

5.1 Unary languages

For unary languages $L \subseteq \{a\}^*$, two-sided derivatives are left derivatives. Thus, a unary nfa is atomic iff it is subatomic.

Example 5.1 (Cyclic unary languages). A unary language L is *cyclic* if its minimal dfa is a cycle [16]. We claim that $\mathrm{ns}(L) = \mathrm{n}\mu(L)$. To see this, let $d := |\mathsf{LD}(L)|$ be the *period* (i.e. number of states) of the minimal dfa. By Fact 1 of [16] (originally from [22]) every state-minimal nfa N accepting L is a disjoint union of cyclic dfas whose periods divide d.[1] Then $|\mathsf{rsc}(N^{\mathsf{r}})| = d$: we have $|\mathsf{rsc}(N^{\mathsf{r}})| \geq d$ since $\mathsf{rsc}(N^{\mathsf{r}})$ is a dfa accepting $L = L^{\mathsf{r}}$ and d is the size of the minimal dfa for L, and $|\mathsf{rsc}(N^{\mathsf{r}})| \leq d$ because after d steps, each cycle will be back in its initial state. Thus N is atomic by Theorem 4.12 and hence subatomic.

We deduce the following result for (not necessarily unary) regular languages:

Theorem 5.2. *If* $\mathsf{syn}(L)$ *is a cyclic group, then* $\mathrm{ns}(L) = \mathrm{n}\mu(L)$.

Proof (Sketch). Suppose that $\mathsf{syn}(L) = \mathsf{tm}(\mathsf{dfa}(L))$ is cyclic. Then there exists $w_0 \in \Sigma^*$ such that the map $\lambda X.w_0^{-1}X \colon \mathsf{LD}(L) \to \mathsf{LD}(L)$ generates $\mathsf{tm}(\mathsf{dfa}(L))$. Fix an alphabet $\Sigma_0 = \{a_0\}$ disjoint from Σ and consider the unary language

$$L_0 := \{\, a_0^n \;:\; n \in \mathbb{N},\ w_0^n \in L \,\} \subseteq \Sigma_0^*.$$

Let $g : \Sigma_0^* \to \Sigma^*$ be the monoid morphism where $g(a_0) := w_0$. Then we have the **JSL**-isomorphism

$$f \colon \mathsf{SLD}(L_0) \xrightarrow{\cong} \mathsf{SLD}(L), \quad f(X^{-1}L_0) := [g[X]]^{-1}L.$$

For each $a \in \Sigma$ choose $n_a \in \mathbb{N}$ such that $a^{-1}K = (w_0^{n_a})^{-1}K$ for all $K \in \mathsf{LD}(L)$. The respective transition endomorphisms of the **JSL**-automata $\mathsf{SLD}(L_0)$ and $\mathsf{SLD}(L)$ determine each other in the sense that the following diagrams commute:

$$\begin{array}{ccc} \mathsf{SLD}(L_0) & \xrightarrow[\cong]{f} & \mathsf{SLD}(L) \\ {\scriptstyle a_0^{-1}(-)}\big\downarrow & & \big\downarrow{\scriptstyle w_0^{-1}(-)} \\ \mathsf{SLD}(L_0) & \xrightarrow[f]{\cong} & \mathsf{SLD}(L) \end{array} \qquad \begin{array}{ccc} \mathsf{SLD}(L_0) & \xrightarrow[\cong]{f} & \mathsf{SLD}(L) \\ {\scriptstyle (a_0^{n_a})^{-1}(-)}\big\downarrow & & \big\downarrow{\scriptstyle a^{-1}(-)} \\ \mathsf{SLD}(L_0) & \xrightarrow[f]{\cong} & \mathsf{SLD}(L) \end{array}$$

Then $\mathrm{ns}(L) = \mathrm{ns}(L_0)$ by Theorem 4.7 and $\mathrm{n}\mu(L) = \mathrm{n}\mu(L_0)$ by Theorem 4.14. Moreover, by Example 5.1 we know that $\mathrm{ns}(L_0) = \mathrm{n}\mu(L_0)$, so the claim follows.

Example 5.3 ($\mathrm{n}\mu(L)$ no larger than Chrobak normal form). A unary nfa is in *Chrobak normal form* [8,13] if it has a single initial state and at most one state with multiple successors, all of which lie in disjoint cycles. We claim that for any nfa N in Chrobak normal form accepting the language L, we have

$$\mathrm{n}\mu(L) \leq |N|,$$

[1] In [16] nfas are restricted to have a single initial state and so are distinguished from unions of dfas; the latter are valid nfas from our perspective.

where $|N|$ denotes the number of states of N. To see this, observe that each state of N up to and including the unique choice state accepts some left derivative of L. The successors of the choice state collectively accept a derivative $u^{-1}L$; this language is cyclic because it is a finite union of cyclic languages. Therefore, by Example 5.1 we may replace the cycles by an atomic nfa accepting $u^{-1}L$, without increasing the number of states. The resulting nfa is atomic.

Since every unary nfa on n states can be transformed into an nfa in Chrobak normal form with $O(n^2)$ states [8, Lemma 4.3], we get:

Corollary 5.4. *If L is a unary regular language, then* $\mathrm{n}\mu(L) = O(\mathrm{ns}(L)^2)$.

5.2 Languages with a canonical state-minimal nfa

There are several natural classes of regular languages for which *canonical* state-minimal nondeterministic acceptors have been identified. We show that these acceptors are actually subatomic. In our arguments, we frequently consider the *length* of a finite semilattice S, i.e. the maximum length n of any ascending chain $s_0 < s_1 < \ldots < s_n$ in S. Note that since every element is uniquely determined by the set of join-irreducibles below it, the length of S is at most $|J(S)|$.

Example 5.5 (Bideterministic and biseparable languages).

(1) A language is called *bideterministic* if it is accepted by a dfa whose reverse is also a dfa. In this case, the minimal dfa is a minimal nfa [34, 38]. Bideterministic languages have been studied in the context of automata learning [2] and coding theory, where they are known as *rectangular codes* [27, 36]. We show that for every bideterministic language L,

$$\mathrm{ns}(L) = \mathrm{n}\mu(L) = |\mathsf{LD}(L)|.$$

To this end, we first note that by [36, Theorem 3.1] a language $L \subseteq \Sigma^*$ is bideterministic iff the left derivatives of L are pairwise disjoint. This implies that $\mathsf{SLD}(L)$ is a boolean algebra with atoms $\mathsf{LD}(L)$. Since the length of a boolean algebra equals the number of atoms (= join-irreducibles), we conclude that for every finite semilattice extension $\mathsf{SLD}(L) \rightarrowtail S$, the semilattice S has length at least $|\mathsf{LD}(L)|$. Thus, $|\mathsf{LD}(L)| \leq |J(S)|$, so any representation ρ extending κ_L or $\kappa_L \circ \mu_L$ satisfies $|\mathsf{LD}(L)| \leq \deg(\rho)$. Hence, $\mathrm{ns}(L) = \mathrm{n}\mu(L) = |\mathsf{LD}(L)|$ by Theorem 4.7 and 4.14. In particular, the minimal dfa of L is a minimal nfa.

(2) A language L is *biseparable* if $\mathsf{SLD}(L)$ is a boolean algebra [28].[2] For every biseparable language L, the *canonical residual automaton* [12], i.e. the nfa N_L of join-irreducibles of the minimal **JSL**-dfa $\mathsf{SLD}(L)$, is a state-minimal nfa; it is subatomic because every state of N_L accepts a derivative of L. This follows exactly as in (1): our argument only used that $\mathsf{SLD}(L)$ is a boolean algebra.

[2] Actually [28] defines biseparability as a property of nfas, and characterizes biseparable nfas as those accepting a language L for which no $\cup$-irreducible left derivative is contained in the union of other $\cup$-irreducible left derivatives. This is equivalent to the lattice $\mathsf{SLD}(L)$ being boolean, i.e. to L being 'biseparable' in our sense.

Example 5.6 (Maximal reachability). A folklore result asserts that if N is an nfa whose accepted language L satisfies $|\mathsf{LD}(L)| = 2^{|N|}$, then N is state-minimal. Since $\mathsf{LD}(L)$ forms the set of states of the minimal dfa for L and $\mathsf{rsc}(N)$ accepts L, we have $\mathsf{rsc}(N) = \mathcal{P}(N)$. It follows the **JSL**-dfa $\mathcal{P}(N)$ is reachable and simple, hence isomorphic to the minimal **JSL**-dfa $\mathsf{SLD}(L)$. This proves that $\mathsf{SLD}(L)$ is a boolean algebra, i.e. L is a biseparable language. We conclude from Example 5.5(2) that $\mathrm{ns}(L) = \mathrm{n}\mu(L) = |N|$ and N_L is a subatomic minimal nfa.

Example 5.7 (BiRFSA and topological languages). So far $\mathsf{SLD}(L)$ has been a boolean algebra. But the argument in Example 5.5 also applies when $\mathsf{SLD}(L)$ is a distributive lattice, noting that the length of a finite distributive lattice is equal to the number of its join-irreducibles [17, Corollary 2.14]. Languages with this property are called *topological* [1]. It thus follows as in Example 5.5(2) that for any topological language L, the canonical residual automaton N_L is subatomic and a state-minimal nfa. Thus, $\mathrm{ns}(L) = \mathrm{n}\mu(L) = |J(\mathsf{SLD}(L))|$.

There is another class of languages where N_L is known to be a state-minimal nfa, the *biRFSA* languages [28]. A language L is called biRFSA if N_L is isomorphic to $(N_{L^{\mathsf{r}}})^{\mathsf{r}}$. Surprisingly, these languages are exactly the topological ones:
(1) *Suppose that L is topological.* Recall that N_L is the nfa of join-irreducibles of the minimal **JSL**-dfa. Thus, it has states $J(\mathsf{SLD}(L))$ and transitions given by $X \xrightarrow{a} Y$ iff $Y \subseteq a^{-1}X$ for $a \in \Sigma$. Moreover, a join-irreducible j is initial iff $j \subseteq L$ and final iff $\varepsilon \in j$. Since the lattice $\mathsf{SLD}(L)$ is distributive, we have a canonical bijection between its join- and meet-irreducibles:

$$\tau\colon J(\mathsf{SLD}(L)) \xrightarrow{\cong} M(\mathsf{SLD}(L)), \quad \tau(j) = \bigcup\{X \in \mathsf{SLD}(L) : j \nsubseteq X\}.$$

Let θ be the unique map making the following diagram commute, where dr_L is the restriction of the isomorphism of Proposition 3.13:

$$\begin{array}{ccc} & J(\mathsf{SLD}(L)) & \\ {\scriptstyle\theta}\swarrow\cong & & \cong\searrow{\scriptstyle\tau} \\ J(\mathsf{SLD}(L^{\mathsf{r}})) & \xrightarrow[\mathrm{dr}_L]{\cong} & M(\mathsf{SLD}(L)) \end{array}$$

One can show θ to be an nfa isomorphism from N_L to $(N_{L^{\mathsf{r}}})^{\mathsf{r}}$. Thus, L is biRFSA.
(2) *Suppose that L is biRFSA.* Then we have a surjective **JSL**-morphism

$$[\mathcal{P}(J(\mathsf{SLD}(L)))]^{\mathsf{op}} \cong \mathcal{P}(J(\mathsf{SLD}(L^{\mathsf{r}}))) \xrightarrow{e_{L^{\mathsf{r}}}} \mathsf{SLD}(L^{\mathsf{r}}) \cong [\mathsf{SLD}(L)]^{\mathsf{op}},$$

where the first isomorphism follows from $N_L \cong (N_{L^{\mathsf{r}}})^{\mathsf{r}}$ and Lemma 3.11, the second isomorphism is given by Proposition 3.13, and $e_{L^{\mathsf{r}}}$ sends $X \subseteq J(\mathsf{SLD}(L^{\mathsf{r}}))$ to $\bigcup X$. The dual of this morphism is the injective **JSL**-morphism

$$m_L\colon \mathsf{SLD}(L) \rightarrowtail \mathcal{P}(J(\mathsf{SLD}(L)))$$

sending $K \in \mathsf{SLD}(L)$ to the set of all $j \in J(\mathsf{SLD}(L))$ with $j \subseteq K$. Note that $e_L \circ m_L = id_{\mathsf{SLD}(Q)}$, showing that $\mathsf{SLD}(L)$ is a retract of $\mathcal{P}(J(\mathsf{SLD}(L)))$. Since **JSL**-retracts of finite distributive lattices are distributive, see e.g. [31, Lemma 2.2.3.15], it follows that $\mathsf{SLD}(L)$ is distributive. Thus, L is topological.

Example 5.8 (Extremal languages). Call a language *extremal* if $\mathsf{SLD}(L)$ has length $|J(\mathsf{SLD}(L))|$ i.e. we have an *extremal lattice* in the sense of Markowsky [29]. Again, the argument of Example 5.5 applies and we get $\mathrm{ns}(L) = \mathrm{n}\mu(L) = |J(\mathsf{SLD}(L))|$. Topological languages are extremal since every distributive lattice is an extremal lattice, although extremal languages need not be topological. Both classes are naturally characterized in terms of the reduced dependency relation:
(1) L is topological iff $\mathcal{DR}_L^j$ is essentially an order relation $\leq_P \subseteq P \times P$ of a finite poset [30, Example 2.2.12].
(2) L is extremal iff $\mathcal{DR}_L^j$ is *upper unitriangularizable* [29, Theorem 11].

The latter means the adjacency matrix of the bipartite graph $\mathcal{DR}_L^j$ can be put in upper triangular form with ones along the diagonal, by permuting rows and columns. An order relation is upper unitriangularizable because it may be extended to a linear order.

6 Conclusion and Future Work

Motivated by the duality theory of deterministic finite automata over semilattices, we introduced a natural class of nondeterministic finite automata called *subatomic nfas* and studied their state complexity in terms of boolean representations of syntactic monoids. Furthermore, we demonstrated that a large body of previous work on state minimization of general nfas actually constructs minimal subatomic ones. There are several directions for future work.

As illustrated by Theorem 4.8, the dependency relation $\mathcal{DR}_L$ forms a useful tool for proving lower bounds on nfas. It is also a key element of the Kameda-Weiner algorithm [26, 37] for minimizing nfas, which rests on computing biclique covers of $\mathcal{DR}_L$. We aim to give an algebraic interpretation of dependency relations based on the representation of finite semilattices by contexts [24], which can be augmented to a categorical equivalence between $\mathbf{JSL}_\mathsf{f}$ and a suitable category of bipartite graphs [31]. Under this equivalence, **JSL**-dfas correspond to *dependency automata*; in particular, the minimal **JSL**-dfa $\mathsf{SLD}(L)$ corresponds to a dependency automaton whose underlying bipartite graph is precisely the dependency relation $\mathcal{DR}_L$. We expect that this observation can lead to a fresh algebraic perspective on the Kameda-Weiner algorithm, as well as a generalization of it computing minimal (sub-)atomic nfas.

On a related note, we also intend to investigate the complexity of the minimization problem for (sub-)atomic nfas. While minimizing general nfas is PSPACE-complete, even if the input automaton is a dfa, we conjecture that the additional structure present in (sub-)atomic acceptors will simplify their minimization to an NP-complete task. First evidence in this direction is provided by Geldenhuys, van der Merve, and van Zijl [14] whose work implies that minimal atomic nfas can be efficiently computed in practice using SAT solvers.

References

1. Adámek, J., Myers, R.S., Urbat, H., Milius, S.: On continuous nondeterminism and state minimality. In: Proc. 30th Conference on the Mathematical Foundations of Programming Semantics (MFPS XXX). vol. 308, pp. 3–23 (2014)
2. Angluin, D.: Inference of reversible languages. J. ACM **29**(3), 741–765 (1982)
3. Arbib, M.A., Manes, E.G.: Adjoint machines, state-behavior machines, and duality. Journal of Pure and Applied Algebra **6**(3), 313–344 (1975)
4. Arbib, M.A., Manes, E.G.: Fuzzy machines in a category. Bulletin of the Australian Mathematical Society **13**(2), 169–210 (1975)
5. Backhouse, R.: Factor theory and the unity of opposites. Journal of Logical and Algebraic Methods in Programming **85**(5, Part 2), 824–846 (2016)
6. Brzozowski, J., Tamm, H.: Theory of átomata. Theoretical Computer Science **539**, 13–27 (2014)
7. Brzozowski, J.A.: Derivatives of regular expressions. J. ACM **11**(4), 481–494 (Oct 1964)
8. Chrobak, M.: Finite automata and unary languages. Theoretical Computer Science **47**, 149–158 (1986)
9. Clemente, L., Mayr, R.: Efficient reduction of nondeterministic automata with application to language inclusion testing. Logical Methods in Computer Science **Volume 15, Issue 1** (2019)
10. Conway, J.H.: Regular Algebra and Finite Machines. Printed in GB by William Clowes & Sons Ltd (1971)
11. De Wulf, M., Doyen, L., Henzinger, T.A., Raskin, J.F.: Antichains: A new algorithm for checking universality of finite automata. In: Ball, T., Jones, R.B. (eds.) Computer Aided Verification. pp. 17–30. Springer (2006)
12. Denis, F., Lemay, A., Terlutte, A.: Residual finite state automata. In: Ferreira, A., Reichel, H. (eds.) STACS 2001: 18th Annual Symposium on Theoretical Aspects of Computer Science Dresden, Germany, February 15–17, 2001 Proceedings. pp. 144–157. Springer Berlin Heidelberg, Berlin, Heidelberg (2001)
13. Gawrychowski, P.: Chrobak normal form revisited, with applications. In: Bouchou-Markhoff, B., Caron, P., Champarnaud, J.M., Maurel, D. (eds.) Implementation and Application of Automata. pp. 142–153. Springer Berlin Heidelberg, Berlin, Heidelberg (2011)
14. Geldenhuys, J., van der Merwe, B., van Zijl, L.: Reducing nondeterministic finite automata with SAT solvers. In: Yli-Jyrä, A., Kornai, A., Sakarovitch, J., Watson, B. (eds.) Finite-State Methods and Natural Language Processing. pp. 81–92. Springer Berlin Heidelberg, Berlin, Heidelberg (2010)
15. Goguen, J.A.: Discrete-time machines in closed monoidal categories. I. J. Comput. Syst. Sci. **10**(1), 1–43 (1975)
16. Gramlich, G.: Probabilistic and nondeterministic unary automata. In: Proc. of Math. Foundations of Computer Science, Springer, LNCS 2747, 2003. pp. 460–469. Springer (2003)
17. Grätzer, G.: General Lattice Theory. Birkhäuser Verlag, 2. edn. (1998)
18. Gruber, H., Holzer, M.: Finding lower bounds for nondeterministic state complexity is hard. In: Ibarra, O.H., Dang, Z. (eds.) Developments in Language Theory: 10th International Conference, DLT 2006, Santa Barbara, CA, USA, June 26-29, 2006. Proceedings. pp. 363–374. Springer Berlin Heidelberg, Berlin, Heidelberg (2006)
19. Horn, A., Kimura, N.: The category of semilattices. Algebra Univ. **1**, 26–38 (1971)

20. Hromkovič, J., Seibert, S., Karhumäki, J., Klauck, H., Schnitger, G.: Communication complexity method for measuring nondeterminism in finite automata. Information and Computation **172**(2), 202–217 (2002), http://www.sciencedirect.com/science/article/pii/S089054010193069X
21. Izhakian, Z., Rhodes, J., Steinberg, B.: Representation theory of finite semigroups over semirings. Journal of Algebra **336**(1), 139–157 (2011)
22. Jiang, T., McDowell, E., Ravikumar, B.: The structure and complexity of minimal nfa's over a unary alphabet. International Journal of Foundations of Computer Science **02**(02), 163–182 (1991)
23. Jiang, T., Ravikumar, B.: Minimal NFA problems are hard. SIAM Journal on Computing **22**(6), 1117–1141 (1993)
24. Jipsen, P.: Categories of algebraic contexts equivalent to idempotent semirings and domain semirings. In: Kahl, W., Griffin, T.G. (eds.) Relational and Algebraic Methods in Computer Science. pp. 195–206. Springer Berlin Heidelberg, Berlin, Heidelberg (2012)
25. Johnstone, P.T.: Stone spaces. Cambridge University Press (1982)
26. Kameda, T., Weiner, P.: On the state minimization of nondeterministic finite automata. IEEE Transactions on Computers **C-19**(7), 617–627 (1970)
27. Kschischang, F.R.: The trellis structure of maximal fixed-cost codes. IEEE Transactions on Information Theory **42**(6), 1828–1838 (1996)
28. Latteux, M., Roos, Y., Terlutte, A.: Minimal NFA and biRFSA languages. RAIRO - Theoretical Informatics and Applications **43**(2), 221–237 (2009)
29. Markowsky, G.: Primes, irreducibles and extremal lattices. Order **9**, 265–290 (09 1992)
30. Myers, R.S.R.: Nondeterministic automata and JSL-dfas. CoRR **abs/2007.06031** (2020), https://arxiv.org/abs/2007.06031
31. Myers, R.S.R.: Representing semilattices as relations. CoRR **abs/2007.10277** (2020), https://arxiv.org/abs/2007.10277
32. Myers, R.S.R., Adámek, J., Milius, S., Urbat, H.: Coalgebraic constructions of canonical nondeterministic automata. Theoretical Computer Science **604**, 81–101 (2015)
33. Pin, J.É.: Mathematical foundations of automata theory (September 2020), available at http://www.liafa.jussieu.fr/~jep/PDF/MPRI/MPRI.pdf
34. Pin, J.E.: On reversible automata. In: Simon, I. (ed.) LATIN '92. pp. 401–416. Springer Berlin Heidelberg, Berlin, Heidelberg (1992)
35. Polák, L.: Syntactic semiring of a language. In: Sgall, J., Pultr, A., Kolman, P. (eds.) Mathematical Foundations of Computer Science 2001: 26th International Symposium, MFCS 2001 Mariánské Lázne, Czech Republic, August 27–31, 2001 Proceedings. pp. 611–620. Springer Berlin Heidelberg, Berlin, Heidelberg (2001)
36. Shankar, P., Dasgupta, A., Deshmukh, K., Rajan, B.: On viewing block codes as finite automata. Theoretical Computer Science **290**(3), 1775–1797 (2003)
37. Tamm, H.: New interpretation and generalization of the Kameda-Weiner method. In: Chatzigiannakis, I., Mitzenmacher, M., Rabani, Y., Sangiorgi, D. (eds.) ICALP 2016, Rome, Italy. LIPIcs, vol. 55, pp. 116:1–116:12. Schloss Dagstuhl - Leibniz-Zentrum für Informatik (2016)
38. Tamm, H., Ukkonen, E.: Bideterministic automata and minimal representations of regular languages. Theoretical Computer Science **328**(1), 135–149 (2004)
39. van Heerdt, G., Moerman, J., Sammartino, M., Silva, A.: A (co)algebraic theory of succinct automata. Journal of Logical and Algebraic Methods in Programming **105**, 112–125 (2019)

A String Diagrammatic Axiomatisation of Finite-State Automata

Robin Piedeleu(✉) and Fabio Zanasi

University College London, London, UK,
{r.piedeleu, f.zanasi}@ucl.ac.uk

Abstract. We develop a fully diagrammatic approach to finite-state automata, based on reinterpreting their usual state-transition graphical representation as a two-dimensional syntax of string diagrams. In this setting, we are able to provide a complete equational theory for language equivalence, with two notable features. First, the proposed axiomatisation is finite— a result which is provably impossible for the one-dimensional syntax of regular expressions. Second, the Kleene star is a derived concept, as it can be decomposed into more primitive algebraic blocks.

Keywords: string diagrams · finite-state automata · symmetric monoidal category · complete axiomatisation

1 Introduction

Finite-state automata are one of the most studied structures in theoretical computer science, with an illustrious history and roots reaching far beyond, in the work of biologists, psychologists, engineers and mathematicians. Kleene [25] introduced regular expressions to give finite-state automata an algebraic presentation, motivated by the study of (biological) neural networks [31]. They are the terms freely generated by the following grammar:

$$e, f ::= e + f \mid ef \mid e^* \mid 0 \mid 1 \mid a \in A \tag{1}$$

Equational properties of regular expressions were studied by Conway [14] who introduced the term *Kleene algebra*: this is an idempotent semiring with an operation $(-)^*$ for iteration, called the (Kleene) star. The equational theory of Kleene algebra is now well-understood, and multiple complete axiomatisations, both for language and relational models, have been given. Crucially, Kleene algebra is not finitely-based: no finite equational theory can appropriately capture the behaviour of the star [35]. Instead, there are purely equational infinitary axiomatisations [28,4] and Kozen's finitary implicational theory [26].

Since then, much research has been devoted to extending Kleene algebra with operations capturing richer patterns of behaviour, useful in program verification. Examples include conditional branching (Kleene algebra with tests [27], and its recent guarded version [37]), concurrent computation (CKA [19,23]), and specification of message-passing behaviour in networks (NetKAT [1]).

S. Kiefer and C. Tasson (Eds.): FOSSACS 2021, LNCS 12650, pp. 469–489, 2021.
https://doi.org/10.1007/978-3-030-71995-1_24

The meta-theory of the formalisms above essentially rests on the same three ingredients: (1) given an operational model (e.g., finite-state automata), (2) devise a syntax (regular expressions) that is sufficiently expressive to capture the class of behaviours of the operational model (regular languages), and (3) find a complete axiomatisation (Kleene algebra) for the given semantics.

In this paper, we open up a direct path from (1) to (3). Instead of thinking of automata as a combinatorial model, we formalise them as a bona-fide (two-dimensional) syntax, using the well-established mathematical theory of *string diagrams* and monoidal categories [36]. This approach lets us axiomatise the behaviour of automata directly, freeing us from the necessity of compressing them down to a one-dimensional notation like regular expressions.

This perspective not only sheds new light on a venerable topic, but has significant consequences. First, as our most important contribution, we are able to provide a *finite and purely equational* axiomatisation of finite-state automata, up to language equivalence. Intriguingly, this does not contradict the impossibility of finding a finite basis for Kleene algebra, as the algebraic setting is different: our result gives a finite presentation as a symmetric monoidal category, while the impossibility result prevents any such presentation to exist as an algebraic theory (in the standard sense). In other words, there is no finite axiomatisation based on terms (*tree*-like structures), but we demonstrate that there is one based on string diagrams (*graph*-like structures).

Secondly, embracing the two-dimensional nature of automata guarantees a strong form of compositionality that the one-dimensional syntax of regular expressions does not have. In the string diagrammatic setting, automata may have multiple inputs and outputs and, as a result, can be decomposed into subcomponents that retain a meaningful interpretation. For example, if we split the automata below left, the resulting components are still valid string diagrams within our syntax, below right:

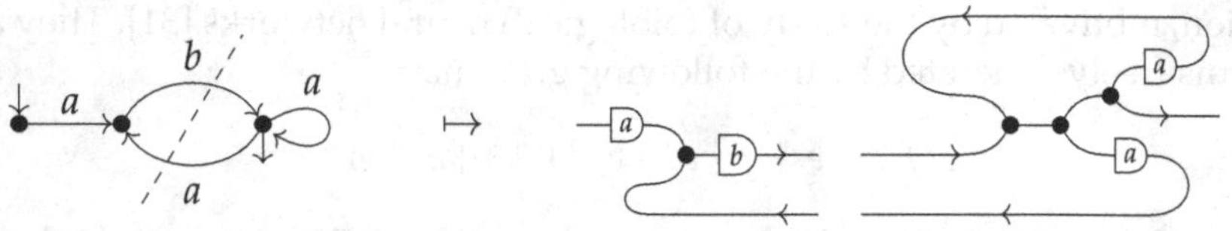

(2)

In line with the compositional approach, it is significant that the Kleene star can be decomposed into more elementary building blocks (which come together to form a feedback loop):

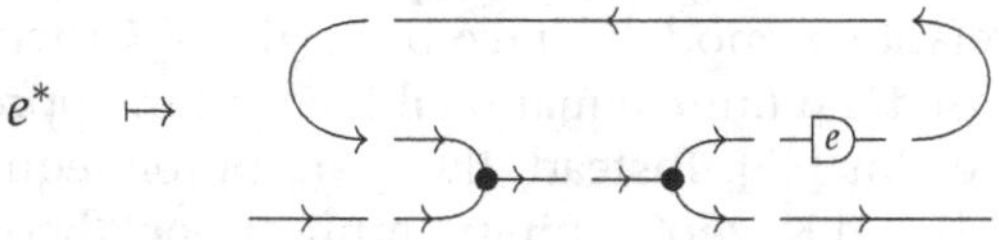

(3)

This opens up for interesting possibilities when studying extensions of Kleene algebra within the same approach— we elaborate on this in Section 6.

Finally, we believe our proof of completeness is of independent interest, as it relies on fully diagrammatic reformulation of Brzozowski's minimisation algorithm [12]. In the string diagrammatic setting, the symmetries of the equational theory give this procedure a particularly elegant and simple form. Because all of the axioms involved in the determinisation procedure come with a dual, a co-determinisation procedure can be defined immediately by simply reversing the former. This reduces the proof of completeness to a proof that determinisation can be performed diagrammatically.

We should also note that this is not the first time that automata and regular languages are recast into a categorical mould. The *iteration theories* [5] of Bloom and Ésik, *sharing graphs* [17] of Hasegawa or *network algebras* [39] of Stefanescu are all categorical frameworks designed to reason about iteration or recursion, that have found fruitful applications in this domain. They are based on a notion of parameterised fixed-point which defines a categorical *trace* in the sense of [22]. While our proposal bears resemblance to (and is inspired by) this prior work, it goes beyond in one fundamental aspect: it is the first to give a *finite* complete axiomatisation of automata up to language equivalence.

A second difference is methodological: our syntax (4) does not feature any primitive for iteration or recursion. In particular, the star is a derived concept, in the sense that it is decomposable into more elementary operations (3). Categorically, our starting point is a compact-closed rather than traced category.

We elaborate on the relation between ours and existing work in Section 6. Omitted proofs can be found in [33].

2 Syntax and semantics

Syntax. We fix an alphabet Σ of letters $a \in \Sigma$. We call Aut_Σ the symmetric strict monoidal category freely generated by the following objects and morphisms:

- three generating objects ▶ ('action'), ▶ ('right') and ◀ ('left') with their identity morphisms depicted respectively as ——, —→— and —←—.
- the following generating morphisms, depicted as *string diagrams* [36]:

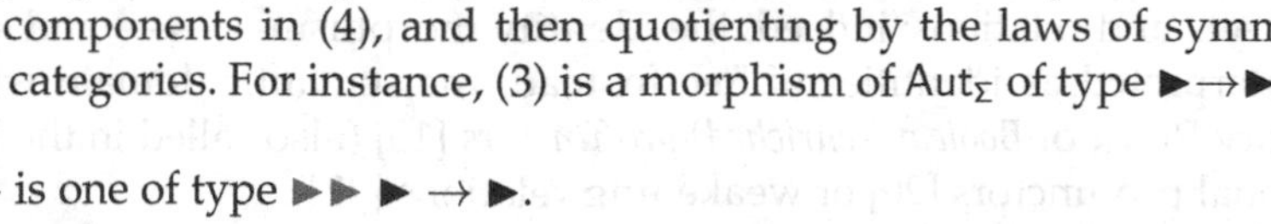
$(a \in \Sigma)$ (4)

Freely generating Aut_Σ from these data (usually called a *symmetric monoidal theory* [42,11]) means that morphisms of Aut_Σ will be the string diagrams obtained by pasting together (by sequential composition and monoidal product in Aut_Σ) the basic components in (4), and then quotienting by the laws of symmetric monoidal categories. For instance, (3) is a morphism of Aut_Σ of type ▶→▶, and

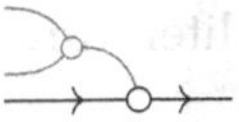 is one of type ▶▶ ▶ → ▶.

Semantics. We first define the semantics for string diagrams simply as a function, and then discuss how to extend it to a functor from Aut_Σ to another category. Our interpretation maps generating morphisms to relations between regular expressions and languages over Σ:

$$[\![\text{——}]\!] = \{((e,e) \mid e \in \mathsf{RegExp}\} \qquad [\![\text{—○—}]\!] = \{(e,e^*) \mid e \in \mathsf{RegExp}\}$$

⟦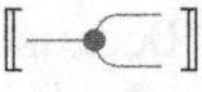⟧ $= \{(e,(e,e)) \mid e \in \mathsf{RegExp}\}$ ⟦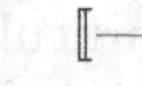⟧ $= \{(e,\bullet) \mid e \in \mathsf{RegExp}\}$

⟦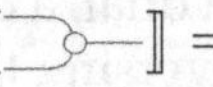⟧ $= \{((e,f),ef) \mid e,f \in \mathsf{RegExp}\}$ ⟦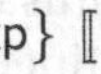⟧ $= \{(\bullet,1)\}$ ⟦⟧ $= \{(\bullet,a)\}$

⟦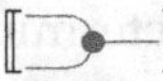⟧ $= \{((e,f),e+f) \mid e,f \in \mathsf{RegExp}\}$ ⟦⟧ $= \{(\bullet,0)\}$

⟦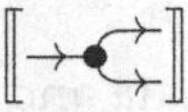⟧ $= \{(L,(K_1,K_2)) \mid L \subseteq K_i,\ i = 1,2 \text{ and } L,K_1,K_2 \subseteq \Sigma^\star\}$

⟦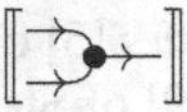⟧ $= \{((L_1,L_2),K) \mid L_i \subseteq K,\ i = 1,2 \text{ and } L_1,L_2,K \subseteq \Sigma^\star\}$

⟦⟧ $= \{(L,\bullet) \mid L \subseteq \Sigma^\star\}$ ⟦⟧ $= \{(\bullet,(L,K)) \mid L \subseteq K \mid L,K \subseteq \Sigma^\star\}$

⟦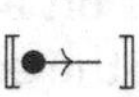⟧ $= \{(\bullet,K) \mid K \subseteq \Sigma^\star\}$ ⟦⟧ $= \{((L,K),\bullet) \mid K \subseteq L \mid L,K \subseteq \Sigma^\star\}$

⟦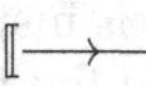⟧ $= \{((L,K),L \subseteq K) \mid L,K \subseteq \Sigma^\star\}$

⟦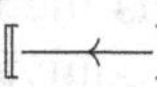⟧ $= \{((L,K),K \subseteq L) \mid L,K \subseteq \Sigma^\star\}$

⟦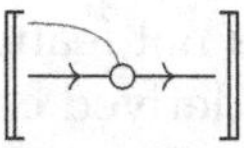⟧ $= \{((e,L),K) \mid L[\![e]\!]_R \subseteq K \text{ and } e \in \mathsf{RegExp},\ L,K \subseteq \Sigma^\star\}$ (5)

In (5), the semantics $[\![e]\!]_R \in 2^{A^*}$ of a regular expression $e \in \mathsf{RegExp}$ is defined inductively on e (see (1)), in the standard way:

$$[\![e+f]\!]_R = [\![e]\!]_R \cup [\![f]\!]_R \qquad [\![ef]\!]_R = \{vw \mid v \in [\![e]\!]_R, w \in [\![f]\!]_R\}$$

$$[\![1]\!]_R = \{\varepsilon\} \qquad [\![0]\!]_R = \varnothing \qquad [\![a]\!]_R = \{a\} \qquad [\![e^*]\!]_R = \bigcup_{n\in\mathbb{N}} [\![e^n]\!]_R$$

where $e^{n+1} := ee^n$ and $e^0 := 1$. The semantics highlights the different roles played by red[1] and black generators. In a nutshell, red generators stand for regular expressions (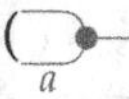 the sum, is 0, 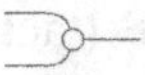 the product,  is 1, 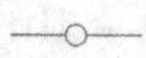 the Kleene star, and  the letters of Σ), and black generators for operations on the set of languages (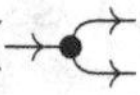 is copy,  is delete,  and  feed back outputs into inputs, in a way made more precise later). These two perspectives, which are usually merged, are kept distinct in our approach and only allowed to communicate via 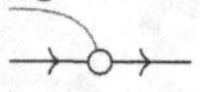, which represents the product action of regular expressions (the red wire) on languages via concatenation on the right.

In order for this mapping to be functorial from Aut_Σ, we now introduce a suitable target semantic category. Interestingly, this will not be the category Rel of sets and relations: indeed, the identity morphisms  and  are not interpreted as identities of Rel. Instead, the semantic domain will be the category $\mathsf{Prof}_{\mathbb{B}}$ of *Boolean(-enriched) profunctors* [15] (also called in the literature relational profunctors [20] or weakening relations [32]).

Definition 1. *Given two preorders* $(X, \leq_X)$ *and* $(Y, \leq_Y)$, *a* Boolean profunctor $R : X \nrightarrow Y$ *is a relation* $R \subseteq X \times Y$ *such that if* $(x,y) \in R$ *and* $x' \leq_X x$, $y \leq_Y y'$ *then* $(x',y') \in R$.

[1] The reader with a greyscale version of the paper should see light grey generators instead.

Preorders and Boolean profunctors form a symmetric monoidal category $\mathsf{Prof}_{\mathbb{B}}$ *with composition given by relational composition. The identity for an object* $(X, \leq_X)$ *is the order relation* $\leq_X$ *itself. The monoidal product is the usual product of preorders.*

The rich features of our diagrammatic language are reflected in the profunctor interpretation. Indeed, the order relation is built into the wires $\rightarrow$ and $\leftarrow$. The two possible directions represent the identities on the ordered set of languages and the same set with the reversed order, respectively. The additional red wire —— represents the set RegExp of regular expressions, with *equality* as the associated order relation.[2] It is clear that all monochromatic generators satisfy the condition of Definition 1. Similarly, the action generator is a Boolean profunctor: if $((e, L), K)$ are such that $L [\![e]\!]_R \subseteq K$ and $L' \subseteq L$, $K \subseteq K'$ then we have $L' [\![e]\!]_R \subseteq L [\![e]\!]_R \subseteq K \subseteq K'$ by monotony of the product of languages. We can conclude that

Proposition 1. $[\![\cdot]\!]$ *defines a symmetric monoidal functor of type* $\mathsf{Aut}_\Sigma \to \mathsf{Prof}_{\mathbb{B}}$.

In particular, because Aut_Σ is free, we can unambiguously assign meaning to any composite diagram from the semantics of its components using composition and the monoidal product in $\mathsf{Prof}_{\mathbb{B}}$:

$$[\![-\boxed{c}-\boxed{d}-]\!] = \{(L,K) \mid \exists M\, (L,M) \in [\![-\boxed{c}-]\!], (M,K) \in [\![-\boxed{d}-]\!]\}$$

$$\left[\!\!\left[\begin{matrix} -\boxed{c_1}- \\ -\boxed{c_2}- \end{matrix} \right]\!\!\right] = \{((L_1,L_2),(K_1,K_2)) \mid (L_i, K_i) \in [\![-\boxed{c_i}-]\!], i = 1,2\}$$

Example 1. We include here a worked out example to show how to compute the behaviour of a composite diagram which, as we will see, represents the action by concatenation of the regular language a^*. We assign variable names to each wire: O to the top wire of the feedback loop, N to the output wire of the action node, and M to the middle wire joining to so that we can compute:

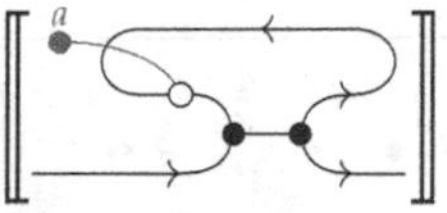

$$\begin{aligned} &= \{(L,K) \mid \exists M,N,O,\ L,N \subseteq M,\ O [\![a]\!]_R \subseteq N,\ M \subseteq O, K\} \\ &= \{(L,K) \mid \exists N,O,\ L,N \subseteq O,\ L,N \subseteq K\ Oa \subseteq N\} \\ &= \{(L,K) \mid \exists O,\ Oa \subseteq O,\ L \subseteq O,\ L,O \subseteq K\}. \end{aligned}$$

Call this diagram d. Since $Oa \subseteq O$ and $L \subseteq O$ is equivalent to $L \cup Oa \subseteq O$, $[\![d]\!] = \{(L,K) \mid \exists O \text{ s.t. } L \cup Oa \subseteq O,\ L,O \subseteq K\}$. Finally, by Arden's lemma [2], La^* is the *least* solution of the language inequality $L \cup Xa \subseteq X$; thus $[\![d]\!] = \{(L,K) \mid \exists O \text{ s.t. } La^* \subseteq O,\ L,O \subseteq K\} = \{(L,K) \mid La^* \subseteq K\}$.

3 Equational theory

In Figure 1 we introduce $=_{KDA}$, the (finite) equational theory of *Kleene Diagram Algebra*, on Aut_Σ. It will be later shown to be *complete* for the given semantics. We explain some salient features of $=_{KDA}$ below.

[2] Note that we can always consider any set with equality as a poset and that, therefore, Rel is a subcategory of $\mathsf{Prof}_{\mathbb{B}}$, but not vice-versa, for the simple reason that the identity relation of an arbitrary poset in $\mathsf{Prof}_{\mathbb{B}}$ is not mapped to the identity relation in Rel.

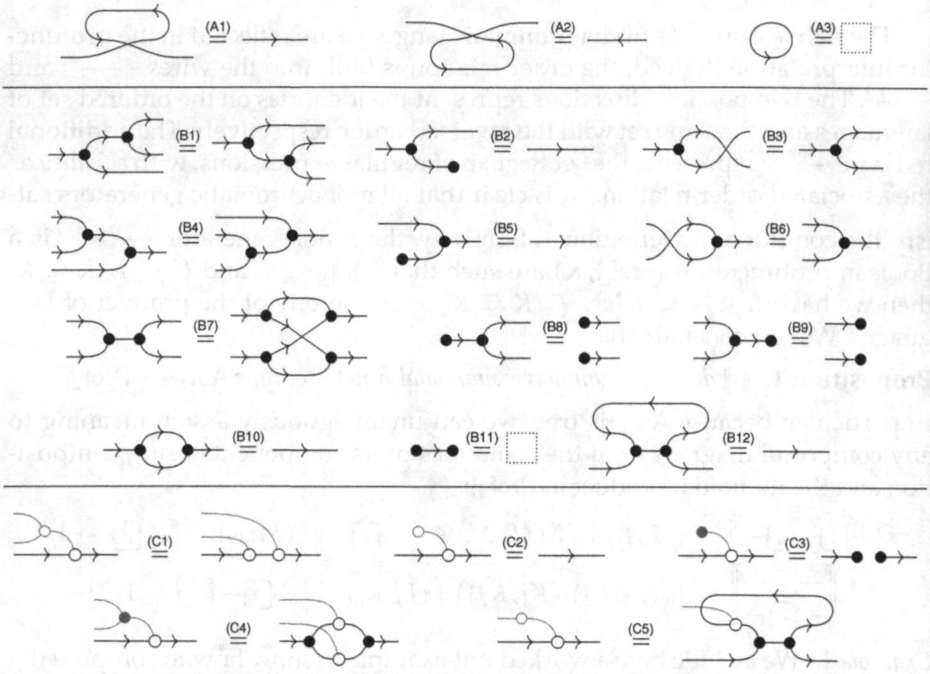

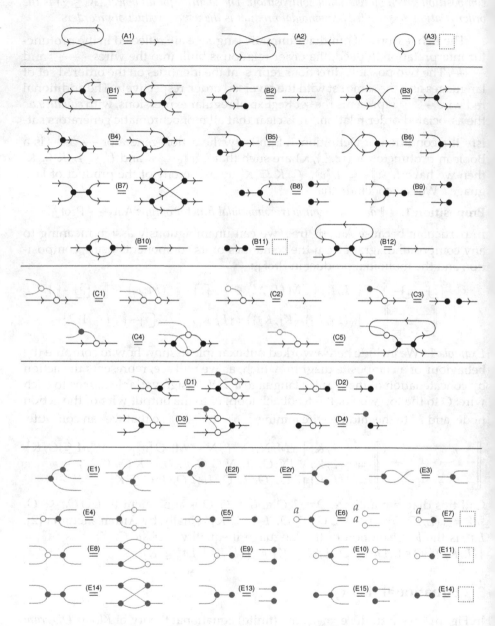

Fig. 1. Equational theory $=_{KDA}$ of Kleene Diagram Algebra.

- (A1)-(A2) relate [diagram] and [diagram], allowing us to bend and straighten wires at will. This makes the full subcategory of Aut_Σ on ▶ and ◀, modulo (A1)-(A2), *compact closed* [24]. (A3) allows us to eliminate isolated loops. Note that the whole category is not compact closed because ▶ has no dual.
- The B block states that [diagram], [diagram] forms a cocommutative comonoid (B1)-(B3), while [diagram], [diagram] form a commutative monoid (B4)-(B6). Moreover, [diagram], [diagram], [diagram], [diagram] form an idempotent bimonoid (B7)-(B11). (B12) allows us to eliminate trivial feedback loops.
- The C block axiomatises the action of regular expressions on languages. These laws mimic the usual definition of the action of a semiring on a set, except for (C5) which is novel and captures the interaction with the Kleene star. Here lies a distinctive feature of our theory: the behaviour of the star is derived from its decomposition as the feedback loop on the right of (C5).
- The D block forces the action to be a comonoid ((D1)-(D2)) and monoid ((D1)-(D2)) homomorphism.
- The E block axiomatises the purely red fragment. Remarkably, these axioms do not describe any of the actual Kleene algebra structure: they just state that [diagram] and [diagram] form a commutative comonoid ((E1)-(E3)) and that all other red generators are comonoid homomorphisms ((E4)-(E15)). This means that the red fragment is actually the *free* (cartesian) algebraic theory (*cf.* [42,11]) on generators [diagram], [diagram], [diagram], [diagram], [diagram], [diagram] ($a \in \Sigma$), where the remaining generators [diagram] and [diagram] act as copy and discard of variables.

Let $=_{KDA}$ be the smallest equational theory containing all equations in Fig. 1. Their *soundness* for the chosen semantics is not difficult to show and, for space reasons, we omit the proof. We now state our *completeness* result, whose proof will be discussed in Section 5.

Theorem 1 (Completeness). *For morphisms d, e in* Aut_Σ *, $d =_{KDA} e$ iff $[\![d]\!] = [\![e]\!]$.*

Remark 1. In the usual approach to the theory of regular languages (e.g. [26]), a completeness result like Theorem 1 is typically proven by first defining a class of models for the algebraic theory, and showing that the standard semantics constitutes the initial/free model. Our proof is different in flavour, but equivalent: taking advantage of the categorical formulation of our diagrammatic syntax and its semantics, we construct an equivalence of categories between our model and the diagrams quotiented by the equations of KDA.

Remark 2. Some axiomatisations of Kleene algebra use a partial order between terms, which can be defined from the idempotent monoid structure: $f \leq e$ iff $e + f = e$. At the semantic level, it corresponds to inclusion of languages. Similarly, using the idempotent bimonoid structure of our equational theory, we can define a partial order on ▶→▶ diagrams: $f \leq e$ iff [diagram] = [diagram]. This partial order structure can also be extended to all morphisms $▶^n \to ▶^m$ by using the vertical composition of n copies of [diagram] and m copies of [diagram] instead.

Remark 3. There are no specific equations relating the atomic actions $\overset{a}{\circ}\!\!-$ ($a \in \Sigma$). This is because, as we study automata, we are interested in the *free* monoid Σ^* over Σ. However, nothing would prevent us from modelling other structures. Free commutative monoids (powers of $\mathbb{N}$), whose rational subsets correspond to semilinear sets [14, Chapter 11] would be of particular interest.

4 Encoding regular expressions and automata

A major appeal of our approach is that both regular expressions and automata can be uniformly represented in the graphical language of string diagrams, and the translation of one into the other becomes an equational derivation in $=_{KDA}$. In fact, we will see there is a close resemblance between automata and the shape of the string diagrams interpreting them — the main difference being that string diagrams are *composable* structures.

In this section we describe how regular expressions (resp. automata) can be encoded as string diagrams, such that their semantics corresponds in a precise way to the languages that they describe (resp. recognise).

In a sense, regular expressions are already part of the graphical syntax, as the red generators: for any regular expression e, one may always construct a 'red' string diagram $(e)\!\!-: 0 \to \blacktriangleright$ such that $[\![(e)\!\!-]\!] = \{(\bullet, e)\}$. However, these alone are meaningless, since their image under the semantics is simply the free term algebra RegExp (see (7)) . They acquire meaning as they *act* on the set of languages over Σ, represented by the black wire.

4.1 From regular expressions to string diagrams

To define these encodings, it is convenient to introduce the following syntactic sugar. We will write $-\!\![e\rangle\!\!-$ for the composite of $(e)\!\!-$ with the action, as defined below left, with the particular case of a letter $a \in \Sigma$ on the right:

(6)

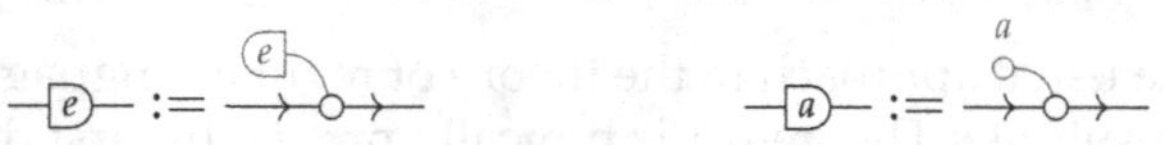

Using this action, we can inductively define an encoding $\langle - \rangle$ of regular expressions into string diagrams of Aut_Σ, as the rightmost diagram for each expression below:

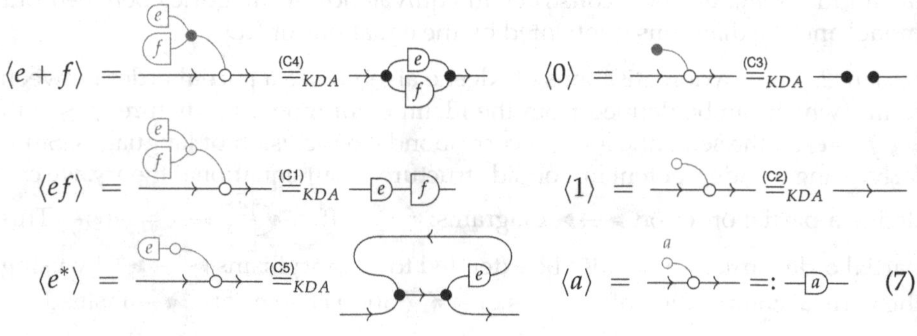
⟨e + f⟩ =
(C4)
=KDA
⟨0⟩ =
(C3)
=KDA
⟨ef⟩ =
(C1)
=KDA
⟨1⟩ =
(C2)
=KDA
⟨e*⟩ =
(C5)
=KDA
⟨a⟩ =
=:
(7)

For example, $\langle ab(a+ab)^* \rangle$ =

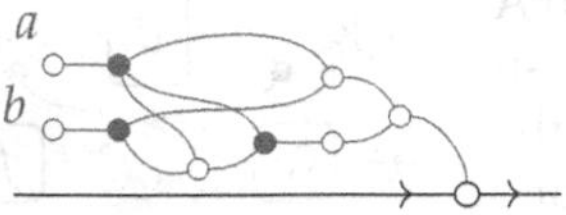

$=_{KDA}$
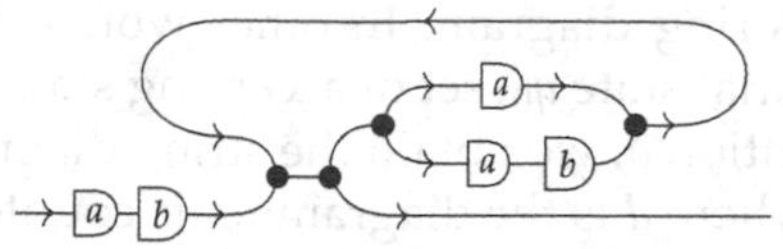

(8)

As expected, the translation preserves the language interpretation of regular expressions in a sense that the following proposition makes precise.

Proposition 2. *For any regular expression e,* $[\![\langle e \rangle]\!] = \{(L,K) \mid [\![e]\!]_R \, L \subseteq K\}$.

4.2 From automata to string diagrams...

Example (8) suggests that the string diagram $\langle e \rangle$ corresponding to a regular expression e looks a lot like a nondeterministic finite-state automaton (NFA) for e. In fact, the translation $\langle - \rangle$ can be seen as the diagrammatic counterpart of Thompson's construction [40] that builds an NFA from a regular expression.

We can generalise the encoding of regular expressions and translate NFA directly into string diagrams, in at least two ways. The first is to encode an NFA as the diagrammatic counterpart of its transition relation. The second is to translate directly its graph representation into the diagrammatic syntax.

Encoding the transition relation. This is a simple variant of the translation of matrices over semirings that has appeared in several places in the literature [29,42].

Let A be an NFA with set of states Q, initial state $q_0 \in Q$, accepting states $F \subseteq Q$ and transition relation $\delta \subseteq Q \times \Sigma \times Q$. We can represent δ as a string diagram d with $|Q|$ incoming wires on the left and $|Q|$ outgoing wires on the right.The left jth port of d is connected to the ith port on the right through an 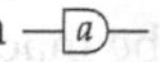 whenever $(q_i, a, q_j) \in \delta$. To accommodate nondeterminism, when the same two ports are connected by several different letters of Σ, we join these using 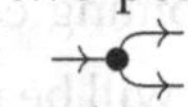 and . When $(q_i, \epsilon, q_j) \in \delta$, the two ports are simply connected via a plain identity wire. If there is no tuple in δ such that $(q_i, a, q_j) \in \delta$ for any a, the two corresponding ports are disconnected.

For example, the transition relation of an NFA with three states and $\delta = \{((q_0, a, q_1), (q_1, b, q_2), (q_2, a, q_1), (q_2, a, q_2))\}$ (disregarding the initial and accepting states for the moment) is depicted on the right. Conversely, given such a diagram, we can recover δ by collecting Σ-weighted paths from left to right ports.

$d =$ 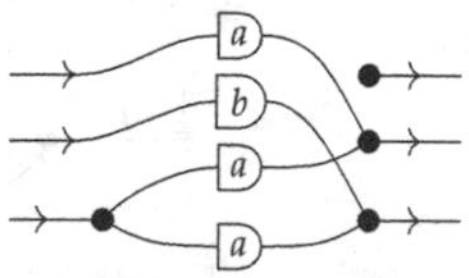

To deal with the initial state, we add an additional incoming wire connected to the right port corresponding to the initial state of the automaton. Similarly, for accepting states we add an additional outgoing wire, connected to the left ports corresponding to each accepting state, via 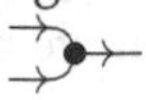 if there is more than one. Finally, we trace out the $|Q|$ wires of the diagrammatic transition relation to obtain the associated string diagram. In other words, for a NFA with initial state q_0, set of accepting states F, transition relation δ, we obtain the string diagram on the right, where d is the diagrammatic counterpart of δ as defined above, e_0 is the injection of a single wire as the first amongst $|Q|$ wires, and f deletes all wires that are not associated to states in F with , and applies to merge them into a single outgoing wire.

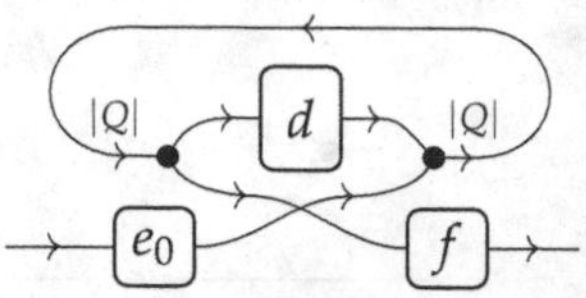

For example, if A with δ as above has initial state q_0 and accepting state $\{q_2\}$, we get the diagram below left; instead, if all states are accepting, we obtain the diagram below right:

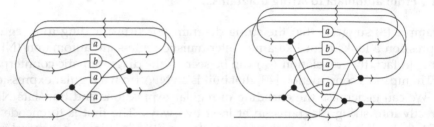

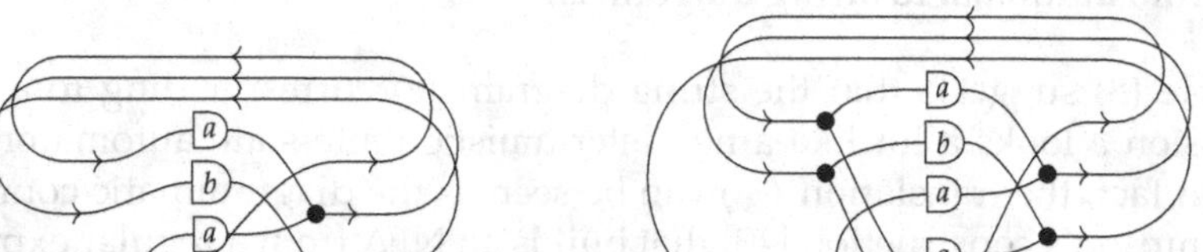

The correctness of this simple translation is justified by a semantic correspondence between the language recognised by a given NFA A and the denotation of the corresponding string diagram.

Proposition 3. *Given an NFA A which recognises the language L, let d_A be its associated string diagram, constructed as above. Then* $[\![d_A]\!] = \{(K, K') \mid LK \subseteq K'\}$.

From graphs to string diagrams. The second way of translating automata into string diagrams mimics more directly the combinatorial representation of automata. The idea (which should be sufficiently intuitive to not need to be made formal here) is, for each state, to use to represent incoming edges, and to represent outgoing edges. As above, labels $a \in A$ will be modelled using . For example, the graph and the associated string diagram corresponding with the NFA above are

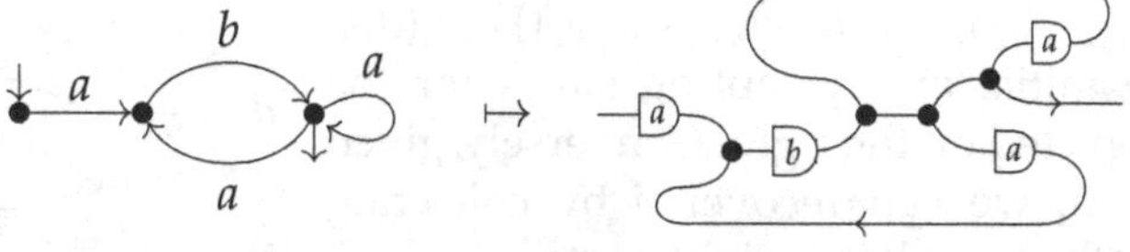

(9)

Note the initial state of the automaton corresponds to the left interface of the string diagram, and the accepting state to the right interface. As before, when there are multiple accepting states, they all connect to a single right interface, via . For example, if we make all states accepting in the automaton above, we get the following diagrammatic representation:

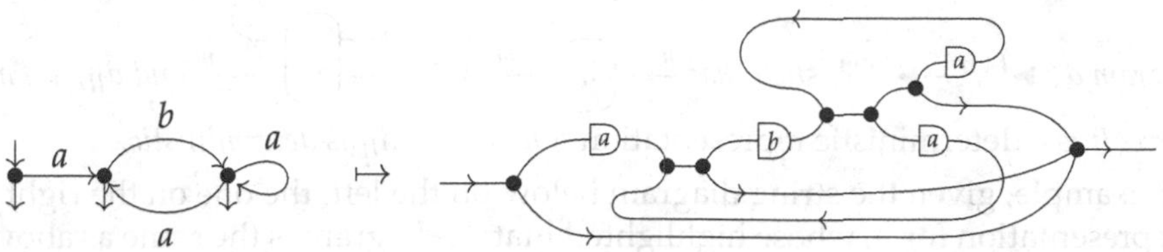

4.3 ...and back

The previous discussion shows how NFAs can be seen as string diagrams of type ▶→▶. The converse is also true: we now show how to extract an automaton from any string diagram d: ▶→▶, such that the language the automaton recognises matches the denotation of d.

In order to phrase this correspondence formally, we need to introduce some terminology. We call *left-to-right* those string diagrams whose domain and codomain contain only ▶, i.e. their type is of the form $▶^n \to ▶^m$. The idea is that, in any such string diagram, the n left interfaces act as *inputs* of the computation, and the m right interfaces act as *outputs*. For instance, (9) is a left-to-right diagram ▶→▶.

A string diagram d is *atomic* if the only red generators occurring in d are of the form $\overset{a}{\circ}$—. By *unfolding* all red components ⓔ— in any left-to-right diagram, using axioms (C1)-(C5), we can prove the following statement.

Proposition 4. *Any left-to-right diagram is* $=_{KDA}$*-equivalent to an atomic one.*

For instance, the string diagram on the left of (8) is $=_{KDA}$-equivalent to the atomic one on the right.

We call *block* of a certain subset of generators a vertical composite of these generators followed by some permutations of the wires.

Definition 2. *A* matrix-diagram *(resp.* generalised matrix-diagram*) is a left-to-right diagram that factors as a block of* →●⊂, →●, *followed by a block of* —ⓐ— *for* $a \in \Sigma$ *(resp.* —ⓔ— *for* $e \in$ RegExp*) and finally, a block of* ⊃●→, ●→.

To each matrix-diagram d we can associate a unique transition relation δ by gathering paths from each input to each output: $(q_i, a, q_j) \in \delta$ if there is —ⓐ— joining the ith input to the jth output.

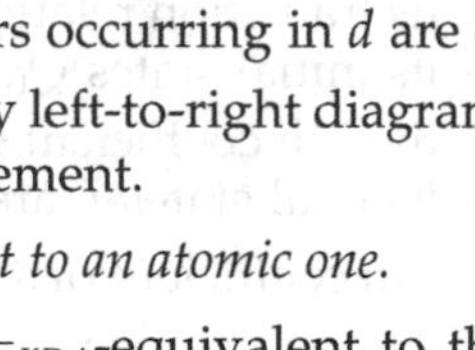

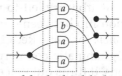

A transition relation is ϵ-*free* if it does not contain the empty word. It is *deterministic* if it is ϵ-free and, for each i and each $a \in \Sigma$ there is at most one j such that $(q_i, a, q_j) \in \delta$. We will apply these terms to matrix-diagrams and the associated transition relation interchangeably. The example of Section 4.2 above, with the three blocks highlighted, is a matrix-diagram. It is ϵ-free but not deterministic since there are two a-labelled transitions starting from the third input.

Given a matrix-diagram $d : ▶^{l+n} \to ▶^{p+m}$, we will write d_{ij}, with $i = l, n$ and $j = p, m$, for the subdiagrams corresponding to the appropriate submatrices.

Definition 3. *For any left-to-right diagram $d : \blacktriangleright^n \to \blacktriangleright^m$, a* representation *is a matrix-diagram $\hat{d} : \blacktriangleright^{l+n} \to \blacktriangleright^{l+m}$, such that*

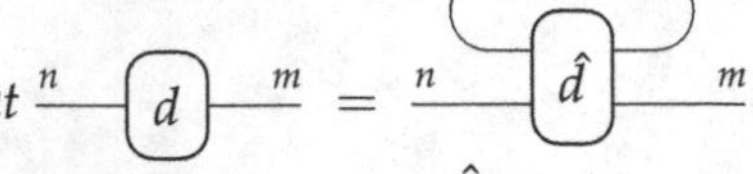

and $\hat{d}_{ll}$, $\hat{d}_{nl}$ are ϵ-free. It is a deterministic representation *if moreover $\hat{d}_{ll}$ is deterministic.*

For example, given the string diagram below on the left, the one on the right is a representation for it, whose highlighted matrix-diagram is the same as above.

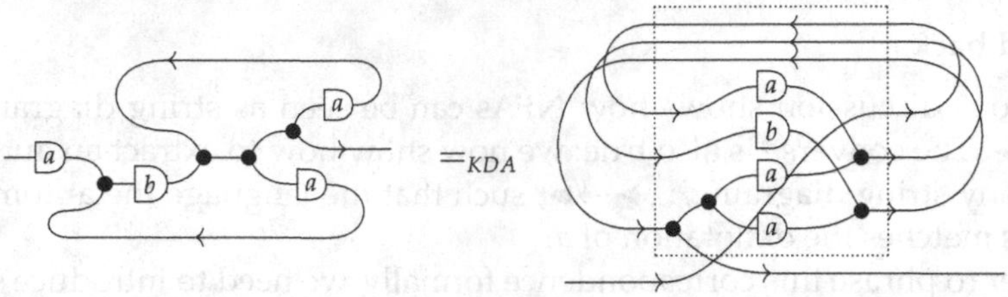

(10)

We will refer to the associated matrix-diagram $\hat{d}$ as the *transition matrix* of a given representation. From a $\blacktriangleright \to \blacktriangleright$ diagram with representation $\hat{d} : \blacktriangleright^{l+1} \to \blacktriangleright^{l+1}$ we can construct an NFA from its transition matrix $\hat{d}$ as follows:

- its state set is $Q = \{q_1, \dots, q_l\}$, i.e., there is one state for each wire of $\hat{d}_{ll}$;
- its transition relation built from $\hat{d}_{ll}$ as described above;
- its initial states Q_0 are those q_i for which there exists an index j such that the ijth coefficient of $\hat{d}_{1l}$ is non-zero (and therefore ϵ);
- its final states F are those q_j for which there exists an index i such that the ijth coefficient of $\hat{d}_{l1}$ is non-zero (and therefore ϵ);

The construction above is the inverse of that of Section 4.2. The link between the constructed automaton and the original string diagram is summarised in the following statement, which is a straightforward corollary of Proposition 3.

Proposition 5. *For a diagram $d : \blacktriangleright \to \blacktriangleright$ with a representation $\hat{d}$, let $A_{\hat{d}}$ be the associated automaton, constructed as above. Then $\hat{L}$ is the language recognised by $A_{\hat{d}}$ iff $[\![d]\!] = \{(K, K') \mid \hat{L}K \subseteq K'\}$.*

The next proposition states that a representation can be extracted from any string diagram.

Proposition 6. *Any left-to-right diagram has a representation.*

We established a correspondence between $\blacktriangleright \to \blacktriangleright$ diagrams and automata. What about arbitrary left-to-right diagrams $\blacktriangleright^n \to \blacktriangleright^m$? To characterise the precise relationship between our syntax and regular expressions we can prove a *Kleene theorem* for Aut_Σ. Recall, from Definition 2 that a *generalised matrix-diagram* is the diagrammatic counterpart of a matrix whose coefficients are regular expressions. It turns out that every left-to-right diagram can be put in this form.

Proposition 7 (Kleene's for Aut_Σ). *Any left-to-right diagram is equal to a generalised matrix diagram.*

As a result, the semantics of a given $\blacktriangleright^n \to \blacktriangleright^m$ diagram is fully characterised by an $m \times n$ array of regular languages.

4.4 Interlude: from regular to context-free languages

It is worth pointing out how a simple modification of the diagrammatic syntax takes us one notch up the Chomsky hierarchy, leaving the realm of regular languages for that of context-free grammars and languages.

Our syntax allows to specify systems of language equations of the form $aX \subseteq Y$. In this context, feedback loops can be interpreted as fixed-points. For example, the automaton below left, and its corresponding string diagram, below right, translate to the system of equations at the center:

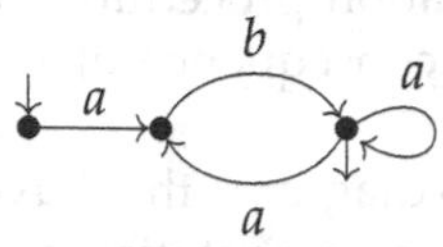

$$\mapsto \begin{cases} \epsilon \subseteq X_0 \\ X_0 a \subseteq X_1 \\ X_1 b \subseteq X_2 \\ X_2 a \subseteq X_1 \\ X_2 a \subseteq X_2 \end{cases} \leftarrow\!\shortmid \tag{11}$$

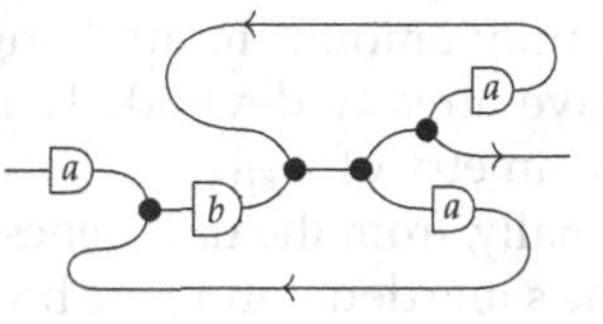

This translation can be obtained by simply labelling each state with a variable and adding one inequality of the form $X_i a \subseteq X_j$ for each a-transition from state i to state j. The system we obtain corresponds very closely to the $[\![-]\!]$-semantics of the associated string diagram.

The distinction between red and black wires can be understood as a type discipline that only allows linear uses of the product of languages. It is legitimate and enlightening to ask what would happen if we forgot about red wires and interpreted the action directly as the product. We would replace the action by a new generator  with semantics $[\![$ 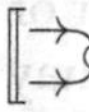 $]\!] = \{((M, L), K) \mid ML \subseteq K\}$.

This would allow us to specify systems of language equations with unrestricted uses of the product on the left of inclusions, e.g. $UVW \subseteq X$. Equations of this form are similar to the production rules (e.g. $X \to UVW$) of context-free grammars and it is well-known that the least solutions of this class of systems are precisely *context-free* languages [14, Chapter 10].

For example we could encode the language $X \to XX \mid (X) \mid \epsilon$ of properly matched parentheses as least solution of the system $\epsilon \subseteq X, (X) \subseteq X, XX \subseteq X$ which gives the diagram displayed on the right.

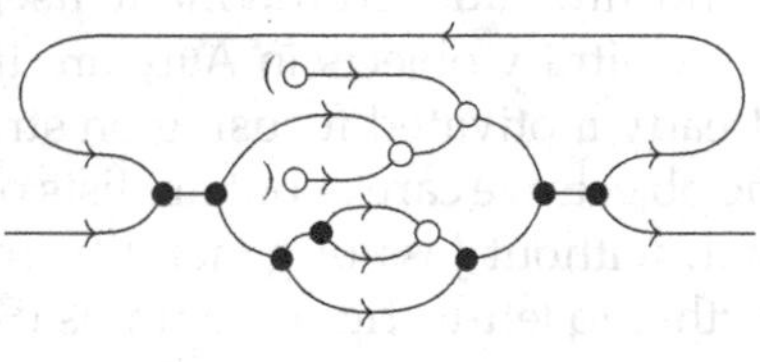

5 Completeness and Determinisation

This section is devoted to prove our completeness result, Theorem 1. We use a normal form argument: more specifically we mimic automata-theoretic results to rewrite every string diagram to a normal form corresponding to a minimal deterministic finite automaton (DFA). We achieve it by implementing Brzozowski's algorithm [12] through diagrammatic equational reasoning. The proof proceeds in three distinct steps.

1. We first show (Section 5.1) how to *determinise* (the representation of) a diagram: this step consists in eliminating all subdiagrams that correspond to nondeterministic transitions in the associated automaton.
2. We use the previous step to implement a *minimisation* procedure (Section 5.2) from which we obtain a minimal representation for a given diagram: this is a representation whose associated automaton is minimal—with the fewest number of states—amongst DFAs that recognise the same language. To do this, we show how the four steps of Brzozowski's minimisation algorithm (reverse; determinise; reverse; determinise) translate into diagrammatic equational reasoning. Note that the first three steps taken together simply amount to applying in reverse the determinisation procedure we have already devised. That this is possible will be a consequence of the symmetry of $=_{KDA}$.
3. Finally, from the uniqueness of minimal DFAs, any two diagrams that have the same denotation are both equal to the same minimal representation and we can derive completeness of $=_{KDA}$.

We will now write equations in $=_{KDA}$ simply as $=$ to simplify notation and say that diagrams c and d are *equal* when $c =_{KDA} d$.

First, we use the symmetries of the equational theory to make simplifying assumptions about the diagrams to consider in the completeness proof.

A few simplifying assumptions. Without loss of generality, the proof we give is restricted to string diagrams with no ▶ in their domain as well as in their codomain. This is simply a matter of convenience: the same proof would work for more general diagrams, that may contain ▶ in their (co)domain, at the cost of significantly cluttering diagrams. Henceforth, one can simply think of the labels for the action $-\boxed{x}-$ as uniquely identifying one open red wire in a diagram. With this convention, two or more occurrences of the same x in a diagram can be seen as connected to the same red wire on the left, via . That we can safely do so is a consequence of the completeness of $=_{KDA}$ restricted to the monochromatic red fragment, itself a consequence of [11, Theorem 6.1].

Arbitrary objects in Aut_Σ are lists of the three generating objects. We have already motivated focusing on string diagrams with no open red wires so that the objects we care about are lists of ▶ and ◀. The following proposition implies that, without loss of generality, for the proof of completeness we can restrict further to left-to-right diagrams (Section 4.2).

Proposition 8. *There is a natural bijection between sets of string diagrams of the form*

where A_i, B_i represent lists of ▶ and ◀.

Proposition 8 tell us that we can always bend the incoming wires to the left and outgoing wires to the right before applying some equations, and recover the original orientation of the wires by bending them into their original place later.

5.1 Determinisation

In diagrammatic terms, a nondeterministic transition of the automaton associated to (a representation of) a given diagram, corresponds to a subdiagram of the form for some $a \in \Sigma$. Clearly, using the definition of := in (6) and the axiom $\overset{(D1)}{=}$, we have = , which will prove to be the engine of our determinisation procedure, along with the fact that any red expression can be copied and deleted. The next two theorems generalise the ability to copy and delete to arbitrary left-to-right diagrams.

Theorem 2. *For any left-to-right diagram* $d : \blacktriangleright^m \to \blacktriangleright^n$*, we have*

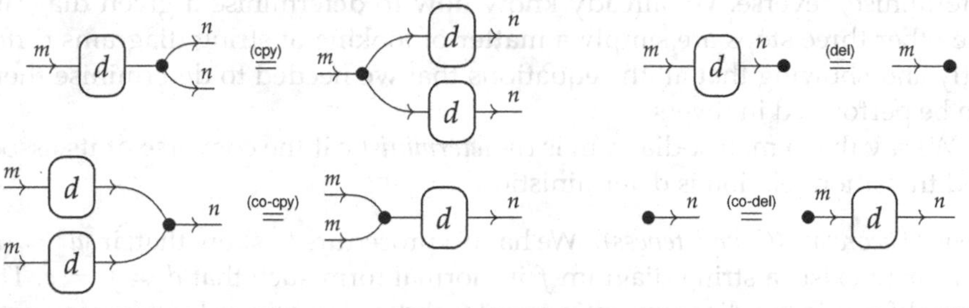

For $d : \blacktriangleright^m \to \blacktriangleright^n$, let d_{ij} be the string diagram of type $\blacktriangleright \to \blacktriangleright$ obtained by composing every input with $\bullet\!\!\rightarrow$ except the ith one, and every output with $\rightarrow\!\!\bullet$ except the jth one. Theorem 2 implies that string diagrams are fully characterised by their $\blacktriangleright \to \blacktriangleright$ subdiagrams.

Corollary 1. *Given* $d, e : \blacktriangleright^m \to \blacktriangleright^n$, $d =_{KDA} e$ *iff* $d_{ij} =_{KDA} e_{ij}$, *for all* $1 \le i \le m$ *and* $1 \le j \le n$.

Thus, we can restrict our focus further to left-to-right $\blacktriangleright \to \blacktriangleright$ diagrams, without loss of generality. We are now able to devise a determinisation procedure for representation of diagrams, which we illustrate below on a simple example.

Proposition 9 (Determinisation). *Any diagram* $\blacktriangleright \to \blacktriangleright$ *has a deterministic representation.*

Example 2.

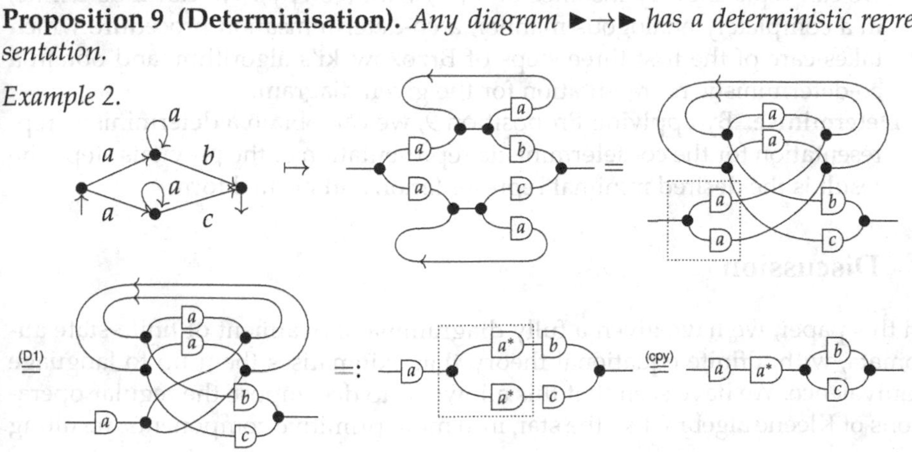

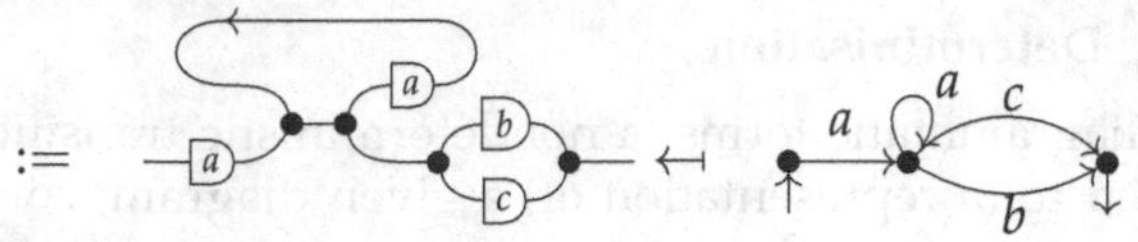

Dealing with useless states. Notice that our deterministic form is *partial* and that the determinisation procedure disregards *useless states*, i.e., parts of a string diagram that do not reach an output wire. None of these contribute to the semantics of the diagram and can be safely eliminated using Theorem 2 (del)-(co-del).

5.2 Minimisation and completeness

As explained above, our proof of completeness is a diagrammatic reformulation of Brzozowski's algorithm which proceeds in four steps: determinise, reverse, determinise, reverse. We already know how to determinise a given diagram. The other three steps are simply a matter of looking at string diagrams differently and showing that all the equations that we needed to determinise them, can be performed in reverse.

We say that a matrix-diagram is *co-deterministic* if the converse of its associated transition relation is deterministic.

Proof (Theorem 1 (Completeness)). We have a procedure to show that, if $[\![d]\!] = [\![e]\!]$, then there exists a string diagram f in normal form such that $d = f = e$. This normal form is the diagrammatic counterpart of the *minimal* automaton associated to d and e. In our setting, it is the deterministic representation equal to d and e with the smallest number of states. This is unique because we can obtain from it the corresponding minimal automaton, which is well-known to be unique. First, given any string diagram we can obtain a representation for it by Proposition 6. Then we obtain a minimal representation by splitting Brzozowski's algorithm in two steps.

1. **Reverse; determinise; reverse.** A close look at the determinisation procedure shows that, at each step, the required laws all hold in reverse. For example, we can replace every instance of (cpy) with (co-cpy). We can thus define, in a completely analogous manner, a co-determinisation procedure which takes care of the first three steps of Brzozowski's algorithm, and obtain a co-deterministic representation for the given diagram.
2. **Determinise.** By applying Proposition 9, we can obtain a deterministic representation for the co-deterministic representation of the previous step. The result is the desired minimal representation and normal form.

6 Discussion

In this paper, we have given a fully diagrammatic treatment of finite-state automata, with a finite equational theory that axiomatises them up to language equivalence. We have seen that this allows us to decompose the regular operations of Kleene algebra, like the star, into more primitive components, resulting in greater modularity. In this section, we compare our contributions with related work, and outline directions for future research.

Traditionally, computer scientists have used *syntax or railroad diagrams* to visualise regular expressions and context-free grammars [41]. These diagrams resemble our very closely but have remained mostly informal More recently, Hinze has treated the single input-output case rigorously as a pedagogical tool to teach the correspondence between finite-state automata and regular expressions [18]. He did not, however, study their equational properties.

Bloom and Ésik's *iteration theories* provide a general categorical setting in which to study the equational properties of iteration for a broad range of structures that appear in programming languages semantics [5]. They are cartesian categories equipped with a parameterised fixed-point operation closely related to the feedback notion we have used to represent the Kleene star. However, the monoidal category of interest in this paper is *compact-closed* (only the full subcategory over ▶ and ◀ to be precise), a property that is incompatible with the existence of categorical products (any category that has both collapses to a preorder [30]). Nevertheless, the subcategory of left-to-right diagrams (Section 4.2) is a (matrix) iteration theory [6], a structure that Bloom and Ésik have used to give an (infinitary) axiomatisation of regular languages [4].

Similarly, Stefanescu's work on *network algebra* provides a unified algebraic treatment of various types of networks, including finite-state automata [39]. In general, network algebras are traced monoidal categories where the product is not necessarily cartesian, and therefore more general than iteration theories. In both settings however, the trace is a global operation, that cannot be decomposed further into simpler components. In our work, on the other hand, the trace can be defined from the compact-closed structure, as was depicted in (3).

Note that the compact closed subcategory in this paper can be recovered from the traced monoidal category of left-to-right diagrams, via the *Int construction* [22]. Therefore, as far as mathematical expressiveness is concerned, the two approaches are equivalent. However, from a methodological point of view, taking the compact closed structure as primitive allows for improved compositionality, as example (2) in the introduction illustrates. Furthermore, the compact closed structure can be finitely presented relative to the theory of symmetric monoidal categories, whereas the trace operation cannot. This matters greatly in this paper, where finding a finite axiomatisation is our main concern.

Finally, the idea of treating regular expressions as a free structure acting on a second algebraic structure also appeared in Pratt's *dynamic algebras*, which axiomatise the propositional fragment of dynamic modal logic [34]. Like our formalism, the variety of dynamic algebras is finitely-based. But they assume more structure: the second algebraic structure is a Boolean algebra.

In all the formalisms we have mentioned, the difficulty typically lies in capturing the behaviour of iteration—whether as the star in Kleene algebra [26,4], or a trace operator [5] in iteration theory and network algebra [39]. The axioms should be coercive enough to force it to be *the least fixed-point* of the language map $L \mapsto \{\epsilon\} \cup LK$. In Kozen's axiomatisation of Kleene algebra [26] for example, this is through (a) the axiom $1+ee^* \leq e^*$ (star is a fixpoint) and (b) the Horn clause $f+ex \leq x \Rightarrow e^*f \leq x$ (star is the least fixpoint). In our work, (a) is a consequence of the unfolding of the star into a feedback loop and can be derived from the other axioms. (b) is more subtle, but can be seen as a consequence of (D1)-(D4) axioms. These allows us to (co)copy and (co)delete arbitrary diagrams (Theorem 2) and we conjecture that this is what forces the star to be a single definite value, not just any fixed-point, but the least one. Making this statement precise is the subject of future work.

The difficulty in capturing the behaviour of fixed-points is also the reason why we decided to work with an additional red wire, to encode the action of regular expressions on the set of languages—without it, global (co)copying and (co)deleting (Theorem 2) cannot be reduced to the local (D1)-(D4) axioms. There is another route, that leads to an infinitary axiomatisation: we could dispense with the red generators altogether and take $-\boxed{a}-$ (for $a \in \Sigma$) as primitive instead, with global axioms to (co)copy and (co)delete arbitrary diagrams. This would pave the way for a reformulation of our work in the context of iteration (matrix) theories, where the ability to (co)copy and (co)delete arbitrary expressions is already built-in. We leave this for future work.

There is an intriguing parallel between our case study and the positive fragment of relation algebra (also known as allegories [16]). Indeed, allegories, like Kleene algebra, do not admit a finite axiomatisation [16]. However, this result holds for standard algebraic theories. It has been shown recently that a structure equivalent to allegories can be given a finite axiomatisation when formulated in terms of string diagrams in monoidal categories [9]. It seems like the greater generality of the monoidal setting—algebraic theories correspond precisely to the particular case of cartesian monoidal categories [11]—allows for simpler axiomatisations in some specific cases. In the future we would like to understand whether this phenomenon, of which now we have two instances, can be understood in a general context.

Lastly, extensions of Kleene Algebra, such as Concurrent Kleene Algebra (CKA) [19,23] and NetKAT [1], are increasingly relevant in current research. Enhancing our theory $=_{KDA}$ to encompass these extensions seems a promising research direction, for two main reasons. First, the two-dimensional nature of string diagrams has been proven particularly suitable to reason about concurrency (see e.g. [7,38]), and more generally about resource exchange between processes (see e.g. [10,13,21,3,8]). Second, when trying to transfer the good meta-theoretical properties of Kleene Algebra (like completeness and decidability) to extensions such as CKA and NetKAT, the cleanest way to proceed is usually in a modular fashion. The interaction between the new operators of the extension and the Kleene star usually represents the greatest challenge to this methodology. Now, in $=_{KDA}$, the Kleene star is decomposable into simpler components (see (3)) and there is only one specific axiom (C5) governing its behaviour. We believe this is a particularly favourable starting point to modularise a meta-theoretic study of CKA and NetKAT with string diagrams, taking advantage of the results we presented in this paper for finite-state automata.

References

1. Anderson, C.J., Foster, N., Guha, A., Jeannin, J.B., Kozen, D., Schlesinger, C., Walker, D.: Netkat: semantic foundations for networks. ACM SIGPLAN Notices **49**(1), 113–126 (2014)
2. Arden, D.N.: Delayed-logic and finite-state machines. In: 2nd Annual Symposium on Switching Circuit Theory and Logical Design (SWCT 1961). pp. 133–151. IEEE (1961)
3. Baez, J.C., Fong, B.: A compositional framework for passive linear networks. Theory & Applications of Categories **33** (2018)
4. Bloom, S.L., Ésik, Z.: Equational axioms for regular sets. Mathematical structures in computer science **3**(1), 1–24 (1993)
5. Bloom, S.L., Ésik, Z.: Iteration theories. Springer (1993)
6. Bloom, S.L., Ésik, Z.: Matrix and matricial iteration theories. Journal of Computer and System Sciences **46**(3), 381–439 (1993)
7. Bonchi, F., Holland, J., Piedeleu, R., Sobociński, P., Zanasi, F.: Diagrammatic algebra: from linear to concurrent systems. In: Proceedings of the 46th Annual ACM SIGPLAN Symposium on Principles of Programming Languages (POPL) (2019)
8. Bonchi, F., Piedeleu, R., Sobociński, P., Zanasi, F.: Graphical affine algebra. In: Proceedings of the 34th Annual ACM/IEEE Symposium on Logic in Computer Science (LICS) (2019)
9. Bonchi, F., Seeber, J., Sobocinski, P.: Graphical conjunctive queries. In: 27th Annual EACSL Conference Computer Science Logic, (CSL). vol. 119 (2018)
10. Bonchi, F., Sobociński, P., Zanasi, F.: The calculus of signal flow diagrams I: linear relations on streams. Information and Computation **252**, 2–29 (2017)
11. Bonchi, F., Sobociński, P., Zanasi, F.: Deconstructing Lawvere with distributive laws. Journal of logical and algebraic methods in programming **95**, 128–146 (2018)
12. Brzozowski, J.A.: Canonical regular expressions and minimal state graphs for definite events. Mathematical theory of Automata **12**(6), 529–561 (1962)
13. Coecke, B., Kissinger, A.: Picturing Quantum Processes - A first course in Quantum Theory and Diagrammatic Reasoning. Cambridge University Press (2017)
14. Conway, J.H.: Regular algebra and finite machines. Courier Corporation (2012)
15. Fong, B., Spivak, D.I.: Seven sketches in compositionality: An invitation to applied category theory. arXiv:1803.05316 (2018)
16. Freyd, P.J., Scedrov, A.: Categories, allegories. Elsevier (1990)
17. Hasegawa, M.: Recursion from cyclic sharing: traced monoidal categories and models of cyclic lambda calculi. In: Proceedings of the Third International Conference on Typed Lambda Calculi and Applications (TLCA). pp. 196–213. Springer (1997)
18. Hinze, R.: Self-certifying railroad diagrams. In: International Conference on Mathematics of Program Construction (MPC). pp. 103–137. Springer (2019)
19. Hoare, C., Möller, B., Struth, G., Wehrman, I.: Concurrent Kleene algebra. In: Proceedings of the 20th International Conference on Concurrency Theory (CONCUR). pp. 399–414. Springer (2009)
20. Hyland, M., Schalk, A.: Glueing and orthogonality for models of linear logic. Theoretical Computer Science **294**(1-2), 183–231 (2003)
21. Jacobs, B., Kissinger, A., Zanasi, F.: Causal inference by string diagram surgery. In: Proceedings of the 22nd International Conference on Foundations of Software Science and Computation Structures (FOSSACS). pp. 313–329. Springer (2019)
22. Joyal, A., Street, R., Verity, D.: Traced monoidal categories. In: Mathematical Proceedings of the Cambridge Philosophical Society. vol. 119, pp. 447–468. Cambridge University Press (1996)

23. Kappé, T., Brunet, P., Silva, A., Zanasi, F.: Concurrent Kleene algebra: Free model and completeness. In: Proceedings of the 27th European Symposium on Programming (ESOP) (2018)
24. Kelly, G.M., Laplaza, M.L.: Coherence for compact closed categories. Journal of Pure and Applied Algebra **19**, 193–213 (1980)
25. Kleene, S.C.: Representation of events in nerve nets and finite automata. Tech. rep., RAND PROJECT AIR FORCE SANTA MONICA CA (1951)
26. Kozen, D.: A completeness theorem for Kleene algebras and the algebra of regular events. Information and Computation **110**(2), 366–390 (1994)
27. Kozen, D.: Kleene algebra with tests. ACM Transactions on Programming Languages and Systems (TOPLAS) **19**(3), 427–443 (1997)
28. Krob, D.: Complete systems of B-rational identities. Theoretical Computer Science **89**(2), 207–343 (1991)
29. Lack, S.: Composing PROPs. Theory and Application of Categories **13**(9), 147–163 (2004)
30. Lambek, J., Scott, P.J.: Introduction to higher-order categorical logic, vol. 7. Cambridge University Press (1988)
31. McCulloch, W.S., Pitts, W.: A logical calculus of the ideas immanent in nervous activity. The bulletin of mathematical biophysics **5**(4), 115–133 (1943)
32. Moshier, A.M.: Coherence for categories of posets with applications. Topology, Algebra and Categories in Logic (TACL) p. 214 (2015)
33. Piedeleu, R., Zanasi, F.: A string diagrammatic axiomatisation of finite-state automata. arXiv:2009.14576 (2020)
34. Pratt, V.: Dynamic algebras as a well-behaved fragment of relation algebras. In: Proceedings of the International Conference on Algebraic Logic and Universal Algebra in Computer Science. pp. 77–110. Springer (1988)
35. Redko, V.N.: On defining relations for the algebra of regular events. Ukrainskii Matematicheskii Zhurnal **16**, 120–126 (1964)
36. Selinger, P.: A survey of graphical languages for monoidal categories. Springer Lecture Notes in Physics **13**(813), 289–355 (2011)
37. Smolka, S., Foster, N., Hsu, J., Kappé, T., Kozen, D., Silva, A.: Guarded Kleene algebra with tests: verification of uninterpreted programs in nearly linear time. Proceedings of the 47th ACM SIGPLAN Symposium on Principles of Programming Languages (POPL) **4**, 1–28 (2020)
38. Sobociński, P., Montanari, U., Melgratti, H., Bruni, R.: Connector algebras for C/E and P/T nets' interactions. Logical Methods in Computer Science **9** (2013)
39. Stefanescu, G.: Network Algebra. Discrete Mathematics and Theoretical Computer Science, Springer London (2000)
40. Thompson, K.: Programming techniques: Regular expression search algorithm. Communications of the ACM **11**(6), 419–422 (1968)
41. Wirth, N.: The programming language pascal. Acta informatica **1**(1), 35–63 (1971)
42. Zanasi, F.: Interacting Hopf Algebras: the theory of linear systems. Ph.D. thesis, Ecole Normale Supérieure de Lyon (2015)

Work-sensitive Dynamic Complexity of Formal Languages*

Jonas Schmidt (✉)[1], Thomas Schwentick[1], Till Tantau[2], Nils Vortmeier[3], and Thomas Zeume[4]

[1] TU Dortmund University, Dortmund, Germany
{jonas2.schmidt,thomas.schwentick}@tu-dortmund.de
[2] Universität zu Lübeck, Lübeck, Germany
tantau@tcs.uni-luebeck.de
[3] University of Zurich, Zurich, Switzerland
nils.vortmeier@uzh.ch
[4] Ruhr University Bochum, Bochum, Germany
thomas.zeume@rub.de

Abstract. Which amount of parallel resources is needed for updating a query result after changing an input? In this work we study the amount of work required for dynamically answering membership and range queries for formal languages in parallel constant time with polynomially many processors. As a prerequisite, we propose a framework for specifying dynamic, parallel, constant-time programs that require small amounts of work. This framework is based on the dynamic descriptive complexity framework by Patnaik and Immerman.

Keywords: Dynamic complexity · work · parallel constant time.

1 Introduction

Which amount of parallel resources is needed for updating a query result after changing an input, in particular if we only want to spend constant parallel time?

In classical, non-dynamic computations, parallel constant time is well understood. Constant time on CRAMs, a variant of CRCW-PRAMs used by Immerman [15], corresponds to constant-depth in circuits, so, to the circuit class AC^0, as well as to expressibility in first-order logic with built-in arithmetic (see, for instance, the books of Immerman [15, Theorem 5.2] and Vollmer [26, Theorems 4.69 and 4.73]). Even more, the amount of work, that is, the overall number of operations of all processors, is connected to the number of variables required by a first-order formula [15, Theorem 5.10].

However, the work aspect of constant parallel time algorithms is less understood for scenarios where the input is subject to changes. To the best of our knowledge, there is only little previous work on constant-time PRAMs in dynamic scenarios. A notable exception is early work showing that spanning trees

* A full version of the paper is available at [21], https://arxiv.org/abs/2101.08735

S. Kiefer and C. Tasson (Eds.): FOSSACS 2021, LNCS 12650, pp. 490–509, 2021.
https://doi.org/10.1007/978-3-030-71995-1_25

and connected components can be computed in constant time by CRCW-PRAMs with $O(n^4)$ and $O(n^2)$ processors, respectively [24].

In an orthogonal line of research, parallel dynamic constant time has been studied from a logical perspective in the dynamic complexity framework by Patnaik and Immerman [20] and Dong, Su, and Topor [7,6]. In this framework, the update of query results after a change is expressed by first-order formulas. The formulas may refer to auxiliary relations, whose updates in turn are also specified by first-order formulas (see Section 3 for more details). The queries maintainable in this fashion constitute the dynamic complexity class DynFO. Such queries can be updated by PRAMs in constant time with a polynomial number of processors. In this line of work, the main focus in recent years has been on proving that queries are in DynFO, and thus emphasised the constant time aspect. It has, for instance, been shown that all context-free languages [11] and the reachability query [5] are in DynFO.

However, if one tries to make the "DynFO approach" for dynamic problems relevant for practical considerations, the work that is needed to carry out the specified updates, hence the *work* of a parallel algorithm implementing them, is a crucial factor. The current general polynomial upper bounds are too coarse. In this paper, we therefore initiate the investigation of more work-efficient dynamic programs that can be specified by first-order logic and that can therefore be carried out by PRAMs in constant time. To do so, we propose a framework for specifying such dynamic, parallel, constant-time programs, which is based on the DynFO framework, but allows for more precise (and better) bounds on the necessary work of a program.

Goal 1.1. *Extend the formal framework of dynamic complexity towards the consideration of parallel work.*

Towards this goal, we link the framework we propose to the CRAM framework in Section 3. In fact, the new framework also takes a somewhat wider perspective, since it does not focus exclusively at one query under a set of change operations, but rather considers dynamic problems that may have several change and query operations (and could even have operations that combine the two). Therefore, from now on we speak about dynamic problems and not about (single) queries.

Goal 1.2. *Find work-efficient* DynFO*-programs for dynamic problems that are known to be in* DynFO *(but whose dynamic programs*[5] *are not competitive, work-wise).*

Ideally we aim at showing that dynamic problems can be maintained in DynFO with sublinear or even polylogarithmic work. One line of attack for this goal is to study dynamic algorithms and to see whether they can be transformed into parallel $\mathcal{O}(1)$-time algorithms with small work. There is a plethora of work

[5] In the field of dynamic complexity the term "dynamic program" is traditionally used for the programs for updating the auxiliary data after a change. The term should not be confused with the "dynamic programming" technique used in algorithm design.

that achieves polylogarithmic sequential update time (even though, sometimes only amortised), see for instance [3,9,12,13]. For many of these problems, it is known that they can be maintained in constant parallel time with polynomial work, e.g. as mentioned above, it has been shown that connectivity and maintenance of regular (and even context-free) languages is in DynFO.

In this paper, we follow this approach for dynamic string problems, more specifically, dynamic problems that allow membership and range queries for regular and context-free languages. Our results can be summarised as follows.

We show in Section 5 that regular languages can be maintained in constant time with $\mathcal{O}(n^\epsilon)$ work for all $\epsilon > 0$ and that for star-free languages even work $\mathcal{O}(\log n)$ can be achieved. These results hold for range and membership queries.

For context-free languages, the situation is not as nice, as we observe in Section 6. We show that subject to a well-known conjecture, we cannot hope for maintaining membership in general context-free languages in DynFO with less than $\mathcal{O}(n^{1.37-\epsilon})$ work. The same statement holds even for the bound $\mathcal{O}(n^{2-\epsilon})$ and "combinatorial dynamic programs". For Dyck languages, that is, sets of well-formed strings of parentheses, we show that this barrier does not apply. Their membership problem can be maintained with $\mathcal{O}(n(\log n)^3)$ work in general, and with polylogarithmic work if there is only one kind of parentheses. By a different approach, range queries can be maintained with work $\mathcal{O}(n^{1+\epsilon})$ in general, and $\mathcal{O}(n^\epsilon)$ for one parenthesis type.

Related work. A complexity theory of incremental time has been developed in [19]. We discuss previous work on dynamic complexity of formal languages in Sections 5 and 6.

2 Preliminaries

Since dynamic programs are based on first-order logic, we represent inputs like graphs and strings as well as "internal" data structures as logical structures.

A *schema* τ consists of a set of relation symbols and function symbols with a corresponding arity. A constant symbol is a function symbol with arity 0. A *structure* $\mathcal{D}$ over schema τ with finite domain D has, for every k-ary relation symbol $R \in \tau$, a relation $R^{\mathcal{D}} \subseteq D^k$, as well as a function $f^{\mathcal{D}} \colon D^k \to D$ for every k-ary function symbol $f \in \tau$. We allow partially defined functions and write $f^{\mathcal{D}}(\bar{a}) = \bot$ if $f^{\mathcal{D}}$ is not defined for $\bar{a}$ in $\mathcal{D}$. Formally, this can be realized using an additional relation that contains the domain of $f^{\mathcal{D}}$. We occasionally also use functions $f^{\mathcal{D}} \colon D^k \to D^\ell$ for some $\ell > 1$. Formally, such a function represents ℓ functions $f_1^{\mathcal{D}}, \dots, f_\ell^{\mathcal{D}} \colon D^k \to D$ with $f^{\mathcal{D}}(\bar{a}) \stackrel{\text{def}}{=} (f_1^{\mathcal{D}}(\bar{a}), \dots, f_\ell^{\mathcal{D}}(\bar{a}))$.

Throughout this work, the structures we consider provide a linear order $\leq$ on their domain D. As we can thus identify D with an initial sequence of the natural numbers, we usually just assume that $D = [n] \stackrel{\text{def}}{=} \{0, \dots, n-1\}$ for some natural number n.

We assume familiarity with first-order logic FO, and refer to [17] for basics of Finite Model Theory. In this paper, unless stated otherwise, first-order formulas *always* have access to a linear order on the domain, as well as compatible functions $+$ and $\times$ that express addition and multiplication, respectively. This holds in particular for formulas in dynamic programs. We use the following "if-then-else" construct: if φ is a formula, and t_1 and t_2 are terms, then $\mathrm{ITE}(\varphi, t_1, t_2)$ is a term. Such a term evaluates to the result of t_1 if φ is satisfied, otherwise to t_2.

Following [11], we encode words of length (at most) n over an alphabet Σ by *word structures*, that is, as relational structures W with universe $\{0, \dots, n-1\}$, one unary relation R_σ for each symbol $\sigma \in \Sigma$ and the canonical linear order $\leq$ on $\{0, \dots, n-1\}$. We only consider structures for which, for every position i, $R_\sigma(i)$ holds for at most one $\sigma \in \Sigma$ and write $W(i) = \sigma$ if $R_\sigma(i)$ holds and $W(i) = \epsilon$ if no such σ exists. We write $\mathrm{word}(W)$ for the word represented by W, that is, the concatenation $w = W(0) \circ \ldots \circ W(n-1)$. As an example, the word structure W_0 with domain $\{0, 1, 2, 3\}$, $W(1) = a$, $W(3) = b$ and $W(0) = W(2) = \epsilon$ represents the string ab. We write $\mathrm{word}(W)[\ell, r]$ for the word $W(\ell) \circ \ldots \circ W(r)$.

Informally, a *dynamic problem* can be seen as a data type: it consists of some underlying structure together with a set Δ of operations. We distinguish between *change operations* that can modify the structure and *query operations* that yield information about the structure, but combined operations could be allowed, as well. Thus, a dynamic problem is characterised by the schema of its underlying structures and the operations that it supports.[6]

In this paper, we are particularly interested in dynamic language problems, defined as follows. Words are represented as word structures W with elementary change operations $\textsc{set}_\sigma(i)$ (with the effect that $W(i)$ becomes σ if it was ϵ before) and $\textsc{reset}(i)$ (with the effect that $W(i)$ becomes ϵ).

For some fixed language L over some alphabet Σ, the dynamic problem $\textsc{RangeMember}(L)$ further supports one query operation $\textsc{range}(\ell, r)$. It yields the result true, if $\mathrm{word}(W)[\ell, r]$ is in L, and otherwise false.

In the following, we denote a word structure W as a sequence $w_0 \ldots w_{n-1}$ of letters with $w_i \in \Sigma \cup \{\epsilon\}$ in order to have an easier, less formal notation. Altogether, the dynamic problem $\textsc{RangeMember}(L)$ is defined as follows.

Problem: $\textsc{RangeMember}(L)$
Input: A sequence $w = w_0 \ldots w_{n-1}$ of letters with $w_i \in \Sigma \cup \{\epsilon\}$
Changes: $\textsc{set}_\sigma(i)$ for $\sigma \in \Sigma$: Sets w_i to σ, if $w_i = \epsilon$
$\textsc{reset}(i)$: Sets w_i to ϵ
Queries: $\textsc{range}(\ell, r)$: Is $w_\ell \circ \cdots \circ w_r \in L$?

In this example, the query $\textsc{range}$ maps (binary) pairs of domain elements to a truth value and thus defines a (binary) relation over the universe of the input word structure. We call such a query *relational.* We will also consider *functional* queries mapping tuples of elements to elements.

Another dynamic problem considered here is $\textsc{Member}(L)$ which is defined similarly as $\textsc{RangeMember}(L)$ but instead of $\textsc{range}$ only has the Boolean query operation $\textsc{member}$ that yields true if $w_0 \circ \ldots \circ w_{n-1} \in L$ holds.

[6] This view is a bit broader than the traditional setting of Dynamic Complexity, where there can be various change operations but usually only one fixed query is supported.

3 Work-sensitive Dynamic Complexity

Since we are interested in the work that a dynamic program does, our specification mechanism for dynamic programs is considerably more elaborated than the one used in previous papers on dynamic complexity. We introduce the mechanism in this section in two steps. First the general form of dynamic programs and then a more pseudo-code oriented syntax. Afterwards, we discuss how these dynamic programs translate into work-efficient constant-time parallel programs.

3.1 The Dynamic Complexity Framework

Our general form of dynamic programs mainly follows [23], but is adapted to the slightly broader view of a dynamic problem as a data type. For a more gentle introduction to dynamic complexity, we refer to [22].

The goal of a *dynamic program* for a dynamic problem Π is to support all its operations Δ. To do so, it stores and updates an auxiliary structure $\mathcal{A}$ over some schema τ_{aux}, over the same domain as the input structure $\mathcal{I}$ for Π.

A (first-order) dynamic program $\mathcal{P}$ consists of a set of (first-order) *update rules* for change operations and *query rules* for query operations. More precisely, a program has one query rule over schema τ_{aux} per query operation that specifies how the (relational) result of that operation is obtained from the auxiliary structure. Furthermore, for each change operation $\delta \in \Delta$, it has one update rule per auxiliary relation or function that specifies the updates after a change based on δ.

A query rule is of the form **on query** $Q(\bar{p})$ **yield** $\varphi_Q(\bar{p})$, where φ_Q is the (first-order) *query formula* with free variables from $\bar{p}$.

An update rule for a k-ary auxiliary relation R is of the form

on change $\delta(\bar{p})$ **update** R **at** $(t_1(\bar{p};\bar{x}),\ldots,t_k(\bar{p};\bar{x}))$ **as** $\varphi^R_\delta(\bar{p};\bar{x})$ **where** $C(\bar{x})$.

Here, φ^R_δ is the (first-order) *update formula*, $t_1,\ldots,t_k$ are first-order terms (possibly using the ITE construct) over τ_{aux}, and $C(\bar{x})$, called a *constraint* for the tuple $\bar{x} = x_1,\ldots,x_\ell$ of variables, is a conjunction of inequalities $x_i \leq f_i(n)$ using functions $f_i\colon \mathbb{N} \to \mathbb{N}$, where n is the size of the domain and $1 \leq i \leq \ell$. We demand that all functions f_i are first-order definable from $+$ and $\times$.

The effect of such an update rule after a change operation $\delta(\bar{a})$ is as follows: the new relation $R^{\mathcal{A}'}$ in the updated auxiliary structure $\mathcal{A}'$ contains all tuples from $R^{\mathcal{A}}$ that are *not* equal to $(t_1(\bar{a};\bar{b}),\ldots,t_k(\bar{a};\bar{b}))$ for any tuple $\bar{b}$ that satisfies the constraints C; and additionally $R^{\mathcal{A}'}$ contains all tuples $(t_1(\bar{a};\bar{b}),\ldots,t_k(\bar{a};\bar{b}))$ such that $\bar{b}$ satisfies C and $\mathcal{A} \models \varphi^R_\delta(\bar{a};\bar{b})$ holds.

Phrased more operationally, an update is performed by enumerating all tuples $\bar{b}$ that satisfy C, evaluating $\varphi^R_\delta(\bar{a};\bar{b})$ on the old auxiliary structure $\mathcal{A}$, and depending on the result adding the tuple $(t_1(\bar{a};\bar{b}),\ldots,t_k(\bar{a};\bar{b}))$ to R (if it was not already present), or removing that tuple from R (if it was present).

Update rules for auxiliary functions are similar, but instead of an update formula that decides whether a tuple of the form $(t_1(\bar{a};\bar{b}),\ldots,t_k(\bar{a};\bar{b}))$ is contained in the updated relation, it features an update term that determines the new function value for a function argument of the form $(t_1(\bar{a};\bar{b}),\ldots,t_k(\bar{a};\bar{b}))$.

We say that $\mathcal{P}$ is a dynamic program for a dynamic problem Π if it supports all its operations and, in particular, always yields correct results for query operations. More precisely, if the result of applying a query operation after a sequence α of change operations on an initial structure $\mathcal{I}_0$ yields the same result as the evaluation of the query rule on the auxiliary structure that is obtained by applying the update rules corresponding to the change operations in α to an initial auxiliary structure $\mathcal{A}_0$. Here, an initial input structure $\mathcal{I}_0$ over some domain D is *empty*, that is, it is a structure with empty relations and with all function values being undefined ($\perp$). The initial auxiliary structure $\mathcal{A}_0$ is over the same domain D as $\mathcal{I}_0$ and is defined from $\mathcal{I}_0$ by some FO-definable initialization.

By DynFO, we denote the class of all dynamic problems that have a dynamic program in the sense we just defined.

3.2 A syntax for work-efficient dynamic programs

In this paper we are particularly interested in dynamic programs that require little work to update the auxiliary structure after every change operation and to compute the result of a query operation. However, since dynamic programs do not come with an execution model, there is no direct way to define, say, when a DynFO-programs has polylogarithmic-work, syntactically.

We follow a pragmatic approach here. We define a pseudo-code-based syntax for *update* and *query procedures* that will be used in place of the update and query *formulas* in rules of dynamic programs. This syntax has three important properties: (1) it is reasonably well readable (as opposed to strict first-order logic formulas), (2) it allows a straightforward translation of rules into proper DynFO-programs, and (3) it allows to associate a "work-bounding function" to each rule and to translate it into a PRAM program with $\mathcal{O}(1)$ parallel time and work bounded by this function.

The syntax of the pseudo-code has similarities with Abstract State Machines [4] and the PRAM-syntax of [16]. For simplicity, we describe a minimal set of syntactic elements that suffice for the dynamic programs in this paper. We encourage readers to have a look at Section 4 for examples of update rules with pseudo-code syntax.

We only spell out a syntax for *update procedures* that can be used in place of the update formula $\varphi_\delta^R(\bar{p};\bar{x})$ of an update rule

on change $\delta(\bar{p})$ **update** R **at** $(t_1(\bar{p};\bar{x}),\ldots,t_k(\bar{p};\bar{x}))$ **as** $\varphi_\delta^R(\bar{p};\bar{x})$ **where** $C(\bar{x})$.

Query procedures are defined similarly, but they can not invoke any change operations for supplementary instances, and their only free variables are from $\bar{p}$.

We allow some compositionality: a dynamic program on some *main instance* can use *supplementary instances* of other dynamic problems and invoke change or query operations of other dynamic programs on those instances. These supplementary instances are declared on a global level of the dynamic program and each has an associated identifier.

Update procedures $P = P_1; P_2$ consist of two parts. In the *initial procedure* P_1 no reference to the free variables from $\bar{x}$ are allowed, but change operations for supplementary instances can be invoked. We require that, for each change operation δ of the main instance and each supplementary instance $\mathcal{S}$, at most one update rule for δ invokes change operations for $\mathcal{S}$.

In the *main procedure* P_2, no change operations for supplementary instances can be invoked, but references to $\bar{x}$ are allowed.

More precisely, both P_1 and P_2 can use (a series of) instructions of the following forms:

– assignments $f(\bar{y}) \leftarrow$ `term` of a function value,
– assignments $R(\bar{y}) \leftarrow$ `condition` of a Boolean value,
– conditional branches **if** `condition` **then** P' **else** P'', and
– parallel branches **for** $z \leq g(n)$ **pardo** P'.

Semantically, here and in the following n always refers to the size of the domain of the main instance. The initial procedure P_1 can further use change invocations `instance.`$\delta(\bar{y})$. However, they are not allowed in the scope of parallel branches. And we recall that in P_1 no variables from $\bar{x}$ can be used.

The main procedure P_2 can further use return statements **return** `condition` or **return** `term`, but not inside parallel branches.

Of course, initial procedures can only have initial procedures P' and P'' in conditional and parallel branches, and analogously for main procedures.

Conditions and terms are defined as follows. In all cases, $\bar{y}$ denotes a tuple of terms and z is a *local variable*, not occurring in $\bar{p}$ or $\bar{x}$. In general, a *term* evaluates to a domain element (or to $\perp$). It is built from

– local variables and variables from $\bar{p}$ and $\bar{x}$,
– function symbols from τ_{aux} and previous function assignments,
– if-then-else terms **if** `condition` **then** `term`$'$ **else** `term`$''$,
– functional queries `instance.`$Q(\bar{y})$, and
– expressions **getUnique**$(z \leq g(n) \mid$ `condition`$)$.

For the latter expression it is required that there is always exactly one domain element $a \leq g(n)$ satisfying `condition`.

A *condition* evaluates to `true` or `false`. It may be

– an atomic formula with relation symbols from τ_{aux} or previous assignments, with terms as above,
– an expression **exists**$(z \leq g(n) \mid$ `condition`$)$,
– a relational query `instance.`$Q(\bar{y})$ with terms $\bar{y}$, and
– a Boolean combination of conditions.

All functions $g\colon \mathbb{N} \rightarrow \mathbb{N}$ in these definitions are required to be FO-definable. For assignments of relations R and functions f we demand that these symbols do *not* appear in τ_{aux}. If an assignment with a head $f(\bar{y})$ or $R(\bar{y})$ occurs in the scope of a parallel branch that binds variable z, then z has to occur as a term y_i

in $\bar{y}$. We further demand that update procedures are well-formed, in the sense that every execution path ends with a return statement of appropriate type.

In our pseudo-code algorithms, we display update procedures $P = P_1; P_2$ with initial procedure P_1 and main procedure P_2 as

on change $\delta(\bar{p})$ **with** P_1
  update R **at** $(t_1(\bar{p}, \bar{x}), \ldots, t_k(\bar{p}, \bar{x}))$, **for all** $C(\bar{x})$, **by:** P_2.

to emphasise that P_1 only needs to be evaluated once for the update of R, and not once for every different value of $\bar{x}$.

In a nutshell, the semantics of an update rule

on change $\delta(\bar{p})$ **update** R **at** $(t_1(\bar{p}; \bar{x}), \ldots, t_k(\bar{p}; \bar{x}))$ **as** P **where** $C(\bar{x})$

is defined as in Subsection 3.1, but $\mathcal{A} \models \varphi_\delta^R(\bar{a}, \bar{b})$ has to be replaced by the condition that P returns true under the assignment $(\bar{p} \mapsto \bar{a}; \bar{x} \mapsto \bar{b})$.

For update rules for auxiliary functions, P returns the new function value instead of a Boolean value.

Since P_1 is independent of $\bar{x}$, in the semantics, it is only evaluated once. In particular, any change invocations are triggered only once.

With Procedural-DynFO-programs we refer to the above class of dynamic update programs. Here and later we will introduce abbreviations as syntactic sugar, for example the sequential loop **for** $z \leq m$ **do** P, where $m \in \mathbb{N}$ needs to be a fixed natural number.

We show next that update and query procedures can be translated into constant-time CRAM programs. Since the latter can be translated into FO-formulas [14, Theorem 5.2], therefore Procedural-DynFO-programs can be translated in DynFO-programs.

3.3 Implementing Procedural-DynFO-programs as PRAMs

We use *Parallel Random Access Machines* (PRAMs) as the computational model to measure the work of our dynamic programs. A PRAM consists of a number of processors that work in parallel and use a shared memory. We only consider *CRAMs*, a special case of Concurrent-Read Concurrent-Write model (CRCW PRAM), i.e. processors are allowed to read and write concurrently from and to the same memory location, but if multiple processors concurrently write the same memory location, then all of them need to write the same value. For an input of size n we denote the *time* that a PRAM algorithm needs to compute the solution as $T(n)$. The *work* $W(n)$ of a PRAM algorithm is the sum of the number of all computation steps of all processors made during the computation. For further details we refer to [14,16].

It is easy to see that Procedural-DynFO programs $\mathcal{P}$ can be translated into $\mathcal{O}(1)$-time CRAM-programs $\mathcal{C}$. To be able to make a statement about (an upper bound of) the work of $\mathcal{C}$, in the full version of this paper we associate a function w with update rules and show that every update rule π can be implemented by a $\mathcal{O}(1)$-time CRAM-program with work $\mathcal{O}(w)$. Likewise for query rules.

In a nutshell, the work of an update procedure mainly depends on the scopes of the (nested) parallel branches and the amount of work needed to query and update the supplementary instances. The work of a whole update rule is then determined by adding the work of the initial procedure once and adding the work of the main procedure for each tuple that satisfies the constraint of the update rule.

4 A simple work-efficient Dynamic Program

In this section we consider a simple dynamic problem with a fairly work-efficient dynamic program. It serves as an example for our framework but will also be used as a subroutine in later sections.

The dynamic problem is to maintain a subset K of an ordered set D of elements under insertion and removal of elements in K, allowing for navigation from an element of D to the next larger and smaller element in K. That is, we consider the following dynamic problem:

Problem: NEXTINK
Input: A set $K \subseteq D$ with canonical linear order $\leq$ on D
Changes: INS(i): Inserts $i \in D$ into K
DEL(i): Deletes $i \in D$ from K
Queries: PRED(i): Returns predecessor of i in K, i.e. $\max\{j \in K \mid i > j\}$
SUCC(i): Returns successor of i in K, i.e. $\min\{j \in K \mid i < j\}$

For the smallest (largest) element the result of a PRED (SUCC) query is undefined, i.e. $\perp$. For simplicity, we assume in the following that D is always of the form $[n]$, for some $n \in \mathbb{N}$.

Sequentially, the changes and queries of NEXTINK can be handled in sequential time $\mathcal{O}(\log \log n)$ [9]. Here we show that the problem also has a dynamic program with parallel time $\mathcal{O}(1)$ and work $\mathcal{O}(\log n)$.

Lemma 4.1. *There is a* DynFO*-program for* NEXTINK with $\mathcal{O}(\log n)$ work per change and query operation.

Proof. The dynamic program uses an ordered binary balanced tree T with leave set $[n]$, and with 0 as its leftmost leaf. Each inner node v represents the interval $S(v)$ of numbers labelling the leafs of the subtree of v. To traverse the tree, the dynamic program uses functions 1st and 2nd that map an inner node to its first or second child, respectively, and a function $\text{anc}(v, j)$ that returns the j-th ancestor of v in the tree.[7] So, $\text{anc}(v, 2)$ returns the parent of the parent of v.

The functions 1st, 2nd and anc are static, that is, they are initialized beforehand and not affected by change operations.

[7] Formally, the $2|D|$ nodes of T can be represented by pairs (a, b) of elements. In our presentation, we disregard these technical issues and use nodes of T just as domain elements.

Algorithm 1 Querying a successor.

1: **on query** $\textsc{succ}(i)$:
2: **if** $\max(T.\text{root}) \leq i$ **then**
3: **return** $\perp$
4: **else**
5: $k \leftarrow$ **getUnique**$(1 \leq k \leq \log(n) \mid \max(T.\text{anc}(i,k)) > i)$
6: $\wedge \max(T.\text{anc}(i, k-1)) \leq i$
7: **return** $\min(T.\text{2nd}(T.\text{anc}(i,k)))$

Algorithm 2 Updating min after an insertion.

1: **on change** $\textsc{ins}(i)$ **update** min **at** $T.\text{anc}(i,k)$, **for all** $k \leq \log n$, **by:**
2: $v \leftarrow T.\text{anc}(i,k)$
3: **if** $\min(v) > i$ **then**
4: **return** i
5: **else**
6: **return** $\min(v)$

The idea of the dynamic program is to maintain, for each node v, the maximal and minimal element in $K \cap S(v)$ (which is undefined if $K \cap S(v) = \emptyset$), by maintaining two functions min and max. It is easy to see that this information can be updated and queries be answered in $\mathcal{O}(\log n)$ time as the tree has depth $\mathcal{O}(\log n)$. For achieving $\mathcal{O}(\log n)$ work and constant time, we need to have a closer look.

Using min and max, it is easy to determine the K-successor of an element $i \in D$: if v is the lowest ancestor of i with $\max(v) > i$, then the K-successor of i is $\min(w)$ for the second child $w \stackrel{\text{def}}{=} \text{2nd}(v)$ of v. Algorithm 1 shows a query rule for the query operation $\textsc{succ}(i)$. The update of these functions is easy when an element i is inserted into K. This is spelled out for min in Algorithm 2. The dynamic program only needs to check if the new element becomes the minimal element in $S(v)$, for every node v that is an ancestor of the leaf i.

Algorithm 3 shows how min can be updated if an element i is deleted from K: if i is the minimal element of K in $S(v)$, for some node v, then $\min(v)$ needs to be replaced by its K-successor, assuming it is in $S(v)$.

It is easy to verify the claimed work upper bounds for $\mathcal{P}$. Querying a successor or predecessor via Algorithm 1 needs $\mathcal{O}(\log n)$ work, since Line 6 requires $\mathcal{O}(\log n)$ and all others require $\mathcal{O}(1)$ work. For maintaining the function min the programs in Algorithms 2 and 3 update the value of $\log n$ tuples, but the work per tuple is constant. In the case of a deletion, Line 3 requires $\mathcal{O}(\log n)$ work but is executed only once. The remaining part consists of $\mathcal{O}(\log n)$ parallel executions of statements, each with $\mathcal{O}(1)$ work.

The handling of max and its work analysis is analogous. □

Algorithm 3 Updating min after a deletion.

```
on change DEL(i)
    with
        s ← SUCC(i)
    update min at T.anc(i, k), for all k ≤ log n, by:
        v ← T.anc(i, k)
        if min(v) ≠ i then
            return min(v)
        else if max(v) = i then
            return ⊥
        else
            return s
```

5 Regular Languages

In this section, we show that the range problem can be maintained with $o(n)$ work for all regular languages and with polylogarithmic work for star-free languages. For the former we show how to reduce the work of a known DynFO-program. For the latter we translate the idea of [9] for maintaining the range problem for star-free languages in $\mathcal{O}(\log \log n)$ sequential time into a dynamic program with $\mathcal{O}(1)$ parallel time.

5.1 DynFO-programs with sublinear work for regular languages

Theorem 5.1. *Let L be a regular language. Then* RANGEMEMBER(L) *can be maintained in DynFO with work $\mathcal{O}(n^\epsilon)$ per query and change operation, for every $\epsilon > 0$.*

The proof of this theorem makes use of the algebraic view of regular languages. For readers not familiar with this view, the basic idea is as follows: for a fixed DFA $\mathcal{A} = (Q, \Sigma, \delta, q_0, F)$, we first associate with each string w a function f_w on Q that is induced by the behaviour of $\mathcal{A}$ on w via $f_w(q) \stackrel{\text{def}}{=} \delta^*(q, w)$, where δ^* is the extension of the transition function δ to strings. The set of all functions $f\colon Q \to Q$ with composition as binary operation is a *monoid*, that is, a structure with an associative binary operation $\circ$ and a neutral element, the identity function. Thus, composing the effect of $\mathcal{A}$ on subsequent substrings of a string corresponds to multiplication of the monoid elements associated with these substrings. The *syntactic monoid* $M(L)$ of a regular language L is basically the monoid associated with its minimal automaton.

It is thus clear that, for the dynamic problem RANGEMEMBER(L) where L is regular, a dynamic program can be easily obtained from a dynamic program for the dynamic problem RANGEEVAL($M(L)$), where RANGEEVAL(M), for finite monoids M, is defined as follows.[8]

[8] We note that, unlike for words, each position always carries a monoid element. However, the empty string of the word case corresponds to the neutral element in

Problem: RANGEEVAL(M)
Input: A sequence $m_0 \dots m_{n-1}$ of monoid elements $m_i \in M$
Changes: $\text{SET}_m(i)$ for $m \in M$: Replaces m_i by m
Queries: $\text{RANGE}(\ell, r)$: $m_\ell \circ \dots \circ m_r$

For the proof of Theorem 5.1 we do not need any insights into monoid theory. However, when studying languages definable by first-order formulas in Theorem 5.3 below, we will make use of a known decomposition result.

From the discussion above it is now clear that in order to prove Theorem 5.1, it suffices to prove the following result.

Proposition 5.2. *Let M be a finite monoid. For every $\epsilon > 0$,* RANGEEVAL(M) *can be maintained in DynFO with work $\mathcal{O}(n^\epsilon)$ per query and change operation.*

Proof sketch. In [11], it was (implicitly) shown that RANGEMEMBER(L) is in DynProp (that is, quantifier-free DynFO), for regular languages L. The idea was to maintain the effect of a DFA for L on $w[\ell, r]$, for each interval (ℓ, r) of positions. This approach can be easily used for RANGEEVAL(M) as well, but it requires a quadratic number of updates after a change operation, in the worst case.

We adapt this approach and only store the effect of the DFA for $\mathcal{O}(n^\epsilon)$ intervals, by considering a hierarchy of intervals of bounded depth.

The first level in the hierarchy of intervals is obtained by decomposing the input sequence into intervals of length t, for a carefully chosen t. We call these intervals *base intervals* of height 1 and their subintervals *special intervals* of height 1. The latter are *special* in the sense that they are exactly the intervals for which the dynamic program maintaines the product of monoid elements. In particular, each base interval of height 1 gives rise to $\mathcal{O}(t^2)$ special intervals of height 1. The second level of the hierarchy is obtained by decomposing the sequence of base intervals of height 1 into sequences of length t. Each such sequence of length t is combined to one base interval of height 2; and each contiguous subsequence of such a sequence is combined to one special interval of height 2. Again, each base interval of height 2 gives rise to $\mathcal{O}(t^2)$ special intervals of height 2. This process is continued recursively for the higher levels of the hierarchy, until only one base interval of height h remains. We refer to Figure 1 for an illustration of this construction.

The splitting factor t is chosen in dependence of n and ϵ such that the height of this hierarchy of special intervals only depends on ϵ and is thus constant for all n. More precisely, we fix $\lambda \stackrel{\text{def}}{=} \frac{\epsilon}{2}$ and $t \stackrel{\text{def}}{=} n^\lambda$. Therefore, $h = \log_t(n) = \frac{1}{\lambda}$.

The idea for the dynamic program is to store the product of monoid elements for each special interval. The two crucial observations are then, that (1) the product of each (not necessary special) interval can be computed with the help of a constant number of special intervals, and (2) that each change operation affects at most t^2 special intervals per level of the hierarchy and thus at most $ht^2 \in \mathcal{O}(n^\epsilon)$ special intervals in total. We refer to the full version for more details. □

the monoid case. In particular, the initial "empty" sequence consists of n copies of the neutral element.

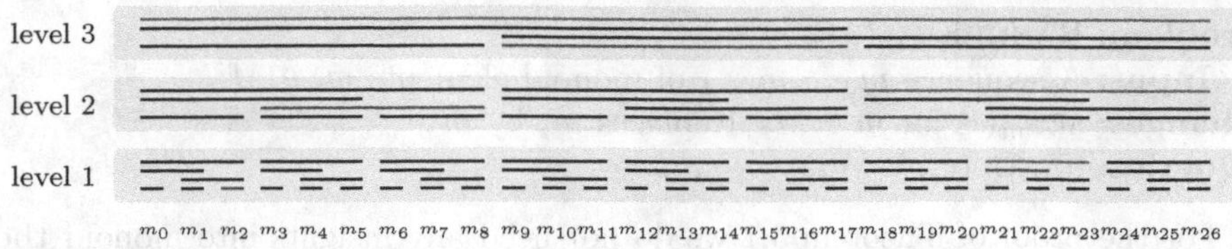

Fig. 1. Illustration of special intervals, for $t = 3$. The special intervals of level 3 are $[0, 9), [9, 18), [18, 27), [0, 18)$ and $[9, 27)$ with base interval $[0, 27)$. The result of a query RANGE(2, 22) can be computed as $\prod_{i=2}^{22} m_i = (m[2, 3) \circ m[3, 9)) \circ m[9, 18) \circ (m[18, 21) \circ m[21, 23))$, illustrated above in blue. The affected base intervals for a change at position 23 are marked in red. E.g., the new product $m'[18, 27)$ can be computed by $m'[18, 27) = m[18, 21) \circ m'[21, 24) \circ m[24, 27)$. As the products are recomputed bottom up, $m'[21, 24)$ is already updated.

5.2 DynFO-programs with polylogarithmic work for star-free languages

Although the work bound of Theorem 5.1 for regular languages is strongly sublinear, one might aim for an even more work-efficient dynamic program, especially, since RANGEMEMBER(L) can be maintained *sequentially* with logarithmic update time for regular languages [9]. We leave it as an open problem whether for every regular language L there is a DynFO-program for RANGEMEMBER(L) with a polylogarithmic work bound. However, we show next that such programs exist for star-free regular languages, in fact they even have a logarithmic work bound. The star-free languages are those that can be expressed by regular expressions that do not use the Kleene star operator but can use complementation.

Theorem 5.3. *Let L be a star-free regular language. Then* RANGEMEMBER(L) *can be maintained in* DynFO *with work $\mathcal{O}(\log n)$ per query and change operation.*

It is well-known that star-free regular languages are just the regular languages that can be defined in first-order logic (without arithmetic!) [18]. Readers might ask why we consider dynamic first-order maintainability of a problem that can actually be *expressed* in first-order logic. The key point is the parallel work here: even though the membership problem for star-free languages can be solved by a parallel algorithm in time $\mathcal{O}(1)$, it inherently requires parallel work $\Omega(n)$.

Proof sketch. The proof uses the well-known connection between star-free languages and group-free monoids (see, e.g., [25, Chapter V.3] and [25, Theorem V.3.2]). It thus follows the approach of [9].

In a nutshell, our dynamic program simply implements the algorithms of the proof of Theorem 2.4.2 in [9]. Those algorithms consist of a constantly bounded number of simple operations and a constantly bounded number of searches for a next neighbour in a set. Since the latter can be done in DynFO with work $\mathcal{O}(\log n)$ thanks to Lemma 4.1, we get the desired result for group-free monoids and then for star-free languages. We refer to the full version for more details. □

6 Context Free Languages

As we have seen in Section 5, range queries to regular languages can be maintained in DynFO with strongly sublinear work. An immediate question is whether context-free languages are equally well-behaved. Already the initial paper by Patnaik and Immerman showed that DynFO can maintain the membership problem for *Dyck languages* D_k, for $k \geq 1$, that is, the languages of well-balanced parentheses expressions with k types of parentheses [20]. It was shown afterwards in [11, Theorem 4.1] that DynFO actually captures the membership problem for all context-free languages and that Dyck languages even do not require quantifiers in formulas (but functions in the auxiliary structure) [11, Proposition 4.4]. These results can easily be seen to apply to range queries as well. However, the dynamic program of [11, Theorem 4.1] uses 4-ary relations and three nested existential quantifiers, yielding work in the order of n^7.

In the following, we show that the membership problem for context-free languages is likely *not* solvable in DynFO with sublinear work, but that the Dyck language D_1 with one bracket type can be handled with polylogarithmic work for the membership problem and work $\mathcal{O}(n^\epsilon)$ for the range problem, and that for other Dyck languages these bounds hold with an additional linear factor n.

6.1 A conditional lower bound for context-free languages

Our conditional lower bound for context-free languages is based on a result from Abboud et al. [2] and the simple observation that the word problem for a language L can be solved, given a dynamic program for its membership problem.

Lemma 6.1. *Let L be a language. If* MEMBER(L) *can be maintained in* DynFO *with work $f(n)$, then the word problem for L can be decided sequentially in time $\mathcal{O}(n \cdot f(n))$.*

The announced lower bound is relative to the following conjecture [1].

Conjecture 6.2 (k-Clique conjecture). For any $\epsilon > 0$, and $k \geq 3$, k-Clique has no algorithm with time bound $\mathcal{O}(n^{(1-\epsilon)\frac{\omega}{3}k})$.

Here, ω is the matrix multiplication exponent [10,27], which is known to be smaller than 2.373 and believed to be exactly two [10,27].

In [2], the word problem for context-free languages was linked to the k-Clique problem as follows.

Theorem 6.3 ([2, Theorem 1.1]). *There is a context free grammar G such that, if the word problem for $L(G)$ can be solved in time $T(n)$, k-Clique can be solved on n node graphs in $\mathcal{O}(T(n^{\frac{k}{3}+1}))$ time, for any $k \geq 3$.*

Putting Lemma 6.1 and Theorem 6.3 together, we get the following result.

Theorem 6.4. *There is a context free grammar G such that, if the membership problem for $L(G)$ can be solved by a* DynFO*-program with work $\mathcal{O}(n^{\omega-1-\epsilon})$, for some $\epsilon > 0$, then the k-Clique conjecture fails.*

The simple proofs of Lemma 6.1 and Theorem 6.4 are presented in the full version.

Thus, we can not reasonably expect any DynFO-programs for general context-free languages with considerable less work than $\mathcal{O}(n^{1.37})$ barring any breakthroughs for matrix multiplication. In fact, for "combinatorial DynFO-programs", an analogous reasoning yields a work lower bound of $\mathcal{O}(n^{2-\epsilon})$.

6.2 On work-efficient dynamic programs for Dyck languages

We next turn to Dyck languages. Clearly, all Dyck languages are deterministic context-free, their word problem can therefore be solved in linear time, and thus the lower bound approach of the previous subsection does not work for them. We first present the DynFO-program with polylogarithmic work for the membership problem of D_1. It basically mimics the sequential algorithm from [8] that maintains D_1 sequentially in time $\mathcal{O}(\log n)$, per change and query operation.

Theorem 6.5. MEMBER(D_1) *can be maintained in* DynFO *with* $\mathcal{O}((\log n)^3)$ *work.*

Proof sketch. Let $\Sigma_1 = \{\langle, \rangle\}$ be the alphabet underlying D_1. The dynamic program uses an ordered binary tree T such that each leaf corresponds to one position from left-to right. A parent node corresponds to the set of positions of its children. We assume for simplicity that the domain is $[n]$, for some number n that is a power of 2. In a nutshell, the program maintains for each node x of T the numbers $\ell(x)$ and $r(x)$ that represent the number of unmatched closing and unmatched opening brackets of the string $\mathrm{str}(x)$ corresponding to x via the leaves of the induced subtree at x. E.g., if that string is $\rangle\langle\rangle\rangle\langle\langle\rangle$ for x, then $\ell(x) = 2$ and $r(x) = 1$. The overall string w is in D_1 exactly if $r(\mathrm{root}) = \ell(\mathrm{root}) = 0$.

In the algorithm of [8], the functions ℓ and r are updated in a bottom-up fashion. However, we will observe that they do not need to be updated sequentially in that fashion, but can be updated in parallel constant time. In the following, we describe how $\mathcal{P}$ can update $\ell(x)$ and $r(x)$ for all ancestor nodes x of a position p, after a closing parenthesis $\rangle$ was inserted at p. Maintaining ℓ and r for the other change operations is analogous.

There are two types of effects that an insertion of a closing parenthesis could have on x: either $\ell(x)$ is increased by one and $r(x)$ remains unchanged, or $r(x)$ is decreased by one and $\ell(x)$ remains unchanged. We denote these effects by the pairs $(+1, 0)$ and $(0, -1)$, respectively.

Table 1 shows how the effect of a change at a position p below a node x with children y_1 and y_2 relates to the effect at the affected child. This depends on whether $r(y_1) \leq \ell(y_2)$ and whether the affected child is y_1 or y_2. A closer inspection of Table 1 reveals a crucial observation: in the upper left and the lower right field of the table, the effect on x is *independent* of the effect on the child (being it y_1 or y_2). That is, these cases induce an effect on x independent of the children. We thus call these cases *effect-inducing*. In the other two fields, the

	p is in str(y_1)	p is in str(y_2)
$r(y_1) \leq \ell(y_2)$	$(+1,0) \rightarrow (+1,0)$ $(0,-1) \rightarrow (+1,0)$	$(+1,0) \rightarrow (+1,0)$ $(0,-1) \rightarrow (0,-1)$
$r(y_1) > \ell(y_2)$	$(+1,0) \rightarrow (+1,0)$ $(0,-1) \rightarrow (0,-1)$	$(+1,0) \rightarrow (0,-1)$ $(0,-1) \rightarrow (0,-1)$

Table 1. The effect on x after a closing parenthesis was inserted at position p. The effects depend on the effect on the children y_1 and y_2 of x: for example, an entry '$(0,-1) \rightarrow (+1,0)$' in the column '$p$ is in str(y_1)' means that if the change operation has effect $(0,-1)$ on y_1 then the change operation has effect $(+1,0)$ on x.

effect on x depends on the effect at the child, but in the simplest possible way: they are just the same. That is the effect at the child is just adopted by x. We call these cases *effect-preserving*. To determine the effect at x it is thus sufficient to identify the highest affected descendant node z of x, where an effect-inducing case applies, such that for all intermediate nodes between x and z only effect-preserving cases apply.

Our dynamic program implements this idea. First it determines, for each ancestor x of the change position p, whether it is effect-inducing and which effect is induced. Then it identifies, for each x, the node z (represented by its height i above p) as the unique effect-inducing node that has no effect-inducing node on its path to x. The node z can be identified with work $\mathcal{O}((\log n)^2)$, as z is one of at most $\log n$ many nodes on the path from x to the leaf of p, and one needs to check that all nodes between x and z are effect-preserving. As the auxiliary relations need to be updated for $\log n$ many nodes, the overall work of $\mathcal{P}$ is $\mathcal{O}((\log n)^3)$. We refer to the full version for more details. □

A work-efficient dynamic program for range queries for D_1 and D_k Unfortunately, the program of Theorem 6.5 does not support range queries, since it seems that one would need to combine the unmatched parentheses of $\log n$ many nodes of the binary tree in the worst case. However, its idea can be combined with the idea of Proposition 5.2, yielding a program that maintains ℓ and r for $\mathcal{O}(n^\epsilon)$ special intervals on a constant number of levels.

In fact, this approach even works for D_k for $k > 1$. Indeed, with the help of ℓ and r, it is possible to identify for each position of an opening parenthesis the position of the corresponding closing parenthesis in $\mathcal{O}(1)$ parallel time with work n^ϵ, and then one only needs to check that they match everywhere. The latter contributes an extra factor $\mathcal{O}(n)$ to the work, for $k > 1$, but can be skipped for $k = 1$.

Theorem 6.6. *For all $\epsilon > 0$, $k > 1$,*

a) RANGEMEMBER(D_1) *can be maintained in* DynFO *with* $\mathcal{O}(n^\epsilon)$ *work, and*
b) RANGEMEMBER(D_k) *can be maintained in* DynFO *with* $\mathcal{O}(n^\epsilon)$ *work per change operation and* $\mathcal{O}(n^{1+\epsilon})$ *work per query operation.*

Proof sketch. In the following we reuse the definition of *special intervals* from the proof of Proposition 5.2 as well as the definition of ℓ and r from the proof of Proposition 6.5. We first describe a dynamic program for RANGEMEMBER(D_1). It maintains ℓ and r for all special intervals, which is clearly doable with $\mathcal{O}(n^\epsilon)$ work per change operation. Similar to the proof of Proposition 5.2, the two crucial observations (justified in the full version) are that (1) a range query can be answered with the help of a constant number of special intervals, and (2) the change operation affects only a bounded number of special intervals per level.

As stated before, the program for RANGEMEMBER(D_k) also maintains ℓ and r, but it should be emphasised that also in the case of several parenthesis types, the definition of these functions ignores the bracket type. With that information it computes, for each opening bracket the position of its matching closing bracket, with the help of ℓ and r, and checks that they match. This can be done in parallel and with work $\mathcal{O}(n^\epsilon)$ per position. We refer to the full version for more details. □

Moderately work-efficient dynamic programs for D_k We now turn to the membership query for D_k with $k > 1$. Again, our program basically mimics the sequential algorithm from [8] which heavily depends on the dynamic problem STRINGEQUALITY that asks whether two given strings are equal.

Problem: STRINGEQUALITY
Input: Two Sequences $u = u_0 \ldots u_{n-1}$ and $v = v_0 \ldots v_{n-1}$ of letters with $u_i, v_i \in \Sigma \cup \{\epsilon\}$
Changes: $\text{SET}_{x,\sigma}(i)$ for $\sigma \in \Sigma, x \in \{u, v\}$: Sets x_i to σ, if $x_i = \epsilon$
$\text{RESET}_x(i)$ for $x \in \{u, v\}$: Sets x_i to ϵ
Queries: EQUALS: Is $u_0 \circ \ldots \circ u_{n-1} = v_0 \circ \ldots \circ v_{n-1}$?

It is easy to show that a linear amount of work is sufficient to maintain STRINGEQUALITY.

Lemma 6.7. STRINGEQUALITY *is in* DynFO *with work* $\mathcal{O}(n)$.

Because of the linear work bound for STRINGEQUALITY our dynamic program for MEMBER(D_k) also has a linear factor in the work bound.

Theorem 6.8. MEMBER(D_k) *is maintainable in* DynFO *with work* $\mathcal{O}(n \log n + (\log n)^3)$ *for every fixed* $k \in \mathbb{N}$.

Proof sketch. The program can be seen as an extension of the program for MEMBER(D_1). As unmatched parentheses are no longer well-defined if we have more than one type of parenthesis the idea of [8] is to maintain the parentheses to the left and right that remain if we reduce the string by matching opening and closing parentheses regardless of their type. To be able to answer MEMBER(D_k), the dynamic program maintains the unmatched parentheses for every node x of a tree spanning the input word, and a bit $M(x)$ that indicates whether the types of the parentheses match properly.

How the unmatched parentheses can be maintained for a node x after a change operation depends on the "segment" of $\mathrm{str}(x)$ in which the change happened and in some cases reduces to finding a node z with a local property on the path from x to the leaf that corresponds to the changed position.

To update $M(x)$ for a node x with children y_1 and y_2 the dynamic program compares the unmatched parentheses to the right of y_1 with the ones to the left of y_2 using STRINGEQUALITY. We refer to the full version for more details. □

Maintaining string equality and membership in D_k for $k > 1$ is even closer related which is stated in the following lemma.

Lemma 6.9. *a) If* STRINGEQUALITY *can be maintained in* DynFO *with work* $W(n)$ *then* MEMBER(D_k) *can be maintained in* DynFO *with work* $\mathcal{O}(W(n) \cdot \log n + (\log n)^3)$, *for each* $k \geq 1$.
b) If MEMBER(D_k) *can be maintained in* DynFO *with work* $W(n)$ *for all* k, *then* STRINGEQUALITY *can be maintained in* DynFO *with work* $\mathcal{O}(W(n))$.

7 Conclusion

In this paper we proposed a framework for studying the aspect of work for the dynamic, parallel complexity class DynFO. We established that all regular languages can be maintained in DynFO with $\mathcal{O}(n^\epsilon)$ work for all $\epsilon > 0$, and even with $\mathcal{O}(\log n)$ work for star-free regular languages. For context-free languages we argued that it will be hard to achieve work bounds lower than $\mathcal{O}(n^{\omega-1-\epsilon})$ in general, where ω is the matrix multiplication exponent. For the special case of Dyck languages D_k we showed that $\mathcal{O}(n \cdot (\log n)^3)$ work suffices, which can be further reduced to $\mathcal{O}(\log^3 n)$ work for D_1. For range queries, dynamic programs with work $\mathcal{O}(n^{1+\epsilon})$ and $\mathcal{O}(n^\epsilon)$ exist, respectively.

We highlight some research directions. One direction is to improve the upper bounds on work obtained here. For instance, it would be interesting to know whether all regular languages can be maintained with polylog or even $\mathcal{O}(\log n)$ work and how close the lower bounds for context-free languages can be matched. Finding important subclasses of context-free languages for which polylogarithmic work suffices is another interesting question. Apart from string problems, many DynFO results concern problems on dynamic graphs, especially the reachability query [5]. How large is the work of the proposed dynamic programs, and are more work-efficient dynamic programs possible?

The latter question also leads to another research direction: to establish further lower bounds. The lower bounds obtained here are relative to strong conjectures. Absolute lower bounds are an interesting goal which seems in closer reach than lower bounds for DynFO without bounds on the work.

References

1. Abboud, A., Backurs, A., Bringmann, K., Künnemann, M.: Fine-grained complexity of analyzing compressed data: Quantifying improvements over decompress-and-

solve. In: Umans, C. (ed.) 58th IEEE Annual Symposium on Foundations of Computer Science, FOCS 2017, Berkeley, CA, USA, October 15-17, 2017. pp. 192–203. IEEE Computer Society (2017). https://doi.org/10.1109/FOCS.2017.26
2. Abboud, A., Backurs, A., Williams, V.V.: If the current clique algorithms are optimal, so is Valiant's parser. SIAM J. Comput. **47**(6), 2527–2555 (2018)
3. Alstrup, S., Husfeldt, T., Rauhe, T.: Dynamic nested brackets. Inf. Comput. **193**(2), 75–83 (2004). https://doi.org/10.1016/j.ic.2004.04.006, https://doi.org/10.1016/j.ic.2004.04.006
4. Börger, E.: Abstract state machines: a unifying view of models of computation and of system design frameworks. Ann. Pure Appl. Log. **133**(1-3), 149–171 (2005). https://doi.org/10.1016/j.apal.2004.10.007
5. Datta, S., Kulkarni, R., Mukherjee, A., Schwentick, T., Zeume, T.: Reachability is in DynFO. J. ACM **65**(5), 33:1–33:24 (2018). https://doi.org/10.1145/3212685
6. Dong, G., Su, J.: First-order incremental evaluation of datalog queries. In: Database Programming Languages (DBPL-4), Proceedings of the Fourth International Workshop on Database Programming Languages - Object Models and Languages, Manhattan, New York City, USA, 30 August - 1 September 1993. pp. 295–308 (1993)
7. Dong, G., Topor, R.W.: Incremental evaluation of datalog queries. In: Database Theory - ICDT'92, 4th International Conference, Berlin, Germany, October 14-16, 1992, Proceedings. pp. 282–296 (1992). https://doi.org/10.1007/3-540-56039-4_48
8. Frandsen, G.S., Husfeldt, T., Miltersen, P.B., Rauhe, T., Skyum, S.: Dynamic algorithms for the Dyck languages. In: Akl, S.G., Dehne, F.K.H.A., Sack, J., Santoro, N. (eds.) Algorithms and Data Structures, 4th International Workshop, WADS '95, Kingston, Ontario, Canada, August 16-18, 1995, Proceedings. Lecture Notes in Computer Science, vol. 955, pp. 98–108. Springer (1995). https://doi.org/10.1007/3-540-60220-8_54
9. Frandsen, G.S., Miltersen, P.B., Skyum, S.: Dynamic word problems. J. ACM **44**(2), 257–271 (1997). https://doi.org/10.1145/256303.256309
10. Gall, F.L.: Powers of tensors and fast matrix multiplication. In: International Symposium on Symbolic and Algebraic Computation, ISSAC '14, Kobe, Japan, July 23-25, 2014. pp. 296–303 (2014)
11. Gelade, W., Marquardt, M., Schwentick, T.: The dynamic complexity of formal languages. ACM Trans. Comput. Log. **13**(3), 19 (2012). https://doi.org/10.1145/2287718.2287719
12. Holm, J., de Lichtenberg, K., Thorup, M.: Poly-logarithmic deterministic fully-dynamic algorithms for connectivity, minimum spanning tree, 2-edge, and biconnectivity. J. ACM **48**(4), 723–760 (2001). https://doi.org/10.1145/502090.502095
13. Holm, J., Rotenberg, E.: Fully-dynamic planarity testing in polylogarithmic time. In: Makarychev, K., Makarychev, Y., Tulsiani, M., Kamath, G., Chuzhoy, J. (eds.) Proccedings of the 52nd Annual ACM SIGACT Symposium on Theory of Computing, STOC 2020, Chicago, IL, USA, June 22-26, 2020. pp. 167–180. ACM (2020). https://doi.org/10.1145/3357713.3384249
14. Immerman, N.: Descriptive complexity. Graduate texts in computer science, Springer (1999). https://doi.org/10.1007/978-1-4612-0539-5
15. Immerman, N.: Descriptive complexity. Springer Science & Business Media (2012)
16. JáJá, J.: An Introduction to Parallel Algorithms. Addison-Wesley (1992)
17. Libkin, L.: Elements of Finite Model Theory. Springer (2004). https://doi.org/10.1007/978-3-662-07003-1
18. McNaughton, R., Papert, S.: Counter-Free Automata. MIT Press (1971)

19. Miltersen, P.B., Subramanian, S., Vitter, J.S., Tamassia, R.: Complexity models for incremental computation. Theor. Comput. Sci. **130**(1), 203–236 (1994). https://doi.org/10.1016/0304-3975(94)90159-7
20. Patnaik, S., Immerman, N.: Dyn-FO: A parallel, dynamic complexity class. In: PODS. pp. 210–221 (1994)
21. Schmidt, J., Schwentick, T., Tantau, T., Vortmeier, N., Zeume, T.: Work-sensitive Dynamic Complexity of Formal Languages. CoRR **abs/2101.08735** (2021), `https://arxiv.org/abs/2101.08735`
22. Schwentick, T., Vortmeier, N., Zeume, T.: Sketches of dynamic complexity. SIGMOD Rec. **49**(2), 18–29 (2020). https://doi.org/10.1145/3442322.3442325
23. Schwentick, T., Zeume, T.: Dynamic Complexity: Recent Updates. SIGLOG News **3**(2), 30–52 (2016). https://doi.org/10.1145/2948896.2948899
24. Sherlekar, D.D., Pawagi, S., Ramakrishnan, I.V.: O(1) parallel time incremental graph algorithms. In: Maheshwari, S.N. (ed.) Foundations of Software Technology and Theoretical Computer Science, Fifth Conference, New Delhi, India, December 16-18, 1985, Proceedings. Lecture Notes in Computer Science, vol. 206, pp. 477–495. Springer (1985). https://doi.org/10.1007/3-540-16042-6_27
25. Straubing, H.: Finite automata, formal logic, and circuit complexity. Birkhauser Verlag (1994)
26. Vollmer, H.: Introduction to circuit complexity: a uniform approach. Springer Science & Business Media (2013)
27. Williams, V.V.: Multiplying matrices faster than Coppersmith-Winograd. In: Karloff, H.J., Pitassi, T. (eds.) Proceedings of the 44th Symposium on Theory of Computing Conference, STOC 2012, New York, NY, USA, May 19 - 22, 2012. pp. 887–898. ACM (2012). https://doi.org/10.1145/2213977.2214056

Learning Pomset Automata*

Gerco van Heerdt[1] (✉), Tobias Kappé[2], Jurriaan Rot[3], and Alexandra Silva[1]

[1] University College London, London, UK
gerco.heerdt@ucl.ac.uk
[2] Cornell University, Ithaca NY, USA
[3] Radboud University, Nijmegen, The Netherlands

Abstract. We extend the L* algorithm to learn bimonoids recognising pomset languages. We then identify a class of pomset automata that accepts precisely the class of pomset languages recognised by bimonoids and show how to convert between bimonoids and automata.

1 Introduction

Automata learning algorithms are useful in automated inference of models, which is needed for verification of hardware and software systems. In *active* learning, the algorithm interacts with a system through tests and observations to produce a model of the system's behaviour. One of the first active learning algorithms proposed was L*, due to Dana Angluin [2], which infers a minimal deterministic automaton for a target regular language. L* has been used in a range of verification tasks, including learning error traces in a program [5]. For more advanced verification tasks, richer automata types are needed and L* has been extended to e.g. input-output [1], register [20], and weighted automata [16]. None of the existing extensions can be used in analysis of concurrent programs.

Partially ordered multisets (pomsets) [13,12] are basic structures used in the modeling and semantics of concurrent programs. Pomsets generalise words, allowing to capture both the sequential and the parallel structure of a trace in a concurrent program. Automata accepting pomset languages are therefore useful to study the operational semantics of concurrent programs—see, for instance, work on concurrent Kleene algebra [17,26,21,24].

In this paper, we propose an active learning algorithm for a class of pomset automata. The approach is algebraic: we consider languages of pomsets recognised by bimonoids [28] (which we shall refer to as pomset recognisers). This can be thought of as a generalisation of the classical approach to language theory of using monoids as word acceptors: bimonoids have an extra operation that models parallel composition in addition to sequential. The two operations give rise to a complex branching structure that makes the learning process non-trivial.

* This work was partially supported by the ERC Starting Grant ProFoundNet (679127) and the EPSRC Standard Grant CLeVer (EP/S028641/1). The authors thank Matteo Sammartino for useful discussions.

S. Kiefer and C. Tasson (Eds.): FOSSACS 2021, LNCS 12650, pp. 510–530, 2021.
https://doi.org/10.1007/978-3-030-71995-1_26

The key observation is that pomset recognisers are tree automata whose algebraic structure satisfies additional equations. We extend tree automata learning algorithms [7,8,31] to pomset recognisers. The main challenge is to ensure that intermediate hypotheses in the algorithm are valid pomset recognisers, which is essential in practical scenarios where the learning process might not run to the very end, returning an approximation of the system under learning. This requires equations of bimonoids to be correctly propagated and preserved in the core data structure of the algorithm—the observation table. The proof of termination, in analogy to $\mathtt{L}^{\star}$, relies on the existence of a canonical pomset recogniser of a language, which is based on its syntactic bimonoid. The steps of the algorithm provide hypotheses that get closer in size to the canonical recogniser.

Finally, we bridge the learning algorithm to pomset automata [21,22] by providing two constructions that enable us to seamlessly move between pomset recognisers and pomset automata. Note that although bimonoids provide a useful formalism to denote pomset languages, which is amenable to the design of the learning algorithm, they enforce a redundancy that is not present in pomset automata: whereas a pomset automaton processes a pomset from left to right in sequence, one letter per branch at a time, a bimonoid needs to be able to take the pomset represented as a binary tree in any way and process it bottom-up. This requirement of different decompositions leading to the same result makes bimonoids in general much larger than pomset automata and hence the latter are, in general, a more efficient representation of a pomset language.

The rest of the paper is organised as follows. We conclude this introductory section with a review of relevant related work. Section 2 contains the basic definitions on pomsets and pomset recognisers. The learning algorithm for pomset recognisers appears in Section 3, including proofs to ensure termination and invariant preservation. Section 4 presents constructions to translate between (a class of) pomset automata and pomset recognisers. We conclude with discussion of further work in Section 5. Omitted proofs appear in the extended version [15].

Related Work. There is a rich literature on adaptations and extensions of $\mathtt{L}^{\star}$ from deterministic automata to various kinds of models, see, e.g., [34,18] for an overview. To the best of our knowledge, this paper is the first to provide an active learning algorithm for pomset languages recognised by finite bimonoids.

Our algorithm learns an algebraic recogniser. Urbat and Schröder [33] provide a very general learning approach for languages recognised by algebras for monads [4,32], based on a reduction to categorical automata, for which they present an $\mathtt{L}^{\star}$-type algorithm. Their reduction gives rise to an infinite alphabet in general, so tailored work is needed for deriving algorithms and finite representations. This can be done for instance for monoids, recognising regular languages, but it is not clear how this could extend to pomset recognisers. We present a direct learning algorithm for bimonoids, which does not rely on any encoding.

Our concrete learning algorithm for bimonoids is closely related to learning approaches for bottom-up tree automata [7,8,31]: pomset languages can be viewed as tree languages satisfying certain equations. Incorporating these equations turned out to be a non-trivial task, which requires additional checks on the observation table during execution of the algorithm.

Conversion between recognisers and automata for a pomset language was first explored by Lodaya and Weil [28,27]. Their results relate the expressive power of these formalisms to *sr-expressions*. As a result, converting between recognisers and automata using their construction uses an sr-expression as an intermediate representation, increasing the resulting state space. Our construction, however, converts recognisers directly to pomset automata, which keeps the state space relatively small. Moreover, Lodaya and Weil work focus on pomset languages of *bounded width*, i.e., with an upper bound on the number of parallel events. In contrast, our conversions work for all recognisable pomset languages (and a suitable class of pomset automata), including those of unbounded width.

Ésik and Németh [9] considered automata and recognisers for *biposets*, i.e., sp-pomsets without commutativity of parallel composition. They equate languages recognised by *bisemigroups* (bimonoids without commutativity or units) with those accepted by *parenthesizing automata*. Our equivalence is similar in structure, but relates a subclass of pomset automata to bimonoids instead. The results in this paper can easily be adapted to learn representations of biposet languages using bisemigroups, and convert those to parenthesizing automata.

2 Pomset Recognisers

Throughout this paper we fix a finite *alphabet* Σ and assume $\square \notin \Sigma$. When defining sets parameterised by a set X, say $\mathsf{S}(X)$, we may use S to refer to $\mathsf{S}(\Sigma)$.

We recall pomsets [12,13], a generalisation of words that model concurrent traces. A *labelled poset* over X is a tuple $\mathbf{u} = \langle S_\mathbf{u}, \leq_\mathbf{u}, \lambda_\mathbf{u} \rangle$, where $S_\mathbf{u}$ is a finite set (the *carrier* of $\mathbf{u}$), $\leq_\mathbf{u}$ is a partial order on $S_\mathbf{u}$ (the *order* of $\mathbf{u}$), and $\lambda_\mathbf{u} \colon S_\mathbf{u} \to X$ is a function (the *labelling* of $\mathbf{u}$). Pomsets are labelled posets up to isomorphism.

Definition 1 (Pomsets). *Let $\mathbf{u}, \mathbf{v}$ be labelled posets over X. An* embedding *of $\mathbf{u}$ in $\mathbf{v}$ is an injection $h : S_\mathbf{u} \to S_\mathbf{v}$ such that $\lambda_\mathbf{v} \circ h = \lambda_\mathbf{u}$ and $s \leq_\mathbf{u} s'$ if and only if $h(s) \leq_\mathbf{v} h(s')$. An* isomorphism *is a bijective embedding whose inverse is also an embedding. We say $\mathbf{u}$ is* isomorphic *to $\mathbf{v}$, denoted $\mathbf{u} \cong \mathbf{v}$, if there exists an isomorphism between $\mathbf{u}$ and $\mathbf{v}$. A* pomset *over X is an isomorphism class of labelled posets over X, i.e., $[\mathbf{v}] = \{\mathbf{u} : \mathbf{u} \cong \mathbf{v}\}$. When $u = [\mathbf{u}]$ and $v = [\mathbf{v}]$ are pomsets, u is a* subpomset *of v when there exists an embedding of $\mathbf{u}$ in $\mathbf{v}$.*

When two pomsets are in scope, we tacitly assume that they are represented by labelled posets with disjoint carriers. We write 1 for the empty pomset. When $\mathsf{a} \in X$, we write a for the pomset represented by the labelled poset whose sole element is labelled by a. Pomsets can be composed in sequence and in parallel:

Definition 2 (Pomset composition). *Let $u = [\mathbf{u}]$ and $v = [\mathbf{v}]$ be pomsets over X. We write $u \parallel v$ for the* parallel composition *of u and v, which is the pomset over X represented by the labelled poset*

$$\mathbf{u} \parallel \mathbf{v} = \langle S_\mathbf{u} \cup S_\mathbf{v},\ \leq_\mathbf{u} \cup \leq_\mathbf{v},\ \lambda_\mathbf{u} \cup \lambda_\mathbf{v} \rangle$$

Similarly, we write $u \cdot v$ *for the* sequential composition *of* u *and* v*, that is, the pomset represented by the labelled poset*

$$\mathbf{u} \cdot \mathbf{v} = \langle S_\mathbf{u} \cup S_\mathbf{v},\ \leq_\mathbf{u} \cup \leq_\mathbf{v} \cup S_\mathbf{u} \times S_\mathbf{v},\ \lambda_\mathbf{u} \cup \lambda_\mathbf{v} \rangle$$

We may elide the dot for sequential composition, for instance writing $\mathtt{ab}$ *for* $\mathtt{a} \cdot \mathtt{b}$.

The pomsets we use can be built using sequential and parallel composition.

Definition 3 (Series-parallel pomsets). *The set of* series-parallel pomsets (sp-pomsets) *over* X*, denoted* $\mathsf{SP}(X)$*, is the smallest set such that* $1 \in \mathsf{SP}(X)$ *and* $\mathtt{a} \in \mathsf{SP}(X)$ *for every* $\mathtt{a} \in X$*, closed under parallel and sequential composition.*

Concurrent systems admit executions of operations that are not only ordered in sequence but also allow parallel branches. An algebraic structure consisting of both a sequential and a parallel composition operation, with a shared unit, is called a *bimonoid*. Formally, its definition is as follows.

Definition 4 (Bimonoid). *A* bimonoid *is a tuple* $\langle M, \odot, \oplus, \mathbf{1} \rangle$ *where*

- M *is a set called the* carrier *of the bimonoid,*
- $\odot$ *is a binary associative operation on* M*,*
- $\oplus$ *is a binary associative and commutative operation on* M*, and*
- $\mathbf{1} \in M$ *is a unit for both* $\odot$ *(on both sides) and* $\oplus$*.*

Bimonoid homomorphisms are defined in the usual way.

Given a set X, the *free bimonoid* [12] over X is $\langle \mathsf{SP}(X), \cdot, \parallel, 1 \rangle$. The fact that it is free means that for every function $f \colon X \to M$ for a given bimonoid $\langle M, \odot, \oplus, \mathbf{1}_M \rangle$ there exists a unique bimonoid homomorphism $f^\sharp \colon \mathsf{SP}(X) \to M$ such that the restriction of $f^\sharp$ to X is f.

Just as monoids can recognise words, bimonoids can recognise pomsets [28]. A bimonoid together with the witnesses of recognition is a *pomset recogniser*.

Definition 5 (Pomset recogniser). *A* pomset recogniser *is a tuple* $\mathcal{R} = \langle M, \odot, \oplus, \mathbf{1}, i, F \rangle$ *where* $\langle M, \odot, \oplus, \mathbf{1} \rangle$ *is a bimonoid,* $i \colon \Sigma \to M$*, and* $F \subseteq M$*. The* language recognised *by* $\mathcal{R}$ *is given by* $\mathcal{L}_\mathcal{R} = \{u \in \mathsf{SP} : i^\sharp(u) \in F\} \subseteq \mathsf{SP}$.

Example 6. Suppose a program consists of a loop, where each iteration runs actions $\mathtt{a}$ and $\mathtt{b}$ in parallel. We can describe the behaviour of this program by

$$\mathcal{L} = \{\mathtt{a} \parallel \mathtt{b}\}^* = \{1, \mathtt{a} \parallel \mathtt{b}, (\mathtt{a} \parallel \mathtt{b}) \cdot (\mathtt{a} \parallel \mathtt{b}), \ldots\}$$

We can describe this language using a pomset recogniser, as follows. Let $M = \{q_\mathtt{a}, q_\mathtt{b}, q_1, q_\bot, \mathbf{1}\}$, and let $\odot$ and $\oplus$ be the operations on M given by

$$q \odot q' = \begin{cases} q & q' = \mathbf{1} \\ q' & q = \mathbf{1} \\ q_1 & q = q' = q_1 \\ q_\bot & \text{otherwise} \end{cases} \qquad q \oplus q' = \begin{cases} q & q' = \mathbf{1} \\ q' & q = \mathbf{1} \\ q_1 & \{q, q'\} = \{q_\mathtt{a}, q_\mathtt{b}\} \\ q_\bot & \text{otherwise} \end{cases}$$

A straightforward proof verifies that $\langle M, \odot, \oplus, \mathbf{1}\rangle$ is a bimonoid.

We set $i(\mathtt{a}) = q_{\mathtt{a}}$, $i(\mathtt{b}) = q_{\mathtt{b}}$, and $F = \{\mathbf{1}, q_1\}$. Now, for $n > 0$:

$$i^\sharp(\underbrace{(\mathtt{a} \parallel \mathtt{b}) \cdots (\mathtt{a} \parallel \mathtt{b})}_{n \text{ times}}) = \underbrace{(i(\mathtt{a}) \parallel i(\mathtt{b})) \odot \cdots \odot (i(\mathtt{a}) \parallel i(\mathtt{b}))}_{n \text{ times}} = \underbrace{q_1 \odot \cdots \odot q_1}_{n \text{ times}} = q_1$$

No other pomsets are mapped to q_1; hence, $\langle M, \odot, \oplus, \mathbf{1}, i, F\rangle$ accepts $\mathcal{L}$.

Example 7. Suppose a program solves a problem recursively, such that the recursive calls are performed in parallel. In that case, the program would either perform the base action $\mathtt{b}$, or some preprocessing action $\mathtt{a}$ followed by running two copies of itself in parallel. This behaviour can be described by the smallest pomset language $\mathcal{L}$ satisfying the following inference rules:

$$\frac{}{\mathtt{b} \in \mathcal{L}} \qquad\qquad \frac{u, v \in \mathcal{L}}{\mathtt{a} \cdot (u \parallel v) \in \mathcal{L}}$$

This language can be described by a pomset recogniser. Let our carrier set be $M = \{q_{\mathtt{a}}, q_{\mathtt{b}}, q_1, q_\bot, \mathbf{1}\}$, and let $\odot$ and $\oplus$ be the operations on M given by

$$q \odot q' = \begin{cases} q & q' = \mathbf{1} \\ q' & q = \mathbf{1} \\ q_{\mathtt{b}} & q = q_{\mathtt{a}}, q' = q_1 \\ q_\bot & \text{otherwise} \end{cases} \qquad q \oplus q' = \begin{cases} q & q' = \mathbf{1} \\ q' & q = \mathbf{1} \\ q_1 & q = q' = q_{\mathtt{b}} \\ q_\bot & \text{otherwise} \end{cases}$$

$\langle M, \odot, \oplus, \mathbf{1}\rangle$ is a bimonoid, $F = \{q_{\mathtt{b}}\}$, and $i \colon \Sigma \to M$ is given by setting $i(\mathtt{a}) = q_{\mathtt{a}}$ and $i(\mathtt{b}) = q_{\mathtt{b}}$. One can then show that $\langle M, \odot, \oplus, \mathbf{1}, i, F\rangle$ accepts $\mathcal{L}$.

Pomset contexts are used to describe the behaviour of individual elements in a pomset recogniser. Formally, the set of pomset contexts over a set X is given by $\mathsf{PC}(X) = \mathsf{SP}(X \cup \{\Box\})$. Here the element $\Box$ acts as a placeholder, where a pomset can be plugged in: given a context $c \in \mathsf{PC}(X)$ and $t \in \mathsf{SP}(X)$, let $c[t] \in \mathsf{SP}(X)$ be obtained by substituting t for $\Box$ in c.

3 Learning Pomset Recognisers

In this section we present our algorithm to learn pomset recognisers from an oracle (*the teacher*) that answers *membership* and *equivalence queries*. A membership query consists of a pomset, to which the teacher replies whether that pomset is in the language; an equivalence query consists of a *hypothesis* pomset recogniser, to which the teacher replies *yes* if it is correct or *no* with a counterexample—a pomset incorrectly classified by the hypothesis—if it is not.

A pomset recogniser is essentially a tree automaton, with the additional constraint that its algebraic structure satisfies the bimonoid axioms. Our algorithm is therefore relatively close to tree automata learning—in particular Drewes and Högberg [7,8]—but there are several key differences: we optimise the algorithm by taking advantage of the bimonoid axioms, and at the same time need to ensure that the hypotheses generated by the learning process satisfy those axioms.

3.1 Observation Table

We fix a target language $\mathcal{L} \subseteq \mathsf{SP}$ throughout this section. As in the original $\mathtt{L}^\star$ algorithm, the state of the learner throughout a run of the algorithm is given by a data structure called the *observation table*, which collects information about $\mathcal{L}$. The table contains rows indexed by pomsets, representing the state reached by the correct pomset recogniser after reading that pomset; and columns indexed by pomset contexts, used to approximately indentify the behaviour of each state. To represent the additional rows needed to approximate the pomset recogniser structure, we use the following definition. Given $U \subseteq \mathsf{SP}$, we define

$$U^+ = \Sigma \cup \{u \cdot v : u, v \in U\} \cup \{u \parallel v : u, v \in U\} \subseteq \mathsf{SP}.$$

Definition 8 (Observation table). *An* observation table *is a pair $\langle S, E\rangle$, with $S \subseteq \mathsf{SP}$ subpomset-closed and $E \subseteq \mathsf{PC}$ such that $1 \in S$ and $\square \in E$. These sets induce the function $\mathsf{row}_{\langle S,E\rangle} : S \cup S^+ \to 2^E$: $\mathsf{row}_{\langle S,E\rangle}(s)(e) = 1 \iff e[s] \in \mathcal{L}$. We often write* row *instead of* $\mathsf{row}_{\langle S,E\rangle}$ *when S and E are clear from the context.*

We depict observation tables, or more precisely row, as two separate tables with rows in S and $S^+ \setminus S$ respectively, see for instance Example 9 below.

The goal of the learner is to extract a *hypothesis* pomset recogniser from the rows in the table. More specifically, the carrier of the underlying bimonoid of the hypothesis will be given by the rows indexed by pomsets in S. The structure on the rows is obtained by transferring the structure of the row labels onto the rows (e.g., $\mathsf{row}(s) \odot \mathsf{row}(t) = \mathsf{row}(s \cdot t)$), but this is not well-defined unless the table satisfies *closedness*, *consistency*, and *associativity*. Closedness and consistency are standard in $\mathtt{L}^\star$, whereas associativity is a new property specific to bimonoid learning. We discuss each of these properties next, also including *compatibility*, a property that is used to show minimality of hypotheses.

The first potential issue is a closedness defect: this is the case when a composed row, indexed by an element of S^+, is not indexed by a pomset in S.

Example 9 (Table not closed). Recall $\mathcal{L} = \{\mathtt{a} \parallel \mathtt{b}\}^*$ from Example 6, and suppose $S = \{1, \mathtt{a}, \mathtt{b}\}$ and $E = \{\square, \mathtt{a} \parallel \square, \square \parallel \mathtt{b}\}$. The induced table is

S	$\square$	$\mathtt{a} \parallel \square$	$\square \parallel \mathtt{b}$
1	1	0	0
a	0	0	1
b	0	1	0

$S^+ \setminus S$	$\square$	$\mathtt{a} \parallel \square$	$\square \parallel \mathtt{b}$
aa	0	0	0
ab	0	0	0
ba	0	0	0
bb	0	0	0
a ∥ a	0	0	0
a ∥ b	1	0	0
b ∥ b	0	0	0

The carrier of the hypothesis bimonoid is $M = \{\mathsf{row}(1), \mathsf{row}(\mathtt{a}), \mathsf{row}(\mathtt{b})\}$, but the composition $\mathsf{row}(\mathtt{a}) \odot \mathsf{row}(\mathtt{a})$ cannot be defined since $\mathsf{row}(\mathtt{aa}) \notin M$.

The absence of the issue described above is captured with *closedness.*

Definition 10 (Closed table). *An observation table* $\langle S, E\rangle$ *is* closed *if for all* $t \in S^{+}$ *there exists* $s \in S$ *such that* $\mathsf{row}(s) = \mathsf{row}(t)$.

Another issue that may occur is that the same row being represented by different index pomsets leads to an inconsistent definition of the structure. The absence of this issue is referred to as *consistency*.

Definition 11 (Consistent table). *An observation table* $\langle S, E\rangle$ *is* consistent *if for all* $s_1, s_2 \in S$ *such that* $\mathsf{row}(s_1) = \mathsf{row}(s_2)$ *we have for all* $t \in S$ *that*

$$\mathsf{row}(s_1 \cdot t) = \mathsf{row}(s_2 \cdot t) \qquad \mathsf{row}(t \cdot s_1) = \mathsf{row}(t \cdot s_2) \qquad \mathsf{row}(s_1 \parallel t) = \mathsf{row}(s_2 \parallel t).$$

Whenever closedness and consistency hold, one can define sequential and parallel composition operations on the rows of the table. However, these operations are not guaranteed to be associative, as we show with the following example.

Example 12 (Table not associative). Consider $\mathcal{L} = \{\mathsf{a}u : u \in \{\mathsf{b}\}^*\}$ over $\Sigma = \{\mathsf{a}, \mathsf{b}\}$, and suppose $S = \{1, \mathsf{a}, \mathsf{b}\}$ and $E = \{\square, \square\mathsf{a}\}$. The induced table is:

	$\square$	$\square$a
1	0	1
a	1	0
b	0	0

	$\square$	$\square$a
aa	0	0
ab	1	0
ba	0	0
bb	0	0
a ∥ a	0	0
a ∥ b	0	0
b ∥ b	0	0

This table does not lead to an associative sequential operation on rows:

$$\begin{aligned}(\mathsf{row}(\mathsf{a}) \odot \mathsf{row}(\mathsf{b})) \odot \mathsf{row}(\mathsf{a}) &= \mathsf{row}(\mathsf{ab}) \odot \mathsf{row}(\mathsf{a}) = \mathsf{row}(\mathsf{a}) \odot \mathsf{row}(\mathsf{a}) = \mathsf{row}(\mathsf{aa}) \\ \neq \mathsf{row}(\mathsf{ab}) = \mathsf{row}(\mathsf{a}) \odot \mathsf{row}(\mathsf{b}) &= \mathsf{row}(\mathsf{a}) \odot \mathsf{row}(\mathsf{ba}) = \mathsf{row}(\mathsf{a}) \odot (\mathsf{row}(\mathsf{b}) \odot \mathsf{row}(\mathsf{a})).\end{aligned}$$

To prevent this issue we enforce the following additional property:

Definition 13 (Associative table). *Let* $\heartsuit \in \{\cdot, \parallel\}$. *An observation table* $\langle S, E\rangle$ *is* $\heartsuit$-associative *if for all* $s_1, s_2, s_3, s_l, s_r \in S$ *with* $\mathsf{row}(s_l) = \mathsf{row}(s_1 \heartsuit s_2)$ *and* $\mathsf{row}(s_r) = \mathsf{row}(s_2 \heartsuit s_3)$ *we have* $\mathsf{row}(s_l \heartsuit s_3) = \mathsf{row}(s_1 \heartsuit s_r)$. *An observation table is* associative *if it is both* $\cdot$*-associative and* $\parallel$*-associative.*

The table from Example 12 is *not* $\cdot$-associative: we have $\mathsf{row}(\mathsf{a}) = \mathsf{row}(\mathsf{ab})$ and $\mathsf{row}(\mathsf{b}) = \mathsf{row}(\mathsf{ba})$ but $\mathsf{row}(\mathsf{aa}) \neq \mathsf{row}(\mathsf{ab})$.

Putting the above definitions of closedness, consistency and associativity of tables together, we have the following result for constructing a hypothesis.

Lemma 14 (Hypothesis). *A closed, consistent and associative table* $\langle S, E\rangle$ *induces a* hypothesis *pomset recogniser* $\mathcal{H} = \langle H, \odot_H, \oplus_H, \mathbf{1}_H, i_H, F_H\rangle$ *where*

$$H = \{\mathsf{row}(s) : s \in S\} \qquad \mathsf{row}(s_1) \odot_H \mathsf{row}(s_2) = \mathsf{row}(s_1 \cdot s_2)$$

$$\mathsf{row}(s_1) \oplus_H \mathsf{row}(s_2) = \mathsf{row}(s_1 \parallel s_2) \qquad \mathbf{1}_H = \mathsf{row}(1) \qquad i_H(\mathsf{a}) = \mathsf{row}(\mathsf{a})$$

$$F_H = \{\mathsf{row}(s) : s \in S, \mathsf{row}(s)(\square) = 1\}.$$

Proof. The operations $\odot_H$ and $\oplus_H$ are well-defined by closedness and consistency, and $\mathbf{1}_H$ is well-defined because $1 \in S$ by the observation table definition. Commutativity of $\oplus_H$ follows from commutativity of $\parallel$, and similarly that $\mathbf{1}_H$ is a unit for both operations follows from 1 being a unit. Associativity follows by associativity of the table (it does *not* follow from $\cdot$ and $\parallel$ being associative: given elements $s_1, s_2, s_3 \in S$, $s_1 \cdot s_2 \cdot s_3$ is not necessarily present in $S \cup S^+$). □

Since a hypothesis is constructed from an observation table $\langle S, E\rangle$ that records for given $s \in S$ and $e \in E$ whether $e[s]$ is accepted by the language or not, one would expect that the hypothesis classifies those pomsets

$$T_{\langle S,E\rangle} = \{e[s] : s \in S, e \in E\}$$

correctly. This is not necessarily the case, as we show in the following example.

Example 15. Consider the language $\mathcal{L}$ from Example 7, and let $S = \{1, \mathtt{b}\}$ and $E = \{\square, \mathtt{a}(\square \parallel \mathtt{b})\}$. The induced table is

	$\square$	$\mathtt{a}(\square \parallel \mathtt{b})$
1	0	0
$\mathtt{b}$	1	1

	$\square$	$\mathtt{a}(\square \parallel \mathtt{b})$
$\mathtt{a}$	0	0
$\mathtt{bb}$	0	0
$\mathtt{b} \parallel \mathtt{b}$	0	0

From this closed, consistent, and associative table we obtain a hypothesis pomset recogniser that satisfies

$$\begin{aligned}(\mathsf{row}(\mathtt{a}) \odot (\mathsf{row}(\mathtt{b}) \oplus \mathsf{row}(\mathtt{b})))(\square) &= (\mathsf{row}(\mathtt{a}) \odot \mathsf{row}(\mathtt{b} \parallel \mathtt{b}))(\square)\\ &= (\mathsf{row}(\mathtt{a}) \odot \mathsf{row}(1))(\square) = \mathsf{row}(\mathtt{a})(\square) = 0 \neq 1\end{aligned}$$

and thus recognises a language that differs from $\mathcal{L}$ on $\mathtt{a} \cdot (\mathtt{b} \parallel \mathtt{b}) \in T_{\langle S,E\rangle}$.

We thus have the following definition, parametric in a subset of $T_{\langle S,E\rangle}$.

Definition 16 (Compatible hypothesis). *A closed, consistent, and associative observation table $\langle S, E\rangle$ induces a hypothesis $\mathcal{H}$ that is X-*compatible *with its table, for $X \subseteq \mathsf{SP}$, if for $x \in X$ we have $x \in \mathcal{L}_{\mathcal{H}} \iff x \in \mathcal{L}$. We say that the hypothesis is* compatible *with its table if it is $T_{\langle S,E\rangle}$-compatible with its table.*

Ensuring hypotheses are compatible with their table will not be a crucial step in proving termination, but plays a key role in ensuring minimality (Section 3.4). This was originally shown by van Heerdt [14] for Mealy machines.

3.2 The Learning Algorithm

We are now ready to introduce our learning algorithm, Algorithm 1. The main algorithm initialises the table to $\langle\{1\}, \{\square\}\rangle$ and starts by augmenting the table to make sure it is closed and associative. We give an example below.

1   $S = \{1\}, E = \{\Box\}$
2   **repeat**
3   **repeat**
4   **while** $\langle S, E \rangle$ is not closed or not associative
5   **if** $\langle S, E \rangle$ is not closed
6   find $t \in S^+$ such that $\mathsf{row}(t) \neq \mathsf{row}(s)$ for all $s \in S$
7                           $S = S \cup \{t\}$
8   **for** $\heartsuit \in \{\cdot, \|\}$
9   **if** $\langle S, E \rangle$ is not $\heartsuit$-associative
10   find $s_1, s_2, s_3, s_l, s_r \in S$ and $e \in E$ such that
                                    $\mathsf{row}(s_l) = \mathsf{row}(s_1 \heartsuit s_2)$,
                                    $\mathsf{row}(s_r) = \mathsf{row}(s_2 \heartsuit s_3)$, and
                                    $\mathsf{row}(s_l \heartsuit s_3)(e) \neq \mathsf{row}(s_1 \heartsuit s_r)(e)$
11   let b be the result of a membership query on $s_1 \heartsuit s_2 \heartsuit s_3$
12   **if** $\mathsf{row}(s_l \heartsuit s_3)(e) \neq b$
13                                     $E = E \cup \{e[\Box \heartsuit s_3]\}$
14   **else**
15                                     $E = E \cup \{e[s_1 \heartsuit \Box]\}$
16   construct the hypothesis $\mathcal{H}$ for $\langle S, E \rangle$
17   **if** $\mathcal{H}$ is not compatible with its table
18   find $s \in S$ and $e \in E$ such that $e[s] \in \mathcal{L}_{\mathcal{H}} \iff e[s] \notin \mathcal{L}$
19                     $E = E \cup \{\textsc{HandleCounterexample}(S, E, e[s], \Box)\}$
20   **until** $\mathcal{H}$ is compatible with its table
21   **if** the teacher replies *no* to $\mathcal{H}$, with a counterexample z
22               $E = E \cup \{\textsc{HandleCounterexample}(S, E, z, \Box)\}$
23   **until** the teacher replies *yes*
24   **return** $\mathcal{H}$

$\textsc{HandleCounterexample}(S, E, z, c)$

1   **if** $z \in S \cup S^+$
2   let $s \in S$ be such that $\mathsf{row}(s) = \mathsf{row}(z)$
3   **if** $c[s] \in \mathcal{L} \iff c[z] \in \mathcal{L}$
4   **return** s
5   **else**
6   **return** c
7   let non-empty $u_1, u_2 \in \mathsf{SP}$ and $\heartsuit \in \{\cdot, \|\}$ be such that $u_1 \heartsuit u_2 = z$
8   $u_1 = \textsc{HandleCounterexample}(S, E, u_1, c[\Box \heartsuit u_2])$
9   **if** $u_1 \notin S$
10   **return** u_1
11 $u_2 = \textsc{HandleCounterexample}(S, E, u_2, c[u_1 \heartsuit \Box])$
12 **if** $u_2 \notin S$
13   **return** u_2
14 **return** $\textsc{HandleCounterexample}(S, E, u_1 \heartsuit u_2, c)$

Algorithm 1: The pomset recogniser learning algorithm.

Example 17 (Fixing closedness and associativity). Consider the table from Example 9, where $\mathsf{row}(\mathsf{aa}) \notin \{\mathsf{row}(1), \mathsf{row}(\mathsf{a}), \mathsf{row}(\mathsf{b})\}$ witnesses a closedness defect. To fix this, the algorithm would add aa to the set S, which means $\mathsf{row}(\mathsf{aa})$ will become part of the carrier of the hypothesis.

Now consider the table from Example 12. Here we found an associativity defect witnessed by $\mathsf{row}(\mathsf{a}) = \mathsf{row}(\mathsf{ab})$ and $\mathsf{row}(\mathsf{b}) = \mathsf{row}(\mathsf{ba})$ but $\mathsf{row}(\mathsf{aa}) \neq \mathsf{row}(\mathsf{ab})$. More specifically, $\mathsf{row}(\mathsf{aa})(\square) \neq \mathsf{row}(\mathsf{ab})(\square)$. Thus, $s_1 = s_3 = s_l = \mathsf{a}$, $s_2 = s_r = \mathsf{b}$, $s_l = \mathsf{a}$, and $e = \square$. A membership query on aba shows $\mathsf{aba} \notin \mathcal{L}$, so $b = 0$. We have $\mathsf{row}(\mathsf{aa})(\square) = 0$, and therefore the algorithm would add the context $\square[\mathsf{a} \cdot \square] = \mathsf{a} \cdot \square$ to E.

Note that the algorithm does not explicitly check for consistency; this is because we actually ensure a stronger property—sharpness [3]—as an invariant (Lemma 25). This property ensures every row indexed by a pomset in S is indexed by exactly one pomset in S (implying consistency):

Definition 18 (Sharp table). *An observation table $\langle S, E\rangle$ is* sharp *if for all $s_1, s_2 \in S$ such that $\mathsf{row}(s_1) = \mathsf{row}(s_2)$ we have $s_1 = s_2$.*

The idea of maintaining sharpness is due to Maler and Pnueli [29].

Once the table is closed and associative, we construct the hypothesis and check if it is compatible with its table. If this is not the case, a witness for incompatibility is a counterexample by definition, so HANDLECOUNTEREXAMPLE is invoked to extract an extension of E, and we return to checking closedness and associativity. Once we obtain a hypothesis that is compatible with its table, we submit it to the teacher to check for equivalence with the target language. If the teacher provides a counterexample, we again process this and return to checking closedness and associativity. Once we have a compatible hypothesis for which there is no counterexample, we return this correct pomset recogniser.

The procedure HANDLECOUNTEREXAMPLE, adapted from [7,8], is provided with an observation table $\langle S, E\rangle$ a pomset z, and a context c and finds a single context to add to E. The main invariant is that $c[z]$ is a counterexample. Recursive calls replace subpomsets from S^+ with elements of S in this counterexample while maintaining the invariant. There are two types of return values: if c is a suitable context, c is returned; otherwise the return value is an element of S that is to replace z. The context c is suitable if $z \in S^+$ and adding c to E would distinguish $\mathsf{row}(s)$ from $\mathsf{row}(z)$, where $s \in S$ is such that currently $\mathsf{row}(s) = \mathsf{row}(z)$. Because S is non-empty and subpomset-closed, if $z \notin S \cup S^+$ it can be decomposed into $z = u_1 \heartsuit u_2$ for non-empty $u_1, u_2 \in \mathsf{SP}$ and $\heartsuit \in \{\cdot, \parallel\}$. We then recurse into u_1 and u_2 to replace them with elements of S and replace z with $u_1 \heartsuit u_2 \in S^+$ in a final recursive call. If $c = \square$, the return value cannot be in S, as we will show in Lemma 25 that these elements are not counterexamples.

Example 19 (Processing a counterexample). Consider $\mathcal{L} = \{\mathsf{a}, \mathsf{aa}, \mathsf{a} \parallel \mathsf{a}\}$, and let $S = \{1, \mathsf{a}\}$ and $E = \{\square\}$. This induces a closed, sharp, and associative table

	$\square$
1	0
a	1

	$\square$
aa	1
a ∥ a	1

Suppose an equivalence query on its pomset recogniser, which rejects only the empty pomset, gives counterexample $z = \mathsf{a} \parallel \mathsf{a} \parallel \mathsf{aa}$. We may decompose z as $(\square \parallel \mathsf{aa})[\mathsf{a} \parallel \mathsf{a}]$, where $\mathsf{a} \parallel \mathsf{a} \in S^+ \setminus S$. Because $\mathsf{row}(\mathsf{a} \parallel \mathsf{a}) = \mathsf{row}(\mathsf{a})$, $(\square \parallel \mathsf{aa})[\mathsf{a}] = \mathsf{a} \parallel \mathsf{aa}$, and $\mathsf{a} \parallel \mathsf{aa} \in \mathcal{L} \iff z \in \mathcal{L}$, we update $z = \mathsf{a} \parallel \mathsf{aa}$ and repeat the process. Now we decompose $z = (\mathsf{a} \parallel \square)[\mathsf{aa}]$. Since $\mathsf{row}(\mathsf{aa}) = \mathsf{row}(\mathsf{a})$, $(\mathsf{a} \parallel \square)[\mathsf{a}] = \mathsf{a} \parallel \mathsf{a}$, and $\mathsf{a} \parallel \mathsf{a} \in \mathcal{L} \iff z \notin \mathcal{L}$, we finish by adding $\mathsf{a} \parallel \square$ to E.

3.3 Termination and Query Complexity

Our termination argument is based on a comparison of the current observation table with the infinite table $\langle \mathsf{SP}, \mathsf{PC} \rangle$. We first show that the latter induces a hypothesis, called the *canonical pomset recogniser* for the language. Its underlying bimonoid is isomorphic to the syntactic bimonoid [28] for the language.

Lemma 20. *$\langle \mathsf{SP}, \mathsf{PC} \rangle$ is a closed, consistent, and associative observation table.*

Definition 21 (Canonical pomset recogniser). *The* canonical pomset recogniser *for $\mathcal{L}$ is the the hypothesis for the observation table $\langle \mathsf{SP}, \mathsf{PC} \rangle$. We denote this hypothesis by $\langle M_\mathcal{L}, \odot_\mathcal{L}, \oplus_\mathcal{L}, \mathbf{1}_\mathcal{L}, i_\mathcal{L}, F_\mathcal{L} \rangle$.*

The comparison of the current table with $\langle \mathsf{SP}, \mathsf{PC} \rangle$ is in terms of the number of distinct rows they hold. In the following lemma we show that the number of the former is bounded by the number of the latter.

Lemma 22. *If $M_\mathcal{L}$ is finite, any observation table $\langle S, E \rangle$ satisfies*

$$|\{\mathsf{row}(s) : s \in S\}| \leq |M_\mathcal{L}|.$$

Proof. Note that $M_\mathcal{L} = \{\mathsf{row}_{\langle \mathsf{SP},\mathsf{PC} \rangle}(s) : s \in S\}$. Given $s_1, s_2 \in S$ such that $\mathsf{row}_{\langle S,E \rangle}(s_1) \neq \mathsf{row}_{\langle S,E \rangle}(s_2)$ we have $\mathsf{row}_{\langle \mathsf{SP},\mathsf{PC} \rangle}(s_1) \neq \mathsf{row}_{\langle \mathsf{SP},\mathsf{PC} \rangle}(s_2)$. This implies $|\{\mathsf{row}(s) : s \in S\}| \leq |M_\mathcal{L}|$. □

An important fact will be that none of the pomsets in S can form a counterexample for the hypothesis of a table $\langle S, E \rangle$. In order to show this we will first show that the hypothesis is always *reachable*, a concept we define for arbitrary pomset recognisers below.

Definition 23 (Reachability). *A pomset recogniser $\mathcal{R} = \langle M, \odot, \oplus, \mathbf{1}, i, F \rangle$ is* reachable *if for all $m \in M$ there exists $u \in \mathsf{SP}$ such that $i^\sharp(u) = m$.*

Our reachability lemma relies on the fact that S is subpomset-closed.

Lemma 24 (Hypothesis reachability). *Given a closed, consistent, and associative observation table $\langle S, E \rangle$, the hypothesis it induces is reachable. In particular, ${i_H}^\sharp(s) = \mathsf{row}(s)$ for any $s \in S$.*

From the above it follows that we always have compatibility with respect to the set of row indices, as we show next.

Lemma 25. *The hypothesis of any closed, consistent, and associative observation table* $\langle S, E \rangle$ *is* S*-compatible.*

Before turning to our termination proof, we show that some simple properties hold throughout a run of the algorithm.

Lemma 26 (Invariant). *Throughout execution of Algorithm 1, we have that* $\langle S, E \rangle$ *is a sharp observation table.*

Proof. Subpomset-closedness holds throughout each run since $\{1\}$ is subpomset-closed and adding a single element of S^+ to S preserves the property.

For sharpness, first note that the initial table is sharp as it only has one row. Sharpness of $\langle S, E \rangle$ can only be violated when adding elements to S. But the only place where this happens is on line 7, and there the new row is unequal to all previous rows, which means sharpness is preserved. □

The preceding results allow us to prove our termination theorem.

Theorem 27 (Termination). *If* $M_{\mathcal{L}}$ *is finite, then Algorithm 1 terminates.*

Proof. First, we observe that fixing a closedness defect by adding a row (line 7) can only happen finitely many times, since, by Lemma 22, the size of $\{\mathsf{row}(s) : s \in S\}$ is bounded by $M_{\mathcal{L}}$.

This means that it suffices to show the following two points:

1. Each iteration of any of the loops starting on lines 2–4 either fixes a closedness defect by adding a row, or adapts E so that $\langle S, E \rangle$ ends up *not* being closed at the end of loop body. In the second case, a closedness defect will be fixed in the following iteration of the inner while loop.
2. The calls to HANDLECOUNTEREXAMPLE terminate.

Combined, these show that the algorithm terminates. For the first point, we treat each of the cases:

- If the table is not closed, we directly find a new row that is taken from the S^+-part of the table and added to the S-part of the table.
- Consider the failure of $\heartsuit$-associativity, for $\heartsuit \in \{\cdot, \parallel\}$, and let $s_1, s_2, s_3, s_l, s_r \in S$ and $e \in E$ be such that $\mathsf{row}(s_l) = \mathsf{row}(s_1 \heartsuit s_2)$, $\mathsf{row}(s_r) = \mathsf{row}(s_2 \heartsuit s_3)$, and $\mathsf{row}(s_l \heartsuit s_3)(e) \neq \mathsf{row}(s_1 \heartsuit s_r)(e)$. Suppose $\mathsf{row}(s_l \heartsuit s_3)(e) \neq b$, with b be the result of a membership query on $s_1 \heartsuit s_2 \heartsuit s_3$. Then $e[\square \heartsuit s_3]$ distinguishes the previously equal rows $\mathsf{row}(s_1 \heartsuit s_2)$ and $\mathsf{row}(s_l)$, so adding it to E creates a closedness defect. The fact that $\mathsf{row}(s_1 \heartsuit s_2)$ cannot remain equal to another row than $\mathsf{row}(s_l)$ is a result of the sharpness invariant.
  Alternatively, $\mathsf{row}(s_l \heartsuit s_3)(e) = b$ means $\mathsf{row}(s_1 \heartsuit s_r)(e) \neq b$, for otherwise we would contradict $\mathsf{row}(s_l \heartsuit s_3)(e) \neq \mathsf{row}(s_1 \heartsuit s_r)(e)$. For similar reasons the context $e[s_1 \heartsuit \square]$ in this case distinguishes the previously equal rows $\mathsf{row}(s_1 \heartsuit s_2)$ and $\mathsf{row}(s_r)$, creating a closedness defect.
- A compatibility defect results in the identification of a counterexample, the handling of which we discuss next.

- Whenever a counterexample is identified, we eventually find a context c, $s \in S$, and $t \in S^+ \setminus S$ such that $\mathsf{row}(t) = \mathsf{row}(s)$ and $c[t] \in \mathcal{L} \iff c[s] \notin \mathcal{L}$. Thus, adding c to E creates a closedness defect.

Termination of HANDLECOUNTEREXAMPLE follows: the first two recursive calls in the procedure replace z with strict subpomsets of z, whereas the last one replaces z with an element of S^+, so no further recursion will happen. □

Query Complexity. We determine upper bounds on the membership and equivalence query numbers of a run of the algorithm in terms of the size of the canonical pomset recogniser $n = |M_{\mathcal{L}}|$, the size of the alphabet $k = |\Sigma|$, and the maximum number of operations (from $\{\cdot, \|\}$, used to compose alphabet symbols) m found in a counterexample. We note that since the number of distinct rows indexed by S is bounded by n and the table remains sharp throughout any run, the final size of S is at most n. Thus, the final size of S^+ is in $\mathcal{O}(n^2 + k)$. Given the initialisation of S with a single element, the number of closedness defects fixed throughout a run is at most $n - 1$. This means that the total number of associativity defects fixed and counterexamples handled (including those resulting from compatibility defects) together is $n - 1$. We can already conclude that the number of equivalence queries posed is bounded by n. Moreover, we know that the final table will have at most n columns, and therefore the total number of cells in that table will be in $\mathcal{O}(n^3 + kn)$.

The number of membership queries posed during a run of the algorithm is given by the number of cells in the table plus the number of queries needed during the processing of counterexamples. Consider the counterexample z that contains the maximum number of operations among those encountered during a run. The first two recursive calls of HANDLECOUNTEREXAMPLE break down one operation, whereas the third is used to execute a base case making two membership queries and does not lead to any further recursion. The number of membership queries made starting from a given counterexample is thus in $\mathcal{O}(m)$. This means the total number of membership queries during the processing of counterexamples is in $\mathcal{O}(mn)$, from which we conclude that the number of membership queries posed during a run is in $\mathcal{O}(n^3 + mn + kn)$.

3.4 Minimality of Hypotheses

In this section we will show that all hypotheses submitted by the algorithm to the teacher are minimal. We first need to define what minimality means. As is the case for DFAs, it is the combination of an absence of unreachable states and of every state exhibiting its own distinct behaviour.

Definition 28 (Minimality). *A pomset recogniser* $\mathcal{R} = \langle M, \odot, \oplus, \mathbf{1}, i, F \rangle$ *is* minimal *if it is reachable and for all* $u, v \in \mathsf{SP}$ *with* $i^\sharp(u) \neq i^\sharp(v)$ *there exists* $c \in \mathsf{PC}$ *such that* $c[u] \in \mathcal{L}_{\mathcal{R}} \iff c[v] \notin \mathcal{L}_{\mathcal{R}}$.

Before proving the main result of this section, we need the following:

Lemma 29. *For all pomset recognisers* $\langle M, \odot, \oplus, \mathbf{1}, i, F\rangle$ *and* $u, v \in \mathsf{SP}$ *such that* $i^\sharp(u) = i^\sharp(v)$ *we have for any* $c \in \mathsf{PC}$ *that* $i^\sharp(c[u]) = i^\sharp(c[v])$.

The minimality theorem below relies on table compatibility, which allows us to distinguish the behaviour of states based on the contents of their rows. Note that the algorithm only submits a hypothesis in an equivalence query if that hypothesis is compatible with its table.

Theorem 30 (Minimality of hypotheses). *A closed, consistent, and associative observation* $\langle S, E\rangle$ *induces a minimal hypothesis if the hypothesis is compatible with its table.*

Proof. We obtain the hypothesis from Lemma 14. Since S is subpomset-closed, we have by Lemma 24 that the hypothesis is reachable. Moreover, for every $s \in S$ we have ${i_H}^\sharp(s) = \mathsf{row}(s)$. Consider $u_1, u_2 \in \mathsf{SP}$ such that ${i_H}^\sharp(u_1) \neq {i_H}^\sharp(u_2)$. Then there exist $s_1, s_2 \in S$ such that $\mathsf{row}(s_1) = {i_H}^\sharp(u_1)$ and $\mathsf{row}(s_2) = {i_H}^\sharp(u_2)$, and we have $\mathsf{row}(s_1) \neq \mathsf{row}(s_2)$. Let $e \in E$ be such that $\mathsf{row}(s_1)(e) \neq \mathsf{row}(s_2)(e)$. We have

$$\begin{aligned}
{i_H}^\sharp(e[u_1]) \in F_H &\iff {i_H}^\sharp(e[s_1]) \in F_H && \text{(Lemma 29)}\\
&\iff e[s_1] \in \mathcal{L}_{\mathcal{H}}\\
&\iff \mathsf{row}(s_1)(e) = 1\\
&\iff \mathsf{row}(s_2)(e) = 0\\
&\iff e[s_2] \notin \mathcal{L}_{\mathcal{H}}\\
&\iff {i_H}^\sharp(e[s_2]) \notin F_H\\
&\iff {i_H}^\sharp(e[u_2]) \notin F_H. && \text{(Lemma 29)} \quad \square
\end{aligned}$$

As a corollary, we find that the canonical pomset recogniser is minimal.

Proposition 31. *The canonical pomset recogniser is minimal.*

4 Conversion to Pomset Automata

Bimonoids are a useful representation of pomset languages because sequential and parallel composition are on an equal footing; in the case of the learning algorithm of the previous section, this helps us treat both operations similarly. On the other hand, the behaviour of a program is usually thought of as a series of actions, some of which involve launching two or more threads that later combine. Here, sequential actions form the basic unit of computation, while fork/join patterns of threads are specified separately. *Pomset automata* [22] encode this more asymmetric model: they can be thought of as non-deterministic finite automata with an additional transition type that brokers forking and joining threads.

In this section, we show how to convert a pomset recogniser to a certain type of pomset automaton, where acceptance of a pomset is guided by its structure; conversely, we show that each of the pomset automata in this class can be represented by a pomset recogniser. Together with the previous section, this establishes that the languages of pomset automata in this class are learnable.

If S is a set, we write $\mathbb{M}(S)$ for the set of *finite multisets* over S. A finite multiset over S is written $\phi = \{\!|s_1, \ldots, s_n|\!\}$.

Definition 32 (Pomset automata). *A* pomset automaton *(PA) is a tuple* $A = \langle Q, I, F, \delta, \gamma\rangle$ *where*

- Q *is a set of* states, *with* $I, F \subseteq Q$ *the* initial *and* accepting states, *and*
- $\delta\colon Q \times \Sigma \to 2^Q$ *the* sequential transition function, *and*
- $\gamma\colon Q \times \mathbb{M}(Q) \to 2^Q$ *the* parallel transition function.

Lastly, for every $q \in Q$ *there are finitely many* $\phi \in \mathbb{M}(Q)$ *such that* $\gamma(q, \phi) \neq \emptyset$.

A finite PA can be represented graphically: every state is drawn as a vertex, with accepting states doubly circled and initial states pointed out by an arrow, while δ-transitions are represented by labelled edges, and γ-transitions are drawn as a multi-ended edge. For instance, in Figure 1a, we have drawn a PA with states q_0 through q_5 with q_5 accepting, and $q_1 \in \delta(q_0, \mathtt{a})$ (among other δ-transitions), while the multi-ended edge represents that $q_2 \in \gamma(q_1, \{\!|q_3, q_4|\!\})$, i.e., q_2 can launch threads starting in q_3 and q_4, which, upon termination, resume in q_2.

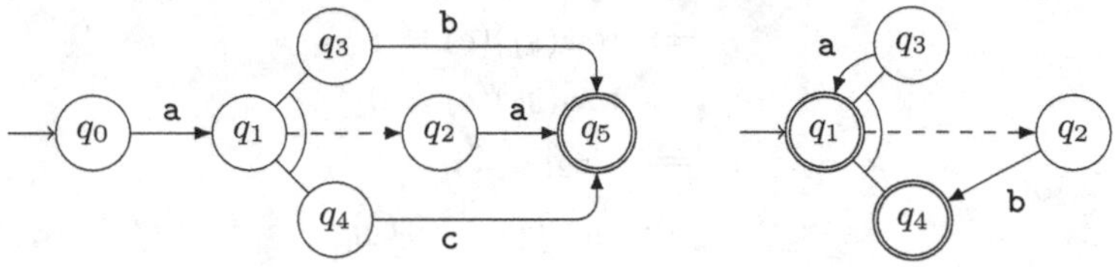

(a) A simple PA. (b) A non-saturated PA.

Fig. 1: Some pomset automata.

The sequential transition function is interpreted as in non-deterministic finite automata: if $q' \in \delta(q, \mathtt{a})$, then a machine in state q may transition to state q' after performing the action $\mathtt{a}$. The intuition to the parallel transition function is that if $q' \in \gamma(q, \{\!|r_1, \ldots, r_n|\!\})$, then a machine in state q may launch threads starting in states r_1 through r_n, and when each of those has terminated succesfully, may proceed in state q'. Note how the representation of starting states in a γ-transition allows for the possibility of launching multiple instances of the same thread, and disregards their order—i.e., $\gamma(q, \{\!|r_1, \ldots, r_n|\!\}) = \gamma(q, \{\!|r_n, \ldots, r_1|\!\})$. This intuition is made precise through the notion of a *run*.

Definition 33 (Run relation). *The* run relation *of a PA* $A = \langle Q, I, F, \delta, \gamma\rangle$, *denoted* $\rightarrow_A$, *is defined as the the smallest subset of* $Q \times \mathsf{SP} \times Q$ *satisfying*

$$\frac{}{q \xrightarrow{1}_A q} \qquad \frac{q' \in \delta(q, \mathtt{a})}{q \xrightarrow{\mathtt{a}}_A q'} \qquad \frac{\forall 1 \leq i \leq n.\ r_i \xrightarrow{u_i}_A r_i' \in F \quad q' \in \gamma(q, \{\!|r_1, \ldots, r_n|\!\})}{q \xrightarrow{u_1 \| \cdots \| u_n}_A q'} \qquad \frac{q \xrightarrow{u}_A q'' \quad q'' \xrightarrow{v}_A q'}{q \xrightarrow{u \cdot v}_A q'}$$

The language accepted *by* A *is* $\mathcal{L}_A = \{u \in \mathsf{SP} : \exists q \in I, q' \in F.\ q \xrightarrow{u}_A q'\}$.

Example 34. If A is the PA from Figure 1a, we can see that $q_3 \xrightarrow{\mathtt{b}}_A q_5$ and $q_4 \xrightarrow{\mathtt{c}}_A q_5$ as a result of the second rule; by the third rule, we find that $q_1 \xrightarrow{\mathtt{b} \parallel \mathtt{c}}_A q_2$. Since $q_2 \xrightarrow{\mathtt{a}} q_5$ and $q_0 \xrightarrow{a}_A q_1$ (again by the second rule), we can conclude $q_0 \xrightarrow{\mathtt{a} \cdot (\mathtt{b} \parallel \mathtt{c}) \cdot \mathtt{a}}_A q_5$ by repeated application of the last rule. The language accepted by this PA is the singleton set $\{\mathtt{a} \cdot (\mathtt{b} \parallel \mathtt{c}) \cdot \mathtt{a}\}$.

In general, finite pomset automata can accept a very wide range of pomset languages, including all context free (pomset) languages [23]. The intuition behind this is that the mechanism of forking and joining encoded in γ can be used to simulate a call stack. For example, the automaton in Figure 1b accepts the strictly context-free language (of words) $\{\mathtt{a}^n \cdot \mathtt{b}^n : n \in \mathbb{N}\}$. It follows that PAs can represent strictly more pomset languages than pomset recognisers. To tame the expressive power of PAs at least slightly, we propose the following.

Definition 35 (Saturation). *We say that* $A = \langle Q, I, F, \delta, \gamma \rangle$ *is* saturated *when for all* $u, v \in \mathsf{SP}$ *with* $u, v \neq 1$*, both of the following are true:*

(i) If $q \xrightarrow{u \cdot v}_A q'$*, then there exists a* $q'' \in Q$ *with* $q \xrightarrow{u}_A q''$ *and* $q'' \xrightarrow{v}_A q'$.
(ii) If $q \xrightarrow{u \parallel v}_A q'$*, then there exist* $r, s \in Q$ *and* $r', s' \in F$ *such that*

$$r \xrightarrow{u}_A r' \qquad s \xrightarrow{v}_A s' \qquad q' \in \gamma(q, \{\!|r, s|\!\})$$

Example 36. Returning to Figure 1, we see that the PA in Figure 1a is saturated, while Figure 1b is not, as a result of the run $q_1 \xrightarrow{\mathtt{a}\cdot\mathtt{a}\cdot\mathtt{b}\cdot\mathtt{b}}_A q_4$, which does not admit an intermediate state q such that $q_1 \xrightarrow{\mathtt{a}\cdot\mathtt{a}}_A q$ and $q \xrightarrow{\mathtt{b}\cdot\mathtt{b}}_A q_4$.

We now have everything in place to convert the encoding of a language given by a pomset recogniser to a pomset automaton. The idea is to represent every element q of the bimonoid by a state which accepts exactly the language of pomsets mapped to q; the transition structure is derived from the operations.

Lemma 37. *Let* $\mathcal{R} = \langle M, \odot, \oplus, \mathbf{1}, i, F \rangle$ *be a pomset recogniser. We construct the pomset automaton* $A = \langle M, F, \{\mathbf{1}\}, \delta, \gamma \rangle$ *(note: we use* F *as the set of initial states) where* $\delta\colon M \times \Sigma \to 2^M$ *and* $\gamma\colon M \times \mathbb{M}(M) \to 2^M$ *are given by*

$$\delta(q, \mathtt{a}) = \{q' : i(\mathtt{a}) \odot q' = q\} \qquad \gamma(q, \phi) = \{q' : (r \oplus r') \odot q' = q,\ \phi = \{\!|r, r'|\!\}\}$$

Then A *is saturated, and* $\mathcal{L}_A = \mathcal{L}_\mathcal{R}$.

Example 38. Let $\langle M, \odot, \oplus, \mathbf{1}, i, F \rangle$ be the pomset recogniser from Example 7. The pomset automaton that arises from the construction above is partially depicted in Figure 2; we have not drawn the state $q_\perp$ and its incoming transitions, or forks into $\mathbf{1}$, to avoid clutter. In this PA, we see that, since $q_\mathtt{a} \odot q_1 = q_\mathtt{b}$ and $i(\mathtt{a}) = q_\mathtt{a}$, we have $q_1 \in \delta(q_\mathtt{b}, \mathtt{a})$. Furthermore, since $(q_\mathtt{b} \oplus q_\mathtt{b}) \odot \mathbf{1} = q_1 \odot \mathbf{1} = q_1$, we also have $\mathbf{1} \in \gamma(q_1, \{\!|q_\mathtt{b}, q_\mathtt{b}|\!\})$. Finally, $q_\mathtt{b}$ is initial, since $F = \{q_\mathtt{b}\}$.

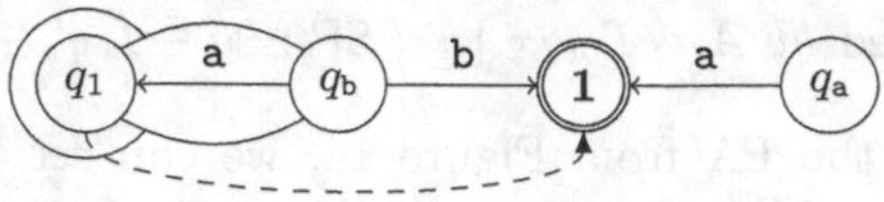

Fig. 2: Part of the PA obtained from the pomset recogniser from Example 7, using the construction from Lemma 37. The state $q_\perp$ (which does not contribute to the language of the automaton) and forks into the state **1** are not pictured.

We have thus shown that the language of any pomset recogniser can be accepted by a finite and saturated PA. In turn, this shows that our algorithm can, in principle, be adapted to work with a teacher that takes a (saturated) PA instead of a pomset recogniser as hypothesis, by simply converting the hypothesis pomset recogniser to an equivalent PA before sending it over.

Conversely, we can show that the transition relations of a saturated PA carry the algebraic structure of a bimonoid, and use that to show that a language recognised by a saturated PA is also recognised by a bimonoid. This shows that our characterisation is "tight", i.e., languages recognised by saturated PAs are precisely those recognised by bimonoids, and hence learnable.

Lemma 39. *Let $A = \langle Q, I, F, \delta, \gamma \rangle$ be a saturated pomset automaton. We can construct a pomset recogniser $\mathcal{R} = \langle M, \odot, \oplus, \mathbf{1}, i, F' \rangle$, where*

$$M = \{ \xrightarrow{u}_A : u \in \mathsf{SP} \} \qquad \xrightarrow{u}_A \odot \xrightarrow{v}_A = \xrightarrow{u \cdot v}_A \qquad \xrightarrow{u}_A \oplus \xrightarrow{v}_A = \xrightarrow{u \parallel v}_A$$

$$i(\mathtt{a}) = \xrightarrow{\mathtt{a}}_A \qquad F' = \{ \xrightarrow{u}_A \in M : \exists q \in I, q' \in F.\ q \xrightarrow{u}_A q' \}$$

Now $\odot$ and $\oplus$ are well-defined, and $\mathcal{R}$ is a pomset recogniser such that $\mathcal{L}_\mathcal{R} = \mathcal{L}_A$.

If A is finite, then so is $\mathcal{R}$, since each of the elements of M is a relation on Q, and there are finitely many relations on a finite set.

In general, the PA obtained from a pomset recogniser may admit runs where the same fork transition is nested repeatedly. Recognisable pomset languages of *bounded width* may be recognised by a pomset recogniser that is *depth-nilpotent* [28], which can be converted into a *fork-acyclic* PA by way of an sr-expression [28,22]. However, this detour via sr-expressions is not necessary: one can adapt Lemma 37 to produce a fork-acyclic PA, when given a depth-nilpotent pomset recogniser. The details are discussed in the full version [15].

We conclude this section by remarking that the minimal pomset recogniser for a bounded-width language is necessarily depth-nilpotent [28]; since our algorithm produces a minimal pomset recogniser, this means that we can also produce a fork-acyclic PA after learning a bounded-width recognisable pomset language.

5 Discussion

To learn DFAs, there are several alternatives to the observation table data structure that reduce the space complexity of the algorithm. Most notable is the *classification tree* [25], which distinguishes individual pairs of words (which for us would be pomsets) at every node rather than filling an entire row for each of them. The TTT algorithm [19] further builds on this and achieves optimal space complexity. Given that we developed the first learning algorithm for pomset languages, we opted for the simplicity of the observation table—optimisations such as those analogous to the aforementioned work are left to future research.

We would like to extend our algorithm to learn recognisers based on arbitrary algebraic theories. One challenge is to ensure that the equations of the theory hold for hypotheses, by generalising our definition of associativity (Definition 13).

Our algorithm can also be specialised to learn languages recognised by commutative monoids. These languages of *multisets* can alternatively be represented as semi-linear sets [30] or described using Presburger arithmetic [11]. While not all languages described this way are recognisable (for instance, the set of multisets over $\Sigma = \{\mathtt{a}, \mathtt{b}\}$ with as many $\mathtt{a}$'s as $\mathtt{b}$'s [28]), it would be interesting to be able to learn at least the fragment representable by commutative monoids, and apply that to one of the domains where semi-linear sets are used.

Our algorithm is limited to learning languages of series-parallel pomsets; there exist pomsets which are not series-parallel, each of which must contain an "N-shape" [12,13,35]. Since N-shapes appear in pomsets that describe message passing between threads, we would like to be able to learn such languages as well. We do not see an obvious way to extend our algorithm to include these pomsets, but perhaps recent techniques from [10] can provide a solution.

Every hypothesis of our algorithm can be converted to a pomset automaton. The final pomset recogniser for a bounded-width language is minimal, and hence depth-nilpotent [28], which means that it can be converted to a fork-acyclic PA. In future work, we would like to guarantee that the same holds for intermediate hypotheses when learning a bounded-width language.

Running two threads in parallel may be implemented by running some initial section of those threads in parallel, followed by running the remainder of those threads in parallel. This interleaving is represented by the *exchange law* [12,13]. One can specialise pomset recognisers to include this interleaving to obtain recognisers of pomset languages closed under subsumption [28], i.e., such that if a pomset u is recognised, then so are all of the "more sequential" versions of u. We would like to adapt our algorithm to learn these types of recognisers, and exploit the extra structure provided by the exchange law to optimise further.

We have shown that recognisable pomset languages correspond to saturated regular pomset languages (Lemmas 37 and 39). One question that remains is whether there is an algorithm that can learn all or at least a larger class of regular pomset languages. Given that pomset automata can accept context-free languages (Figure 1b), we wonder if a suitable notion of context-free grammars for pomset languages could be identified. Clark [6] showed that there exists a subclass of context-free languages that can be learned via an adaptation of $\mathtt{L}^\star$. Arguably, this adaptation learns recognisers with a monoidal structure and reverses this structure to obtain a grammar. An extension of this work to pomset languages might lead to a learning algorithm that learns more PAs.

References

1. Aarts, F., Vaandrager, F.W.: Learning I/O automata. In: CONCUR. pp. 71–85 (2010). https://doi.org/10.1007/978-3-642-15375-4_6
2. Angluin, D.: Learning regular sets from queries and counterexamples. Inf. Comput. **75**(2), 87–106 (1987). https://doi.org/10.1016/0890-5401(87)90052-6
3. Barlocco, S., Kupke, C.: Angluin learning via logic. In: LFCS. LNCS, vol. 10703, pp. 72–90. Springer (2018). https://doi.org/10.1007/978-3-319-72056-2_5
4. Bojanczyk, M.: Recognisable languages over monads. In: DLT. pp. 1–13 (2015). https://doi.org/10.1007/978-3-319-21500-6_1
5. Chapman, M., Chockler, H., Kesseli, P., Kroening, D., Strichman, O., Tautschnig, M.: Learning the language of error. In: ATVA. pp. 114–130 (2015). https://doi.org/10.1007/978-3-319-24953-7_9
6. Clark, A.: Distributional learning of some context-free languages with a minimally adequate teacher. In: ICGI. pp. 24–37 (2010). https://doi.org/10.1007/978-3-642-15488-1_4
7. Drewes, F., Högberg, J.: Learning a regular tree language from a teacher. In: DLT. pp. 279–291 (2003). https://doi.org/10.1007/3-540-45007-6_22
8. Drewes, F., Högberg, J.: Query learning of regular tree languages: How to avoid dead states. Theory Comput. Syst. **40**, 163–185 (2007). https://doi.org/10.1007/s00224-005-1233-3
9. Ésik, Z., Németh, Z.L.: Higher dimensional automata. J. Autom. Lang. Comb. **9**(1), 3–29 (2004). https://doi.org/10.25596/jalc-2004-003
10. Fahrenberg, U., Johansen, C., Struth, G., Thapa, R.B.: Generating posets beyond $\mathbb{N}$. In: RAMiCS. pp. 82–99 (2020). https://doi.org/10.1007/978-3-030-43520-2_6
11. Ginsburg, S., Spanier, E.H.: Bounded ALGOL-like languages. Trans. Am. Math. Soc. **113**(2), 333–368 (1964). https://doi.org/10.2307/1994067
12. Gischer, J.L.: The equational theory of pomsets. Theor. Comput. Sci. **61**, 199–224 (1988). https://doi.org/10.1016/0304-3975(88)90124-7
13. Grabowski, J.: On partial languages. Fundam. Inform. **4**(2), 427 (1981)
14. van Heerdt, G.: Efficient Inference of Mealy Machines. Bachelor's thesis, Radboud University (2014), https://www.cs.ru.nl/bachelors-theses/2014/Gerco_van_Heerdt___4167503___Efficient_Inference_of_Mealy_Machines.pdf
15. van Heerdt, G., Kappé, T., Rot, J., Silva, A.: Learning pomset automata (2021), to appear on arXiv.
16. van Heerdt, G., Kupke, C., Rot, J., Silva, A.: Learning weighted automata over principal ideal domains. In: FOSSACS. pp. 602–621 (2020). https://doi.org/10.1007/978-3-030-45231-5_31
17. Hoare, T., Möller, B., Struth, G., Wehrman, I.: Concurrent Kleene algebra. In: Proc. Concurrency Theory (CONCUR). pp. 399–414 (2009). https://doi.org/10.1007/978-3-642-04081-8_27
18. Howar, F., Steffen, B.: Active automata learning in practice - an annotated bibliography of the years 2011 to 2016. In: Machine Learning for Dynamic Software Analysis. pp. 123–148 (2018). https://doi.org/10.1007/978-3-319-96562-8_5
19. Isberner, M., Howar, F., Steffen, B.: The TTT algorithm: A redundancy-free approach to active automata learning. In: RV. LNCS, vol. 8734, pp. 307–322. Springer (2014). https://doi.org/10.1007/978-3-319-11164-3_26
20. Isberner, M., Howar, F., Steffen, B.: The open-source learnlib - A framework for active automata learning. In: CAV. pp. 487–495 (2015). https://doi.org/10.1007/978-3-319-21690-4_32

21. Kappé, T., Brunet, P., Luttik, B., Silva, A., Zanasi, F.: Brzozowski goes concurrent - A Kleene theorem for pomset languages. In: CONCUR. pp. 25:1–25:16 (2017). https://doi.org/10.4230/LIPIcs.CONCUR.2017.25
22. Kappé, T., Brunet, P., Luttik, B., Silva, A., Zanasi, F.: Equivalence checking for weak bi-Kleene algebra (2018), https://arxiv.org/abs/1807.02102, under submission
23. Kappé, T., Brunet, P., Luttik, B., Silva, A., Zanasi, F.: On series-parallel pomset languages: Rationality, context-freeness and automata. J. Log. Algebr. Meth. Program. **103**, 130–153 (2019). https://doi.org/10.1016/j.jlamp.2018.12.001
24. Kappé, T., Brunet, P., Silva, A., Zanasi, F.: Concurrent Kleene algebra: Free model and completeness. In: ESOP. pp. 856–882 (2018). https://doi.org/10.1007/978-3-319-89884-1_30
25. Kearns, M.J., Vazirani, U.V.: An Introduction to Computational Learning Theory. MIT press (1994)
26. Laurence, M.R., Struth, G.: Completeness theorems for bi-Kleene algebras and series-parallel rational pomset languages. In: Proc. Relational and Algebraic Methods in Computer Science (RAMiCS). pp. 65–82 (2014). https://doi.org/10.1007/978-3-319-06251-8_5
27. Lodaya, K., Weil, P.: A Kleene iteration for parallelism. In: FSTTCS. pp. 355–366 (1998). https://doi.org/10.1007/978-3-540-49382-2_33
28. Lodaya, K., Weil, P.: Series-parallel languages and the bounded-width property. Theoretical Computer Science **237**(1), 347–380 (2000). https://doi.org/10.1016/S0304-3975(00)00031-1
29. Maler, O., Pnueli, A.: On the learnability of infinitary regular sets. Inf. Comput. **118**, 316–326 (1995). https://doi.org/10.1006/inco.1995.1070
30. Parikh, R.: On context-free languages. J. ACM **13**(4), 570–581 (1966). https://doi.org/10.1145/321356.321364
31. Sakakibara, Y.: Learning context-free grammars from structural data in polynomial time. Theor. Comput. Sci. **76**(2-3), 223–242 (1990). https://doi.org/10.1016/0304-3975(90)90017-C
32. Urbat, H., Adámek, J., Chen, L., Milius, S.: Eilenberg theorems for free. In: MFCS. pp. 43:1–43:15 (2017). https://doi.org/10.4230/LIPIcs.MFCS.2017.43
33. Urbat, H., Schröder, L.: Automata learning: An algebraic approach. In: LICS. pp. 900–914 (2020). https://doi.org/10.1145/3373718.3394775
34. Vaandrager, F.W.: Model learning. Commun. ACM **60**(2), 86–95 (2017). https://doi.org/10.1145/2967606
35. Valdes, J., Tarjan, R.E., Lawler, E.L.: The recognition of series parallel digraphs. SIAM J. Comput. **11**(2), 298–313 (1982). https://doi.org/10.1137/0211023

The Structure of Sum-Over-Paths, its Consequences, and Completeness for Clifford*

Renaud Vilmart(✉)

Université Paris-Saclay, ENS Paris-Saclay, Inria, CNRS, LMF, 91190, Gif-sur-Yvette, France
vilmart@lsv.fr

Abstract. We show that the formalism of "Sum-Over-Path" (SOP), used for symbolically representing linear maps or quantum operators, together with a proper rewrite system, has the structure of a dagger-compact PROP. Several consequences arise from this observation:
– Morphisms of SOP are very close to the diagrams of the graphical calculus called ZH-Calculus, so we give a system of interpretation between the two
– A construction, called the discard construction, can be applied to enrich the formalism so that, in particular, it can represent the quantum measurement.
We also enrich the rewrite system so as to get the completeness of the Clifford fragments of both the initial formalism and its enriched version.

Keywords: Categorical Quantum Mechanics · Dagger-Compact PROP · Sum-Over-Paths · Clifford Fragment · Normal Form · Rewriting · Discard Construction · Verification.

1 Introduction

The "Sum-Over-Paths" (SOP) formalism [1] was introduced in order to perform verification on quantum circuits. It is inspired by Feynman's notion of path-integrals, and can be conceived as a discrete version of it.

The core idea here is to represent unitary transformations in a symbolic way, so as to be able to simplify the term, which would for instance accelerate its evaluation. To do so, the formalism comes equipped with a rewrite system, which reduces any term into an equivalent one.

As pure quantum circuits (which represent unitary maps) can easily be mapped to an SOP morphism, one can try and perform verification: given a specification $\mathcal{S}$ and another SOP morphism t obtained from a circuit supposed to implement the specification, we can compute the term $\mathcal{S}\circ t^\dagger$ and try to reduce it to the identity. In a very similar way, one can check whether two quantum circuits implement the same unitary map.

* This work was made during a Postdoc funded by the project PIA-GDN/Quantex. Proofs can be found at arXiv:2003.05678

S. Kiefer and C. Tasson (Eds.): FOSSACS 2021, LNCS 12650, pp. 531–550, 2021.
https://doi.org/10.1007/978-3-030-71995-1_27

The rewrite system is known to be complete for Clifford unitary maps, i.e. in the Clifford fragment of quantum mechanics, the term obtained from $t_1 \circ t_2^\dagger$ will reduce to the identity iff t_1 and t_2 represent the same unitary map. Moreover, this reduction terminates in time polynomial in the size of the SOP term (itself related to the size of the quantum circuit), and still performs well outside the Clifford fragment.

Lately, the SOP formalism has been used for efficient verification of optimisation strategies such as [4,12], as well as for specification of quantum circuits [6].

In this paper, we are interested in extensions of the formalism. We first focus on its categorical structure, and show that arbitrary terms already go beyond the representation of unitary maps. We then turn to extending the formalism to encompass mixed quantum processes. In both cases, we show a completeness result for their respective Clifford fragment.

In Section 2, we explain in details the structure of †-compact PROP, which we show in Section 3 to be shared by **SOP**.

Because the formalism is no longer restricted to unitary maps, we argue that it could benefit from a slight redefinition, which is done in Section 4.

Another "family" of categories that share this structure is the family of graphical languages for quantum computation: ZX-Calculus, ZW-Calculus and ZH-Calculus [3,7,8]. All three formalisms represent morphisms of **Qubit** using diagrams, and come with equational theories, proven to be complete for the whole category [3,11,19], i.e. whenever two diagrams represent the same morphism of **Qubit**, the first can be turned into the other using only the equational theory.

In Section 5, we present interpretations between the respective Clifford fragments of the ZH-calculus and **SOP**, in a slightly different way than in [14,15], partly thanks to our redefinition of sums-over-paths.

In Section 6, we realise that the original rewrite system of **SOP** is not enough for the completeness of the Clifford fragment of **Qubit**. We hence enrich the set of rules so as to get the completeness in this restriction.

In Section 7, we enrich the whole formalism using the discard construction [5], so as to be able to represent completely positive maps, as well as the operator of partial trace. Again, one can consider the Clifford fragment of this formalism. We give a new set of rewrite rules, and show that it makes the fragment complete.

2 Background

2.1 PROPs and String Diagrams

The first kind of category we will be interested in is the *PROP* [13,20]. A PROP **C** is a strict symmetric monoidal category (SMC) [16,18] generated by a single object, or equivalently, whose objects form $\mathbb{N}$. Hence the morphisms of **C** are of the form $f : n \to m$. They can be composed sequentially $(. \circ .)$ or in parallel $(. \otimes .)$, and they satisfy the following axioms:

$$f \circ (g \circ h) = (f \circ g) \circ h \qquad f \otimes (g \otimes h) = (f \otimes g) \otimes h$$

$$id_m \circ f = f = f \circ id_n \qquad id_0 \otimes f = f = f \otimes id_0$$
$$(f_2 \circ f_1) \otimes (g_2 \circ g_1) = (f_2 \circ g_2) \circ (f_1 \otimes g_1)$$

The category is also equipped with a particular family of morphisms $\sigma_{n,m} : n+m \to m+n$. Intuitively, these allow morphisms to swap places. They satisfy additional axioms:

$$\sigma_{n,m+p} = (id_m \otimes \sigma_{n,p}) \circ (\sigma_{n,m} \otimes id_p) \qquad \sigma_{n+m,p} = (\sigma_{n,p} \otimes id_m) \circ (id_n \otimes \sigma_{m,p})$$
$$\sigma_{m,n} \circ \sigma_{n,m} = id_{n+m} \qquad (id_p \otimes f) \circ \sigma_{n,p} = \sigma_{m,p} \circ (f \otimes id_p)$$

2.2 †-Compact PROPs

Some PROPs can have additional structure, such as a compact-closed structure, or a †-functor.

A †-PROP **C** is a PROP together with an involutive, identity-on-objects functor $(.)^\dagger : \mathbf{C}^{\mathrm{op}} \to \mathbf{C}$ compatible with $(. \otimes .)$. That is, for every morphism $f : n \to m$, there is a morphism $f^\dagger : m \to n$ such that $f^{\dagger\dagger} = f$. It behaves with the compositions by $(f \circ g)^\dagger = g^\dagger \circ f^\dagger$ and $(f \otimes g)^\dagger = f^\dagger \otimes g^\dagger$. Finally, we have $\sigma_{n,m}^\dagger = \sigma_{m,n}$.

A †-compact PROP has two particular families of morphisms: $\eta_n : 0 \to 2n$ and $\epsilon_n : 2n \to 0$. These are dual by the †-functor: $\eta_n^\dagger = \epsilon_n$. They satisfy the following axioms:

$$(\epsilon_n \otimes id_n) \circ (id_n \otimes \eta_n) = id_n = (id_n \otimes \epsilon_n) \circ (\eta_n \otimes id_n)$$
$$\sigma_{n,n} \circ \eta_n = \eta_n \qquad \eta_{n+m} = (id_n \otimes \sigma_{n,m} \otimes id_m) \circ (\eta_n \otimes \eta_m)$$

In this context, one can define the transpose operator of a morphism f as:

$$f^t := (\epsilon_m \otimes id_n) \circ (id_m \otimes f \otimes id_n) \circ (id_m \otimes \eta_m)$$

One can check that, thanks to the axioms of †-compact PROP, $(f \circ g)^t = g^t \circ f^t$, $(f \otimes g)^t = f^t \otimes g^t$, and $f^{tt} = f$.

We can then compose $(.)^t$ and $(.)^\dagger$: $\overline{(.)} := (.)^{\dagger t}$. Again using the axioms of †-compact PROP, one can check that $(.)^{\dagger t} = (.)^{t\dagger}$.

2.3 Example: Qubit

The usual example of a strict symmetric †-compact monoidal category is **FHilb**, the category whose objects are finite dimensional Hilbert spaces, and whose morphisms are linear maps between them. It is not, however, a PROP, as it is not generated by a single object.

One subcategory of **FHilb** that *is* a PROP, though, is **Qubit**, the subcategory of **FHilb** generated by the object $\mathbb{C}^2$, considered as the object 1. A morphism $f : n \to m$ of **Qubit** is hence a linear map from $\mathbb{C}^{2^n}$ to $\mathbb{C}^{2^m}$. $(. \circ .)$ is then the usual composition of linear maps, and $(. \otimes .)$ is the usual tensor product of linear maps. One can check that the first set of axioms is satisfied.

This is not enough to conclude that **Qubit** is a PROP. We still need to define a family of morphisms $\sigma_{n,m}$. In the Dirac notation, given a basis $\mathcal{B}$ of $\mathbb{C}^2$, we can define $\sigma_{n,m}$ as $\sigma_{n,m} := \sum_{(\boldsymbol{x},\boldsymbol{y})\in\mathcal{B}^n\times\mathcal{B}^m} |\boldsymbol{y},\boldsymbol{x}\rangle\langle\boldsymbol{x},\boldsymbol{y}|$. One can then check that all the axioms of PROPs are satisfied.

Qubit is not only a PROP, but also †-compact. Indeed, first, given a morphism:

$$f = \sum_{(\boldsymbol{x},\boldsymbol{y})\in\mathcal{B}^n\times\mathcal{B}^m} a_{\boldsymbol{x},\boldsymbol{y}} |\boldsymbol{y}\rangle\langle\boldsymbol{x}|$$

we can define its dagger $f^\dagger := \sum_{(\boldsymbol{x},\boldsymbol{y})\in\mathcal{B}^n\times\mathcal{B}^m} \overline{a_{\boldsymbol{x},\boldsymbol{y}}} |\boldsymbol{x}\rangle\langle\boldsymbol{y}|$, which is the usual definition of the dagger for linear maps.

Its compact structure can be given by $\eta_n := \sum_{\boldsymbol{x}\in\mathcal{B}^n} |\boldsymbol{x},\boldsymbol{x}\rangle$, which implies $\epsilon_n = \eta_n^\dagger = \sum_{\boldsymbol{x}\in\mathcal{B}^n} \langle\boldsymbol{x},\boldsymbol{x}|$. One can check that all the axioms of †-compact PROPs are satisfied.

Since **Qubit** is †-compact, we can define the transpose $(.)^t$ which happens to be the usual transpose of linear maps, and the conjugate $\overline{(.)}$, which again is the usual conjugation in linear maps over $\mathbb{C}$.

There is a subcategory of **Qubit** that is of importance: **Stab**. It is the smallest †-compact subcategory of **Qubit** (the compact structure is preserved) that contains:

- $|0\rangle : 0 \to 1$
- $H := \frac{1}{\sqrt{2}}(|0\rangle\langle 0| + |0\rangle\langle 1| + |1\rangle\langle 0| - |1\rangle\langle 1|) : 1 \to 1$
- $S := |0\rangle\langle 0| + i\,|1\rangle\langle 1| : 1 \to 1$
- $CZ := |00\rangle\langle 00| + |01\rangle\langle 01| + |10\rangle\langle 10| - |11\rangle\langle 11| : 2 \to 2$

3 The Category SOP

3.1 SOP as a PROP

The point of the Sum-Over-Paths formalism [1], is to *symbolically* manipulate morphisms written in a form akin to the Dirac notation. Reasoning on symbolic terms allow us to detect where a term can be simplified to a "smaller" one, or to give a specification on a term.

A morphism of the category will be of the form:
$|\boldsymbol{x}\rangle \mapsto s \sum_{\boldsymbol{y}\in V^k} e^{2i\pi P(\boldsymbol{x},\boldsymbol{y})} |\boldsymbol{Q}(\boldsymbol{x},\boldsymbol{y})\rangle$ where:

- $\boldsymbol{x} = x_1, \ldots, x_n$ is the input signature, it is a list of variables
- V is a set of variables (hence $\boldsymbol{y}$ is a collection of these variables)
- P is a multivariate polynomial, instantiated by the variables $\boldsymbol{x}$ and $\boldsymbol{y}$
- $\boldsymbol{Q} = Q_1, \ldots, Q_m$ is the output signature, it is a multivariate, multivalued boolean polynomial
- s is a real scalar

We may denote V_f a subset of the variables V used in f. Then by default, if V_f and V_g are used in the same term, we consider that $V_f \cap V_g = \varnothing$. To distinguish the two sum operators (the one in P and the one in $\boldsymbol{Q}$), we can denote the one in the output signature $\boldsymbol{Q}$ as $\oplus$. Moreover, it will sometimes be necessary to immerse one of the boolean polynomials Q_i in the polynomial P. We hence define $\widehat{Q_i}$ inductively as $\widehat{x} = x$ for a variable x, $\widehat{pq} = \widehat{p}\widehat{q}$ and $\widehat{p \oplus q} = \widehat{p} + \widehat{q} - 2\widehat{p}\widehat{q}$.

Definition 1 (SOP). **SOP** *is defined as the PROP where, given a set of variables V:*

- *Identity morphisms are $id_n : |\boldsymbol{x}\rangle \mapsto |\boldsymbol{x}\rangle$*
- *Morphisms $f : n \to m$ are of the form $f : |\boldsymbol{x}\rangle \mapsto s \sum\limits_{\boldsymbol{y} \in V^k} e^{2i\pi P(\boldsymbol{x},\boldsymbol{y})} |\boldsymbol{Q}(\boldsymbol{x},\boldsymbol{y})\rangle$ where $s \in \mathbb{R}$, $\boldsymbol{x} \in V^n$, $P \in \mathbb{R}[X_1, \dots, X_{n+k}]/(1, X_i^2 - X_i)$, and $\boldsymbol{Q} \in (\mathbb{F}_2[X_1, \dots, X_{n+k}])^m$*
- *Composition is obtained as*
$$f \circ g := |\boldsymbol{x}_g\rangle \mapsto s_f s_g \sum_{\substack{\boldsymbol{y}_f \in V_f^{k_f} \\ \boldsymbol{y}_g \in V_g^{k_g}}} e^{2i\pi(P_g + P_f[\boldsymbol{x}_f \leftarrow \widehat{\boldsymbol{Q}_g}])} |\boldsymbol{Q}_f[\boldsymbol{x}_f \leftarrow \boldsymbol{Q}_g]\rangle$$
- *Tensor product is obtained as*
$$f \otimes g := |\boldsymbol{x}_f \boldsymbol{x}_g\rangle \mapsto s_f s_g \sum_{\substack{\boldsymbol{y}_f \in V_f^{k_f} \\ \boldsymbol{y}_g \in V_g^{k_g}}} e^{2i\pi(P_g + P_f)} |\boldsymbol{Q}_f \boldsymbol{Q}_g\rangle$$
- *The symmetric braiding is $\sigma_{n,m} : |\boldsymbol{x}_1, \boldsymbol{x}_2\rangle \mapsto |\boldsymbol{x}_2, \boldsymbol{x}_1\rangle$*

The polynomial P is called the *phase polynomial*, as it appears in the morphism in $e^{2i\pi \cdot}$. Because of this, we consider the polynomial modulo 1. We also consider the polynomial quotiented by $X^2 - X$ for all its variables X, as these variables are to be evaluated in $\{0, 1\}$, so we consider $X^2 = X$.

Notice that the definition of the identities does not directly fit the description of the morphisms. However, we can rewrite it as $|\boldsymbol{x}\rangle \mapsto |\boldsymbol{x}\rangle = |\boldsymbol{x}\rangle \mapsto 1 \sum\limits_{\boldsymbol{y} \in V^0} e^{2i\pi 0} |\boldsymbol{x}\rangle$. Hence, when we sum over a single element, we may forget the sum operator, and when the phase polynomial is 0, we may not write it. Notice by the way that $id_0 = |\rangle \mapsto |\rangle$. Indeed, $|\rangle$ is absolutely valid, it represents an empty register.

Example 1. We can give the **SOP** version of the usual quantum gates:

$$R_Z(\alpha) := |x\rangle \mapsto e^{2i\pi \frac{\alpha x}{2\pi}} |x\rangle \qquad CNot := |x_1, x_2\rangle \mapsto |x_1, x_1 \oplus x_2\rangle$$
$$H := |x\rangle \mapsto \frac{1}{\sqrt{2}} \sum_{y \in V} e^{2i\pi \frac{xy}{2}} |y\rangle \qquad CZ := |x_1, x_2\rangle \mapsto e^{2i\pi \frac{x_1 x_2}{2}} |x_1, x_2\rangle$$

Example 2. Let us derive the operation $(id \otimes H) \circ CNot$:

$(id \otimes H) \circ CNot$

$$= \left(|x_1, x_2\rangle \mapsto \frac{1}{\sqrt{2}} \sum_{y \in V} e^{2i\pi \frac{x_2 y}{2}} |x_1, y\rangle \right) \circ \left(|x_1, x_2\rangle \mapsto |x_1, x_1 \oplus x_2\rangle \right)$$

$$= |x_1, x_2\rangle \mapsto \frac{1}{\sqrt{2}} \sum_{y \in V} e^{2i\pi \frac{(x_1 + x_2 - 2x_1 x_2) y}{2}} |x_1, y\rangle$$

where $x_1 + x_2 - 2x_1 x_2 = \widehat{x_1 \oplus x_2}$.

The previous definition contains a claim: that **SOP** is a PROP. To be so, one has to check all the axioms of PROPs. One has to be careful when doing so. Indeed, the sequential composition $(.\circ.)$ induces a substitution. Hence, one has to check all the axioms in the presence of a "context", that is, one has to show that the axioms can be applied *locally.*

If an axiom states $t_1 \to t_2$, one should ideally check that $A \circ (id_n \otimes t_1 \otimes id_m) \circ B \to A \circ (id_n \otimes t_2 \otimes id_m) \circ B$ for any "before" morphism B and any "after" morphism A. However, this can be easily reduced to checking that $A \circ t_1 \circ B \to A \circ t_2 \circ B$.

In the case of the axioms of PROPs, this can further be reduced to showing the axioms without context, as neither id_n nor $\sigma_{n,m}$ introduce variables or phases. For the other axioms, however, the context will have to be taken into account. A fairly straightforward but tedious verification gives that, indeed, **SOP** is a PROP.

3.2 From SOP to Qubit

To check the soundness of what we are going to do in the following, it may be interesting to have a way of interpreting morphisms of **SOP** as morphisms of **Qubit**.

Definition 2. *The functor* $[\![.]\!] : \mathbf{SOP} \to \mathbf{Qubit}$ *is defined as being identity on objects, and such that*

$$\left[\!\!\left[|\boldsymbol{x}\rangle \mapsto s \sum_{\boldsymbol{y} \in V^k} e^{2i\pi P(\boldsymbol{x},\boldsymbol{y})} |Q(\boldsymbol{x},\boldsymbol{y})\rangle \right]\!\!\right] := s \sum_{(\boldsymbol{x},\boldsymbol{y}) \in \{0,1\}^n \times \{0,1\}^k} e^{2i\pi P(\boldsymbol{x},\boldsymbol{y})} |Q(\boldsymbol{x},\boldsymbol{y})\rangle\langle \boldsymbol{x}|$$

Example 3. The interpretation of H is as intended the Hadamard gate:

$$[\![H]\!] = \frac{1}{\sqrt{2}} \sum_{x,y \in \{0,1\}} e^{2i\pi \frac{xy}{2}} |y\rangle\langle x| = \frac{1}{\sqrt{2}} (|0\rangle\langle 0| + |0\rangle\langle 1| + |1\rangle\langle 0| - |1\rangle\langle 1|)$$

Proposition 1. *The interpretation* $[\![.]\!]$ *is a* PROP-functor, *meaning:*
i) $[\![.\circ.]\!] = [\![.]\!] \circ [\![.]\!]$, *ii)* $[\![.\otimes.]\!] = [\![.]\!] \otimes [\![.]\!]$, *iii)* $[\![\sigma_{n,m}]\!] = \sigma_{n,m}$

3.3 SOP as a †-Compact PROP

Towards a Compact Structure. It is tempting to try and adapt the compact structure of **Qubit** to **SOP**. To do so, we can first define $\eta_n := |\rangle \mapsto \sum_{\boldsymbol{y}\in V^n} |\boldsymbol{y},\boldsymbol{y}\rangle$. However, we cannot as easily define ϵ_n. To do so, we need to put the phase polynomial to use: $\epsilon_n := |\boldsymbol{x}_1,\boldsymbol{x}_2\rangle \mapsto \frac{1}{2^n}\sum_{\boldsymbol{y}\in V^n} e^{2i\pi\frac{\boldsymbol{x}_1\cdot\boldsymbol{y}+\boldsymbol{x}_2\cdot\boldsymbol{y}}{2}}|\rangle$.

One can easily check that $[\![\epsilon_n]\!] = \epsilon_n$. We can also easily check that the axioms of †-compact PROP where ϵ_n does not appear, such as $\sigma_{n,n}\circ\eta_n = \eta_n$ and $(id_n\otimes\sigma_{n,m}\otimes id_m)\circ(\eta_n\otimes\eta_m) = \eta_{n+m}$ are satisfied.

However, the equation $(\epsilon_n\otimes id_n)\circ(id_n\otimes\eta_n) = id_n = (id_n\otimes\epsilon_n)\circ(\eta_n\otimes id_n)$ is not satisfied, as:

$$(\epsilon_n\otimes id_n)\circ(id_n\otimes\eta_n) = |\boldsymbol{x}\rangle \mapsto \frac{1}{2}\sum_{\boldsymbol{y}_1,\boldsymbol{y}_2\in V^n} e^{2i\pi\frac{\boldsymbol{x}\cdot\boldsymbol{y}_2+\boldsymbol{y}_1\cdot\boldsymbol{y}_2}{2}}|\boldsymbol{y}_1\rangle \neq id_n$$

The fact that we have $(\epsilon_n\otimes id_n)\circ(id_n\otimes\eta_n)\neq id_n$ while its interpretation in **Qubit** holds, hints at a way to *rewrite* the first term as the second.

An Equational Theory. A rewrite strategy is given in [1], and we show in Figure 1 the rules we are going to use in the paper. Each rewrite rule contains a condition, which usually ensures that a variable (the one we want to get rid of) does not appear in some polynomials. We hence use Var as the operator that gets all the variables from a sequence of polynomials. For simplicity, the input signature is omitted, as well as the parameters in the polynomials.

$$\sum_{\boldsymbol{y}} e^{2i\pi P}|\boldsymbol{Q}\rangle \underset{y_0\notin\mathrm{Var}(P,\boldsymbol{Q})}{\longrightarrow} 2\sum_{\boldsymbol{y}\setminus\{y_0\}} e^{2i\pi P}|\boldsymbol{Q}\rangle \tag{Elim}$$

$$\sum_{\boldsymbol{y}} e^{2i\pi\left(\frac{y_0}{2}(y_0'+\widehat{Q_2})+R\right)}|\boldsymbol{Q}\rangle \underset{\substack{y_0\notin\mathrm{Var}(R,Q_2,\boldsymbol{Q})\\ y_0'\notin\mathrm{Var}(Q_2)}}{\longrightarrow} 2\sum_{\boldsymbol{y}\setminus\{y_0,y_0'\}} e^{2i\pi\left(R[y_0'\leftarrow\widehat{Q_2}]\right)}|\boldsymbol{Q}[y_0'\leftarrow Q_2]\rangle \tag{HH}$$

$$\sum_{\boldsymbol{y}} e^{2i\pi\left(\frac{y_0}{4}+\frac{y_0}{2}\widehat{Q_2}+R\right)}|\boldsymbol{Q}\rangle \underset{y_0\notin\mathrm{Var}(Q_2,R,\boldsymbol{Q})}{\longrightarrow} \sqrt{2}\sum_{\boldsymbol{y}\setminus\{y_0\}} e^{2i\pi\left(\frac{1}{8}-\frac{1}{4}\widehat{Q_2}+R\right)}|\boldsymbol{Q}\rangle \tag{ω}$$

Fig. 1. Rewrite strategy $\underset{\text{Clif}}{\longrightarrow}$.

$\underset{\text{Clif}}{\longrightarrow}$ denotes the rewrite system formed by the three rules (Elim), (HH) and (ω). $\overset{*}{\underset{\text{Clif}}{\longrightarrow}}$ is the transitive closure of the rewrite system. Notice that all the rules remove at least one variable from the morphism, so we know $\underset{\text{Clif}}{\longrightarrow}$ terminates.

When the rules are not oriented, we get an equivalence relation on the morphisms of **SOP**. We denote this equivalence $\underset{\text{Clif}}{\sim}$.

We denote $\mathbf{SOP}/\underset{\text{Clif}}{\sim}$ the category $\mathbf{SOP}$ quotiented by the equivalence relation $\underset{\text{Clif}}{\sim}$.

It is to be noticed that:

Proposition 2. *For any rule r of $\underset{\text{Clif}}{\longrightarrow}$ and $t_1, t_2 \in \mathbf{SOP}$:*

$$t_1 \underset{r}{\longrightarrow} t_2 \Longrightarrow \begin{cases} A \circ t_1 \circ B \underset{r}{\longrightarrow} A \circ t_2 \circ B & \text{for all } A \text{ and } B \text{ composable} \\ A \otimes t_1 \otimes B \underset{r}{\longrightarrow} A \otimes t_2 \otimes B & \text{for all } A \text{ and } B \end{cases}$$

This obviously generalises to $\underset{\text{Clif}}{\sim}$.

This result allows us to forget about the context in the rewriting process.

The newly obtained category $\mathbf{SOP}/\underset{\text{Clif}}{\sim}$ is still a PROP. It even has a compact structure, as the last necessary axiom is now derivable:

$$(\epsilon \otimes id) \circ (id \otimes \eta) = |x\rangle \mapsto \frac{1}{2} \sum_{y_1, y_2 \in V} e^{2i\pi\left(\frac{y_1 y_2}{2} + \frac{x y_2}{2}\right)} |y_1\rangle \underset{\text{(HH)}}{\longrightarrow} |x\rangle \mapsto |x\rangle = id$$

and similarly for $(id \otimes \epsilon) \circ (\eta \otimes id) = id$.

†-Functor for SOP. To show that $\mathbf{SOP}/\underset{\text{Clif}}{\sim}$ is †-compact, we lack a notion of †-functor $\mathbf{SOP}$.

Remember that we defined $\overline{(.)}$ as $(.)^{\dagger t}$. Since we have a compact structure, we can already define the functor $(.)^t$. Thanks to the new equivalence relation $\underset{\text{Clif}}{\sim}$, this functor is involutive. Hence, we have $(.)^\dagger = \overline{(.)}^t$. An appropriate definition of the conjugation can be given:

Definition 3. *The conjugation is defined as:*

$$\overline{|\boldsymbol{x}\rangle \mapsto s_f \sum e^{2i\pi P_f} |\boldsymbol{Q}_f\rangle} := |\boldsymbol{x}\rangle \mapsto s_f \sum e^{-2i\pi P_f} |\boldsymbol{Q}_f\rangle$$

By combination of $(.)^t$ this gives a definition of $(.)^\dagger$. These three functors are the expected ones:

Proposition 3. $[\![(.)^t]\!] = [\![.]\!]^t \;,\quad [\![\overline{(.)}]\!] = \overline{[\![.]\!]} \;,\quad [\![(.)^\dagger]\!] = [\![.]\!]^\dagger$

We can finally prove the wanted result:

Theorem 1. $\mathbf{SOP}/\underset{\text{Clif}}{\sim}$ *is a †-compact PROP.*

4 Redefinition of SOP

In **Qubit**, and hence in **SOP**, because the strutures are †-compact, it may feel unnatural to have an asymmetry between inputs and outputs of the process. Why not have morphisms of the form $f = s\sum_{\boldsymbol{y}} e^{2i\pi P}\,|\boldsymbol{O}\rangle\langle\boldsymbol{I}|$? In this case, we have to change the definition of the composition, which has for consequence that the **SOP** morphisms do not form a category. However, it is a category when quotiented by $\underset{\text{Clif}}{\sim}$. This is the reason why we did not define **SOP** like this at first, although it greatly simplifies the notions of compact structure and †-functor.

We now redefine **SOP**, and will use this new definition in the rest of the paper:

Definition 4 (SOP). *We redefine* **SOP** *as the collection of objects* $\mathbb{N}$ *and morphisms between them:*

- *Identity morphisms are* $id_n : \sum_{\boldsymbol{y}\in V^n} |\boldsymbol{y}\rangle\langle\boldsymbol{y}|$
- *Morphisms* $f : n \to m$ *are of the form* $f : s\sum_{\boldsymbol{y}\in V^k} e^{2i\pi P(\boldsymbol{y})}\,|\boldsymbol{O}(\boldsymbol{y})\rangle\langle\boldsymbol{I}(\boldsymbol{y})|$ *where* $s \in \mathbb{R}$, $P \in \mathbb{R}[X_1,\ldots,X_k]/(1, X_i^2 - X_i)$, $\boldsymbol{O} \in (\mathbb{F}_2[X_1,\ldots,X_k])^m$ *and* $\boldsymbol{I} \in (\mathbb{F}_2[X_1,\ldots,X_k])^n$
- *Composition is obtained as* $f \circ g := \frac{s_f s_g}{2^{|\boldsymbol{I}_f|}} \sum_{\substack{\boldsymbol{y}_f,\boldsymbol{y}_g\\ \boldsymbol{y}\in V^m}} e^{2i\pi\left(P_g+P_f+\frac{\boldsymbol{O}_g\cdot\boldsymbol{y}+\boldsymbol{I}_f\cdot\boldsymbol{y}}{2}\right)}\,|\boldsymbol{O}_f\rangle\langle\boldsymbol{I}_g|$
- *Tensor product is obtained as* $f \otimes g := s_f s_g \sum_{\boldsymbol{y}_f,\boldsymbol{y}_g} e^{2i\pi(P_g+P_f)}\,|\boldsymbol{O}_f\boldsymbol{O}_g\rangle\langle\boldsymbol{I}_f\boldsymbol{I}_g|$
- *The symmetric braiding is* $\sigma_{n,m} = \sum_{\boldsymbol{y}_1,\boldsymbol{y}_2} |\boldsymbol{y}_2,\boldsymbol{y}_1\rangle\langle\boldsymbol{y}_1,\boldsymbol{y}_2|$
- *The compact structure is* $\eta_n = \sum_{\boldsymbol{y}} |\boldsymbol{y},\boldsymbol{y}\rangle\langle|$ *and* $\epsilon_n = \sum_{\boldsymbol{y}} |\rangle\langle\boldsymbol{y},\boldsymbol{y}|$
- *The †-functor is given by:* $f^\dagger := s\sum_{\boldsymbol{y}} e^{-2i\pi P}\,|\boldsymbol{I}\rangle\langle\boldsymbol{O}|$
- *The functor* $[\![.]\!]$ *is defined as:* $[\![f]\!] := s\sum_{\boldsymbol{y}\in\{0,1\}^k} e^{2i\pi P(\boldsymbol{y})}\,|\boldsymbol{O}(\boldsymbol{y})\rangle\langle\boldsymbol{I}(\boldsymbol{y})|$

As announced, this is not a category, as $id \circ id = \frac{1}{2}\sum_{\boldsymbol{y}} e^{2i\pi\frac{y_1+y_2}{2}y_3}\,|y_2\rangle\langle y_1| \neq \sum_y |y\rangle\langle y| = id$. This problem is solved by reintroducing the rewrite rules, adapted to the new formalism. In the following, references to the rewrite rules are to their adapted version.

The results given for the previous formalisation can easily be adapted. In particular:

Proposition 4. $\mathbf{SOP}/\underset{\text{Clif}}{\sim}$ *is a †-compact PROP, and* $[\![.]\!]$ *is a †-compact PROP-functor.*

Remark 1. When building a **SOP**-morphism t from a circuit (or a diagram as we will show in the following) in this formalism, provided the complexity of the gates is bounded (e.g. in the gateset $\langle H, R_Z(\alpha), CNot\rangle$), the resulting t is always of size $O(d \times n)$ where n is the size of the register, and d the *depth* of the circuit (and for a diagram in $O(G \times a)$ where G is the number of generators and a the maximum arity of these generators). This contrasts with the first definition of **SOP**, where the size of the constructed **SOP** term gets exponential in general.

5 SOP and Graphical Languages

The sum-over-paths formalism was initially intended to be used for isometries. As such, it was given a weak form of completeness – as we will discuss in the next section. However, if transforming a quantum circuit – that describes an isometry – into an **SOP** morphism is easy, the converse, transforming a **SOP** morphism into a circuit is not. And actually, all **SOP** morphisms do not represent an isometry. For instance, the morphism ϵ_1 described above is not an isometry. An even smaller example is $\sum_y |\rangle\langle y|$ which is a valid **SOP** morphism, but clearly does not represent an isometry.

Monoidal categories, and subsequently PROPs, have the benefit of having a nice graphical representation, using string diagrams. The fact that **SOP** is one hints at another (family) of language(s) more suited for representing it: the Z$*$-Calculi: ZX, ZW and ZH [7,8,10,3]. These are all †-compact graphical languages, that have an interpretation in **Qubit**, and are universal for **Qubit**. This means that any morphism of **Qubit** can be represented as a morphism of either of these 3 languages.

The language that happens to be the closest to **SOP** is the ZH-Calculus. This is the one we are going to present in the following. However, bear in mind that, as we have semantics-preserving functors between any two of these three languages, one can do the same work with ZX and ZW-Calculi.

The link between the sum-over-paths formalism and the ZH-Calculus was first shown in [14,15]. We give here a slightly different but equivalent presentation, that in particular uses the fact that we altered the formalism of **SOP**, and we will focus this presentation to the Clifford fragment, as it is sufficient for the scope of the present article, although a more general presentation could be given (see the previous two references, or the longer version of the present article).

5.1 The Cliffrord Fragment of the ZH-Calculus

$\mathbf{ZH}_{\mathrm{Clif}}$ is a PROP whose morphisms are composed (sequentially $(. \circ .)$ or in parallel $(. \otimes .)$) from the generators 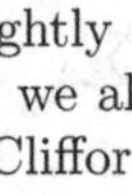 , 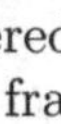 , 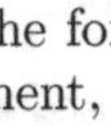 and  ; where $\alpha \in \frac{\pi}{2}\mathbb{Z}$ and $s \in \langle \sqrt{2}, e^{i\frac{\pi}{4}} \rangle$ the multiplicative group freely generated by $\sqrt{2}$ and $e^{i\frac{\pi}{4}}$.

$\mathbf{ZH}_{\mathrm{Clif}}$ is made a †-compact PROP, which means it also has the symmetric structure $\sigma_{n,m}$ :: 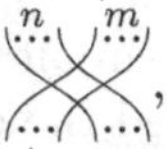, the compact structure $\Big(\eta_n$ :: 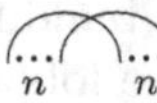, ϵ_n :: $\Big)$, and a †-functor $(.)^\dagger : \mathbf{ZH}_{\mathrm{Clif}}^{\mathrm{op}} \to \mathbf{ZH}_{\mathrm{Clif}}$.

For convenience, we define two additional spiders:

$:=$ and ¬ $:=$

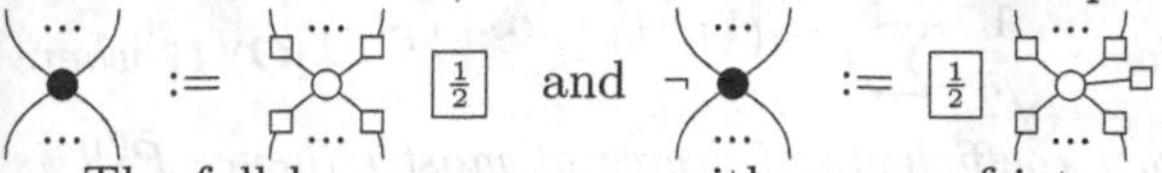

The full language comes with a way of interpreting the morphisms as morphisms of **Qubit**, and whose restriction to $\mathbf{ZH}_{\mathrm{Clif}}$ maps to **Stab**. The standard interpretation $\llbracket . \rrbracket : \mathbf{ZH}_{\mathrm{Clif}} \to \mathbf{Stab}$ is a †-compact-PROP-functor, defined as:

$$\left\llbracket \text{\includegraphics{}} \right\rrbracket = |0^m\rangle\langle 0^n| + |1^m\rangle\langle 1^n|, \qquad \left\llbracket \text{\includegraphics{}} \right\rrbracket = \sum_{x,y \in \{0,1\}} (-1)^{xy} |y\rangle\langle x|,$$

$$\left\llbracket \boxed{e^{i\alpha}} \right\rrbracket = |0\rangle + e^{i\alpha}|1\rangle, \qquad \left\llbracket \boxed{s} \right\rrbracket = s$$

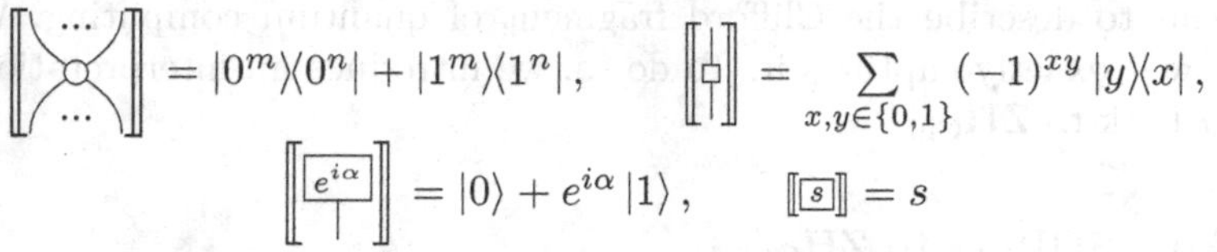

Notice that we used the same symbol for two different functors: the two interpretations $\llbracket . \rrbracket : \mathbf{SOP} \to \mathbf{Qubit}$ and $\llbracket . \rrbracket : \mathbf{ZH}_{\mathrm{Clif}} \to \mathbf{Stab}$. It should be clear from the context which one is to be used.

The language is universal for **Stab**:

Proposition 5. $\llbracket . \rrbracket : \mathbf{ZH}_{\mathrm{Clif}} \to \mathbf{Stab}$ *is onto, i.e.*

$$\forall f \in \mathbf{Stab}, \ \exists D_f \in \mathbf{ZH}_{\mathrm{Clif}}, \ \ \llbracket D_f \rrbracket = f$$

Since it is not a 1-to-1 correspondence, the language comes with an equational theory, which in particular gives the axioms for a †-compact PROP. We will not present it here.

5.2 From $\mathbf{ZH}_{\mathrm{Clif}}$ to SOP

We show in this section how any $\mathbf{ZH}_{\mathrm{Clif}}$ morphism can be turned into a **SOP** morphism in a way that preserves the semantics. We define $[.]^{\mathrm{sop}} : \mathbf{ZH}_{\mathrm{Clif}} \to$ **SOP** as the †-compact PROP-functor such that:

$$\left[\text{\includegraphics{}} \right]^{\mathrm{sop}} := \sum_y |y,\ldots,y\rangle\langle y,\ldots,y| \qquad \left[\text{\includegraphics{}} \right]^{\mathrm{sop}} := \sum_{y_0,y_1} e^{2i\pi\frac{y_0 y_1}{2}} |y_0\rangle\langle y_1|$$

$$\left[\boxed{e^{i\alpha}} \right]^{\mathrm{sop}} := \sum_y e^{2i\pi\frac{\alpha}{2\pi}y} |y\rangle \qquad \left[\boxed{\rho e^{i\theta}} \right]^{\mathrm{sop}} := \rho \sum_\emptyset e^{2i\pi\frac{\theta}{2\pi}} |\rangle\langle| \quad \text{for } \rho e^{i\theta} \in \langle\sqrt{2}, e^{i\frac{\pi}{4}}\rangle$$

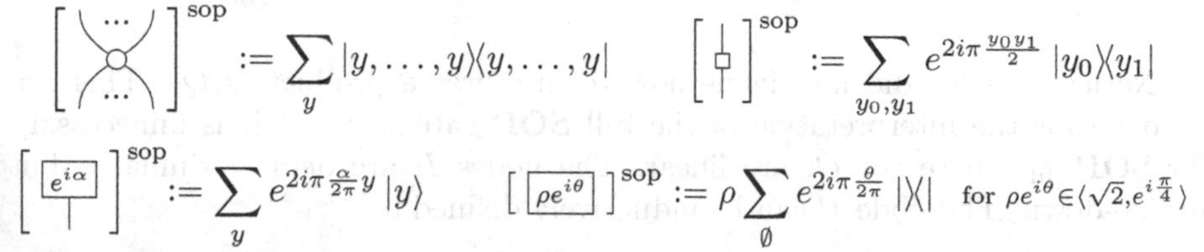

This interpretation can be extended to the full graphical language. It preserves the semantics:

Proposition 6. $\llbracket [.]^{\mathrm{sop}} \rrbracket = \llbracket . \rrbracket$.

5.3 The Clifford Fragment of SOP

Since $\mathbf{ZH}_{\mathrm{Clif}}$ is universal for **Stab**, the Clifford fragment of **Qubit**, and since we have an interpretation $[.]^{\mathrm{sop}} : \mathbf{ZH}_{\mathrm{Clif}} \to$ **SOP** that preserves the semantics, we can define $\mathbf{SOP}_{\mathrm{Clif}}$ as the the image of $\mathbf{ZH}_{\mathrm{Clif}}$ by $[.]$. This gives a characterisation of the fragment:

Definition 5. $\mathbf{SOP}_{\mathrm{Clif}}$ *is the subPROP of* $\mathbf{SOP}$ *with the same objects, and whose morphisms are of the form* $\frac{1}{\sqrt{2}^p}\sum e^{2i\pi\left(\frac{1}{8}P^{(0)}+\frac{1}{4}P^{(1)}+\frac{1}{2}P^{(2)}\right)}|\boldsymbol{O}\rangle\langle\boldsymbol{I}|$ *where* $P^{(i)}$ *is a polynomial with integer coefficients of degree at most* i *(hence* $P^{(0)}$ *is in fact merely an integer); and where all the* O_i *and* I_i *are linear.*

It is an easy check that $[\mathbf{ZH}_{\mathrm{Clif}}]^{\mathrm{sop}} \subseteq \mathbf{SOP}_{\mathrm{Clif}}$, so $\mathbf{SOP}_{\mathrm{Clif}}$ has enough morphisms to describe the Clifford fragment of quantum computing. We can even show it exactly captures it. To do so, we introduce an interpretation from $\mathbf{SOP}_{\mathrm{Clif}}$ back to $\mathbf{ZH}_{\mathrm{Clif}}$.

5.4 From $\mathbf{SOP}_{\mathrm{Clif}}$ to $\mathbf{ZH}_{\mathrm{Clif}}$

We define $[.]^{\mathrm{ZH}} : \mathbf{SOP}_{\mathrm{Clif}} \to \mathbf{ZH}_{\mathrm{Clif}}$ on arbitrary $\mathbf{SOP}_{\mathrm{Clif}}$ morphisms as:

$$\left[s\sum_{\boldsymbol{y}} e^{2i\pi P}\,|O_1,\ldots,O_m\rangle\langle I_1,\ldots,I_n|\right]^{\mathrm{ZH}} :=$$

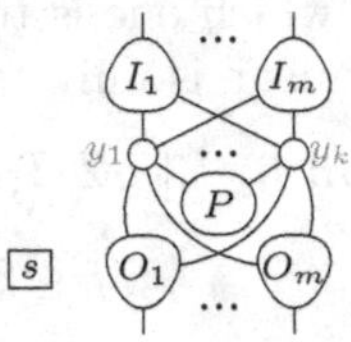

where the row of Z-spiders represents the variables $y_1,\cdots,y_k$.

The inputs of O_i are linked to $y_1,\cdots,y_k$. The nodes O_i can be inductively defined as:

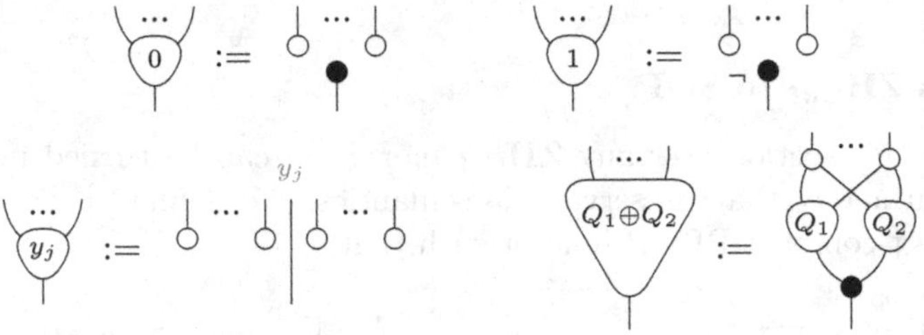

Notice that we did not define how to interpret a product Q_1Q_2. This can be done for the interpretation of the full $\mathbf{SOP}$ category, but it is unnecessary for $\mathbf{SOP}_{\mathrm{Clif}}$ where the O_i are linear. The nodes I_i are defined similarly, but upside-down. The node P can be inductively defined as:

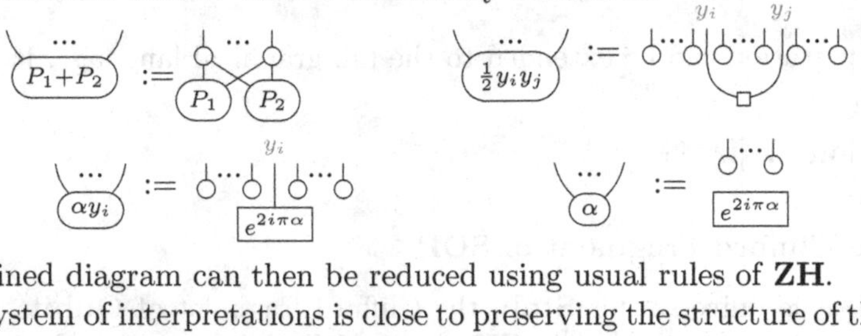

The obtained diagram can then be reduced using usual rules of $\mathbf{ZH}$.

The system of interpretations is close to preserving the structure of the terms:

Proposition 7. $\left[[.]^{\mathrm{ZH}}\right]^{\mathrm{sop}} \underset{\mathrm{Clif}}{\sim} (.)$

Corollary 1. $[\![[.]^{\mathrm{ZH}}]\!] = [\![.]\!]$.

This result allows us to prove $\mathbf{SOP}_{\mathrm{Clif}}$ does capture the Clifford fragment of quantum mechanics:

Proposition 8. $[\![.]\!] : \mathbf{SOP}_{\mathrm{Clif}} \to \mathbf{Stab}$, *the restriction of the standard interpretation to* $\mathbf{SOP}_{\mathrm{Clif}}$ *is onto* **Stab**.

6 A Complete Rewrite System for Clifford

In [1], where the rewrite rules are introduced, the author gives a notion of completeness for Clifford *unitaries*, that we will refer to in the following as "weak completeness":

Proposition 9 (Weak Completeness for Clifford Unitaries). *Given two terms* t_1, t_2 *of* $\mathbf{SOP}_{\mathrm{Clif}}$ *such that* $[\![t_i]\!] \circ [\![t_i]\!]^\dagger = id = [\![t_i]\!]^\dagger \circ [\![t_i]\!]$, *we have:*

$$t_1 \circ t_2^\dagger \xrightarrow[\mathrm{Clif}]{*} id \iff [\![t_1]\!] = [\![t_2]\!]$$

In practice, this is sufficient for deciding the equivalence of two Clifford quantum circuits, as they are represented as unitary morphisms of $\mathbf{SOP}_{\mathrm{Clif}}$. However, in our case, where we deal with more than unitaries, we cannot use this trick. Instead, we aim at a result like "$t_1 \xrightarrow{*} t \xleftarrow{*} t_2 \iff [\![t_1]\!] = [\![t_2]\!]$". In other words, we want a rewrite system that will transform any term of $\mathbf{SOP}_{\mathrm{Clif}}$ into a unique normal form. However, the rewrite system $\xrightarrow[\mathrm{Clif}]{}$ is not enough for this:

Lemma 1. $\xrightarrow[\mathrm{Clif}]{}$ *is not confluent in* $\mathbf{SOP}_{\mathrm{Clif}}$.

To address this problem, we propose to add three rewrite rules to the previously presented ones. These new rewrite rules are shown in Figure 2.

$$\sum e^{2i\pi(P)} |O_1, ..., \underbrace{y_0 \oplus O_i'}_{O_i}, ..., O_m\rangle\langle \boldsymbol{I}| \underset{y_0 \notin \mathrm{Var}(O_1,...,O_{i-1},O_i') \,\wedge\, O_i' \neq 0}{\longrightarrow} \sum e^{2i\pi\left(P[y_0 \leftarrow \widehat{O_i}]\right)} \left(|\boldsymbol{O}\rangle\langle \boldsymbol{I}|\right)[y_0 \leftarrow O_i] \quad \text{(ket)}$$

$$\sum e^{2i\pi(P)} |\boldsymbol{O}\rangle\langle I_1, ..., \underbrace{y_0 \oplus I_i'}_{I_i}, ..., I_m| \underset{y_0 \notin \mathrm{Var}(\boldsymbol{O},I_1,...,I_{i-1},I_i') \,\wedge\, I_i' \neq 0}{\longrightarrow} \sum e^{2i\pi\left(P[y_0 \leftarrow \widehat{I_i}]\right)} \left(|\boldsymbol{O}\rangle\langle \boldsymbol{I}|\right)[y_0 \leftarrow I_i] \quad \text{(bra)}$$

$$s \sum_{\boldsymbol{y}} e^{2i\pi\left(\frac{y_0}{2} + R\right)} |\boldsymbol{O}\rangle\langle \boldsymbol{I}| \underset{(R \neq 0 \text{ or } \boldsymbol{OI} \neq \boldsymbol{0}) \,\wedge\, y_0 \notin \mathrm{Var}(R,\boldsymbol{O},\boldsymbol{I})}{\longrightarrow} \sum_{y_0} e^{2i\pi\left(\frac{y_0}{2}\right)} |0, ..., 0\rangle\langle 0, ..., 0| \quad \text{(Z)}$$

Fig. 2. Together with those of $\xrightarrow[\mathrm{Clif}]{}$, these rules constitute the rewrite system $\xrightarrow[\mathrm{Clif+}]{}$.

The last rule (Z) describes what happens for a term that represents the linear map 0. Rule (bra) is simply the continuation of (ket). They explain how to operate suitable changes of variables.

Proposition 10. *The rewrite system* $\underset{\text{Clif+}}{\longrightarrow}$ *terminates.*

Not only does this rewrite system terminate, it is confluent in $\mathbf{SOP}_{\text{Clif}}$ and the induced equivalence relation $\underset{\text{Clif+}}{\sim}$ is complete for Clifford. The plan to prove this is by showing that any morphism of $\mathbf{SOP}_{\text{Clif}}$ reduces to a normal form that is unique, up to α-conversion (upcoming Thm. 2). To get there, we first need a few intermediary results.

Lemma 2. *Any morphism of* $\mathbf{SOP}_{\text{Clif}}$ *reduces by* $\underset{\text{Clif+}}{\longrightarrow}$ *to a morphism of the form* $\frac{1}{\sqrt{2}^p}\sum e^{2i\pi P}\,|\boldsymbol{O}\rangle\langle\boldsymbol{I}|$ *where:*

- $\text{Var}(P) \subseteq \text{Var}(\boldsymbol{O},\boldsymbol{I})$ *or* $P = \frac{y_0}{2}$ *where* $y_0 \notin \text{Var}(\boldsymbol{O},\boldsymbol{I})$
- $O_i = \begin{cases} \text{either } y_k \text{ or} \\ c \oplus \bigoplus\limits_{y\in\text{Var}(O_1,\ldots,O_{i-1})} c_y y \quad \textit{where } c, c_y \in \{0,1\} \end{cases}$
- $I_i = \begin{cases} \text{either } y_k \text{ or} \\ c \oplus \bigoplus\limits_{y\in\text{Var}(\boldsymbol{O},I_1,\ldots,I_{i-1})} c_y y \quad \textit{where } c, c_y \in \{0,1\} \end{cases}$

To start with, we deal with the case where the term represents the null map.

Proposition 11. *Let* t *be a morphism of* $\mathbf{SOP}_{\text{Clif}}$ *such that* $[\![t]\!] = 0$. *Then:*

$$t \underset{\text{Clif+}}{\overset{*}{\longrightarrow}} \sum_{y_0} e^{2i\pi\frac{y_0}{2}}\,|0,\ldots,0\rangle\langle 0,\ldots,0|$$

Corollary 2. *If a morphism* $t = \frac{1}{\sqrt{2}^r}\sum e^{2i\pi P}\,|\boldsymbol{O}\rangle\langle\boldsymbol{I}|$ *of* $\mathbf{SOP}_{\text{Clif}}$ *is irreducible such that* $\text{Var}(P) \subseteq \text{Var}(\boldsymbol{O},\boldsymbol{I})$, *then* $[\![t]\!] \neq 0$.

Before moving on to the completeness by normal forms theorem, we need a result for the uniqueness of the phase polynomial:

Lemma 3. *Let* P_1 *and* P_2 *be two polynomials of* $\mathbb{R}[X_1,\ldots,X_k]/(1,X^2-X)$. *We have* $\left(\forall \boldsymbol{x} \in \{0,1\}^k,\ P_1(\boldsymbol{x}) = P_2(\boldsymbol{x})\right) \implies (P_1 = P_2)$

Theorem 2. *Let* t_1*, and* t_2 *be two morphisms of* $\mathbf{SOP}_{\text{Clif}}$ *such that* $[\![t_1]\!] = [\![t_2]\!]$. *Then, there exists* t *in* $\mathbf{SOP}_{\text{Clif}}$ *such that* $t_1 \underset{\text{Clif+}}{\overset{*}{\longrightarrow}} t \underset{\text{Clif+}}{\overset{*}{\longleftarrow}} t_2$, *up to* α*-conversion.*

This result is not totally surprising, since, as exposed by [15], the rules of $\underset{\text{Clif}}{\longrightarrow}$ are generalisations of the so-called pivoting and local complementation which can be used to reduce any Clifford ZX (or ZH)-diagram into a *pseudo*-normal form [9,2] there, a diagram can have several different but equivalent pseudo-normal form. The rules introduced to get $\underset{\text{Clif+}}{\longrightarrow}$ are simply here to further rewrite terms in pseudo-normal form into terms in proper (unique) normal form.

Corollary 3. *The equality of morphisms in* $\mathbf{SOP}_{\text{Clif}}/\underset{\text{Clif+}}{\sim}$ *is decidable in time polynomial in the size of the phase polynomial and in the combined size of the ket/bra polynomials.*

Although the set of rules is confluent in $\mathbf{SOP}_{\text{Clif}}$, it is not in **SOP**:

Lemma 4 (Non-confluence). *The rewrite systems* $\underset{\text{Clif}}{\longrightarrow}$ *and* $\underset{\text{Clif+}}{\longrightarrow}$ *are not confluent in* **SOP**.

7 SOP with Discards

We want in this section to extend **SOP** to be able to express the larger formalism of mixed quantum operators. The discard construction can be used for that purpose, as well as for extending the rewrite system for the Clifford fragment. We finally leverage the previous completeness theorem to get a similar result in this extension.

7.1 The Discard Construction on SOP

In [5], a construction is given to extend any †-compact PROP for *pure* quantum mechanics to another †-compact PROP for quantum mechanics with environment. This new formalism can also be understood as the previous one, but where on top of it, one can discard the qubits. Because **SOP** fits the requirements, the construction can be applied to it.

First, we have to create the subcategory $\mathbf{SOP}_{\text{iso}}$ of **SOP** that contains all its isometries. The objects of the new category are the same, and its morphisms are $\{f \in \mathbf{SOP} \mid [\![f^\dagger \circ f]\!] = id\}$.

These are important, as the isometries are exactly the pure quantum operators that can be discarded. The next step in the construction does just that. We perform the affine completion of $\mathbf{SOP}_{\text{iso}}$, that is, for every object n, we add a new morphism $!_n : n \to 0$, and we impose that $! \circ f = !$ for any f in the new category, that we denote $\mathbf{SOP}^!_{\text{iso}}$. We also need to impose that $!_n \otimes !_m = !_{n+m}$ and $!_0 = id_0$.

Finally, the category $\mathbf{SOP}^{\ground}$ is obtained as the following pushout in the category of SMCs, where the arrows are the inclusion functors:

$$\begin{array}{ccc} \mathbf{SOP}_{\text{iso}} & \longrightarrow & \mathbf{SOP} \\ \downarrow & & \downarrow \\ \mathbf{SOP}^!_{\text{iso}} & \longrightarrow & \mathbf{SOP}^{\ground} \end{array}$$

We write the new morphisms in the form $s \sum_{y \in V^k} e^{2i\pi P(y)} |\boldsymbol{O}(y)\rangle !\boldsymbol{D}(y) \langle \boldsymbol{I}(y)|$ where the additional $\boldsymbol{D}$ is a set of multivariate polynomials of $\mathbb{F}_2$. The fact that it is a set, and not a list, already captures some rules on the discard: first permuting qubits and then discarding them is equivalent to discarding them right away. Similarly, copying data and discarding the copies is equivalent to discarding the data right away.

Pure morphisms are those such that $\boldsymbol{D} = \{\}$. In those, no qubits are discarded. We hence easily induce usual morphisms such as H and CZ in the new formalism.

The new morphisms $!_n$ are given by: $!_n := \sum_{\boldsymbol{y}\in V^n} |\rangle!\{y_1,\dots,y_n\}\langle y_1,\dots,y_n|$

In the new formalism, the compositions are obtained exactly like previously, where the resulting set of discarded polynomial is the union of the other two.

It might be useful to be able to give an interpretation to the morphisms of the new formalism. To do so, we use the CPM construction [17] to map morphisms of $\mathbf{SOP}^{\ground}$ to morphisms of $\mathbf{SOP}$.

Definition 6. *The map* CPM : $\mathbf{SOP}^{\ground} \to \mathbf{SOP}$ *is defined as:*

$$s\sum_{\boldsymbol{y}} e^{2i\pi P}\,|\boldsymbol{O}\rangle!\boldsymbol{D}\,\langle\boldsymbol{I}| \;\mapsto$$
$$\frac{s^2}{2^{|\boldsymbol{D}|}}\sum_{\boldsymbol{y}_1,\boldsymbol{y}_2,\boldsymbol{y}} e^{2i\pi\left(P(\boldsymbol{y}_1)-P(\boldsymbol{y}_2)+\frac{\boldsymbol{D}(\boldsymbol{y}_1)\cdot\boldsymbol{y}+\boldsymbol{D}(\boldsymbol{y}_2)\cdot\boldsymbol{y}}{2}\right)}\,|\boldsymbol{O}(\boldsymbol{y}_1),\boldsymbol{O}(\boldsymbol{y}_2)\rangle\langle\boldsymbol{I}(\boldsymbol{y}_1),\boldsymbol{I}(\boldsymbol{y}_2)|$$

We can now define a standard interpretation of $\mathbf{SOP}^{\ground}$-morphisms as:

Definition 7. *The standard interpretation* $[\![.]\!]$ *of* $\mathbf{SOP}^{\ground}$ *is defined as* $[\![.]\!] := [\![\mathrm{CPM}(.)]\!]$.

Again, it is easy to transform any morphism of $\mathbf{SOP}^{\ground}$ in $\mathbf{ZH}^{\ground}$ and vice-versa:

$$\left[s\sum_{\boldsymbol{y}\in V^k} e^{2i\pi P(\boldsymbol{y})}\,|\boldsymbol{O}(\boldsymbol{y})\rangle!\boldsymbol{D}(\boldsymbol{y})\,\langle\boldsymbol{I}(\boldsymbol{y})|\right]^{\mathrm{ZH}} :=$$

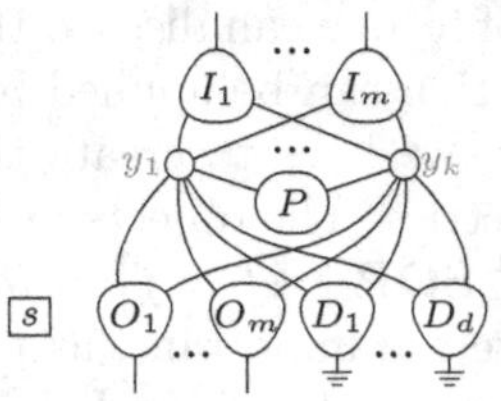

and $[\ground]^{\mathrm{sop}} = !_1$.

7.2 SOP with Discards for Clifford

The discard construction can be applied to the subcategory $\mathbf{SOP}_{\mathrm{Clif}}$. We end up with a new category $\mathbf{SOP}^{\ground}_{\mathrm{Clif}}$, such that the following diagram, whose arrows are inclusions, commutes:

$$\begin{array}{ccc}\mathbf{SOP}_{\mathrm{Clif}} & \longrightarrow & \mathbf{SOP}\\ \downarrow & & \downarrow\\ \mathbf{SOP}^{\ground}_{\mathrm{Clif}} & \longrightarrow & \mathbf{SOP}^{\ground}\end{array}$$

Following the characterisation of $\mathbf{SOP}_{\mathrm{Clif}}$ morphisms, we determine that all the morphisms of $\mathbf{SOP}^{\ground}_{\mathrm{Clif}}$ are of the form: $\frac{1}{\sqrt{2}^p}\sum e^{2i\pi\left(\frac{1}{8}P^{(0)}+\frac{1}{4}P^{(1)}+\frac{1}{2}P^{(2)}\right)}\,|\boldsymbol{O}\rangle!\boldsymbol{D}\,\langle\boldsymbol{I}|$ where $p\in\mathbb{Z}$, where $P^{(i)}$ is a polynomial with integer coefficients and of degree at most i, and where the polynomials of $\boldsymbol{O}, \boldsymbol{D}$ and $\boldsymbol{I}$ are linear.

The rewrite system presented previously can obviously be adapted to the new formalism (when there is a substitution, it has to be applied in $!\boldsymbol{D}$ as well).

On top of that, the condition that makes $\mathbf{SOP}^!_{\text{iso}}$ terminal can be translated as a meta rule which sadly is not easy to apply. Thankfully, the last part of [5] is devoted to showing that this big meta rule can sometimes be replaced by a few small ones. The idea is that, in some cases (in particular in the Clifford fragment), all the isometries can be generated from a finite set of generators. In particular, it is enough to impose the following equations:

$$e^{i\alpha} = 1 \qquad !_1 \circ |0\rangle = 1 \qquad !_1 \circ H = ! \qquad !_1 \circ S = !_1 \qquad !_2 \circ CZ = !_2$$

Based on this, we can give an updated set of rewrite rules fit for the introduction of $\ground$. Due to the size of this rewrite system, we do not provide it here, but it can be found in the extended version of this paper. The rewrite system is denoted $\underset{\text{Clif}\,\ground}{\longrightarrow}$ and induces a equivalence relation $\underset{\text{Clif}\,\ground}{\sim}$. Notice that we can extend CPM to $\text{CPM} : \mathbf{SOP}^{\ground}/\underset{\text{Clif}\,\ground}{\sim} \to \mathbf{SOP}/\underset{\text{Clif+}}{\sim}$, which makes it a functor.

Proposition 12. *The rewrite system* $\underset{\text{Clif}\,\ground}{\longrightarrow}$ *terminates.*

We aim to prove a similar result to that of the $\ground$-free Clifford fragment, that is that the new rewrite system rewrites any morphism of the Clifford fragment into a unique normal form. The idea here it to make use of the previous result.

Lemma 5. *Any non-null morphism of* $\mathbf{SOP}^{\ground}_{\text{Clif}}$ *can be reduced to:*

$$\frac{1}{\sqrt{2}^p}\sum_{\boldsymbol{y},\boldsymbol{y_d}} e^{2i\pi\left(\frac{1}{4}P^{(1)}(\boldsymbol{y})+\frac{1}{2}P^{(2)}(\boldsymbol{y},\boldsymbol{y_d})\right)} \left|\boldsymbol{O}(\boldsymbol{y},\boldsymbol{y_d})\right\rangle !\{\boldsymbol{y_d}\} \left\langle \boldsymbol{I}(\boldsymbol{y},\boldsymbol{y_d})\right| \text{ where:}$$

- *polynomials of* $\boldsymbol{O}$ *and* $\boldsymbol{I}$ *are linear*
- *the set of discarded polynomials is reduced to a set of variables* $\{\boldsymbol{y_d}\}$
- $P^{(1)}$ *and* $P^{(2)}$ *have no constants*
- *no monomial of* $P^{(2)}$ *uses only variables of* $\boldsymbol{y_d}$
- $\{\boldsymbol{y_d}\} \subseteq \text{Var}(\boldsymbol{O},\boldsymbol{I})$
- $\text{Var}(P^{(1)},P^{(2)}) \subseteq \text{Var}(\boldsymbol{O},\boldsymbol{I},\boldsymbol{D})$ *or* $P = \frac{y_0}{2}$ *with* $y_0 \notin \text{Var}(\boldsymbol{O},\boldsymbol{I},\boldsymbol{D})$.

Corollary 4. *Any morphism of* $\mathbf{SOP}^{\ground}_{\text{Clif}}$ *eventually reduces to a morphism of the form given in Lem. 5.*

Lemma 6. *Any morphism* t *of* $\mathbf{SOP}^{\ground}_{\text{Clif}}$ *such that* $[\![t]\!] = 0$ *reduces to:*

$$\sum_{y_0} e^{2i\pi\left(\frac{y_0}{2}\right)} |0,\cdots,0\rangle !\{\} \langle 0,\cdots,0|$$

Corollary 5. *If* $t \in \mathbf{SOP}^{\ground}_{\text{Clif}}$ *is terminal with* $\text{Var}(P) \subseteq \text{Var}(\boldsymbol{O},\boldsymbol{D},\boldsymbol{I})$, *then* $[\![t]\!] \neq 0$.

Definition 8. *We define* $\overline{\mathbf{SOP}^{\ground}_{\text{Clif}}}$ *as the set of morphisms of* $\mathbf{SOP}^{\ground}_{\text{Clif}}$ *in the form given in Lem. 5. We define the function* F *on* $\overline{\mathbf{SOP}^{\ground}_{\text{Clif}}}$ *such that, for any morphism* $t = \frac{1}{\sqrt{2}^p}\sum_{\boldsymbol{y},\boldsymbol{y_d}} e^{2i\pi P(\boldsymbol{y},\boldsymbol{y_d})} |\boldsymbol{O}(\boldsymbol{y},\boldsymbol{y_d})\rangle !\{\boldsymbol{y_d}\} \langle \boldsymbol{I}(\boldsymbol{y},\boldsymbol{y_d})|$ *of* $\overline{\mathbf{SOP}^{\ground}_{\text{Clif}}}$*:*

$$F(t) := \frac{1}{\sqrt{2}^{2p}}\sum_{\boldsymbol{y},\boldsymbol{y'},\boldsymbol{y_d}} e^{2i\pi\left(P(\boldsymbol{y},\boldsymbol{y_d})-P(\boldsymbol{y'},\boldsymbol{y_d})\right)} \left|\boldsymbol{O}(\boldsymbol{y},\boldsymbol{y_d}),\boldsymbol{O}(\boldsymbol{y'},\boldsymbol{y_d})\right\rangle\!\left\langle\boldsymbol{I}(\boldsymbol{y},\boldsymbol{y_d}),\boldsymbol{I}(\boldsymbol{y'},\boldsymbol{y_d})\right|$$

This new functor F can be seen as a simplified CPM construction, applicable only for terms that are already simplified (in the form of Lem. 5).

Proposition 13. *For any* $t \in \overline{\mathbf{SOP}^{⏚}_{\text{Clif}}}$, $F(t) \underset{\text{Clif+}}{\sim} \text{CPM}(t)$.
This implies $[\![F(.)]\!] = [\![\text{CPM}(.)]\!]$.

Definition 9. *We define a function* G *on some morphisms of* $\mathbf{SOP}_{\text{Clif}}$ *that have an appropriate form. Let* $t = \frac{1}{\sqrt{2}^{2p}} \sum_{\boldsymbol{y}} e^{2i\pi P} |\boldsymbol{O}_1, \boldsymbol{O}_2\rangle\langle \boldsymbol{I}_1, \boldsymbol{I}_2|$ *with* $|\boldsymbol{O}_1| = |\boldsymbol{O}_2|$ *and* $|\boldsymbol{I}_1| = |\boldsymbol{I}_2|$. *Let us partition* $\boldsymbol{y}$ *into:* $\{\boldsymbol{y}_d\} := \{\boldsymbol{y}\} \setminus \text{Var}(\boldsymbol{O}_1 \oplus \boldsymbol{O}_2, \boldsymbol{I}_1 \oplus \boldsymbol{I}_2)$, $\{\boldsymbol{y}_1\} := \text{Var}(\boldsymbol{O}_1, \boldsymbol{I}_1) \setminus \{\boldsymbol{y}_d\}$ *and* $\{\boldsymbol{y}_2\} := (\{\boldsymbol{y}\} \setminus \{\boldsymbol{y}_1\}) \setminus \{\boldsymbol{y}_d\}$. *If* $|\boldsymbol{y}_1| = |\boldsymbol{y}_2|$ *and if there exists a unique bijection* $\delta : \{\boldsymbol{y}_2\} \to \{\boldsymbol{y}_1\}$ *such that:* $(\boldsymbol{O}_1 \oplus \boldsymbol{O}_2, \boldsymbol{I}_1 \oplus \boldsymbol{I}_2)[\boldsymbol{y}_2 \leftarrow \delta(\boldsymbol{y}_2)] = \boldsymbol{0}$, *then* $G(t)$ *is defined, and:*

$$G(t) := \frac{1}{\sqrt{2}^{p}} \sum_{\boldsymbol{y}_1, \boldsymbol{y}_d} e^{-2i\pi P[\boldsymbol{y}_1 \leftarrow \boldsymbol{0}][\boldsymbol{y}_2 \leftarrow \delta(\boldsymbol{y}_2)]} \big(|\boldsymbol{O}_2\rangle !\{\boldsymbol{y}_d\} \langle \boldsymbol{I}_2| \big)[\boldsymbol{y}_1 \leftarrow \boldsymbol{0}][\boldsymbol{y}_2 \leftarrow \delta(\boldsymbol{y}_2)]$$

The function G is designed to be an inverse of F for morphisms where it is defined, while at the same being impervious to some rewrite rules.

Proposition 14. *Let* t *be terminal with* $\underset{\text{Clif ⏚}}{\longrightarrow}$, *and* t' *such that* $F(t) \underset{\text{Clif+}}{\overset{*}{\longrightarrow}} t'$. *Then,* $G(F(t))$ *and* $G(t')$ *exist, and* $G(F(t)) = G(t')$.

Theorem 3. *Let* t_1 *and* t_2 *be two morphisms of* $\mathbf{SOP}^{⏚}_{\text{Clif}}$ *such that* $[\![t_1]\!] = [\![t_2]\!]$. *If* t_1' *and* t_2' *are terminal such that* $t_1 \underset{\text{Clif ⏚}}{\overset{*}{\longrightarrow}} t_1'$ *and* $t_2 \underset{\text{Clif ⏚}}{\overset{*}{\longrightarrow}} t_2'$, *then* $t_1' = t_2'$ *up to* α*-conversion.*

Remark 2. Interestingly, the previous proposition and theorem show that the simplification of a term of $\mathbf{SOP}^{⏚}_{\text{Clif}}$ can be operated in the "pure" setting, and then G can be used to retrieve the normal form.

Corollary 6. *The equality of morphisms in* $\mathbf{SOP}^{⏚}_{\text{Clif}} / \underset{\text{Clif ⏚}}{\sim}$ *is decidable in time polynomial in the size of the phase polynomial and in the combined size of the ket/bra/discarded polynomials.*

References

1. Amy, M.: Towards large-scale functional verification of universal quantum circuits. In: Selinger, P., Chiribella, G. (eds.) Proceedings of the 15th International Conference on Quantum Physics and Logic, Halifax, Canada, 3-7th June 2018. Electronic Proceedings in Theoretical Computer Science, vol. 287, pp. 1–21 (2019). https://doi.org/10.4204/EPTCS.287.1
2. Backens, M.: The ZX-calculus is complete for stabilizer quantum mechanics. In: New Journal of Physics. vol. 16, p. 093021. IOP Publishing (Sep 2014). https://doi.org/10.1088/1367-2630/16/9/093021, https://doi.org/10.1088%2F1367-2630%2F16%2F9%2F093021

3. Backens, M., Kissinger, A.: ZH: A complete graphical calculus for quantum computations involving classical non-linearity. In: Selinger, P., Chiribella, G. (eds.) Proceedings of the 15th International Conference on Quantum Physics and Logic, Halifax, Canada, 3-7th June 2018. Electronic Proceedings in Theoretical Computer Science, vol. 287, pp. 23–42 (2019). https://doi.org/10.4204/EPTCS.287.2
4. de Beaudrap, N., Bian, X., Wang, Q.: Fast and effective techniques for t-count reduction via spider nest identities (2020)
5. Carette, T., Jeandel, E., Perdrix, S., Vilmart, R.: Completeness of Graphical Languages for Mixed States Quantum Mechanics. In: Baier, C., Chatzigiannakis, I., Flocchini, P., Leonardi, S. (eds.) 46th International Colloquium on Automata, Languages, and Programming (ICALP 2019). Leibniz International Proceedings in Informatics (LIPIcs), vol. 132, pp. 108:1–108:15. Schloss Dagstuhl–Leibniz-Zentrum fuer Informatik, Dagstuhl, Germany (2019). https://doi.org/10.4230/LIPIcs.ICALP.2019.108, http://drops.dagstuhl.de/opus/volltexte/2019/10684
6. Chareton, C., Bardin, S., Bobot, F., Perrelle, V., Valiron, B.: A deductive verification framework for circuit-building quantum programs (2020)
7. Coecke, B., Duncan, R.: Interacting quantum observables: Categorical algebra and diagrammatics. New Journal of Physics **13**(4), 043016 (Apr 2011). https://doi.org/10.1088/1367-2630/13/4/043016, https://doi.org/10.1088%2F1367-2630%2F13%2F4%2F043016
8. Coecke, B., Kissinger, A.: The compositional structure of multipartite quantum entanglement. In: Automata, Languages and Programming, pp. 297–308. Springer Berlin Heidelberg (2010). https://doi.org/10.1007/978-3-642-14162-1_25, https://doi.org/10.1007%2F978-3-642-14162-1_25
9. Duncan, R., Perdrix, S.: Pivoting makes the ZX-calculus complete for real stabilizers. In: Coecke, B., Hoban, M. (eds.) Proceedings of the 10th International Workshop on Quantum Physics and Logic, Castelldefels (Barcelona), Spain, 17th to 19th July 2013. Electronic Proceedings in Theoretical Computer Science, vol. 171, pp. 50–62 (2014). https://doi.org/10.4204/EPTCS.171.5
10. Hadzihasanovic, A.: A diagrammatic axiomatisation for qubit entanglement. In: 2015 30th Annual ACM/IEEE Symposium on Logic in Computer Science. pp. 573–584 (Jul 2015). https://doi.org/10.1109/LICS.2015.59
11. Hadzihasanovic, A., Ng, K.F., Wang, Q.: Two complete axiomatisations of pure-state qubit quantum computing. In: Proceedings of the 33rd Annual ACM/IEEE Symposium on Logic in Computer Science. pp. 502–511. LICS '18, ACM, New York, NY, USA (2018). https://doi.org/10.1145/3209108.3209128, http://doi.acm.org/10.1145/3209108.3209128
12. Kissinger, A., van de Wetering, J.: Reducing T-count with the ZX-calculus (2019)
13. Lack, S.: Composing PROPs. In: Theory and Applications of Categories. vol. 13, pp. 147–163 (2004), http://www.tac.mta.ca/tac/volumes/13/9/13-09abs.html
14. Lemonnier, L.: Relating high-level frameworks for quantum circuits. Master's thesis, Radbound University (2019), https://www.cs.ox.ac.uk/people/aleks.kissinger/papers/lemonnier-high-level.pdf
15. Lemonnier, L., van de Wetering, J., Kissinger, A.: Hypergraph simplification: Linking the path-sum approach to the zh-calculus (2020), arXiv:2003.13564
16. Mac Lane, S.: Categories for the Working Mathematician, vol. 5. Springer Science & Business Media (2013)
17. Selinger, P.: Dagger compact closed categories and completely positive maps. Electronic Notes in Theoretical Computer Science **170**, 139–163 (Mar

2007). https://doi.org/10.1016/j.entcs.2006.12.018, https://doi.org/10.1016%2Fj.entcs.2006.12.018
18. Selinger, P.: A survey of graphical languages for monoidal categories. In: New Structures for Physics, pp. 289–355. Springer (2010)
19. Vilmart, R.: A near-minimal axiomatisation of zx-calculus for pure qubit quantum mechanics. In: 2019 34th Annual ACM/IEEE Symposium on Logic in Computer Science (LICS). pp. 1–10 (June 2019). https://doi.org/10.1109/LICS.2019.8785765
20. Zanasi, F.: Interacting Hopf Algebras – the theory of linear systems. Ph.D. thesis, Université de Lyon (2015), http://www.zanasi.com/fabio/#/publications.html

A Quantified Coalgebraic van Benthem Theorem

Paul Wild (✉) and Lutz Schröder[*]

Friedrich-Alexander-Universität Erlangen-Nürnberg, Erlangen, Germany
{paul.wild,lutz.schroeder}@fau.de

Abstract. The classical van Benthem theorem characterizes modal logic as the bisimulation-invariant fragment of first-order logic; put differently, modal logic is as expressive as full first-order logic on bisimulation-invariant properties. This result has recently been extended to two flavours of quantitative modal logic, viz. fuzzy modal logic and probabilistic modal logic. In both cases, the quantitative van Benthem theorem states that every formula in the respective quantitative variant of first-order logic that is bisimulation-invariant, in the sense of being nonexpansive w.r.t. behavioural distance, can be approximated by quantitative modal formulae of bounded rank. In the present paper, we unify and generalize these results in three directions: We lift them to full coalgebraic generality, thus covering a wide range of system types including, besides fuzzy and probabilistic transition systems as in the existing examples, e.g. also metric transition systems; and we generalize from real-valued to quantale-valued behavioural distances, e.g. nondeterministic behavioural distances on metric transition systems; and we remove the symmetry assumption on behavioural distances, thus covering also quantitative notions of simulation.

Keywords: Modal logic · Quantale · Fuzzy logic · Coalgebra · Behavioural distance · Modal characterization.

1 Introduction

Modal logic takes part of its popularity from the fact that it specifies transition systems at what for many purposes may be regarded as the right level of granularity; that is, it is invariant under the standard process-theoretic notion of *bisimulation* in the sense that bisimilar states satisfy the same modal formulae. There are two quite different well-known converses to this elementary property, which both witness the *expressiveness* of modal logic: By the *Hennessy-Milner theorem* [29], states in finitely branching systems that satisfy the same modal formulae are bisimilar, and by the *van Benthem theorem*, every first-order definable bisimulation-invariant property is expressible by a modal formula. Since modal logic embeds into first-order logic, the latter result may be phrased as saying that modal logic is the bisimulation-invariant fragment of first-order logic.

* Work of both authors forms part of the DFG project *Probabilistic description logics as a fragment of probabilistic first-order logic* (SCHR 1118/6-2)

S. Kiefer and C. Tasson (Eds.): FOSSACS 2021, LNCS 12650, pp. 551–571, 2021.
https://doi.org/10.1007/978-3-030-71995-1_28

In the two-valued setting, there has been increased recent interest in variants and generalizations of this result (e.g. [54,14,52,22,55,1])

For quantitative systems, it has long been realized (e.g. [26,15,10]) that quantitative notions of process equivalence, generally referred to as *behavioural metrics* (although they are in general only *pseudo*metrics, as distinct but equivalent states have distance zero), are often more appropriate than two-valued bisimilarity. In particular, while two-valued notions of process equivalence just flag small deviations between systems as inequivalence, behavioural metrics can provide more fine-grained information on the degree of similarity of systems. Behavioural metrics are correspondingly used, e.g., in verification [25], differential privacy [13], and conformance testing of hybrid systems [36].

In the same way that two-valued modal logic constitutes a natural specification language for two-valued transition systems, quantitative systems correlate to quantitative modal logics. In this context, bisimulation invariance is read as *nonexpansiveness* w.r.t. behavioural distance, i.e. two states differ on a modal formula at most by their behavioural distance; we refer to this property as *behavioural nonexpansiveness*. Notably, van Breugel and Worrell [10] prove a Hennessy-Milner type theorem for a quantitative probabilistic modal logic: They show that on compact state spaces, the formulae of the logic lie dense in the space of behaviourally nonexpansive state properties, which implies that behavioural distance and logical distance coincide.

In the present paper, we are mainly interested in the other converse to behavioural nonexpansiveness, i.e. in *quantitative van Benthem theorems*. In previous work with Pattinson and König, we have established such theorems for quantitative modal logics of fuzzy [57] and probabilistic [58] transition systems. In the quantitative setting, these theorems take the form of approximability properties, and state that every behaviourally nonexpansive quantitative first-order property is approximable by quantitative modal formulae *of bounded rank*. The latter qualification is in fact the key content of the respective theorems – without it, approximability is closer in flavour to Hennessy-Milner-type theorems, which apply to arbitrary rather than just first-order definable properties (although one should note additionally that our van Benthem theorems do not assume compactness of the state space).

Our present contribution is to unify and generalize these results in three directions: First, we allow for full *coalgebraic generality*, i.e. we cover system types subsumed under the paradigm of *universal coalgebra* [49]. Besides the fuzzy and probabilistic systems featuring in the previous concrete instances of our result, this includes a wide range of weighted, game-based, and preferential systems; for illustration, we concentrate on the (comparatively simple) case of *metric transition systems* [3,20] in the presentation. Second, we generalize from real-valued to *quantale-valued* metrics (e.g. [24,33]). Using the unit interval quantale, we recover our previous results on real-valued logics as special cases. Beyond this, quantales in particular provide support for what may be termed *metrics with effects*; we illustrate this on a notion of *convex-nondeterministic behavioural distance* on metric transition systems, where the behavioural distance gives an interval of possible real-valued distances. Lastly, we remove the assumption that distances need to be symmetric, so that we cover also notions of quantitative simulation. At this level of generality, we prove both a Hennessy-Milner type theorem stating coincidence of logical and behavioural distance, effectively generalizing the existing coalgebraic quantitative Hennessy-Milner theorem [37] to quantale-valued distances; and, as our main result, a quantitative van Benthem theorem stating that all behaviourally non-expansive first-order properties can be modally approximated in bounded rank.

Related Work There is a substantial body of work on two-valued modal characterization theorems, e.g. for logics with frame conditions [14], coalgebraic modal logics [52], fragments of XPath [12,22,1], neighbourhood logic [28], modal logic with team semantics [38], modal μ-calculi (within monadic second order logics) [35,19], PDL (within weak chain logic) [11], modal first-order logics [6,54], and two-dimensional modal logics with an $S5$-modality [55]. We are not aware of quantitative modal characterization theorems other than the mentioned ones for fuzzy and probabilistic modal logics [57,58]. Prior to the quantitative Hennessy-Milner theorems mentioned above [10,37], Hennessy-Milner theorems have been established for *two-valued* logics and two-valued bisimilarity over quantitative systems, e.g. on probabilistic transition systems [39,16,17]. There is work on Hennessy-Milner theorems for certain Heyting-valued modal logics [21,18]; since Heyting algebras are quantales but often fail to meet a continuity assumption needed in our generic Hennessy-Milner theorem, we do not claim to subsume these results.

2 Preliminaries

We briefly recall basic definitions and examples on quantales and universal coalgebra, and fix some data needed throughout the paper. We need some elementary category theory, see, e.g., [2].

Quantales are order-algebraic structures that serve as objects of truth values in suitable multi-valued logics, and also support a useful notion of generalized (pseudo-)metric space (e.g. [24,33,32]). Our arguments will rely on a certain amount of epsilontics, and hence require more specifically the use of *value quantales* [24].

We recall some basic order and lattice theory. A *complete lattice* is a partially ordered set $(V, \leq)$ having all suprema $\bigvee A$ for $A \subseteq V$, equivalently all infima $\bigwedge A$. We denote binary meets and joins by $\wedge$ and $\vee$, respectively. Given $x, y \in V$, we say that x is *well above* y, and write $x \gg y$, if whenever $y \geq \bigwedge A$ for some $A \subseteq V$, then $x \geq a$ for some $a \in A$. A complete lattice $(V, \leq)$ is *completely distributive* if all joins in V distribute over all meets, equivalently all meets distribute over all joins [46]. Another equivalent characterization is that $(V, \leq)$ is completely distributive iff

$$y = \bigwedge \{x \in V \mid x \gg y\}$$

for every $y \in V$ [47].

In the definition of value quantale, we follow Flagg [24] in dualizing the usual continuity condition for quantales in order to avoid having to reverse the order when moving between the general development and basic examples such as the unit interval; deviating from his terminology, we emphasize this by the prefix 'co-':

Definition 2.1 ((Value) co-quantales). A *(commutative) co-quantale* $\mathcal{V}$ is a complete lattice $(V, \leq)$ equipped with a commutative monoid structure $(0, \oplus)$ that is *meet-continuous*:

$$a \oplus \bigwedge_{i\in I} b_i = \bigwedge_{i\in I}(a \oplus b_i).$$

A co-quantale $\mathcal{V}$ is a *value co-quantale* [24] if 0 is the bottom element of $\mathcal{V}$ and moreover $(V, \leq)$ is a *value distributive lattice*, i.e. a completely distributive complete lattice such that $|V| > 1$ and for all $x, y \in V$, $x, y \gg 0$ implies $x \wedge y \gg 0$. Correspondingly, we denote the greatest element of $\mathcal{V}$ by 1.

(Dually, in a *quantale* the operation $\oplus$ is required to be *join-continuous*.) By meet-continuity, we obtain a further binary operator $\ominus$ on a co-quantale $\mathcal{V}$ by adjunction, defined by

$$a \ominus b \leq v \quad \text{iff} \quad a \leq b \oplus v$$

(equivalently, $a \ominus b = \bigwedge\{v \mid a \leq b \oplus v\}$). The operator $\ominus$ is sometimes called the *internal hom* of $\mathcal{V}$ [7]. Moreover, in a value co-quantale, we have that for each $\varepsilon \gg 0$, there exists $\delta \gg 0$ such that $2 \cdot \delta := \delta \oplus \delta \leq \varepsilon$ [24, Theorem 2.9]. This allows for proofs where an error bound $\varepsilon \gg 0$ needs to be split up into multiple smaller parts.

A simple example of a value co-quantale is the unit interval $[0, 1]$ with the usual ordering, with truncated addition $a \oplus b = \min(a + b, 1)$ as the monoid structure. Correspondingly, the $\ominus$ operation is truncated subtraction $a \ominus b = \max(a - b, 0)$. We have $a \gg b$ iff $a > b$. We will give further examples in Section 3.

Universal Coalgebra serves as a unified framework for many types of state-based systems [49], such as nondeterministic, probabilistic, alternating, game-based, or weighted systems. It is based on encapsulating the system type as a *functor* T, for our purposes on the category Set of sets and functions; such a T assigns to each set X a set TX, thought of as a type of structured collections over X, and to each map $f\colon X \to Y$ a map $Tf\colon TX \to TY$, respecting identities and composition. A *T-coalgebra* (A, α) consists of a set A of *states* and a *transition map* $\alpha\colon A \to TA$, thought of as assigning to each state a structured collection of successors. Taking T to be the *covariant powerset functor* $\mathcal{P}$, which assigns to each set X its powerset $\mathcal{P}X$, we obtain relational transition systems as T-coalgebras. As a further example, the *(discrete) subdistribution functor* $\mathcal{S}$ assigns to each set X the set $\mathcal{S}X$ of discrete probability subdistributions μ on X (i.e. $\mu(X_0) = \mu(X) \leq 1$ for some countable subset $X_0 \subseteq X$), and to each map $f\colon X \to Y$ the image measure function (i.e. $\mathcal{S}f(\mu)(B) = \mu(f^{-1}[B])$ for $B \subseteq Y$).

$\mathcal{S}$-coalgebras are probabilistic transition systems (or Markov chains) with possible deadlock: They assign to each state a subdistribution over possible successor states, with the gap of the total probability to 1 interpreted as the probability of deadlock. Additional instances are seen in Example 4.4. For the remainder of the paper, we *fix a set functor T and require that $T\emptyset$ is nonempty* (hence our use of subdistributions instead of distributions in the examples). Moreover, we require w.l.o.g. that T is *standard*, i.e. preserves subset inclusions [5].

3 Quantale-Valued Distances and Lax Extensions

A *$\mathcal{V}$-valued relation* between sets A and B is a map $R\colon A \times B \to V$, which we also denote by $R\colon A \nrightarrow B$. For fixed A and B, we order the $\mathcal{V}$-valued relations between A and B pointwise: $R_1 \leq R_2 \iff \forall a \in A, b \in B.\ R_1(a,b) \leq R_2(a,b)$. We compose relations $R\colon A \nrightarrow B$ and $S\colon B \nrightarrow C$ using the monoid operation on $\mathcal{V}$:

$$(R;S)(a,c) = \bigwedge\{R(a,b) \oplus S(b,c) \mid b \in B\}.$$

Given a function $f\colon A \to B$ and $\varepsilon \in V$, the *ε-graph* $\mathrm{Gr}_{\varepsilon,f}$ is the relation

$$\mathrm{Gr}_{\varepsilon,f}(a,b) = \begin{cases} \varepsilon, & \text{if } f(a) = b; \\ 1, & \text{otherwise.} \end{cases}$$

We also write $\mathrm{Gr}_f = \mathrm{Gr}_{0,f}$ and, in case of the identity function, $\Delta_{\varepsilon,X} = \mathrm{Gr}_{\varepsilon,\mathsf{id}_X}$ and $\Delta_X = \Delta_{0,X}$.

Definition 3.1 ($\mathcal{V}$-continuity space). Let X be a set and let $d\colon X \nrightarrow X$. The pair (X,d) is a *$\mathcal{V}$-continuity space* [24] if $d \leq \Delta_X$ and $d \leq d;d$, or equivalently, if for all $x,y,z \in X$,

$$d(x,x) = 0 \qquad \text{and} \qquad d(x,z) \leq d(x,y) \oplus d(y,z).$$

The *dual* of (X,d) is the $\mathcal{V}$-continuity space (X,d^*) where $d^*(x,y) = d(y,x)$. The *symmetrization* of (X,d) is the space (X,d^s) with $d^s(x,y) = d(x,y) \vee d^*(x,y)$. We say that (X,d) is *symmetric* if $d = d^*$.

Remark 3.2. Recall that omission of the metric symmetry axiom $d(x,y) = d(y,x)$ is standardly designated by the prefix 'quasi-' and omission of the anti-symmetry axiom $d(x,y) = 0 \Rightarrow x = y$ by the prefix 'pseudo-'; thus, continuity spaces could be termed *generalized pseudo-quasimetric spaces*, and symmetric continuity spaces *generalized pseudometric spaces.*

The co-quantale $\mathcal{V}$ itself is made into a $\mathcal{V}$-continuity space $(V, d_{\mathcal{V}})$ using the operator $\ominus$:

$$d_{\mathcal{V}}(a,b) = a \ominus b.$$

For any set A, the *supremum distance* between $\mathcal{V}$-valued maps $f,g\colon A \to V$ is

$$d_{\mathcal{V}}^{\vee}(f,g) = \bigvee_{a \in A} d_{\mathcal{V}}(f(a), g(a)).$$

The usual notion of nonexpansive map generalizes as expected:

Definition 3.3 (Nonexpansive maps). A map $f\colon X \to Y$ between $\mathcal{V}$-continuity spaces (X, d_1) and (Y, d_2) is *nonexpansive* if $d_2(f(x), f(y)) \leq d_1(x, y)$ for all $x, y \in X$. We denote the space of nonexpansive maps between (X, d_1) and (Y, d_2) by $(X, d_1) \to_1 (Y, d_2)$. In the special case of nonexpansive $\mathcal{V}$-valued maps we write $\mathsf{Pred}(X, d) = (X, d) \to_1 (V, d_\mathcal{V})$.

Ultimately we are interested in defining and reasoning about *behavioural distances.* Generally speaking, a behavioural distance is a $\mathcal{V}$-continuity space defined on the carrier of a T-coalgebra $\alpha\colon A \to TA$ in such a way that the behaviour defined by the coalgebra map α is incorporated into the distance values of states in A. This is accomplished using *relation liftings*, which lift $\mathcal{V}$-valued relations giving distances between states to those giving distances between successor structures of states. We specifically generalize the notion of nonexpansive lax extension [56] to the quantale-valued case:

Definition 3.4 (Lax Extension). A *nonexpansive lax extension* of T is a mapping L that maps $\mathcal{V}$-valued relations $R\colon A \times B \to V$ to relations $LR\colon TA \times TB \to V$ and satisfies the following axioms:

$$\begin{aligned}
&(L1)\quad R_1 \leq R_2 \implies LR_1 \leq LR_2\\
&(L2)\quad L(R;S) \leq LR; LS\\
&(L3)\quad L\mathrm{Gr}_f \leq \mathrm{Gr}_{Tf}\\
&(L4)\quad L\Delta_{\varepsilon,A} \leq \Delta_{\varepsilon,TA}
\end{aligned}$$

for all $R, R_1, R_2\colon A \nrightarrow B$, $S\colon B \nrightarrow C$, $f\colon A \to B$ and $\varepsilon \in V$.

(The notion of *lax extension*, given by axioms $(L1)$–$(L3)$, is standard, e.g. [31]; the axiom $(L4)$, introduced in [56], guarantees nonexpansiveness w.r.t. the supremum metric as shown in Lemma 3.6.)

Lemma 3.5. *If L is a lax extension of T and (A, d) is a $\mathcal{V}$-continuity space, then so is (TA, Ld).*

Lemma 3.6. *If L is a nonexpansive lax extension of T, then L is in fact nonexpansive w.r.t. the supremum metric. That is, for $R_1, R_2\colon A \nrightarrow B$ we have $d_\mathcal{V}^\vee(LR_1, LR_2) \leq d_\mathcal{V}^\vee(R_1, R_2)$.*

Proof. We have $d_\mathcal{V}^\vee(R_1, R_2) \leq \varepsilon \iff R_1 \leq R_2; \Delta_\varepsilon$. Using $(L1)$, $(L2)$ and $(L4)$, we have $LR_1 \leq L(R_2; \Delta_\varepsilon) \leq LR_2; L\Delta_\varepsilon \leq LR_2; \Delta_\varepsilon$, so $d_\mathcal{V}^\vee(LR_1, LR_2) \leq \varepsilon$. □

For technical purposes, we will be interested in a generalized version of total boundedness (recall that a standard metric space is compact iff it is complete and totally bounded):

Definition 3.7 (Total boundedness). Let (X, d) be a $\mathcal{V}$-continuity space. For $\varepsilon \gg 0$, we write $B_\varepsilon^s(x) = \{y \in X \mid d^s(x, y) \leq \varepsilon\}$ for the *(symmetric) ball* of radius ε around $x \in X$. A *finite ε-cover* of (X, d) is a choice of finitely many $x_1, \ldots, x_n \in X$ such that $X = \bigcup_{i=1}^n B_\varepsilon^s(x_i)$. We say that (X, d) is *totally bounded* if X has a finite ε-cover for each $\varepsilon \gg 0$.

Remark 3.8. Note that use of the symmetrization d^s is essential in the above definition; e.g. in the unit interval, with $d(x,y) = x \ominus y$, the set $\{y \mid d(0,y) \leq \varepsilon\}$ is the whole space, so 0 alone would form an ε-cover of $[0,1]$ if we replaced d^s with d.

Moreover, our main result involves a generalization of the standard notion of density:

Definition 3.9 (Density). Let (X,d) be a $\mathcal{V}$-continuity space. A subset $Y \subseteq X$ is *dense* if for every $x \in X$ and $\varepsilon \gg 0$ there exists $y \in Y$ such that $d^s(x,y) \leq \varepsilon$.

Assumption 3.10. Throughout the paper, we *fix a value co-quantale $\mathcal{V}$ that is totally bounded as a $\mathcal{V}$-continuity space.* Moreover, we fix a dense subset $\mathcal{V}_0 \subseteq \mathcal{V}$ for use as a set of truth constants in the relevant logics, with a view to keeping the syntax countable in the central examples. (The technical development, on the other hand, does not require $\mathcal{V}_0$ to be countable, so we can always take $\mathcal{V}_0 = \mathcal{V}$.)

Example 3.11 ((Value) co-quantales).

1. The set $2 = \{0,1\}$, with $0 \leq 1$ and with binary join as the monoid structure, is a value co-quantale [24], and of course totally bounded. 2-Continuities d are just preorders, with y being above x if $d(x,y) = 0$ (!); symmetric 2-continuities are equivalence relations. Notice that $0 \gg 0$ in 2. The $\ominus$ operator is given by $a \ominus b = 1$ iff $a = 1$ and $b = 0$.
2. The dual of every *locale* (e.g. [8]), in particular the set of closed subsets of any topological space, forms a co-quantale, with binary join as the monoid structure. However, locales are not in general value co-quantales. The dual $\Omega(R)$ of the *free* locale over a set R, described as the lattice of downclosed systems of finite subsets of R (ordered by reverse inclusion of such set systems), does form a value co-quantale [24], and is totally bounded [30]. $\Omega(R)$-continuity spaces are known as *structure spaces* [30,24].
3. The unit interval $[0,1]$ is totally bounded. $[0,1]$-Continuity spaces coincide with 1-bounded pseudo-quasimetric spaces, and symmetric $[0,1]$-continuity spaces with 1-bounded pseudometric spaces in the standard sense (cf. Remark 3.2).
4. *Convex-nondeterministic distances:* The set $\mathcal{I}$ of nonempty closed subintervals (i.e. finitely generated nonempty convex subsets) of $[0,1]$, written in the form $[a,b]$ with $a \leq b$, ordered by $[a,b] \leq [c,d]$ iff $a \leq c$ and $b \leq d$, and equipped with truncated Minkowski addition $[a,b] \oplus [c,d] = [a \oplus c, b \oplus d]$ (with $\oplus$ on $[0,1]$ defined as in the previous item), is a totally bounded value co-quantale. We write $\{\!\![a,b\}\!\!\} = [a, \max(a,b)]$. We have $[a,b] \gg 0 = [0,0]$ iff $a > 0$, and $[a,b] \ominus [c,d] = \{\!\![a \ominus c, b \ominus d\}\!\!\}$, again with $\ominus$ on $[0,1]$ described as in the previous item. We can think of an $\mathcal{I}$-continuity space as assigning to each pair of points a nondeterministic distance, given as an interval of possible distances.

4 Quantale-Valued Modal and Predicate Logics

We next introduce the main objects of study, quantale-valued coalgebraic modal and predicate logics. They will feature modalities interpreted using a quantitative version of *predicate liftings* [45,50,51]. Predicate liftings take their name from the fact that they lift predicates on a base set X to predicates on the set TX (where T is our globally fixed functor representing the system type according to Section 2). We work with *$\mathcal{V}$-valued predicates*, which are organized in the *contravariant $\mathcal{V}$-powerset* functor $\mathcal{Q}$ given on sets X by $\mathcal{Q}X = X \to V$ and on functions $f\colon X \to Y$ by $\mathcal{Q}f(g) = g \circ f$ (that is, $\mathcal{Q}$ is a functor $\mathsf{Set}^{\mathrm{op}} \to \mathsf{Set}$ where $\mathsf{Set}^{\mathrm{op}}$ is the opposite category of Set). In keeping with the prevalent reading in fuzzy and probabilistic logics (where, typically, $\mathcal{V} = [0,1]$), we read $0 \in \mathcal{V}$ as 'false' and $1 \in \mathcal{V}$ as 'true' (opposite choices are also found in the literature, e.g. in modal logics for metric transition systems [3], where $0 \in [0,1]$ is interpreted as 'true'). Predicate liftings can have arbitrary finite arities [50]. For brevity, we restrict the presentation to unary modalities and predicate liftings; generalizing to higher arities requires only more indexing.

Definition 4.1. A *($\mathcal{V}$-valued) predicate lifting* is a natural transformation $\lambda\colon \mathcal{Q} \to \mathcal{Q} \circ T$, i.e. a family of maps $\lambda_X\colon \mathcal{Q}X \to \mathcal{Q}TX$, indexed over all sets X, such that $\lambda_Y(f)(Th(t)) = \lambda_X(f \circ h)(t)$ for all $f\colon Y \to V$, $h\colon X \to Y$, $t \in TX$.

Definition 4.2. Let λ be a predicate lifting.
1. λ is *monotone* if for all sets X and all $f, g \in \mathcal{Q}X$ with $f \leq g$ we have $\lambda_X(f) \leq \lambda_X(g)$.
2. λ is *nonexpansive* if for all sets X and all $f, g \in \mathcal{Q}X$ we have $d^{\vee}_{\mathcal{V}}(\lambda_X(f), \lambda_X(g)) \leq d^{\vee}_{\mathcal{V}}(f, g)$.

For the remainder of the paper, we *fix a set Λ of monotone and nonexpansive predicate liftings*, which, by abuse of notation, we also use as modalities in the syntax. A basic example is the $\Diamond$ modality of quantitative probabilistic modal logic [10], which denotes expected probability (in the next transition step) and corresponds to a predicate lifting for the (sub-)distribution functor $\mathcal{S}$ (Section 2); see Example 4.4.2 for details. The generic **syntax** of *($\mathcal{V}$-valued) quantitative coalgebraic modal logic* is then given by the grammar

$$\varphi, \psi ::= c \mid \varphi \oplus c \mid \varphi \ominus c \mid \varphi \wedge \psi \mid \varphi \vee \psi \mid \lambda\varphi \qquad (c \in V_0, \lambda \in \Lambda).$$

The operators $\oplus$, $\ominus$, $\vee$, $\wedge$ denote co-quantale operations, the meaning of λ is determined by the associated predicate lifting. As usual, the *rank* of a formula φ is the maximal nesting depth of modalities λ in φ. We denote the set of all modal formulae by $\mathcal{L}^{\Lambda}$ and the set of formulae of rank at most n by $\mathcal{L}^{\Lambda}_n$.

Formally, the **semantics** is defined by assigning to each formula φ and each T-coalgebra $\alpha\colon A \to TA$ the *extension* $[\![\varphi]\!]_\alpha\colon A \to V$, or just $[\![\varphi]\!]$, of φ over α, recursively defined by

$$[\![\varphi \oplus c]\!](a) = [\![\varphi]\!](a) \oplus c \qquad [\![\varphi \ominus c]\!](a) = [\![\varphi]\!](a) \ominus c$$
$$[\![\varphi \wedge \psi]\!](a) = [\![\varphi]\!](a) \wedge [\![\psi]\!](a) \qquad [\![\varphi \vee \psi]\!](a) = [\![\varphi]\!](a) \vee [\![\psi]\!](a)$$
$$[\![c]\!](a) = c \qquad [\![\lambda\varphi]\!](a) = \lambda_A([\![\varphi]\!])(\alpha(a))$$

Remark 4.3. Fuzzy logics differ widely in their interpretation of propositional connectives (e.g [41]). In our modal syntax, we necessarily restrict to nonexpansive operations, in order to ensure nonexpansiveness w.r.t. behavioural distance later; this is typical of characteristic logics for behavioural distances (such as quantitative probabilistic modal logic [10]). The logic hence does not include binary $\oplus$ or $\ominus$ (in the above syntax, we insist that one of the arguments is a constant). In terminology usually applied to $\mathcal{V} = [0,1]$, we thus allow *Zadeh* connectives (such as $\vee$, $\wedge$) but not *Łukasiewicz* connectives, so for $\mathcal{V} = [0,1]$, the above version of quantitative coalgebraic modal logic is essentially the Zadeh fragment of Łukasiewicz fuzzy coalgebraic modal logic [51].

The syntax does not include negation $1 \ominus (-)$; if $\mathcal{V}$ satisfies the De Morgan laws (e.g. these hold in $[0,1]$), Λ is closed under *duals* $1 \ominus (\lambda(1 \ominus (-)))$, and V_0 is closed under negation (i.e. $c \in V_0$ implies $1 \ominus c \in V_0$), then negation can be defined via negation normal forms as usual.

As the ambient predicate logic of the above modal logic, we use *($\mathcal{V}$-valued) quantitative coalgebraic predicate logic*, a quantitative variant of two-valued coalgebraic predicate logic [40]. Its **syntax** is given by

$$\varphi, \psi ::= c \mid x = y \mid \varphi \oplus c \mid \varphi \ominus c \mid \varphi \wedge \psi \mid \varphi \vee \psi \mid \exists x.\varphi \mid \forall x.\varphi \mid x\lambda\lceil y \colon \varphi\rceil$$

where $c \in V_0$, $\lambda \in \Lambda$, and x, y come from a fixed supply Var of (individual) variables. The reading of $x\lambda\lceil y \colon \varphi\rceil$ is the modalized truth degree (according to λ) to which the successors y of a state x satisfy φ; e.g. with $\Diamond$ as above, $x\Diamond\lceil y \colon \varphi\rceil$ is the expected truth value of φ at a random successor y of x. The **semantics** over (A, α) as above is given by $\mathcal{V}$-valued maps $[\![\varphi]\!]_\alpha$, or just $[\![\varphi]\!]$, that are defined on valuations $\kappa \colon \mathsf{Var} \to A$. The interesting clauses in the definition are

$$[\![\exists x.\varphi]\!](\kappa) = \bigvee_{a \in A} [\![\varphi]\!](\kappa[x \mapsto a]) \qquad [\![\forall x.\varphi]\!](\kappa) = \bigwedge_{a \in A} [\![\varphi]\!](\kappa[x \mapsto a])$$

$$[\![x\lambda\lceil y \colon \varphi\rceil]\!](\kappa) = \lambda_A([\![\varphi]\!](\kappa[y \mapsto \cdot]))(\alpha(\kappa(x)))$$

(where $\kappa[y \mapsto a]$ maps y to a and otherwise behaves like κ, and by $[\![\varphi]\!](\kappa[y \mapsto \cdot])$ we mean the predicate that maps a to $[\![\varphi]\!](\kappa[y \mapsto a])$). Moreover, equality is crisp, i.e. $[\![x = y]\!](\kappa)$ is 1 if $\kappa(x) = \kappa(y)$, and 0 otherwise.

Example 4.4. We discuss some instances of the above framework.

1. *Fuzzy modal logic:* Take T to be the *covariant* $\mathcal{V}$-valued powerset functor, i.e. $TX = X \to V$ and $Tf(A)(y) = \bigvee\{A(x) \mid f(x) = y\}$ for $f : X \to Y$. We think of $A \in TX$ as a $\mathcal{V}$-valued fuzzy subset of X; we say that A is *crisp* if $A(x) \in \{0,1\}$ for all x. Put $\Lambda = \{\Diamond\}$ where $\Diamond_X(A)(B) = \bigvee\{A(x) \wedge B(x) \mid x \in X\}$ for $A \in \mathcal{Q}X$, $B \in TX$. Then T-coalgebras are equivalent to fuzzy Kripke frames, which consist of a set X and a fuzzy relation $R \colon X \times X \to V$, and $\Diamond$ is the natural fuzzification of the standard diamond modality. Fuzzy propositional atoms from a set At

can be added by passing to the functor that maps a set X to $\mathcal{Q}(\mathsf{At}) \times TX$. Instantiating to $\mathcal{V} = [0,1]$, we obtain a basic modal logic of fuzzy relations, or in description logic terminology *Zadeh fuzzy $\mathcal{ALC}$* [53]. The corresponding instance of quantitative coalgebraic predicate logic is essentially the Zadeh fragment of Novak's Łukasiewicz fuzzy first order logic [43].

2. *Probabilistic modal logic:* As indicated in Section 2, coalgebras for the subdistribution functor $\mathcal{S}$ are probabilistic transition systems (with possible deadlock). We take $\mathcal{V} = [0,1]$ and $\Lambda = \{\Diamond\}$, interpreted by the predicate lifting

$$\Diamond_X(A)(\mu) = \mathbb{E}_\mu(A) \qquad \text{for } \mu \in \mathcal{S}X$$

where $\mathbb{E}_\mu(A)$ denotes the expected value of $A(x)$ when x is distributed according to μ. The induced instance of quantitative coalgebraic modal logic is *(quantitative) probabilistic modal logic* [10], which may be seen as a quantitative variant of two-valued probabilistic modal logic [39], and embeds into the probabilistic μ-calculus [34,42]. Propositional atoms are treated analogously as in the previous item (and indeed probabilistic modal logic is trivial without them). The ambient quantitative probabilistic first-order logic arising as the corresponding instance of quantitative coalgebraic predicate logic is a quantitative variant of Halpern's type-1 (i.e. statistical) probabilistic first-order logic [27].

3. *Metric modal logic:* In their simplest form, *metric transition systems* [3] are just transition systems in which states are labelled in a metric space S (numerous variants exist, e.g. with states themselves forming a metric space or with transitions labelled in a metric space [9]). We work with a generalized version where (S, d_S) is a $\mathcal{V}$-continuity space. Metric transition systems are then coalgebras for the functor TX given on sets by $TX = S \times \mathcal{P}X$. We take $\Lambda = \{\Diamond\} \cup S$. We interpret Λ using predicate liftings

$$\Diamond_X(A)(s,B) = \bigvee\{A(x) \mid x \in B\} \qquad r_X(A)(s,B) = d_S(s,r)$$

for $A \in \mathcal{Q}X$, $(s,B) \in TX$, $r \in S$. Note that $r \in S$ ignores its argument A, so is effectively a nullary modality. Note also that as per our interpretation of truth values, this nullary modality is read as distinctness from r; in case $\mathcal{V} = [0,1]$, the degree of equality to r can be expressed as $1 \ominus r$. The induced instance of coalgebraic modal logic is related to characteristic logics for branching-time behavioural distances on metric transition systems [3,9].

4. *Convex-nondeterministic metric modal logic:* We continue to consider metric transition systems as recalled in the previous item, reusing the designators T, S, d_S, and taking $\mathcal{V} = [0,1]$ for simplicity. Recall the value co-quantale $\mathcal{I}$ of nonempty closed subintervals of $[0,1]$ from Example 3.11.4. We turn the predicate liftings for $r \in S$ defined in the previous item into $\mathcal{I}$-valued predicate liftings by prolonging them along the inclusion $\iota\colon [0,1] \hookrightarrow \mathcal{I}$, given by $\iota(a) = [a,a]$. We define an $\mathcal{I}$-valued predicate lifting M for T, where $\mathcal{I}$ is the value quantale of closed intervals introduced in Example 3.11.4, by

$$M_X(A)(s,B) = [\,\bigwedge\{\pi_1(A(x)) \mid x \in B\},\ \bigvee\{\pi_2(A(x)) \mid x \in B\}]$$

where $\pi_i\colon \mathcal{I} \to [0,1]$ denote the evident *projections* $\pi_1([a,b]) = a$, $\pi_2([a,b]) = b$. That is, M returns the range of truth values that A takes on B.

5 Behavioural Distance and Quantitative Bisimulation Invariance

The behavioural distance between states of a coalgebra $\alpha\colon A \to TA$ is defined as a least fixpoint that arises from an iterative process: Initially, at depth 0, all states are thought of as equivalent and their distance is therefore 0. In order to increase the depth of the behavioural distance from n to $n+1$, we lift the depth-n distance on A to a the set TA of successor structures. Formally, this is accomplished using the following quantale-valued version of the coalgebraic Kantorovich lifting [4,56]:

Definition 5.1 (Kantorovich lifting). Let A and B be sets and $R\colon A \nrightarrow B$.

1. A pair (f, g) of functions $f\colon A \to V$, $g\colon B \to V$ is *R-nonexpansive* if $f(a) \ominus g(b) \leq R(a, b)$ for all $a \in A$, $b \in B$.
2. The *Kantorovich lifting* of R is the relation $K_\Lambda(R)\colon TA \nrightarrow TB$ given by

$$K_\Lambda(R)(t_1, t_2) = \bigvee\{\lambda_A(f)(t_1) \ominus \lambda_B(g)(t_2) \mid \lambda \in \Lambda, (f, g) \text{ } R\text{-nonexpansive}\}.$$

(Here, Λ is the set of modalities fixed in Section 4.) Generalizing [56, Theorem 5.6], we have:

Lemma 5.2. *The Kantorovich lifting is a nonexpansive lax extension.*

Example 5.3 (Kantorovich liftings).

1. For $\mathcal{V} = [0, 1]$ and $\mathcal{V}$-valued *fuzzy modal logic* with $\Lambda = \{\Diamond\}$ (i.e. for simplicity without propositional atoms; cf. Example 4.4.1), the Kantorovich lifting $K_\Lambda(R)$ of a $\mathcal{V}$-valued relation $R\colon X \nrightarrow Y$ coincides with an asymmetric generalized Hausdorff lifting; i.e.

$$K_\Lambda(R)(A, B) = \bigvee_{x \in X} \bigwedge_{y \in Y} ((A(x) \ominus B(y)) \vee (A(x) \wedge R(x, y)))$$

for $A \in TX = X \to V$, $B \in TY$. (Obtaining a similar description for general $\mathcal{V}$ remains an open problem.) In particular, on crisp sets A, B, the symmetrization $K_\Lambda(R)^s$ is the usual Hausdorff lifting $K_\Lambda(R)^s(A, B) = \max(\bigvee_{A(x)=1} \bigwedge_{B(y)=1} R(x, y), \bigvee_{B(y)=1} \bigwedge_{A(x)=1} R(x, y))$.

2. For *probabilistic modal logic* (Example 4.4.2), the restriction of K_Λ to distributions coincides, by definition, with the usual (symmetric) Kantorovich-Wasserstein lifting (e.g. [10]). On subdistributions, one obtains an asymmetric variant, whose symmetrization then coincides with the standard one.

3. For $\mathcal{V}$-valued *metric modal logic* (Example 4.4.3), with $\Lambda = \{\Diamond\} \cup S$, we similarly obtain a $\mathcal{V}$-valued (asymmetric) Hausdorff distance

$$K_\Lambda(R)((s, A), (t, B)) = d(s, t) \vee \bigvee_{x \in A} \bigwedge_{y \in B} R(x, y)$$

on $(s, A) \in TX = S \times \mathcal{P}(X)$, $(t, B) \in TY$, and $R\colon X \nrightarrow Y$; a characterization that in this case holds for unrestricted $\mathcal{V}$.

4. *Convex-nondeterministic metric modal logic:* The $\mathcal{I}$-valued Kantorovich lifting induced by the set $\Lambda = \{M\} \cup S$ of modalities on metric transition systems, with notation as in Examples 3.11.4 and 4.4.4, is given by

$$K_\Lambda(R)((s,A),(t,B)) = \iota(d(s,t)) \vee \\ \{\!\!\{\bigvee_{y\in B} \bigwedge_{x\in A} \pi_1(R(x,y)), \bigvee_{x\in A} \bigwedge_{y\in B} \pi_2(R(x,y))\}\!\!\}$$

on $(s,A) \in TX = S \times \mathcal{P}(X)$, $(t,B) \in TY$, and $R\colon X \nrightarrow Y$ (recall that the π_i are the projections $\mathcal{I} \to [0,1]$, and $\iota\colon [0,1] \to \mathcal{I}$ denotes the evident injection).

For purposes of lifting $\mathcal{V}$-continuity structures as relations, nonexpansive pairs can be replaced with the more familiar notion of nonexpansive map:

Lemma 5.4. *Let (A,d) be a $\mathcal{V}$-continuity space and let (f,g) be d-nonexpansive. Put $h(b) = \bigvee_{a\in A} f(a) \ominus d(a,b)$. Then $f \leq h \leq g$ and $h \in \mathsf{Pred}(A,d)$.*

By monotonicity of predicate liftings we get the following alternative formulation for the Kantorovich lifting of a $\mathcal{V}$-continuity structure:

Lemma 5.5. *Let (A,d) be $\mathcal{V}$-continuity space. Then for all $t_1, t_2 \in TA$*

$$K_\Lambda(d)(t_1,t_2) = \bigvee\{\lambda_A(h)(t_1) \ominus \lambda_A(h)(t_2) \mid \lambda \in \Lambda, h \in \mathsf{Pred}(A,d)\}.$$

Using the Kantorovich lifting, we can now define a sequence of behavioural distances between states a, b in a T-coalgebras $\alpha\colon A \to TA, \beta\colon B \to TB$:

$$d_0^K(a,b) = 0 \quad d_{n+1}^K(a,b) = K_\Lambda(d_n^K)(\alpha(a),\beta(b)) \quad d_\omega^K(a,b) = \bigvee_{n<\omega} d_n^K(a,b).$$

By general fixed point theory, the continuation of this ordinal-indexed sequence past ω eventually stabilizes, that is, there exists some ordinal γ such that $d_{\gamma+1}^K = d_\gamma^K$. The arising least fixed point is the unbounded *behavioural distance* d^K, alternatively given by

$$d^K = \bigwedge\{d \mid d = K_\Lambda(d) \circ (\alpha \times \beta)\}.$$

These behavioural distances lead to an appropriate generalization of the notion of bisimulation invariance. A family f of $\mathcal{V}$-valued predicates f_α indexed over T-coalgebras $\alpha\colon A \to TA$ – such as the extension of a modal formula or of a first-order formula with a single free variable – is said to be *behaviourally nonexpansive* if it is nonexpansive with respect to behavioural distance d^K, i.e. if for all coalgebras $\alpha\colon A \to TA, \beta\colon B \to TB$ and all $a \in A, b \in B$,

$$f_\alpha(a) \ominus f_\beta(b) \leq d^K(a,b). \tag{1}$$

Similarly, f is *depth-n behaviourally nonexpansive* for finite depth n if f is nonexpansive with respect to depth-n behavioural distance d_n^K.

To match these notions to the classical setting, consider the binary co-quantale 2. In the general case, the above notion of behavioural nonexpansiveness should then be thought of as preservation under simulation: States a, b have (asymmetric) distance 0 if b simulates a, and in this case, (1) stipulates that if f is true at a, then f is also true at b.

Example 5.6. The behavioural distance arising from the Kantorovich lifting of metric modal logic (Example 5.3.3) is a *simulation distance*. The value $d^K(a,b)$ quantifies the degree to which traces starting at b simulate traces starting at a, where the distance from one trace to another is the supremum over the distances at all time steps.

On the other hand, there are many cases where the behavioural distance d^K is symmetric. If $\mathcal{V} = [0,1]$ and the set Λ is closed under duals (Remark 4.3), then we have that $K_\Lambda(R^*) = K_\Lambda(R)^*$ for all R and therefore d^K is symmetric [56]. Concretely, if we put $\Box_X(A) = 1 \ominus \Diamond_X(1 \ominus A)$, then in the case of fuzzy modal logic (Example 4.4.1) we have $\Box_X(A)(B) = \bigwedge\{(1 \ominus B(x)) \vee A(x) \mid x \in X\}$ and in the case of probabilistic modal logic (Example 4.4.2) we have $\Box_X(A)(\mu) = \mathbb{E}_\mu(A) \oplus (1 \ominus \mu(X))$, and in both cases $\Lambda = \{\Diamond, \Box\}$ yields a symmetric distance.

In these symmetric cases distance 0 determines a notion of *bi*similarity, and behavioural nonexpansiveness amounts to the standard notion of bisimulation invariance. Thus, the following straightforward lemma generalizes both bisimulation invariance of modal logic and preservation of positive modal logic (with only diamond modalities) under simulation:

Lemma 5.7. *All modal formulae are behaviourally nonexpansive, and all modal formulae of rank at most n are depth-n behaviourally nonexpansive.*

As expected, coalgebra morphisms preserve behaviour on the nose:

Lemma 5.8. *Let $\alpha\colon A \to TA$ and $\beta\colon B \to TB$ be coalgebras and $h\colon A \to B$ a coalgebra morphism, that is $Th \circ \alpha = \beta \circ h$. Then $d^{K,s}(a, h(a)) = 0$ for all $a \in A$.*

Another way to define distances between states of a coalgebra is in terms of the modal formulae:

Definition 5.9 (Logical distance). Let a, b be states in coalgebras $\alpha\colon A \to TA, \beta\colon B \to TB$. We define

$$d_n^L(a,b) = \bigvee\{[\![\varphi]\!](a) \ominus [\![\varphi]\!](b) \mid \varphi \in \mathcal{L}_n^\Lambda\}$$
$$d^L(a,b) = \bigvee\{[\![\varphi]\!](a) \ominus [\![\varphi]\!](b) \mid \varphi \in \mathcal{L}^\Lambda\}$$

The relationship between fixpoint-based distances d^K and logical distances d^L is at the heart of the study of behavioural nonexpansiveness and modal expressiveness. For instance, Lemma 5.7 can equivalently be expressed by the inequalities $d^L \leq d^K$ and $d_n^L \leq d_n^K, n < \omega$. In Section 6, we investigate the converse inequalities.

6 Modal Approximation

We now establish our first contribution, a quantitative coalgebraic Hennessy-Milner theorem. To this end, we first need to pin down the exact relationship of the two families of distances at finite depth.

Theorem 6.1. *Let the set Λ of monotone and nonexpansive predicate liftings from Section 4 be finite and let (A, α) be a coalgebra. For all $n < \omega$:*

1. *We have $d_n^K = d_n^L =: d_n$*
2. *The space (A, d_n) is totally bounded.*
3. *The set $\mathcal{L}_n^\Lambda$ is a dense subset of $\mathsf{Pred}(A, d)$.*

Remark 6.2. The need for assuming that the set Λ of modalities is finite is specific to quantitative Hennessy-Milner theorems (and implicitly present also in the existing $[0,1]$-valued version of the theorem [37]), and not needed in the two-valued case [45,50]. It relates to the total boundedness claim in Theorem 6.1, and features also in the van Benthem theorem, where in fact it is needed also in the two-valued case [52]; indeed, proofs of the original van Benthem theorem start by assuming, in that case w.l.o.g., that there are only finitely many propositional atoms and relational modalities. In our running examples, only the ones featuring metric transition systems are affected by this assumption; indeed, for our theorems to apply to such systems, the space of labels needs to be finite.

Theorem 6.1 is proven by induction on n and most of Section 6 is devoted to the inductive step (the base case $n = 0$ is immediate from $d_0^K = d_0^L = 0$). We fix a coalgebra $\alpha\colon A \to TA$ and an integer $n > 1$ and assume as the inductive hypothesis that the three items of Theorem 6.1 have already been proven for all $m < n$. We show Item 1 in Lemma 6.3, Item 2 in Lemma 6.6, and Item 3 in Lemma 6.7.

Lemma 6.3. *We have $d_n^K = d_n^L$ on A.*

Proof (sketch). We use the alternative formula for the Kantorovich lifting as given in Lemma 5.5. By Item 3 of the inductive hypothesis, and because the predicate liftings are nonexpansive, the maps $\lambda(f) \circ \alpha$ with $f \in \mathsf{Pred}(A, d_{n-1})$ can be approximated using formula expansions $[\![\lambda\psi]\!]$ with $\psi \in \mathcal{L}_{n-1}^\Lambda$. □

Having shown that $d_n^K = d_n^L$, from now on we simply use d_n to denote both. To show that d_n is totally bounded, we make use of the following version of the Arzelà-Ascoli theorem [23, Theorem 4.13].

Lemma 6.4 (Arzelà-Ascoli). *Let (X, d_1) and (Y, d_2) be totally bounded $\mathcal{V}$-continuity spaces. Then the space $(X, d_1) \to_1 (Y, d_2)$ is also totally bounded.*

Using Lemma 6.4, we show that the Kantorovich lifting preserves total boundedness; this generalizes a previous result for the case $\mathcal{V} = [0,1]$ [37, Proposition 29], which in turn generalizes [57, Lemma 5.6].

Lemma 6.5. *If the set Λ of predicate liftings is finite and (X, d) is a totally bounded $\mathcal{V}$-continuity space, then $(TX, K_\Lambda(d))$ is totally bounded.*

The following is now an easy consequence:

Lemma 6.6. *The space (A, d_n) is totally bounded.*

Finally, we show that the modal formulae up to depth n form a dense subspace of the space of all nonexpansive properties:

Lemma 6.7. *Let $f \in \mathsf{Pred}(A, d_n)$ be a nonexpansive map and let $\varepsilon \gg 0$. Then there exists some modal formula $\varphi \in \mathcal{L}_n^\Lambda$ such that $d_{\mathcal{V}}^{\vee,s}(f, [\![\varphi]\!]) \leq \varepsilon$.*

Proof (sketch). We use the fact that for all $x, y \in A$

$$f(x) = \bigwedge_{y \in A} d_n(x, y) \oplus f(y) = \bigwedge_{y \in A} (\bigvee_{\gamma \in \mathcal{L}_n^\Lambda} [\![\gamma]\!](x) \ominus [\![\gamma]\!](y)) \oplus f(y).$$

The latter term can be approximated using formulae of $\mathcal{L}_n^\Lambda$, where the infimum over y and the supremum over γ are made finite using ε-covers of A and $\mathcal{L}_n^\Lambda$. □

Having shown that behavioural distance and logical distance coincide at all finite depths, we are now equipped to prove our first main result, a version of the Hennessy-Milner theorem stating that behavioural distance and logical distance coincide not only at finite depths (Theorem 6.1.1), but in fact also at unbounded depth. In general, this equivalence of distances can only be expected to hold if the functor T in question is *finitary*, or admits approximation by a finitary subfunctor [56]. The functor T is finitary if for all sets X and all $t \in TX$ there exists a finite subset $Y \subseteq X$ such that $t = Ti(s)$ for some $s \in TY$, where $i \colon Y \to X$ is set inclusion. Examples of finitary functors include the *finite powerset functor* $\mathcal{P}_\omega X = \{Y \subseteq X \mid Y \text{ finite}\}$ and the *finite subdistribution functor* $\mathcal{S}_\omega$ which maps a set X to the set of finitely supported probability subdistributions on X. König and Mika-Michalski [37] prove a quantitative coalgebraic Hennessy-Milner theorem for the case of the co-quantale $[0, 1]$. We generalize their result as follows:

Definition 6.8. We say that the value co-quantale $\mathcal{V}$ is *continuous from below* if for every monotone increasing sequence $(a_n)_{n<\omega}$ in $\mathcal{V}$ and every $\varepsilon \gg 0$, there exists some n such that $a_n \oplus \varepsilon \geq \bigvee_{n<\omega} a_n$.

This condition essentially allows the use of epsilontic arguments also for joins of increasing sequences, while value co-quantales in general allow this only for meets. It holds in all our running examples.

Theorem 6.9 (Quantified Hennessy-Milner theorem). *Let Λ be a finite set of monotone and nonexpansive predicate liftings, let T be a finitary functor and let $\mathcal{V}$ be a totally bounded value co-quantale that is continuous from below. Then we have $d^K = d^L$.*

Proof (sketch). Because $\mathcal{V}$ is continuous from below, we have $K_\Lambda(d_\omega^K) = \bigvee_{n<\omega} K_\Lambda(d_n^K)$ on finite sets, and as T is finitary, this also holds for all sets. This implies that $d_\omega^K = d_{\omega+1}^K = d^K$, so that

$$d^K = \bigvee_{n<\omega} d_n^K = \bigvee_{n<\omega} d_n^L = d^L. \qquad \square$$

Besides examples already covered by the $[0, 1]$-valued version of the theorem [37], this result instantiates, e.g., to a quantitative Hennessy-Milner theorem for convex-nondeterministic metric modal logic (Example 4.4.4).

7 Locality and Modal Characterization

We proceed to establish our main result, the quantitative coalgebraic van Benthem theorem. The main tool in the proof of this result is a notion of *locality*, which characterizes formulae that only depend on the structure of the model in some neighbourhood of the state under consideration. This poses a challenge when it comes to coalgebraic models, as these need not come with a built-in graph structure that could be used to define what it means for two states to be neighbouring. To solve this, we make use of a technique based on *supported* coalgebras that has previously been used in the proof of a two-valued coalgebraic van Benthem theorem [52].

Recall from Section 2 that we assume $T\emptyset \neq \emptyset$. We fix an element $\bot \in T\emptyset$, and for each set A put $\bot_A = Ti(\bot)$, where $i\colon \emptyset \to A$ is the empty map.

Definition 7.1 (Support). Let A be a set. We say that a set $B \subseteq A$ is a *support* of $t \in TA$ if $t \in TB$. A *supported coalgebra* is a coalgebra $\alpha\colon A \to TA$ together with a map $\mathrm{supp}_\alpha\colon A \to \mathcal{P}A$ such that $\mathrm{supp}_\alpha(a)$ is a support of $\alpha(a)$ for every $a \in A$.

Every coalgebra can be supported because we can always put $\mathrm{supp}_\alpha(a) = A$ for all $a \in A$. Supporting a coalgebra equips it with a graph structure:

Definition 7.2 (Neighbourhood). Let $\mathcal{A} = (A, \alpha, \mathrm{supp}_\alpha)$ be a supported coalgebra.

1. The *Gaifman graph* of $\mathcal{A}$ is the undirected graph with vertex set A and edge set $\{\{a, b\} \mid b \in \mathrm{supp}_\alpha(a)\}$.
2. For any $a, b \in A$, the *Gaifman distance* $D_{\mathrm{supp}}(a, b)$ is the least number of steps to get from a to b in the Gaifman graph (or ∞, if no path from a to b exists).
3. The *radius-k neighbourhood* of a state $a \in A$ is the set $U^k(a) = \{b \in A \mid D_{\mathrm{supp}}(a, b) \leq k\}$.

For any $k < \omega$ and any state a in a supported coalgebra $\mathcal{A} = (A, \alpha, \mathrm{supp}_\alpha)$, we can define a supported coalgebra $\mathcal{A}_a^k = (U^k(a), \alpha^k, \mathrm{supp}_{\alpha^k})$ on the radius-k neighbourhood of a. The coalgebra map $\alpha^k\colon U^k(a) \to T(U^k(a))$ is given by $\alpha^k(b) = \alpha(b)$ if $\mathrm{supp}_\alpha(b) \subseteq U^k(a)$ and $\alpha^k(b) = \bot_A$ otherwise. We note that the latter case only occurs for states on the edge of $U^k(a)$, that is when $D_{\mathrm{supp}}(a, b) = k$. Note that $\bot_A$ has empty support by construction, so that we can put $\mathrm{supp}_{\alpha^k}(b) = \emptyset$ in this latter case and $\mathrm{supp}_{\alpha^k}(b) = \mathrm{supp}_\alpha(b)$ otherwise.

Using the neighbourhood around a state and the coalgebra structure defined on it, we can now define our notion of locality:

Definition 7.3. A formula φ is *k-local* if we have $[\![\varphi]\!]_\alpha(a) = [\![\varphi]\!]_{\alpha^k}(a)$ for all supported coalgebras $\mathcal{A} = (A, \alpha, \mathrm{supp}_\alpha)$ and all $a \in A$.

Lemma 7.4. *For every supported coalgebra $\mathcal{A} = (A, \alpha, \mathrm{supp}_\alpha)$, $k < \omega$ and $a \in A$, we have $d_k^{K,s}(a, a) = 0$, where the first a lives in $\mathcal{A}$ and the second in $\mathcal{A}_a^k$.*

A key step in the proof is the following locality result, which in similar form appears also in proofs of the classical van Benthem theorem [44], and is proved, in our case, by a game-theoretic method that is related to classical Ehrenfeucht-Fraïssé games:

Lemma 7.5. *Let $\varphi(x)$ be a behaviourally nonexpansive formula with* $\mathsf{qr}(\varphi) \leq n$. *Then φ is k-local for $k = 3^n$.*

Proof (sketch). Consider a spoiler-duplicator game over n rounds, where both players place a pebble every round and the second player needs to maintain the invariant that if there are m rounds remaining the radius 3^m neighbourhoods around the pebbles need to be isomorphic. One can show that this invariant guarantees equivalence on formulae of rank at most m.

We use this game to prove for every supported coalgebra $\mathcal{A}$ that φ has the same value on $\mathcal{A}$ and $\mathcal{A}_a^k$. Nonexpansiveness of φ is used to extend the two coalgebras in such a way that the duplicator always has a suitable response. □

We next show that every nonexpansive formula that is local is also nonexpansive at some finite depth. We make use of an unravelling construction, where a coalgebra is enlarged so that the successors of every state in the unravelling (as given by the support relation) form a tree.

Definition 7.6 (Unravelling). The *unravelling* of a supported coalgebra $\mathcal{A} = (A, \alpha, \mathrm{supp}_\alpha)$ is the supported coalgebra $\mathcal{A}^* = (A^+, \alpha^*, \mathrm{supp}_{\alpha^*})$, where A^+ is the set of nonempty sequences over A and for $a_1 \dots a_n \in A^+$ we have $\alpha^\star(a_1 \dots a_n) = Tf(\alpha(a_n))$ and $\mathrm{supp}_{\alpha^*}(a_1 \dots a_n) = f[\mathrm{supp}_\alpha(a_n)]$, where $f\colon A \to A^+, a \mapsto a_1 \dots a_n a$.

Lemma 7.7. *For every supported coalgebra $\mathcal{A} = (A, \alpha, \mathrm{supp}_\alpha)$ and every $a \in A$, we have $d^{K,s}(a, a) = 0$, where the first a lives in $\mathcal{A}$ and the second in $\mathcal{A}^*$.*

The mentioned nonexpansiveness at finite depth follows:

Lemma 7.8. *Let φ be behaviourally nonexpansive and k-local. Then φ is also depth-k behaviourally nonexpansive.*

Proof (sketch). By the assumptions on φ we may pass from any supported coalgebra to the radius-k neighbourhood in the unravelling, which is shaped like a tree of depth k. Between any two such tree structures we have $d_k^K = d^K$, as their behaviour past depth k is fully characterized by the default value $\bot \in T\emptyset$. □

The target result then follows by combining the above lemmas with Theorem 6.1 and a final chain argument that allows us to detach the technical development from the choice of a fixed coalgebra:

Theorem 7.9 (Quantified van Benthem theorem). *Let Λ be a finite set of monotone and nonexpansive predicate liftings, let T be a standard functor with $T\emptyset \neq \emptyset$, and let $\mathcal{V}$ be a totally bounded value co-quantale. Then for every behaviourally nonexpansive formula φ of quantitative coalgebraic predicate logic with quantifier rank at most n and every $\varepsilon \gg 0$ there exists a modal formula $\psi \in \mathcal{L}^\Lambda$ such that for all coalgebras $\alpha\colon A \to TA$ and all $a \in A$, $d^s_{\mathcal{V}}([\![\varphi]\!]_\alpha(a), [\![\psi]\!]_\alpha(a)) \leq \varepsilon$ and the modal rank of ψ is bounded by 3^n.*

Proof (sketch). Using the final chain $(T^n 1)_{n<\omega}$, where 1 is a singleton set, we can construct a coalgebra (Z, ζ) such that for all (A, α) and all φ, ψ we have $d_{\mathcal{V}}^{\vee,s}([\![\varphi]\!]_\alpha, [\![\psi]\!]_\alpha) \le d_{\mathcal{V}}^{\vee,s}([\![\varphi]\!]_\zeta, [\![\psi]\!]_\zeta)$.

As φ is behaviourally nonexpansive, we get that it is also depth-k behaviourally nonexpansive for $k = 3^{\mathsf{qr}(\varphi)}$ by Lemmas 7.5 and 7.8, and by Theorem 6.1.3 for every $\varepsilon \gg 0$ there is $\psi \in \mathcal{L}_k^\Lambda$ such that $d_{\mathcal{V}}^{\vee,s}([\![\varphi]\!]_\zeta, [\![\psi]\!]_\zeta) \le \varepsilon$. □

To our best knowledge, the only previously known instances of this result in the real-valued setting are the ones for $[0, 1]$-valued fuzzy modal logic [57] and for quantitative probabilistic modal logic [58]. In the two-valued setting, we cover a previous coalgebraic van Benthem result [52] by instantiating to $\mathcal{V} = 2$, and in fact obtain an additional asymmetric version, characterizing fragments that are preserved under simulation. In our running examples, we obtain new concrete van Benthem theorems for $[0, 1]$-valued metric modal logic (Example 4.4.3) and convex-nondeterministic metric modal logic (Example 4.4.4). We cover, by default, the asymmetric case (to be thought of as characterizing fragments that are preserved under quantitative simulation) and, in the cases $\mathcal{V} = [0, 1]$ and $\mathcal{V} = 2$, also the symmetric case (to be thought of as characterizing fragments that are invariant under bisimulation).

8 Conclusions

We have established a highly general quantitative version of van Benthem's modal characterization theorem, stating that given a value quantale $\mathcal{V}$ that is totally bounded and continuous from below, all state properties, in a given type of quantitative systems, that are nonexpansive w.r.t. $\mathcal{V}$-valued behavioural distance and expressible in $\mathcal{V}$-valued coalgebraic (first-order) predicate logic can be approximated by $\mathcal{V}$-valued modal formulae of bounded rank. A key technical tool in the proof are versions of the classical Arzela-Ascoli and Stone-Weierstraß theorems for totally bounded quantale-valued (pseudo-quasi-)metric spaces. Coalgebraic generality implies that this result not only subsumes existing quantitative van-Benthem type theorems for fuzzy [57] and probabilistic [58] systems, but we also obtain new results, e.g. for metric transition systems. Via the additional parametrization over a value quantale, we moreover obtain, e.g., a van Benthem theorem for convex-nondeterministic behavioural distance ('states x, y have distance between a and b') on metric transition systems. Our result complements previous coalgebraic results for two-valued logics [52]. We do leave some open problems, in particular to determine whether the main result can be sharpened to exact modal expressibility instead of approximability, and to obtain a quantitative modal characterization over finite models, in generalization of Rosen's finite-model variant of van Benthem's theorem [48].

Acknowledgements We wish to thank Barbara König for valuable discussions.

References

1. Abriola, S., Descotte, M., Figueira, S.: Model theory of XPath on data trees. Part II: Binary bisimulation and definability. Inf. Comput. (2017), in press
2. Adámek, J., Herrlich, H., Strecker, G.: Abstract and Concrete Categories. Wiley Interscience (1990), available as *Reprints Theory Appl. Cat.* 17 (2006), pp. 1-507
3. de Alfaro, L., Faella, M., Stoelinga, M.: Linear and branching system metrics. IEEE Trans. Software Eng. **35**(2), 258–273 (2009)
4. Baldan, P., Bonchi, F., Kerstan, H., König, B.: Coalgebraic behavioral metrics. Log. Methods Comput. Sci. **14**(3) (2018)
5. Barr, M.: Terminal coalgebras in well-founded set theory. Theoret. Comput. Sci. **114**, 299–315 (1993)
6. van Benthem, J.: Correspondence theory. In: Gabbay, D., Guenthner, F. (eds.) Handbook of Philosophical Logic, vol. 3, pp. 325–408. Springer (2001)
7. Bilkova, M., Kurz, A., Petrisan, D., Velebil, J.: Relation lifting, with an application to the many-valued cover modality. Logical Methods in Computer Science **Volume 9, Issue 4** (Oct 2013)
8. Borceux, F.: Handbook of Categorical Algebra: Volume 3, Sheaf Theory. Cambridge University Press (1994)
9. van Breugel, F.: A behavioural pseudometric for metric labelled transition systems. In: Abadi, M., de Alfaro, L. (eds.) Concurrency Theory, CONCUR 2005. LNCS, vol. 3653, pp. 141–155. Springer (2005)
10. van Breugel, F., Worrell, J.: A behavioural pseudometric for probabilistic transition systems. Theor. Comput. Sci. **331**, 115–142 (2005)
11. Carreiro, F.: PDL is the bisimulation-invariant fragment of weak chain logic. In: Logic in Computer Science, LICS 2015. pp. 341–352. IEEE (2015)
12. ten Cate, B., Fontaine, G., Litak, T.: Some modal aspects of XPath. J. Appl. Non-Classical Log. **20**, 139–171 (2010)
13. Chatzikokolakis, K., Gebler, D., Palamidessi, C., Xu, L.: Generalized bisimulation metrics. In: Concurrency Theory, CONCUR 2014. LNCS, vol. 8704, pp. 32–46. Springer (2014)
14. Dawar, A., Otto, M.: Modal characterisation theorems over special classes of frames. In: Logic in Computer Science, LICS 05. pp. 21–30. IEEE Computer Society (2005)
15. Desharnais, J., Gupta, V., Jagadeesan, R., Panangaden, P.: Metrics for labelled Markov processes. Theor. Comput. Sci. **318**, 323–354 (2004)
16. Desharnais, J., Edalat, A., Panangaden, P.: Bisimulation for labelled markov processes. Inf. Comput. **179**(2), 163–193 (2002)
17. Doberkat, E.: Stochastic Coalgebraic Logic. Monographs in Theoretical Computer Science. An EATCS Series, Springer (2009)
18. Eleftheriou, P., Koutras, C., Nomikos, C.: Notions of bisimulation for Heyting-valued modal languages. J. Logic Comput. **22**(2), 213–235 (2012)
19. Enqvist, S., Seifan, F., Venema, Y.: Monadic second-order logic and bisimulation invariance for coalgebras. In: Logic in Computer Science, LICS 2015. IEEE (2015)
20. Fahrenberg, U., Legay, A., Thrane, C.: The quantitative linear-time–branching-time spectrum. In: Foundations of Software Technology and Theoretical Computer Science, FSTTCS 2011. LIPIcs, vol. 13, pp. 103–114. Schloss Dagstuhl – Leibniz-Zentrum für Informatik (2011)
21. Fan, T.: Fuzzy bisimulation for Gödel modal logic. IEEE Trans. Fuzzy Sys. **23**, 2387–2396 (Dec 2015)

22. Figueira, D., Figueira, S., Areces, C.: Model theory of XPath on data trees. Part I: Bisimulation and characterization. J. Artif. Intell. Res. (JAIR) **53**, 271–314 (2015)
23. Flagg, B., Kopperman, R.: Continuity spaces: Reconciling domains and metric spaces. Theoretical Computer Science **177**(1), 111 – 138 (1997)
24. Flagg, R.: Quantales and continuity spaces. Algebra Univ. **37**(3), 257–276 (1997)
25. Gavazzo, F.: Quantitative behavioural reasoning for higher-order effectful programs: Applicative distances. In: Logic in Computer Science, LICS 2018. pp. 452–461. ACM (2018)
26. Giacalone, A., Jou, C., Smolka, S.: Algebraic reasoning for probabilistic concurrent systems. In: Programming concepts and methods, PCM 1990. pp. 443–458. North-Holland (1990)
27. Halpern, J.Y.: An analysis of first-order logics of probability. Artif. Intell **46**, 311–350 (1990)
28. Hansen, H., Kupke, C., Pacuit, E.: Neighbourhood structures: Bisimilarity and basic model theory. Log. Meth. Comput. Sci. **5**(2) (2009)
29. Hennessy, M., Milner, R.: Algebraic laws for non-determinism and concurrency. J. ACM **32**, 137–161 (1985)
30. Henriksen, M., Kopperman, R.: A general theory of structure spaces with applications to spaces of prime ideals. Alg. Univ. **28**(3), 349–376 (1991)
31. Hofmann, D.: Topological theories and closed objects. Adv. Math. **215**(2), 789 – 824 (2007)
32. Hofmann, D., Reis, C.: Probabilistic metric spaces as enriched categories. Fuzzy Sets Sys. **210**, 1–21 (2013)
33. Hofmann, D., Waszkiewicz, P.: Approximation in quantale-enriched categories. Topol. Appl. **158**(8), 963–977 (2011)
34. Huth, M., Kwiatkowska, M.: Quantitative analysis and model checking. In: Logic in Computer Science, LICS 1997. pp. 111–122. IEEE (1997)
35. Janin, D., Walukiewicz, I.: Automata for the modal μ-calculus and related results. In: Mathematical Foundations of Computer Science, MFCS 1995. LNCS, vol. 969, pp. 552–562. Springer (1995)
36. Khakpour, N., Mousavi, M.R.: Notions of conformance testing for cyber-physical systems: Overview and roadmap (invited paper). In: Concurrency Theory, CONCUR 2015. LIPIcs, vol. 42, pp. 18–40. Schloss Dagstuhl – Leibniz-Zentrum für Informatik (2015)
37. König, B., Mika-Michalski, C.: (Metric) Bisimulation Games and Real-Valued Modal Logics for Coalgebras. In: Schewe, S., Zhang, L. (eds.) Concurrency Theory, CONCUR 2018. LIPIcs, vol. 118, pp. 37:1–37:17. Schloss Dagstuhl – Leibniz-Zentrum für Informatik (2018)
38. Kontinen, J., Müller, J., Schnoor, H., Vollmer, H.: A van Benthem theorem for modal team semantics. In: Computer Science Logic, CSL 2015. LIPIcs, vol. 41, pp. 277–291. Schloss Dagstuhl - Leibniz-Zentrum für Informatik (2015)
39. Larsen, K., Skou, A.: Bisimulation through probabilistic testing. Inform. Comput. **94**, 1–28 (1991)
40. Litak, T., Pattinson, D., Sano, K., Schröder, L.: Coalgebraic predicate logic. In: Czumaj, A., Mehlhorn, K., Pitts, A., Wattenhofer, R. (eds.) Automata, Languages, and Programming, ICALP 2012. LNCS, vol. 7392, pp. 299–311. Springer (2012)
41. Lukasiewicz, T., Straccia, U.: Managing uncertainty and vagueness in description logics for the semantic web. J. Web Sem. **6**(4), 291–308 (2008)
42. Morgan, C., McIver, A.: A probabilistic temporal calculus based on expectations. In: Groves, L., Reeves, S. (eds.) Formal Methods Pacific, FMP 1997. Springer (1997)

43. Novák, V.: First-order fuzzy logic. Stud. Log. **46**, 87–109 (1987)
44. Otto, M.: Elementary proof of the van Benthem-Rosen characterisation theorem. Tech. Rep. 2342, Department of Mathematics, Technische Universität Darmstadt (2004)
45. Pattinson, D.: Expressive logics for coalgebras via terminal sequence induction. Notre Dame J. Formal Log. **45**, 19–33 (2004)
46. Raney, G.: Completely distributive complete lattices. Proc. AMS **3**(5), 677–680 (1952)
47. Raney, G.: A subdirect-union representation for completely distributive complete lattices. Proc. AMS **4**(4), 518–522 (1953)
48. Rosen, E.: Modal logic over finite structures. J. Logic, Language and Information **6**(4), 427–439 (1997)
49. Rutten, J.: Universal coalgebra: A theory of systems. Theoret. Comput. Sci. **249**, 3–80 (2000)
50. Schröder, L.: Expressivity of coalgebraic modal logic: The limits and beyond. Theoret. Comput. Sci. **390**, 230–247 (2008)
51. Schröder, L., Pattinson, D.: Description logics and fuzzy probability. In: Walsh, T. (ed.) Int. Joint Conf. Artificial Intelligence, IJCAI 2011. pp. 1075–1081. AAAI (2011)
52. Schröder, L., Pattinson, D., Litak, T.: A van Benthem/Rosen theorem for coalgebraic predicate logic. J. Log. Comput. **27**(3), 749–773 (2017)
53. Straccia, U.: A fuzzy description logic. In: Artificial Intelligence, AAAI 1998. pp. 594–599. AAAI Press / MIT Press (1998)
54. Sturm, H., Wolter, F.: First-order expressivity for S5-models: Modal vs. two-sorted languages. J. Philos. Logic **30**, 571–591 (2001)
55. Wild, P., Schröder, L.: A characterization theorem for a modal description logic. In: Int. Joint Conf. Artificial Intelligence, IJCAI 2017. pp. 1304–1310. ijcai.org (2017)
56. Wild, P., Schröder, L.: Characteristic logics for behavioural metrics via fuzzy lax extensions. In: Concurrency Theory, CONCUR 2020. LIPIcs, vol. 171, pp. 27:1–27:23. Schloss Dagstuhl – Leibniz-Zentrum für Informatik (2020)
57. Wild, P., Schröder, L., Pattinson, D., König, B.: A van Benthem theorem for fuzzy modal logic. In: Logic in Computer Science, LICS 2018. pp. 909–918. ACM (2018)
58. Wild, P., Schröder, L., Pattinson, D., König, B.: A modal characterization theorem for a probabilistic fuzzy description logic. In: Kraus, S. (ed.) International Joint Conference on Artificial Intelligence, IJCAI 2019. pp. 1900–1906. ijcai.org (2019)

Author Index

Printed in the United States
by Baker & Taylor Publisher Services

Printed in the United States
by Baker & Taylor Publisher Services